# Appendix A.2

## UPPER PERCENTILES
## OF STUDENT *t* DISTRIBUTIONS

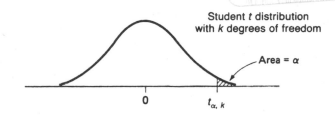

Student *t* distribution
with *k* degrees of freedom

Area = α

0     $t_{\alpha, k}$

$\alpha$

| df | 0.20 | 0.15 | 0.10 | 0.05 | 0.025 | 0.01 | 0.005 |
|----|------|------|------|------|-------|------|-------|
| 1 | 1.376 | 1.963 | 3.078 | 6.3138 | 12.706 | 31.821 | 63.657 |
| 2 | 1.061 | 1.386 | 1.886 | 2.9200 | 4.3027 | 6.965 | 9.9248 |
| 3 | 0.978 | 1.250 | 1.638 | 2.3534 | 3.1825 | 4.541 | 5.8409 |
| 4 | 0.941 | 1.190 | 1.533 | 2.1318 | 2.7764 | 3.747 | 4.6041 |
| 5 | 0.920 | 1.156 | 1.476 | 2.0150 | 2.5706 | 3.365 | 4.0321 |
| 6 | 0.906 | 1.134 | 1.440 | 1.9432 | 2.4469 | 3.143 | 3.7074 |
| 7 | 0.896 | 1.119 | 1.415 | 1.8946 | 2.3646 | 2.998 | 3.4995 |
| 8 | 0.889 | 1.108 | 1.397 | 1.8595 | 2.3060 | 2.896 | 3.3554 |
| 9 | 0.883 | 1.100 | 1.383 | 1.8331 | 2.2622 | 2.821 | 3.2498 |
| 10 | 0.879 | 1.093 | 1.372 | 1.8125 | 2.2281 | 2.764 | 3.1693 |
| 11 | 0.876 | 1.088 | 1.363 | 1.7959 | 2.2010 | 2.718 | 3.1058 |
| 12 | 0.873 | 1.083 | 1.356 | 1.7823 | 2.1788 | 2.681 | 3.0545 |
| 13 | 0.870 | 1.079 | 1.350 | 1.7709 | 2.1604 | 2.650 | 3.0123 |
| 14 | 0.868 | 1.076 | 1.345 | 1.7613 | 2.1448 | 2.624 | 2.9768 |
| 15 | 0.866 | 1.074 | 1.341 | 1.7530 | 2.1315 | 2.602 | 2.9467 |
| 16 | 0.865 | 1.071 | 1.337 | 1.7459 | 2.1199 | 2.583 | 2.9208 |
| 17 | 0.863 | 1.069 | 1.333 | 1.7396 | 2.1098 | 2.567 | 2.8982 |
| 18 | 0.862 | 1.067 | 1.330 | 1.7341 | 2.1009 | 2.552 | 2.8784 |
| 19 | 0.861 | 1.066 | 1.328 | 1.7291 | 2.0930 | 2.539 | 2.8609 |
| 20 | 0.860 | 1.064 | 1.325 | 1.7247 | 2.0860 | 2.528 | 2.8453 |
| 21 | 0.859 | 1.063 | 1.323 | 1.7207 | 2.0796 | 2.518 | 2.8314 |
| 22 | 0.858 | 1.061 | 1.321 | 1.7171 | 2.0739 | 2.508 | 2.8188 |
| 23 | 0.858 | 1.060 | 1.319 | 1.7139 | 2.0687 | 2.500 | 2.8073 |
| 24 | 0.857 | 1.059 | 1.318 | 1.7109 | 2.0639 | 2.492 | 2.7969 |
| 25 | 0.856 | 1.058 | 1.316 | 1.7081 | 2.0595 | 2.485 | 2.7874 |
| 26 | 0.856 | 1.058 | 1.315 | 1.7056 | 2.0555 | 2.479 | 2.7787 |
| 27 | 0.855 | 1.057 | 1.314 | 1.7033 | 2.0518 | 2.473 | 2.7707 |
| 28 | 0.855 | 1.056 | 1.313 | 1.7011 | 2.0484 | 2.467 | 2.7633 |
| 29 | 0.854 | 1.055 | 1.311 | 1.6991 | 2.0452 | 2.462 | 2.7564 |
| 30 | 0.854 | 1.055 | 1.310 | 1.6973 | 2.0423 | 2.457 | 2.7500 |
| 31 | 0.8535 | 1.0541 | 1.3095 | 1.6955 | 2.0395 | 2.453 | 2.7441 |
| 32 | 0.8531 | 1.0536 | 1.3086 | 1.6939 | 2.0370 | 2.449 | 2.7385 |
| 33 | 0.8527 | 1.0531 | 1.3078 | 1.6924 | 2.0345 | 2.445 | 2.7333 |
| 34 | 0.8524 | 1.0526 | 1.3070 | 1.6909 | 2.0323 | 2.441 | 2.7284 |
| ∞ | 0.84 | 1.04 | 1.28 | 1.64 | 1.96 | 2.33 | 2.58 |

SOURCE: *Scientific Tables,* 6th ed. (Basel, Switzerland: J.R. Geigy, 1962), pp. 32–33.

# Statistics

# Statistics

**Richard J. Larsen**
*Vanderbilt University*

**Morris L. Marx**
*University of West Florida*

*Prentice Hall , Englewood Cliffs, New Jersey 07632*

**Library of Congress Cataloging-in-Publication Data**

Larsen, Richard J.
  Statistics.

  Includes bibliographical references (p. 771)
  1. Statistics.    I. Marx, Morris L.    II. Title.
QA276.12.L364    1990    519.5    89–25557
ISBN 0–13–844085–9

Interior and cover design: Suzanne Behnke
Editorial development: Raymond Mullaney
Production: Nicholas Romanelli
Illustrator: Ron Weickart, Network Graphics
Manufacturing buyer: Paula Massenaro
Cover photo: Larry Dale Gordon/Image Bank

STATISTICS by Richard J. Larsen and Morris L. Marx

Printed in the United States of America
10 9 8 7 6 5 4 3 2 1

ISBN 0-13-844085-9

Prentice-Hall International (UK) Limited, *London*
Prentice-Hall of Australia Pty. Limited, *Syndey*
Prentice-Hall Canada Inc., *Toronto*
Prentice-Hall Hispanoamericana, S.A., *Mexico*
Prentice-Hall of India Private Limited, *New Delhi*
Prentice-Hall of Japan, Inc., *Tokyo*
Simon & Schuster Asia Pte. Ltd., *Singapore*
Editora Prentice-Hall do Brasil, Ltda., *Rio de Janeiro*
Prentice Hall, Englewood Cliffs, *New Jersey*

# Contents

## 10  THE TWO-SAMPLE PROBLEM  448

## 11  THE PAIRED-DATA PROBLEM  483

## 12  REGRESSION–CORRELATION DATA  504

## 17 MULTIPLE REGRESSION     722

## BIBLIOGRAPHY     771

## APPENDIX TABLES     780

## GLOSSARY     797

## ANSWERS TO ODD-NUMBERED QUESTIONS     807

## INDEX     823

# Preface

*"The time has come," the Walrus said*
*"To talk of many things:*
*Of shoes—and ships—and sealing wax*
*Of cabbages—and kings.*
*And why the sea is boiling hot*
*And whether pigs have wings."*

Lewis Carroll's famous passage from *Alice Through the Looking-Glass* was an invitation for a dinner conversation extended to a group of unwitting oysters. In spirit, it describes pretty well what lies ahead for us. Like the Walrus's agenda, statistics is a subject made up of a little bit of this and a little bit of that. If every discipline we might study, from archaeology and astronomy to urban studies and zoology, were written up in a grand, academic Betty Crocker cookbook, statistics would be somewhere in the middle of the mulligan stew chapter.

To learn statistics, in other words, requires that we *do* talk about many things. By its very nature, statistics is both elegantly conceptual and eminently practical. Theory and applications walk side by side, hand in hand. On a first reading, the enormous range of topics that needs to be discussed may seem overwhelming. Don't be discouraged. Once the objectives of statistics become clearer, the fact that we must deal with everything from mathematical cabbages to mathematical kings is precisely what gives the subject its charm.

Painted with the broadest of brushes, statistical applications in the real world tend to proceed along one of two lines: either we seek to characterize what has *already* happened (for example, constructing a graph to show the proportion of never-married women in the United States as a function of their age) or we try to predict what *might* happen (How likely is another *Challenger* disaster within the next five years?). Methods for addressing problems of the first kind are known collectively as *descriptive statistics*. These are covered in the first three chapters. *Inferential statistics*, by which we mean procedures for making predictions and forming generalizations, are the subject of Chapters 7 through 17. Sandwiched between descriptive statistics and inferential statistics are three chapters that introduce the basis principles of *probability*.

Many statistics books target specific audiences by emphasizing certain topics or drawing examples from a single discipline. A trip to the library will reveal any number of texts advertising themselves to be statistics for biologists, statistics for economists, statistics for engineers, statistics for the social sciences, and so on. Not to be outdone, this book also has a theme, but the conceptual dock to which we are moored is not an area of application. *Our theme is data identification.*

By virtue of the way in which observations are collected, every set of data has a certain intrinsic structure, irrespective of whether the measurements come from an engineering experiment or a psychology experiment. One set of data, for example, might represent what a statistician calls a "paired-data" problem; another might be classified as a "regression–correlation" problem. Fortunately, the number of such structures, or models, is comfortably small: the vast majority of data sets likely to be encountered belong to one of only seven prototypes. Why are these models so important? Because each requires a different kind of statistical analysis. For a statistician, learning to distinguish a paired-data problem from a regression–correlation problem is every bit as important as a physician's ability to diagnose whether a patient complaining of stomach pains has a bleeding ulcer or acute appendicitis.

Chapter 2 is devoted entirely to data identification. Each of the seven basic models is first described in general terms and then illustrated with several examples. Later chapters reinforce that same idea by pursuing the consequences of a data set's structure. What kind of descriptive techniques, for example, are appropriate for paired-data? How are the objectives of inferential statistics expressed when the data have a regression–correlation format?

Case studies are featured prominently in every chapter. These are somewhat expanded discussions of data sets taken from actual experiments. Read them carefully. Think about what the experimenter was trying to accomplish. See if you agree with the way the data were analyzed. Would you have approached the problem the same way? Quite often, more statistics can be learned by reflecting on the details of a specific experiment than by pouring over the assumptions of a formally stated theorem.

We hope you enjoy this introduction to probability and statistics. That parts of it will be useful is a foregone conclusion. Whether the future finds you conducting your own research, passing judgment on the work of others, or simply coping as an informed citizen with an increasingly numerical world, there will be times when knowing something about statistical methodology will come in handy. And who knows, a little knowledge and a little luck may help us avoid the fate that befell the oysters in Lewis Carroll's poem—they ended up as walrus food!

We would like to express our gratitude to all the researchers whose data we have used for examples. Specific acknowledgments appear in the bibliography. We appreciate the efforts of Tom Whipple, who served as Development Editor, and, for their detailed comments, criticisms, and suggestions, we thank the reviewers:

Ernest A. Blaisdell, Jr., Emory University

James C. Curl, Modesto Junior College

Frank A. Gunnip, Oakland Community College

Gordon W. Hoagland, Ricks College

John T. Kontogianes, Tulsa Junior College

Phillip McGill, Illinois Central College

John D. Neff, Georgia Institute of Technology

William Notz, Ohio State University

Lawrence Riddle, Emory University

Finally, we would like to express our sincere thanks to Bob Sickles and the entire editorial production and design staff at Prentice Hall.

RJL
Nashville, TN

MLM
Pensacola, FL

## SUPPLEMENTS FOR STUDENTS AND INSTRUCTORS

To ensure that this text is as effective a learning and teaching tool as possible, a complete package of innovative supplements is available for students and instructors.

The *Annotated Instructor's Edition (AIE)*, prepared by the authors, contains the complete student text, with numerous on-page annotations for the instructor (answers, teaching tips, cautions, explanations), and an introduction on how to use the AIE.

The *Instructor's Solutions Manual*, by David Welch of Southern Methodist University, contains complete solutions to every exercise in the text.

The *Test Item File* contains approximately 1,000 questions, referenced to chapter number and section. Also available is a microcomputer testing service—the *Bibliogem Test Generator*—that allows the instructor to access questions from the Test Item File, edit them if necessary, and then prepare and print out customized tests on an IBM personal computer.

There is also a *Transparency Pack* that includes both two-color acetates and masters.

The *Study Guide and Student's Solutions Manual*, also by David Welch, contains complete solutions to every sixth exercise and offers the student helpful hints, and problem-solving strategies.

An especially helpful supplement for the student is *How to Study Statistics*. Written by the authors, it contains strategies, suggestions, and hints for learning and achieving success in statistics. This supplement is provided free with each new copy of the text.

A number of supplements are available to enhance the use of software in the statistics course. *Mystat* software for both the IBM and Macintosh and a *Mystat Manual* are supplied free to adopters of the text. Also available are manuals on SAS, SPSS, and Minitab that show how these materials can be used effectively in an introductory statistics course.

# Statistics

# *1* *Why Statistics Is Good for You*

*Thou shalt not answer questionnaires*
*Or quizzes upon world affairs,*
*Nor with compliance*
*Take any test.*
*Thou shalt not sit with statisticians nor commit*
*A social science.*

*W.H. Auden*

# 1.1

## THE GOAL

When training for his 1946 heavyweight title fight with Billy Conn, Joe Louis fired a memorable warning at his quick-footed challenger. "He can run," said Joe, "but he can't hide." The context may be different, but the wisdom in that trenchant bit of advice applies to us as well. As citizens of a computerized and information-rich society, we are all facing—like Joe Louis's foe—an adversary from which flight is futile and hiding is hopeless. Learning how to cope with that adversary is what this book is about.

*Our battle is with data.* Think about it. When was the last time you watched a news broadcast or read a magazine like *Time* or *Newsweek* and were not expected to interpret at least one table, chart or graph? How often have you tried to read articles intended for "general" audiences only to find that the author assumed that you knew the meanings of *standard deviation, correlation coefficient, regression analysis, sampling error,* or *statistical significance*? Like it or not, we live in a time and a place where the road of persuasion has taken a decidedly numerical turn.

The extent of our fascination with quantification is quite amazing. Traditionally numerical fields such as engineering and the physical sciences are not the only affected areas. The federal government, for example, routinely evaluates the state of the economy by using a variety of stock market indicators and fiscal barometers; candidates for major elective offices rely heavily on polls to profile constituencies and to monitor changes in political momentum; health officials notify us of potential medical hazards by publishing the latest "risk" probabilities associated with exposure to certain carcinogens and mutagens; even Cupid has become fair game for the slings and arrows of numerical scrutiny—the entire cover of a recent issue of Newsweek (114) was a graph showing a college-educated woman's chances of getting married, for the first time, as a function of her age.

Our point is not that quantification is bad—quite the contrary. Numbers strongly influence our thinking and the way we make decisions, and this influence is likely to continue; indeed, it may be irreversible. But not every set of data or every analysis based on that data is meaningful. A result derived from a formula or appearing on a computer printout has no guarantee of validity. Quantification misadventures (better known as numerical garbage) are becoming an increasingly bothersome form of scientific litter. Therefore, learning how to collect, analyze, and interpret data *correctly* is well worth your time. The subject that embraces these various concerns, and the object of our attention in this book, is ***statistics***.

statistics—The numbers recorded or information collected. In other contexts, it refers to all the procedures and mathematical properties associated with the collection, analysis, and interpretation of data.

## 1.2
## THE GAME PLAN

The last paragraph in Section 1.1 summarizes our intentions: (1) to develop our powers of numerical persuasion and (2) to protect ourselves from numerical abuse, we want to learn how to collect, analyze, and interpret data. This objective seems reasonable enough, but it doesn't tell us *how* to accomplish this minor academic miracle. Where do we start? What skills do we need along the way? How should we organize our pursuit to maximize our chances of success?

The key to our game plan, and a major theme of the book, is recognizing data models. Put more simply, we must be able to distinguish one form of data from another. Most data can be put into one of seven different categories or *models*. These categories are established according to basic properties of how the data are collected. A set of cholesterol levels, for example, from a nutrition experiment may have the same fundamental properties (that is, belong to the same data model) as mileage estimates recorded in an EPA study of two kinds of catalytic converters.

**data models**—All data have certain fundamental properties relating to the nature, number, and type of variables recorded. Most data sets fall into one of seven categories or models.

Why is it so important to identify a data set's underlying model? Because each of these seven models must be analyzed in a different way; its objectives are different, and its formulas are different. Any "diagnostic" errors up front will come back to haunt us—indeed, our solution is guaranteed to be wrong! It makes good sense, therefore, to learn something about data identification in general before we get enmeshed in the details of specific procedures.

Chapter 2, using 17 case studies, discusses data classification by defining the characteristics of the seven different types and illustrating each with several sets of real-world applications. Chapter 2 accomplishes another important objective. The data used as examples of data models allow us to introduce informally certain definitions and concepts that are pursued formally in other chapters. Getting a quick look at some of these notions now will prepare us for a more detailed examination later.

With Chapter 2 putting us into a classification frame of mind, we should not be surprised that methods for *presenting* data are similarly categorized. **Descriptive statistics** is the name given to the potpourri of procedures for summarizing and displaying numerical information. Some procedures are graphical, some are numerical, and a few are a little bit of both. All of them, though, should look comfortably familiar: they are the graphs, charts, and numbers that show up everywhere from textbooks to magazines to the six o'clock news. Chapter 3 retraces the seven data models profiled in Chapter 2 and explains the basic graphical and numerical conventions commonly associated with each one.

**descriptive statistics**—The set of procedures for summarizing and displaying data.

The most important mathematical tool for understanding statistics is probability. Using probability, statisticians perform their most valuable service—drawing inductive inferences. The material is difficult, but the range of problems it allows us to address is quite impressive. We will talk about encoding messages, testing for rare blood diseases, looking for flying saucers, reading Braille, solving crimes, teaching chimpanzees to read, playing baseball, attacking Klingons, investigating ESP, ordering hamburgers, getting arrested for drunk driving, and writing screenplays. These, of course, are in addition to what you would expect to use probability for, playing poker and shooting craps!

Four chapters may seem like a lot of time to spend on probability when our ultimate objective is analyzing data, but it really isn't. Statistical inference is rooted entirely in probability. You cannot accurately interpret experimental results without the conceptual framework in Chapters 4 through 7.

Beginning with Chapter 8, our attention turns unrelentingly statistical. Most of the chapters are now keyed to one of the seven data models. As in Chapters 2 and 3, our presentation is decidely data-oriented. Formulas appropriate for each model are introduced with as much generality as feasible, but the major emphasis is on learning how to apply those formulas to actual data.

Listing the topics in Chapters 8 through 16 would not serve any purpose because the terminology is unfamiliar. Suffice it to say that what we learn there can be applied to many different experimental situations—a *great* many.

Indicative of the scope and the power of this material are the following questions, all of which arise within various case studies:

1.  Are the point spreads quoted for NFL games accurate?
2.  Is it likely that the Etruscan civilization was native to Italy?
3.  Does it make sense to argue that certain rectangles are esthetically more pleasing than others?
4.  How effective are bats at catching insects?
5.  How precise are geological dates estimated by radioactive decay?
6.  Was Quintus Curtius Snodgrass a pen name of Mark Twain?
7.  Are women more likely than men to suffer from frequent nightmares?
8.  Does aspirin affect the rate at which a person's blood clots?
9.  What is the functional relationship between the velocity of a galaxy and its distance?
10. Does the full moon induce aberrant behavior?

Quite clearly, our pursuit of statistics will take us to many exotic ports of call. Trivia aficionado or not, there should be something of interest here for everyone.

# 1.3

## SIGMA NOTATION

The mathematical notation used in statistics differs a bit from the way symbols are defined in algebra. In algebra, problems often involve only one variable, which we denote by the letter $x$. ("If it takes three rock stars 15 minutes to demolish a hotel suite, how long $(x)$ would it take their five-piece band to trash an entire floor?") When two variables are involved, we denote one by $x$ and the other by $y$, and so on. Invariably, the number of variables needed for a problem always ran out long before the alphabet was used up.

In statistics, however, we could run out of letters of the alphabet. For reasons that will soon be readily apparent, we need a different variable name for each measurement recorded. Assigning a new letter to each measurement is obviously inadequate if more than 26 observations are taken. A much better approach, regardless of the number of measurements involved, is to represent each observation by a letter that has a numerical *subscript* that represents the *order* in which the data point was recorded. For example, suppose a psychologist recruits five student volunteers for a learning experiment and records the IQ of each student. Her notebook might have an entry like that in Table 1.1. Since the 134 recorded for student DF is the first IQ listed, its variable name is $x_1$. Likewise, the 120 for ML is $x_2$, and so on. (In equation form, we would write $x_1 = 134$.)

Subscript notation obviously solves our problem of depleting the supply of variable names; if a set of data contains 1000 entries, we simply denote them $x_1, x_2, \ldots, x_{1000}$. But we still have a *second* notation problem. As we will see in the next paragraph, some situations need a notation to represent an "arbitrary" observation. Subscripted numbers will not work for that purpose. As soon as we associate a numerical subscript with $x$, say $x_{17}$, the observation is no longer arbitrary. Subscripted *letters*, on the other hand, camouflage an observation's identity. In general, we use the symbol $x_i$ to denote an "arbitrary" observation.

Subscripts and symbols for arbitrary observations are important for our purposes because of the mathematical operation of summation, which figures prominently in what lies ahead. Many statistical problems require that (1) the observations recorded be added together or that (2) certain quantities computed from the observations be added together. For example, suppose the psychologist who collected the data in Table 1.1 wanted to know the *average* IQ of the five students in her study. The answer, 122.8, is a sum divided by a constant:

$$\text{Average IQ} = \frac{134 + 120 + 140 + 105 + 115}{5}$$

$$= \frac{614}{5} = 122.8$$

**TABLE 1.1**

| Student | IQ |
|---------|-----|
| DF | 134 |
| ML | 120 |
| CW | 140 |
| SH | 105 |
| EJ | 115 |

A shorthand notation for the five-term summation

$$134 + 120 + 140 + 105 + 115$$

would be convenient. Definition 1.1 introduces a symbol, $\sum$ (capital Greek sigma), that we use, in general, to denote the sum of $n$ terms.

---

**DEFINITION 1.1**

Let $x_1, x_2, \ldots, x_n$ be a set of $n$ observations. The symbol

$$\sum_{i=1}^{n} x_i$$

represents the sum of all measurements from $x_1$ to $x_n$:

$$\sum_{i=1}^{n} x_i = x_1 + x_2 + \cdots + x_n$$

More generally, if $f(x_i)$ is any function of $x_i$, then

$$\sum_{i=1}^{n} f(x_i) = f(x_1) + f(x_2) + \cdots + f(x_n)$$

---

**sigma notation ($\sum$)**—A mathematical shorthand for representing the sum of a set of observations. Given $x_1, x_2, \ldots, x_n$,

$$\sum_{i=1}^{n} x_i = x_1 + x_2 + \cdots + x_n$$

The notation applies to *functions* of the $x_i$'s as well. For example,

$$\sum_{i=1}^{n} x_i^2 = x_1^2 + x_2^2 + \cdots + x_n^2$$

Example 1.1 works through several applications of this **summation** (or **sigma**) **notation**. Pay attention to the distinction between **(b)** and **(d)**. We will often encounter both expressions in the same formula—keeping them straight is important.

---

*EXAMPLE 1.1*    An experimenter records four observations

$$x_1 = 3, \quad x_2 = 1, \quad x_3 = 0, \quad \text{and} \quad x_4 = 1$$

Compute the following sums:

(a) $\sum_{i=1}^{4} x_i$    (b) $\sum_{i=1}^{4} x_i^2$    (c) $\sum_{i=1}^{4} (x_i - 2)^2$    (d) $\left( \sum_{i=1}^{4} x_i \right)^2$

(Note: In the terminology of Definition 1.1, $f(x_i) = x_i^2$ for (b) and $f(x_i) = (x_i - 2)^2$ for (c).)

*Solution*    (a) $\sum_{i=1}^{4} x_i = x_1 + x_2 + x_3 + x_4 = 3 + 1 + 0 + 1 = 5$

(b) $\sum_{i=1}^{4} x_i^2 = x_1^2 + x_2^2 + x_3^2 + x_4^2 = 3^2 + 1^2 + 0^2 + 1^2 = 11$

(c) $\displaystyle\sum_{i=1}^{4} (x_i - 2)^2 = (x_1 - 2)^2 + (x_2 - 2)^2 + (x_3 - 2)^2 + (x_4 - 2)^2$

$$= (3 - 2)^2 + (1 - 2)^2 + (0 - 2)^2 + (1 - 2)^2$$

$$= 1 + 1 + 4 + 1 = 7$$

(d) $\displaystyle\left(\sum_{i=1}^{4} x_i\right)^2 = (x_1 + x_2 + x_3 + x_4)^2 = (3 + 1 + 0 + 1)^2 = 5^2 = 25$

---

**QUESTIONS**

**1.1**  Inventions, whether simple or complex, can take a long time to become marketable. Minute Rice, for example, was "invented" in 1931 but didn't appear on grocery shelves until 1949. Listed below are the conception dates and the realization dates for eight well-known products (176). Computed for each product and shown in the last column is the interval $x_i$ between its year of conception and its year of realization. The interval associated with the automatic transmission, for example, is

$$x_1 = 1946 - 1930 = 16$$

| Product | Year of Conception | Year of Realization | $x_i$ = interval |
|---|---|---|---|
| Automatic transmission | 1930 | 1946 | 16 |
| Ball-point pen | 1938 | 1945 | 7 |
| Instant coffee | 1934 | 1956 | 22 |
| Photography | 1782 | 1838 | 56 |
| Roll-on deodorant | 1948 | 1955 | 7 |
| Television | 1884 | 1947 | 63 |
| Xerox copying | 1935 | 1950 | 15 |
| Zipper | 1883 | 1913 | 30 |

Find

(a) $\displaystyle\sum_{i=1}^{8} x_i$

(b) The average time between the year of conception and the year of realization for the eight products.

**1.2**  A "statistic" frequently used to evaluate baseball pitchers is their strikeout ratio (i.e., the average number of batters they strike out in nine innings). According to official records compiled through 1980 (132), the all-time major league leader in that category is Nolan Ryan, who registered 3109 strikeouts in 2926 innings, an average of 1.062 (= 3109/2926) strikeouts per inning or 9.6 strikeouts every nine innings. Listed in the accompanying table are the five best career strikeout ratios in the history of baseball:

| Pitcher | Nine-Inning Strikeout Ratio, $x_i$ |
|---|---|
| Nolan Ryan | 9.6 |
| Sandy Koufax | 9.3 |
| Sam McDowell | 8.8 |
| J. R. Richard | 8.4 |
| Bob Veale | 8.0 |

(a) Compute $\sum\limits_{i=1}^{5} x_i$ and (b) use it to find the average nine-inning strikeout ratio for these five pitchers.

1.3   Let $x_1 = 2.0$, $x_2 = 4.0$, $x_3 = 4.0$, and $x_4 = 1.0$. Find

(a) $\sum\limits_{i=1}^{4} x_i$
(b) $\sum\limits_{i=1}^{4} x_i^2$
(c) $\left( \sum\limits_{i=1}^{4} x_i \right)^2$

(d) $\sum\limits_{i=1}^{4} (x_i - 2)^2$
(e) $\left( \sum\limits_{i=1}^{4} (x_i - 2) \right)^2$
(f) $\left( \sum\limits_{i=1}^{4} (x_i - 2)^2 \right)^2$

1.4   Let $x_1 = -1$, $x_2 = 4$, $x_3 = -2$, $x_4 = 5$, and $x_5 = 3$. Find

(a) $\sum\limits_{i=1}^{5} x_i$
(b) the average of the $x_i$'s
(c) $\sum\limits_{i=1}^{5} x_i^2$

(d) $\left( \sum\limits_{i=1}^{5} x_i \right)^2$
(e) $\left( \sum\limits_{i=1}^{5} x_i^2 \right)^2$
(f) $\sqrt{ \sum\limits_{i=1}^{5} x_i^2 }$

(g) $\left( \sum\limits_{i=1}^{5} x_i^2 - \left( \sum\limits_{i=1}^{5} x_i \right)^2 \right)^2$

1.5   If $x_1 = 2$, $x_2 = 6$, and $x_3 = 5$, express the following sums in sigma notation:

(a) $2^2 + 6^2 + 5^2$

(b) $(2 + 6 + 5)^3$

(c) $(2 - 1)^2 + (6 - 1)^2 + (5 - 1)^2$

(d) $\dfrac{1}{4} \cdot 2 + \dfrac{1}{4} \cdot 6 + \dfrac{1}{4} \cdot 5$

(e) $\dfrac{1}{2}(2 - 4)^2 + \dfrac{1}{2}(6 - 4)^2 + \dfrac{1}{2}(5 - 4)^2$

1.6   Let $x_1$ and $x_2$ be any two measurements. What must be true in order that

$$\sum\limits_{i=1}^{2} x_i^2 = \left( \sum\limits_{i=1}^{2} x_i \right)^2$$

1.7 For a given area the male-female ratio is defined as

$$\frac{\text{Number of males}}{\text{Number of females}} \times 100$$

In the mid-1960s, the overall male-female ratio in the United States was 97.0, indicating a deficit in the number of males. At the same time, however, there was a surplus of *single* males. Listed below are the 1966 male-female ratios for single persons, aged 14 and over, in the four geographical regions of the United States (121):

| Region | Male-Female Ratio (singles, age 14+), $x_i$ |
|---|---|
| Northeast | 109.9 |
| North central | 122.7 |
| South | 124.5 |
| West | 149.1 |
| Total U.S. | 123.4 |

(a) Compute

$$\frac{1}{4} \sum_{i=1}^{4} x_i$$

the average of the first four ratios shown in the table.

(b) Why doesn't the average you computed equal 123.4, the figure given at the bottom of the table for the entire United States?

### Sigma Notation for Two Variables

One minor extension of sigma notation occurs when *two* different kinds of measurements are made on each subject. For instance, suppose the psychologist of our earlier example had recorded SATs in addition to IQs. Let $x_i$ denote the $i$th subject's IQ, and $y_i$ the $i$th subject's SAT, where $i = 1, 2, 3, 4, 5$. The results might look like Table 1.2.

Now, suppose we wanted to multiply each subject's IQ by his or her SAT and then add the products. We would write

$$\sum_{i=1}^{5} x_i y_i = (134)(1540) + (120)(1250) + (140)(1460) + (105)(1120)$$

$$+ (115)(1220)$$

$$= 818{,}660$$

**TABLE 1.2**

| Subject | IQ ($x_i$) | SAT ($y_i$) |
|---|---|---|
| DF | 134 | 1540 |
| ML | 120 | 1250 |
| CW | 140 | 1460 |
| SH | 105 | 1120 |
| EJ | 115 | 1220 |

More generally, if $f(x_i, y_i)$ is any function of $x_i$ and $y_i$, then

$$\sum_{i=1}^{5} f(x_i, y_i) = f(x_1, y_1) + f(x_2, y_2) + \cdots + f(x_n, y_n)$$

Example 1.2 looks at several expressions where the quantities being added are functions of two variables. We will see terms such as those in parts (a) and (b) especially often.

---

***EXAMPLE 1.2***   An experimenter collects the set of $x$ and $y$ measurements shown in Table 1.3. Compute

(a) $\displaystyle\sum_{i=1}^{4} x_i y_i$    (b) $\left(\displaystyle\sum_{i=1}^{4} x_i\right)\left(\displaystyle\sum_{i=1}^{4} y_i\right)$    (c) $\displaystyle\sum_{i=1}^{4} x_i^2 y_i^2$    (d) $\left(\displaystyle\sum_{i=1}^{4} x_i y_i\right)^2$

**TABLE 1.3**

| Subject | $x_i$ | $y_i$ |
|---------|-------|-------|
| 1 | $2\,(=x_1)$ | $0\,(=y_1)$ |
| 2 | $4\,(=x_2)$ | $1\,(=y_2)$ |
| 3 | $3\,(=x_3)$ | $1\,(=y_3)$ |
| 4 | $1\,(=x_4)$ | $3\,(=y_4)$ |

*Solution*   (a) $\displaystyle\sum_{i=1}^{4} x_i y_i = x_1 y_1 + x_2 y_2 + x_3 y_3 + x_4 y_4$

$$= 2 \cdot 0 + 4 \cdot 1 + 3 \cdot 1 + 1 \cdot 3 = 10$$

(b) $\left(\displaystyle\sum_{i=1}^{4} x_i\right)\left(\displaystyle\sum_{i=1}^{4} y_i\right) = (x_1 + x_2 + x_3 + x_4)(y_1 + y_2 + y_3 + y_4)$

$$= (2 + 4 + 3 + 1)(0 + 1 + 1 + 3)$$
$$= 10 \cdot 5 = 50$$

(c) $\displaystyle\sum_{i=1}^{4} x_i^2 y_i^2 = x_1^2 y_1^2 + x_2^2 y_2^2 + x_3^2 y_3^2 + x_4^2 y_4^2$

$$= 2^2 \cdot 0^2 + 4^2 \cdot 1^2 + 3^2 \cdot 1^2 + 1^2 \cdot 3^2 = 34$$

(d) $\left(\displaystyle\sum_{i=1}^{4} x_i y_i\right)^2 = (x_1 y_1 + x_2 y_2 + x_3 y_3 + x_4 y_4)^2$

$$= (2 \cdot 0 + 4 \cdot 1 + 3 \cdot 1 + 1 \cdot 3)^2$$
$$= (10)^2 = 100$$

**QUESTIONS**

1.8 If scientists could be nominated for a Researcher's Hall of Fame, a deserving candidate for "most unusual experiment" would be Wayne Rowley, who wanted to know how far mosquitoes can fly without stopping. Rowley attached a mosquito to a rotating arm with a very fine wire (Figure 1.1).

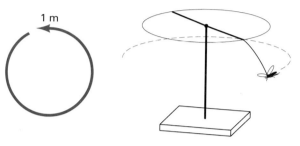

1 m

**FIG. 1.1**

The dimension of the arm was such that a complete revolution by the mosquito covered 1 meter (m). Poking the mosquito encouraged it to start flying. When the mosquito stopped, a technician poked it again and the mosquito usually resumed its flight, albeit grudgingly. Eventually, exhaustion set in, and no amount of prodding had any effect.

Two variables were recorded for each mosquito: age $x_i$ (in weeks) and distance $y_i$ (in thousands of meters) flown until exhausted. The results are listed in the table in the left margin (140). Compute

| Age (weeks), $x_i$ | Distance Flown (1000s of meters), $y_i$ |
|---|---|
| 2 | 11.6 |
| 3 | 6.8 |
| 1 | 12.6 |
| 4 | 9.2 |

(a) $\displaystyle\sum_{i=1}^{4} x_i$   (b) $\displaystyle\sum_{i=1}^{4} x_i^2$   (c) $\displaystyle\sum_{i=1}^{4} y_i$

(d) $\displaystyle\sum_{i=1}^{4} y_i^2$   (e) $\left(\displaystyle\sum_{i=1}^{4} x_i\right)\left(\displaystyle\sum_{i=1}^{4} y_i\right)$   (f) $\displaystyle\sum_{i=1}^{4} x_i y_i$

1.9 Fossil records indicate that "extinction episodes" have occurred throughout earth's history. By definition, extinction episodes are geologically short intervals in which numerous species have vanished. Probably the best known episode occurred about 65 million years ago when dinosaurs became extinct. Until recently, gradual climatic changes have been the accepted explanation for these die-offs. But scientists have now discovered two striking facts that may shed some new light on the question. First, the separation times between these mass extinctions are astonishingly regular, about 26 million years. Second, the sediment that formed during each extinction episode shows an abnormally high concentration of iridium, an element frequently associated with comets.

Putting these two discoveries together has led to the highly controversial *nemesis theory*, which proposes that some extraterrestrial mechanism— perhaps a companion star to our sun—periodically perturbs the orbits of some of the comets surrounding the solar system and induces them to strike the earth. The enormous amount of dust resulting from the impacts then produces a nuclear winter scenario, which decimates certain species. Shown in the table are the estimated times of the four most recent extinction events. (165) (Notice the similarity in the gaps between episodes 1 and 2, 2 and 3, and 3 and 4.) Compute

| Extinction Episode, $x_i$ | Millions of Years before Present, $y_i$ |
|---|---|
| 1 | 11 |
| 2 | 38 |
| 3 | 65 |
| 4 | 91 |

(a) $\sum_{i=1}^{4} x_i$    (b) $\sum_{i=1}^{4} x_i^2$       (c) $\sum_{i=1}^{4} y_i$

(d) $\sum_{i=1}^{4} y_i^2$    (e) $\left(\sum_{i=1}^{4} x_i\right)\left(\sum_{i=1}^{4} y_i\right)$    (f) $\sum_{i=1}^{4} x_i y_i$

1.10   An experimenter collects a set of five $(x_i, y_i)$ observations: $(3, 1)$, $(2, 2)$, $(1, 0)$, $(4, 3)$, and $(-1, -2)$. Compute

(a) $\sum_{i=1}^{5} x_i y_i$       (b) $\sum_{i=1}^{5} x_i^2 y_i^2$

(c) $\left(\sum_{i=1}^{5} x_i y_i\right)^2$       (d) $\sum_{i=1}^{5} (x_i - 3)(y_i + 2)$

1.11   For the data

| $x_i$ | $y_i$ |
|---|---|
| 4 | 2 |
| 5 | 1 |
| 3 | 2 |

express the following sums in sigma notation:

(a) $(4 + 5 + 3)(2 + 1 + 2)$       (b) $4 \cdot 2 + 5 \cdot 1 + 3 \cdot 2$

(c) $(4 + 5 + 3)^2(2^2 + 1^2 + 2^2)$       (d) $(4^2 \cdot 2^2 + 5^2 \cdot 1^2 + 3^2 \cdot 2^2)^2$

1.12   Three observations, $x_i$, $y_i$, and $z_i$, are recorded on each of four subjects:

| Subject | $x_i$ | $y_i$ | $z_i$ |
|---|---|---|---|
| DF | 3 | 1 | 4 |
| ML | 2 | 0 | 1 |
| SH | 2 | 2 | 1 |
| NL | 4 | 3 | 2 |

Compute

(a) $\sum_{i=1}^{4} x_i y_i z_i$       (b) $\left(\sum_{i=1}^{4} x_i\right)\left(\sum_{i=1}^{4} y_i\right)\left(\sum_{i=1}^{4} z_i\right)$

(c) $\displaystyle\sum_{i=1}^{4} x_i^2 y_i^3 z_i$  (d) $\displaystyle\sum_{i=1}^{4} (x_i + y_i + z_i)^2$

(e) $\displaystyle\left(\sum_{i=1}^{4} (x_i + y_i + z_i)^2\right)^2$

1.13  Male dragonflies spend time near ponds for the sole purpose of meeting female dragonflies. How long the male dragonfly stays depends on his estimated chance of "success," which in turn depends on the number of males he is competing against. If the density of male dragonflies is high, a typical male is likely to get discouraged and leave. Tabled on the left are the arrival rates ($x_i$) and the average durations of stay ($y_i$) for male dragonflies recorded during six periods of observation near a small pond (data modified from 17). Notice the "inverse" relationship: large values of $x_i$ tend to be associated with small values of $y_i$ and small values of $x_i$ tend to be associated with large values of $y_i$. Compute

| Arrival Rate (males/hr), $x_i$ | Average Duration of Stay (min), $y_i$ |
|---|---|
| 25.0 | 6.8 |
| 11.0 | 10.1 |
| 4.5 | 20.8 |
| 8.8 | 15.4 |
| 10.4 | 9.3 |
| 5.0 | 12.3 |

(a) $\displaystyle\sum_{i=1}^{6} x_i$  (b) $\displaystyle\sum_{i=1}^{6} x_i^2$  (c) $\displaystyle\sum_{i=1}^{6} y_i$

(d) $\displaystyle\sum_{i=1}^{6} y_i^2$  (e) $\displaystyle\left(\sum_{i=1}^{6} x_i\right)\left(\sum_{i=1}^{6} y_i\right)$  (f) $\displaystyle\sum_{i=1}^{6} x_i y_i$

(g) $\displaystyle\sqrt{\sum_{i=1}^{6} x_i y_i}$  (h) $\displaystyle\sqrt{\sum_{i=1}^{6} x_i^2 y_i^2}$  (i) $\displaystyle\sqrt{\left(\sum_{i=1}^{6} x_i y_i\right)^2}$

# 1.4

## A BRIEF HISTORY (Optional)

The picture of statistics presented in textbooks is, unfortunately, a single snapshot, one recent photograph from a rather thick family album. What we see is the subject as it is now, with no hint of what it was 100 years ago, much less why it got started in the first place. For those interested in matters historical (or those who wish to procrastinate), we offer in this section a glimpse at what statistics used to be.

### Sources of Statistics: A Summary

Historians generally agree that, as a subject, statistics began to take definite shape in the middle of the nineteenth century. What triggered its emergence was the union of three different "sciences," each of which had been developing along more or less independent lines (173).

The first of these sciences, what the Germans called *Staatenkunde*, involved the collection of comparative information on the history, resources, and military prowess of nations. Although efforts in this direction peaked in the seventeenth and eighteenth centuries, the concept was hardly new: Aristotle had done something similar in the fourth century B.C. Of the three movements this one had the least influence on the development of modern statistics, but it did contribute some terminology: the word *statistics*, itself, first arose in connection with studies of this type.

The second movement, known as *political arithmetic*, was defined by one of its early proponents as "the art of reasoning by figures, upon things relating to government." Of more recent vintage than Staatenkunde, political arithmetic's roots were in seventeenth-century England. Making population estimates and constructing mortality tables were two of the problems it frequently dealt with. In spirit, political arithmetic was similar to what is now called *demography*.

The third movement, originating in seventeenth-century France, was the formulation of a *calculus of probability*, which was a collection of mathematical theorems and techniques for problems involving uncertainty. Initially, probability theory developed in response to certain specific problems posed by gamblers. More general applications soon followed.

**STAATENKUNDE: THE COMPARATIVE DESCRIPTION OF STATES**  The need for gathering information on the customs and resources of nations has been obvious since antiquity. Aristotle is credited with the first major effort toward that objective: his *Politeiai*, written in the fourth century B.C., contained detailed descriptions of some 158 different city-states. Unfortunately, the thirst for knowledge that led to the *Politeiai* fell victim to the intellectual drought of the Dark Ages, and almost 2000 years elapsed before any similar projects of like magnitude were undertaken.

The subject resurfaced during the Renaissance, and the Germans showed the most interest. They not only gave it a name, *Staatenkunde*, meaning the comparative description of states, but they were also the first (in 1660) to incorporate the subject into a university curriculum. A leading figure in the German movement was Gottfried Achenwall, who taught at the University of Göttingen during the middle of the eighteenth century. Among Achenwall's claims to fame is that he was the first to use the word *statistics* in print. It appeared in the preface of his 1749 book *Abriss der Statswissenschaft der heutigen vornehmsten europaishen Reiche und Republiken.* (The word comes from the Italian root *stato*, meaning state, implying that a statistician is someone concerned with government affairs.) As terminology, it seems to have been well-received: for almost 100 years the word *statistics* continued to be associated with the comparative description of states. In the middle of the nineteenth century, though, the term was redefined, and statistics became the new name for what had previously been called political arithmetic.

How important was the work of Achenwall and his predecessors to the development of statistics? That would be difficult to say. To be sure, their contributions were more indirect than direct. They left no methodology and no general theory. But they did point out the need for collecting accurate data and, perhaps more importantly, reinforced the notion that something complex—even as complex as an entire nation—can be effectively studied by gathering information on its component parts. Thus, they were lending important support to the then-growing belief that *induction*, rather than *deduction*, was a more sure-footed path to scientific truth.

POLITICAL ARITHMETIC  In the sixteenth century the English government began to compile records, called ***bills of mortality***, on a parish-to-parish basis, showing numbers of deaths and their underlying causes. Their motivation largely stemmed from the plague epidemics that had periodically ravaged Europe in the not-too-distant past and were threatening to become a problem in England. Certain government officials, including the very influential Thomas Cromwell, felt that these bills would prove invaluable in helping to control the spread of an epidemic. At first, the bills were published only occasionally, but by the early seventeenth century they had become a weekly institution.[1]

Figure 1.2 (141) shows a portion of a bill that appeared in London in 1665. The gravity of the plague epidemic is strikingly apparent when we look at the numbers at the top: out of 97,306 deaths, 68,596 (over 70%) were caused by the plague. The breakdown of certain other afflictions, though they caused fewer deaths, raise some interesting questions. What happened, for example, to the 23 people who were "frighted" or to the 397 who suffered from "rising of the lights"?

Among the faithful readers of the bills was London merchant John Graunt. Graunt not only read the bills, he studied them intently. He looked for patterns, computed death rates, devised ways of estimating population sizes, and even set up a primitive life table. His results were published in the 1662 treatise *Natural and Political Observations upon the Bills of Mortality*. This work was a landmark: Graunt had launched the twin sciences of vital statistics and demography, and, although the name came later, it also signaled the beginning of political arithmetic. (Graunt did not have to wait long for accolades; in the year his book was published, he was elected to the prestigious Royal Society of London.)

High on the list of innovations that made Graunt's work unique were his objectives. Not content simply to describe a situation, although he was adept at doing so, Graunt often sought to go beyond his data and make generalizations (or, in current statistical terminology, draw *inferences*). Having been blessed with this particular turn of mind, he almost certainly qualifies as the world's first statistician. All Graunt really lacked was the probability theory that would have enabled him to frame his inferences more mathematically. That theory, though, was just beginning to unfold, several hundred miles away in France.

Other seventeenth-century writers were quick to follow through on Graunt's ideas. William Petty's *Political Arithmetick* was published in 1690, although it was probably written some 15 years earlier. (It was Petty who

---

[1] An interesting account of the bills of mortality is given in Daniel Defoe's *A Journal of the Plague Year*, which purportedly chronicles the London plague outbreak of 1665.

The bill for the year—A General Bill for this present year, ending the 19 of December, 1665, according to the Report made to the King's most excellent Majesty, by the Co. of Parish Clerks of Lond., &c.— gives the following summary of the results; the details of the several parishes we omit, they being made as in 1625, except that the out-parishes were now 12:—

| | |
|---|---|
| Buried in the 27 Parishes within the walls ... ... ... ... | 15,207 |
| Whereof of the plague ... ... ... ... ... ... ... | 9,887 |
| Buried in the 16 Parishes without the walls ... ... ... | 41,351 |
| Whereof of the plague ... ... ... ... ... ... | 28,838 |
| At the Pesthouse, total buried ... ... ... ... ... ... | 159 |
| Of the plague ... ... ... ... ... ... ... | 156 |
| Buried in the 12 out-Parishes in Middlesex and surrey ... | 18,554 |
| Whereof of the plague ... ... ... ... ... ... | 21,420 |
| Buried in the 5 Parishes in the City and Liberties of Westminster | 12,194 |
| Whereof the plague ... ... ... ... ... ... | 8,403 |
| The total of all the christenings ... ... ... ... ... | 9,967 |
| The total of all the burials this year ... ... ... ... | 97,306 |
| Whereof of the plague ... ... ... ... ... ... | 68,596 |

| | | | | | |
|---|---|---|---|---|---|
| Abortive and Stillborne ... | 617 | Griping in the Guts ... | 1,288 | Palsie ... ... ... ... | 30 |
| Aged ... ... ... ... | 1,545 | Hang'd & made away themselved | 7 | Plague ... ... ... ... | 68,596 |
| Ague & Feaver ... ... | 5,257 | Headmould shot and mould fallen | 14 | Plannet ... ... ... ... | 6 |
| Appolex and Suddenly ... | 116 | Jaundice ... ... ... ... | 110 | Plurisie ... ... ... ... | 15 |
| Bedrid ... ... ... ... | 10 | Impostume ... ... ... ... | 227 | Poysoned ... ... ... ... | 1 |
| Blasted ... ... ... ... | 5 | Kill by several accidents ... | 46 | Quinsie ... ... ... ... | 35 |
| Bleeding ... ... ... ... | 16 | King's Evill ... ... ... | 86 | Rickets ... ... ... ... | 535 |
| Cold & Cough ... ... ... | 68 | Leprosie ... ... ... ... | 2 | Rising of the Lights ... ... | 397 |
| Collick & Winde ... ... | 134 | Lethargy ... ... ... ... | 14 | Rupture ... ... ... ... | 34 |
| Comsumption & Tissick... | 4,808 | Livergrown ... ... ... | 20 | Scurry ... ... ... ... | 105 |
| convulsion & Mother ... | 2,036 | Bloody Flux, Scowring & Flux | 18 | Shingles & Swine Pox ... | 2 |
| Distracted ... ... ... | 5 | Burnt and Scalded ... ... | 8 | Sores, Ulcers, Broken and | |
| Dropsie & Timpany ... ... | 1,478 | Calenture ... ... ... ... | 3 | Bruised Limbs ... ... | 82 |
| Drowned ... ... ... ... | 50 | Cancer, Cangrene & Fistula | 56 | Spleen ... ... ... ... | 14 |
| Executed ... ... ... ... | 21 | Canker and Thrush ... ... | 111 | Spotted Feaver & Purples ... | 1,929 |
| Flox & Smallpox ... ... | 655 | Childbed ... ... ... ... | 625 | Stopping of the Stomach ... | 332 |
| Found Dead in streets, fields, &c. | 20 | Chrisomes and Infants ... | 1,258 | Stone and Stranguary ... | 98 |
| French Pox ... ... ... | 86 | Meagrom and Headach ... | 12 | Surfe ... ... ... ... | 1,251 |
| Frighted ... ... ... ... | 23 | Measles ... ... ... ... | 7 | Teeth & Worms ... ... ... | 2,614 |
| Gout & Sciatica ... ... | 27 | Murthered & Shot ... ... | 9 | Vomiting ... ... ... ... | 51 |
| Grief ... ... ... ... | 46 | Overlaid & Starved ... ... | 45 | Wenn ... ... ... ... | 8 |

| | | | | | | |
|---|---|---|---|---|---|---|
| Christened.—Males | ... | 5,114. | Females ... | 4,853. | In all ... | 9,967 |
| Buried.—Males | ... | 58,569. | Females ... | 48,737. | In all ... | 97,306 |
| Of the Plague | | ... ... ... ... ... ... ... | | | | 68,596 |
| Increase in the Burials in the 130 Parishes and the Pesthouse this year | | | | | | 79,009 |
| Increase of he Plague in the 130 Parishes and the Pesthouse this year | | | | | | 68,590 |

**FIG. 1.2**

gave the movement its name.) Perhaps even more significant were the contributions of Edmund Halley (of "Halley's comet" fame). Principally an astronomer, he also dabbled in political arithmetic, and in 1693 wrote *An Estimate of the Degrees of the Mortality of Mankind, drawn from Curious Tables of the Births and Funerals at the city of Breslaw; with an attempt to ascertain the Price of Annuities upon Lives.* (Book titles were longer then!) Halley shored up, mathematically, the efforts of Graunt and others to construct an accurate mortality table. In doing so, he laid the foundation for the important theory of annuities. Today all life insurance companies base their premium schedules on methods similar to Halley's. (The first company to follow his lead was The Equitable, founded in 1765.)

For all its initial flurry of activity, political arithmetic did not fare particularly well in the eighteenth century, at least in terms of having its methodology fine-tuned. Still, the second half of the century did see some notable achievements for improving the quality of the data bases: several countries, including the United States in 1790, established a periodic census. To some extent, answers to the questions that interested Graunt and his followers had to be deferred until the theory of probability could develop just a little bit more.

THE CALCULUS OF PROBABILITY  The last, and perhaps most important, of the three movements that coalesced in the nineteenth century to form the foundation for modern statistics was the rise of a rigorous theory of probability. At their most rudimentary levels the notions of probability and chance go back a long way. Egyptian pharaohs, for example, played games with primitive dice fashioned from the heel bones of sheep. Still, the first attempts to formulate an actual theory, or *calculus*, of probability are much more recent, with 1654 being the date usually quoted by historians.

What happened in 1654 was rather curious. The story begins with Antoine Gombaud, better known as the Chevalier de Mere, a minor French nobleman with a penchant for gambling. While pursuing his avocation, de Mere wanted to know how the stakes should be divided in a game interrupted before its completion where one player holds an advantage. He turned to his countryman Blaise Pascal for help. Though only in his twenties, Pascal was already one of the most prominent mathematicians on the Continent. Intrigued by de Mere's question, Pascal worked out a solution and sent it to Pierre Fermat, another distinguished French mathematician, for his comments. Fermat responded graciously and asked some questions of his own. Thus began a correspondence that lasted nearly a year: out of it came several basic principles that we now recognize as the first steps toward a comprehensive theory of probability.

Three years later, Christiaan Huygens, a Dutch scientist best remembered for his work in optics, published the first treatise devoted solely to probability, *De Ratiociniis in Ludo Alea* (Calculations in Games of Chance). Although *De Ratiociniis* did manage to arouse some interest in probability, the subject did not take off immediately. Most mathematicians were too busy taking sides in the Newton-Leibniz controversy to get involved with probability. (Newton in England and Leibniz in Germany had codiscovered the calculus in the 1660s, but their approaches were somewhat different, and the mathematical community was trying to decide which to pursue.) Nevertheless, progress was made. Bernoulli's *Ars Conjectandi* (The Art of Conjecture) was published posthumously in 1713, the familiar bell-shaped curve first appeared in 1733, and in 1812 Laplace released his profoundly influential *Theorie Analytique des Probabilities*. By then, probability had come of age.

### Quetelet: The Catalyst

With political arithmetic furnishing the data, and many of the questions, and the theory of probability holding out the promise of rigorous answers, the birth of statistics was at hand. All that was needed was a catalyst—someone to bring the two together. Several individuals served with distinction in that capacity. Karl Friedrich Gauss, the superb German mathematician and astronomer, was especially helpful in showing how statistical concepts could be useful in the physical sciences. Similar efforts in France were made by Laplace. But the man who perhaps best deserves the title of "matchmaker" was a Belgian, Adolphe Quetelet.

Quetelet was a mathematician, astronomer, physicist, sociologist, anthropologist, and poet. One of his passions was collecting data, and he was fascinated by the apparent regularity and predictability of social phenomena. In commenting on the nature of criminal tendencies, he once wrote (66):

> Thus we pass from one year to another with the sad perspective of seeing the same crimes reproduced in the same order and calling down the same punishments in the same proportions. Sad condition of humanity! . . . We might enumerate in advance how many individuals will stain their hands in the blood of their fellows, how many will be forgers, how many will be poisoners, almost we can enumerate in advance the births and deaths that should occur. There is a budget which we pay with a frightful regularity; it is that of prisons, chains and the scaffold.

Given such an orientation, it was not surprising that Quetelet would see in probability theory an elegant means for expressing human behavior. For much of the nineteenth century he vigorously championed the cause of statistics, and as a member of more than 100 learned societies his influence was enormous. When he died in 1874, statistics had been brought to the brink of its modern era.

### Statistics in the 1980s

Statistics has flourished in the last 100 years, and the subject today bears little resemblance to the methodology that Quetelet so fondly extolled. But now is not the time to comment on these advances; we will do that later as the occasion demands. Suffice it to say that it was once again the British, in the tradition of John Graunt, who were at the forefront of statistical innovation when the subject entered its modern era near the turn of the twentieth century.

**REVIEW QUESTIONS**

**1.14**  Listed below are the eight foods lowest in calories, as reported by the U.S. Department of Agriculture (176).

| Rank | Food | Calories/100 grams |
|------|------|--------------------|
| 1 | Water | 0 |
| 2 | Club soda | 0 |
| 3 | Coffee | 1 |
| 4 | Tea | 2 |
| 5 | Jerusalem artichoke | 7 |
| 6 | Sour pickle | 10 |
| 7 | Sauerkraut juice | 10 |
| 8 | Dill pickle | 11 |

If $x_i$ denotes the calories/100 grams in the $i$th food, compute

(a) $\displaystyle\sum_{i=1}^{8} x_i$     (b) $\displaystyle\sum_{i=1}^{8} x_i^2$     (c) $\left(\displaystyle\sum_{i=1}^{8} x_i\right)^2$

**1.15**  The sum of the first $n$ integers is

$$\frac{n(n+1)}{2}$$

The sum of the squares of the first $n$ integers is

$$\frac{n(n+1)(2n+1)}{6}$$

Write those two statements as equations using sigma notation. Verify that the statements are correct for $n = 3$ and $n = 5$.

**1.16**  During their six home games last season, State University's football team fumbled 2, 0, 3, 1, 4, and 3 times, respectively. If $x_i$ denotes the number of fumbles made in the $i$th game, compute

(a) $\displaystyle\sum_{i=1}^{6} (x_i - 2)$     (b) $\displaystyle\sum_{i=1}^{6} x_i - 2$     (c) $\displaystyle\sum_{i=1}^{6} (x_i - 2)^2$     (d) $\displaystyle\sum_{i=1}^{6} x_i^2 - 2^2$

**1.17**  In words, what does $\displaystyle\sum_{i=1}^{n} (2i - 1)$ represent?

**1.18**  For the following set of data,

| $x_i$ | $y_i$ |
|-------|-------|
| 2 | 1 |
| −1 | 2 |
| 0 | 1 |
| −2 | 0 |

which is larger,

$$\left( \sum_{i=1}^{4} x_i \right) \left( \sum_{i=1}^{4} y_i \right) \quad \text{or} \quad \sum_{i=1}^{4} x_i y_i?$$

1.19   The brain weights (in ounces) of five famous people are listed below (76).

| | Brain Weight, $x_i$ |
|---|---|
| Lord Byron | 82 |
| Marilyn Monroe | 51 |
| Leon Trotsky | 56 |
| Howard Hughes | 49 |
| Ivan Turgenev | 74 |

Let $\bar{x}$ denote the average of the five weights.

(a) Write a formula for $\bar{x}$ using sigma notation.

(b) Compute $\bar{x}$.

1.20   In general, which, if any, of the following statements are true?

(a) $\displaystyle \sum_{i=1}^{n} (x_i + y_i) = \sum_{i=1}^{n} x_i + \sum_{i=1}^{n} y_i$   (b) $\displaystyle \sum_{i=1}^{n} x_i y_i = \left( \sum_{i=1}^{n} x_i \right) \left( \sum_{i=1}^{n} y_i \right)$

(c) $\displaystyle \frac{\sum_{i=1}^{n} x_i^2}{\sum_{i=1}^{n} x_i} = \sum_{i=1}^{n} x_i$   (d) $\displaystyle \sum_{i=1}^{n} \frac{x_i^2}{x_i} = \sum_{i=1}^{n} x_i$

1.21   Use sigma notation to summarize the following expressions:

(a) $x_1 y_1^2 + x_2 y_2^2 + x_3 y_3^2$   (b) $1 \cdot x_1 + 2 \cdot x_2 + 3 \cdot x_3 + 4 \cdot x_4$

(c) $(x_1 + y_1)^2 + (x_2 + y_2)^2 + \cdots + (x_n + y_n)^2$

1.22   Let $c$ be any non-zero constant. In general, which of the following statements are true?

(a) $\displaystyle c \sum_{i=1}^{n} x_i = \sum_{i=1}^{n} c x_i$   (b) $\displaystyle \sum_{i=1}^{n} x_i^c = \left( \sum_{i=1}^{n} x_i \right)^c$   (c) $\displaystyle \sum_{i=1}^{n} \frac{x_i}{c} = \frac{\sum_{i=1}^{n} x_i}{c}$

1.23   Subscripts can be used to index *arrays* of numbers, as well as *sets* of numbers. Define $x_{ij}$ as the entry in the $i$th row and $j$th column of a table of numbers. For the following data,

| | | COLUMN ($j$) | | |
|---|---|---|---|---|
| | | 1 | 2 | 3 |
| ROW ($i$) | 1 | 3.0 | 1.5 | 2.0 |
| | 2 | 1.3 | 6.4 | 4.7 |

find

(a) $x_{13}$    (b) $x_{31}$    (c) $\sum\limits_{j=1}^{3} x_{1j}$    (d) $\sum\limits_{i=1}^{2} x_{i2}$

1.24   The table below gives the numbers of commercials broadcast per hour by two television stations. Data were collected on five different occasions.

| Time Period | NUMBER OF COMMERCIALS | |
| --- | --- | --- |
| | Station A ($x_i$) | Station B ($y_i$) |
| March 21, 9–10 AM | 3 | 4 |
| April 4, 5–6 PM | 4 | 4 |
| April 27, 1–2 AM | 8 | 6 |
| May 15, 3–4 PM | 5 | 7 |
| June 4, 10–11 PM | 3 | 3 |

Let $d_i - y_i - x_i$. Compute the average of (a) the $d_i$'s and (b) the $d_i^2$'s.

1.25   Suppose the causes of death listed in Figure 1.2 are numbered consecutively from 1 (for Abortive and Stillborne) to 62 (for Wenn). Let $x_i$ denote the number of deaths due to the $i$th cause. Use sigma notation to write a formula for the proportion of total deaths due to "Griping in the Guts."

1.26   The symbol $x_{(i)}$ is often defined to mean the $i$th ordered value in a set of $n$ observations. For example, if a data set consists of three observations, $x_1 = 16$, $x_2 = 6$, and $x_3 = 12$, then $x_{(1)} = 6$, $x_{(2)} = 12$, and $x_{(3)} = 16$. For the following sample,

| Subject | $x_i$ |
| --- | --- |
| DP | 18.1 |
| SH | 14.2 |
| BB | 17.6 |
| LM | 15.4 |

compute (a) $x_2 + x_3$ and (b) $x_{(2)} + x_{(3)}$.

1.27   Each semester the Registrar at State University publishes the average grade (on a 4.0 scale) given out by each academic department. Results for six departments are summarized below.

| Department | Average Grade, $x_i$ |
| --- | --- |
| Economics | 2.6 |
| Mathematics | 3.0 |
| Political Science | 3.1 |
| Physics | 2.7 |
| Fine Arts | 3.2 |
| Molecular Biology | 3.1 |

Suppose the average for the entire university is 3.06. Let

$$y_i = \begin{cases} 1 & \text{if } x_i \geq 3.06 \\ 0 & \text{if } x_i < 3.06 \end{cases}$$

Compute

(a) $\sum_{i=1}^{6} y_i$    (b) $\sum_{i=1}^{6} y_i^2$    (c) $\left( \sum_{i=1}^{6} y_i \right)^2$

1.28    Listed below are the periods of revolution $(x)$ and the mean distances from the sun $(y)$ for the nine planets in the solar system.

| Planet | $x_i$ (years) | $y_i$ (astronomical units) |
|--------|-----------|------------------------|
| Mercury | 0.241 | 0.387 |
| Venus | 0.615 | 0.723 |
| Earth | 1.000 | 1.000 |
| Mars | 1.881 | 1.524 |
| Jupiter | 11.86 | 5.203 |
| Saturn | 29.46 | 9.54 |
| Uranus | 84.01 | 19.18 |
| Neptune | 164.8 | 30.06 |
| Pluto | 248.4 | 39.52 |

Kepler's Third Law states that "the squares of the periods of the planets are proportional to the cubes of their mean distance from the Sun." Check the veracity of that statement by computing $x_i^2/y_i^3$ for $i = 1, 2, \ldots, 9$. If a new planet is discovered with a period of revolution of 5.2 years, what would we expect its mean distance from the Sun to be (3)?

1.29    Listed below are the points scored in a marksmanship contest by the five members of State's rifle team. Compute the team's average score. If $\bar{x}$ denotes the team average, for which shooter is $(x_i - \bar{x})^2$ the largest?

| Shooter | Points Scored, $x_i$ |
|---------|----------------------|
| Tom | 62 |
| Kyle | 81 |
| Randi | 70 |
| Mike | 84 |
| Beth | 75 |

1.30    If $f(x_i) = x_i + x_i^2$, compute $\sum_{i=1}^{3} f(x_i)$ for $x_1 = 2$, $x_2 = 5$, and $x_3 = -1$. What value would you associate with $f\left( \sum_{i=1}^{3} x_i \right)$?

1.31   Use sigma notation to represent the following sums:

(a) $(x_1 + 1) + (x_2 + 2) + \cdots + (x_n + n)$

(b) $x_1^0 + x_2^1 + \cdots + x_n^{n-1}$

1.32   Listed below are the number of copies sold $(x)$ and the number of years since publication $(y)$ for five diet books.

| Book | No. of Copies Sold, $x_i$ | No. of Years Since Publication, $y_i$ |
|------|---------------------------|----------------------------------------|
| 1 | 20,000 | 3 |
| 2 | 10,000 | 2 |
| 3 | 55,000 | 4 |
| 4 | 5,000 | 1 |
| 5 | 5,000 | 5 |

Compute

(a) $\sum_{i=1}^{5} x_i$ (b) $\sum_{i=1}^{5} y_i$ (c) $\sum_{i=1}^{5} x_i y_i$ (d) $\left( \sum_{i=1}^{5} x_i \right) \left( \sum_{i=1}^{5} y_i \right)$.

1.33   As part of an assignment to collect a set of data, Lisa counts the number of times in each of six lectures that her statistics teacher doesn't seem to know what he's taking about. Her results are summarized below.

| Class Period | Number of Times Incoherent, $x_i$ |
|--------------|-----------------------------------|
| Monday, Week 1 | 3 |
| Wednesday, Week 1 | 4 |
| Friday, Week 1 | 2 |
| Monday, Week 2 | 1 |
| Wednesday, Week 2 | 2 |
| Friday, Week 2 | 6 |

Compute

(a) the sum of the $x_i$'s;

(b) the sum of the squares of the $x_i$'s;

(c) the square of the sum of the $x_i$'s.

1.34   Cindy, a senior, has kept track of the number of Valentine's Day cards that she received during her four years at college. If the pattern shown on the next page continues, what should be the value of $x_6$, the number of cards she receives after two years of graduate school? (Hint: Compute $2^{i-1} - 1$ for years 1, 2, 3, and 4.)

| Year | No. of Cards in Year $i$, $x_i$ |
|------|--------------------------------|
| 1 | 0 |
| 2 | 1 |
| 3 | 3 |
| 4 | 7 |

1.35  For the conception to realization intervals in Question 1.1, compute the quantity $s$, where

$$s = \sqrt{\frac{n \sum_{i=1}^{n} x_i^2 - \left( \sum_{i=1}^{n} x_i \right)^2}{n(n-1)}}$$

(In Chapter 3 we shall learn that $s$ is the *sample standard deviation* of the $x_i$'s. Its numerical value reflects the extent to which the $x_i$'s are not all the same.)

1.36  The *product* of a set of quantities is denoted by the symbol $\prod$. More specifically,

$$\prod_{i=1}^{n} f(x_i) = f(x_1) \cdot f(x_2) \cdot \ \cdots \ \cdot f(x_n)$$

If $f(x_i) = x_i^2$ and $x_1 = 3$, $x_2 = 4$, and $x_3 = 1$, compute $\prod_{i=1}^{3} f(x_i)$.

1.37  Refer to Question 1.36. Write the following products using $\prod$ notation:

(a) $x_1 \cdot x_2 \cdot \ \cdots \ \cdot x_n$

(b) $(x_1 - 2)(x_2 - 2) \cdot \ \cdots \ \cdot (x_n - 2)$

(c) $1 \cdot 2 \cdot 3 \cdot \ \cdots \ \cdot n$

1.38  If a data set consists of the numbers $x_1, x_2, \ldots, x_n$, where $n$ is even, what do the sums

$$\sum_{i=1}^{n/2} x_{2i-1} \qquad \text{and} \qquad \sum_{i=1}^{n/2} x_{2i}$$

represent? What would be another expression for

$$\sum_{i=1}^{n/2} x_{2i-1} + \sum_{i=1}^{n/2} x_{2i}?$$

# 2 Data Models

*A statistician is a specialist
who assembles figures and then leads them astray.*

*Anonymous*

### INTRODUCTION

A statistician treats data like a doctor treats patients. Both must first diagnose the problem. That means the statistician must recognize the underlying "structure" of the data. (Fortunately, distinguishing types of data is usually much easier than figuring out why a two-month-old baby cries all night. What greatly simplifies matters is that the number of structures that needs to be considered is small. By defining only seven, we can set up a classification system that is adequate for most of the experimental situations likely to be encountered. This chapter characterizes and illustrates these seven data structures, or models.

Two concepts hold the key to classifying data: sample and population. What they represent in a given situation determines the category to which the data belong. A *sample* is a set of measurements. The number of measurements in that set is the *sample size*. For example, imagine the football trainer for Swampwater Tech weighing in the three freshman noseguards his alma mater has illegally recruited. The entry in his records might look like Table 2.1. Here, the sample (of size 3) is the set of numbers 285, 290, and 280. The sample subjects—that is, what or whom the measurements are made on—are Bubba, Joe Bob, and Billy Ray. Using the subscript notation of Section 1.3, we might let $y_i$ denote the weight of the $i$th recruit, $i = 1, 2, 3$, in which case $y_1 = 285$, $y_2 = 290$, and $y_3 = 280$.

In many statistical applications our primary concern is not with what *did* happen but with what *might have* happened. The sample is what did happen. What might have happened is a set of hypothetical measurements called the *population*. Presumably, the sample "represents" the population (in ways we will discuss in future chapters). For the data in Table 2.1 the population is the weights associated with the entire set of Swampwater noseguards illegally recruited past, present, and future. The sample values of 285, 290, and 280 recorded for Bubba, Joe Bob, and Billy Ray are three members of that population.

In an experiment there might be one population or several. The samples might be composed of a single kind of measurement, or the information might be multifaceted. The subjects themselves could influence the formation of the samples. Combinations of these options produce the seven data models profiled in this Chapter:

| Model I | One-sample data |
| Model II | Two-sample data |
| Model III | Paired data |
| Model IV | Regression–correlation data |
| Model V | Categorical data |
| Model VI | $k$-sample data |
| Model VII | Randomized block data |

**sample**—A set of measurements chosen (or selected or drawn) from a population of all possible such measurements.

**TABLE 2.1**

| Name | Weight |
|---|---|
| Bubba | 285 |
| Joe Bob | 290 |
| Billy Ray | 280 |

**population**—The (usually) hypothetical set of measurements from which the observed sample is presumed to have come.

This chapter will not make you an instant expert in diagnosing data, any more than a home medical dictionary makes a parent a physician. But it will give you a very useful preview of several important ideas and teach you to look at data more critically. At the end of the book, after these themes have resurfaced several times, we still may not be experts, but at least we won't kill the patient!

| QUESTIONS | | |
|---|---|---|

2.1 Choosing samples that accurately reflect a targeted population is an important skill for experimenters to develop. We will have many occasions to address this problem from various perspectives. For now it may be helpful to think about ways *not* to choose a sample. For instance, if we wanted data on the average daily amount of alcohol drunk by young adults, it would not be a good idea to interview students leaving a Friday night fraternity party (or, worse yet, students unable to leave a Friday night fraternity party!). Or if we wanted to estimate the average height of Americans, it would not make much sense to measure the starters at an NBA all-star game or the rank-and-file members attending the National Little People Convention. How would you *not* collect data to investigate

(a) the average IQ of all college freshmen?

(b) the academic reputation of the college or university you attend?

(c) your own personal significance and potential usefulness as a leader of the free world?

2.2 Cereal boxes attached to a conveyor belt are filled automatically at the rate of 2000 per hour. The production supervisor intends to check 20 of the boxes each hour to make certain they are not being overfilled or underfilled. What are some clearly inappropriate ways to choose the 20?

2.3 Imagine drawing two chips from an urn that contains four chips, numbered 1, 2, 3, and 4 (Figure 2.1). A typical sample is the two chips (1, 3). Notice that the population in this situation (the set of all possible samples of size 2 is

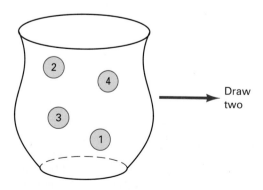

**FIG. 2.1**

easy to identify. By inspection, it contains six pairs: $(1, 2), (1, 3), (1, 4), (2, 3),$ $(2, 4),$ and $(3, 4)$. Suppose three chips are to be drawn from an urn containing five chips, numbered 1, 2, 3, 4, and 5. List the samples that constitute the population.

2.4   The last four times you were in the supermarket express lane, the person in front of you had an unmarked item that required a price check. Having nothing better to do with your time, you measured the resulting delays (in minutes) and recorded 2.5, 4.0, 5.5, and 3.5.

(a) What is the sample, and what is the population associated with your "experiment?" What is the sample *size*?

(b) In the future how long would you expect to wait if the person in front of you showed up with some unmarked broccoli?

2.5   An aide to the governor is seeking information on local support for a bill pending in the state legislature that would legalize horse racing as a means of generating tax revenues. As they are leaving a baseball game, she asks 10 adults the question, "Are you in favor of legalizing horse racing in our state?" and receives the following responses:

| Subject | Response | Subject | Response |
|---------|----------|---------|----------|
| 1 | Yes | 6 | Yes |
| 2 | Yes | 7 | No |
| 3 | No | 8 | Yes |
| 4 | Yes | 9 | Yes |
| 5 | No | 10 | Yes |

(a) What population is represented by these 10 responses? Is the sample likely to be biased? In which direction?

(b) If you were willing to assume that this sample represents the population you wish to consider, what percentage would you say indicates the level of support for legalizing horse racing?

# 2.2

## ONE-SAMPLE DATA

**one-sample data**—A single sample of size $n$ is observed, $y_1, y_2, \ldots, y_n$. The $y_i$'s measure (1) a single set of conditions or (2) the effects of one particular treatment.

The simplest of the seven data models is the **one-sample problem**. Figure 2.2 shows its sample–population structure. The heavy dots symbolize the entire group of subjects or conditions (real or hypothetical) under consideration.

Associated with the population is a **treatment**, by which we mean either a condition applied to a subject or an attribute characteristic of a subject. A flu vaccine, for example, given to $n$ adults participating in a

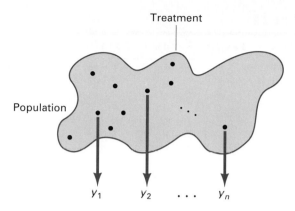

Treatment

Population

$y_1$ $y_2$ $\cdots$ $y_n$

**FIG. 2.2**

**treatment**—Any identifiable condition or trait whose effect on the measured variable is being studied.

clinical trial, is a treatment. "Mexican-American" becomes a treatment when an economist studies family income and focuses on that particular ethnic group. Reflecting the effect of a treatment will be some measurement $y$. In the clinical trial, $y$ might be a subject's antibody level two weeks after receiving the vaccine; for the economist, $y$ is the annual income, in dollars, of a Mexican-American family.

In general, it will be impossible to determine the actual population of $Y$ values; that is, we are physically unable to measure the response of each potential subject. Instead, we observe a *sample* of $Y$ values, $y_1$, $y_2, \ldots, y_n$. Our objective is to use the information in the sample (the $y_i$'s) to make an inference about the population.

## CASE STUDY 2.1

*"Toxic Torts"*

"Toxic torts," in the opinion of many legal experts, will dominate litigation in the 1990s in much the same way that civil rights cases flourished during the 1960s. Toxic torts refer to lawsuits filed by individuals who have come in contact with hazardous substances and developed medical problems that might be due to that contact. An office worker, for example, may allege that the asbestos in the ceiling above her desk caused the lung cancer from which she now suffers.

In general, such claims are difficult to support or refute. The evidence is seldom the smoking-gun variety, and *proving* causality is impossible (in contrast to traffic accidents, where who did what to whom, and when, are more easily established). The study described here typifies the kind of data that can bring about toxic tort litigation.

Producing a certain detergent causes workers to receive prolonged exposure to *Bacillus subtilis* enzyme, a substance that may be harmful to the respiratory system. As a preliminary study to assess the severity of the

**TABLE 2.2**

| Subject | FEV$_1$/VC | Subject | FEV$_1$/VC |
|---------|-----------|---------|-----------|
| RH | 0.61 | WS | 0.78 |
| RB | 0.70 | RV | 0.84 |
| MB | 0.63 | EN | 0.83 |
| DM | 0.76 | WD | 0.82 |
| WB | 0.67 | FR | 0.74 |
| RB | 0.72 | PD | 0.85 |
| BF | 0.64 | EB | 0.73 |
| JT | 0.82 | PC | 0.85 |
| PS | 0.88 | RW | 0.87 |
| RB | 0.82 | | |

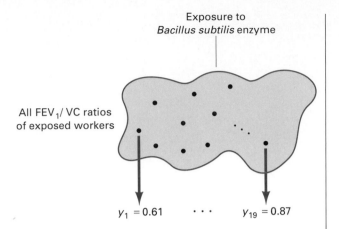

All FEV$_1$/ VC ratios
of exposed workers

Exposure to
*Bacillus subtilis* enzyme

$y_1 = 0.61$   $\cdots$   $y_{19} = 0.87$

**FIG. 2.3**

problem (if there is a problem), the airflow rates[1] of 19 workers routinely exposed to *B. subtilis* enzyme were measured (154) and tabulated (Table 2.2).

These data offer a clear example of the one-sample structure; that is, they show a single sample representing a single population. The sample (of size 19) is the set of numbers 0.61, 0.70, . . . , 0.87. The population is the set of FEV$_1$/VC ratios for all workers receiving similar exposures to *B. subtilis* (Figure 2.3).

Among healthy adults not exposed to enzymes of this sort, the typical value (or norm) for an FEV$_1$/VC ratio is 0.80. Airflow rates less than that value are interpreted as a sign of respiratory impairment. Here, the average airflow rate for the 19 workers listed in Table 2.2 is less than 0.80:

$$\text{Average airflow rate} = \frac{0.61 + 0.70 + \cdots + 0.87}{19} = 0.77$$

The question is, does the sample average of 0.77 constitute enough evidence for us to conclude, in general, that exposure to *B. subtilis* enzyme will lower a person's airflow rate? Would a jury be convinced?

Conceptually, there are two explanations (or hypotheses) for why the sample average is less than the norm. The first explanation says that *B. subtilis* is *not* harmful, and that the difference between 0.77 and 0.80 is due entirely to chance. The second explanation is that *B. subtilis is* harmful, as indicated by the sample average being less than 0.80. Choosing between conflicting explanations is a common problem faced by experimenters. The solution requires a statistical technique known as *hypothesis testing*, which we formally introduce in Chapter 8.

---

[1] Airflow rate is defined to be the ratio of a person's forced expiratory volume (FEV$_1$) to his vital capacity, (VC). Vital capacity is the maximum volume of air a person can exhale after taking as deep a breath as possible. FEV$_1$ is the maximum volume of air a person can exhale in 1 s.

## CASE STUDY 2.2

*Young Scientists' Discoveries*     Great discoveries in science are often made by young persons, a marked contrast to the stereotype of the white-haired, absent-minded professor. Newton developed the calculus and formulated the theory of gravitation at the age of 23. Einstein showed that $E = mc^2$ and proposed the special theory of relativity when he was only 26.

Table 2.3 lists 12 famous scientific discoveries, all major breakthroughs in terms of the impact they had on prevailing theories, and the ages of the scientists who made them. The sample is the set of these particular scientists' ages. The population is defined as the set of scientists' ages associated with all great discoveries (184).

**TABLE 2.3**

| Discovery | Discoverer | Date | Age |
|---|---|---|---|
| Earth goes around sun | Copernicus | 1543 | 40 |
| Telescope, basic laws of astronomy | Galileo | 1600 | 34 |
| Principles of motion, gravitation, calculus | Newton | 1665 | 23 |
| Nature of electricity | Franklin | 1746 | 40 |
| Burning is uniting with oxygen | Lavoisier | 1774 | 31 |
| Earth evolved by gradual processes | Lyell | 1830 | 33 |
| Evidence for natural selection controlling evolution | Darwin | 1858 | 49 |
| Field equations for light | Maxwell | 1864 | 33 |
| Radioactivity | Curie | 1896 | 34 |
| Quantum theory | Planck | 1901 | 43 |
| Special theory of relativity, $E = mc^2$ | Einstein | 1905 | 26 |
| Mathematical foundations for quantum theory | Schroedinger | 1926 | 39 |

QUESTIONS

2.6    Making a general statement about the "average" age at which great scientific discoveries are made would be misleading if that average were a function of the time in which the scientist lived. Make a graph of "date" versus "age" for the discoveries in Table 2.3. Put "date" on the horizontal axis. Are there any apparent trends? Does it seem reasonable to talk about "age" as a phenomenon independent of time?

2.7    In the home the amount of radiation emitted by a color television set is not harmful. But in a department store 15 or 20 sets may be on in a relatively confined area. To see what levels of radiation are produced under these conditions, readings were taken at 10 different department stores, each having at least five sets turned on in their display areas. The figure shown for each store is an average radiation level, in milliroentgens per hour (mR/hr) based on readings taken at several locations in each area (82).

| Location | Net mR/hr |
|----------|-----------|
| 1 | 0.40 |
| 2 | 0.48 |
| 3 | 0.60 |
| 4 | 0.15 |
| 5 | 0.50 |
| 6 | 0.80 |
| 7 | 0.50 |
| 8 | 0.36 |
| 9 | 0.16 |
| 10 | 0.89 |

(a)  To what population do these 10 observations refer?

(b)  Why do these data represent the one-sample model?

(c)  A recommended exposure limit to this type of radiation has been set by the National Council on Radiation Protection at 0.5 mR/hr. What proportion of department stores do you estimate have TV sales areas with radiation levels exceeding the NCRP recommended limit?

2.8   For reasons probably best left to the imagination, an experimenter wanted to know whether vampire bats prefer blood at room temperature (23°C) or at the body temperature of cattle (38.5°C). Eight vampire bats in a cage were allowed equal access to two blood-filled feeding tubes. The temperature of the blood in one tube was 23°C; in the other tube, 38.5°C. Listed below are the percentages of total blood consumed from the tube at 38.5° (20).

| Bat | % Blood Consumed at 38.5°C |
|-----|-----------------------------|
| 1 | 37.5 |
| 2 | 80.0 |
| 3 | 37.5 |
| 4 | 37.5 |
| 5 | 35.0 |
| 6 | 47.5 |
| 7 | 47.5 |
| 8 | 45.0 |

What subjects make up the population represented by these data?

2.9   Fabrics washed without bleach or without any special sanitizing additives might look clean and smell clean and yet not be "bacteriologically clean." Swatches cut from 10 colored napkins, commerically laundered and ironed, were soaked in water, shredded with a blender, and poured over Petri dishes filled with a growth medium. Listed in the accompanying table are the estimated bacteria counts (per square inch of fabric) recorded 48 hr later (115).

| Napkin | Bacteria/in² |
|--------|-------------|
| 1 | 1,600,000 |
| 2 | 96,500 |
| 3 | 407,000 |
| 4 | 185,000 |
| 5 | 34,400 |
| 6 | 5,200 |
| 7 | 33,300 |
| 8 | 21,000 |
| 9 | 3,740,000 |
| 10 | 259,000 |

Why do these data qualify as a one-sample problem? What population is represented?

2.10 Not surprisingly, people who jump out of airplanes flying at 10,000 ft undergo rapid increases in their heart rates. Eleven Navy parachutists were electronically monitored as they made such a jump. Among the measurements recorded were their maximum heart rates (133):

| Subject | Maximum Heart Rate (beats/min) |
|---------|-------------------------------|
| HWP | 160 |
| SCK | 184 |
| SR | 173 |
| TSF | 176 |
| KMM | 168 |
| BHO | 156 |
| RHR | 160 |
| KAW | 160 |
| REA | 174 |
| JGW | 140 |
| AWB | 166 |

Explain the distinction between a sample and a population as it applies to these data.

## Bernoulli Variables

The $y_i$'s that we have seen thus far (the airflow rates in Case Study 2.1 and the scientists' ages in Case Study 2.2) could have taken on many possible values. In Table 2.2, for example, the entries range from 0.61 to 0.88; the ages in Table 2.3 vary from 23 to 49. Not all variables have that property. In some problems each $y_i$ has only two possible values, 0 or 1. For example, a pollster trying to estimate the popularity of a political candidate might interview $n$ people and record $y_1, y_2, \ldots, y_n$, where $y_i$ is the number of votes person $i$ intends to cast for the candidate. Clearly, each $y_i$ must be

0 or 1:

$$y_i = \begin{cases} 1 & \text{if person } i \text{ does support candidate} \\ 0 & \text{if person } i \text{ does not support candidate} \end{cases} \tag{2.1}$$

**Bernoulli variable**—A variable that *for each subject* can have only one of two possible values, 0 or 1.

Measurements having the 0–1 structure of Equation 2.1 are called ***Bernoulli variables***. We will see them often.

---

### CASE STUDY 2.3

*Flower Pollination*

Nature has given flowers that are pollinated by bees various biological mechanisms for insuring that the job gets done properly. Many, for example, have strongly scented pollen; others are brightly colored. What might be another device, one certainly not so obvious, is the set of radiating lines sometimes found on petals. It has been suggested that these lines may serve as "nectar guides," leading bees from the outside of the flower to the inside where the pollen is stored. Bernoulli variables figured prominently in an experiment designed to test that theory (53).

First, a group of honeybees was trained to feed off a special table. The top of the table consisted of two panes of glass; pressed between the panes were some irregularly shaped paper flowers approximately 3 cm in diameter (Figure 2.4a). Drops of sugar syrup were placed on the tabletop above each paper flower. Released near the table, the bees ate the syrup, presumably associating the food with the paper flowers.

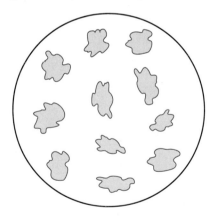 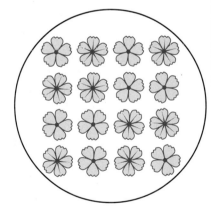

(a) Training Table        (b) Test Table

**FIG. 2.4**

After the bees were accustomed to the training table, the experimenter removed the syrup and replaced the original flowers with a grid of 16 new ones, eight with lines and eight without (Figure 2.4b). Reintroduced to the area, the bees were now carefully watched, and the number landing on each pattern was counted. When the buzzing subsided, the researcher's tally showed that 107 bees had landed on the test table, 66 on flowers with lines and 41 on flowers without lines.

Does the "66 versus 41" imply that bees do have a genuine affinity for flowers with lines? Maybe. If bees are not influenced by nectar guides, we would expect approximately half of the 107 landings to have been on flowers with lines and half on flowers without lines (since the test table had equal numbers of each). What we observed was that 66 out of 107, or 62%, of the landings were on flowers with lines. Are the 12% additional "line" landings a large enough increase to support the nectar guide theory, or is 12% a small enough number that we can write if off to chance? We will find out in Chapter 8.

In problems involving Bernoulli variables, like this one, the underlying structure of the data can sometimes be misread. Keep in mind that we need to focus on the $y_i$'s actually recorded, not on quantities computed from those $y_i$'s. Here, for example, the "data" are not the fact that 66 out of 107 bees landed on flowers with lines. What was actually recorded by the experimenter was a sample of size 107, where each $y_i$ was either 0 or 1:

$$y_i = \begin{cases} 1 & \text{if } i\text{th bee lands on flower with lines} \\ 0 & \text{if } i\text{th bee lands on flower without lines} \end{cases}$$

(Adding the 107 $y_i$'s together gives the value 66; that is, $\sum_{i=1}^{107} y_i = 66$.) Viewed in this context, the experiment clearly belongs to the one-sample model. The (single) population being represented by the 107 $y_i$'s is the set of 0's and 1's associated with all bees who might be trained and then tested in this same fashion.

---

**QUESTIONS**

**2.11**   Notice that the flowers with lines in Figure 2.4a are positioned in such a way that exactly two appear in each row and in each column of the test table. Why did the experimenter do that? (Hint: Suppose you were doing agricultural research that involved the comparison of two varieties of wheat. To get some data, you subdivide a large field into a 4 × 4 grid analogous to Figure 2.4b with the intention of planting one variety in eight of the "plots" and the second variety in the remaining eight. Why would it *not* be a good idea to assign the varieties according to the pattern shown below:

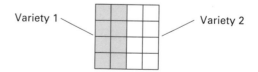

**2.12**   In certain countries, handwriting analysis is widely used for assessing an individual's personality traits. One way to validate the technique is to demonstrate that an expert can distinguish one personality type from another, solely on the basis of the way an individual dots his $i$'s, crosses his $t$'s, and so on. In one such study a handwriting expert was given 10 folders, each containing handwriting samples of two persons, one "normal" and the other schizophrenic. Her objective was to indicate which was which. Listed in the table are the results (112):

| Folder No. | Made Correct Identification? |
|------------|------------------------------|
| 1 | Yes |
| 2 | Yes |
| 3 | No |
| 4 | Yes |
| 5 | No |
| 6 | No |
| 7 | Yes |
| 8 | Yes |
| 9 | Yes |
| 10 | No |

(a) Define the $y_i$'s associated with these 10 observations.

(b) What population is represented here?

(c) Intuitively, does it appear that the expert has demonstrated the ability to distinguish a normal person's writing from that of a schizophrenic?

2.13 Which professional football conference is better, the NFC or the AFC? In the first 20 Super Bowls, the NFC won 8 and the AFC won 12:

| Super Bowl | Score | Winner |
|------------|-------|--------|
| I | Green Bay 35, Kansas City 10 | NFC |
| II | Green Bay 33, Oakland 14 | NFC |
| III | New York 16, Baltimore 7 | AFC |
| IV | Kansas City 23, Minnesota 7 | AFC |
| V | Baltimore 16, Dallas 13 | AFC |
| VI | Dallas 24, Miami 3 | NFC |
| VII | Miami 14, Washington 7 | AFC |
| VIII | Miami 24, Minnesota 7 | AFC |
| IX | Pittsburg 16, Minnesota 6 | AFC |
| X | Pittsburg 21, Dallas 17 | AFC |
| XI | Oakland 32, Minnesota 14 | AFC |
| XII | Dallas 27, Denver 10 | NFC |
| XIII | Pittsburg 35, Dallas 31 | AFC |
| XIV | Pittsburg 31, LA Rams 19 | AFC |
| XV | Oakland 27, Philadelphia 10 | AFC |
| XVI | San Francisco 26, Cincinnati 21 | NFC |
| XVII | Washington 27, Miami 17 | NFC |
| XVIII | LA Raiders 38, Washington 9 | AFC |
| XIX | San Francisco 38, Miami 16 | NFC |
| XX | Chicago 46, New England 10 | NFC |

Expressing these results as a percentage, we could say that the NFC won 40% of all the championship games.

(a) Define a set of Bernoulli variables appropriate for these data. What does $y_9$ equal?

(b) Express the 40% in terms of the Bernoulli variables in part (a).

(c) What population is represented by these data?

2.14   Data that originally have a "non-Bernoulli" structure can always be reduced to a set of Bernoulli variables by defining $y_i$ to be 1 if the original $i$th observation exceeds a certain value and letting $y_i$ be 0 otherwise. For example, the $FEV_1/VC$ ratios in Table 2.2 could be *dichotomized* according to whether each one was less than the norm for persons not having any lung dysfunction:

$$y_i = \begin{cases} 1 & \text{if } i\text{th subject's } FEV_1/VC > 0.80 \\ 0 & \text{if } i\text{th subject's } FEV_1/VC \leqslant 0.80 \end{cases}$$

How might the radiation data in Question 2.7 be similarly reduced to a set of Bernoulli variables? (Hint: Look at Question 2.7(c)).

2.15   Do female cockroaches look for Mr. Right? A recent study (  ) focused on mating preferences exhibited by female cockroaches when the latter were each presented with four males from which to choose. Before vying for the affections of the lady in waiting, the four males in each group had their behavior carefully monitored, and the dominant member of the foursome was identified. A total of 39 matings occurred. The following tally was kept of whether the female chose the dominant male in a group:

| Mating Number | With Dominant Male? | Mating Number | With Dominant Male? |
|---|---|---|---|
| 1 | No | 21 | Yes |
| 2 | Yes | 22 | Yes |
| 3 | Yes | 23 | Yes |
| 4 | No | 24 | No |
| 5 | No | 25 | No |
| 6 | Yes | 26 | Yes |
| 7 | No | 27 | No |
| 8 | Yes | 28 | Yes |
| 9 | No | 29 | Yes |
| 10 | No | 30 | No |
| 11 | Yes | 31 | No |
| 12 | No | 32 | Yes |
| 13 | No | 33 | No |
| 14 | No | 34 | No |
| 15 | Yes | 35 | Yes |
| 16 | No | 36 | Yes |
| 17 | Yes | 37 | No |
| 18 | No | 38 | No |
| 19 | No | 39 | Yes |
| 20 | Yes | | |

(a) Define the $y_i$'s appropriate for this experiment.

(b) If dominance is not a trait valued by female cockroaches, how many of the 39 observations would we have expected to be a yes?

# 2.3

## TWO-SAMPLE DATA

two-sample data—Two sets of independent samples,

$$x_1, x_2, \ldots, x_n \quad \text{and} \quad y_1, y_2, \ldots, y_m$$

representing treatment X and treatment Y are observed. The objective is to use the $x_i$'s and $y_i$'s to compare the effects of the two treatments.

Most real-world experiments are more complicated than the one-sample format described in Section 2.2. Researchers must often compare several different treatments or several kinds of subjects (making it necessary to define several different populations). The simplest of these comparative models is the *two-sample problem*.

The data for a two-sample problem consist of two samples,

$$x_1, x_2, \ldots, x_n \quad \text{and} \quad y_1, y_2, \ldots, y_m$$

The two sample sizes, $n$ and $m$, are not necessarily the same. Two variations are possible: (1) the $x_i$'s and $y_i$'s are measurements taken on similar subjects receiving different treatments (Figure 2.5a), or (2) the $x_i$'s and $y_i$'s are measurements taken on dissimilar subjects observed under the same conditions (Figure 2.5b).

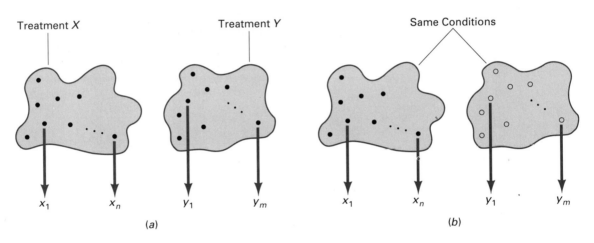

**FIG. 2.5**

An example of the first variation is comparing two different teaching techniques (method X and method Y) for preparing students to take the SAT. The $x_i$'s and $y_i$'s are the SAT scores earned by the participants in the two groups. An example of the second variation is a study designed to see whether men and women respond similarly to a new Coca-Cola advertising campaign. The "condition" is the advertisement; "men" and "women" are the treatments. The $x_i$'s (for men) and $y_i$'s (for women) might be scores on a questionnaire (On a scale of 1 to 10, how would you rate the commercial?), or they might be actual amounts of Coca-Cola purchased over a specific time.

### CASE STUDY 2.4

*Silver Percentage in Coins*     In 1965 a silver shortage in the United States prompted Congress to authorize the minting of silverless dimes and quarters. They also recommended that the silver content of half-dollars be reduced from 90% to 40%. Historically, fluctuations in the amount of rare metals found in coins are not uncommon. The data in Table 2.4 show a possible shift in the silver content of a Byzantine coin minted on two separate occasions during the reign of Manuel I (1143–1180).

**TABLE 2.4**

| Early Coinage, $x_i$ (% Ag) | Late Coinage, $y_i$ (% Ag) |
|---|---|
| 5.9 | 5.3 |
| 6.8 | 5.6 |
| 6.4 | 5.5 |
| 7.0 | 5.1 |
| 6.6 | 6.2 |
| 7.7 | 5.8 |
| 7.2 | 5.8 |
| 6.9 | |
| 6.2 | |
| Average: 6.7 | Average: 5.6 |

The 16 coins in the combined sample, 9 belonging to the earlier minting and 7 to the later, were part of a large cache discovered not long ago in Cyprus. The measurements in Table 2.4 were obtained by dissolving chips from a specimen in a 50% nitric acid solution, titrating that solution with sodium chloride until all the silver chloride precipitated out, and weighing the precipitate (71).

The two treatments in this study refer to *when* a coin was minted ("early" versus "late"). The sample of nine $x_i$'s represents the population of silver percentages in all the coins minted during the early years of Manuel's reign; the seven $y_i$'s represent a second population, the silver percentages in all the coins minted during the later years. These data belong to the first type of two-sample problem (Figure 2.5a).

Statistically, the objective here is to decide whether the average silver percentage *in the two populations* is different (the averages in the two *samples* are 6.7 for the $x_i$'s and 5.6 for the $y_i$'s). Once again, we need to rely on the principles of hypothesis testing to tell us whether the difference between 6.7 and 5.6 is (1) so large that it "proves" that the average silver percentage shifted during Manuel's reign or (2) so small that it could easily have happened by chance even if no population shift had occurred.

**CASE STUDY 2.5**

*First Walk for Baby*    Baby books are filled with "first" dates—when the baby first crawled, or first stood up, or said his or her first word. Surprisingly, our knowledge of these developmental stages is quite limited. We know their norms but little else. Most medical researchers have simply not pursued the various factors and conditions that may influence a child's early growth. The following study is an exception.

Zelazo and his colleagues (194) studied one of the most important of these firsts: when a baby first walks alone, typically around age 14 months. The Zelazo study hypothesizes that infants given a special series of exercises would walk much earlier.

Twelve one-week-old male infants from middle-class and upper-middle-class families were used as subjects. All were white. The birth order of the infants and the ages of the parents were kept as similar as possible to guard against possible biases.

Two groups, each of size 6, were randomly chosen. The first, or active exercise, group received four 3-minute sessions of special walking and placing exercises each day, beginning with the second week and lasting through the eighth week. Infants in the second, or passive exercise, group received the same overall amount of daily exercise but were not given the walking and placing training. This second group acted as a control. In each instance, the exercises were conducted in the child's home by the parents. After eight weeks the program was discontinued.

Table 2.5 shows the ages, in months, when the children first walked by themselves. This study is a two-sample problem of the first type—the subjects are all similar, but the treatments are different. The $x_i$'s represent the population of first walking times that would be generated if all children were given these special exercises; the $y_i$'s are a sample from the control population.

**TABLE 2.5 First Walking Times (months)**

| Active Exercise Group, $x_i$ | Passive Exercise Group, $y_i$ |
|---|---|
| 9.00 | 11.00 |
| 9.50 | 10.00 |
| 9.75 | 10.00 |
| 10.00 | 11.75 |
| 13.00 | 10.50 |
| 9.50 | 15.00 |

Are the exercises effective? Probably. We will have to wait until Chapter 10 for a more definitive answer, but there does appear to be evidence of a trend: the $x_i$'s, in general, have shifted to the left of the $y_i$'s. Children receiving the special exercises, in other words, tend to walk sooner.

QUESTIONS

**2.16** Children not receiving any special exercise are expected to begin walking on their own at age 14 months. Taking that as given, how would you account for the fact that five of the six infants in the control group developed markedly faster than the norm?

**2.17** A person exposed to an infectious agent, either by contact or by vaccination, normally develops antibodies to that agent. These antibodies are proteins carried in the blood that can inactivate the infectious agents. Presumably, the severity of an infection is related to the number of antibodies produced. The degree of antibody response is indicated by saying that the person's blood serum has a certain *titer*, with higher titers indicating greater concentration of antibodies. Listed below are the titers of 22 persons involved in a tularemia epidemic in Vermont. Eleven were quite ill, whereas 11 were asymptomatic (19).

| SEVERELY ILL | | ASYMPTOMATIC | |
|---|---|---|---|
| Subject | Titer | Subject | Titer |
| 1 | 640 | 12 | 10 |
| 2 | 80 | 13 | 320 |
| 3 | 1280 | 14 | 320 |
| 4 | 160 | 15 | 320 |
| 5 | 640 | 16 | 320 |
| 6 | 640 | 17 | 80 |
| 7 | 1280 | 18 | 160 |
| 8 | 640 | 19 | 10 |
| 9 | 160 | 20 | 640 |
| 10 | 320 | 21 | 160 |
| 11 | 160 | 22 | 320 |

(a) Define what $x_i$ and $y_i$ represent for these data.

(b) Is this a two-sample problem of the first type or the second type?

(c) Do the magnitudes of the observations support the general statement that titer levels and disease severity are related?

| Nonfilter, $x_i$ | Charcoal Filter, $y_i$ |
|---|---|
| 8 | 21 |
| 7 | 37 |
| 11 | 24 |
| 8 | 27 |
| 9 | 19 |
| 8 | 14 |

**2.18** The effectiveness of charcoal filters on cigarettes was investigated with an experiment involving protozoa (180). *Paramecium aurelia* were suspended in a hanging drop inside a smoke chamber. Every 60 s a 6-s puff of smoke was drawn through the chamber. The movements of the paramecia were watched through a stereomicroscope. The variable recorded was the length of time from the start of the experiment to when the last paramecium died. The experiment was replicated 12 times. On six of those occasions the smoke came from a nonfilter cigarette; on the other six occasions the smoke came from a charcoal filter cigarette. The two sets of survival times (in minutes) are shown in the accompanying table on the left.

(a) Briefly describe the populations represented by the $x_i$'s and $y_i$'s.

(b) Would it have made sense to do this experiment using the one-sample structure—that is, collect data using only the charcoal filter? Explain.

2.19 Because of its association with a variety of serious disorders, caffeine is a closely monitored food additive. The measurements in the accompanying table were made using high-performance liquid chromatography. They show the caffeine content (grams per 100 grams of dry matter) in 12 brands of instant coffee, 8 spray-dried and 4 freeze-dried (167).

| Spray-dried | Freeze-dried |
|-------------|--------------|
| 4.8 | 3.7 |
| 4.0 | 3.4 |
| 3.8 | 2.8 |
| 4.3 | 3.7 |
| 3.9 | |
| 4.6 | |
| 3.1 | |
| 3.7 | |

(a) Define $x_i$ and $y_i$. What populations do they represent?

(b) Is this a two-sample problem of the first type or the second type?

(c) From what we have seen thus far, what is the first numerical step that should be taken in comparing the caffeine contents of spray-dried and freeze-dried coffee?

2.20 In forensic medicine or after a bad accident, identifying the sex of a body can be quite difficult. In some cases dental structure is a useful criterion, because teeth remain in good condition long after other tissues have deteriorated. Furthermore, studies have shown that female teeth and male teeth have different physical and chemical characteristics. X-ray penetration of tooth enamel, for instance, differs for men and women. Listed in the table in the left margin are spectropenetration gradients for eight female teeth and eight male teeth (54). These numbers are measures of the rate of change in the amount of x-ray penetration through a 500-micron section of tooth enamel, first at a wavelength of 600 nanometers (nm) and then at 400 nm.

| Male, $x_i$ | Female, $y_i$ |
|-------------|---------------|
| 4.9 | 4.8 |
| 5.4 | 5.3 |
| 5.0 | 3.7 |
| 5.5 | 4.1 |
| 5.4 | 5.6 |
| 6.6 | 4.0 |
| 6.3 | 3.6 |
| 4.3 | 5.0 |

(a) Is this a two-sample problem of the first type or the second type?

(b) Ideally, what numerical characteristics would you like the $x_i$'s and $y_i$'s to have, assuming you intend to use spectropenetration gradients for identification purposes?

## Two-Sample Problems with Bernoulli Variables

It is not uncommon in two-sample problems for the $x_i$'s and $y_i$'s to be Bernoulli variables. Case Study 2.6 shows what that type of data look like.

## CASE STUDY 2.6

*Diet and Arteriosclerosis*     To test whether a controlled diet can retard the process of arteriosclerosis, researchers monitored 846 randomly chosen persons for eight years. Half were instructed to eat only certain foods; the other half could eat whatever they wanted. At the end of eight years the numbers of individuals in each group who had died of myocardial or cerebral infarctions were tallied. The results are shown in Table 2.6 (183).

**TABLE 2.6**

|  | Diet Group | Control Group |
|---|---|---|
| Died of infarction | 66 | 93 |
| Did not die of infarction | 357 | 330 |
| Total | 423 | 423 |
| Deaths due to infarction | 15.6% | 22.0% |

Notice that Table 2.6 summarizes the information collected; it does not list the original data as Tables 2.4 and 2.5 did. Here, the actual data consisted of two different samples of Bernoulli variables, each of size 423. If we let

$$x_i = \begin{cases} 1 & \text{if } i\text{th person in diet group died of infarction} \\ 0 & \text{if } i\text{th person in diet group did not die of infarction} \end{cases}$$

and

$$y_i = \begin{cases} 1 & \text{if } i\text{th person in control group died of infarction} \\ 0 & \text{if } i\text{th person in control group did not die of infarction} \end{cases}$$

then

$$\sum_{i=1}^{423} x_i = \text{total number in diet group who died of infarction}$$

$$= 66$$

and

$$\sum_{i=1}^{423} y_i = \text{total number in control group who died of infarction}$$

$$= 93$$

(Figure 2.6).

Based on the two previous examples, the analysis appropriate for the data of Table 2.6 should be clear. If diet has no effect on a person's chances of dying from an infarction, we would expect the two percentages listed in the bottom row of the table to be "similar." Do 15.6 and 22.0 fall into that category? Perhaps. Until we learn how to do the kind of formal analysis

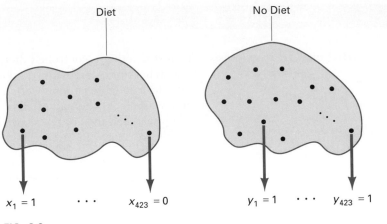

**FIG. 2.6**

introduced in Chapter 10, any attempt to draw an inference on the basis of the similarity, or dissimilarity, between 15.6 and 22.0 would be nothing but pure speculation.

QUESTIONS

**2.21** Intuitively, what role do you think sample sizes play in making inferences about observed differences between proportions? In Table 2.6, the 15.6 and 22.0 were based on sample sizes of 423 and 423, respectively. Might our interpretation of those figures change if they had each been based on 200 observations? 2000 observations?

**2.22** Surveys traditionally show that men and women in the marketplace, when performing similar kinds of work, derive different levels of job satisfaction. (Among the explanations offered for why women feel more fulfilled is that they tend to have lower expectations.) In one study 18 female employees and 29 male employees of a moderate-sized manufacturing plant were asked to indicate whether their job satisfaction was high or low. Their responses are listed below (192).

| Female | | | Male | | | |
|---|---|---|---|---|---|---|
| High | High | High | Low | High | High | Low |
| High | Low | High | High | Low | Low | Low |
| Low | High | High | High | Low | High | High |
| High | Low | High | Low | Low | Low | High |
| High | High | Low | High | Low | Low | Low |
| Low | Low | High | Low | High | High | Low |
| | | | Low | Low | Low | Low |
| | | | High | | | |

(a) Define $x_i$ and $y_i$ for these data. What populations do they represent?

(b) Summarize these data in a table similar to the one in Case Study 2.6.

(c) Compute the percentage of females and the percentage of males who report a high level of job satisfaction.

2.23 Water witching, the practice of using the movements of a forked twig to locate underground water (or minerals), dates back over 400 years. Its first detailed description appears in Agricola's *De re Metallica*, published in 1556. That water witching works remains a belief widely held among rural people in Europe and throughout the Americas. (In 1960 the number of "active" water witches in the United States was estimated to be more than 20,000 (174).) Reliable evidence supporting or refuting water witching is hard to find. Personal accounts of isolated successes or failures tend to be strongly biased by the attitude of the observer. The data below show the outcomes of all the wells dug in Fence Lake, New Mexico, where 29 "witched" wells and 32 "nonwitched" wells were sunk. Recorded for each well was whether it proved to be successful (S) or unsuccessful (U).

| Witched Wells | | | | |
|---|---|---|---|---|
| S | S | S | S | U |
| U | S | S | S | S |
| S | U | S | U | S |
| S | S | S | S | S |
| S | S | S | S | S |
| U | S | S | S | |

| Nonwitched Wells | | | | |
|---|---|---|---|---|
| U | S | S | S | S |
| S | S | S | S | S |
| S | U | S | S | S |
| S | U | S | S | S |
| S | S | S | U | S |
| U | S | S | S | S |
| S | S | | | |

(a) Define $x_i$ and $y_i$ and describe the populations they represent.

(b) Carry out the necessary summation to put the data into a tabular format.

(c) From these data does it appear that witching helps to locate potentially successful wells?

2.24 If flying saucers are a genuine phenomena, it would follow that the nature of sightings (that is, their physical characteristics) should be similar in different parts of the world. A prominent UFO investigator compiled a listing of 91 sightings reported in Spain and 1117 reported elsewhere. Among the information recorded was whether the saucer was on the ground or hovering. His data are summarized below (178).

| | In Spain | Not in Spain |
|---|---|---|
| Saucer on ground | 53 | 705 |
| Saucer hovering | 38 | 412 |

(a) Define $x_i$ and $y_i$ for these data. What populations do they represent? What are the sample sizes?

(b) How might the nature of the sightings in a given area be reduced to a single number?

2.25   Suicide rates in the United States tend to be much higher for men than for women, at all ages. That pattern, however, may not extend to all professions. Death certificates obtained for the 3637 members of the American Chemical Society who died between April 1948 and July 1967 revealed that 106 of the 3522 male deaths were suicides and 13 of the 115 female deaths were suicides (76).

(a) Clarify the structure of these data by defining $x_i$ and $y_i$. What did the data look like as they were being recorded?

(b) Express these numbers as entries in a table. What percentages need to be computed as a first step in analyzing the results?

# 2.4

## PAIRED DATA

**independent data**—Two variables, x and y, are independent if the value of one does not help to predict the value of the others.

**dependent (data)**—Two variables, x and y, are dependent if the conditions under which $x_i$ is measured are deliberately "matched" to be similar to the conditions under which $y_i$ is measured.

**paired data**—Two dependent observations, representing treatment X and treatment Y, are observed on each of n pairs. The set of within-pair response differences, $d_1 = y_1 - x_1,$   $d_2 = y_2 - x_2,$   ..., $d_n = y_n - x_n$ becomes the basis for comparing the treatments.

Implicit in the two-sample model is the assumption that the $x_i$'s and $y_i$'s are *independent*. In other words, knowing the value of $x_i$ doesn't help us predict the value of $y_i$. Look back at Table 2.5. The value recorded for $x_1$ was 9.00. Does that figure give us any clue that $y_1$ is 11.00? No.

The details will be deferred until Chapter 11, but there are sometimes compelling reasons to deliberately choose the subjects so that $x_1$ and $y_1$ are related, $x_2$ and $y_2$ are related, and so on. Samples with this property are *dependent* or *paired*.

Figure 2.7 shows schematically how pairing is done. We try to match up, one-on-one, each subject scheduled to be given treatment X with a "similar" subject slated for treatment Y. (What "similar" means depends on the context.) Later, when the results are analyzed, we compute the within-pair response differences,

$$d_i = y_i - x_i, \qquad i = 1, 2, \ldots, n$$

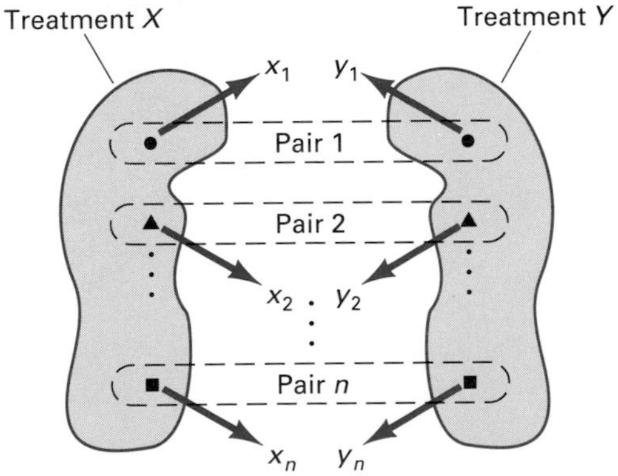

**FIG. 2.7**

and use those numbers rather than the original $x_i$'s and $y_i$'s to compare the two treatments. (If the treatments are comparable, we expect the $d_i$'s, on the average, to be zero.)

Paired-data experiments have exactly the same objective as two-sample experiments: both are trying to compare the effect of treatment $X$ with the effect of treatment $Y$. *How* they do that is where the experiments differ. The crucial distinction is in the way the data are collected. Look closely at the structure of the next two case studies. Notice how the relationship between the $x_i$'s and $y_i$'s differs from what was true in Section 2.3, where the (independent) $x_i$'s and $y_i$'s were collected according to the two-sample format.

## CASE STUDY 2.7

**Running Base Paths**

Baseball rules allow a batter considerable leeway in how he is permitted to run from home plate to second base. Two possibilities are the narrow-angle and wide-angle paths diagrammed in Figure 2.8. To compare the two paths, investigators held time trials with 22 players (185). *Each player ran both paths.* Recorded for each runner was the time it took to go from a point 35 feet from home plate to a point 15 ft from second base.

The original data are shown in the second and third columns of Table 2.7. The two paths are the treatments. The 22 players, each having run both paths, are acting as 22 "pairs." Can we presume that a given $x_i$ and $y_i$ are dependent?

Definitely. If someone is fast (player 14), both of his times will tend to be low ($x_{14} = 5.00$, $y_{14} = 4.95$). Conversely, someone slow (player 22) will have trouble getting from home to second no matter how he rounds first base ($x_{22} = 6.30$, $y_{22} = 6.25$).

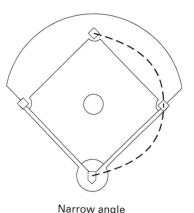

Narrow angle

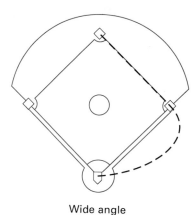

Wide angle

**FIG. 2.8**

**TABLE 2.7  Times (s) Required to Round First Base**

| Player | Narrow-Angle, $x_i$ | Wide-Angle, $y_i$ | $d_i = y_i - x_i$ |
|--------|--------------------|--------------------|--------------------|
| 1  | 5.50 | 5.55 | 0.05 |
| 2  | 5.70 | 5.75 | 0.05 |
| 3  | 5.60 | 5.50 | $-0.10$ |
| 4  | 5.50 | 5.40 | $-0.10$ |
| 5  | 5.85 | 5.70 | $-0.15$ |
| 6  | 5.55 | 5.60 | 0.05 |
| 7  | 5.40 | 5.35 | $-0.05$ |
| 8  | 5.50 | 5.35 | $-0.15$ |
| 9  | 5.15 | 5.00 | $-0.15$ |
| 10 | 5.80 | 5.70 | $-0.10$ |
| 11 | 5.20 | 5.10 | $-0.10$ |
| 12 | 5.55 | 5.45 | $-0.10$ |
| 13 | 5.35 | 5.45 | 0.10 |
| 14 | 5.00 | 4.95 | $-0.05$ |
| 15 | 5.50 | 5.40 | $-0.10$ |
| 16 | 5.55 | 5.50 | $-0.05$ |
| 17 | 5.55 | 5.35 | $-0.20$ |
| 18 | 5.50 | 5.55 | 0.05 |
| 19 | 5.45 | 5.25 | $-0.20$ |
| 20 | 5.60 | 5.40 | $-0.20$ |
| 21 | 5.65 | 5.55 | $-0.10$ |
| 22 | 6.30 | 6.25 | $-0.05$ |

The fourth column lists the within-pair response differences, $d_i = y_i - x_i$. Most of these numbers are negative, suggesting that runners take more time to get to second base when they use the narrow-angle path. If that impression were supported by a formal hypothesis test it would make sense for a coach to recommend the wide-angle path.

## CASE STUDY 2.8

*Treating Fear of Heights*    Acrophobia (fear of heights) can be treated by *contact desensitization* or by *demonstration participation*. Using the first method, the therapist demonstrates a task that is difficult to do for someone with acrophobia, such as looking over a ledge or standing on a ladder. The therapist then guides the subject through the task while maintaining physical contact. In the second treatment method the therapist tries to talk the subject through the task without touching him.

These two techniques were compared in a study involving 10 volunteers, all of whom had severe acrophobia (137). Investigators realized at the outset, however, that the affliction was much more incapacitating in some subjects than in others, and that this heterogeneity might compromise the therapy comparison. To prevent that from happening, data were collected according to a paired-data format, where the pairs were defined by the severity of a subject's acrophobia. Each subject was given the height avoidance test (HAT), a series of 44 tasks related to ladder climbing. "Points" were received for each task successfully completed. On the basis of their final scores, the 10 volunteers were divided into five pairs (A, B, C, D, and E). The two subjects in pair A had the lowest scores (that is, the most severe acrophobia), those in pair B had the second lowest scores, and so on.

Each therapy was then assigned at random to one of the two subjects in each pair. When the counseling sessions were over, the subjects retook the HAT. Table 2.8 lists the *changes* in their scores (score after therapy − score before therapy).

**TABLE 2.8 HAT Score Changes**

| Pair | Contact Desensitization, $x_i$ | Demonstration Participation, $y_i$ | $d_i = y_i - x_i$ |
|------|-------------------------------|-----------------------------------|-------------------|
| A | 8 | 2 | −6 |
| B | 11 | 1 | −10 |
| C | 9 | 12 | 3 |
| D | 16 | 11 | −5 |
| E | 24 | 19 | −5 |

With paired data we focus on the $d_i$'s rather than the original $x_i$'s and $y_i$'s. For this experiment the $d_i$'s represent the population of differences in HAT score improvements, where a difference is defined to be a subject's demonstration participation improvement ($y_i$) *minus* his or her contact desensitization improvement ($x_i$). If the two treatments were equally effective, we would expect some $d_i$'s to be positive, some to be negative, and their average to be close to 0. The last column of Table 2.8 shows that these $d_i$'s tend to be negative, implying that contact desensitization is a more effective therapy than demonstration participation.

Notice that the self-pairing approach that worked so well in Case Study 2.7 would not be a good idea here, since the first therapy given to a subject would bias our estimate of the effectiveness of the second. As an alternative strategy, the experimenter has quite properly turned to a pretest as a means of identifying *pairs* of subjects having similar severities of acrophobia.

QUESTIONS

2.26 How would the two baserunning paths described in Case Study 2.7 be compared if the experimenter carried out the study with a two-sample model? Which strategy—paired data or independent data—seems more reasonable?

2.27 Every automobile, foreign or domestic, sold as a new car is assigned an EPA mileage estimate based on the car's gas consumption as measured by a laboratory dynamometer that simulates actual driving conditions. Are those estimates accurate? Maybe not. The following set of data shows a comparison of EPA estimates with actual mileages obtained by professional drivers (112).

| Car | EPA, $x_i$ | Actual, $y_i$ | $d_i = y_i - x_i$ |
|---|---|---|---|
| AMC Pacer | 17 | 14 | $-3$ |
| Buick Skyhawk | 19 | 15 | $-4$ |
| Cadillac Seville | 13 | 11 | $-2$ |
| Chevrolet Vega | 19 | 17 | $-2$ |
| Datsun B-210 | 27 | 27 | 0 |
| Datsun 610 | 22 | 22 | 0 |
| Dodge Charger | 13 | 10 | $-3$ |
| Dodge Colt | 20 | 20 | 0 |
| Dodge Dart | 17 | 13 | $-4$ |
| Fiat 128 | 20 | 19 | $-1$ |
| Ford Granada | 12 | 11 | $-1$ |
| Ford Mustang | 18 | 15 | $-3$ |
| Ford Pinto | 18 | 16 | $-2$ |
| Lincoln Continental | 10 | 8 | $-2$ |
| Mercury Comet | 16 | 13 | $-3$ |
| Oldsmobile Cutlass | 15 | 13 | $-2$ |
| Plymouth Fury | 11 | 10 | $-1$ |
| Pontiac Grand LeMans | 12 | 10 | $-2$ |
| Toyota Corolla | 21 | 20 | $-1$ |
| Volkswagen Rabbit | 24 | 24 | 0 |

(a) What population do the $d_i$'s represent?

(b) Look down the column of $d_i$'s. It seems quite obvious that the EPA estimates, in general, are higher than the actual mileages. How can the magnitude of that discrepancy be estimated?

2.28 Dust bathing is a common activity among birds living on the ground. Its function, though, and the factors that trigger or inhibit it are a mystery. Whether light plays a role was investigated in an experiment with domestic Japanese quail. Five birds were kept uncaged in a windowless room. The floor had a layer of wood chips and sawdust. Each bird was watched under two different lighting conditions. For condition A the room was illuminated 14 hr a day; for condition B, 10 hr a day. Shown in the table on the left is the average percentage of daylight hours that each bird spent dust bathing (159).

| Bird | Condition A | Condition B |
|---|---|---|
| 1 | 2.3 | 5.8 |
| 2 | 4.1 | 4.5 |
| 3 | 3.6 | 3.2 |
| 4 | 3.5 | 3.0 |
| 5 | 2.6 | 2.8 |

(a) Compute the "within-bird" response differences. Does a quail's tendency to dust bathe seem much affected by whether it receives 14 hr of light a day or 10?

(b) How would an experimenter compare conditions A and B within the framework of a two-sample model?

| Location | 6TA-204 | Rosner |
|----------|---------|--------|
| El Centro | 16.2 | 16.6 |
| Five Points | 15.6 | 16.7 |
| Davis | 16.9 | 16.8 |
| Tulelake | 15.9 | 16.7 |

**2.29**  Triticale is a hybrid cereal grain derived from wheat and rye. Only recently cultivated, its nutritional properties are largely unknown. Among the efforts to learn more about triticale was an experiment that compared the protein content in two of its varieties, 6TA-204 and Rosner. Both were grown at four widely separated sites in central California. The table on the left shows the average protein content (dry-weight percentage) measured for the eight triticale crops (142).

(a) Compute the $d_i$'s for these data. What population do they represent?

(b) Why are these data considered to be paired?

**2.30**  Two manufacturing processes are available for annealing copper tubing, the primary difference being in the temperature required. The critical response variable is the resulting tensile strength. In an experiment to compare the methods, 15 pieces of tubing were cut into pairs. One piece from each pair was randomly selected to be annealed at a moderate temperature, the other piece at a high temperature. The two tensile strengths (in tons/in$^2$) for each pair are tabulated below.

| Pair | Moderate Temperature | High Temperature |
|------|---------------------|------------------|
| 1 | 16.5 | 16.9 |
| 2 | 17.6 | 17.2 |
| 3 | 16.9 | 17.0 |
| 4 | 15.8 | 16.1 |
| 5 | 18.4 | 18.2 |
| 6 | 17.5 | 17.7 |
| 7 | 17.6 | 17.9 |
| 8 | 16.1 | 16.0 |
| 9 | 16.8 | 17.3 |
| 10 | 15.8 | 16.1 |
| 11 | 16.8 | 16.5 |
| 12 | 17.3 | 17.6 |
| 13 | 18.1 | 18.4 |
| 14 | 17.9 | 17.2 |
| 15 | 16.4 | 16.5 |

(a) Define $x_i$, $y_i$, and $d_i$ for these data.

(b) In general, what would you infer if the $d_i$'s for a set of data were all close to zero but the $x_i$'s and $y_i$'s took on a wide range of values?

# 2.5

## REGRESSION–CORRELATION DATA

"Comparison" was the watchword for the two-sample and paired-data problems. Whether the $x_i$'s and $y_i$'s were independent or dependent, the question being asked was the same: Is the population associated with treatment $X$ the same as the population associated with treatment $Y$? In this section we look at data having a much different structure, and we use them to answer a different set of questions.

Suppose that each of $n$ subjects is measured for two dissimilar traits. For example, the $X$ trait might be a person's systolic blood pressure, and the $Y$ trait that person's serum cholesterol level. Part of the data might look like Table 2.9. Notice that the units of these two measurements are different, so it makes no sense to compare the average value of the $x_i$'s with the average value of the $y_i$'s. On the other hand, it is reasonable to ask whether there is any functional relationship between the variables. Do people with high blood pressures, for example, also tend to have high cholesterol levels?

**TABLE 2.9**

| Name | Blood Pressure (mm Hg), $x_i$ | Cholesterol Level (mg/100 mL), $y_i$ |
|---|---|---|
| DF | 120 | 204.6 |
| ML | 118 | 184.4 |
| CW | 130 | 216.3 |
| EJ | 110 | 174.7 |

**regression–correlation data**—Two (dissimilar) measurements, $x$ and $y$, are taken on each of $n$ subjects. The objective is to quantify the relationship evidenced by the resulting set of $n$ points,

$$(x_1, y_1), \quad (x_2, y_2), \quad \ldots, \quad (x_n, y_n)$$

**linear regression**—A kind of regression–correlation data where the relationship between the $x$ and $y$ measurements can be satisfactorily approximated by a function of the form $y = a + bx$.

**nonlinear regression**—Regression–correlation data that can *not* be adequately approximated by a function of the form $y = a + bx$.

Quantifying the way in which two or more dissimilar variables are related is the primary objective of the ***regression–correlation problem***. Depending on how many variables are involved and what assumptions are made, there are many variations of this particular model but we will defer those subtleties. Figure 2.9 shows the only version we need to be aware of until we get to Chapter 12: two (dissimilar) measurements, $x_i$ and $y_i$, are taken on each of $n$ subjects. The responses of subject $i$ are denoted $(x_i, y_i)$, so the entire sample can be written

$$(x_1, y_1), \quad (x_2, y_2), \quad \ldots, \quad (x_n, y_n)$$

Usually the first step in analyzing regression–correlation data is finding the equation of the line or curve that "best" describes the relationship between the two variables. The case studies concluding this section show three examples of these equations. The first two describe ***linear regressions***; the third, a ***nonlinear regression***.

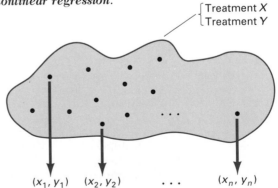

**FIG. 2.9**

*Radioactive Contamination*

The oil embargo of 1973 raised some very serious questions about energy policies in the United States. One of the most controversial is whether nuclear reactors should assume a more prominent role in the production of electric power. Those in favor point to the efficiency of the reactors and to the availability of nuclear material; those against warn of nuclear incidents and emphasize the health hazards of low-level radiation.

Since nuclear power is relatively new, there is not an abundance of past experience on which to draw. One notable exception though, is a West Coast reactor that has been in continuous operation for 30 years. What happened there is what environmentalists fear will be a recurrent problem if the role of nuclear reactors is expanded.

Since World War II, plutonium for use in atomic weapons has been produced at a government facility in Hanford, Washington. One of the major safety problems encountered there has been the storage of radioactive wastes. Over the years, significant quantities of radioactive wastes, including strontium 90 and cesium 137, have leaked from their open-pit storage areas into the nearby Columbia River, which flows along the Washington-Oregon border and eventually empties into the Pacific Ocean.

To measure the health consequences of this contamination, experimenters calculated an index of exposure for each of the nine Oregon counties having frontage on either the Columbia River or the Pacific Ocean. Among the factors included in the index were the county's stream distance from Hanford and the average distance of the county's population from any water frontage. As a second (dissimilar) variable, the cancer mortality rate was determined for each of the counties.

Table 2.10 shows the index of exposure and the cancer mortality rate (deaths per 100,000) for the nine Oregon counties affected (44). Higher index values represent higher levels of contamination.

A simple graph of the data (Figure 2.10) suggests that mortality rate

**FIG. 2.10**

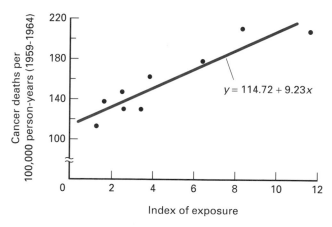

$y = 114.72 + 9.23x$

**TABLE 2.10**

| County | Index of Exposure, $x_i$ | Cancer Mortality per 100,000, $y_i$ |
|---|---|---|
| Umatilla | 2.49 | 147.1 |
| Morrow | 2.57 | 130.1 |
| Gilliam | 3.41 | 129.9 |
| Sherman | 1.25 | 113.5 |
| Wasco | 1.62 | 137.5 |
| Hood River | 3.83 | 162.3 |
| Portland | 11.64 | 207.5 |
| Columbia | 6.41 | 177.9 |
| Clatsop | 8.34 | 210.3 |

($y$) and index of exposure ($x$) vary *linearly*; that is, the relationship between $x$ and $y$ can be approximated by a function of the form $y = a + bx$. Finding the numerical values for $a$ and $b$ that position the line so it best fits the data is an important problem in an area of statistics known as *regression analysis*. Here, the optimal line, based on methods described in Chapter 3, has the equation $y = 114.72 + 9.23x$.

## CASE STUDY 2.10

*The Expanding Universe*   One of the most startling and profound scientific revelations of the twentieth century was the discovery in 1929 by the American astronomer Edwin Hubble that the universe is expanding. Hubble's announcement shattered forever the ancient belief that the heavens are basically in a state of cosmic equilibrium. Actually galaxies are receding from each other at fantastic velocities. If $y$ is a galaxy's recession velocity relative to any other galaxy and $x$ is its distance from that other galaxy, Hubble's law states that

$$y = bx \qquad (2.2)$$

where $b$ is known as *Hubble's constant*. (The $a$ in Case Study 2.9 has no relevance in Equation 2.2. You may recall from high school algebra that in straight-line equations $a$ is the $y$ intercept (the value of $y$ when $x = 0$). Here, according to the Big Bang hypothesis, both $x$ and $y$ were zero when the universe began. Therefore, $a$ must be zero.)

Table 2.11 gives distance and velocity measurements made on 11 galactic clusters (26). A graph of $y_i$ versus $x_i$ shows that the data agree very well with the linear relationship predicted by Equation 2.2 (Figure 2.11). The dashed line on the graph is the one that "best" fits the 11 points: it

**TABLE 2.11**

| Cluster | Distance, $x_i$ (millions of light years) | Velocity, $y_i$ (thousands of miles/s) |
|---|---|---|
| Virgo | 22 | 0.75 |
| Pegasus | 68 | 2.4 |
| Perseus | 108 | 3.2 |
| Coma Berenices | 137 | 4.7 |
| Ursa Major No. 1 | 255 | 9.3 |
| Leo | 315 | 12.0 |
| Corona Borealis | 390 | 13.4 |
| Gemini | 405 | 14.4 |
| Boötes | 685 | 24.5 |
| Ursa Major No. 2 | 700 | 26.0 |
| Hydra | 1100 | 38.0 |

**FIG. 2.11**

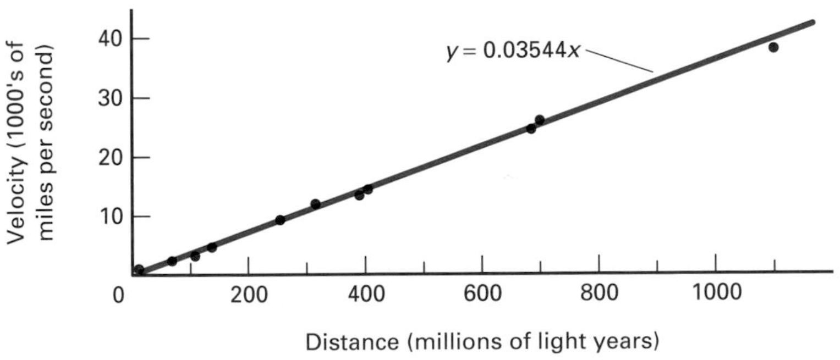

has the equation $y = 0.03544x$. (Hubble's constant, estimated from these data to be 0.03544, is an extremely significant number to cosmologists. When we take its reciprocal and convert its units to time, it becomes an estimate of the age of the universe.)

## CASE STUDY 2.11

*Nonlinear Relationships*  Most of the regression–correlation data that we will see involve linear relationships, which can be described by the equations $y = bx$ or $y = a + bx$. Exceptions certainly occur, one of the most common being relationships of the form

$$y = ax^b \tag{2.3}$$

Depending on the value of $b$, data following Equation 2.3 can exhibit a variety of patterns (Figure 2.12). In this study we look at a nonlinear relationship for which $b$ is between 0 and 1.

Among mammals the relationship between the age at which an animal develops locomotion and the age at which it begins to play has been widely studied. Listed in Table 2.12 are typical onset times for locomotion and for play in 11 different species (26). When graphed, the data show a pattern for which $y = ax^b$ appears to be an appropriate model (Figure 2.13). Using

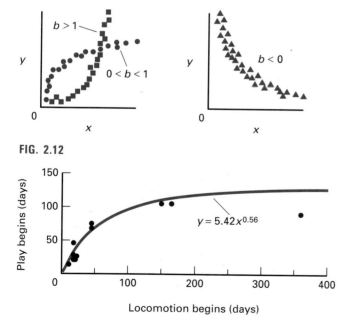

**FIG. 2.12**

**FIG. 2.13**

**TABLE 2.12**

| Species | Locomotion Begins, $x_i$ (days) | Play Begins, $y_i$ (days) |
|---|---|---|
| *Homo sapiens* | 360 | 90 |
| *Gorilla gorilla* | 165 | 105 |
| *Felis catus* | 21 | 21 |
| *Canis familaris* | 23 | 26 |
| *Rattus norvegicus* | 11 | 14 |
| *Turdus merula* | 18 | 28 |
| *Macaca mulatta* | 18 | 21 |
| *Pan troglodytes* | 150 | 105 |
| *Saimiri sciurens* | 45 | 68 |
| *Cercocebus alb.* | 45 | 75 |
| *Tamiasciureus hud.* | 18 | 46 |

formulas from Chapter 12, we find that $a = 5.42$ and $b = 0.56$, so the equation of the dashed curve shown on the graph is $y = 5.42x^{0.56}$. The only data point not falling close to the curve is the one for humans, $(x_1, y_1) = (360, 90)$.

---

**QUESTIONS**

2.31  Suppose a mammal not included among the 11 listed in Table 2.12 began locomotion at the age of 125 days. When would we expect it to begin playing?

2.32  Numerous studies have shown that the diversity of bird species in a given area tends to be linearly related to various measures of plant diversity. That latter measure may reflect variation in foliage *heights*, or it may reflect the *mixture* of plant types. The data given below belong to the second category; they show the relationship between bird species diversity and plant cover diversity in 13 desert-type habitats (99).

| Area | Plant Cover Diversity, $x_i$ | Bird Species Diversity, $y_i$ |
|------|------|------|
| 1 | 0.90 | 1.80 |
| 2 | 0.76 | 1.36 |
| 3 | 1.67 | 2.92 |
| 4 | 1.44 | 2.61 |
| 5 | 0.20 | 0.42 |
| 6 | 0.16 | 0.49 |
| 7 | 1.12 | 1.90 |
| 8 | 1.04 | 2.38 |
| 9 | 0.48 | 1.24 |
| 10 | 1.33 | 2.80 |
| 11 | 1.10 | 2.41 |
| 12 | 1.56 | 2.80 |
| 13 | 1.15 | 2.16 |

(a) Graph these data. Put plant cover diversity on the horizontal axis. Does the relationship look linear?

(b) Using methods described in Chapter 3, we can approximate the relationship here by the linear equation $y = 0.26 + 1.70x$. Does that equation overestimate or underestimate the bird-species diversity in area 7?

2.33  Listed in the table are the average weights of varsity football players at the University of Texas for selected years from the turn of the century to the middle 1960s (data slightly modified from (98)).

| Year | Average Weight (lb) |
|------|------|
| 1905 | 164 |
| 1919 | 163 |
| 1932 | 181 |
| 1945 | 192 |
| 1955 | 194 |
| 1965 | 199 |

(a) A graph of average weight versus year (Figure 2.14) shows that the six points could be described fairly well by the equation $y = a + bx$. Would such a model be reasonable over a wider range of $x$ values? Explain.

(b) The straight line that best fits the six points in Figure 2.14 has the equation $y = -1108.0 + 0.666x$. Assuming the model is valid for all

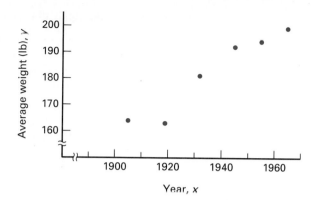

**FIG. 2.14**

values of $x$, what should we expect the average University of Texas football player to weigh in the year 3000? Does your answer to part (**b**) change your answer to part (**a**)?

2.34   Estimating directly the number of ants that populate a colony can be very difficult. Part of the problem is the sheer magnitude of the numbers involved: hundreds of thousands of ants can be in a colony. Regression techniques provide biologists with an indirect and easier approach to the problem. Studies have shown that certain ants in a colony are assigned foraging duties, which necessitate that they come and go from the colony on a regular basis. Furthermore, if $y$ is the colony size and $x$ is the number of ants that forage, the relationship between $y$ and $x$ has the form $y = ax^b$. Therefore, by counting the (much smaller) number of ants leaving a colony to forage, we can estimate the total colony size. Listed below are the actual colony sizes ($y$) and foraging sizes ($x$) for 15 colonies of red wood ants (*Formica polyctena*) (90).

| Colony | Foraging Size, $x_i$ | Colony Size, $y_i$ |
|--------|--------------------|------------------|
| P1B | 45 | 280 |
| P2 | 74 | 222 |
| P3 | 118 | 288 |
| P4 | 70 | 601 |
| P5 | 220 | 1,205 |
| P6 | 823 | 2,769 |
| P7 | 647 | 2,828 |
| C1 | 446 | 3,229 |
| C2 | 765 | 3,762 |
| C3 | 338 | 7,551 |
| C4 | 611 | 8,834 |
| C5 | 4,119 | 12,584 |
| C6 | 850 | 12,605 |
| F1B | 11,600 | 34,661 |
| F2 | 64,512 | 139,043 |

(a) Graph the data. Put foraging size on the horizontal axis.

(b) Using methods described in Chapter 12, we can approximate the $xy$ relationship for these 15 points by the equation $y = 12.88x^{0.86}$. If 400 ants were counted on a foraging trail, how many ants should we estimate to be in the entire colony?

2.35 *Suggestibility* is a psychological term that refers to a person's acceptance of external influences. Its relationship with chronological age was the focus of an experiment. The subjects, ages 6 to 18, were all students at a private academy. Each student was shown a set of slides that required a series of yes/no responses. The experimenter tried to "intimidate" the students into giving wrong answers. For example, a slide might show two geometrical figures of equal area, and the subject would be instructed to decide whether the figures had (1) different areas or (2) the same area. The examiner, however, would preface the slide with a comment that most students are able to distinguish a difference in area. The more suggestible students then would tend to answer (1) even though their uncoerced response might have been (2). After the results were tallied, each age group was assigned a "suggestibility score." Large positive scores represented the greatest amount of suggestibility; large negative scores, the least amount. Figure 2.15 shows the 13 data points (101).

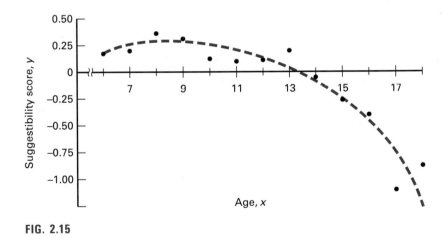

**FIG. 2.15**

(a) These numbers can be described quite nicely by a quadratic model,

$$y = -0.81 + 0.26x - 0.015x^2$$

(the dashed curve in Figure 2.15). What suggestibility score does the model predict for 12-year olds? What age would we estimate for a student whose suggestibility score is $-0.60$?

(b) Is the model in part (a) appropriate for older subjects? What do you think the relationship between suggestibility and age would be like for $x$ values between 20 and 40?

# 2.6

## CATEGORICAL DATA

**qualitative variable**—A variable whose possible values are not numerical, but descriptive.

**quantitative variable**—A variable whose possible values are numerical.

**categorical data**—Two dissimilar variables, $x$ and $y$, are recorded on each subject. The range of possible $x$ responses is reduced to a set of $r$ categories; the range of possible $y$ responses is reduced to a set of $c$ categories. The number of observations belonging to each ($x$ category, $y$ category) combination is reported.

All the data we have seen thus far have been numerical, or *quantitative*. In many situations, however, the observed responses are *qualitative*. A geneticist might investigate the relative prevalence of eye color and records either brown, blue, or black for each subject; a sociologist surveying race and religion might record White, Black, Hispanic, and Other for the first variable and Catholic, Protestant, Jewish, and Other for the second. Such problems, in which the variables are classified rather than measured, are examples of *categorical data*.

Our contact with categorical data will be limited to problems where two (dissimilar) traits, $x$ and $y$, are recorded for each of $n$ subjects. Values for each $x_i$ will belong to exactly one of $r$ categories, where $r \geq 2$; values for each $y_i$ will belong to exactly one of $c$ categories, where $c \geq 2$ (Figure 2.16). The objective in these problems is understanding the relationship between the two traits. In particular, is a subject's $x$ value independent of its $y$ value?

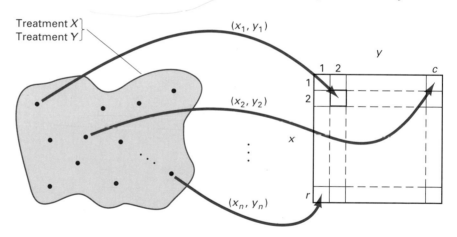

**FIG. 2.16**

**contingency table**—A grid of $r$ rows and $c$ columns for summarizing categorical data. Entries are the numbers of observations belonging to each ($x$ category, $y$ category) combination.

When a set of categorical data is compiled, it generates a *contingency table* of $r$ rows and $c$ columns. The entry in, say, the second row and fourth column is simply the number of subjects whose $x$ trait belongs to the second category for $x$ and whose $y$ trait belongs to the fourth category for $y$. The *sum* of all the entries is the sample size $n$, since each ($x_i, y_i$) belongs to exactly one location in the grid. Methods for analyzing contingency tables are the subject of Chapter 13.

*An Early Antiseptic*

Until almost the end of the nineteenth century, the mortality associated with surgical operations, even minor ones, was extremely high. The major problem was infection. The germ theory as a model for disease transmission was still unknown, so there was no concept of sterilization, and many patients died from postoperative complications.

The major breakthrough so desperately needed came when the British physician Joseph Lister began reading about work done by Louis Pasteur. In a series of classic experiments, Pasteur had demonstrated the role that yeasts and bacteria play in fermentation.

Lister conjectured that human infections might have a similar organic origin; if so, they could be controlled by certain chemicals. To test his theory, he began using carbolic acid as an operating room disinfectant. The results were dramatic, but resistance to change was so strong that it took almost 10 years for his ideas to be accepted.

Seventy-five patients, all needing amputations, "volunteered" for Lister's study. Two traits, each having two categories, were recorded for each subject:

$x$ trait:  Did patient survive? $\begin{cases} \text{No} \\ \text{Yes} \end{cases}$

$y$ trait:  Was patient given carbolic acid? $\begin{cases} \text{No} \\ \text{Yes} \end{cases}$

Altogether, four different $(x_i, y_i)$'s were possible:

(No, No)
(No, Yes)
(Yes, No)
(Yes, Yes)

Table 2.13 summarizes the final disposition of the 75 observations (183).

**TABLE 2.13**

|  |  | CARBOLIC ACID USED? | |
|---|---|---|---|
|  |  | Yes | No |
| **PATIENT LIVED?** | No | 6 | 16 |
|  | Yes | 34 | 19 |

If the $x$ and $y$ traits are independent, the proportion of patients who died and had received carbolic acid should be roughly the same as the proportion of patients who died and had not received the disinfectant. Here, those two proportions are 6/40 ($=15\%$) and 16/35 ($=46\%$), a difference that was large enough to convince Lister that the $x$ and $y$ traits were *not* independent. The formal procedure for assessing the statistical significance of the difference between 6 out of 40 and 16 out of 35 is called a ***chi-square test***. We will learn about it in Chapter 13.

---

### CASE STUDY 2.13

*Delinquency and Birth Order*

Are birth order and juvenile delinquency related? Are oldest children, for example, more predisposed to a life of crime than are youngest children, or vice versa? A group of 1154 girls (116) attending public high school were given a questionnaire that measured the degree to which each had exhibited delinquent behavior in terms of criminal acts or immoral conduct. After tallying the results and agreeing on their criteria, the researchers categorized some 111 of the girls as "delinquent." Everyone in the initial sample was also asked to indicate her birth order, and each girl was listed as being (1) the oldest, (2) in between, (3) the youngest, or (4) an only child.

The two (dissimilar) traits being observed here, delinquency and birth order, have two and four categories, respectively. Table 2.14 shows the frequencies associated with the eight possible $(x_i, y_i)$'s. As must be the case, the sum of the entries is the total sample size, 1154.

Intuitively, what should be true of the frequencies in Table 2.14 if delinquency and birth order are independent? The proportions of delinquents in the four birth-order categories should be similar. Are they? No. Oldest children, for example, have a much lower delinquency rate than "only" children—24 out of 474 ($=5.1\%$) versus 23 out of 93 ($=24.7\%$). For whatever reasons delinquency and birth order are *not* independent, a conclusion we will reach more formally later.

**TABLE 2.14**

|  |  | BIRTH ORDER | | | |
|---|---|---|---|---|---|
|  |  | Oldest | In Between | Youngest | Only Child |
| DELINQUENT? | Yes | 24 | 29 | 35 | 23 |
|  | No | 450 | 312 | 211 | 70 |

**2.36**   Categorical data with the structure that each trait allows only two possible responses can also be analyzed as a two-sample model of the type shown in Case Study 2.6. Define a set of $x_i$'s and $y_i$'s that would reformulate the Lister data in Table 2.13 as a two-sample problem. (Ultimately, having two "choices" for classifying data of this sort is irrelevant; whatever conclusions one model implies, the other implies also.)

**2.37**   A phenomenon often studied in ESP research is the "sheep-and-goat" effect—the tendency for persons (sheep) who believe in extrasensory perception (ESP) to perform better on ESP tests than those who do not believe (goats). In one experiment volunteers were given a questionnaire to determine their attitudes toward ESP. The questions were all similar to the following two:

1.   Have you ever had a feeling in advance that you are going to receive a particular letter on a particular day?

2.   When you were participating in some game, have you ever felt compelled to bet on a certain result and won the bet?

On the basis of a subject's responses, he or she was classified as (1) believing in ESP, (2) not believing in ESP, or (3) undecided. All subjects were then given a precognition test consisting of 500 blank squares. Each subject was to predict which of the five standard ESP symbols ($+$, $\star$, $\bigcirc$, $\ggg$, $\square$) would later be entered (by a random generator) into each of the 500 squares. Taking the test were 52 "sheep" and 11 "goats." The scores (numbers of correct predictions) of these 63 subjects were classified according to whether they were above or below the average of all persons taking the test. The results are summarized below (144).

|  | Sheep | Goats |
|---|---|---|
| Above Average | 24 | 1 |
| Below Average | 28 | 10 |

(a)   Define $(x_i, y_i)$ for these data. What does $n$ equal?

(b)   Do the numbers lend any support to the theory that ESP attitude and precognition ability are related?

**2.38**   High blood pressure is one of the major contributors to coronary heart disease. A related issue is whether there is a relationship between the blood pressures of children and those of their fathers. (If such a relationship exists, it might be possible to use one group to screen for high-risk individuals in the other.) One study addressing that question involved 92 eleventh graders, 47 males and 45 females, and their fathers. Blood pressures for the children and the fathers were categorized as belonging to the lower, middle, or upper third of their respective distributions. The resulting frequencies are tabulated below (79).

|  |  | CHILD'S BLOOD PRESSURE | | |
|  |  | Lower Third | Middle Third | Upper Third |
|---|---|---|---|---|
| **FATHER'S BLOOD PRESSURE** | Lower Third | 14 | 11 | 8 |
| | Middle Third | 11 | 11 | 9 |
| | Upper Third | 6 | 10 | 12 |

(a) Define $(x_i, y_i)$.

(b) What would the frequencies in this contingency table look like if, in general, a child was strongly predisposed to having a blood pressure comparable to his or her father's?

2.39  Does a mouse's environment early in life affect its aggressiveness later in life? Judge for yourself. Two groups of mice were tested: the mice in one group had been raised by their natural mothers, the other mice by foster mice. (A foster mouse was a female whose own litter had been removed shortly after birth.) When a mouse was three months old, it was placed in a box partitioned into two compartments. On the other side of the partition was another mouse, one that had had no previous contact with the "test" mouse. The partition was then removed, and the behavior of the two mice was observed for the next 6 minutes. Eventually, 307 mice were monitored under similar conditions. Of the 167 mice raised by their natural mothers, 27 began fighting with the mouse on the other side of the partition. Among the remaining 140 mice, each of which had been raised by a foster mouse, 47 began fighting (74).

(a) Put these data into a contingency table having two rows and two columns. Define $(x_i, y_i)$.

(b) Does it appear that environment and aggressiveness are independent? Explain.

2.40  Market researchers often gather survey information by telephone. Ideally, the numbers dialed should be random, thus affording every subscriber an equal chance of being included in the sample. In practice, though, doing a survey that way would be very costly because of the sizable number of nonproductive calls resulting from dialing invalid numbers. The obvious remedy is to sample from the telephone directory, but that raises another problem: unlisted numbers would not be included. Nationally, more than 15% of all telephone numbers are unlisted, and the percentage is much higher in certain locales (in Los Angeles, for example, more than 35% of all telephone numbers are unlisted). Clearly, the directory method is unreliable unless the factors being studied are independent of whether or not the telephone is listed. The following responses (slightly modified) were obtained in a survey of 1000 subscribers by Pacific Bell (135).

|  | Listed Number | Unlisted Number |
|---|---|---|
| **Own Home** | 628 | 146 |
| **Rent** | 172 | 54 |

(a) Define $(x_i, y_i)$.

(b) Suppose a political poll is to be taken in California to assess the popularity of a proposition to reduce property taxes. One possible strategy is to choose the sample to be interviewed from the telephone directory. How does the information here provide input into the feasibility of that approach?

# 2.7

## k-SAMPLE DATA

Recall the sample-population structure that characterizes two-sample data. Two independent samples,

$$x_1, x_2, \ldots, x_n \quad \text{and} \quad y_1, y_2, \ldots, y_m$$

are observed. The first set measures the effect of treatment $X$, the second set the effect of treatment $Y$. The objective is to use the $x_i$'s and $y_i$'s to make inferences about how treatment $X$ compares with treatment $Y$. In this section we want to extend that idea to include experiments where $k$ treatments (or treatment levels) are to be compared, where $k$ is some integer greater than 2.

**k-sample data**—Samples of sizes $n_1, n_2, \ldots, n_k$ $(k > 2)$ are chosen to represent treatment 1, treatment 2, ..., treatment $k$, respectively. All observations are independent.

Data for a **k-sample problem** consist of $n_1$ observations taken under treatment 1, $n_2$ taken under treatment 2, ..., and $n_k$ taken under treatment $k$. The total sample size is $n$, where $n = n_1 + n_2 + \cdots + n_k$. *All samples are assumed to be independent* (Figure 2.17).

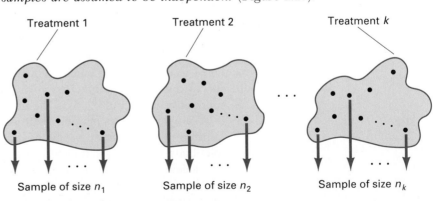

FIG. 2.17

Associated with each treatment is a population. The usual objective in a *k*-sample problem is to test that all *k* of those populations are identical. Despite their similarities, we distinguish two-sample data from *k*-sample data because the mathematical methods used to analyze the two models are entirely different.

## CASE STUDY 2.14

*Declination and Magnetic Field*

Many attempts have been made to document the directional changes over time in the earth's magnetic field. In one study (163) samples were taken of three lava flows from volcanic eruptions of Mount Etna in 1669, in 1780, and in 1865. In each instance the magnetic field of the molten lava aligned itself with the earth's magnetic field as it prevailed at that time. When the lava cooled and hardened, the magnetic field was "captured" and its direction remained fixed.

Three blocks of lava, each weighing approximately 1 kg, were collected from each flow. The measurement recorded for each block was the declination (in degrees) of the magnetic field as it existed when the lava originally cooled. Table 2.15 summarizes the findings.

The *k* = 3 populations being represented here are the magnetic field declinations for all the lava blocks associated with the 1669, the 1780, and the 1865 eruptions. Using these nine sample values to make inferences about the averages for those three populations requires a technique known as the *analysis of variance*, which we develop in Chapter 14. Notice that the sample values in Table 2.15 are clearly independent: the 57.8 listed under 1669, for example, is not linked in any way with the 57.9 under 1780 or the 52.7 under 1865.

**TABLE 2.15  Declination of Magnetic field**

| 1669 | 1780 | 1865 |
|------|------|------|
| 57.8 | 57.9 | 52.7 |
| 60.2 | 55.2 | 53.0 |
| 60.3 | 54.8 | 49.4 |

## CASE STUDY 2.15

*Sodium Nitrate and N-nitrosopyrrolidine*

Sodium nitrite has been used for many years as a curing agent for bacon. Until recently it was thought to be harmless. But now it appears that, during frying, sodium nitrite induces the formation of *N*-nitrosopyrrolidine (NPYR), a suspected carcinogen. To assess the seriousness of the problem from the consumer's standpoint, researchers performed a study (151) where the variable measured was the amount of NPYR, in parts per billion (ppb), recovered after frying a slice of bacon. Four commercially available brands were studied, and three slices from each brand were tested. The 12 resulting NPYR determinations are shown in Table 2.16.

Notice that the *k* describing this set of data is 4, that being the number of populations under comparison. All the bacon slices produced by brand 1, for example, would yield, upon frying, a population of NPYR values. Three members of that population are the 20, 40, and 18 in column 1 of Table 2.16.

**TABLE 2.16 NPYR Recovered from Bacon (ppb)**

|          | Brand 1 | Brand 2 | Brand 3 | Brand 4 |
|----------|---------|---------|---------|---------|
|          | 20      | 75      | 15      | 25      |
|          | 40      | 25      | 30      | 30      |
|          | 18      | 21      | 21      | 31      |
| Average: | 26.0    | 40.3    | 22.0    | 28.7    |

The consumer wants to know whether certain brands, in general, are more carcinogenic than others. The bottom row in Table 2.16 shows the average amounts of NPYR found in each sample. They range from a low of 22.0 for brand 3 to a high of 40.3 for brand 2. Should a consumer conclude that Brand 3 is safer than Brand 2? We will find out in Chapter 14.

QUESTIONS

2.41 Look again at the 12 entries in Table 2.16. Is there anything "questionable" about those figures? If you were the statistician handed those numbers, what might you want to ask the experimenter?

2.42 Japanese fishing villages have suffered two widely publicized epidemics of acute mercury poisoning. As a result, mercury pollution has become a hotly debated ecological issue all over the world. Much of the mercury released into the environment originates as a by-product of coal combustion and other industrial processes. It is not dangerous until it falls into large bodies of water, where microorganisms change it into methyl mercury, an organic form highly toxic to humans. Edible fish then become the intermediaries. They ingest and absorb the methyl mercury, making themselves a source of contamination. To get some idea of the seriousness of the problem, a research team recorded the mercury uptake (in ppm) found in 12 walleyed pike caught in Lake Erie. Three different ages were singled out to serve as treatment levels (125).

| Younger Than 1 Year | Yearlings | 2 Years and Older |
|---------------------|-----------|-------------------|
| 0.60                | 0.75      | 1.03              |
| 0.64                | 0.92      | 0.67              |
| 0.62                | 0.93      | 0.78              |
| 0.44                | 0.75      | 0.98              |

(a) What populations are being represented here? Can we assume that the observations are independent?

(b) Just by looking at the three samples, do you think that the populations they represent might be different?

2.43  The predatory behavior of the octopus has not often been studied outside a laboratory environment. In one investigation, though, six *Octopus cyanea* were captured in their native Indo-Pacific Ocean and taken to large coral-enclosed holding ponds in the Hawaiian Islands, where they were tagged and carefully watched. Crabs are the main staple in an octopus's diet. The following data show the distances traveled (in meters) by each octopus in three crab-hunting forages (188).

| | | OCTOPUS | | | |
|---|---|---|---|---|---|
| 1 | 2 | 3 | 4 | 5 | 6 |
| 5 | 6 | 34 | 61 | 21 | 31 |
| 11 | 21 | 69 | 15 | 24 | 40 |
| 18 | 12 | 24 | 23 | 91 | 18 |

(a)  What populations do these data represent? Is it reasonable to assume that the observations are independent? Why?

(b)  If the numbers of measurements recorded for each octopus were all different, would this still be classified as a *k*-sample problem?

2.44  Geologists always want to know why one water well has better quality, productivity, and longevity than another. Are these well-to-well differences random, or are they the product of specific geological factors? The question is obviously practical. If the differences do have assignable causes, future wells should be restricted to areas that are geologically favorable. Eighty water wells in central Pennsylvania were studied to see whether the type of rock in which a well is located affects its yield. Each well was dug in one of five types of rock: upper sandy dolomite, nittany dolomite, bellefonte dolomite, limestone, and shale. The productivities of the 80 wells were ranked from 1 (lowest) to 80 (highest) as follows (155).

| Upper Sandy Dolomite | | Nittany Dolomite | | Bellefonte Dolomite | | Limestone | Shale |
|---|---|---|---|---|---|---|---|
| 9 | 61 | 12 | 57 | 1 | 30 | 5 | 8 |
| 18 | 62 | 24 | 67 | 2 | 32 | 13 | 10 |
| 28 | 63 | 29 | 72 | 3 | 34 | 21 | 14 |
| 41 | 65 | 33 | 77 | 4 | 36 | 22 | 19 |
| 43 | 68 | 39 | 78 | 6 | 37 | 27 | 23 |
| 44 | 70 | 46 | 79 | 7 | 38 | 31 | 25 |
| 45 | 71 | 49 | 80 | 11 | 40 | 42 | 26 |
| 50 | 73 | 53 | | 15 | 48 | 54 | 35 |
| 58 | 74 | | | 16 | 51 | 56 | 47 |
| 59 | 75 | | | 17 | 52 | 66 | 55 |
| 60 | 76 | | | 20 | 64 | 69 | |

(a) Notice that the numbers recorded here are ranks, not actual productivities. If location has no influence on a well's productivity, what would we expect to be true for the average ranks associated with the five different rock types?

(b) Do the data suggest that rock type does influence a well's productivity?

# 2.8

## RANDOMIZED BLOCK DATA

**randomized block data**—$k$ ($>2$) dependent observations, representing treatment 1, treatment 2, . . . , treatment $k$, are observed in each of $n$ blocks. The objective is to compare the $k$ treatments.

**block**—A group of similar subjects or similar experimental conditions to which the entire set of treatments being compared is applied.

The $k$-sample problem generalizes the two-sample problem by allowing more treatments to be compared. In the same spirit the *randomized block problem*, the last of our seven data structures, generalizes the paired-data problem. Conceptually $k$-sample data differ from randomized block data in the way the sample subjects are chosen: observations in $k$-sample data are *independent*, observations in randomized block data are *dependent*.

Figure 2.18 shows the structure of randomized block data: $k$ treatment levels are compared, where $k > 2$. The subjects are grouped into $n$ **blocks**, $k$ subjects to a block. Each treatment level is applied to exactly one subject in a block. In general, the blocks are defined according to whatever criterion is appropriate for the situation. We want the $k$ subjects in any block to be as similar as possible. (A block in randomized block data plays the same role for $k$ treatment levels that a pair in paired data plays for two treatment levels.)

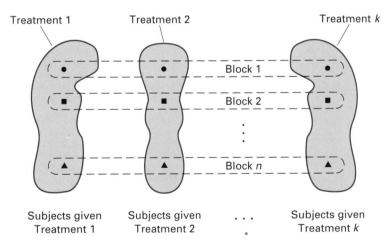

**FIG. 2.18**

As in the paired-data structure, blocks can be defined in many ways. If each subject, for example, is given all $k$ treatments, then each of those $n$ subjects becomes a block (recall Case Study 2.7). In animal experiments litters are often used as blocks, meaning that each of $k$ littermates is given a different treatment. Time and place are also frequently used criteria for matching subjects.

*Lunar Influence*
*on Humans*

In folklore the full moon is often portrayed as something sinister, an evil force possessing the power to control our behavior. Many prominent writers and philosophers have shared that belief (10). Milton, in *Paradise Lost*, refers to

> Demoniac frenzy, moping melancholy
> And moon-struck madness.

and Sir William Blackstone, the renowned eighteenth-century barrister, defined a lunatic as

> one who hath . . . lost the use of his reason and
> who hath lucid intervals, sometimes enjoying his
> senses and sometimes not, and that frequently
> depending upon changes of the moon.

The possibility of lunar phases influencing human affairs has supporters in the scientific community. Studies by reputable medical researchers have attempted to link the "Transylvania effect," as lunar influence has come to be known, with a variety of ills, including disorderly conduct, pyromania, and suicide. Here we look at still another context in which this phenomenon might be expected to occur. Table 2.17 shows the admission rates to the emergency room of a Virginia mental health clinic before, during, and after the 12 full moons from August 1971 to July 1972.

Here time is used as the blocking criterion: the three measurements in August are one block, the three in September are a second block, and so on. Are the observations in a given month dependent? Definitely. All the admission rates in August, for example, are low, regardless of lunar phase; all in April, on the other hand, are high. For whatever reason, time of year is definitely a factor affecting admission rates.

**TABLE 2.17 Admission Rates (patients/day)**

| Month | Before Full Moon | During Full Moon | After Full Moon |
|-------|-----------------|------------------|-----------------|
| August | 6.4 | 5.0 | 5.8 |
| September | 7.1 | 13.0 | 9.2 |
| October | 6.5 | 14.0 | 7.9 |
| November | 8.6 | 12.0 | 7.7 |
| December | 8.1 | 6.0 | 11.0 |
| January | 10.4 | 9.0 | 12.9 |
| February | 11.5 | 13.0 | 13.5 |
| March | 13.8 | 16.0 | 13.1 |
| April | 15.4 | 25.0 | 15.8 |
| May | 15.7 | 13.0 | 13.3 |
| June | 11.7 | 14.0 | 12.8 |
| July | 15.8 | 20.0 | 14.5 |
| Averages: | 10.9 | 13.3 | 11.4 |

Notice the last line in Table 2.17, giving the three averages for the 12 months. As the Transylvania hypothesis predicts, the admission rate *is* higher during the full moon. The question that needs to be addressed, though, is whether the differences between 10.9, 13.3, and 11.4 could reasonably have arisen by chance alone, or must we conclude that the phases of the moon *do* have an effect? Phrased in more technical jargon, are the differences *statistically significant*? What do you think?

## CASE STUDY 2.17

*The Ponzo Illusion*

Hypnotic age regression is a procedure whereby an adult is instructed under hypnosis to return to an earlier chronological age. Experiments have shown convincingly that this process can recapture *behavioral* patterns of the earlier age, but less is known about its ability to reinstate *perceptual* patterns. Among the efforts to investigate the latter was an interesting experiment using the Ponzo illusion in Figure 2.19 (126).

When the vertical lines in the figure are equal in length, most people perceive the rightmost one as being shorter. Furthermore, studies have shown that the strength of the illusion is a function of age: it becomes increasingly more pronounced from childhood through adolescence; in the late teens it levels off.

The Ponzo illusion was shown to eight college students. These students were then regressed under hypnosis to age nine and to age five. At each age their perceptions of the illusion were measured. If hypnosis can recapture perceptual patterns, the illusion strengths at these three ages should be considerably different.

The Ponzo illusion was drawn on a set of 21 cards. On each card the leftmost line was 4 in long; the rightmost line varied from card to card, ranging from $2\frac{1}{2}$ to 5 inches in increments of $\frac{1}{8}$ in. The cards were shown to each subject in random order. In each instance the subject was instructed to point to the longer line. If the subject perceived the rightmost line to be longer when it was actually $4\frac{6}{8}$ in but shorter when it was $4\frac{5}{8}$ in, the strength of the illusion was defined to be

$$\frac{5}{8} + \frac{1}{2}\left(\frac{6}{8} - \frac{5}{8}\right) = 0.69 \text{ in}$$

Results for the eight subjects are listed in Table 2.18.

The three treatments being compared are the mental states awake, regressed to age nine, and regressed to age five. Are the three observations in a given row dependent? Yes, because all were made on the same subject, so all are reflecting, to some extent, that subject's general psychological makeup. Because they were given all three treatments, the subjects are acting as blocks. The key to interpreting these data is the three treatment averages, 0.48, 0.39, and 0.44, shown at the bottom of Table 2.18. We will

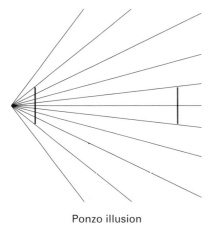

Ponzo illusion

**FIG. 2.19**

**TABLE 2.18  Illusion Stengths**

| Subject | Awake | Regressed to Age 9 | Regressed to Age 5 |
|---------|-------|--------------------|--------------------|
| 1 | 0.81 | 0.69 | 0.56 |
| 2 | 0.44 | 0.31 | 0.44 |
| 3 | 0.44 | 0.44 | 0.44 |
| 4 | 0.56 | 0.44 | 0.44 |
| 5 | 0.19 | 0.19 | 0.31 |
| 6 | 0.94 | 0.44 | 0.69 |
| 7 | 0.44 | 0.44 | 0.44 |
| 8 | 0.06 | 0.19 | 0.19 |
| Averages: | 0.48 | 0.39 | 0.44 |

learn in Chapter 15 what the differences among those averages imply about the effectiveness of hypnotic age regression in reinstating a person's perception of the Ponzo illusion.

---

QUESTIONS

2.45  In Case Study 2.17 what two general "hypotheses" could be offered to explain the differences among the three illusion strength averages 0.48, 0.39, and 0.44? (Hint: Reread the conclusion of Case Study 2.16).

2.46  Historically, agricultural research provided many early applications for the randomized block structure. The blocks were geographically contiguous fields. Soil quality, drainage conditions, and so on were presumably fairly homogeneous within each field. Typical of that kind of experiment is the following data, where five different levels of potash were applied to cotton being grown in three different fields. The variable recorded was the Pressley strength index of the cotton fibers (27).

| | POUNDS OF POTASH PER ACRE | | | | |
|-------|------|------|------|------|------|
| Field | 36 | 54 | 72 | 108 | 144 |
| 1 | 7.62 | 8.14 | 7.76 | 7.17 | 7.46 |
| 2 | 8.00 | 8.15 | 7.73 | 7.57 | 7.68 |
| 3 | 7.93 | 7.87 | 7.74 | 7.80 | 7.21 |

(a) Why are the observations considered dependent rather than independent?

(b) What numbers might we compute that would reflect the inherent similarity or dissimilarity of the fields?

2.47  Accurate laboratory methods for determining the percentage content of solids in heterogeneous foods are important for maintaining quality control and complying with government regulations. In one study (42), three techniques were compared for measuring the percent solids in uncreamed cottage cheese. All three methods were used on each of nine different samples.

| | PERCENT SOLIDS DETERMINED | | |
|---|---|---|---|
| Sample | Method 1 | Method 2 | Method 3 |
| 1 | 17.766 | 17.707 | 17.677 |
| 2 | 19.167 | 19.222 | 19.224 |
| 3 | 19.071 | 19.113 | 19.092 |
| 4 | 19.636 | 19.646 | 19.520 |
| 5 | 28.703 | 28.784 | 28.704 |
| 6 | 19.072 | 19.131 | 19.024 |
| 7 | 18.382 | 18.490 | 18.362 |
| 8 | 18.576 | 18.675 | 18.530 |
| 9 | 18.756 | 18.870 | 18.637 |

(a) Is there anything about these numbers to suggest that the samples themselves can be quite variable?

(b) Can these data be used to infer which of the three methods is best? Explain.

2.48   Male cockroaches can be very antagonistic toward other male cockroaches. Encounters may be fleeting or quite spirited, the latter often resulting in missing antennae and broken wings. A study was done to see whether cockroach density has any effect on the frequency of serious altercations. Ten groups of four male cockroaches (*Byrsotria fumigata*) were each subjected to three levels of density: high, intermediate, and low. (In the low-density environment, the four cockroaches were placed in a 1-m² cage. The two higher density levels were achieved by making the cage smaller.) Listed below are the numbers of "serious" encounters per minute that were observed (13).

| Group | High | Intermediate | Low |
|---|---|---|---|
| 1 | 0.30 | 0.11 | 0.12 |
| 2 | 0.20 | 0.24 | 0.28 |
| 3 | 0.17 | 0.13 | 0.20 |
| 4 | 0.25 | 0.36 | 0.15 |
| 5 | 0.27 | 0.20 | 0.31 |
| 6 | 0.19 | 0.12 | 0.16 |
| 7 | 0.27 | 0.19 | 0.20 |
| 8 | 0.23 | 0.08 | 0.17 |
| 9 | 0.37 | 0.18 | 0.18 |
| 10 | 0.29 | 0.20 | 0.20 |
| Averages: | 0.25 | 0.18 | 0.20 |

(a) Why are the three observations in a given row dependent?

(b) If density has no effect on a cockroach's aggressiveness, what would we expect to be true of the column averages? Would a density effect have any influence on differences among row averages?

# 2.9
## *SUMMARY*

Identifying the underlying statistical structure in a set of data is an exercise in asking the right questions. Are the observations qualitative or quantitative? How many treatment groups are being compared? Are the measurement units similar or dissimilar? Are the samples dependent or independent? At most, it will take the answers to those four questions to classify a set of data into one of the seven models defined in Sections 2.2 through 2.8.

In general, recognizing data models is fairly easy. The questions are few, and the answers are almost automatic. Does it matter in which *order* the questions are asked? No, but some orders are more "natural" than others. The flowchart in Figure 2.20 gives a sequence that works well.

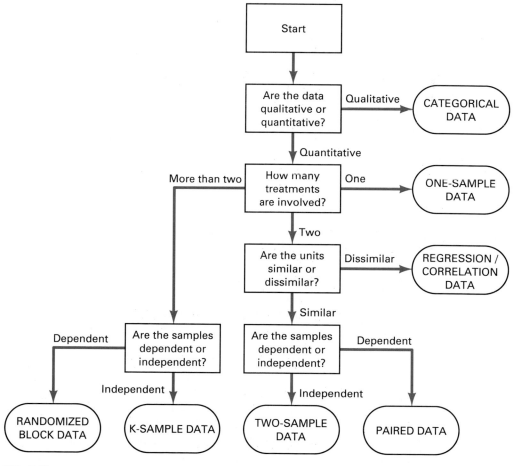

**FIG. 2.20**

*EXAMPLE 2.1*   A civil rights group plans to file suit against a city because of its alleged noncompliance with a school busing order. The black-white ratio in this community is 20 : 80, and all the public schools are expected to reflect that ratio as closely as possible. Table 2.19 lists the actual percentages of black students in two randomly selected sets of schools. The first set comprises schools located in predominantly black neighborhoods, the second set comprises schools located in predominantly white neighborhoods. What data structure do these numbers represent?

**TABLE 2.19 Percentage of Black Students**

| Schools in Black Neighborhoods | Schools in White Neighborhoods |
|---|---|
| 36 | 21 |
| 28 | 14 |
| 41 | 11 |
| 32 | 30 |
| 46 | 29 |
| 39 | 6 |
| 24 | 18 |
| 32 | 25 |
| 45 | 23 |

Look at the flowchart in Figure 2.20. Are these 18 observations qualitative or quantitative? *Quantitative*. How many treatments are involved? *Two*. Are the units similar or dissimilar? *Similar*. Are the samples dependent or independent? *Independent*. According to the flowchart, then, the percentages in Table 2.19 qualify as a set of *two-sample data*.

*EXAMPLE 2.2*   Debates about the arms race and the level of our military preparedness have been a staple of political oratory for many, many years. The nature of the weapons changes as technology progresses, but the basic arguments remain pretty much the same. In the 1960s, long before the "Star Wars" initiative was conceived, the buildup in intercontinental ballistic missiles (ICBMs) was a major controversy of this type. Table 2.20 shows the rapid growth of the American ICBM force from 1960 to 1969 (7). To which of the seven data models do these data belong? *Regression–correlation*. To see why, look again at the questions in the flowchart in Figure 2.20. Are the data qualitative or quantitative? *Quantitative*. How many treatments are involved? *Two*. Are the units similar or dissimilar? *Dissimilar*. The answers to those questions are all we need: according to the flowchart, the data belong to the regression–correlation model.

**TABLE 2.20**

| Year | Number of ICBMs |
|---|---|
| 1960 | 18 |
| 1961 | 63 |
| 1962 | 294 |
| 1963 | 424 |
| 1964 | 834 |
| 1965 | 854 |
| 1966 | 904 |
| 1967 | 1,054 |
| 1968 | 1,054 |
| 1969 | 1,054 |

**QUESTIONS**

*Identify the data structure (one-sample, two-sample, etc.) that each of the following hypothetical experiments represents.*

**2.49** The production of a certain organic chemical requires ammonium chloride. The manufacturer can obtain the ammonium chloride in one of three forms: powdered, moderately ground, and coarse. To see if the consistency of the $NH_4Cl$ is itself a factor that needs to be considered, the manufacturer decides to run the reaction seven times with each form of ammonium chloride. The resulting yields (in pounds) are as follows:

| Powdered $NH_4Cl$ | Moderately Ground $NH_4Cl$ | Coarse $NH_4Cl$ |
|---|---|---|
| 146 | 150 | 141 |
| 152 | 144 | 138 |
| 149 | 148 | 142 |
| 161 | 155 | 146 |
| 158 | 154 | 139 |
| 154 | 148 | 137 |
| 149 | 150 | 145 |

**2.50** A sports publicist for State University has assembled records on the winning 100-m times posted by athletes at S.U. during the school's five most recent track and field decathlons. The purpose was to see if the winning times are getting noticeably shorter.

| Year | Winning Time (s) |
|---|---|
| 1983 | 10.4 |
| 1984 | 10.3 |
| 1985 | 10.3 |
| 1986 | 10.4 |
| 1987 | 10.2 |

**2.51** In Eastern Europe a study was done on 50 people bitten by rabid animals. Twenty victims were given the standard Pasteur treatment, while the other 30 were given the Pasteur treatment in addition to one or more doses of antirabies gamma globulin. Nine of those given the standard treatment survived; twenty survived in the gamma globulin group.

**2.52** As part of an affirmative-action litigation, records were produced showing the average salaries earned by white, black, and Hispanic workers in a large manufacturing plant. Three different departments were selected at random for the comparison. The entries shown are average annual salaries, in thousands of dollars.

|  | White | Black | Hispanic |
|---|---|---|---|
| Department 1 | 20.2 | 19.8 | 19.9 |
| Department 2 | 20.6 | 19.0 | 19.2 |
| Department 3 | 19.7 | 20.0 | 18.4 |

**2.53** The amount of corrosion sustained by underground pipes can be gauged by measuring the maximum pit depth found in a randomly selected foot of pipe. Six such pieces are inspected; the recorded pit depths (in inches) are 0.0039, 0.0041, 0.0038, 0.0044, 0.0040, and 0.0036.

**2.54** A new liquid-protein diet is tested on five adult males. Listed below are their weights (in lb) (1) prior to starting the diet and (2) three weeks later.

| Subject | Weight Before Diet | Weight After Diet |
|---|---|---|
| DF | 150 | 143 |
| ML | 195 | 190 |
| CW | 188 | 185 |
| SH | 197 | 191 |
| NL | 200 | 204 |

**2.55** A sample of 400 voters is divided into four groups (A, B, C, D) on the basis of income. The 100 voters having the lowest annual earnings are put into group A, the 100 with the next lowest incomes are put into group B, and so on. Members in each group are asked to assess the competence of their newly elected mayor. Their opinions are summarized in the following 2 × 4 table.

|  |  | A | B | C | D |
|---|---|---|---|---|---|
| Perceived to be Capable | Yes | 54 | 45 | 73 | 69 |
|  | No | 46 | 55 | 27 | 31 |

**2.56** A time-study engineer is assigned the problem of comparing the efficiencies of two different machines that measure the shear strength of polyester fibers. He randomly divides 12 garment workers into two groups. The first group measures the shear strength by using machine A; the second group uses machine B. The numbers shown in the table are the 12 completion times (in seconds).

| Machine A | Machine B |
|-----------|-----------|
| 220 | 247 |
| 235 | 223 |
| 214 | 215 |
| 197 | 219 |
| 206 | 207 |
| 214 | 236 |

2.57   The admissions counselor at State University is trying to decide if he should recommend a professional review course to high school seniors planning to retake the SAT. Ten students in State's applicant pool retook the test last year. Five had completed the review course, and five had not. Students who had taken the course registered gains of 30, 50, 45, 60, and 35 points; those not taking the review course improved by gains of 30, 25, 35, 15, and 40 points.

2.58   Three different catalytic converters being tested by an automobile manufacturer meet government standards, but they may not give the same mileage. Twelve identical new cars are selected for the test. Each type of converter is installed on four cars, which are then driven over a specially engineered track that simulates city driving. The resulting mileage estimates are listed below.

| Converter A | Converter B | Converter C |
|-------------|-------------|-------------|
| 21.6 | 22.8 | 23.9 |
| 23.2 | 25.6 | 24.6 |
| 20.5 | 24.7 | 23.8 |
| 21.7 | 24.1 | 21.5 |

2.59   In a small clinical trial designed to assess the value of a new tranquilizer on psychoneurotic adults, each patient was given a week's treatment with the drug and a week's treatment with a placebo. At the end of each week the patient was asked to complete a questionnaire, on the basis of which he or she was assigned an anxiety score (with possible values from 0 to 30). High scores correspond to high states of anxiety.

| Patient | Drug | Placebo |
|---------|------|---------|
| HW | 19 | 22 |
| EP | 11 | 18 |
| JS | 14 | 17 |
| JT | 17 | 19 |
| AD | 23 | 22 |
| DM | 11 | 12 |

2.60   Four air-to-surface missile launchers are tested for accuracy. The same artillery crew fires four rounds with each launcher, each round consisting of 20 missiles. A hit is scored if the missile lands within 10 yd of its target. The table below gives the number of hits registered in each round.

| Launcher 1 | Launcher 2 | Launcher 3 | Launcher 4 |
|------------|------------|------------|------------|
| 13 | 15 | 9 | 12 |
| 11 | 16 | 11 | 12 |
| 10 | 18 | 10 | 9 |
| 14 | 17 | 8 | 11 |

2.61   A chemist is trying to determine the pH of a catalyst by using a standard titration procedure. She replicates the experiment four times and gets readings of 6.4, 6.2, 6.3, and 6.3.

2.62   The relationship between a student's class ranking in high school (expressed as a percentile) and first-year college grade point average (GPA) was investigated by pulling the records of a random sample of six currently enrolled sophomores. The results are summarized below.

| Student | Class Rank (percentile) | First-Year GPA |
|---------|-------------------------|----------------|
| FJ | 85 | 3.12 |
| LL | 91 | 3.34 |
| CK | 76 | 2.96 |
| CF | 72 | 2.38 |
| NL | 96 | 3.62 |
| CW | 80 | 2.63 |

2.63   A pharmaceutical company is testing two new drugs for increasing the blood-clotting ability of hemophiliacs. Six subjects are chosen for the study and randomly divided into two groups of size 3. The first group is given drug A; the second group, drug B. The response variable in each case is the subject's prothrombin time, a number that reflects the time it takes for a clot to form. The results (in seconds) for group A are 32.6, 46.7, and 81.2; for group B, 25.9, 33.6, and 35.1.

2.64   Marketing executives for a major TV network are comparing the popularities of three news broadcasts, all aired at the same time. Nielsen ratings for the three programs were determined for each of five large cities, chosen at random.

| City | Network 1 | Network 2 | Network 3 |
|------|-----------|-----------|-----------|
| A | 9.4 | 8.6 | 14.3 |
| B | 10.6 | 11.2 | 12.1 |
| C | 13.1 | 12.6 | 13.2 |
| D | 15.8 | 15.6 | 14.0 |
| E | 14.9 | 14.3 | 11.8 |

2.65   An investigation was made of 107 fatal poisonings of children. Each death was caused by one of three drugs. In each instance the investigator asked how the child received the fatal overdose. Responsibility for the accidents was assessed according to the breakdown shown below.

| | Drug A | Drug B | Drug C |
|------|--------|--------|--------|
| Child Responsible | 10 | 10 | 18 |
| Parent Responsible | 10 | 14 | 10 |
| Another Person Responsible | 4 | 18 | 13 |

2.66   A paint manufacturer is experimenting with a new latex derivative that may improve the chalkiness of its house paint. The problem is that the additive might also affect the tint, thereby requiring the company to rework its blending procedures. As a test, a quality control engineer draws a sample from each of seven batches of Osage Orange. Each sample is then split into two aliquots, with the latex derivative being added to one of the two. Both samples are then examined with a spectroscope. The results are shown below in standardized lumen units (if the tint of the aliquot was exactly right, it would register a reading of 1.00).

| Batch | Osage Orange | Osage Orange + Additive |
|-------|--------------|-------------------------|
| 1 | 1.10 | 1.06 |
| 2 | 1.05 | 1.02 |
| 3 | 1.08 | 1.17 |
| 4 | 0.98 | 1.21 |
| 5 | 1.01 | 1.01 |
| 6 | 0.96 | 1.23 |
| 7 | 1.02 | 1.19 |

REVIEW QUESTIONS

2.67   According to the package label, a bag of cement is supposed to weigh at least 50 lb. Suppose the weights of five bags chosen at random are found to be 51.2, 49.6, 50.4, 48.6, and 49.3 lb, respectively. Define a set of Bernoulli variables whose average represents the likelihood of a customer being cheated.

**2.68** Suppose you intend to set up a taste test for the purpose of comparing Coca Cola with Pepsi Cola. Describe how you would carry out the study using (a) a two-sample format and (b) a paired-data format.

**2.69** Would the following be classified as paired data or correlation–regression data? Explain.

| Subject | $x_i$ (years) | $y_i$ (blood pressure) |
|---------|---------------|------------------------|
| TK | $x_1$ | $y_1$ |
| ND | $x_2$ | $y_2$ |
| BW | $x_3$ | $y_3$ |
| JB | $x_4$ | $y_4$ |

**2.70** Three men (A, B, and C) and three women (X, Y, and Z) have been nominated for an annual award recognizing the fraternity member and the sorority member doing the most community service. List the combinations making up the population of possible winners.

**2.71** Suppose you intend to study the effect of caffeine on the aggressiveness of rats. A control group of 10 rats will be fed a standard diet while a second group of 10 rats will be fed a high-caffeine diet. The 20 rats from which the two groups will be formed are currently housed in a community cage. In choosing the particular rats to be fed the high-caffeine diet, would it be acceptable simply to reach into the cage and pull out a sample of 10? Why or why not?

**2.72** A chemist is titrating a nitric acid and silver solution with sodium chloride for the purpose of precipitating silver chloride. She repeats the process several times, on each occasion using the amount of silver chloride precipitated, $y_i$, as a way of estimating the amount of silver in the original solution. Is the population represented by the $y_i$'s real or hypothetical?

**2.73** The six subjects described below have volunteered for a paired-data study designed to compare two diet plans. How would you divide the six into three groups of two?

| Subject | Age | Sex | Height | Weight |
|---------|-----|-----|--------|--------|
| BM | 23 | M | 6'2" | 240 |
| CK | 26 | M | 5'8" | 180 |
| SH | 37 | F | 5'2" | 145 |
| RR | 23 | M | 5'6" | 165 |
| JS | 21 | M | 5'11" | 225 |
| CF | 43 | F | 5'4" | 165 |

**2.74** Two hundred registered voters—120 Democrats and 80 Republicans—are asked to give an overall "positive" or "negative" rating to the incumbent president's foreign policy record. What data structure would the responses represent?

2.75  The figure below shows a straight line, $y = a + bx$, drawn through a set of $n$ points. The dashed lines represent the vertical distances of each point from the regression line. Let $L$ denote the sum of the squares of the distances represented by the dashed lines. Using sigma notation, write a formula for $L$. (Hint: The coordinates of the $i$th point are $(x_i, y_i)$ and the height of the line when $x = x_i$ is $a + bx_i$.)

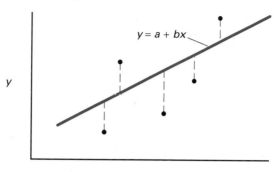

2.76  Three memory-improving techniques (A, B, and C) are taught to each of three subjects (DP, CW, and JH). Tests are given after each technique is learned. Subject DP's scores are 60, 70, and 90; CW's are 80, 80, and 70; JH's are 60, 40, and 70. What data structure is being represented?

2.77  An advertising firm wants to know whether men respond differently than women to a certain commercial. Three men and three women are shown the same 60-second video and are asked to rank its appeal on a scale of 1 to 10. The three men give it scores of 6, 8, and 7; the three women give it scores of 6, 3, and 5. What data structure is being represented?

2.78  The two-sample data shown here summarize the responses of six subjects to two treatments, $X$ and $Y$.

| Treatment X | Treatment Y |
|:---:|:---:|
| 3 | 4 |
| 1 | 3 |
| 2 | 2 |

Would it make sense to graph these data in the format shown in the figure? Why or why not?

2.79  What data model has the same relationship to randomized block data that the two-sample model has to $k$-sample data?

**2.80**   When they are presented in tabular form, what distinguishes $k$-sample data from randomized block data?

**2.81**   Why are categorical data said to be "qualitative" when the entries in contingency tables are always numerical?

**2.82**   A sample of 60 high school teachers was surveyed to see whether the size of the school they graduated from has any relationship to the level of satisfaction they derive from their careers. Of the forty teachers who graduated from a university, 30 were satisfied with their careers and 10 were not. Responding to the same question, there were 15 "satisfied" and 5 "unsatisfied" among the 20 teachers who had graduated from a four-year college. What data structure do these numbers represent?

**2.83**   Suppose $x_1, x_2, \ldots, x_n$ represent the IQ scores recorded for a sample of $n$ first-graders. How might the $x_i$'s be reduced to a set of Bernoulli variables?

**2.84**   What do two-sample data and paired data have in common? How are they different?

**2.85**   A test is given to six 10-year-old boys and six 10-year-old girls that measures their ability to relate to their peers. The scores for the boys are 76, 83, 80, 64, 73, and 79; for the girls, 86, 81, 74, 84, 92, and 71. Are these two-sample data or paired data? Why?

**2.86**   What is the difference between an $x_i$ and a $y_i$ that are dependent and an $x_i$ and a $y_i$ that are independent?

**2.87**   A market analyst is trying to construct a profile of adults who watch the network news. Each of 100 respondents is asked two questions: (a) How often do you watch a network news program? (b) How much formal education have you had? As summarized in the table, which data model do the responses represent?

|  |  | FREQUENCY OF WATCHING | | |
|---|---|---|---|---|
|  |  | Regularly | Often | Seldom |
|  | High School | 12 | 10 | 13 |
| EDUCATION | Some College | 13 | 9 | 8 |
|  | College Degree | 13 | 12 | 10 |

**2.88**   A developmental psychologist is studying the relationship between a child's age $(x)$ and his or her score on a manual dexterity test $(y)$. The $(x_i, y_i)$'s recorded for a sample of five children were (6, 82), (7, 85), (6, 76), (5, 62), and (8, 92). What data structure do these numbers represent?

**2.89**   Recall Case Study 2.1. Suppose there was concern that men might be affected differently by the enzyme than women. How might an experiment be set up using the paired data format that would test that hypothesis?

Assume that workers assigned to a similar task would receive similar exposures.

2.90 Do the following results represent categorical data or randomized block data, or do we need additional information about the Treatments and the Methods before making an identification? Explain.

|  |  | TREATMENT | | |
|---|---|---|---|---|
|  |  | A | B | C |
|  | 1 | 2.6 | 15.1 | 8.0 |
| METHOD | 2 | 11.2 | 20 9 | 10.3 |
|  | 3 | 4.6 | 18.2 | 11.9 |

2.91 What do regression–correlation data and categorical data have in common? How are they different?

# 3 Descriptive Statistics

*There never were in the world
two opinions alike, any more than two hairs or two grains;
the most universal quality is diversity.*

M. de Montaigne

### INTRODUCTION

*Recognizing* the underlying structure in a data set is the first step in a statistical analysis. *Summarizing* those data is the second. The results of an experiment cannot be effectively used unless they are displayed so that their interpretation is unambiguous. There are many ways to highlight a sample's information: some are graphical, others are numerical, a few are a little of both. Collectively, these techniques are known as ***descriptive statistics***. Using the seven data models, we will learn which descriptive procedures can be used with each kind of data.

This material is extremely useful. Our initial "picture" of a set of data is important; not only does it give us insight about what the numbers might imply, but it also suggests how to analyze them formally. For example, we will discuss a procedure for finding the equation of the straight line that best fits a set of regression–correlation data (Figures 2.10 and 2.11). But the procedure makes sense only if the data points show a linear relationship. How do we know that the points $(x_i, y_i)$ are linearly related? *We graph them.* Many of the mathematical analyses we will study are based on assumptions that can be checked by what we learn here. Descriptive statistics is useful to anyone who will ever have to write a report that involves the presentation of numerical information. In our data-oriented society that includes practically everybody.

Over the years a kind of grammar—a set of "do's" and "don'ts"—associated with displaying data has evolved. It may be unfair, but the perceived validity and significance of numerical results are inextricably linked with their presentation. In short, an experimenter's credibility suffers if he fails to summarize data in formats consistent with accepted practice. This chapter shows you when and how to use those formats.

**descriptive statistics**—Procedures for summarizing and displaying data. They may be graphical (e.g., a histogram) or numerical (e.g., a standard deviation).

## 3.2

### DESCRIPTIVE STATISTICS FOR ONE-SAMPLE DATA

We begin by looking at *graphical* descriptive statistics for the one-sample problem. We assume that the sample consists of $n$ observations $y_1, y_2, \ldots, y_n$. If $n$ is "small," meaning less than 15, the standard graphical format is a graph showing the points plotted along an imaginary vertical line. The vertical axis is scaled according to the units of the measurements; the horizontal axis is unmarked.

## CASE STUDY 3.1

*Lead Poisoning from Wine*

Studies have documented the medical risks associated with the consumption of heavy metals, especially lead. One source of this problem is table wines produced from vineyards where lead arsenate is used as an insecticide to control caterpillars.

In Australia the maximum residue limit (MRL) for lead in table wine is set by the government at 0.2 mg/L. Lead contents, however, are not indicated on wine bottles, nor are vineyards required by law to stay below the MRL. Hence, the actual amounts of lead consumed by the public are largely unknown.

Handson (65) tested eight different red table wines, all produced by Australian vineyards using lead arsenate. Table 3.1 lists the results. How might we show these data graphically? The sample size is small ($n = 8$), so the format in Figure 3.1 is the only option we have. Notice the scale on the vertical axis and the absence of any scale on the horizontal axis. (After reviewing the findings—that five out of eight wines exceeded the MRL, the Australian government banned the use of lead arsenate.)

### TABLE 3.1

| Winery | Lead Content (mg/L), $x_i$ |
|--------|-----------------------------|
| A | 0.09 |
| B | 0.22 |
| C | 0.44 |
| D | 0.06 |
| E | 0.27 |
| F | 0.47 |
| G | 0.39 |
| H | 0.10 |

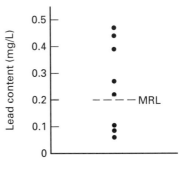

**FIG. 3.1**

QUESTIONS

**3.1** Recall the discussion in Case Study 2.2 of when scientists make their greatest discoveries.

(a) Graph the "ages" listed in Table 2.3 using the format in Figure 3.2

(b) Suppose that poets tend to produce their best work at age 30. Does the latter figure seem inconsistent with the data you just graphed?

**3.2** The *track length* of a storm (the distance it covers while maintaining a certain minimum wind velocity and precipitation intensity) is an important parameter in a storm's "profile." Listed below are the track lengths recorded for eight severe hailstorms that occurred in New England over a five-year period (56). Graph the data. Draw a line on the graph indicating the *average* track length.

| Date of Storm | Track Length (km) |
| --- | --- |
| 6 June 1961 | 16 |
| 30 June 1961 | 160 |
| 1 July 1964 | 95 |
| 1 July 1964 | 65 |
| 5 August 1964 | 30 |
| 10 August 1965 | 26 |
| 13 August 1965 | 26 |
| 7 June 1966 | 24 |

3.3   Gorillas are not the solitary creatures that many people perceive them to be: they live in groups whose average size is about 16, which usually includes 3 adult males, 6 adult females, and 7 "youngsters." Also, they stay surprisingly close to one another; the "diameter" of a group rarely exceeds 200 ft at any moment. Listed below are the sizes of 10 groups of mountain gorillas observed in the volcanic highlands of the Albert National Park in the Congo (147). Summarize these results with an appropriate graphical format.

| Group | Number of Members |
| --- | --- |
| I | 8 |
| II | 19 |
| III | 5 |
| IV | 24 |
| V | 11 |
| VI | 20 |
| VII | 18 |
| VIII | 21 |
| IX | 27 |
| XI | 16 |

## Frequency Distribution

As the number of observations increases, the helpfulness and appropriateness of a graph like Figure 3.2 diminish. *Representing* data is not always enough. When $n$ is large ($n \geq 15$), we need a format for *summarizing* data. For example, suppose a researcher has collected the 50 observations shown in Table 3.2. What impression do those data convey? Probably not much of anything. The sheer number of observations has overwhelmed our ability to discern any underlying patterns. Now look at the summary of those data

**TABLE 3.2**

| 20 | 43 | 44 | 17 | 46 | 31 | 35 | 60 | 23 | 35 |
| 31 | 45 | 44 | 41 | 35 | 37 | 43 | 33 | 38 | 29 |
| 33 | 35 | 69 | 55 | 48 | 12 | 39 | 37 | 31 | 41 |
| 28 | 36 | 32 | 36 | 37 | 43 | 25 | 46 | 86 | 45 |
| 56 | 20 | 35 | 48 | 39 | 74 | 53 | 58 | 43 | 36 |

in Table 3.3, where the original 50 observations have been grouped into eight classes. Now the data pattern is abundantly clear: the observations begin in the 10–19 range, rise quickly to a maximum in the 30–39 range, and then tail off slowly.

A listing that shows data grouped into classes and gives the number of $y_i$'s belonging to each of those classes is called a *frequency distribution*. In Table 3.3 the classes are the intervals

$$10\text{–}19, \quad 20\text{–}29, \quad \ldots, \quad 80\text{–}89$$

The frequencies are obtained by inspection; for example, two of the original $y_i$'s (12, 17) fall into the first class, six (20, 20, 23, 25, 28, 29) are contained in the second, and so on. *For one-sample problems where $n \geq 15$, constructing a frequency distribution is the first step to take in summarizing the data.*

The rules for setting up a frequency distribution are not entirely objective. Two people given the same set of $y_i$'s could each construct a different frequency distribution, each equally valid. On the other hand, the rules are not entirely subjective. Four widely adhered to guidelines restrict our options in defining classes:

1. The classes should be defined so that each observation belongs to one and only one class. (It would have been incorrect, for example, to define the first two classes as 10–20 and 20–30.)

2. Each class should have the same width. (By definition, the *class width* is the difference between successive *lower class limits*. In Table 3.3 the lower class limits are the numbers, 10, 20, 30, and so on. Therefore, the class width is

$$10 \, ( = 20 - 10 = 30 - 20 = \ldots)$$

3. The *number* of classes should not be less than 5 or greater than 15. (If we define fewer than five classes, we would probably be guilty of oversummarizing—too much detail would be lost. More than 15 classes usually undersummarizes, leaving us not much better off than when we started.)

4. The lower class limits should be multiples of 5 or 10, if possible. (This rule is the most arbitrary, but its purpose is to ensure that classes are defined using intervals to which the reader can easily

**frequency distribution**—A listing that shows the numbers of observations falling into each of several classes.

**TABLE 3.3**

| Class | Frequency |
|-------|-----------|
| 10–19 | 2 |
| 20–29 | 6 |
| 30–39 | 20 |
| 40–49 | 14 |
| 50–59 | 4 |
| 60–69 | 2 |
| 70–79 | 1 |
| 80–89 | 1 |
| | 50 |

relate. For example, the intervals

$$12-21, \quad 22-31, \quad 32-41, \quad \ldots$$

comply with rules 1, 2, and 3 in terms of the data given in Table 3.2, but they would still be a poor choice of classes. We are simply not accustomed to dealing with intervals having such spacings. Given that a class width of 10 is deemed appropriate, there is no reason to use anything other than

$$10-19, \quad 20-29, \quad 30-39, \quad \ldots,$$

or maybe       $5-14, \quad 15-24, \quad 25-34, \quad \ldots.)$

After we define the classes, we need to specify the second column in a frequency distribution. Three variations are in common use: a class's *frequency* (as in Table 3.3), its *relative frequency*, or its *percentage frequency*.

**relative frequency**—The frequency of a class divided by the total number of observations in the sample.

By definition the **relative frequency** of a class is the proportion of the entire sample that belongs to that class. Thus, the relative frequency of the 10–19 class is 2/50, or 0.04; the relative frequency of the 20–29 class is 6/50, or 0.12. Note that the sum of all the relative frequencies will always be 1 except for possible roundoff error. (Why?) The **percentage frequency** of a class is simply its relative frequency multiplied by 100. The sum of all the percentage frequencies will always equal 100 (except for possible round-off errors). Table 3.4 shows the three sets of frequencies that could have been listed for the classes in Table 3.3.

**TABLE 3.4**

| Class | Frequency | Relative Frequency | Percentage Frequency |
|-------|-----------|--------------------|----------------------|
| 10–19 | 2 | 0.04 | 4 |
| 20–29 | 6 | 0.12 | 12 |
| 30–39 | 20 | 0.40 | 40 |
| 40–49 | 14 | 0.28 | 28 |
| 50–59 | 4 | 0.08 | 8 |
| 60–69 | 2 | 0.04 | 4 |
| 70–79 | 1 | 0.02 | 2 |
| 80–89 | 1 | 0.02 | 2 |
|       | 50 | 1.00 | 100 |

If our sole objective is to describe a single set of data, any of the three options in Table 3.4 is equally effective. On the other hand, if we want to *compare* the patterns exhibited by several sets of data, then relative frequency or percentage frequency should be used because they prevent differences in sample sizes from affecting the relative shapes of the distributions.

**CASE STUDY 3.2**

*Decision Making:*
*Morality or*
*Expediency?*

How do people make decisions? What factors loom large when it comes to choosing between plan A and plan B? Researchers in this area hypothesize that decision makers can be assigned positions along a *motivation scale*. At one end of the scale are the moralists, individuals who always do what they believe to be right regardless of the consequences. At the other end sit the expedients, persons whose decisions reflect more pragmatic concerns (peer group pressure, short-range gain, desire to be liked). Most decision makers would fall somewhere between these extremes, sometimes acting out of righteousness, sometimes out of expediency.

What does the *distribution* of decision makers along this motivation scale look like? Is it uniform? Are moderate positions more common? Do people tend to favor one extreme over the other? In a study with 106 volunteers (60) each subject was presented with 37 different conflict situations. A subject received a point for every proposed resolution that seemed to be motivated by moral considerations. The 106 totals (ranging from a possible low of 0 to a possible high of 37) are listed in Table 3.5.

**TABLE 3.5**

| | | | | | | |
|---|---|---|---|---|---|---|
| 13 | 12 | 11 | 19 | 24 | 2 | 13 |
| 17 | 15 | 2 | 17 | 15 | 7 | 15 |
| 13 | 27 | 4 | 16 | 13 | 9 | 5 |
| 8 | 19 | 4 | 17 | 12 | 5 | 28 |
| 7 | 23 | 13 | 13 | 6 | 21 | 20 |
| 10 | 6 | 10 | 7 | 17 | 18 | 19 |
| 10 | 2 | 13 | 9 | 27 | 17 | 14 |
| 21 | 9 | 19 | 12 | 3 | 18 | 11 |
| 18 | 11 | 25 | 11 | 10 | 12 | 14 |
| 17 | 5 | 14 | 30 | 7 | 15 | 4 |
| 19 | 18 | 11 | 19 | 1 | 13 | 8 |
| 15 | 20 | 4 | 4 | 14 | 13 | 10 |
| 15 | 24 | 14 | 11 | 22 | 15 | 7 |
| 23 | 15 | 12 | 18 | 16 | 6 | 23 |
| 12 | 14 | 23 | 18 | 10 | 25 | 18 |
| | | | | | | 24 |

**TABLE 3.6**

| Score | Frequency |
|---|---|
| 0–4 | 10 |
| 5–9 | 16 |
| 10–14 | 33 |
| 15–19 | 29 |
| 20–24 | 12 |
| 25–29 | 5 |
| 30–34 | 1 |
| | 106 |

The smallest observation is 1; the largest, 30. Using the four guidelines on page 88, we see that a good set of classes is 0–4, 5–9, 10–14, . . . , 30–34. Table 3.6 shows the corresponding frequency distribution. Using the entries in the second column, we can address the questions raised earlier. The overall distribution, for example, is decidedly *not* uniform, in the sense that moderate scores are showing up much more frequently than extreme scores. There is also a slight tendency for low scores (the expedients) to outnumber high scores (the moralists), but we would not consider this distribution to be markedly asymmetric.

**CASE STUDY 3.3**

*Amorous Toads*    Male toads often have trouble distinguishing other male toads from female toads, a state of affairs that can lead to awkward moments during mating season. When male toad A inadvertently makes inappropriate romantic overtures to male toad B, the latter emits a short call known as a release chirp. Researchers measured the release chirps of 15 male toads innocently caught up in misadventures of the heart. Table 3.7 shows the lengths of those chirps (18).

**TABLE 3.7**

| Toad | Length of Release Chirp (s) |
|------|------------------------------|
| 1 | 0.11 |
| 2 | 0.06 |
| 3 | 0.06 |
| 4 | 0.06 |
| 5 | 0.11 |
| 6 | 0.08 |
| 7 | 0.08 |
| 8 | 0.10 |
| 9 | 0.06 |
| 10 | 0.06 |
| 11 | 0.15 |
| 12 | 0.16 |
| 13 | 0.11 |
| 14 | 0.10 |
| 15 | 0.07 |

**TABLE 3.8**

| Length of Release Chirp (s) | Relative Frequency |
|------------------------------|---------------------|
| 0.06–0.07 | 0.40 |
| 0.08–0.09 | 0.13 |
| 0.10–0.11 | 0.33 |
| 0.12–0.13 | 0.00 |
| 0.14–0.15 | 0.07 |
| 0.16–0.17 | 0.07 |
| | 1.00 |

In defining classes for these data, we need to accommodate values as small as 0.06 and as large as 0.16. One set of lower and upper class limits that meets our four guidelines would be 0.06–0.07, 0.08–0.09, . . . , 0.16–0.17. Table 3.8 shows the corresponding relative frequency distribution. There are six release chirps, for example, in the 0.06–0.07 class; that class's relative frequency is then 6/15, or 0.40. The frequencies here are showing a distinctly *skewed* pattern—short release chirps occur much more frequently than long release chirps.

---

QUESTIONS

**3.4**    Make a frequency distribution for the $FEV_1/VC$ ratios described in Case Study 2.1 and listed in Table 2.2. Show the frequency and the relative frequency associated with each class.

**3.5**    The following table shows the 20 U.S. chemical companies with the highest reported assets in 1977. The company's average annual growth rate applies to the period from 1971 through 1977 (57). Make a frequency distribution for the growth rates and show the percentage frequency for each class.

| Company | Rank | Annual Growth Rate (%) | Company | Rank | Annual Growth Rate (%) |
|---|---|---|---|---|---|
| Dow | 1 | 16.4 | Diamond Shamrock | 11 | 17.1 |
| DuPont | 2 | 10.9 | Hercules | 12 | 11.2 |
| Union Carbide | 3 | 13.1 | Dart | 13 | 9.5 |
| Monsanto | 4 | 12.4 | Stauffer | 14 | 18.9 |
| 3M | 5 | 12.4 | Olin | 15 | 1.2 |
| W. R. Grace | 6 | 10.1 | Air Products | 16 | 14.7 |
| Allied Chemical | 7 | 9.7 | Rohm & Haas | 17 | 8.3 |
| FMC | 8 | 11.8 | Ethyl | 18 | 7.3 |
| Celanese | 9 | 3.2 | Airco | 19 | 6.6 |
| Williams | 10 | 17.3 | Koppers | 20 | 12.0 |

3.6 Mauna Loa, a 14,000-ft mountain in Hawaii, is one of the most active volcanoes in the world. It averages one eruption every 3.6 years. Listed below are the intervals (in months) between 37 consecutive eruptions of Mauna Loa from 1832 to 1950 (96). Define a set of classes for these data and construct the corresponding frequency distribution. Show the frequency and the relative frequency associated with each class.

| | | | | | |
|---|---|---|---|---|---|
| 126 | 73 | 3 | 6 | 37 | 23 |
| 73 | 23 | 2 | 65 | 94 | 51 |
| 26 | 21 | 6 | 68 | 16 | 20 |
| 6 | 18 | 6 | 41 | 40 | 18 |
| 41 | 11 | 12 | 38 | 77 | 61 |
| 26 | 3 | 38 | 50 | 91 | 12 |

3.7 In a nongeriatric population the number of platelets (per cubic millimeter) in a person's blood will typically range from 140,000 to 440,000. Whether those norms are also appropriate for the elderly prompted a recent hematologic study that used 24 female patients at a rest home as subjects. The following platelet counts were recorded (162). Put these data into classes. Show the frequency associated with each class.

| | | | |
|---|---|---|---|
| 125,000 | 170,000 | 250,000 | 270,000 |
| 144,000 | 184,000 | 176,000 | 100,000 |
| 220,000 | 200,000 | 170,000 | 160,000 |
| 180,000 | 180,000 | 280,000 | 240,000 |
| 270,000 | 220,000 | 110,000 | 176,000 |
| 280,000 | 176,000 | 188,000 | 176,000 |

3.8 An occupational hazard of airline pilots is the hearing loss due to the high noise levels they are exposed to. To document the magnitude of the problem, researchers (89) measured the cockpit noise levels of 18 commercial aircraft. The results (in decibels) are listed below. Define a set of classes and construct a frequency distribution.

| | | |
|---|---|---|
| 74 | 77 | 80 |
| 82 | 82 | 85 |
| 80 | 75 | 75 |
| 72 | 90 | 87 |
| 73 | 83 | 86 |
| 83 | 83 | 80 |

**3.9**  A baseball player's strikeout ratio is computed by dividing career at bats by career strikeouts. Mickey Mantle, for example, had 8102 at bats during his career and struck out 1710 times, giving him a strikeout ratio of 4.73 ($=8102/1710$). Listed below are the 20 major leaguers who were the easiest to strike out (132). (In contrast, the player hardest to strike out was Joe Sewell, who fanned only 114 times in 7132 at bats, a ratio of 62.6.) Construct a percentage frequency distribution for these data.

| Player | Strike-Out Ratio | Player | Strike-Out Ratio |
|---|---|---|---|
| Dave Nicholson | 2.48 | Frank Howard | 4.44 |
| Dave Kingman | 3.37 | Deron Johnson | 4.51 |
| Richie Allen | 3.82 | Jimmy Wynn | 4.66 |
| Mike Schmidt | 3.96 | Mickey Mantle | 4.73 |
| Reggie Jackson | 3.97 | Harmon Killebrew | 4.80 |
| Bobby Bonds | 4.06 | Lee May | 4.85 |
| Donn Clendenon | 4.06 | Bob Allison | 4.87 |
| Rick Monday | 4.09 | Doug Rader | 4.90 |
| Willie Stargell | 4.10 | Wally Post | 4.92 |
| Woodie Held | 4.25 | Tony Perez | 5.22 |

**3.10**  Shown below is a chronology of the 30 most destructive earthquakes, worldwide, from the late 1800s to the middle 1900s. Also listed are the numbers of casualties attributed to each one (68). Summarize the number of deaths with a frequency distribution.

| Year | Place | Deaths | Year | Place | Deaths |
|---|---|---|---|---|---|
| 1897 | India | 1,500 | 1948 | Japan | 5,000 |
| 1898 | Japan | 22,000 | 1949 | Ecuador | 6,000 |
| 1906 | Chile | 1,500 | 1950 | India | 1,500 |
| 1906 | USA | 500 | 1953 | Turkey | 1,200 |
| 1907 | Jamaica | 1,400 | 1954 | Algeria | 1,600 |
| 1908 | Italy | 160,000 | 1956 | Afghanistan | 2,000 |
| 1915 | Italy | 30,000 | 1957 | Iran | 2,500 |
| 1920 | China | 180,000 | 1957 | Iran | 1,400 |
| 1923 | Japan | 143,000 | 1957 | Mongolia | 1,200 |
| 1930 | Italy | 1,500 | 1960 | Chile | 5,700 |
| 1932 | China | 70,000 | 1960 | Morocco | 12,000 |
| 1935 | Baluchistan | 60,000 | 1962 | Iran | 12,000 |
| 1939 | Chile | 30,000 | 1963 | Formosa | 100 |
| 1939 | Turkey | 40,000 | 1963 | Yugoslavia | 1,000 |
| 1946 | USA | 150 | 1964 | USA | 114 |

### The Histogram

Does our summary of a set of $y_i$'s end with the construction of their frequency distribution? Not necessarily. For visual impact it may be desirable to display the data graphically. The two most common formats are the *histogram* and the *frequency polygon*.

Mark off the horizontal axis using the same classes that appeared in the frequency distribution. Above each class, draw a bar with height equal to the frequency associated with that class. The resulting graph is a *histogram*. Figure 3.2 shows the histogram of the motivation score data from Table 3.6.

**histogram**—A graphical format for grouped data: the frequency of a class is represented by the height of a bar. The vertical axis of a histogram can be scaled in terms of either frequency, relative frequency, or percentage frequency.

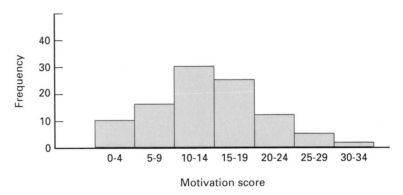

**FIG. 3.2**

If we had used relative frequency or percentage frequency on the vertical axis, the *shape* of the histogram would still be the same. Figure 3.3 shows the relative frequency version of the motivation score histogram. (The relative frequency of the 0–4 class is 10/106, or 0.094; for the 5–9 class, 16/106, or 0.151; and so on).

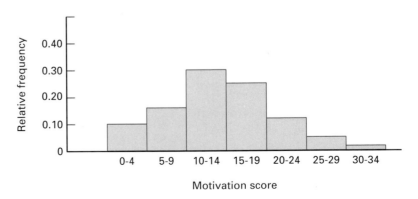

**FIG. 3.3**

## CASE STUDY 3.4

*Hurricane Tracking*   One of the most important characteristics of a distribution is its shape, and one of the most frequently encountered shapes is Figure 3.2. Distributions in which class frequencies are largest in the center and fall off *symmetrically* in the two tails are **bell-shaped**. For theoretical reasons this configuration is extremely important, and we will say much more about it later. A distribution is *not* bell-shaped if one tail is much longer than the other. In that case the distribution is said to be **skewed**. The hurricane precipitation levels described in this study are an example of the latter.

**bell-shaped histogram**—A histogram in which class frequencies are largest in the center and taper off symmetrically in the two tails.

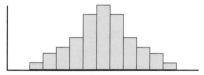

Although hurricanes generally strike only the eastern and southern coastal regions of the United States, they do occasionally sweep inland before completely dissipating. Records kept by the U.S. Weather Bureau from 1900 to 1969 document that 36 hurricanes roared across the Appalachians. Table 3.9 lists the maximum 24-hr precipitation levels measured for those 36 storms while they were over the mountains (63).

**TABLE 3.9**

| Year | Name | Location | Maximum Precipitation (in) | Year | Name | Location | Maximum Precipitation (in) |
|------|------|----------|-----|------|------|----------|-----|
| 1969 | Camille | Tye River, Va. | 31.00 | 1932 | | Caesars Head, S.C. | 4.75 |
| 1968 | Candy | Hickley, N.Y. | 2.82 | 1932 | | Rockhouse, N.C. | 6.85 |
| 1965 | Betsy | Haywood Gap, N.C. | 3.98 | 1929 | | Rockhouse, N.C. | 6.25 |
| 1960 | Brenda | Cairo, N.Y. | 4.02 | 1928 | | Roanoke, Va. | 3.42 |
| 1959 | Gracie | Big Meadows, Va. | 9.50 | 1928 | | Caesars Head, S.C. | 11.80 |
| 1957 | Audrey | Russels Point, Ohio | 4.50 | 1923 | | Mohonk Lake, N.Y. | 0.80 |
| 1955 | Connie | Slide Mt., N.Y. | 11.40 | 1923 | | Wappingers Falls, N.Y. | 3.69 |
| 1954 | Hazel | Big Meadows, Va. | 10.71 | 1920 | | Landrum, S.C. | 3.10 |
| 1954 | Carol | Eagles Mere, Pa. | 6.31 | 1916 | | Altapass, N.C. | 22.22 |
| 1952 | Able | Bloserville 1-N, Pa. | 4.95 | 1916 | | Highlands, N.C. | 7.43 |
| 1949 | | North Ford #1, N.C. | 5.64 | 1915 | | Lookout Mt., Tenn. | 5.00 |
| 1945 | | Crossmore, N.C. | 5.51 | 1915 | | Highlands, N.C. | 4.58 |
| 1942 | | Big Meadows, Va. | 13.40 | 1912 | | Norcross, Ga. | 4.46 |
| 1940 | | Rhodhiss Dam, N.C. | 9.72 | 1906 | | Horse Cove, N.C. | 8.00 |
| 1939 | | Caesars Head, S.C. | 6.47 | 1902 | | Sewanee, Tenn. | 3.73 |
| 1938 | | Hubbardston, Mass. | 10.16 | 1901 | | Linville, N.C. | 3.50 |
| 1934 | | Balcony Falls, Va. | 4.21 | 1900 | | Marrobone, Ky. | 6.20 |
| 1933 | | Peekamoose, N.Y. | 11.60 | 1900 | | St. Johnsbury, Vt. | 0.67 |

A workable set of classes for these data are the intervals 0–3.99, 4.00–7.99, 8.00–11.99, and so on (Table 3.10). Figure 3.4 shows the corresponding histogram. Is the pattern skewed? Definitely. Thirty-three of the distribution's 36 observations occur in the first three classes. What few data points remain trail off very gradually to the right.

**TABLE 3.10**

| Maximum Precipitation (in) | Frequency |
|---|---|
| 0– 3.99 | 9 |
| 4.00– 7.99 | 16 |
| 8.00–11.99 | 8 |
| 12.00–15.99 | 1 |
| 16.00–19.99 | 0 |
| 20.00–23.99 | 1 |
| 24.00–27.99 | 0 |
| 28.00–31.99 | 1 |
| | 36 |

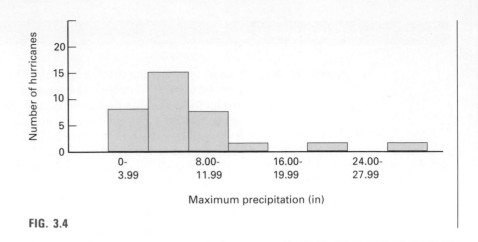

**FIG. 3.4**

QUESTIONS

3.11 Listed below are the 20 largest U.S. industrial corporations ranked according to 1968 revenues. Next to each is the number of persons they employ (109). Construct a frequency distribution for the employee figures and draw the corresponding histogram.

| Company | Rank | Employees (1000s) |
|---|---|---|
| General Motors | 1 | 728 |
| Standard Oil (of New Jersey) | 2 | 150 |
| Ford | 3 | 394 |
| General Electric | 4 | 375 |
| Chrysler | 5 | 216 |
| Mobil | 6 | 79 |
| IBM | 7 | 222 |
| Texaco | 8 | 79 |
| Gulf | 9 | 58 |
| U.S. Steel | 10 | 197 |
| Western Electric | 11 | 169 |
| Standard Oil (of California) | 12 | 47 |
| DuPont | 13 | 112 |
| Shell | 14 | 38 |
| RCA | 15 | 128 |
| McDonnell Douglas | 16 | 140 |
| Standard Oil (of Indiana) | 17 | 45 |
| Westinghouse | 18 | 132 |
| Boeing | 19 | 142 |
| Swift | 20 | 48 |

3.12 Draw a histogram for the data in Question 3.5. Put percentage frequency on the vertical axis.

3.13  Sixty-eight free-speed measurements (in feet per second) were recorded for traffic on the New Jersey Turnpike (40). Free speeds are speeds observed for cars traveling in uncongested lanes, meaning they represent, to some extent, the driver's *desired* speed. Summarize these data with a histogram.

| 91 | 97 | 90 | 113 | 80 | 93 | 87 |
|-----|-----|-----|-----|-----|-----|-----|
| 102 | 95 | 86 | 101 | 80 | 82 | 81 |
| 82 | 94 | 102 | 95 | 91 | 97 | 106 |
| 83 | 76 | 107 | 102 | 72 | 104 | 105 |
| 92 | 101 | 89 | 84 | 97 | 101 | 76 |
| 92 | 106 | 86 | 88 | 98 | 86 | 90 |
| 104 | 105 | 92 | 82 | 99 | 86 | 95 |
| 108 | 90 | 97 | 75 | 108 | 89 | 107 |
| 96 | 93 | 98 | 79 | 91 | 82 | 78 |
| 104 | 99 | 83 | 80 | 92 | | |

3.14  Draw the relative frequency histogram for the Mauna Loa eruption intervals in Question 3.6.

3.15  Synovial fluid is the clear, viscid secretion that lubricates joints and tendons. For some conditions its hydrogen ion concentration (pH) has diagnostic significance. In healthy adults the average pH for synovial fluid is 7.39. Listed below are the synovial pH values for fluids drawn from the knees of 44 patients with various arthritic conditions (166). Construct a set of classes for these data and draw the corresponding relative frequency histogram.

| 7.02 | 7.26 | 7.31 | 7.14 | 7.45 | 7.32 | 7.21 |
|------|------|------|------|------|------|------|
| 7.35 | 7.25 | 7.24 | 7.20 | 7.39 | 7.40 | 7.33 |
| 7.32 | 7.35 | 7.34 | 7.41 | 7.28 | 6.99 | 7.28 |
| 7.33 | 7.38 | 7.32 | 7.77 | 7.34 | 7.10 | 7.35 |
| 7.15 | 7.20 | 7.34 | 7.12 | 7.22 | 7.30 | 7.24 |
| 7.36 | 7.09 | 7.32 | 6.95 | 7.35 | 7.36 | 6.60 |
| 7.29 | 7.31 | | | | | |

### The Frequency Polygon

**frequency polygon**—A descriptive technique that presents graphically the information contained in a frequency distribution. The frequency of a class is represented by the height of a point drawn above the midpoint of that class. Successive points are connected with straight lines. The vertical axis can be scaled in terms of either frequency, relative frequency, or percentage frequency.

A second common graphical format is the *frequency polygon*. Along the horizontal axis we mark off the *midpoint* of each class. (The class midpoint is the average of its lower and upper class limits. If a class is defined to be 5–9, for example, the midpoint is $(5 + 9)/2 = 7$. Above the midpoint of each class we plot a point at a height equal to the *frequency* of that class. We then connect the points with straight lines. Figure 3.5 is the frequency polygon for the motivation data in Table 3.6. We could have scaled the vertical axis using either frequency, relative frequency, or percentage frequency.

For representing a single set of data, the histogram format and frequency polygon format are equivalent: they provide the same information and create the same visual effect. In situations, though, where several sets of data are to be *compared* on the same set of axes, the frequency polygon

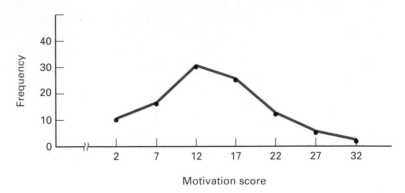

**FIG. 3.5**

format is much more useful. Superimposing two or more histograms inevitably results in a cluttered-looking graph; frequency polygons, on the other hand, can be superimposed with little or no loss in clarity.

## CASE STUDY 3.5

*Mark Twain or Quintus Curtius Snodgrass?*

Mark Twain's role in the Civil War has long been a subject of historical debate. Some scholars have suggested that he was a Confederate deserter; others maintain that his loyalty to the South was unswerving. Twain himself refused to shed any light on the matter. In his later years he put off biographers who pressed him on the point with the warning, "When I was younger, I could remember anything, whether it happened or not; but now I am getting old, and soon I shall remember only the latter."

Curiously enough, an important piece of evidence brought forth in support of Twain's military achievements is a set of 10 letters, signed Quintus Curtius Snodgrass, that appeared in the *New Orleans Daily Crescent* in 1861. The letters purported to chronicle the author's adventures as a member of the Louisiana militia. Historians generally agree that the accounts are true, but there is no record of anyone of that name. Adding to the mystery is the fact that the style of the letters bears unmistakable traces (to some critics) of the humor and irony that made Mark Twain so famous.

In the past, incidents of disputed authorship have not been uncommon. Several tracts in the Bible, for example, have been attributed by different scholars to different sources. Speculation has persisted for several hundred years that some of Shakespeare's works were written by Sir Francis Bacon. And whether it was Alexander Hamilton or James Madison who wrote certain of the Federalist papers is still an open question.

Efforts to unravel controversies of this type often rely heavily on historical clues, but not always. A distinctly statistical approach can be taken. Studies have shown that writers leave verbal "fingerprints"—an author will use roughly the same proportion of, say, three-letter words in something he writes this year as he did in whatever he wrote last year. But the proportion of three-letter words that author A consistently uses will very likely be different than the proportion of three-letter words that author B uses. Theoretically, then, by constructing a word-length frequency count for essays known to be written by Mark Twain and comparing that to a similar count made for the Snodgrass letters, we can assess the likelihood that the two authors are the same.

Table 3.11 shows the word-length distributions for three sets of essays known to be the work of Mark Twain (16). Of the 1885 words making up sample 1, for example, 74 are one-letter words (giving a relative frequency of 74/1885 = 0.039), 349 are two-letter words, and so on. Figure 3.6 on page 100 superimposes the relative frequency polygons for the three Twain samples. (Note that the classes are single values rather than intervals.)

**TABLE 3.11**

| Word Length | SAMPLE 1 | | SAMPLE 2 | | SAMPLE 3 | |
|---|---|---|---|---|---|---|
| | Frequency | Relative Frequency | Frequency | Relative Frequency | Frequency | Relative Frequency |
| 1 | 74 | 0.039 | 312 | 0.051 | 116 | 0.039 |
| 2 | 349 | 0.185 | 1.146 | 0.188 | 496 | 0.167 |
| 3 | 456 | 0.242 | 1,394 | 0.228 | 673 | 0.226 |
| 4 | 374 | 0.198 | 1.177 | 0.193 | 565 | 0.190 |
| 5 | 212 | 0.113 | 661 | 0.108 | 381 | 0.128 |
| 6 | 127 | 0.067 | 442 | 0.072 | 249 | 0.084 |
| 7 | 107 | 0.057 | 367 | 0.060 | 185 | 0.062 |
| 8 | 84 | 0.045 | 231 | 0.038 | 125 | 0.042 |
| 9 | 45 | 0.024 | 181 | 0.030 | 94 | 0.032 |
| 10 | 27 | 0.014 | 109 | 0.018 | 51 | 0.017 |
| 11 | 13 | 0.007 | 50 | 0.008 | 23 | 0.008 |
| 12 | 8 | 0.004 | 24 | 0.004 | 8 | 0.003 |
| 13+ | 9 | 0.005 | 12 | 0.002 | 8 | 0.005 |
| | 1,885 | 1.000 | 6,106 | 1.000 | 2,974 | 1.003 |

Two observations can be drawn. First, the similarity of the three polygons attests to the constancy of Twain's writing style: the frequencies with which he used words of certain lengths remained pretty much the same from essay to essay. Second, the graph gives us a definitive word-length pattern, a standard for Twain against which we can compare the letters of Snodgrass. Are the two men the same? Answer Question 3.16 and judge for yourself.

---

**QUESTIONS**

**3.16** Tabulated below is the word-length distribution for the 13,175 words in the 10 Snodgrass letters. Sketch the relative frequency polygon for these data on the three Twain polygons in Figure 3.6. How well do the four polygons agree? Does it seem likely that Twain and Snodgrass were the same person?

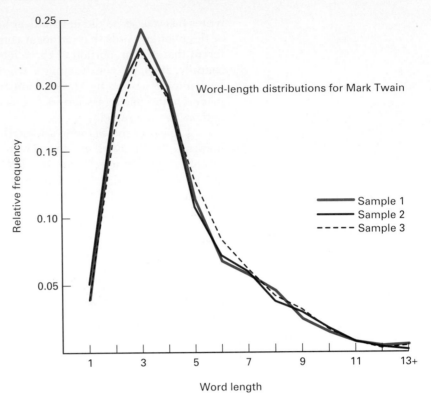

**FIG. 3.6**

| Word Length | Frequency | Relative Frequency |
|:---:|:---:|:---:|
| 1 | 424 | 0.032 |
| 2 | 2,685 | 0.204 |
| 3 | 2,752 | 0.209 |
| 4 | 2,302 | 0.175 |
| 5 | 1,431 | 0.109 |
| 6 | 992 | 0.075 |
| 7 | 896 | 0.068 |
| 8 | 638 | 0.048 |
| 9 | 465 | 0.035 |
| 10 | 276 | 0.021 |
| 11 | 152 | 0.011 |
| 12 | 101 | 0.008 |
| 13+ | 61 | 0.005 |
| | 13,175 | 1.000 |

3.17   (a) Was it necessary in Figure 3.6 to scale the vertical axis with relative frequency rather than frequency? Would percentage frequency have worked?

(b) Was it necessary to construct more than one polygon for Twain? Could we have made any inference on the basis of one polygon for Twain and one polygon for Snodgrass?

(c) In a paternity suit certain medical information (for example, blood type and length of pregnancy) can be introduced to prove the defendant is not the child's father. But there is no way that such information can prove that the defendant is the child's father. Is it likely that statistical tests of authorship are similarly one-directional? Explain.

3.18 Draw a frequency polygon for the cockpit noise levels in Question 3.8.

3.19 Summarize the strikeout ratios in Question 3.9 with a percentage frequency polygon.

3.20 In 1969 an extensive survey of American college and university professors revealed that their political preferences were closely tied to their fields of interest. A sociologist, for example, reacts to public issues quite differently than an engineer, even though the two might be identical with respect to all the usual political indices. Summarized below are the political leanings of faculty members in agriculture, medicine, law, and social science (95). Superimpose a set of percentage frequency polygons based on these data. Use "very liberal," "liberal," and so on, as the classes.

| Political Leaning | Agri. | Med. | Law | Soc. Sci. |
|---|---|---|---|---|
| Very liberal | 3% | 13% | 23% | 34% |
| Liberal | 10 | 25 | 28 | 30 |
| Moderate | 16 | 20 | 18 | 17 |
| Conservative | 30 | 24 | 18 | 13 |
| Very conservative | 42 | 19 | 14 | 7 |

## *Measures of Central Tendency*

Summarizing a set of one-sample data typically begins with one of the formats we just covered. When $n$ is small ($<15$), we draw a simple point graph (Figure 3.1). For larger sample sizes we choose among the frequency distribution, the histogram, or the frequency polygon. All of these formats give us an overview of the entire distribution.

Any further summarizing of the data necessarily takes a more numerical turn and has a narrower focus. In particular, we want to define numbers that reflect specific features of a distribution. Two especially significant features are *location* and *dispersion*. By **location**, we mean the "position" of the data along the horizontal axis; **dispersion** refers to the extent to which the observations are spread out—that is, how variable they are.

Figure 3.7a shows two distributions that differ with respect to location (but have the same dispersion). The data sets in Figure 3.7b have different dispersions but the same location. In the rest of this section we will look

**location**—The position or center of a distribution. Location is measured by the sample mean. If the distribution is markedly skewed, location is more appropriately measured by the sample median.

**dispersion**—The amount of scatter in a set of $y_i$'s.

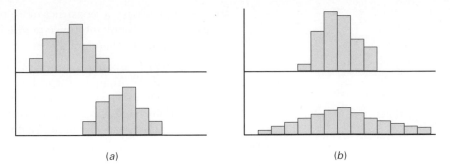

(a)                                                                      (b)

**FIG. 3.7**

at ways of measuring and interpreting these two critically important properties.

THE SAMPLE MEAN   Whether we realize it or not, we have all had considerable experience in computing numbers that measure location. Suppose your grades for three chemistry exams were 85, 90, and 51 (you had two other tests the same day, and your cat died). How would you summarize numerically the level of your performance? By computing the *average* of your three exam grades:

$$\text{Exam average} = \frac{85 + 90 + 51}{3} = 75.3$$

In statistics we call the average of a set of numbers the **sample mean**.

---

**DEFINITION 3.1    The Sample Mean**

Let $y_1, y_2, \ldots, y_n$ be a set of $n$ observations. The **sample mean** $\bar{y}$ is the average of the $y_i$'s:

$$\bar{y} = \frac{1}{n} \sum_{i=1}^{n} y_i \tag{3.1}$$

---

**CASE STUDY 3.6**

*A Mosquito's*   *Aedes aegypti* is the scientific name of the mosquito that transmits yellow
*Feast*   fever. Although no longer a major health problem in the West, for almost 200 years yellow fever was probably the most devastating communicable disease in the United States.

Controlling any communicable disease requires that we understand its *vector* (in this case, the mosquito). Jones and Pilitt (83) investigated the length of time it takes the *Aedes* mosquito to complete a feeding. For five young *A. aegypti* females the experiment was a mosquito's dream come true. They were allowed to bite an exposed human forearm, undisturbed and unswatted. The resulting bite durations are recorded in Table 3.12. *On the average*, how long did it take these mosquitoes to complete their repast?

Let $y_i$, $i = 1, 2, 3, 4, 5$, denote the bite duration for the $i$th mosquito. What happens on the average is measured by the sample mean. Applying Definition 3.1,

$$\bar{y} = \frac{1}{n} \sum_{i=1}^{n} y_i$$

$$- \frac{1}{5} (176.0 + 202.9 + 315.0 + 374.6 + 352.5)$$

$$= 284.2 \text{ seconds}$$

Mosquitoes may be lacking in other social graces, but they can't be accused of eating and running: it takes them an average of almost 5 minutes to complete a single bite!

**TABLE 3.12**

| Mosquito | Bite Duration (s) |
|----------|-------------------|
| 1 | 176.0 |
| 2 | 202.9 |
| 3 | 315.0 |
| 4 | 374.6 |
| 5 | 352.5 |

## QUESTIONS

**3.21** Look again at Case Study 2.2. Compute the average age of the scientists making the 12 discoveries listed in Table 2.3.

**3.22** Let $y_1, y_2, \ldots, y_{10}$ denote the set of radiation levels measured at the 10 department stores described in Question 2.7. Compute $\bar{y}$.

**3.23** *Phainopepla nitens* is a small black bird indigenous to the western and southwestern United States. As part of a study investigating the energy expenditures of these birds, estimates were made of the number of hours per day that they spend in flight. Listed below are the values obtained for six of the birds living in the Santa Monica Mountains (178). Compute the sample mean.

| | |
|------|------|
| 3.77 | 3.13 |
| 3.34 | 3.92 |
| 4.20 | 3.52 |

**3.24** If a set of data appears in histogram form, we cannot compute the sample mean because the exact $y_i$'s are unknown. However, we can approximate $\bar{y}$ by assuming that each observation in a given class is equal to the midpoint of that class. For example, suppose we were given the breakdown of six observations shown in the table on the left.

If we assumed that the two observations in the first class equaled 4.5 and the four in the second class equaled 14.5, the "mean" would be

$$\frac{4.5 + 4.5 + 14.5 + 14.5 + 14.5 + 14.5}{6} = 11.2$$

| Class | Midpoint | Frequency |
|-------|----------|-----------|
| 0–9 | 4.5 | 2 |
| 10–19 | 14.5 | 4 |
| | | 6 |

Because of the original form of the data, an average computed in this fashion is called a **grouped mean**. In general, if classes 1, 2, . . . , $k$ have midpoints $m_1, m_2, \ldots, m_k$ and frequencies $f_1, f_2, \ldots, f_k$ where

$$f_1 + f_2 + \cdots + f_k = n$$

then the grouped mean $\bar{y}_g$ is defined as

$$\bar{y}_g = \frac{1}{n} \sum_{i=1}^{k} f_i m_i$$

Compute the grouped mean for the motivation score data in Case Study 3.2.

3.25  Listed below are a recent year's traffic death rates (per 100 million motor vehicle miles) for each of the 50 states (103). Make a histogram of these observations, using as classes 2.0–2.9, 3.0–3.9, and so on. Compute $\bar{y}_g$ (see Question 3.24) and compare it with $\bar{y}$.

| | | | | | |
|---|---|---|---|---|---|
| Ala | 6.4 | La | 7.1 | Ohio | 4.5 |
| Alaska | 8.8 | Maine | 4.6 | Okla | 5.0 |
| Ariz | 6.2 | Mass | 3.5 | Ore | 5.3 |
| Ark | 5.6 | Md | 3.9 | Pa | 4.1 |
| Cal | 4.4 | Mich | 4.2 | RI | 3.0 |
| Colo | 5.3 | Minn | 4.6 | SC | 6.5 |
| Conn | 2.8 | Miss | 5.6 | SD | 5.4 |
| Del | 5.2 | Mo | 5.6 | Tenn | 7.1 |
| Fla | 5.5 | Mont | 7.0 | Tex | 5.2 |
| Ga | 6.1 | NC | 6.2 | Utah | 5.5 |
| Hawaii | 4.7 | ND | 4.8 | Va | 4.5 |
| Idaho | 7.1 | Neb | 4.4 | Vt | 4.7 |
| Ill | 4.3 | Nev | 8.0 | WVa | 6.2 |
| Ind | 5.1 | NH | 4.6 | Wash | 4.3 |
| Iowa | 5.9 | NJ | 3.2 | Wisc | 4.7 |
| Kans | 5.0 | NM | 8.0 | Wy | 6.5 |
| Ky | 5.6 | NY | 4.7 | | |

THE SAMPLE MEDIAN  Although Definition 3.1 carries no explicit restrictions, we do not always use $\bar{y}$ to measure location. In some situations the sample mean is misleading and should be avoided. For example, consider the set of observations 6, 4, 5, 6, 100. For those numbers $\bar{y} = 24.2$, yet that average is nowhere near any of the $y_i$'s (see Figure 3.8). The reason is that the one extremely large observation dominates $\sum_{i=1}^{5} y_i$ and inflates $\bar{y}$.

In general, the sample mean is most useful when it coincides with the bulk of a distribution (see, for example, Figure 3.9). When that happens, $\bar{y}$ represents not only the location of the distribution but also the value of a "typical" observation. On the other hand, for a distribution that is sharply skewed (see Figure 3.10), $\bar{y}$ will be heavily influenced by the extremely large (or extremely small) observations in the tail and will move away from where most of the $y_i$'s lie. The sample mean is then less representative of the data, and its numerical value could be deceptive.

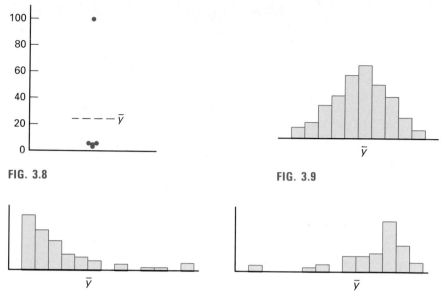

FIG. 3.8

FIG. 3.9

FIG. 3.10

Another measure of location (or central tendency) is available and preferred, for skewed distributions—the *sample median* $\tilde{y}$.

---

**DEFINITION 3.2   The Sample Median**

For a set of $n$ observations $y_1, y_2, \ldots, y_n$ the **sample median** $\tilde{y}$ is

1. The middle observation (in terms of magnitude) if $n$ is odd.
2. The average of the two middle observations (in terms of magnitude) if $n$ is even.

---

*EXAMPLE 3.1*   An experimenter records the following sample of size 5. Find $\tilde{y}$.

| Observation Number | $y_i$ |
|:---:|:---:|
| 1 | 25 |
| 2 | 17 |
| 3 | 160 |
| 4 | 34 |
| 5 | 33 |

*Solution*    Arrange the $y_i$'s from smallest to largest:

$$17 < 25 < 33 < 34 < 160$$

Since $n$ is odd, the sample median is the middle observation: $\tilde{y} = 33$. Notice that $\tilde{y}$ is unaffected by the magnitude of the largest observation. Even though the 160 would have greatly inflated (and distorted) $\bar{y}$, it does not unduly affect $\tilde{y}$. Indeed, $\tilde{y}$ would be the same if the largest value had been 160,000,000.

---

*EXAMPLE 3.2*    Consider the following hypothetical sample of size 6. Find $\tilde{y}$.

| Observation Number | $x_i$ |
|:---:|:---:|
| 1 | 48 |
| 2 | 60 |
| 3 | 71 |
| 4 | 4 |
| 5 | 64 |
| 6 | 49 |

*Solution*    Arranging these numbers from smallest to largest shows that the middle two observations are 49 and 60:

$$4 < 48 < 49 < 60 < 64 < 71$$

Therefore,
$$\tilde{y} = \frac{49 + 60}{2} = 54.5$$

---

**CASE STUDY 3.7**

*Survival Times*    Survival time is a widely used index for measuring the effectiveness of a medical treatment or the severity of a serious disease. It follows that computing an "average" survival time is a reasonable way to summarize a set of $n$ such observations. But what kind of average should we use? Experience has shown that distributions of survival times are frequently skewed, so median survival times tend to be more informative than mean survival times.

Table 3.13 lists the survival times of 11 acute leukemia patients after they received bone marrow transplants (58). Here the only option we have for describing location is the median survival time, because one of the subjects was still alive when the study ended. Not knowing Y.O.'s exact survival time makes it impossible to compute $\bar{y}$, but the 451+ causes no problems in determining $\tilde{y}$.

**TABLE 3.13**

| Patient | Post-Transplant Survival Time (days) |
|---------|--------------------------------------|
| J.B. | 131 |
| Y.O. | 451+ |
| S.A. | 90 |
| L.P. | 332 |
| J.J. | 33 |
| M.P. | 66 |
| D.V. | 92 |
| J.T. | 47 |
| W.E. | 32 |
| T.W. | 215 |
| P.W. | 75 |

Arranging the observations from smallest to largest gives

$$32 < 33 < 47 < 66 < 75 < 90 < 92 < 131 < 215 < 352 < 451+$$

Since $n$ is odd, we look for the middle observation; so $\tilde{y} = 90$ days.

---

**QUESTIONS**

**3.26** Find the sample median for the bacteria count data in Question 2.9. Would $\bar{y}$ or $\tilde{y}$ be the better measure of location for these data? Why?

**3.27** Compute $\tilde{y}$ for the mosquito bite data in Case Study 3.6.

**3.28** Tree roots can lengthen anywhere from 1 mm/day to more than 5 cm/day, depending on the type of tree, the season, and the amount of daylight (roots grow more at night than they do during the day). Listed below are elongations recorded during daylight hours for six *Prunus avium* roots (88). Find the sample median for these data.

| Root Number | Growth (mm) |
|-------------|-------------|
| 1 | 3.9 |
| 2 | 2.3 |
| 3 | 3.4 |
| 4 | 3.3 |
| 5 | 4.0 |
| 6 | 3.6 |

**3.29** One other measure of location is occasionally used. In a set of $n$ observations the *sample mode* $y_m$ is the $y$ value that occurs most often. For example, if a sample of size 7 is the set of numbers 3.6, 2.8, 2.8, 4.3, 5.6, 2.8, and 4.3, the sample mode is 2.8. Compute $\bar{y}$, $\tilde{y}$, and $y_m$ for the parachute data in Question 2.10.

**3.30**   Dental structure provides an effective criterion for classifying certain fossils. Listed below are the third molar lengths of nine baboons belonging to the genus *Papio* (106).

| Specimen | Third Molar Length (mm) |
|----------|-------------------------|
| MCZ   8466 | 8.3 |
| MCZ   29790 | 8.5 |
| MCZ   10570 | 8.6 |
| MCZ   11395 | 7.8 |
| MCZ   29791 | 8.1 |
| MCZ   29792 | 8.1 |
| MCZ   26472 | 7.2 |
| MCZ   29789 | 8.2 |
| MCZ   22752 | 8.8 |

(a) Compute $\bar{y}$, $\tilde{y}$, and $y_m$.

(b) Recently a baboon skull of unknown origin was discovered in a cave in Angola. Its third molar length was 9.0 mm. Is it likely that this baboon was a member of the genus *Papio*?

**THE SAMPLE STANDARD DEVIATION** Having dealt at length with the measurement of location, we now turn our attention to the second of the two properties of distributions mentioned earlier—*dispersion*. Conceptually, dispersion is the extent to which all the $y_i$'s in a sample are not all the same. (If they are all the same, the dispersion is zero.) In general, distributions whose $y_i$'s are numerically similar have a small dispersion; those whose $y_i$'s vary over a broad range have a large dispersion (recall Figure 3.7).

The question is, what kind of number can we define to reflect the scatter in a set of points (in the same way that we defined $\bar{y}$ to quantify location)? It might seem reasonable to measure dispersion by computing the average deviation of each of the points from $\bar{y}$ that is, by forming the quotient

$$D = \frac{\sum_{i=1}^{n}(y_i - \bar{y})}{n} \tag{3.2}$$

Equation 3.2 might look like a good idea, but it won't work. Why not? Because it doesn't measure dispersion! No matter how spread out or how concentrated the data points are, $D$ is *always* zero, since

$$\frac{1}{n}\sum_{i=1}^{n}(y_i - \bar{y}) = \frac{1}{n}\sum_{i=1}^{n}y_i - \frac{1}{n}\sum_{i=1}^{n}\bar{y} = \bar{y} - \frac{n\bar{y}}{n} = \bar{y} - \bar{y} = 0$$

In effect, the negative and positive deviations from $\bar{y}$ cancel each other.

To keep $D$ from summing to zero, we can square each deviation before performing the addition. That is, let

$$D' = \frac{1}{n} \sum_{i=1}^{n} (y_i - \bar{y})^2 \tag{3.3}$$

Equation 3.3 does reflect the scatter in a set of points, but a few additional modifications give an even better measure of dispersion. Note that, as a consequence of squaring $y_i - \bar{y}$, we have also squared the units of $D'$: if the $y_i$'s are given in pounds, for example, $D'$ is in terms of pounds squared. For obvious reasons, we would prefer to express the $y_i$'s and the number measuring the dispersion of those $y_i$'s on the same scale. Taking the square root of $D'$ will achieve that objective: that is,

$$\sqrt{D'} = \sqrt{\frac{1}{n} \sum_{i=1}^{n} (y_i - \bar{y})^2} \tag{3.4}$$

will have the same units as the $y_i$'s.

Equation 3.4 is almost what we want. For a technical reason beyond our concern, it is standard practice to use $n - 1$ rather than $n$ in the denominator. The resulting expression is called the *sample standard deviation s*.

---

**DEFINITION 3.3   The Sample Standard Derivation**

For a set of $n$ observations $y_1, y_2, \ldots, y_n$ the dispersion of the $y_i$'s is measured by the **sample standard deviation s**, where

$$s = \sqrt{\frac{1}{n-1} \sum_{i=1}^{n} (y_i - \bar{y})^2} \tag{3.5}$$

---

**CASE STUDY 3.8**

*Armadillos*   When the Spanish invaded the New World, they encountered a strange, rodentlike creature unlike anything they had seen before. Giving it a name was a challenge, but they finally settled on "armadillo," meaning "little armored thing."

Appearance, as it turns out, is not the only thing strange about armadillos. Some of their behavioral traits are also a bit unusual. They are, for example, inordinately fond of sleeping. Van Twyver and Allison (170) recorded the number of hours slept per day for six different nine-banded armadillos. The results are shown in Table 3.14. Find the corresponding sample standard deviation.

**TABLE 3.14**

| Armadillo | Hours Slept per Day, $y_i$. |
|-----------|------------------------------|
| CE | 17.4 |
| JE | 18.5 |
| CM | 20.0 |
| FR | 16.1 |
| SS | 15.3 |
| TR | 17.2 |

**TABLE 3.15**

| $y_i$ | $y_i - 17.4$ | $(y_i - 17.4)^2$ |
|-------|--------------|-------------------|
| 17.4 | 0 | 0 |
| 18.5 | 1.1 | 1.21 |
| 20.0 | 2.6 | 6.76 |
| 16.1 | −1.3 | 1.69 |
| 15.3 | −2.1 | 4.41 |
| 17.2 | −0.2 | 0.04 |
| | | 14.11 |

Note, first of all, that the sample mean for these data is 17.4 hours:

$$\bar{y} = \frac{1}{6} \sum_{i=1}^{6} y_i = \frac{1}{6}(17.4 + 18.5 + \cdots + 17.2) = 17.4 \text{ hr}$$

Having found $\bar{y}$, we can compute $\sum_{i=1}^{n} (y_i - \bar{y})^2$: Table 3.15 shows the details. From Equation 3.5,

$$s = \sqrt{\frac{14.11}{6-1}} = \sqrt{2.82} = 1.7 \text{ hr}$$

QUESTIONS

**3.31** Pick any two sets of four $y_i$'s. Compute the corresponding $\bar{y}$'s. Verify that $\sum_{i=1}^{4} (y_i - \bar{y}) = 0$ in both cases (which is what the short argument after Equation 3.2 proves in general).

**3.32** For the vampire bat blood consumption percentages given in Question 2.8, $\bar{y} = 45.9$. Use the method in Table 3.15 to find the sample standard deviation for those eight observations.

**3.33** Compute the sample standard deviation for the five winning 100-m times in Question 2.50.

**3.34** The grouped mean $\bar{y}_g$ was defined in Question 3.24. A similar approximation can be made for the sample standard deviation when the data are given as a frequency distribution rather than as a set of individual $y_i$'s. If classes $1, 2, \ldots, k$ have midpoints $m_1, m_2, \ldots, m_k$ and frequencies $f_1, f_2, \ldots, f_k$ (where $f_1 + f_2 + \cdots + f_k = n$), then the grouped standard deviation $s_g$ is defined as

$$s_g = \sqrt{\frac{\sum_{i=1}^{n} f_i(m_i - \bar{y}_g)^2}{n-1}}$$

Compute $s_g$ for the motivation scores in Case Study 3.2. Use the data in Table 3.6.

A COMPUTING FORMULA FOR $s$ In practice, we never use Equation 3.5 to calculate $s$. The sequence of steps it requires—taking a difference, squaring the difference, adding the squares—is awkward, even for a calculator. A much easier expression to use is the *computing formula for s* given in Theorem 3.1. Algebraically, the two formulas are equivalent, but we leave the proof of that statement to Question 3.40.

---

**THEOREM 3.1**   Computing Formula for $s$

The sample standard deviation $s$ for a set of observations $y_1, y_2, \ldots, y_n$ can be computed from the formula

$$s = \sqrt{\frac{n \sum_{i=1}^{n} y_i^2 - \left( \sum_{i=1}^{n} y_i \right)^2}{n(n-1)}} \qquad (3.6)$$

---

*EXAMPLE 3.3*    For the data in Case Study 3.8, Table 3.16 shows how to form the sums we need to find $s$ by using Equation 3.6. Substituting

$$\sum_{i=1}^{6} y_i = 104.5 \quad \text{and} \quad \sum_{i=1}^{6} y_i^2 = 1834.15$$

**TABLE 3.16**

| $y_i$ | $y_i^2$ |
|-------|---------|
| 17.4 | 302.76 |
| 18.5 | 342.25 |
| 20.0 | 400.00 |
| 16.1 | 259.21 |
| 15.3 | 234.09 |
| 17.2 | 295.84 |
| 104.5 | 1834.15 |

into Equation 3.6 gives

$$s = \sqrt{\frac{6(1834.15) - (104.5)^2}{6(5)}} = \sqrt{2.82} = 1.7 \text{ hr}$$

which is the same answer we found by using Equation 3.5.

---

QUESTIONS

3.35    For the 10 radiation levels in Question 2.7,

$$\sum_{i=1}^{10} y_i = 4.84 \quad \text{and} \quad \sum_{i=1}^{10} y_i^2 = 2.8602$$

Use Equation 3.6 to find $s$.

**3.36**   Find the sample standard deviation for the $FEV_1/VC$ ratios in Table 2.2. Note that

$$\sum_{i=1}^{19} y_i = 14.56 \quad \text{and} \quad \sum_{i=1}^{19} y_i^2 = 11.2904$$

**3.37**   Use Equation 3.6 to find the sample standard deviation for the 12 ages at discovery in Case Study 2.2.

**3.38**   Compute the sample standard deviation of the nine fossil molar lengths in Question 3.30.

**3.39**   The formula for the grouped standard deviation given in Question 3.34 can be simplified in the same way that Theorem 3.1 simplified the formula for the ungrouped standard deviation. Specifically, if classes $1, 2, \ldots, k$ have midpoints $m_1, m_2, \ldots, m_k$ and frequencies $f_1, f_2, \ldots, f_k$ (where $f_1 + f_2 + \cdots + f_k = n$), then

$$s_g = \sqrt{\frac{n \sum_{i=1}^{k} f_i m_i^2 - \left( \sum_{i=1}^{k} f_i m_i \right)^2}{n(n-1)}}$$

Compute $s_g$ for the motivation scores in Table 3.6 of Case Study 3.2. Compare your answer with your result from Question 3.34.

**3.40**   Prove that the two formulas for $s$ (Equations 3.5 and 3.6) are algebraically equivalent. Hint: Begin by writing $(y_i - \bar{y})^2$ as $y_i^2 - 2y_i\bar{y} + \bar{y}^2$ and remember that

$$\sum_{i=1}^{n} \bar{y} = n\bar{y}$$

**INTERPRETING THE STANDARD DEVIATION** Computing a sample standard deviation is one problem; *interpreting* that number is quite another. What does it mean, for example, to say that the sample standard deviation of the armadillo sleep durations is 1.7 hr? Is that figure large or small, good or bad? Is there some inference we can draw from that information? Unfortunately, the questions have no easy answers. We all have an intuitive feeling for what the sample mean represents, but the sample standard deviation is difficult to pin down. The best we can do at this point is to make three statements, the origin of which will have to be deferred until Chapter 6:

**STATEMENT 1:** If the $y_i$'s have a bell-shaped histogram (see Case Study 3.4), approximately 68% of the data points lie in the interval from $\bar{y} - s$ to $\bar{y} + s$.

**STATEMENT 2:** If the $y_i$'s have a bell-shaped histogram, approximately 95% of the data points lie in the interval from $\bar{y} - 2s$ to $\bar{y} + 2s$.

**STATEMENT 3:** If the $y_i$'s have a bell-shaped histogram, *almost all* the data points lie in the interval from $\bar{y} - 3s$ to $\bar{y} + 3s$.

Each statement relates the value of $s$ to the notion of dispersion, albeit in a roundabout way. If, for example, $\bar{y}$ and $s$ for a set of data were 10.0 and 2.0, respectively, then Statement 1 says that approximately 68% of the distribution is in the interval (8, 12), since $\bar{y} - s = 8.0$ and $\bar{y} + s = 12.0$. In other words, a standard deviation of 2.0 implies that the variation among the $y_i$'s is such that we can expect approximately 68% of the observations to lie within 2.0 units of $\bar{y}$. Figure 3.11 illustrates the first two statements graphically.

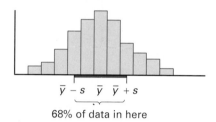

**FIG. 3.11**

---

*EXAMPLE 3.4*   Figure 3.2 is the histogram for the 106 $y_i$'s in Table 3.5. Since the distribution is roughly bell-shaped, the three statements about the standard deviation should apply. Verify that they do.

*Solution*   We first find $\bar{y}$ and $s$. Adding the $y_i$'s and the $y_i^2$'s gives

$$\sum_{i=1}^{106} y_i = 1464 \quad \text{and} \quad \sum_{i=1}^{106} y_i^2 = 24{,}670$$

Therefore,
$$\bar{y} = \frac{1464}{106} = 13.8$$

and
$$s = \sqrt{\frac{(106)(24{,}670) - (1464)^2}{(106)(105)}} = 6.5$$

According to Statement 1, approximately 68% of the observations should lie in the interval from 7.3 ($= 13.8 - 6.5$) to 20.3 ($= 13.8 + 6.5$) or, since the observations are whole numbers, from 8 to 20, inclusive. How many actually did fall in that range? From Table 3.5, we find that exactly 69 entries had values between 8 and 20, inclusive. Therefore, the actual percentage of observations in the $\bar{y} - s$ to $\bar{y} + s$ range is $69/106 = 65\%$, a figure close to the 68% predicted by Statement 1.

Statement 2 predicts that approximately 95% of the observations lie in the interval

$$13.8 - 2(6.5) = 0.8 \quad \text{to} \quad 13.8 + 2(6.5) = 26.8$$

or from 1 to 26. All but four observations from Table 3.5 are in that range.

The actual proportion, therefore, of observations in the interval from $\bar{y} - 2s$ to $\bar{y} + 2s$ is 102/106 (or 96%), a figure very close to Statement 2's predicted value of 95%.

---

QUESTIONS

**3.41**   Does Statement 3 hold for the motivation score data analyzed in Example 3.4?

**3.42**   Verify Statements 1, 2, and 3 for the traffic fatality data given in Question 3.25. Note that

$$\sum_{i=1}^{50} y_i = 266.5 \quad \text{and} \quad \sum_{i=1}^{50} y_i^2 = 1499.91$$

**3.43**   Suppose the sample mean and the sample standard deviation for a set of bell-shaped data are 160 and 10, respectively. What percentage of the observations would we expect to be

(a) between 150 and 170?       (b) between 140 and 180?

(c) between 150 and 180?       (d) less than 170?

(e) greater than 190?

**3.44**   Approximately 43% of the observations in a bell-shaped distribution lie in the interval from $\bar{y}$ to $\bar{y} + 1.5s$; another 43% lie in the interval from $\bar{y} - 1.5s$ to $\bar{y}$.

(a) What percentage of the distribution would we expect to find in the interval from $\bar{y} - s$ to $\bar{y} + 1.5s$? in the interval from $\bar{y} - 1.5s$ to $\bar{y} + 2s$?

(b) If a set of data has $\bar{y} = 28$ and $s = 4$, what proportion of the observations are likely to be in the interval from 22 to 32? from 34 to 36?

**3.45**   Consider the following set of 40 observations, all rounded off to the nearest integer. Compute $\bar{y}$ and $s$. What proportion of the sample is within one standard deviation of the mean? Why is the actual percentage of observations in the interval from $\bar{y} - s$ to $\bar{y} + s$ so different from the 68% predicted by Statement 1?

| | | | | |
|---|---|---|---|---|
| 1 | 8 | 1 | 12 | 1 |
| 6 | 1 | 12 | 12 | 4 |
| 2 | 12 | 4 | 1 | 11 |
| 11 | 7 | 10 | 1 | 5 |
| 1 | 12 | 11 | 8 | 9 |
| 10 | 3 | 11 | 2 | 12 |
| 12 | 12 | 12 | 10 | 2 |
| 3 | 1 | 1 | 9 | 2 |

### Relative Standing

Histograms, means, medians, and standard deviations all characterize the nature of the $y_i$'s. The histogram pictures the overall pattern exhibited by the data, the mean and the median estimate the average observation, and the standard deviation measures dispersion. But none of these techniques tells us *where* an observation falls in a distribution. Knowing the latter—that is, knowing an observation's relative standing—is sometimes absolutely essential.

Suppose you earned a 620 on the verbal portion of the SAT. Could you expect to make good grades in English class? Maybe yes and maybe no. If almost everyone else in your class has verbal SATs from 750 to 800, then a 620, relatively speaking, is marginal at best and you will probably struggle just to pass. On the other hand, if most of your classmates have verbal SATs less than 500, then you will be the pacesetter. The point is that any observation can have entirely different implications, depending on where that observation falls relative to the rest of the distribution.

How do we indicate relative standing? Probably the most common way is to express the observation as a *percentile*. By definition, a **percentile** is the value that exceeds a certain proportion of the entire sample (or population). For example, if 5% of all the students in your English class have verbal SATs *less than or equal to 620*, then 620 is the fifth percentile (see Figure 3.12*a*). On the other hand, if 90% of your classmates have scores *less than or equal to 620*, then 620 is the 90th percentile (see Figure 3.12*b*).

**percentile**—A number having the property that it equals or exceeds a certain percentage of all the observations in a distribution.

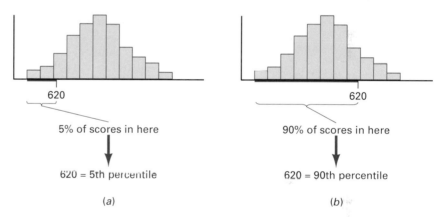

(a)

(b)

**FIG. 3.12**

Notice how much better we can put the 620 in perspective when we know the percentile it represents. Being told only that our score was 620 leaves unanswered the crucial question: is 620 a "good" score or a "bad" score? With the percentile information, we can answer that question: if 620 is the 5th percentile, then 620 is a low score; if 620 is the 90th percentile, it's a high score.

Certain percentiles are given special names: the 25th percentile is often called the **first quartile**; the 75th percentile is the **third quartile**. The **second quartile** (or the 50th percentile) is the median (see Figure 3.13*a*). Percentiles that are multiples of 10 are sometimes called **deciles**: the 10th percentile, for example, is the **first decile**, the 20th percentile is the **second decile**, and so on (see Figure 3.13*b*).

**quartile**—Percentiles that are multiples of 25.

**decile**—Percentiles that are multiples of 10.

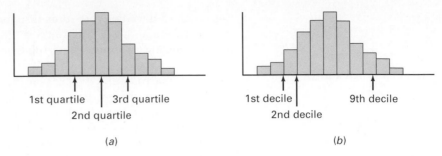

1st quartile | 3rd quartile

2nd quartile

1st decile | 9th decile

2nd decile

(a)

(b)

FIG. 3.13

## CASE STUDY 3.9

*Fingerprint Facts*

Diversity, wrote the philosopher Montaigne, is the "most universal quality." There may be no more familiar example of that statement than fingerprints. Formed during the beginning of the second trimester, the delicate swirls and ridges that make up a fingerprint leave each individual with a unique identity. No two sets of fingerprints are the same (even in "identical" twins), and their original pattern never changes.

Despite their manifest differences, fingerprints do have certain things in common. Toward the end of the last century, Sir Francis Galton classified all fingerprints into three generic types: the whorl, the loop, and the arch (Figure 3.14). A few years later, Sir Edward Richard Henry, a pioneer criminologist who later became Commissioner of Scotland Yard, refined Galton's system to include eight generic types. The Henry system, as it came to be known, revolutionized the way criminals are identified. It is still used by the FBI.

Not unexpectedly, the classification of fingerprints inspired a form of fortune telling (known more elegantly as dactylomancy). According to those who believe in such things, a person "having whorls on all fingers is restless, vacillating, doubting, sensitive, clever, eager for action, and inclined to crime." A "mixture of loops and whorls signifies a neutral character, a person who is kind, obedient, truthful, but often undecided and impatient" (33).

Many characteristics, in addition to those proposed by Galton and Henry, can be used to distinguish fingerprints. One is the *ridge count*. Look again at Figure 3.14. The loop pattern shows a point, known as the *triradius*, where the three ridge systems meet. A straight line drawn from the triradius to the center of the loop crosses a number of ridges (11 in Figure 3.15). If the number of crossings is determined for each finger and those numbers are added together, the resulting sum is the *ridge count*.

Table 3.17 is a frequency distribution of the ridge counts found for 825 males (33). What ridge count is exceeded by 75% of the sample? That is, what is the 25th percentile of the ridge count distribution?

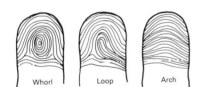

Whorl     Loop     Arch

FIG. 3.14

FIG. 3.15

**TABLE 3.17**

| Ridge Count | Frequency |
|-------------|-----------|
| 0–19 | 10 |
| 20–39 | 12 |
| 40–59 | 24 |
| 60–79 | 40 |
| 80–99 | 73 |
| 100–119 | 100 |
| 120–139 | 90 |
| 140–159 | 117 |
| 160–179 | 139 |
| 180–199 | 100 |
| 200–219 | 67 |
| 220–239 | 36 |
| 240–259 | 10 |
| 260–279 | 4 |
| 280–299 | 3 |
| | 825 |

**TABLE 3.18**

| Ridge Count | Cumulative Frequency | % Cumulative Frequency |
|-------------|----------------------|------------------------|
| ≤19 | 10 | 1.2 |
| ≤39 | 22 | 2.7 |
| ≤59 | 46 | 5.6 |
| ≤79 | 86 | 10.4 |
| ≤99 | 159 | 19.3 |
| ≤119 | 259 | 31.4 |
| ≤139 | 349 | 42.3 |
| ≤159 | 466 | 56.5 |
| ≤179 | 605 | 73.3 |
| ≤199 | 705 | 85.4 |
| ≤219 | 772 | 93.6 |
| ≤239 | 808 | 97.9 |
| ≤259 | 818 | 99.2 |
| ≤279 | 822 | 99.6 |
| ≤299 | 825 | 100.0 |

**percentage frequency**—One hundred times the frequency of a class divided by the total number of observations in the sample.

When we estimate percentiles, it helps to reexpress the frequency distribution as a cumulative frequency distribution. Table 3.18 is the percentage cumulative frequency distribution based on the numbers in Table 3.17. Notice how the classes have been redefined. Each one includes all the observations *less than or equal to its upper limit*, hence the name *cumulative frequency*. We compute the **percentage cumulative frequency** by dividing the total sample size (825) into the cumulative frequency and multiplying by 100. For example, the percentage cumulative frequency for the ≤99 class is

$$\frac{159}{825} \times 100 = 19.3\%$$

If one of the cumulative percentage frequencies is exactly 25.0, then the 25th percentile, $y_{(25)}$, would equal the upper limit for that class. (Why?) In general, though, life is not that simple, and we have to make an approximation. Here 19.3% of the observations are less than or equal to 99, and 31.4% are less than or equal to 119. It follows that $y_{(25)}$ is between 99 and 119 (see Figure 3.16). A more explicit statement would necessarily be an approximation based on interpolation.

Figure 3.16 suggests how to proceed. We would expect $y_{(25)}$ to be proportionally the same distance from 99 that 25 is from 19.3. That is,

$$\frac{x}{119 - 99} = \frac{25 - 19.3}{31.4 - 19.3}$$

| 99 | $y_{(25)}$ | 119 |
|----|-----------|-----|
| ‖ | ‖ | ‖ |
| 19.3rd percentile | 25th percentile | 31.4th percentile |

**FIG. 3.16**

Solving this equation gives

$$x(31.4 - 19.3) = (119 - 99)(25 - 19.3)$$

$$12.1x = 20(5.7)$$

$$x = 9.4$$

Therefore, an estimate of the 25th percentile of the ridge count distribution is

$$y_{(25)} = 99 + 9.4 = 108.4$$

---

QUESTIONS

**3.46**  Estimate the 65th percentile of the ridge count distribution as summarized in Table 3.18.

**3.47**  In general, which is numerically larger, the third quartile or the seventh decile?

**3.48**  Suppose the distribution for a certain variable is bell-shaped. Will $y_{(51)} - y_{(49)}$ be less than, equal to, or greater than $y_{(3)} - y_{(1)}$? Explain.

**3.49**  Construct a percentage cumulative frequency distribution for the motivation scores based on the information in Table 3.6. Estimate (a) the fourth decile, (b) the third quartile, (c) the 97th percentile.

**3.50**  The *interquartile range Q* is defined to be the difference between the third quartile and the first quartile:

$$Q = y_{(75)} - y_{(25)}$$

Is $Q$ a measure of location or dispersion? Explain.

# 3.3

## DESCRIPTIVE STATISTICS FOR TWO-SAMPLE DATA

Formats for summarizing and displaying two-sample data are very similar to those for one-sample data. Observations $x_1, x_2, \ldots, x_n$ will denote the sample representing treatment X, and $y_1, y_2, \ldots, y_m$ the sample representing treatment Y. In general, what we do depends on the sizes of $n$ and $m$ and on whether we want to present the data graphically or numerically.

If $n$ and $m$ are small ($<15$), the standard graphical format is a plot of the two sets of points along two imaginary vertical lines (as in Figure 3.1). Although simple, such graphs are quite useful in pointing out shifts in location or dispersion. For data sets where $n$ and $m$ are larger, histograms or frequency polygons are more effective formats. Of course, regardless of the magnitude of $n$ and $m$, we can always summarize each sample's location and dispersion numerically by computing its mean and standard deviation.

## CASE STUDY 3.10

***Crustacean Chemistry***

Male fiddler crabs solicit attention from the opposite sex by standing in front of their burrows and waving their claws at the females who walk by. If a female likes what she sees, she pays the male a brief visit in his burrow. If everything goes well and the crustacean chemistry is just right, she will stay a little longer and mate.

In what may be a ploy to lessen the risk of spending the night alone, some of the males build elaborate mud domes over their burrows. Zucker (196) observed each of 12 courting males for 10 to 20 min. Five males had constructed a dome; seven had not. Table 3.19 shows the percentages of time each of the 12 spent waving to passing females.

**TABLE 3.19**

| % OF TIME SPENT WAVING TO FEMALES | |
|---|---|
| Males with Domes, $x_i$ | Males without Domes, $y_i$ |
| 100.0 | 76.4 |
| 58.6 | 84.2 |
| 93.5 | 96.5 |
| 83.6 | 88.8 |
| 84.1 | 85.3 |
| | 79.1 |
| | 83.6 |

Figure 3.17 shows the corresponding graph. There seems to be no strong indication here that the presence or absence of a dome has any demonstrable effect on the *average* amount of time a male crab spends waving at female crabs. But, for whatever reason, the *dispersion* of the $x_i$'s is somewhat larger than that of the $y_i$'s.

These last two impressions should be reflected in the means and standard deviations of the two samples. Adding the $x_i$'s and $x_i^2$'s for the males with domes gives

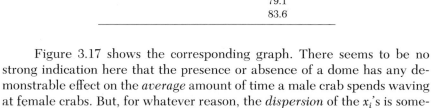

$$\sum_{i=1}^{5} x_i = 419.8 \quad \text{and} \quad \sum_{i=1}^{5} x_i^2 = 36{,}237.98$$

Thus,

$$\bar{x} = \frac{419.8}{5} = 84.0$$

and

$$s_X = \sqrt{\frac{5(36{,}237.98) - (419.8)^2}{5(4)}} = 15.7$$

Similarly, for males without domes,

$$\sum_{i=1}^{7} y_i = 593.9 \quad \text{and} \quad \sum_{i=1}^{7} y_i^2 = 50{,}646.15$$

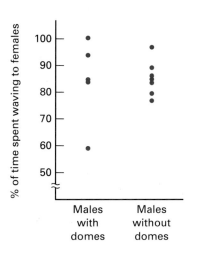

**FIG. 3.17**

Therefore,
$$\bar{y} = \frac{593.9}{7} = 84.8$$

and
$$s_Y = \sqrt{\frac{7(50{,}646.15) - (593.9)^2}{7(6)}} = 6.6$$

The numerical descriptive statistics, then, do agree with Figure 3.17: the means for the two samples are quite similar, but the standard deviation of the $x_i$'s is more than twice the standard deviation of the $y_i$'s.

## CASE STUDY 3.11

*Parent-Child Relationships*

One of the standard personality inventories used by psychologists is the thematic apperception test (TAT), which consists of a series of pictures of everyday experiences. The subject is asked to examine the pictures and to invent a story about each one. Interpreted properly, the content of those stories can provide valuable insights into the subject's mental well-being.

In order for this test to have any credibility, it must be able to distinguish one personality type from another. That is, the distribution of TAT scores achieved by individuals having personality type $X$ must differ with respect to shape, location, dispersion, or a combination of all three from the distribution of scores characteristic of individuals having personality type $Y$. In this study that idea is carried one step further. Two groups of children are involved, one normal and one schizophrenic, but the question is whether their *mothers* will score differently on the TAT (181).

Forty mothers participated, 20 of whom had schizophrenic children. Each mother was shown the same set of 10 pictures, and the stories they made up were categorized according to the parent-child relationship exhibited. The data in Table 3.20 give the number of those stories (out of 10)

**TABLE 3.20 Number of Stories (out of 10) Showing a Positive Parent-Child Relationship**

| NORMAL CHILDREN | | | | SCHIZOPHRENIC CHILDREN | | | |
|---|---|---|---|---|---|---|---|
| Mother | Score, $x_i$ | Mother | Score, $x_i$ | Mother | Score, $y_i$ | Mother | Score, $y_i$ |
| 1 | 8 | 11 | 2 | 21 | 2 | 31 | 0 |
| 2 | 4 | 12 | 1 | 22 | 1 | 32 | 2 |
| 3 | 6 | 13 | 1 | 23 | 1 | 33 | 4 |
| 4 | 3 | 14 | 4 | 24 | 3 | 34 | 2 |
| 5 | 1 | 15 | 3 | 25 | 2 | 35 | 3 |
| 6 | 4 | 16 | 3 | 26 | 7 | 36 | 3 |
| 7 | 4 | 17 | 2 | 27 | 2 | 37 | 0 |
| 8 | 6 | 18 | 6 | 28 | 1 | 38 | 1 |
| 9 | 4 | 19 | 3 | 29 | 3 | 39 | 2 |
| 10 | 2 | 20 | 4 | 30 | 1 | 40 | 2 |

showing a *positive* parent-child relationship—one where the mother was clearly capable of interacting with her child in a flexible, open-minded way.

Figure 3.18 shows these data as two histograms. Does the graph tell us anything? Definitely. The histogram for mothers of schizophrenic children is noticeably shifted to the left of the histogram for mothers of normal children, implying that mothers of schizophrenic children are more likely to be experiencing a strained parent-child relationship.

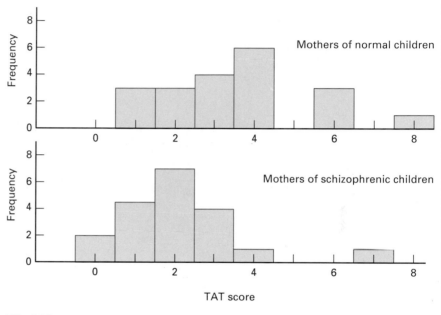

**FIG. 3.18**

## CASE STUDY 3.12

*The Measure of a Man*

Scientific journals often summarize two-sample data by a format that is graphical and numerical at the same time. Figure 3.19, which appears at the end of the data we are about to describe, is one such example.

According to recent speculation, height and life expectancy may be related, with short people enjoying longer lives than tall people. Death certificates of short and tall baseball players and short and tall boxers appear to bear out that contention (146). Another source of data on which to check out the theory is former U.S. presidents. Table 3.21 lists the 31 presidents who have died of natural causes.

Let $x_i$ denote the age at death for the $i$th short president, $i = 1, 2, \ldots,$ 5. Let $y_i$ denote the age at death for the $i$th tall president, $i = 1, 2, \ldots, 26.$

**TABLE 3.21**

| SHORT PRESIDENTS (<5'8") | | | TALL PRESIDENTS (≥5'8") | | |
|---|---|---|---|---|---|
| President | Height | Age | President | Height | Age |
| Madison | 5'4" | 85 | W. Harrison | 5'8" | 68 |
| Van Buren | 5'6" | 79 | Polk | 5'8" | 53 |
| B. Harrison | 5'6" | 67 | Taylor | 5'8" | 65 |
| J. Adams | 5'7" | 90 | Grant | 5'8½" | 63 |
| J. Q. Adams | 5'7" | 80 | Hayes | 5'8½" | 70 |
| | | | Truman | 5'9" | 88 |
| | | | Fillmore | 5'9" | 74 |
| | | | Pierce | 5'10" | 64 |
| | | | A. Johnson | 5'10" | 66 |
| | | | T. Roosevelt | 5'10" | 60 |
| | | | Coolidge | 5'10" | 60 |
| | | | Eisenhower | 5'10" | 78 |
| | | | Cleveland | 5'11" | 71 |
| | | | Wilson | 5'11" | 67 |
| | | | Hoover | 5'11" | 90 |
| | | | Monroe | 6' | 73 |
| | | | Tyler | 6' | 71 |
| | | | Buchanan | 6' | 77 |
| | | | Taft | 6' | 72 |
| | | | Harding | 6' | 57 |
| | | | Jackson | 6'1" | 78 |
| | | | Washington | 6'2" | 67 |
| | | | Arthur | 6'2" | 56 |
| | | | F. Roosevelt | 6'2" | 63 |
| | | | L. Johnson | 6'2" | 64 |
| | | | Jefferson | 6'2½" | 83 |

For the short presidents,

$$\sum_{i=1}^{5} x_i = 401 \quad \text{and} \quad \sum_{i=1}^{5} x_i^2 = 32{,}455$$

Hence,

$$\bar{x} = \frac{401}{5} = 80.2 \text{ years}$$

and

$$s_X = \sqrt{\frac{5(32{,}455) - (401)^2}{5(4)}} = 8.6 \text{ years}$$

For the tall presidents,

$$\sum_{i=1}^{26} y_i = 1798 \quad \text{and} \quad \sum_{i=1}^{26} y_i^2 = 126{,}508$$

Thus,

$$\bar{y} = \frac{1798}{26} = 69.2 \text{ years}$$

and

$$s_Y = \sqrt{\frac{26(126{,}508) - (1798)^2}{26(25)}} = 9.3 \text{ years}$$

Our objective is to incorporate all this information into a graphical format.

Note that $s_X$ and $s_Y$ measure the dispersions *of the observations*. But the $x_i$'s and $y_i$'s are not the only variables in an experiment. If we compute quantities *from* those observations, such as $\bar{x}$ and $\bar{y}$, then those quantities should also be considered variables. However, for reasons we will defer until Chapter 7, the dispersion of a mean is *less* than the dispersion of the observations that went into that mean. In particular, the standard deviation of a mean (also known as the **standard error of the mean**) is equal to the standard deviation of the observations *divided by the square root of the sample size*. Here, the two standard errors of the mean are $s_X/\sqrt{n}$ and $s_Y/\sqrt{m}$.

**standard error of the mean**—A number that reflects the dispersion that would be observed among a set of sample means if each mean was based on a different random sample.

A common way to present two-sample data is to plot the two means as points and draw an interval around each one representing one standard error above that value and one standard error below. The *width* of these intervals indicates the "precision" of $\bar{x}$ and $\bar{y}$: the shorter the interval, the more confidence we have that if the experiment were repeated the "new" sample mean would be close to the "old" sample mean.

Figure 3.19 shows the two intervals

$$\left(\bar{x} - \frac{s_X}{\sqrt{n}}, \bar{x} + \frac{s_X}{\sqrt{n}}\right) = \left(80.2 - \frac{8.6}{\sqrt{5}}, 80.2 + \frac{8.6}{\sqrt{5}}\right)$$

$$= (76.4 \text{ years}, 84.0 \text{ years})$$

and

$$\left(\bar{y} - \frac{s_Y}{\sqrt{m}}, \bar{y} + \frac{s_Y}{\sqrt{m}}\right) = \left(69.2 - \frac{9.3}{\sqrt{26}}, 69.2 + \frac{9.3}{\sqrt{26}}\right)$$

$$= (67.4 \text{ years}, 71.0 \text{ years}).$$

The fact that the two intervals fail to overlap suggests that the *difference* between the two sample means, 69.2 years versus 80.2 years, may be *statistically significant*. We will learn what that means in precise terms in Chapter 10. In not-so-precise terms, it suggests that the data support the theory that short people live longer than tall people.

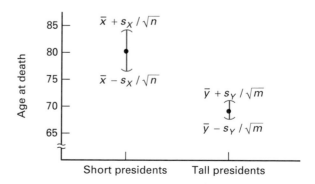

**FIG. 3.19**

**3.51** A parameter used to evaluate myocardial function is the end diastolic volume (EDV). Listed in the table are the EDVs (in mL/mm$^2$) recorded for eight persons with normal cardiac function and for six with constrictive pericarditis (173).

| Normal, $x_i$ | Constrictive Pericarditis, $y_i$ |
|---|---|
| 62 | 24 |
| 60 | 56 |
| 78 | 42 |
| 62 | 74 |
| 49 | 44 |
| 67 | 28 |
| 80 | |
| 48 | |

(a) Summarize the location and dispersion of these two samples by computing $\bar{x}$, $\bar{y}$, $s_X$, and $s_Y$.

$$\sum_{i=1}^{8} x_i = 506 \qquad \sum_{i=1}^{8} x_i^2 = 32{,}966$$

$$\sum_{i=1}^{6} y_i = 268 \qquad \sum_{i=1}^{6} y_i^2 = 13{,}672$$

(b) Draw a graph of these data in the format of Figure 3.17. Indicate the location of $\bar{x}$ and $\bar{y}$.

**3.52** Nod-swimming in male ducks is a highly ritualized behavioral trait. The term refers to the rapid back-and-forth movement of a duck's head. It frequently occurs during courtship displays and occasionally occurs when the duck is approached by another male perceived to have higher status. It *may* depend on the duck's "race." In an experiment investigating the latter (91), two sets of green-winged teals, American and European, were photographed for several days. Listed below are the frequencies (per 10,000 frames of film) with which each bird initiated the nod-swimming motion.

| American Male | Frequency, $x_i$ | European Male | Frequency, $y_i$ |
|---|---|---|---|
| 07 | 14.6 | 8 | 3.6 |
| 59 | 28.8 | 4 | 8.2 |
| 14 | 19.1 | 6 | 7.8 |
| 58 | 23.1 | 18 | 27.5 |
| 10 | 50.3 | 3 | 7.0 |
| 6 | 35.7 | 15 | 19.7 |
| | | 13 | 17.0 |
| | | 1 | 3.5 |
| | | 12 | 13.3 |
| | | 18 | 12.4 |
| | | 2 | 19.0 |
| | | 17 | 14.1 |

(a) Compute the sample mean and the sample standard deviation for both sets of data.

$$\sum_{i=1}^{6} x_i = 171.6 \qquad \sum_{i=1}^{6} x_i^2 = 5745.60$$

$$\sum_{i=1}^{12} y_i = 159.5 \qquad \sum_{i=1}^{12} y_i^2 = 2567.05$$

(b) Draw a point graph for these data in the style of Figure 3.17. Does the graph reflect the values for $\bar{x}$, $\bar{y}$, $s_X$, and $s_Y$ found in part (a)?

**3.53** Do male fruit flies spend as much time preening themselves as female fruit flies do? Make histograms for the following two sets of data and judge for yourself. The numbers listed are the average times spent (in seconds) for a cleaning bout. The subjects were 15 male fruit flies (*Drosophila melanogaster*) and 15 female fruit flies (30). Put the two histograms on the same graph, using the format in Figure 3.18.

| Male Times, $x_i$ | | | Female Times, $y_i$ | | |
|---|---|---|---|---|---|
| 2.3 | 1.9 | 3.3 | 3.7 | 5.4 | 2.2 |
| 2.9 | 2.2 | 1.3 | 11.7 | 2.8 | 2.4 |
| 2.2 | 2.4 | 2.1 | 4.0 | 2.8 | 2.0 |
| 1.2 | 2.0 | 2.7 | 2.8 | 2.4 | 2.4 |
| 2.3 | 1.9 | 1.2 | 2.9 | 10.7 | 3.2 |

**3.54** Another part of the *Phainopepla nitens* energy expenditure study cited in Question 3.23 compared the perching times of birds living in the Colorado Desert with the perching times of birds living in the Santa Monica Mountains. Listed below are the percentages of a day's activities spent in perching by 17 birds living in the desert and 15 birds living in the mountains (178). Summarize these data by drawing two relative frequency histograms. Use the format in Figure 3.18. Describe the differences and the similarities between the distributions.

| % Time Perching, $x_i$ (desert birds) | | % Time Perching, $y_i$ (mountain birds) | |
|---|---|---|---|
| 83.0 | 85.6 | 74.0 | 70.8 |
| 90.1 | 84.9 | 74.1 | 40.7 |
| 85.4 | 85.0 | 35.2 | 37.9 |
| 90.7 | 39.8 | 34.1 | 31.8 |
| 53.4 | 38.5 | 30.2 | 63.0 |
| 36.8 | 72.6 | 65.1 | 58.0 |
| 71.1 | 73.9 | 54.5 | 54.0 |
| 68.9 | 70.9 | 48.7 | |
| 66.4 | | | |

**3.55** Compute $\bar{x}$, $\bar{y}$, $s_X$, and $s_Y$ for the silver-content data in Case Study 2.4. Graph the intervals

$$\left(\bar{x} - \frac{s_X}{\sqrt{n}}, \bar{x} + \frac{s_X}{\sqrt{n}}\right) \quad \text{and} \quad \left(\bar{y} - \frac{s_Y}{\sqrt{m}}, \bar{y} + \frac{s_Y}{\sqrt{m}}\right)$$

using the format in Figure 3.19.

$$\sum_{i=1}^{9} x_i = 60.7 \qquad \sum_{i=1}^{9} x_i^2 = 411.75$$

$$\sum_{i=1}^{7} y_i = 39.3 \qquad \sum_{i=1}^{7} y_i^2 = 221.43$$

3.56    Graph the caffeine-content data of Question 2.19. Follow the steps outlined in Case Study 3.12.

# 3.4

## DESCRIPTIVE STATISTICS FOR PAIRED DATA

Descriptive techniques for paired data are a little different than what we have seen so far, because of the dependence between $x_i$ and $y_i$ (see Section 2.4). Whether we opt for a graphical format or a numerical format, the specific relationship of $x_i$ to $y_i$ (due to their membership in pair $i$) must be emphasized.

We must pay attention to the one-to-one relationship between $x_i$ and $y_i$ even when simply writing down the data. *Tables of paired data must have at least three columns.* The second and third columns are the $x_i$'s and $y_i$'s; the first column indicates the nature of the pairing. Look again at Tables 2.7 and 2.8. In the former the first column shows that each player is being paired with himself, that is, he runs both paths. In Table 2.8 the pairs are subjects who scored similarly on the height avoidance pretest. The two in pair A, for example, had the lowest scores; those in pair E, the highest scores.

Because the $x_i$'s and $y_i$'s are dependent, graphical formats for paired data are limited. When the number of pairs is large, there are no practical ways to show the one-to-one relationship between the $x_i$'s and $y_i$'s. For small $n$, though, we can use the format in Case Study 3.13.

## CASE STUDY 3.13

*Cerebral Circulation*    One reason cited for the mental deterioration so often seen in the very elderly is the reduction in cerebral blood flow that accompanies the aging process. If that speculation is correct, it would follow that drugs capable of dilating blood vessels and stimulating arterial circulation might be able to retard mental senility.

Eleven residents of a rest home volunteered for a study where they received daily doses of cyclandelate, a widely used vasodilator. At the beginning of the experiment, radioactive tracers were used to determine each subject's *mean circulation time* (MCT, the number of seconds it takes blood to travel from the carotid artery to the jugular vein). After four months on the drug each subject's MCT was measured again. The results before and after are listed in Table 3.22 (5).

**TABLE 3.22  Mean Circulation Time**

| Subject | Before, $x_i$ | After, $y_i$ |
|---------|--------|-------|
| J.B.    | 15     | 13    |
| M.B.    | 12     | 8     |
| A.B.    | 12     | 12.5  |
| M.B.    | 14     | 12    |
| J.L.    | 13     | 12    |
| S.M.    | 13     | 12.5  |
| M.M.    | 13     | 12.5  |
| S.McA.  | 12     | 14    |
| A.McL.  | 12.5   | 12    |
| F.S.    | 12     | 11    |
| P.W.    | 12.5   | 10    |

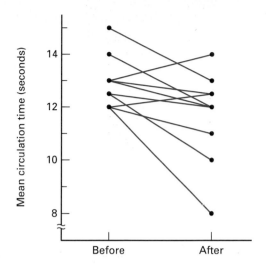

**FIG. 3.20**

Figure 3.20 is the corresponding graph. The lines connect the two MCTs recorded for a given subject. The "lowest" line on the graph, for example, shows the response pattern for subject M.B., whose MCT dropped from a "before" of 12 s to an "after" of 8 s. Cyclandelate seems to have accomplished what the researchers had predicted: most of the lines are sloping downward, indicating that cerebral circulation generally improved.

**QUESTIONS**

**3.57**  Research has shown that speech patterns can be used to measure the severity of certain mental illnesses. A speech pattern frequently used is the type-token ratio (TTR), which is the number of different words a person uses divided by the total number of words spoken. To see whether TTRs can distinguish between normal people and schizophrenics, researchers defined the following set of four pairs, where the matching was based on ethnic group, sex, age, and education (all factors that affect speech patterns):

| | SCHIZOPHRENICS | | | | | NORMALS | | | |
|------|-----------------|-----|------|--------------|------|-----------------|-----|------|--------------|
| Pair | Ethnic Group | Sex | Age | Educ. (yr) | Pair | Ethnic Group | Sex | Age | Educ. (yr) |
| 1 | Negro   | M | 27 | 10 | 1 | Negro   | M | 30 | 9  |
| 2 | Negro   | F | 40 | 10 | 2 | Negro   | F | 32 | 10 |
| 3 | Italian | F | 44 | 8  | 3 | Italian | F | 40 | 6  |
| 4 | Jewish  | M | 24 | 12 | 4 | Jewish  | M | 23 | 14 |

Each subject was asked to talk about any topic whatsoever for 30 min. Listed below are the TTRs computed from those eight monologues (64). Graph these data in the format shown in Figure 3.20. Does the pattern of lines suggest that the TTR would be useful as a diagnostic tool for treating schizophrenia?

| Pair | Schizophrenic ($x_i$) | Normal ($y_i$) |
|------|----------------------|----------------|
| 1 | 0.329 | 0.332 |
| 2 | 0.322 | 0.349 |
| 3 | 0.288 | 0.294 |
| 4 | 0.343 | 0.360 |

3.58 Graph the HAT data described in Case Study 2.8. In general, why would it be incorrect to use the format of Figure 3.20 on data like the first-walking times shown in Case Study 2.5?

### Within-Pair Responses

As pointed out in Section 2.4, the key to analyzing paired data is the set of *within-pair* response differences

$$d_i = y_i - x_i \quad (i = 1, 2, \ldots, n)$$

Those are the numbers that measure the relative effects of treatments $X$ and $Y$. If the $d_i$'s are all close to zero, for example, the reasonable inference would be that the two treatments, on the average, elicit essentially the same response. In general, whatever we do numerically to summarize a set of paired data must necessarily focus on the $d_i$'s.

---

**CASE STUDY 3.14**

---

*Hypnosis and ESP*  Several research projects in extrasensory perception (ESP) have examined the possibility that hypnosis may help strengthen a person's ESP. The obvious way to test such a hypothesis is with a self-paired experiment: the ESP of an unhypnotized subject is compared to that subject's ESP when hypnotized. To test this idea, Casler (24) used a sample of 15 college students. Each student was tested 200 times: 100 times while awake and 100 while hypnotized. On each occasion the subject was asked to identify a card on which a "sender" was concentrating. Each of five different cards was equally likely to be the one "sent." Out of 100 trials, then, a subject with no ESP would be expected to get one fifth of the total, or 20, correct. Table 3.23 shows the results.

The last column in Table 3.23 lists each of the within-pair response differences. A *positive* difference means a subject scored *higher* while hypnotized. Think of those 15 numbers as a set of one-sample data. Let $\bar{d}$ and $s_d$ denote their sample mean and sample standard deviation, respectively.

**TABLE 3.23 Number of Correct Responses (out of 100)**

| Student | Sender and Student in Waking State, $x_i$ | Sender and Student in Hypnotic State, $y_i$ | $d_i = y_i - x_i$ |
|---------|-------------------------------------------|---------------------------------------------|-------------------|
| 1  | 18 | 25 | +7  |
| 2  | 19 | 20 | +1  |
| 3  | 16 | 26 | +10 |
| 4  | 21 | 26 | +5  |
| 5  | 16 | 20 | +4  |
| 6  | 20 | 23 | +3  |
| 7  | 20 | 14 | −6  |
| 8  | 14 | 18 | +4  |
| 9  | 11 | 18 | +7  |
| 10 | 22 | 20 | −2  |
| 11 | 19 | 22 | +3  |
| 12 | 29 | 27 | −2  |
| 13 | 16 | 19 | +3  |
| 14 | 27 | 27 | 0   |
| 15 | 15 | 21 | +6  |

Since

$$\sum_{i=1}^{15} d_i = 43 \qquad \text{and} \qquad \sum_{i=1}^{15} d_i^2 = 363$$

we can use Equations 3.1 and 3.6 to obtain

$$\bar{d} = \frac{1}{n}\sum_{i=1}^{n} d_i = \frac{43}{15} = 2.9$$

and

$$s_d = \sqrt{\frac{n\sum_{i=1}^{n} d_i^2 - \left(\sum_{i=1}^{n} d_i\right)^2}{n(n-1)}} = \sqrt{\frac{15(363) - (43)^2}{15(14)}} = 4.1$$

What do these numbers imply? If $\bar{d}$ were *exactly* zero, we would certainly have to infer that hypnosis has no effect. What is not so obvious is how much greater than zero must $\bar{d}$ be before we can reasonably argue that ESP does enhance a person's ability. In particular, does a $\bar{d}$ of $+2.9$ constitute statistical "proof" that hypnosis strengthens ESP? We will need to defer our response until Chapter 12, where we learn how to interpret $s_d$.

---

**QUESTIONS**

3.59  Depth perception is a life-or-death ability for lambs living in rugged mountain terrain. How quickly a lamb develops that faculty may depend on the amount of time it spends with its ewe. Thirteen sets of lamb twins were the subjects of an experiment that addressed that question. One member of each set was left with its mother; the other was removed immediately after birth.

Once every hour, the lambs were placed on a simulated "cliff," part of which included a platform of glass. If the lamb placed its feet on the glass, it "failed" the test, since that would have been equivalent to walking off the cliff. The data below are the trial numbers when the lambs first learned not to walk on the glass—that is, when they first developed depth perception (92). Find $d_i$ for each pair and compute $\bar{d}$ and $s_d$. Does it appear that a lamb's early environment has any effect on its ability to learn depth perception?

| | NUMBER OF TRIALS TO LEARN DEPTH PERCEPTION | |
|---|---|---|
| Pair | Mothered $d$, $x_i$ | Unmothered, $y_i$ |
| 1 | 2 | 3 |
| 2 | 3 | 11 |
| 3 | 5 | 10 |
| 4 | 3 | 5 |
| 5 | 2 | 5 |
| 6 | 1 | 4 |
| 7 | 1 | 2 |
| 8 | 5 | 7 |
| 9 | 3 | 5 |
| 10 | 1 | 4 |
| 11 | 7 | 8 |
| 12 | 3 | 12 |
| 13 | 5 | 7 |

3.60    Compute the within-pair sample mean and sample standard deviation for the HAT score changes listed in Table 2.8. Do the numerical magnitudes of $\bar{d}$ and $s_d$ seem to correspond with the graph of these data asked for in Question 3.58.

3.61    If the average within-pair response difference is small, can we assume that the standard deviation of the within-pair differences will also be small? If $\bar{d}$ is large, does it follow that $s_d$ will be large?

# 3.5

## DESCRIPTIVE STATISTICS FOR REGRESSION-CORRELATION DATA

Suppose $(x_1, y_1), (x_2, y_2), \ldots, (x_n, y_n)$ is a set of regression-correlation data, meaning that $x_i$ and $y_i$ are (dissimilar) measurements made on the $i$th subject's $X$ trait and $Y$ trait, respectively. Recall the example at the beginning of Section 2.5. The $X$ trait there was a person's blood pressure (expressed in mm Hg) and the $Y$ trait was that person's cholesterol level (expressed in mg/100 mL). How should such data be summarized? Since the reason for collecting regression-correlation data is to learn something about the *relationship* between $X$ and $Y$, we begin by graphing that relationship; that is, we plot $x_i$ versus $y_i$. Such a graph is called a ***scatter diagram***, examples of which we have already seen in Figures 2.9, 2.10, and 2.11.

**scatter diagram**—The basic graphical descriptive technique for regression-correlation data. The $n$ observations are plotted as points in the $xy$ plane.

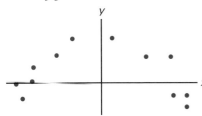

In general, relationships are classified according to the appearance of their scatter diagrams. For example, if the $(x_i, y_i)$'s show a basically straight-line pattern (one that can be represented by an equation of the form $y = a + bx$), the relationship is *linear*. Case Studies 2.9 and 2.10 belong to that category. If the $(x_i, y_i)$'s can *not* be adequately described by a straight line, the relationship is *nonlinear*. Some of the most frequently encountered equations associated with nonlinear relationships are

$$y = ax^b, \qquad y = ae^{bx}, \qquad y = a + \frac{b}{x}, \quad \text{and} \quad y = a + bx + cx^2$$

That different scatter diagram patterns are associated with different mathematical expressions motivates what is probably the most fundamental question in connection with regression-correlation data: given a set of $(x_i, y_i)$'s, which equation fits them best? Should we use a model of the form $y = a + bx$, for example, or is the relationship better described by

$$y = ax^b? \quad \text{or} \quad y = a + bx + cx^2?$$

or something entirely different?

These questions are difficult to answer in complete generality, too difficult for us to deal with here. We can, however, handle a simpler problem that still leads to a useful descriptive technique. Suppose we draw the scatter diagram for a set of points, and, from the graph, it seems quite obvious that the relationship has the form $y = a + bx$. How should we choose values for $a$ and $b$? What is the equation, in other words, of the best *straight* line through a given set of points?

Before answering that, let's review what high school algebra says about straight lines and their equations. Any straight line can be written in the form $y = a + bx$. The coefficient $b$ is the *slope* of the line and represents the amount that $y$ changes when $x$ increases by 1. If the line is horizontal, $b = 0$; if the line slopes upward, $b > 0$; if the line slopes downward, $b < 0$. The coefficient $a$ is the $y$ intercept, that is; the value of $y$ when $x = 0$.

Suppose we wish to graph the equation $y = -1.0 + 2.5x$. Since two points determine a straight line, we need to choose two values for $x$ and solve for the corresponding values of $y$.

If $x = 0$, $\qquad\qquad\qquad\qquad y = -1.0 + 2.5(0) = -1.0$

if $x = 2$, $\qquad\qquad\qquad\qquad y = -1.0 + 2.5(2) = \quad 4.0$

Figure 3.21 shows the points $(0, -1.0)$ and $(2, 4.0)$ plotted in the $xy$ plane. The line passing through those points has the equation $y = -1.0 + 2.5x$.

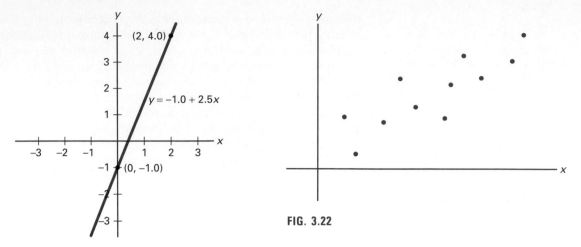

**FIG. 3.21**

**FIG. 3.22**

We now return to our earlier question. We are given a set of $n$ points that have a scatter diagram with a linear pattern (Figure 3.22). Which straight line best describes the relationship between $X$ and $Y$? The key word here is *best*: what does that word mean in this particular mathematical context? Consider these same linearly related points and imagine superimposing on their scatter diagram an *arbitrary* straight line $y = a + bx$ (Figure 3.23). How might we measure the extent to which the $i$th point is *not* "fit" by that line? By computing the vertical distance from the point to the line. Since the "height" of the $i$th point is $y_i$ and the "height" of the line when $x = x_i$ is $a + bx_i$, it follows that the vertical distance (or deviation) of the $i$th point from the line is the difference between these two heights:

Deviation of $(x_i, y_i)$ from $y = a + bx$
$$= y_i - (a + bx_i)$$

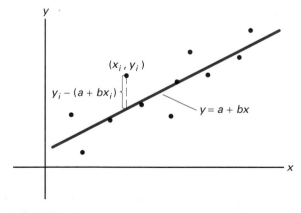

**FIG. 3.23**

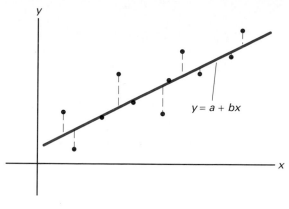

**FIG. 3.24**

Similar deviations could be constructed for *all* the points (Figure 3.24). Obviously, as the orientation of the line changes (that is, as *a* and *b* change), the values of $y_i - (a + bx_i)$, $i = 1, 2, \ldots, n$ will also change.

---

### LEAST SQUARES CRITERION

$y = a + bx$ is the *best* line describing the linear relationship between X and Y if it minimizes the sum of the squared deviations of the points from the line.

---

That is, $y = a + bx$ is the straight line that best describes the $(x_i, y_i)$'s if its values for *a* and *b* are such that

$$\sum_{i=1}^{n} (y_i - (a + bx_i))^2$$

is as small as possible.

Figure 3.25 shows two straight lines—$y = -1.0 + 1.0x$ and $y = 1.5 + 0.5x$—superimposed over three points, (2, 1), (3, 4), and (4, 2). Which line provides a better approximation of the *xy* relationship? Table 3.24 shows the two corresponding sets of deviations and their squares. Since $5.5 < 6$, the line $y = 1.5 + 0.5x$ provides a better fit than does the line $y = -1.0 + 1.0x$.

These computations suggest the obvious question that we need to confront: what is the equation of the line that gives the *smallest* possible value for the sum of the squared deviations? The answer, in the form of equations for *a* and *b*, is given in Theorem 3.2. Its proof requires calculus and will have to be omitted. (Note: To distinguish between the "best" line and an

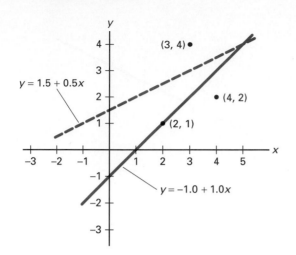

**FIG. 3.25**

**TABLE 3.24**

| $(x_i, y_i)$ | $y_i - (-1.0 + 1.0x_i)$ | $(y_i - (-1.0 + 1.0x_i))^2$ |
|---|---|---|
| $(2, 1)$ | 0 | 0 |
| $(3, 4)$ | 2 | 4 |
| $(4, 2)$ | $-1$ | 1 |
| | | 6 |

| $(x_i, y_i)$ | $y_i - (1.5 + 0.5x_i)$ | $(y_i - (1.5 + 0.5x_i))^2$ |
|---|---|---|
| $(2, 1)$ | $-1.5$ | 2.25 |
| $(3, 4)$ | 1 | 1 |
| $(4, 2)$ | $-1.5$ | 2.25 |
| | | 5.5 |

arbitrary line, we will begin denoting the $y$ intercept and the slope of the best line by the symbols $\hat{a}$ and $\hat{b}$.)

---

**THEOREM 3.2   The Least Squares Line**

Let $(x_1, y_1), (x_2, y_2), \ldots, (x_n, y_n)$ be a set of $n$ points. The straight line $y = \hat{a} + \hat{b}x$ that minimizes the sum of the squared deviations from the points to the line is called the **least squares line** (or **regression line**). It has slope

$$\hat{b} = \frac{n \sum_{i=1}^{n} x_i y_i - \left( \sum_{i=1}^{n} x_i \right)\left( \sum_{i=1}^{n} y_i \right)}{n \sum_{i=1}^{n} x_i^2 - \left( \sum_{i=1}^{n} x_i \right)^2} \tag{3.7}$$

and $y$ intercept

$$\hat{a} = \frac{1}{n} \left( \sum_{i=1}^{n} y_i - \hat{b} \sum_{i=1}^{n} x_i \right) \tag{3.8}$$

---

*EXAMPLE 3.5*   For the points $(2, 1), (3, 4)$, and $(4, 2)$ we showed that the line $y = 1.5 + 0.5x$ fits that set of data better than $y = -1.0 + 1.0x$. But $y = 1.5 + 0.5x$, itself, might not be all that good. Use Theorem 3.2 to find the *best* line describing the relationship between the $(x_i, y_i)$'s.

*Solution*  Note that

$$\sum_{i=1}^{3} x_i = 2 + 3 + 4 = 9$$

$$\sum_{i=1}^{3} x_i^2 = 2^2 + 3^2 + 4^2 = 29$$

$$\sum_{i=1}^{3} y_i = 1 + 4 + 2 = 7$$

$$\sum_{i=1}^{3} x_i y_i = 2 \cdot 1 + 3 \cdot 4 + 4 \cdot 2 = 22$$

So, by Equations 3.7 and 3.8,

$$\hat{b} = \frac{3(22) - 9(7)}{3(29) - (9)^2} = \frac{1}{2} \quad \text{and} \quad \hat{a} = \frac{7 - \frac{1}{2}(9)}{3} = \frac{5}{6}$$

Therefore, the least squares line has the equation $y = \frac{5}{6} + \frac{1}{2}x$.

Table 3.25 shows the corresponding sum of squared deviations. Notice that the figure at the bottom of column 3 is less than either sum in Table 3.24. What the theorem guarantees is that 4.16 is a lower limit; in other words, *no* straight line through these same three points can possibly have a smaller sum of squared deviations.

**TABLE 3.25**

| $(x_i, y_i)$ | $y_i - (\frac{5}{6} + \frac{1}{2}x_i)$ | $(y_i - (\frac{5}{6} + \frac{1}{2}x_i))^2$ |
|---|---|---|
| (2, 1) | −5/6 | 0.69 |
| (3, 4) | 10/6 | 2.78 |
| (4, 2) | −5/6 | 0.69 |
| | | 4.16 |

---

## CASE STUDY 3.15

*Working Women*  Among the problems faced by women seeking to reenter the workforce and compete with men is that the years spent away from the job raising children can erode skills and allow backgrounds to become outdated. Women with such gaps in their employment histories can find it difficult to keep their careers on track.

Roberts (138) conducted a survey that focused on the relationship between the chances that medical technologists have of getting rehired and the number of years they have been inactive in the profession. Specifically, administrators at 67 randomly selected hospitals throughout the United

States were asked to indicate their willingness to rehire med techs who had been away from the field for $x$ years. The results are summarized in Table 3.26.

**TABLE 3.26**

| Years of Inactivity, $x$ | % of Hospitals Willing to Hire, $y$ |
|---|---|
| $\frac{1}{2}$ | 100 |
| $1\frac{1}{2}$ | 94 |
| 4 | 75 |
| 8 | 44 |
| 13 | 28 |
| 18 | 17 |

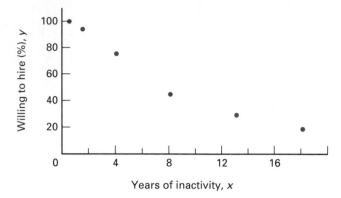

**FIG. 3.26**

The scatter diagram for these data (Figure 3.26) shows a distinctly linear relationship. Find the regression line $y = \hat{a} + \hat{b}x$ that best describes these six points.

Table 3.27 shows the computations necessary to generate the sums that appear in the formulas for $\hat{a}$ and $\hat{b}$. Substituting these values into Equations 3.7 and 3.8 we find that

$$\hat{b} = \frac{6(1513) - (45)(358)}{6(575.5) - (45)^2} \qquad \hat{a} = \frac{358 - (-4.9)(45)}{6}$$

$$= -4.9 \qquad\qquad\qquad = 96.4$$

The line satisfying the least squares criterion, therefore, is

$$y = 96.4 - 4.9x.$$

Notice that the slope is negative, which is not surprising. As $x$ (the number of years a woman is out of the workforce) increases, we would expect $y$ (her likelihood of being rehired) to decrease. The numerical value

**TABLE 3.27**

| $x_i$ | $y_i$ | $x_i^2$ | $x_i y_i$ |
|---|---|---|---|
| 0.5 | 100 | 0.25 | 50 |
| 1.5 | 94 | 2.25 | 141 |
| 4 | 75 | 16 | 300 |
| 8 | 44 | 64 | 352 |
| 13 | 28 | 169 | 364 |
| 18 | 17 | 324 | 306 |
| 45 | 358 | 575.5 | 1513 |

of $\hat{b}$ tells us even more: a slope of $-4.9$ implies that a med tech's chances of being rehired drop by 4.9% for every year spent away from the job.

To graph the regression line, we simply choose any two values for $x$ and solve for $y$. For $x = 0$,

$$y = 96.4 - 4.9(0) = 96.4$$

For $x = 10$,

$$y = 96.4 - 4.9(10) = 47.4$$

Figure 3.27 shows the equation $y = 96.4 - 4.9x$ superimposed over the original scatter diagram.

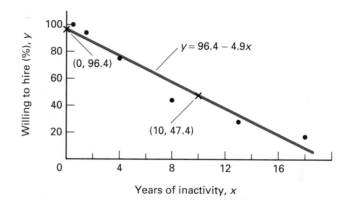

**FIG. 3.27**

**3.62** Child development studies often focus more on the *relationship* between different activities than on the activities themselves. The reason is that observations made on isolated behavioral traits (such as fighting) are so dependent on the experimental "environment" that comparisons from study to study are difficult to draw. *Patterns* of behavior, on the other hand, that involve pairs of activities are thought to be a bit more reproducible. Two such activities are running and laughing. In the table below are observations made on the frequencies with which children in a playground engaged in those activities. The subjects were five males, ages three to five (84).

| Child | No. of Times Seen Running, $x_i$ | No. of Times Seen Laughing, $y_i$ |
|-------|---------------------------------|-----------------------------------|
| A | 3 | 3 |
| B | 9 | 7 |
| C | 7 | 8 |
| D | 6 | 3 |
| E | 4 | 5 |

(a) Graph the data.

(b) Find the equation of the regression line.

(c) Plot your answer to part (b) on the graph in part (a).

(d) Does it seem safe to conclude that children who run more, laugh more?

3.63 Draw the scatter diagram and find the regression line for the mosquito data described in Question 1.8. Plot the line on the scatter diagram. How far might we expect a $2\frac{1}{2}$-week-old mosquito to fly?

3.64 The aging of whisky in charred oak barrels brings about chemical changes that enhance the whisky's taste and darken its color. Shown in the adjacent table is the change in a whisky's proof as a function of the number of years aged (149).

| Age (years), $x_i$ | Proof, $y_i$ |
|---|---|
| 0 | 104.6 |
| 0.5 | 104.1 |
| 1 | 104.4 |
| 2 | 105.0 |
| 3 | 106.0 |
| 4 | 106.8 |
| 5 | 107.7 |
| 6 | 108.7 |
| 7 | 110.6 |
| 8 | 112.1 |

(a) Draw the scatter diagram for these data.

(b) Find the equation of the regression line.

$$\sum_{i=1}^{10} x_i = 36.5 \qquad \sum_{i=1}^{10} x_i^2 = 204.25$$

$$\sum_{i=1}^{10} y_i = 1070.0 \qquad \sum_{i=1}^{10} x_i y_i = 3973.35$$

(c) Show the regression line on the scatter diagram.

(d) If the age-proof relationship continued past the period covered by the data, what proof would we expect a 10-year-old whisky to have?

3.65 Crickets make their chirping sound by rapidly sliding one wing cover back and forth over the other. Biologists have long been aware that there is a linear relationship between temperature and the frequency with which a cricket chirps, although the slope and $y$ intercept of the relationship vary from species to species. Listed on the next page are 15 frequency–temperature observations recorded for the striped ground cricket, *Nemobius fasciatus fasciatus* (124).

| Observation Number | Chirps per Second, $x_i$ | Temperature (°F), $y_i$ |
|:---:|:---:|:---:|
| 1 | 20.0 | 88.6 |
| 2 | 16.0 | 71.6 |
| 3 | 19.8 | 93.3 |
| 4 | 18.4 | 84.3 |
| 5 | 17.1 | 80.6 |
| 6 | 15.5 | 75.2 |
| 7 | 14.7 | 69.7 |
| 8 | 17.1 | 82.0 |
| 9 | 15.4 | 69.4 |
| 10 | 16.2 | 83.3 |
| 11 | 15.0 | 79.6 |
| 12 | 17.2 | 82.6 |
| 13 | 16.0 | 80.6 |
| 14 | 17.0 | 83.5 |
| 15 | 14.4 | 76.3 |

(a) Plot the data and find the regression line.

$$\sum_{i=1}^{15} x_i = 249.8 \qquad \sum_{i=1}^{15} x_i^2 = 4200.56$$

$$\sum_{i=1}^{15} y_i = 1200.6 \qquad \sum_{i=1}^{15} x_i y_i = 20,127.47$$

(b) The existence of a relationship between $x$ and $y$ raises the possibility of using a cricket as a thermometer. Suppose we time the chirps of a striped ground cricket and record a frequency of 18 chirps per second. What is the estimated temperature?

3.66   Verify that the least squares line for the weight-versus-year data given in Question 2.33 is $y = -1108.0 + 0.666x$, where $x$ represents the year and $y$ is the average weight in pounds of a University of Texas football player.

3.67   In some situations the experimenter knows that the regression line should go through the origin—that is, $y$ should be zero when $x$ is zero. The model being fit in such cases is $y = bx$ rather than $y = a + bx$. When we know from physical constraints that $a$ should be zero, the formula for the estimated slope is

$$\hat{b} = \frac{\sum_{i=1}^{n} x_i y_i}{\sum_{i=1}^{n} x_i^2}$$

An example of a phenomenon where $a$ is necessarily zero is the galactic recession data described in Case Study 2.10. Verify that the regression line summarizing the data in Table 2.11 has the equation $y = 0.03544x$.

$$\sum_{i=1}^{11} x_i^2 = 2,685,141 \qquad \sum_{i=1}^{11} x_i y_i = 95,161.2$$

3.68   For a given car, studies have shown that fuel economy $z$ (in gal/mile) is related to average speed $s$ (in miles/hr) according to the equation

$$z = k_1 + \frac{k_2}{s}$$

Furthermore, the constant $k_1$ reflects the amount of fuel consumed per mile for the purpose of overcoming the car's rolling friction. As such, $k_1$ should be approximately a linear function of $w$, the car's weight. Since a weightless car would produce no friction, the appropriate model relating $w$ and $k_1$ is

$$k_1 = bw$$

For nine different cars the following values were determined for $w$ and $k_1$ (43). Plot the data. Find $\hat{b}$ and graph the function $k_1 = \hat{b}w$ on your scatter diagram.

| Vehicle | Weight, $w$ (lb) | $k_1$ (gal/mile) |
|---|---|---|
| Standard size | 4980 | 0.0474 |
| Standard size | 5051 | 0.0402 |
| Intermediate size | 3792 | 0.0362 |
| Small import | 2277 | 0.0194 |
| Luxury | 5474 | 0.0518 |
| Subcompact station wagon | 2833 | 0.0307 |
| Subcompact | 3620 | 0.0384 |
| Intermediate size | 5100 | 0.0365 |
| Subcompact | 2353 | 0.0248 |

### The Sample Correlation Coefficient

If a scatter diagram shows that sets of $(x_i, y_i)$'s are linearly related, then finding the equation of the least squares straight line $y = \hat{a} + \hat{b}x$ is certainly the first step to take to describe mathematically the $xy$ relationship. Having such an equation is not only a convenient way of summarizing the data, it also helps us make predictions about $y$ if we are interested in certain values for $x$.

Look again at Case Study 3.15. Suppose a med tech has been out of work for five years. What are her chances of reentering the profession? Even though $x = 5$ is not in the original data (Table 3.26), we can nevertheless *estimate* her chances of future employment by substituting $x = 5$ into the equation for the least squares line. Specifically, if $x = 5$,

$$y = 96.4 - 4.9(5) = 71.9$$

implying that a woman in her situation can expect to have a 71.9% chance of being rehired (see Figure 3.28).

But not everything we need to know about a linear relationship is contained in the equation of the least squares line. A second question remains to be addressed. Figure 3.29 shows two sets of points that have the same regression line. Would we consider those two relationships to be en-

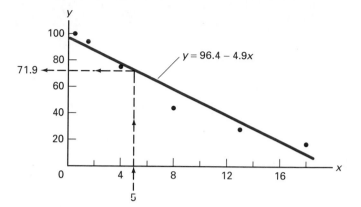

**FIG. 3.28**

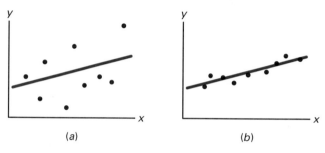

(a)                                          (b)

**FIG. 3.29**

tirely comparable? No. Despite the fact that both scatter diagrams produce the same $y$ intercept and the same slope, the picture we get of the two relationships is quite different. In Figure 3.29*a* the points have a *weak* linear relationship due to their considerable scatter from the regression line; in Figure 3.29*b* the deviations of the $(x_i, y_i)$'s from $y = \hat{a} + \hat{b}x$ are all very small, meaning the relationship is *strong*.

To an experimenter the strength of a linear relationship is crucial information. When the points look like Figure 3.29*b*, predictions about the value of $y$ for a given value of $x$ can be made with a good deal of confidence. On the other hand, predictions based on weak relationships are subject to a good deal of error. *Finding a number that measures the strength of a linear relationship is the second contribution that descriptive statistics makes to the analysis of regression-correlation data.*

Figure 3.30 shows two sets of points: one with a weak linear relationship, the other with a strong linear relationship. Superimposed on each scatter diagram are the least squares line

$$y = \hat{a} + \hat{b}x$$

and the horizontal line

$$y = \bar{y} \qquad \text{where} \quad \bar{y} = \frac{1}{n} \sum_{i=1}^{n} y_i$$

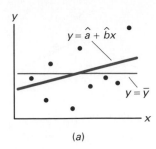 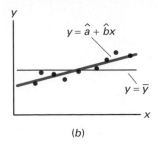

(a)                              (b)

**FIG. 3.30**

For the set of points with a weak fit, the sum of the squared deviations of the points from the regression line will not be that much smaller than the sum of the squared deviations of the points from the line $y = \bar{y}$. That is,

$$\frac{\sum_{i=1}^{n} (y_i - (\hat{a} + \hat{b}x_i))^2}{\sum_{i=1}^{n} (y_i - \bar{y})^2}$$

will be a number not much smaller than 1. On the other hand, for the set of points with a strong fit,

$$\sum_{i=1}^{n} (y_i - \bar{y})^2 \quad \text{will be much larger than} \quad \sum_{i=1}^{n} (y_i - (\hat{a} + \hat{b}x_i))^2$$

the result being that

$$\frac{\sum_{i=1}^{n} (y_i - (\hat{a} + \hat{b}x_i))^2}{\sum_{i=1}^{n} (y_i - \bar{y})^2}$$

will be close to zero. The fact that the numerical values of the two ratios just cited can distinguish between weak and strong linear relationships motivates the next definition.

---

**DEFINITION 3.4   The Sample Correlation Coefficient**

Let $(x_1, y_1)$, $(x_2, y_2)$, . . . , $(x_n, y_n)$ be a set of $n$ measurements made on an $X$ trait and on a $Y$ trait, respectively. The strength of the linear relationship between $X$ and $Y$ is measured by the **sample correlation coefficient r**, where

$$r = \pm \sqrt{1 - \frac{\sum_{i=1}^{n} (y_i - (\hat{a} + \hat{b}x_i))^2}{\sum_{i=1}^{n} (y_i - \bar{y})^2}} \tag{3.9}$$

If $\hat{b} > 0$, then $r > 0$; if $\hat{b} < 0$, then $r < 0$.

---

We can deduce several things about the magnitude of $r$.

1. No matter what the scatter diagram for a set of points looks like, it can be shown that $-1 \leqslant r \leqslant +1$.
2. If the points all fall exactly on a straight line with a positive slope, then $r = +1$. (Why?)
3. If the points all fall on a straight line with a negative slope, then $r = -1$.
4. If there is absolutely no relationship between the $x_i$'s and the $y_i$'s and the points are just randomly scattered over the $xy$ plane, then $r = 0$.

Linear relationships with strengths between these extremes will likewise have $r$'s with correspondingly intermediate values. Figure 3.31 shows six sets of 10 points representing a variety of straight-line patterns. The weakest relationship has $r = -0.06$, the strongest $r = 0.96$.

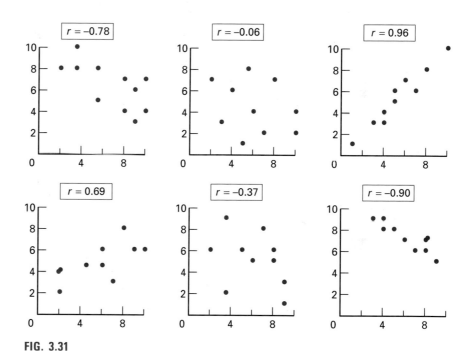

**FIG. 3.31**

In practice, we never use Equation 3.9 for computational purposes. The latter's only function is to show how the correlation coefficient measures the strength of a linear relationship (by taking the ratio of two different sums of squared deviations). To compute $r$ we use Equation 3.10. Although it looks entirely different from Equation 3.9, the two expressions are equivalent, but Equation 3.10 is much easier to use.

---

**THEOREM 3.3   A Formula for Computing $r$**

The sample correlation coefficient $r$ is computed by the formula

$$r = \frac{n \sum_{i=1}^{n} x_i y_i - \left( \sum_{i=1}^{n} x_i \right)\left( \sum_{i=1}^{n} y_i \right)}{\sqrt{n \sum_{i=1}^{n} x_i^2 - \left( \sum_{i=1}^{n} x_i \right)^2} \sqrt{n \sum_{i=1}^{n} y_i^2 - \left( \sum_{i=1}^{n} y_i \right)^2}} \qquad (3.10)$$

---

## CASE STUDY 3.16

*Duck Crossings*   When two closely related species are crossed, the progeny will tend to have physical traits that lie somewhere "between" the corresponding traits of the parents. But are behavioral traits similarly "mixed"? Does a hybrid whose physical traits favor one parent tend to behave like that same parent?

One attempt to answer these questions was an experiment done with mallard and pintail ducks (153). Eleven males were studied; all were second-generation crosses. A rating scale was devised that measured the extent to which the plumage of each duck resembled the plumage of the first generation's parents. A score of 0 indicated that the hybrid had the same appearance (phenotype) as a pure mallard; a score of 20 meant that the hybrid looked like a pintail. Similarly, certain behavioral traits were quantified, and a second scale was constructed that ranged from 0 (completely mallard-like) to 15 (completely pintail-like).

**TABLE 3.28**

| Male | Plumage Index, $x$ | Behavioral Index, $y$ |
|------|--------------------|-----------------------|
| R | 7 | 3 |
| S | 13 | 10 |
| D | 14 | 11 |
| F | 6 | 5 |
| W | 14 | 15 |
| K | 15 | 15 |
| U | 4 | 7 |
| O | 8 | 10 |
| V | 7 | 4 |
| J | 9 | 9 |
| L | 14 | 11 |

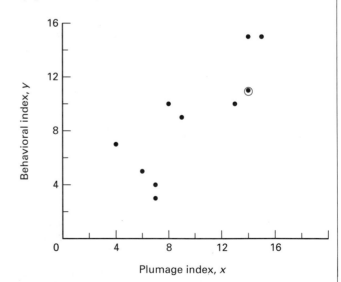

**FIG. 3.32**

Table 3.28 lists the *plumage index* $(x)$ and the *behavioral index* $(y)$ for the 11 hybrids studied. Graphed, the data show a linear relationship (Figure 3.32). We measure the strength of this relationship by computing the sample correlation coefficient $r$.

Table 3.29 lists the sums that are needed for Equation 3.10. Thus,

$$r = \frac{11(1141) - (111)(100)}{\sqrt{11(1277) - (111)^2}\sqrt{11(1072) - (100)^2}} = 0.82$$

**TABLE 3.29**

| $x_i$ | $y_i$ | $x_i^2$ | $x_i y_i$ | $y_i^2$ |
|-------|-------|---------|-----------|---------|
| 7     | 3     | 49      | 21        | 9       |
| 13    | 10    | 169     | 130       | 100     |
| 14    | 11    | 196     | 154       | 121     |
| 6     | 5     | 36      | 30        | 25      |
| 14    | 15    | 196     | 210       | 225     |
| 15    | 15    | 225     | 225       | 225     |
| 4     | 7     | 16      | 28        | 49      |
| 8     | 10    | 64      | 80        | 100     |
| 7     | 4     | 49      | 28        | 16      |
| 9     | 9     | 81      | 81        | 81      |
| 14    | 11    | 196     | 154       | 121     |
| 111   | 100   | 1277    | 1141      | 1072    |

(Does a correlation coefficient of 0.82 seem "about right" for these data? Look at the scatter diagram in Figure 3.32 and compare it with the relationships in Figure 3.31.)

---

QUESTIONS

3.69    Graph the nemesis data given in Question 1.9. Use Equation 3.10 to compute the sample correlation coefficient. Do the appearance of your scatter diagram and the magnitude of $r$ seem to agree?

3.70    It is important to remember that $r$ measures the strength of the *linear* relationship between $x$ and $y$. It says nothing about the presence or strength of any other kind of relationship. For the following set of points show that $r$ is zero. Does it follow that $x$ and $y$ are unrelated? (Hint: Plot the data and superimpose the function $y = x^2$.)

| $x_i$ | $y_i$ |
|-------|-------|
| $-2$  | 4     |
| 0     | 0     |
| 1     | 1     |
| 2     | 4     |
| $-1$  | 1     |

**3.71**   An experimenter collects a set of regression-correlation data

$$(x_1, y_1), \quad (x_2, y_2), \quad \ldots, \quad (x_n, y_n)$$

What is the first step she needs to take to analyze those results? Explain.

**3.72**   Compute the sample correlation coefficient for the plant-cover-diversity/bird-species-diversity data in Question 2.32.

$$\sum_{i=1}^{13} x_i = 12.91 \qquad \sum_{i=1}^{13} x_i^2 = 15.6171$$

$$\sum_{i=1}^{13} y_i = 25.29 \qquad \sum_{i=1}^{13} y_i^2 = 57.8103$$

$$\sum_{i=1}^{13} x_i y_i = 29.8762$$

**3.73**   Correlation coefficients for two sets of 10 points are computed to be $+0.65$ and $-0.65$, respectively. Which set of data has the stronger linear relationship?

**3.74**   (a)  Quantify the strength of the linear relationship evidenced in the tethered mosquito data in Question 1.8.

   (b)  Draw the data's scatter diagram. Is your value for $r$ consistent with the appearance of the graph? Is the *sign* of $r$ appropriate?

# 3.6

## DESCRIPTIVE STATISTICS FOR CATEGORICAL DATA

Summarizing categorical data is easy. There are no graphical formats and only the most rudimentary numerical formats to learn. In effect, we just need to display the data in tabular form (see Section 2.6). Recall that the data consist of $n$ pairs of *qualitative* observations

$$(x_1, y_1), \quad (x_2, y_2), \quad \ldots, \quad (x_n, y_n)$$

Each $x_i$ is assigned to one of $r$ classes, each $y_i$ to one of $c$ classes. As *points*, therefore, each $(x_i, y_i)$ belongs to one of the $rc$ classes formed by making a grid of $r$ rows and $c$ columns (Figure 2.16). Descriptive statistics for categorical data begins and ends with the construction of a *contingency table*, which is simply a tally of the number of $(x_i, y_i)$'s belonging to each of these $rc$ classes.

In many situations the $r$ classes and the $c$ classes into which the $x_i$'s and $y_i$'s are ultimately put are immediately apparent from the nature of the data. Other times, however, the situation is more subjective, and the statistician is the one who decides how the classes should be defined. Case Study 3.17 typifies the former; Case Study 3.18 the latter.

## CASE STUDY 3.17

*Are Good Patients Religious?*

If asked, "What makes a good patient?" we would probably respond by reciting a list of *personality* traits (trust, a positive attitude, willingness to follow orders, etc.). But Vincent (172) showed that a better answer might be a list of *biological* and *cultural* traits.

Fifty-seven elderly persons visiting a clinic on an outpatient basis were each labeled as a "complier" or a "noncomplier" according to whether the patient followed his or her doctor's orders about taking medication. Factors that may influence the likelihood of a patient following directions were the subject of Vincent's inquiry. One factor considered was religion. Among the patients, 17 were Catholics and 40 were Protestants. Ten Catholics and 15 Protestants were compliers.

Notice that two *qualitative* pieces of information are being recorded on each subject. The $X$ trait is whether the patient complies with the doctor's orders. As a variable, $X$ has two classes: either the patient is a complier or the patient is a noncomplier. Religion is the $Y$ trait. For these subjects only two classes for $Y$ need to be defined: Catholic and Protestant.

**TABLE 3.30**

|  | Catholic | Protestant |
|---|---|---|
| Compliers | 10 | 15 |
| Noncompliers | 7 | 25 |
| % Compliers | 58.8 | 37.5 |

Table 3.30 is the contingency table summarizing the information recorded. The two classes for $X$ and the two classes for $Y$ generate a grid with four $(= 2 \cdot 2)$ classes for the 57 $(x_i, y_i)$'s. What do the four entries in the contingency table suggest? That maybe there *is* a connection between religion and being a good patient. Ten of the 17 Catholics, or 58.8%, proved to be good patients; only 37.5% of the Protestants were. At this point we can only speculate as to what the magnitude of that difference implies. We will learn in Chapter 12 how to tell whether the difference between 58.8% and 37.5% is large enough for us to conclude *statistically* that Catholics are better patients than Protestants. (With the application of this same criterion of compliance versus noncompliance, Vincent's data also suggest that whites are better patients than nonwhites and women are better patients than men.)

## CASE STUDY 3.18

*Suicide Rates*   Ever since Emile Durkheim published *Suicide: A Study in Sociology* in 1897, the phenomenon of suicide has been widely studied as a means of evaluating the levels of stress prevailing in a society. Over the years, suicide rates have been shown to follow some definite patterns. They are much higher for males than for females and for whites than for blacks. They are high for the elderly, the unmarried, and the childless. They are low in times of war and high in times of peace.

What causes these patterns is still largely unknown. Alienation, though, is undoubtedly one of the factors high on the list, which might explain why suicide rates are higher in urban areas than in smaller communities. It also suggests that cities whose populations are more transient should have higher suicide rates than do cities whose populations are more settled.

Young and Schmid (190) put the latter hypothesis to a test by collecting information on the suicide rate and the transiency level for residents of 25 major American cities. Transiency was measured in terms of a *mobility index*, with lower index values corresponding to populations that are more transient. What they found is presented in Table 3.31.

**TABLE 3.31**

| City | Suicides per 100,000, $x_i$ | Mobility Index, $y_i$ | City | Suicides per 100,000, $x_i$ | Mobility Index, $y_i$ |
|---|---|---|---|---|---|
| New York | 19.3 | 54.3 | Washington | 22.5 | 37.1 |
| Chicago | 17.0 | 51.5 | Minneapolis | 23.8 | 56.3 |
| Philadelphia | 17.5 | 64.6 | New Orleans | 17.2 | 82.9 |
| Detroit | 16.5 | 42.5 | Cincinnati | 23.9 | 62.2 |
| Los Angeles | 23.8 | 20.3 | Newark | 21.4 | 51.9 |
| Cleveland | 20.1 | 52.2 | Kansas City | 24.5 | 49.4 |
| St. Louis | 24.8 | 62.4 | Seattle | 31.7 | 30.7 |
| Baltimore | 18.0 | 72.0 | Indianapolis | 21.0 | 66.1 |
| Boston | 14.8 | 59.4 | Rochester | 17.2 | 68.0 |
| Pittsburg | 14.9 | 70.0 | Jersey City | 10.1 | 56.5 |
| San Francisco | 40.0 | 43.8 | Louisville | 16.6 | 78.7 |
| Milwaukee | 19.3 | 66.2 | Portland | 29.3 | 33.2 |
| Buffalo | 13.8 | 67.6 | | | |

As given, these measurements are not categorical, and we could summarize them by using the techniques appropriate for regression-correlation data. Any set of regression-correlation data, though, can be transformed into categorical data by superimposing a set of classes over the original $(x_i, y_i)$'s. Why would we do that? To make the association (or lack of association) between the two variables a little easier to see.

A common way of "reducing" data from a regression-correlation format to a categorical format is to *dichotomize* the scales of each of the two variables. The $X$ scale, for example, is replaced by two classes: "less than $\bar{x}$" and "greater than $\bar{x}$." Similarly, the $Y$ scale is simplified to "less than $\bar{y}$" and "greater than $\bar{y}$." For the 25 observations in Table 3.31,

$$\bar{x} = \frac{19.3 + 17.0 + \cdots + 29.3}{25} = 20.8$$

and

$$\bar{y} = \frac{54.3 + 51.5 + \cdots + 33.2}{25} = 56.0$$

Table 3.32 shows the frequencies with which the 25 $(x_i, y_i)$'s in Table 3.31 fall into the four categories created by the two $X$ classes and the two $Y$ classes. New York, for example, with its original measurements of (19.3, 54.3), is one of the three observations listed in the lower left-hand corner of the contingency table (since $19.3 < 20.8$ and $54.3 < 56.0$). Boston, on the other hand, is one of the 11 cities assigned to the lower right-hand corner.

**TABLE 3.32**

|  |  | MOBILITY INDEX | |
|---|---|---|---|
|  |  | Low ($<56.0$) | High ($>56.0$) |
| SUICIDE RATE | High ($>20.8$) | 7 | 4 |
|  | Low ($<20.8$) | 3 | 11 |
|  | % "High Suicide" | 70.0 | 26.7 |

Notice how the table effectively highlights the relationship between mobility index and suicide rate. Of the cities with a low mobility index (that is, with a high transiency level), 70% show a high suicide rate; in contrast, only 26.7% of the cities with a low transiency level have a high suicide rate. At first glance, then, these data would seem to support the contention that alienation is a significant factor in the incidence of suicide.

---

**QUESTIONS**

**3.75**  Three of the most important predators on the Serengeti plain in East Africa are the cheetah, the lion, and the wild dog. One of their favorite meals is Thompson's gazelle, but the latter is quite fast and can often outrun its attackers. Extensive observations on these life-and-death struggles have shown that out of 56 attacks initiated by cheetahs, 54.5% were successful (from the cheetah's standpoint, not the gazelle's!). Lions, on the other hand, prevailed in 26% of 417 encounters, and wild dogs gave chase 47 times and brought down their prey on 49% of those occasions (41). Summarize these results with a contingency table.

**3.76**   Certain viral infections contracted during pregnancy can cause birth defects. Among the most dangerous of these defects are rubella infections, better known as German measles. The timing of the infection—that is, when it occurs during the pregnancy—may also have an effect. In a study seeking to relate these two factors, researchers cross-classified 578 pregnancies, all involving a rubella infection. Each birth was judged "normal" or "abnormal," the latter category consisting of all abortions, stillbirths, deaths within two years, and children with birth defects. The timing of the infection was also reduced to two classes: the first three months or the last six months of the pregnancy. Of the 202 pregnancies complicated by an early infection, 59 resulted in abnormal births. Of the 376 women who contracted German measles during the last six months of their pregnancies, 349 experienced a normal delivery (49).

(a)  Put these data into a tabular format.

(b)  Compute whatever percentages you think would be helpful in highlighting the relationship between type of birth and time of infection.

**3.77**   Compute the average chirps per second and the average temperature for the data in Question 3.65. Dichotomize the two sets of readings in the manner described in Case Study 3.18 and enter the resulting frequencies in a contingency table.

**3.78**   Listed below are data collected on 19 patients with advanced nephritis. Heart weight and blood pressure were recorded for each subject (123). Put these data into a contingency table having two rows and two columns. Do heart weight and blood pressure seem to be related? Explain.

| Patient | Heart Weight (oz), $x_i$ | Blood Pressure (mm Hg), $y_i$ |
|---|---|---|
| 1 | 14.3 | 179 |
| 2 | 14.0 | 141 |
| 3 | 27.3 | 197 |
| 4 | 18.2 | 214 |
| 5 | 15.8 | 221 |
| 6 | 10.9 | 115 |
| 7 | 14.0 | 132 |
| 8 | 14.6 | 202 |
| 9 | 15.1 | 197 |
| 10 | 27.2 | 258 |
| 11 | 10.3 | 151 |
| 12 | 13.8 | 175 |
| 13 | 17.0 | 210 |
| 14 | 8.0 | 120 |
| 15 | 24.1 | 192 |
| 16 | 12.7 | 125 |
| 17 | 16.3 | 249 |
| 18 | 19.9 | 235 |
| 19 | 22.1 | 234 |
| | 315.6 | 3547 |

# 3.7

## DESCRIPTIVE STATISTICS FOR k-SAMPLE DATA

In Section 2.7 we saw that *k*-sample data have the same fundamental structure as two-sample data, the only difference being that *k* independent samples are being compared rather than two. We should not be surprised, then, to find that any of the numerical or graphical formats used for two-sample data can easily be extended to the general *k*-sample problem.

If the data are presented in tabular form (Figure 3.33), we typically list each sample's mean ($\bar{y}$) and standard error ($s/\sqrt{n}$), in the form $\bar{y} \pm s/\sqrt{n}$, under each treatment column. If the number of points in each sample is fairly small, the data are graphed as in Figure 3.17. We can also use the format in Figure 3.19, where each sample's mean and standard error are depicted graphically. Case Study 3.19 shows all of these options applied to the same set of data.

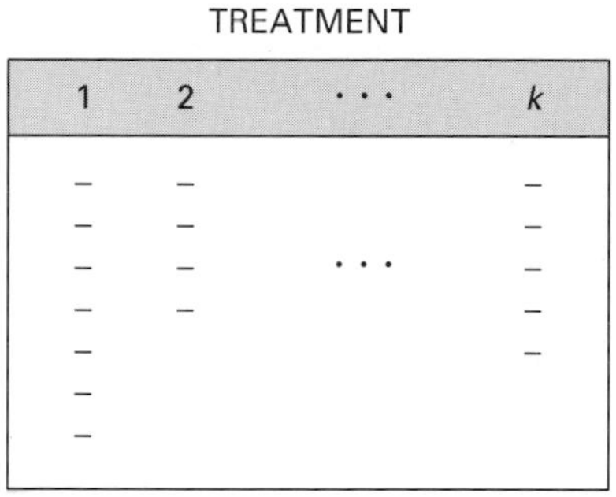

TREATMENT

| 1 | 2 | $\cdots$ | k |
|---|---|---|---|
| – | – | | – |
| – | – | | – |
| – | – | $\cdots$ | – |
| – | – | | – |
| – | | | |
| – | | | |

**FIG. 3.33**

### CASE STUDY 3.19

***Binding of Antibiotics***

A certain fraction of antibiotics injected into the bloodstream is bound to serum proteins. This phenomenon has considerable pharmacological importance, because as the extent of the binding increases the systemic uptake of the drug decreases. And as the drug's uptake decreases, so does its effectiveness as a medication.

Ziv and Sulman (195) investigated this problem by determining the binding percentages characteristic of five widely used antibiotics. Table 3.33 shows the four measurements made on each drug. The last row lists $\bar{y} \pm s/\sqrt{n}$ for each antibiotic. For example, the $n = 4$ observations recorded for penicillin G give

$$\sum_{i=1}^{4} y_i = 29.6 + 24.3 + 28.5 + 32.0 = 114.4$$

**TABLE 3.33**

| Penicillin G | Tetracycline | Streptomycin | Erythromycin | Chloramphenicol |
|:---:|:---:|:---:|:---:|:---:|
| 29.6 | 27.3 | 5.8 | 21.6 | 29.2 |
| 24.3 | 32.6 | 6.2 | 17.4 | 32.8 |
| 28.5 | 30.8 | 11.0 | 18.3 | 25.0 |
| 32.0 | 34.8 | 8.3 | 19.0 | 24.2 |
| $28.6 \pm 1.6$ | $31.4 \pm 1.6$ | $7.8 \pm 1.2$ | $19.1 \pm 0.9$ | $27.8 \pm 2.0$ |

and $\qquad \sum_{i=1}^{4} y_i^2 = (29.6)^2 + (24.3)^2 + (28.5)^2 + (32.0)^2 = 3302.9$

It follows that

$$\bar{y} = \frac{114.4}{4} = 28.6$$

and $\qquad s = \sqrt{\frac{4(3302.9) - (114.4)^2}{4(3)}} = 3.2$

Therefore, the sample mean plus or minus one standard error reduces to $28.6 \pm 3.2/\sqrt{4}$, or $28.6 \pm 1.6$.

Figure 3.34 shows how the observations in their original form would be graphed. If our intention is to emphasize the "precision" of each sample mean, we would graph the intervals $\bar{y} \pm s/\sqrt{n}$, as shown in Figure 3.35. The extent to which some of these intervals fail to overlap—most notably, the ones for streptomycin and erythromycin—suggests strongly that there are significant differences in the binding percentages characteristic of these five antibiotics.

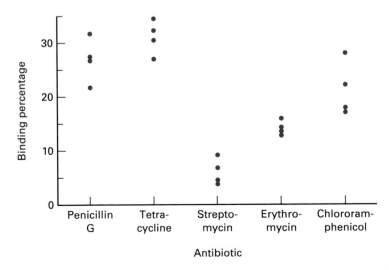

**FIG. 3.34**

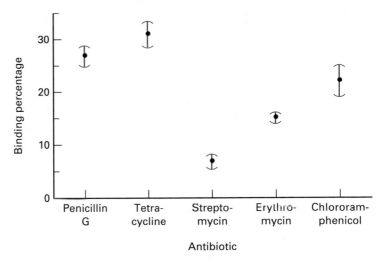

**FIG. 3.35**

**3.79** Use the format of Figure 3.34 to graph the magnetic field data in Case Study 2.14. Compute the sample mean and the sample standard deviation for each group and draw a graph similar to Figure 3.35. Does it appear that the direction of earth's magnetic field has changed?

$$\text{For 1669,} \quad \sum_{i=1}^{3} y_i = 178.3 \quad \sum_{i=1}^{3} y_i^2 = 10{,}600.97$$

$$\text{For 1780,} \quad \sum_{i=1}^{3} y_i = 169.9 \quad \sum_{i=1}^{3} y_i^2 = 9402.49$$

$$\text{For 1865,} \quad \sum_{i=1}^{3} y_i = 155.1 \quad \sum_{i=1}^{3} y_i^2 = 8026.65$$

**3.80** Graph the intervals $\bar{y} \pm s/\sqrt{n}$ for the sodium nitrite data described in Case Study 2.15.

$$\text{For brand 1,} \quad \sum_{i=1}^{3} y_i = 78 \quad \sum_{i=1}^{3} y_i^2 = 2324$$

$$\text{For brand 2,} \quad \sum_{i=1}^{3} y_i = 121 \quad \sum_{i=1}^{3} y_i^2 = 6691$$

$$\text{For brand 3,} \quad \sum_{i=1}^{3} y_i = 66 \quad \sum_{i=1}^{3} y_i^2 = 1566$$

$$\text{For brand 4,} \quad \sum_{i=1}^{3} y_i = 86 \quad \sum_{i=1}^{3} y_i^2 = 2486$$

**3.81** Three different biofeedback techniques were tested on a group of 12 hypertensives. Four subjects were chosen at random and given procedure S, five were given procedure L, and three were given procedure N. Each treatment was an audiovisual presentation; differences among S, L, and N were basically related to length rather than content. Changes in heart rates (before — after) were the measurements recorded (in beats per minute). Summarize these data numerically and graphically. Does it look like the procedures are all equally effective?

| Procedure S | Procedure L | Procedure N |
|:-----------:|:-----------:|:-----------:|
| 10 | 4 | 0 |
| 8 | 5 | 6 |
| 6 | 6 | 3 |
| 7 | 5 | |
| | 2 | |

3.82    Suppose an experiment involved three treatment groups with 75 independent observations being collected in each sample. What graphical techniques could be used to summarize those responses? (Hint: Is the format of Figure 3.34 practical for sample sizes as large as 75?)

# 3.8
## DESCRIPTIVE STATISTICS FOR RANDOMIZED BLOCK DATA

In Section 3.7 we saw that $k$-sample data are summarized and graphed by using the same techniques and formats that we use for two-sample data. Both kinds of data have the same basic structure, so the similarities in dealing with them should not have been totally unexpected. How we summarize randomized block data, however, does *not* carry directly over from what we did with paired data, even though they have the same basic structure.

In general, randomized block data are not graphed; they are simply tabulated. Included with the original observations are the treatment averages and sometimes the block averages (see Figure 2.18). Ultimately, the magnitudes of the differences among the treatment averages are what we use to decide whether the treatments are all equally effective. The significance of the differences among the block averages will become apparent when we discuss the analysis of variance in Chapter 15.

## CASE STUDY 3.20

*Rat Poison Recipes*    Normally, rat poison is made by mixing its active chemical ingredients with ordinary cornmeal. In many urban areas, though, rats can find food that they prefer to cornmeal, so the poison goes untouched. One solution is to make the cornmeal more appealing to the rodent palate by adding food supplements such as peanut butter or meat. That works, but the cost is high, and natural additives tend to spoil quickly. Hulbert and Krumbiegel (75) investigated whether artificial food supplements might be a viable alternative.

The study took place in Milwaukee, Wisconsin, and consisted of five two-week surveys. For each survey approximately 3200 baits were placed around garbage storage areas; 800 were plain cornmeal, another 800 were cornmeal mixed with butter-vanilla flavoring, a third 800 were cornmeal flavored with artificial roast beef, and a final 800 were cornmeal supplemented with an additive that tasted like bread. The baits were placed side by side so that rats would always have equal access to all four flavors.

After two weeks the sites were inspected, and the percentages of baits that had been eaten were recorded. Then a different set of locations in the same general area was selected, and the experiment was repeated for another two weeks. Altogether, the study was replicated five times.

Table 3.34 lists the bait-acceptance percentages for the four poisons. Notice that the observations are dependent (that is, the data are blocked) because all four observations associated with a given survey shared all the same parameters (weather conditions, availability of other food, activity around the bait sites, etc.) that characterized that particular survey.

**TABLE 3.34 Percentage of Baits Eaten**

| Survey | Plain | Butter-Vanilla | Roast Beef | Bread | Average |
|--------|-------|----------------|------------|-------|---------|
| 1 | 13.8 | 11.7 | 14.0 | 12.6 | 13.0 |
| 2 | 12.9 | 16.7 | 15.5 | 13.8 | 14.7 |
| 3 | 25.9 | 29.8 | 27.8 | 25.0 | 27.1 |
| 4 | 18.0 | 23.1 | 23.0 | 16.9 | 20.2 |
| 5 | 15.2 | 20.2 | 19.0 | 13.7 | 17.0 |
| Average: | 17.2 | 20.3 | 20.0 | 16.4 | |

The bottom row in the table lists the *average* bait-acceptance percentages for the four flavors. Are the differences shown there great enough for us to conclude that rats prefer certain flavors? Maybe yes and maybe no. The inference to be drawn from these data is certainly not as obvious as what we encountered in Case Study 3.19, which is all the more reason that we need to learn the formal analysis in Chapter 15.

The rightmost column in Table 3.34 lists the block averages. What do they suggest?—that the conditions under which the data were collected varied considerably from survey to survey. The rats were especially hungry, for example, during survey 3 and especially finicky during survey 1. Fortunately, the method we will learn for comparing *treatment* averages is unaffected by differences that may exist among *block* averages.

QUESTIONS

**3.83**   Compute the treatment averages and the block averages for the potash data in Question 2.46. Would you conclude on the basis of the differences among the treatment averages that the amount of potash applied to a cotton plant has an effect on the Pressley strength index of its fibers?

**3.84**   *Blood doping* refers to the practice of injecting additional red blood cells into an athlete's blood stream, thereby increasing the blood's oxygen-carrying capacity and boosting the athlete's endurance. Whether performance actually is enhanced, though, remains a controversial question: different studies have reached different conclusions. In one investigation of the practice, a group of six well-conditioned long-distance runners were timed on three separate occasions while competing in 10,000-m races. For one of the runs the subjects were given no injection whatsoever; on another occasion they were given a placebo; on the third occasion they were given a 400-mL injection of their own red blood cells that had been collected (and frozen) earlier. The order of the treatments were randomized. Shown below are the times (in minutes) recorded for each subject under each condition (15).

| | TIMES TO COMPLETE 10,000-m RUN | | |
|---|---|---|---|
| Subject | Base Line | After Placebo | After RBCs |
| 1 | 34.03 | 34.53 | 33.03 |
| 2 | 32.85 | 32.70 | 31.55 |
| 3 | 33.50 | 33.62 | 32.33 |
| 4 | 32.52 | 31.23 | 31.20 |
| 5 | 34.15 | 32.85 | 32.80 |
| 6 | 33.77 | 33.05 | 33.07 |

(a) Summarize these data by computing the appropriate averages.

(b) Do the differences here among the block averages imply anything about the three treatment levels? Do the differences among the treatment averages imply anything about the similarity of the subjects?

**3.85**   Why do the results of the experiments referred to in Questions 3.83 and 3.84 qualify as randomized block data rather than categorical data?

## *3.9*
## *COMPUTER NOTES*

The descriptive statistics discussed in this chapter are part of the computational routines of most computer statistical software. Very good packages for both mainframe and personal computers are available. We have chosen SAS and Minitab to provide a sense of doing statistics by computer. Although both are more than sufficient for the statistical needs of most users, SAS is full-scale, versatile, and flexible, whereas Minitab has sacrificed some power for ease of use and is often the choice of beginners.

Figure 3.36 reproduces the computer screen from a Minitab session to produce descriptive statistics for the data of Case Study 3.2. The work was done *interactively*; in other words, the computer responded with calculations and prompts for more commands while the user was at the keyboard. The data are named C1, for reference, and entered. The command for a histogram follows. Its parameters are the interval midpoints and their width. A frequency distribution is also provided.

```
SET C1
 13  17  13   8   7  10  10  21  18  17  19  15  15  23  12
 12  15  27  19  23   6   2   9  11   5  18  20  24  15  14
 11   2   4   4  13  10  13  19  25  14  11   4  14  12  23
 19  17  16  17  13   7   9  12  11  30  19   4  11  18  18
 24  15  13  12   6  17  27   3  10   7   1  14  22  16  10
  2   7   9   5  21  18  17  18  12  15  13  13  15   6  25
 13  15   5  28  20  19  14  11  14   4   8  10   7  23  18  24
END
HISTOGRAM C1, FIRST MIDPOINT 2.5, WIDTH 5

C1

MIDDLE OF        NUMBER OF
 INTERVAL        OBSERVATIONS
   2.50             10    **********
   7.50             16    ****************
  12.50             33    *********************************
  17.50             29    *****************************
  22.50             12    ************
  27.50              5    *****
  32.50              1    *
```

**FIG. 3.36**

```
DESCRIBE C1

                    C1
N                  106
MEAN             13.81
MEDIAN           13.50
TMEAN            13.72
STDEV             6.51
SEMEAN            0.63
MAX              30.00
MIN               1.00
Q3               18.00
Q1                9.75
```

**FIG. 3.37**

Next, the computer was asked to give the one-sample statistics for the data set (see Figure 3.37). SEMEAN is the standard error of the mean. A new statistic, TMEAN, is the *trimmed mean*, which the program calculates by deleting the bottom 5% and the top 5% of the data and averaging what remains. The trimmed mean cancels the effect of very large or very small observations. Q1 is the first quartile, and Q3 the third.

Next, we examine the SAS printout for the same data. This output was obtained in *batch mode*: all instructions and data were put in a file that was submitted for processing. The user is not involved once the program is submitted. Figure 3.38 shows a histogram for the data of Case Study 3.11. Figure 3.39 is the relevant portion of the resulting printout.

```
DATA ONE;
    INPUT SCORE;
CARDS;

(The data are entered here, one observation per line.)

PROC CHART;
    VBAR SCORE / MIDPOINTS = 2.49 TO 32.49 BY 5;
```

**FIG. 3-38**

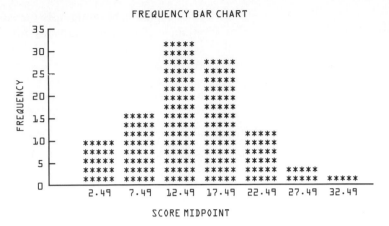

**FIG. 3.39**

To obtain a histogram equivalent to Figure 3.2, the midpoints for the SAS histogram command differ from those used for the Minitab version. The routines treat the end points of the interval defining the frequency distribution differently. For example, in Minitab we asked for an interval centered at 2.5 of width 5. The program considers this the interval $[0, 5)$, which corresponds to the meaning of the 0–4 range in Table 3.6. Since SAS would look at the same interval as $(0, 5]$, we must use the SAS interval $(-0.01, 4.99]$ to get the same frequency distribution.

The basic one-variable statistics are obtained through a SAS procedure known as PROC UNIVARIATE. Figure 3.40 gives a portion of the output from PROC UNIVARIATE. (Some of the given statistical values are not of interest to us at this time.) Try to match the SAS labels with the corresponding Minitab indicators.

## UNIVARIATE

VARIABLE=SCORE

| MOMENTS | | | | QUANTILES (DEF=4) | | | | |
|---|---|---|---|---|---|---|---|---|
| N | 106 | SUM WGTS | 106 | 100% | MAX 30 | | 99% | 29.86 |
| MEAN | 13.8113 | SUM | 1464 | 75% | Q3 18 | | 95% | 25 |
| STD DEV | 6.51023 | VARIANCE | 42.3831 | 50% | MED 13.5 | | 90% | 23 |
| SKEWNESS | 0.168361 | KURTOSIS | -0.433537 | 25% | Q1 9.75 | | 10% | 4.7 |
| USS | 24670 | CSS | 4450.23 | 0% | MIN 1 | | 5% | 3.35 |
| CV | 47.1369 | STD MEAN | 0.63233 | | | | 1% | 1.07 |
| T:MEAN=0 | 21.842 | PROB > \|T\| | 0.0001 | | RANGE 29 | | | |
| SGN RANK | 2835.5 | PROB > \|S\| | 0.0001 | | Q3-Q1 8.25 | | | |
| NUM ¬= 0 | 106 | MODE 13 | | | | | | |

**FIG. 3.40**

Routines are available to calculate the other statistics in Chapter 3, but we will examine them in later chapters.

# 3.10
## SUMMARY

Chapter 3 has presumably taught us two important lessons. First, that the *structure* of a set of data dictates to a great extent the options we have in drawing graphs and computing numerical summaries. We tabulate and display regression–correlation data, for example, entirely differently from two-sample data.

Second, descriptive statistics can be enormously helpful in isolating the important information contained in a set of measurements. Recall, for instance, how nicely the histogram in Figure 3.2 highlighted what was nothing but a list of numbers in Table 3.5. Or how sharply the contingency table in Case Study 3.18 delineated the relationship between transiency levels and suicide rates. Indeed, *any* statistical analysis, no matter how sophisticated it eventually becomes, should begin with the graphs and computations covered in this chapter.

The key word here is "begins." Much of this material has necessarily been a prelude to themes we have yet to compose. We have learned, for example, how to compute standard deviations and standard errors, and we can see that they measure dispersion. But why, exactly, is dispersion so important? Until we come face to face with the problem of hypothesis testing in Chapters 8 and 9, there is really no good way to answer that. Other ideas introduced in this chapter will be similarly fleshed out as our background permits and the need arises.

REVIEW QUESTIONS

3.86  Listed below are the percentages of votes for the Republican gubernatorial candidate reported by 20 rural precincts in a recent election.

| | | | |
|---|---|---|---|
| 25 | 49 | 72 | 23 |
| 21 | 68 | 54 | 28 |
| 47 | 25 | 28 | 38 |
| 36 | 29 | 51 | 37 |
| 26 | 33 | 21 | 31 |

(a) Construct a set of classes for these data and draw the corresponding histogram.

(b) The average for the 20 precincts is 37.1. If these results are typical, can we expect that approximately 68% of all rural precincts will report Republican percentages within one standard deviation of 37.1? Why or why not?

3.87 An experimenter collects the following set of regression data:

| $x_i$ | $y_i$ | $x_i$ | $y_i$ |
|---|---|---|---|
| 3 | 3 | 7 | 2 |
| 5 | 3 | 3 | 4 |
| 6 | 3 | 5 | 4 |
| 1 | 0 | 7 | 1 |
| 1 | 2 | 1 | 1 |

Is it reasonable to say that the $xy$-relationship is described by the equation $y = 1.68 + 0.16x$? Explain.

3.88 What, if anything, is wrong with the two graphs pictured below?

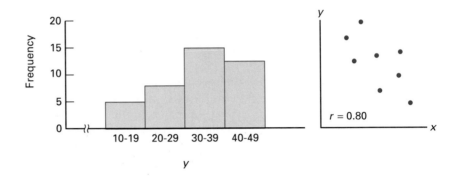

3.89 The 20 observations below are summarized by the frequency distribution shown to the right. Compute $\bar{y}$ for the 20 observations and compare it with the grouped mean, $\bar{y}_g$, computed from the frequency distribution (see Question 3.24). Describe a situation for which $\bar{y}$ will equal $\bar{y}_g$.

| 10 | 15 | 17 | 16 |
|---|---|---|---|
| 16 | 18 | 4 | 9 |
| 16 | 19 | 8 | 12 |
| 2 | 10 | 13 | 13 |
| 12 | 5 | 18 | 16 |

| Class | Frequency |
|---|---|
| 0–3 | 1 |
| 4–7 | 2 |
| 8–11 | 4 |
| 12–15 | 5 |
| 16–19 | 8 |
| | 20 |

3.90 An engineering firm is studying the insulating efficiency of a new kind of awning. For each of six months, heating and cooling costs are reported for two similar houses, on one of which has been installed the special awnings. The results are summarized below. Graph the data.

| Month | With Awnings | Without Awnings |
|---|---|---|
| April | $30 | $35 |
| May | 28 | 27 |
| July | 60 | 75 |
| August | 65 | 80 |
| December | 112 | 110 |
| January | 120 | 115 |

3.91 Two mathematics achievement tests, Math I and Math II, can be taken as part of the SAT exam. The latter is more difficult and covers a more advanced set of topics. Suppose a student takes both tests and is sent the following summary of his performance:

| | Score | Percentile |
|---|---|---|
| Math I | 700 | 92 |
| Math II | 740 | 86 |

Why does a higher score on the more difficult test earn a lower percentile? Has a mistake been made?

3.92 A quality control engineer is concerned about variation in the amount of cereal in boxes that are advertised to be 12 oz. She weighs four boxes chosen at random and gets values of 11.4, 12.1, 12.2, and 12.6. Based on those four observations, what would be a reasonable estimate for the lightest "12 oz." box she is likely to encounter?

3.93 There are seven charter members in a local Star Trek Fan Club. Listed below are the numbers of times each has seen the "Trouble with Tribbles" episode.

| Member | Number of viewings, $y_i$ |
|---|---|
| AW | 3 |
| PR | 2 |
| NM | 7 |
| LN | 2 |
| JC | 9 |
| BB | 4 |
| MS | 5 |

Compute the corresponding sample mean, sample median, and sample standard deviation.

**3.94**    Let $y_{(i)}$ denote the $i$th percentile of a distribution. If

$$y_{(3)} - y_{(1)} = 2, \qquad y_{(51)} - y_{(49)} = 10, \quad \text{and} \quad y_{(99)} - y_{(97)} = 2$$

what might the distribution of the $y_i$'s look like?

**3.95**    Shown below are graphs of two sets of $k$-sample data. For each treatment, the intervals drawn extend from $\bar{y} - s/\sqrt{n}$ to $\bar{y} + s/\sqrt{n}$. For which experiment, if either, are the three treatments more likely to be equal? Explain. Assume that $n$ is the same for all six samples.

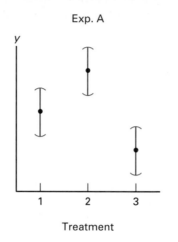
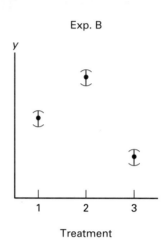

**3.96**    As part of its MBA recruiting brochure, a university lists the following starting salaries (in thousands) for 25 of last year's graduates. Summarize the data by drawing a relative frequency polygon.

| | | | | |
|---|---|---|---|---|
| $26.5 | 43.0 | 30.5 | 24.8 | 31.4 |
| 32.3 | 38.9 | 29.5 | 30.6 | 27.4 |
| 34.0 | 26.4 | 38.7 | 35.2 | 25.9 |
| 22.8 | 25.3 | 21.4 | 29.7 | 56.2 |
| 31.5 | 26.9 | 26.7 | 26.8 | 24.3 |

**3.97**    Gadgets, Inc., has offered a kitchen novelty item at a variety of prices. Shown below are the profits that resulted at each price.

| Time Period | Unit Price, $x_i$ | Total Profits, $y_i$ |
|---|---|---|
| Fall, '87 | $1.50 | $2000 |
| Spring, '88 | $1.25 | $1700 |
| Fall, '88 | $1.75 | $2500 |
| Spring, '89 | $1.85 | $2600 |

(a) Graph the data.

(b) Find the least squares regression line.

(c) Plot your answer to part (b) on the graph in part (a).

**3.98** Display the data in the following scatter diagram as a contingency table having two rows and two columns. (Hint: Define the rows and columns of the table in terms of $\bar{y}$ and $\bar{x}$, respectively.

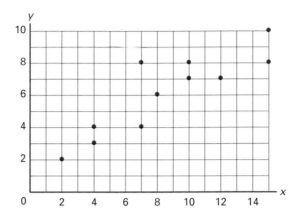

**3.99** The graph below contains a set of two-sample data. What does its appearance suggest about the shapes of the two populations from which the data have been drawn?

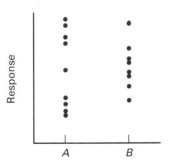

**3.100** Consider the following set of one-sample data:

| Subject, $i$ | Response, $y_i$ |
| --- | --- |
| 1 | 4.0 |
| 2 | 2.6 |
| 3 | 4.1 |
| 4 | 5.2 |

What, if anything, is wrong with presenting those data in a more visual format by graphing $y_i$ versus $i$, as shown below?

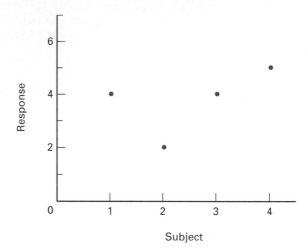

3.101   As part of a study investigating the impact of children on professional growth, family size ($x$) and number of promotions ($y$) were recorded for each of five female lawyers.

| Lawyer | Number of Children, $x_i$ | Number of Promotions, $y_i$ |
|--------|---------------------------|------------------------------|
| CF | 2 | 1 |
| DP | 0 | 4 |
| SH | 4 | 1 |
| LM | 1 | 4 |
| CW | 2 | 1 |

(a) Graph the data.

(b) Find the equation of the regression line, $y = \hat{a} + \hat{b}x$.

(c) Plot your answer to part (b) on the graph drawn for part (a).

3.102   For the data in Question 3.101, compute

$$L = \sum_{i=1}^{5} [y_i - (\hat{a} + \hat{b}x_i)]^2$$

Suppose $y = c + dx$ is any line other than $y = \hat{a} + \hat{b}x$. Define $L'$ by the equation,

$$L' = \sum_{i=1}^{5} [y_i - (c + dx_i)]^2$$

What mathematical relationship exists between $L$ and $L'$?

3.103   Freshman SAT scores at Miskatonic Tech have a bell-shaped distribution with a sample mean of 1200 and a sample standard deviation of 250. If 600 students are in the freshman class, how many would we expect to have SAT's above 1450?

3.104   The scores below are the results of an on-the-job stress test given to thirty air traffic controllers.

| 129 | 141 | 154 | 135 | 137 | 146 |
|-----|-----|-----|-----|-----|-----|
| 138 | 130 | 128 | 143 | 123 | 151 |
| 131 | 131 | 136 | 152 | 133 | 140 |
| 148 | 144 | 136 | 149 | 159 | 135 |
| 156 | 142 | 121 | 148 | 140 | 126 |

(a) Construct a frequency distribution.

(b) Draw the relative frequency histogram based on your answer to part (a).

(c) Would you consider these data to be bell-shaped?

3.105   When does the sample size dictate the way in which a set of data should be graphed?

3.106   Refer to Case Study 2.2. Graph the ages at which the twelve scientists made their greatest discoveries.

3.107   What, if anything, is wrong with the following graph?

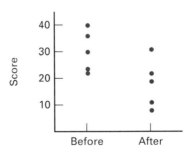

3.108   Compute the sample mean, the sample median, and the sample standard deviation for the maximum heart rates in Question 2.10. If similar measurements were taken on another set of 11 Navy parachutists, would you be surprised if one of the observations in that second group was equal to 206? Explain.

3.109   Construct a percentage cumulative frequency distribution for the $FEV_1/VC$ ratios in Case Study 2.1. Begin by making a frequency distribution using 0.60–0.64, 0.65–0.69, and so on, as classes.

3.110   For any set of regression-correlation data, it can be proved that the line $y = \hat{a} + \hat{b}x$ goes through the point $(\bar{x}, \bar{y})$. Verify that property for the dragonfly data in Question 1.13.

# 4 Combinatorics and Combinatorial Probability

*They are called wise who put things in their right order.*

**St. Thomas Aquinas**

## 4.1

### A BRIEF HISTORY

No one knows where or when the notion of chance first arose; it fades into our prehistory. Nevertheless, evidence linking early humans with devices for generating random events is plentiful: archaeological digs, for example, throughout the ancient world consistently turn up a curious overabundance of *astragali*, the heel bones of sheep and other vertebrates. Why should the frequencies of these bones be so disproportionately high? One could hypothesize that our forebearers were fanatical foot fetishists, but two other explanations seem more plausible: the bones were used for religious ceremonies *and for gambling*.

Astragali have six sides but are not symmetrical (Figure 4.1). Those found in excavations typically have their sides numbered or engraved. For many ancient civilizations, astragali were the primary mechanism through which oracles solicited the opinions of their gods. In Asia Minor, for example, it was customary in divination rites to roll, or *cast*, five astragali. Each possible configuration was associated with the name of a god and carried with it the sought after advice. An outcome of (1, 3, 3, 4, 4), for instance, was said to be the throw of the saviour Zeus, and its appearance was taken as a sign of encouragement (35):

Sheep *astragalus*

**FIG. 4.1**

> One one, two threes, two fours
> The deed which thou meditatest, go do it boldly.
> Put thy hand to it. The gods have given thee
>     favorable omens
> Shrink not from it in thy mind, for no evil
>     shall befall thee.

A (4, 4, 4, 6, 6), on the other hand, the throw of the child-eating Cronos, would send everyone scurrying for cover:

> Three fours and two sixes. God speaks as follows.
> Abide in thy house, nor go elsewhere,
> Lest a ravening and destroying beast come nigh thee.
> For I see not that this business is safe. But bide
>     thy time.

Gradually, over thousands of years, astragali were replaced by dice, and the latter became the most common means for generating random events. Pottery dice have been found in Egyptian tombs built before 2000 B.C.; by the time the Greek civilization was in full flower, dice were everywhere. (*Loaded* dice have also been found. Mastering the mathematics of probability would prove to be a formidable task for our ancestors, but they quickly learned how to cheat!)

Our lack of historical records blurs the distinction initially drawn between divination ceremonies and recreational gaming. Among more recent societies, though, gambling emerged as a distinct entity, and its popularity was irrefutable. The Greeks and Romans were consummate gamblers, as were the early Christians (86).

Rules for many of the Greek and Roman games have been lost, but we can recognize the lineage of certain modern diversions in what was played during the Middle Ages. The most popular dice game of that period was called *hazard*, the name deriving from the Arabic *al zhar*, which means "a die." Hazard is thought to have been brought to Europe by soldiers returning from the Crusades; its rules are much like those of our modern day craps. Cards were first introduced in the fourteenth century and immediately gave rise to a game known as *Primero*, an early form of poker. Board games, such as backgammon, were also popular during this period.

Given this rich tapestry of games and the obsession with gambling that characterized so much of the Western world, it may seem more than a little puzzling that a formal study of probability was not undertaken sooner than it was. As we will see shortly, the first instance of anyone *conceptualizing* probability, in terms of a mathematical model, occurred in the sixteenth century. That means that more than 2000 years of dice games, card games, and board games passed by before someone finally had the insight to write down even the simplest of probabilistic abstractions.

Historians generally agree that, as a subject, probability got off to a rocky start because of its incompatibility with two of the most dominant forces in the evolution of our Western culture, Greek philosophy and early Christian theology. The Greeks were comfortable with the notion of chance (something the Christians were not), but it went against their nature to suppose that random events could be quantified in any useful fashion. They believed that any attempt to reconcile mathematically what *did* happen with what *should have* happened was, in their phraseology, an improper juxtaposition of the "earthly plane" with the "heavenly plane."

Making matters worse was the antiempiricism that permeated Greek thinking. Knowledge, to them, was not something that should be derived by experimentation. It was better to reason out a question logically than to search for its explanation in a set of numerical observations. Together, these two attitudes had a deadening effect: the Greeks had no motivation to think about probability in any abstract sense, nor were they faced with the problems of interpreting data that might have pointed them in the direction of a probability calculus.

If the prospects for the study of probability were dim under the Greeks, they became even worse when Christianity broadened its sphere of influence. The Greeks and Romans at least accepted the *existence* of chance. They believed their gods to be either unable or unwilling to get involved in matters so mundane as the outcome of the roll of a die. Cicero writes:

Nothing is so uncertain as a cast of dice, and yet there is no one who plays often who does not make a Venus-throw[1] and occasionally twice and thrice in succession. Then are we, like fools, to prefer to say that it happened by the direction of Venus rather than by chance?

For the early Christians, though, there was no such thing as chance: every event that happened, no matter how trivial, was perceived to be a direct manifestation of God's deliberate intervention. In the words of St. Augustine:

> Nos eas causas quae dicuntur fortuitae . . . non dicimus
>   nullas, sed latentes; easque tribuimus vel veri Dei . . .
> (We say that those causes that are said to be by chance
>   are not non-existent but are hidden, and we attribute
>   them to the will of the true God . . .)

Taking Augustine's position makes the study of probability moot, and it makes a probabilist a heretic. Not surprisingly, nothing of significance was accomplished in the subject for the next fifteen hundred years.

It was in the sixteenth century that probability, like a mathematical Lazarus, arose from the dead. Orchestrating its resurrection was one of the most eccentric figures in the entire history of mathematics, Gerolamo Cardano. By his own admission, Cardano personified the worst and the best—the Jekyll and the Hyde—of the Renaissance man. He was born in 1501 in Pavia. Facts about his personal life are difficult to verify. He wrote an autobiography, but his penchant for lying raises doubts about much of what he says. Whether true or not, though, his "one-sentence" self-assessment paints an interesting portrait (117):

> Nature has made me capable in all manual work, it has given me the spirit of a philosopher and ability in the sciences, taste and good manners, voluptuousness, gaiety, it has made me pious, faithful, fond of wisdom, meditative, inventive, courageous, fond of learning and teaching, eager to equal the best, to discover new things and make independent progress, of modest character, a student of medicine, interested in curiosities and discoveries, cunning, crafty, sarcastic, an initiate in the mysterious lore, industrious, diligent, ingenious, living only from day to day, impertinent, contemptuous of religion, grudging, envious, sad, treacherous, magician and sorcerer, miserable, hateful, lascivious, obscene, lying, obsequious, fond of the prattle of old men, changeable, irresolute, indecent, fond of women, quarrelsome, and because of the conflicts between my nature and soul I am not understood even by those with whom I associate most frequently.

---

[1] When rolling four astragali, each of which is numbered on *four* sides, a Venus-throw was having each of the four numbers appear.

Formally trained in medicine, Cardano's interest in probability derived from his addiction to gambling. His love of dice and cards was so all-consuming that he is said to have once sold all his wife's possessions just to get table stakes! Fortunately, something positive came out of Cardano's obsession. He began looking for a mathematical model that would describe, in some abstract way, the outcome of a random event. What he eventually formalized is now called the *classical definition of probability*: if the total number of possible outcomes, all equally likely, associated with some action is $n$ and if $m$ of those $n$ result in the occurrence of some given event, then the probability of that event is $m/n$. If a fair die is rolled, there are $n = 6$ possible outcomes. If the event "outcome is greater than or equal to 5" is the one in which we are interested, then $m = 2$ (the outcomes 5 and 6) and the probability of the event is 2/6, or 1/3 (see Figure 4.2).

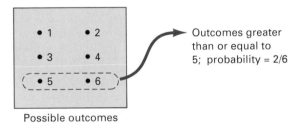

Possible outcomes

**FIG. 4.2**

Cardano had tapped into the most basic principle in probability. The model he discovered may seem trivial in retrospect, but it represented a giant step forward: his was the first recorded instance of anyone computing a *theoretical*, as opposed to an empirical, probability. Still, the actual impact of Cardano's work was minimal. He wrote a book in 1525, but its publication was delayed until 1663. By then, the focus of the Renaissance, as well as interest in probability, had shifted from Italy to France.

The date cited by many historians (those who are not Cardano supporters) as the "beginning" of probability is 1654. In Paris a well-to-do gambler, the Chevalier de Mere, asked several prominent mathematicians, including Blaise Pascal, a series of questions, the best-known of which was the *problem of points*:

> Two people, A and B, agree to play a series of fair games until one person has won six games. They each have wagered the same amount of money, the intention being that the winner will be awarded the entire pot. But suppose, for whatever reason, the series is prematurely terminated, at which point A has won five games and B three. How should the stakes be divided?

[The correct answer is that A should receive seven eighths of the total amount wagered. See if you can figure out why. (Hint: Suppose the contest were resumed. What scenarios would lead to A's being the first person to win six games?)]

Pascal was intrigued by de Mere's questions and shared his thoughts with Pierre Fermat, a Toulouse civil servant and probably the most brilliant mathematician in Europe. Fermat graciously replied, and from the now famous Pascal-Fermat correspondence came not only the solution to the problem of points but the foundation for more general results. More significantly, news of what Pascal and Fermat were working on spread quickly. Others got involved, of whom the best known was the Dutch scientist and mathematician, Christiaan Huygens. The delays and the indifference that plagued Cardano a century earlier were not going to happen again.

Best remembered for his work in optics and astronomy, Huygens, early in his career, was intrigued by the problem of points. In 1657 he published *De Ratiociniis in Aleae Ludo* (Calculations in Games of Chance), a very significant work, far more comprehensive than anything Pascal and Fermat had done, and for almost 50 years the standard "textbook" in the theory of probability. Huygens, of course, has supporters who feel that *he* should be credited as the founder of probability.

Almost all the mathematics of probability was still waiting to be discovered. What Huygens wrote was only the humblest of beginnings, a set of 14 propositions bearing little resemblance to the topics we teach today. But the foundation was there. Probability, as a subject, was at last on firm footing.

# 4.2

## A FUNDAMENTAL PRINCIPLE: THE MULTIPLICATION RULE

The probability problems that Cardano, Pascal, Fermat, and Huygens first struggled with were all conceptually much the same. Each invoked the $m/n$ argument: if an action, or game, can lead to $n$ equally likely outcomes and if $m$ of those outcomes satisfy a certain condition, then the probability of that condition is $m/n$.

Finding $n$ and $m$ sometimes is easy and sometimes is not. If the action being considered is something simple like rolling a die, we can get $n$ and $m$ by direct enumeration. But suppose we want to know how many 5-card poker hands can be dealt from a 52-card deck? Or how many sets of six numbers can be picked to win the Ohio lottery? The answers to these last two questions are 2,598,960 and 3,838,380, respectively—numbers much too large to be obtained by simply listing them (If we wrote down one poker hand per minute around the clock, it would take us five years to list all $2\frac{1}{2}$ million hands!)

Must we be limited, then, to simple games where $n$ and $m$ are small? No. A branch of mathematics known as ***combinatorics*** deals with this problem. Combinatorics examines the structure of the outcomes associated with

combinatorics—The branch of mathematics concerned with the selection, arrangement, and enumeration of ordered and unordered sequences.

a situation and allows us to count $n$ and $m$ *systematically*. The basic principles of counting are presented in Sections 4.3 and 4.4. In Section 4.5 we return to situations where it makes sense to compute a probability by taking the ratio of *favorable* outcomes to *total* outcomes.

We begin with a simple notion at the very heart of combinatorics. It leads to a result for counting the number of different ways to perform an ordered sequence of events. Identifying and characterizing such sequences are the keys to solving a great many problems.

Imagine a freshman planning his weekend festivities. His Friday night and Saturday night options are listed in Table 4.1. (To simplify matters, we are making the admittedly unrealistic assumption that the weekend starts on Friday rather than on Thursday . . . .) Clearly, his weekend entertainment will be one of *six* possible sequences. He can follow up any of his three Friday night activities with either of the two possibilities for Saturday (see Figure 4.3). By the same reasoning, if there were *five* things he, might do Sunday night, his set of possible

Friday–Saturday–Sunday

sequences would increase to 30 ($= 6 \cdot 5$), because each of the six sequences in Figure 4.3 could precede any of his five options for Sunday. Extending this argument to sequences of $k$ components gives us the most basic result in combinatorics, the **multiplication rule**.

**TABLE 4.1**

| Friday Night | Saturday Night |
| --- | --- |
| 1. Sleep | 1. Party |
| 2. Go to rock concert | 2. Do laundry |
| 3. Read | |

| Friday | Saturday | Weekend Sequence |
| --- | --- | --- |
| Sleep | Party<br>Laundry | 1. (sleep, party)<br>2. (sleep, laundry) |
| Rock concert | Party<br>Laundry | 3. (rock concert, party)<br>4. (rock concert, laundry) |
| Read | Party<br>Laundry | 5. (read, party)<br>6. (read, laundry) |

**FIG. 4.3**

---

**THEOREM 4.1**　Multiplication Rule

If step 1 can be performed in $n_1$ ways, step 2 in $n_2$ ways, . . . , and step $k$ in $n_k$ ways, the ordered sequence

$$(\text{step } 1, \quad \text{step } 2, \quad \ldots, \quad \text{step } k)$$

can be performed in

$$n_1 \cdot n_2 \cdots n_k \quad \text{ways.}$$

*EXAMPLE 4.1*   When access to a computer file needs to be restricted, authorized users are assigned a *password*, a sequence of letters and numbers with a prescribed format. Only if the password is keyed in can a person log on to the system. How many passwords are possible if each must be in the form

letter   letter   number   number

*Solution*   We can consider a password as an ordered sequence of four ($k = 4$) steps. Steps 1 and 2 can each be "performed" in 26 ways, steps 3 and 4 in 10 ways. It follows, then, that the total number of passwords is the product

$$26 \cdot 26 \cdot 10 \cdot 10 = 67,600.$$

**tree diagram**—A graphical device for picturing the formation of an ordered sequence.

The ***tree diagram*** in Figure 4.4 graphically depicts the multiplication rule for this situation. Any admissible password is necessarily an ordered sequence of four consecutive "branches" (branches are denoted by heavy lines; note, for example, the branches producing the password AA09). Study the diagram until it is clear to you that the number of different sequences

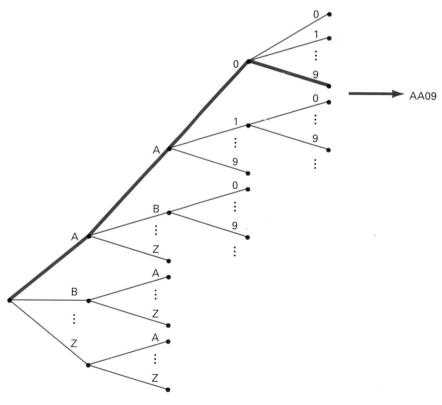

**FIG. 4.4**

of four consecutive branches is

$$n_1 \cdot n_2 \cdot n_3 \cdot n_4 = 26 \cdot 26 \cdot 10 \cdot 10$$

---

**EXAMPLE 4.2**   Against her better judgment, Nancy has agreed to babysit all day Saturday for an obnoxious set of twins. Her only hope of remaining sane is to let them do what they enjoy most—watch slasher movies. The local video store has *Chainsaw Massacre* (parts I and II), *Halloween* (parts I, II, and III), and 15 assorted movies about maniacs terrorizing sororities. How many sets of three videos can she select if she wants to choose one from each "category"?

*Solution*   The number of triple features from which Nancy must choose is a direct application of Theorem 4.1. With 2 choices available in the first category, 3 in the second, and 15 in the third, she has her pick of $2 \cdot 3 \cdot 15 = 90$ different *sets* of three videos, where each set contains exactly one movie of each type.

---

**EXAMPLE 4.3**   When they were first introduced, postal zip codes were five-digit numbers, theoretically ranging from 00000 to 99999. (Actually the lowest zip code was 00601 for San Juan, Puerto Rico, the highest was 99950 for Ketchikan, Alaska, and not all intermediate numbers were used.) An additional four digits have recently been added, so each zip code is now a nine-digit number.

(a) If all possible nine-digit numbers were used, how many different zip codes could be written?

(b) How many different zip codes are at least as large as 60000−0000, have 7 as their third digit, and end on an even number?

*Solution*   (a) Since each digit can be assigned any of 10 different values (0 through 9), the total number of possible nine-digit zip codes is

$$\underbrace{10 \cdot 10 \cdots 10}_{9 \text{ factors}} = 10^9 = 1 \text{ billion}$$

(b) Counting the number of zip codes restricted by certain constraints is conceptually no different from what we did in part (**a**). We simply need to readjust the $n_i$'s. For the zip code to be at least as large as

$$60000-0000$$

the first digit must be a 6, 7, 8, or 9, giving us *four* choices. The third digit is fixed at 7, so our options at that step are reduced to *one*. Finally, if the last digit is to be even, it must be one of *five* possible values: 0, 2, 4, 6, or 8. Figure 4.5 shows the nine $n_i$'s. By the multiplication rule the total number of zip codes complying with the conditions imposed on the first, third, and last digits is

$$4 \cdot 10 \cdot 1 \cdot 10 \cdot 10 \cdot 10 \cdot 10 \cdot 10 \cdot 5 = 20,000,000$$

Options:

| (4) | (10) | (1) | (10) | (10) | (10) | (10) | (10) | (5) |
|-----|------|-----|------|------|------|------|------|-----|
| 1 | 2 | 3 | 4 | 5 | 6 | 7 | 8 | 9 |

Zip code digits

**FIG. 4.5**

---

*EXAMPLE 4.4*  A restaurant offers a choice of 4 appetizers, 14 entrees, 6 desserts, and 5 beverages. In how many ways can a diner "design" his evening meal

(a) if he is hungry enough to elect one option from each category?

(b) if he intends to order only three courses?

(Consider the beverage as a "course.")

*Solution*  (a) Think of a meal as an ordered sequence of menu selections, where $n_1 = 4$, $n_2 = 14$, $n_3 = 6$, and $n_4 = 5$. The total number of different four-course meals is the product $4 \cdot 14 \cdot 6 \cdot 5 = 1680$.

(b) There are four types of three-course meals:

> (appetizer, entree, dessert)
> (appetizer, entree, beverage)
> (appetizer, dessert, beverage)
> (entree, dessert, beverage).

According to the multiplication rule, the numbers of different meals within each type are

$$4 \cdot 14 \cdot 6 = 336$$
$$4 \cdot 14 \cdot 5 = 280$$
$$4 \cdot 6 \cdot 5 = 120$$
$$14 \cdot 6 \cdot 5 = 420$$

respectively. Therefore, the number of possible three-course meals is

$$336 + 280 + 120 + 420 = 1156.$$

---

*EXAMPLE 4.5*  Alphonse Bertillon, a nineteenth-century French criminologist, developed an identification system based on 11 anatomical variables (height, head width, ear length, etc.) that presumably remained essentially unchanged during an individual's adult life. The range of each variable was divided into three subintervals: small, medium, and large. A person's *Bertillon configuration* was an ordered sequence of 11 letters, say

$$s, s, m, m, l, s, l, s, s, m, s$$

where a letter indicated the individual's "size" relative to that variable. In New York City will there necessarily be two adults with the same Bertillon configuration?

*Solution*    Viewed as an ordered sequence, a Bertillon configuration is an 11-step classification system with three options available at each step. It follows that $3^{11}$, or 177,147, distinct sequences are possible. Therefore, any city with at least 177,148 adults would necessarily have at least two residents with the same pattern. (The limited number of possibilities generated by Bertillon's variables proved to be one of its major weaknesses. Still, it was widely used in Europe for criminal identification before the development of finger-printing.)

---

**EXAMPLE 4.6**    In 1824 Louis Braille invented what would eventually become the standard alphabet for the blind. Based on an earlier form of "night writing" used by the French army for reading battlefield communiques in the dark, Braille's system replaced each written character with a six-dot matrix:

where certain dots were raised, the choice depending on the character being transcribed. The letter *e*, for example, has two raised dots and is written

Punctuation marks, common words, suffixes, and so on, also have specified dot patterns. In all, how many different characters can be enciphered in Braille?

*Solution*    Think of the dots as six distinct steps, numbered 1 to 6 (see Figure 4.6). In forming a Braille letter, we have two options for each dot: we can raise it or *not* raise it. Viewed in that context the letter *e* corresponds to the six-step sequence (raise, don't raise, don't raise, don't raise, raise, don't raise). The number of such sequences, with $k = 6$ and $n_1 = n_2 = \cdots = n_6 = 2$, is $2^6$. However, one of the $2^6$ configurations has *no* raised dots, a pattern of no use to a blind person. Figure 4.7 shows the entire 63-character ($= 2^6 - 1$) Braille alphabet.

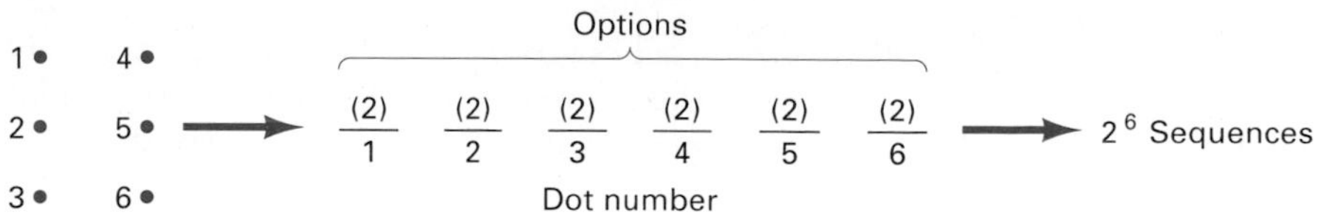
Options
1 4
2 5
3 6
(2) (2) (2) (2) (2) (2)
1 2 3 4 5 6
Dot number
$2^6$ Sequences

FIG. 4.6

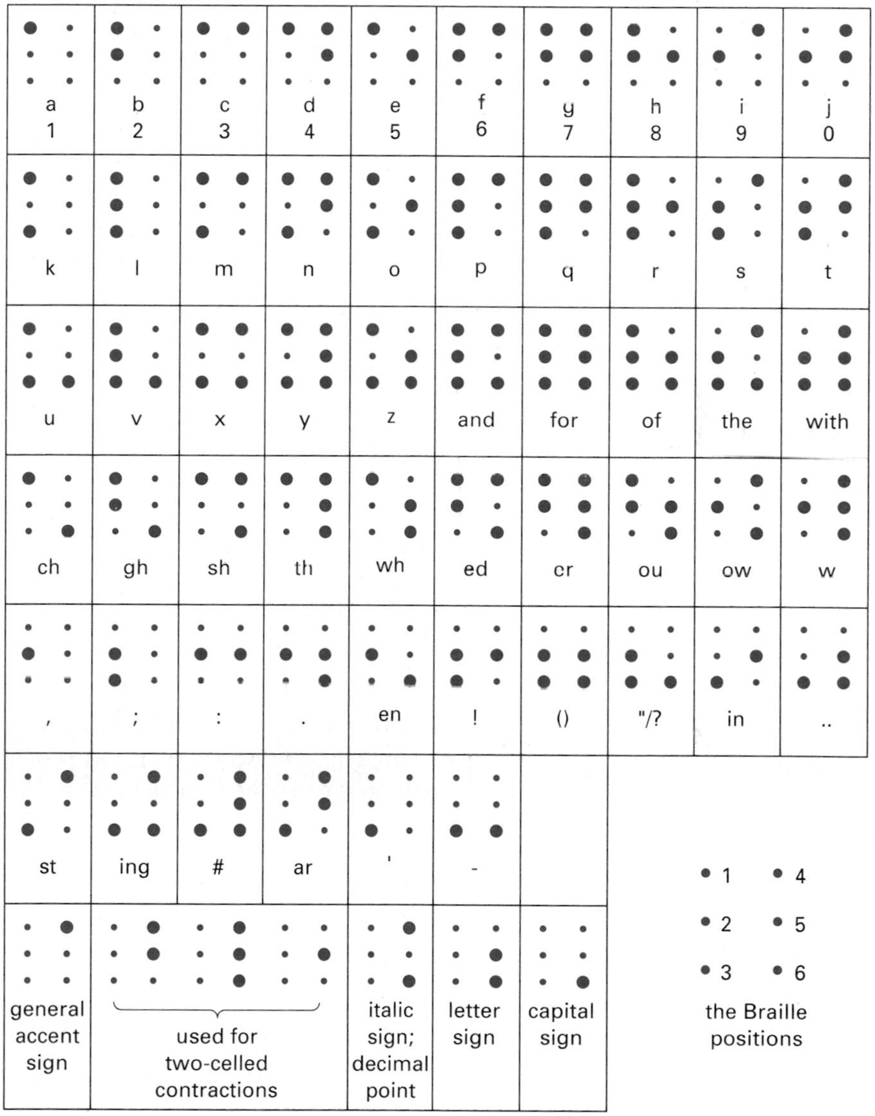
a 1
b 2
c 3
d 4
e 5
f 6
g 7
h 8
i 9
j 0
k
l
m
n
o
p
q
r
s
t
u
v
x
y
z
and
for
of
the
with
ch
gh
sh
th
wh
ed
er
ou
ow
w
,
;
:
.
en
!
()
"/?
in
..
st
ing
#
ar
'
-
general accent sign
used for two-celled contractions
italic sign; decimal point
letter sign
capital sign
1 4
2 5
3 6
the Braille positions

FIG. 4.7

*EXAMPLE 4.7*  Hamburger Heaven features a condiment bar where customers personalize their burgers. Eight toppings are available. Jamie insists on eating there every weekday. How long can she go without making the same kind of hamburger twice?

*Solution*  Notice the similarity between the structure of this problem and that of Example 4.6. Here, each topping corresponds to one of the *eight* steps involved in customizing a hamburger. Furthermore, each step can be performed in two ways: mustard, for example, can be added or not added. By the multiplication rule the number of possible hamburgers is $2^8 = 256$, just four short of the number of weekdays in a year.

---

**QUESTIONS**

**4.1**  A coded message from a CIA operative to his Russian KGB counterpart is to be sent in the form $Q\ 4\ E\ T$, where the first and last entries must be consonants, the second entry an integer 1 through 9, and the third entry one of six vowels. How many different ciphers can be transmitted?

**4.2**  A coin is tossed three times. How many different sequences of heads and tails are possible? Draw the corresponding tree diagram.

**4.3**  Suppose that three Democrats and four Republicans enter their respective party primaries in preparation for the next presidential campaign. How many different matchups are possible in the general election? Draw the corresponding tree diagram.

**4.4**  An octave contains 13 distinct notes (on a piano, 5 black keys and 8 white keys). How many different 8-note melodies within a single octave can be written if only the white keys are used? only the black keys?

**4.5**  A businesswoman is scheduled to fly from Atlanta to Philadelphia to Boston. She has her choice of five flights connecting Atlanta with Philadelphia and six connecting Philadelphia with Boston.

(a) How many different flight plans can she arrange from Atlanta to Boston via Philadelphia?

(b) How many different round-trips can she schedule?

**4.6**  In baseball how many ways can the three bases (first, second, and third) be occupied? (Hint: Recall the solution given in Example 4.6.)

**4.7**  Consider again the video options available to Nancy in Example 4.2. How many three-movie matinees could she schedule if the *order* in which the films were shown made a difference?

**4.8**  How many terms will be included in the expansion of

$$(a + b + c)(d + e + f)(x + y + u + v + w)$$

(Hint: Consider the simpler problem of counting the number of terms in the expansion of $(a + b)(x + y)$. Use the answer to suggest a general solution.)

**4.9**  Each letter in the English alphabet is represented by international Morse code as a series of dots and dashes: the letter *a*, for example, is signaled as · —. How many different letters can be transcribed by using dot-dash sequences of "length" 3 or less?

**4.10**  The combination locks on the post office boxes at Arkham Tech have two dials, each marked off with 16 divisions (see the figure). To open a box, a person first turns the leftmost dial in a certain direction for two revolutions and then stops on a mark. The right dial is set in a similar fashion, after having been turned in a certain direction for two revolutions. Will this system provide enough combinations to accommodate Arkham's 1015 undergraduates?

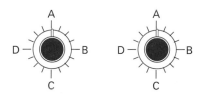

**4.11**  A chemical engineer wishes to observe the effect of temperature, pressure, and catalyst concentration on the yield resulting from a certain reaction. If she chooses to include two different temperatures, three pressures, and two levels of catalyst, how many different runs must she make in order to try each temperature-pressure-catalyst combination exactly twice?

**4.12**  A word puzzle found in many newspapers has 20 sentences, each requiring that a word be filled in. For each sentence two choices are provided, only one of which is correct. How many entries must a person submit before one of the solutions is necessarily a winner?

**4.13**  In the famous science fiction story by Arthur C. Clarke, "The Nine Billion Names of God," a computer firm is hired by the lamas in a Tibetan monastery to write a program to generate all possible names of God. For reasons never divulged, the lamas believe that all such names are nine or fewer letters in length. If no letter combinations are ruled inadmissible, is the "nine billion" in the story's title a large enough number to accommodate all the possibilities?

# 4.3
## PERMUTATIONS

We saw in Section 4.2 that the ordered sequence model has surprisingly wide applicability. Phenomena as diverse as computer passwords, Braille letters, and hamburgers can all be counted by properly invoking the multiplication rule. Another combinatorial model that we use frequently involves the formulation of ordered sequences whose components are selected, without replacement, from a finite set of elements. Such arrangements are called *permutations*. For example,

$$1\,2\,0, \quad 1\,0\,2, \quad 2\,0\,1, \quad 2\,1\,0, \quad 0\,2\,1, \quad 0\,1\,2$$

are permutations of the first three digits.

---

*EXAMPLE 4.8*   Starting with a set of four elements (the letters $A$, $B$, $C$, and $D$), in how many different ways can we arrange *three* of them in a row? Assume that no letter is to be used more than once.

*Solution*   A typical arrangement here is easy to visualize: $DAC$, for example, is one possibility. Enumerating the entire such sequence is an exercise in the multiplication rule. Think of the formation of a permutation as a three-step sequence (see Figure 4.8). We can put four different letters in the first position, three in the second position (the letters *not* used in the first position), and two in the third (the letters not used in the first or second positions). How many sequences can then be formed? By Theorem 4.1 the answer is 24 ($= 4 \cdot 3 \cdot 2$). The tree diagram in Figure 4.9 shows how each of those 24 *permutations of length* 3 is generated.

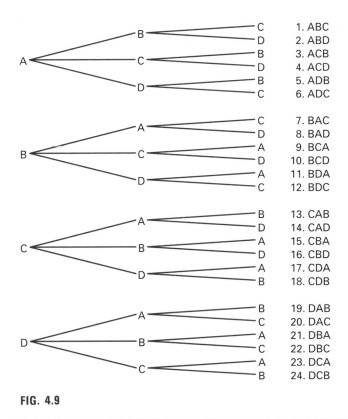

Choices:   $\dfrac{(4)}{1}$   $\dfrac{(3)}{2}$   $\dfrac{(2)}{3}$

Positions

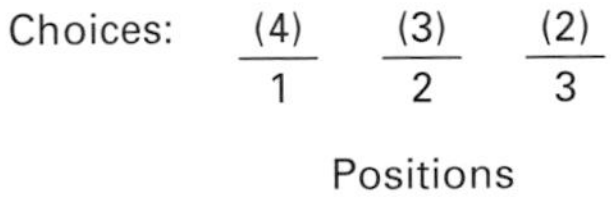

**FIG. 4.8**

**FIG. 4.9**

QUESTIONS      4.14    How many ordered sequences of length 3 could be formed if each of the four elements could be used any number of times?

### The Permutation Rule

Generalizing the question asked in Example 4.8 is not difficult: given $n$ distinct elements, in how many ways can we form permutations of length $r$? The answer follows readily from Theorem 4.1 but it requires one piece of notation that we have not encountered thus far. If $n$ is any positive integer, we will define $n!$ (read "$n$ factorial") to be the product of all positive integers less than or equal to $n$. That is,

$$n! = n(n-1)(n-2)\cdots(2)(1)$$

Also, we define $0!$ to be 1. Thus,

$$6! = 6(6-1)(6-2)(6-3)(6-4)(6-5) = 6 \cdot 5 \cdot 4 \cdot 3 \cdot 2 \cdot 1 = 720$$

---

**THEOREM 4.2**    Permutation Rule

The number of ordered arrangements (or **permutations**) of length $r$ that can be formed from a set of $n$ distinct elements, with no element being used more than once, is $_nP_r$, where

$$_nP_r = n(n-1)(n-2)\cdots(n-r+1) = \frac{n!}{(n-r)!}$$

---

*Proof*   Any of the given $n$ objects may occupy the first position in the arrangement, $n-1$ the second (we cannot reuse whatever appears in the first position), any of $n-2$ the third, and so on; the number of choices available for filling the $r$th position will be $n-(r-1) = n-r+1$ (see Figure 4.10). The result then follows from the multiplication rule: there will be a total of

$$n(n-1)(n-2)\cdots(n-r+1)$$

ordered arrangements of length $r$.

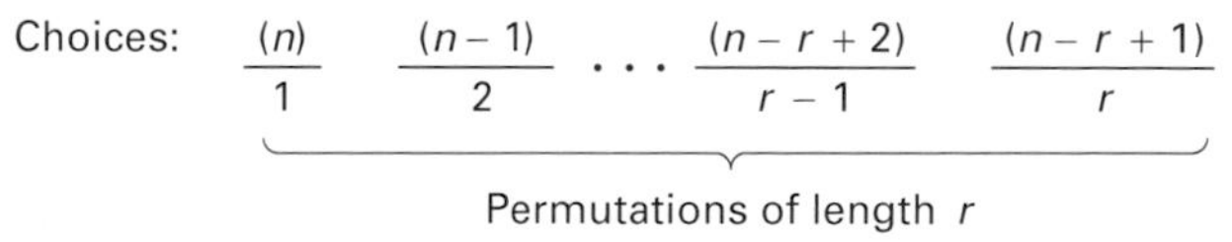

**FIG. 4.10**

Factorial notation allows the answer to be written in a slightly more compact form. By dividing $n!$ by $(n-r)!$, we can cancel all terms in $n!$ less than $n-r+1$:

$$\frac{n!}{(n-r)!} = \frac{n(n-1)\cdots(n-r+1)(n-r)(n-r-1)\cdots 1}{(n-r)(n-r-1)\cdots 1}$$

$$= n(n-1)\cdots(n-r+1)$$

Therefore, the number of permutations of length $r$, that can be formed from a set of $n$ distinct objects is

$$\frac{n!}{(n-r)!} \quad \blacksquare$$

---

**COROLLARY 4.1**

The number of ways to permute an entire set of $n$ distinct objects is $n!$ $(= {}_nP_n)$.

---

**EXAMPLE 4.9**   The crew of *Apollo 17* consisted of a pilot, a copilot, and a geologist. Suppose that NASA had actually trained nine aviators and four geologists as candidates for the flight. How many different crews could they have assembled?

*Solution*   Figure 4.11 shows the crew represented as a three-step sequence. Notice that choosing the pilot and the copilot is analogous to forming a permutation of length 2 from a set of nine distinct elements. By Theorem 4.2

$$_9P_2 = \frac{9!}{7!} = 9 \cdot 8 = 72$$

Furthermore, to each of those 72 pilot/copilot selections, NASA could have assigned any of four geologists. The number of possible crews, then, is $9 \cdot 8 \cdot 4 = 288$.

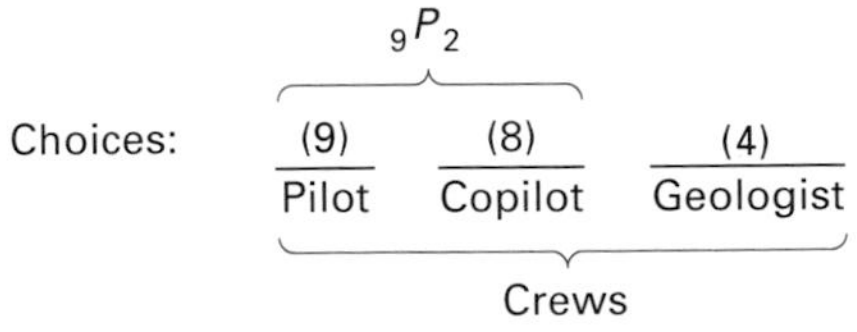

Choices:   (9)        (8)        (4)
          Pilot     Copilot    Geologist
                    Crews

**FIG. 4.11**

*EXAMPLE 4.10*  The nine members of the music faculty baseball team, the Mahler Maulers, are so versatile that each person can play any of the nine positions. (Their versatility derives from their total ineptness—they all play each position equally poorly.) In how many different ways can the Maulers take the field?

*Solution*  Think of the players as being the letters *A* through *I*. "Taking the field" means that those nine letters have been assigned in some fashion to the nine available positions. Figure 4.12 shows one possibility, where *B* is the pitcher, *F* is the catcher, *A* is the first baseman, and so on.

| B | F | A | C | E | D | I | G | H |
|---|---|---|---|---|---|---|---|---|
| p | c | 1b | 2b | 3b | ss | rf | cf | lf |

**FIG. 4.12**

In general, the pitcher could have been any one of nine players, the catcher any of the remaining 8, the first baseman any of the remaining 7, and so on (see Figure 4.13). Since each assignment of players to positions is a permutation of length 9, and the number of players available to permute is also 9, the number of different teams follows from Corollary 4.1: for $n = 9$,

$$_nP_n = {_9}P_9 = 9! = 9 \cdot 8 \cdot 7 \cdot 6 \cdot 5 \cdot 4 \cdot 3 \cdot 2 \cdot 1$$

meaning that 362,880 Mahler variations could be fielded.

Choices:

| (9) | (8) | (7) | (6) | (5) | (4) | (3) | (2) | (1) |
|---|---|---|---|---|---|---|---|---|
| p | c | 1b | 2b | 3b | ss | rf | cf | lf |

Possible teams

**FIG. 4.13**

*EXAMPLE 4.11*  Years ago—long before Rubik's cube fever had become epidemic—puzzles were much simpler. A popular combinatorial-related diversion was a 4 × 4 grid consisting of 15 movable plastic squares and one empty space. The object was to maneuver as quickly as possible an arbitrary configuration (Figure 4.14a) into a specific pattern (Figure 4.14b). In how many different ways can we initially position the 15 tiles?

(a)

(b)

**FIG. 4.14**

*Solution*    Let the empty space be tile number 16 and imagine the four rows of the grid laid end to end to make a 16-digit sequence. Each permutation of that sequence corresponds to a different pattern for the grid. Therefore, by Corollary 4.1, the number of ways to arrange the tiles is

$$_{16}P_{16} = 16! = 20,922,789,888,000$$

To put the latter in perspective, the number of ways to permute the tiles in this puzzle is more than 50 times greater than the number of stars in the Milky Way!

---

*EXAMPLE 4.12*    A new horror movie, *Friday the 13th*, *Part X* stars Jason's great-grandson as a psychotic trying to dismember eight camp counselors (four men and four women).

(a) How many scenarios can the scriptwriters devise, assuming they intend for Jason to dispatch all the men before going after any of the women?

(b) How many scripts are possible if the only restriction put on Jason is that he save Muffy for last?

*Solution*    (a) Like many combinatorial problems, this one cannot be solved by a single application of a single rule. Instead, we need to reduce what is being asked for to a series of simpler problems. The solution requires that we use Corollary 4.1 twice and the multiplication rule once.

Suppose the male counselors are denoted A, B, C, and D, and the female counselors W, X, Y, and Z. Among the admissible plots would be the sequence pictured in Figure 4.15, where B is done in first, then D, and so on. The men, if they are to be restricted to the first four positions, can still be permuted in $_4P_4 = 4!$ ways. The same number of rearrangements can be found for the women. Furthermore, the plot in its entirety can be thought of as a two-step sequence: first the men are eliminated, then the women. Since 4! ways are available to do the former and 4! the latter, the total number of different scripts, by the multiplication rule, is $4! \cdot 4! = 576$.

|  | Men |  |  |  | Women |  |  |
|---|---|---|---|---|---|---|---|
| B | D | A | C | Y | Z | W | X |
| 1 | 2 | 3 | 4 | 5 | 6 | 7 | 8 |

Order of killing

**FIG. 4.15**

(b) If the only condition to be met is that Muffy be dealt with last, then the number of admissible scripts is simply $_7P_7 = 7! = 5040$, that being the number of ways to permute the other seven counselors.

*EXAMPLE 4.13 (Optional)*     A deck of 52 cards is shuffled and dealt face up in a row. In how many different ways can the aces be adjacent?

*Solution*     A standard problem-solving technique in combinatorics is to picture the structure of a typical admissible arrangement and to let that picture dictate what needs to be done to count *all* such arrangements. Figure 4.16 shows the basic format that we need to consider here: the four aces appear as a "clump" somewhere between or around the 48 non-aces.

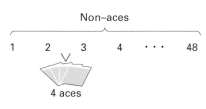

FIG. 4.16

Notice that there are 49 "spaces" that could be occupied by the four aces (in front of the first non-ace, between the first and second non-aces, . . . , after the 48th non-ace). Furthermore, once the four aces are assigned to one of the 49 positions, they can still be permuted in 4! ways. Similarly, the 48 non-aces can be arranged in 48! ways.

Putting these partial enumerations together requires an application of the multiplication rule. The physical formation of an admissible arrangement of the 52 cards is a three-step sequence:

1. Choose a position for the four aces.
2. Permute the aces.
3. Permute the 48 non-aces.

Therefore, the number of ways we can deal 52 cards *and keep the four aces together* is

$$49 \cdot 4! \cdot 48! \approx 1.46 \times 10^{64}$$

*EXAMPLE 4.14 (Optional)*     Four Nigerians (A, B, C, D), three Chinese ($\#$, $*$, $\&$), and two Greeks ($\alpha$, $\beta$) are lined up at the box office to buy tickets for the World's Fair.

(a) In how many ways can they position themselves if the Nigerians are to hold the first four places in line, the Chinese the next three, and the Greeks the last two?

(b) How many arrangements are possible if members of the same nationality must stay together?

(c) How many arrangements are possible if the two Greeks are to be separated by exactly three people?

(d) How many different queues can be formed?

*Solution*     (a) Figure 4.17 shows a typical admissible arrangement. Notice the similarity between this problem and Example 4.12. Restricted to the first four positions, the Nigerians can still be permuted in $_4P_4 = 4!$ ways. Similarly, the Chinese and the Greeks can be permuted in $_3P_3 = 3!$ and $_2P_2 = 2!$ ways, respectively. By the multiplication rule the number of admissible arrangements is

$$4! \cdot 3! \cdot 2 = 288$$

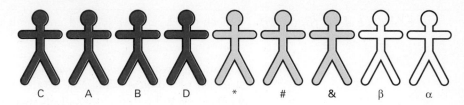

**FIG. 4.17**

(b) Part of the total being asked for here has already been counted—the 288 permutations enumerated in part (a). But now there are other possibilities, queues where the Nigerians, say, are not necessarily first. Arrangements like Figure 4.18 must also be included. By Corollary 4.1 the three nationalities can be permuted in $_3P_3 = 3! = 6$ ways:

$$
\begin{array}{l}
\text{(Nigerians, Chinese, Greeks)}\\
\text{(Nigerians, Greeks, Chinese)}\\
\text{(Chinese, Nigerians, Greeks)}\\
\text{(Chinese, Greeks, Nigerians)}\\
\text{(Greeks, Nigerians, Chinese)}\\
\text{(Greeks, Chinese, Nigerians)}
\end{array}
$$

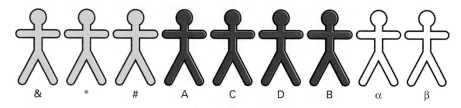

**FIG. 4.18**

Since each of these permutations has 288 distinct arrangements, it follows from the multiplication rule that the number of queues having members of the same nationality grouped together is

$$6 \cdot 288 = 1728$$

(c) One possibility is to put $\alpha$ in position 1 and $\beta$ in position 5 (see Figure 4.19). The remaining seven positions can then be filled in 7! ways (why?).

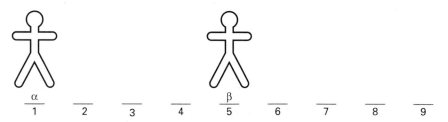

**FIG. 4.19**

Another set of 7! arrangements are generated by sliding $\alpha$ and $\beta$ down to positions 2 and 6, respectively (see Figure 4.20).

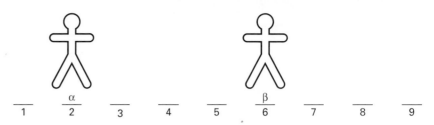

**FIG. 4.20**

By inspection, $\alpha$ and $\beta$ can be located in five different places: $(1, 5)$, $(2, 6)$, $(3, 7)$, $(4, 8)$, and $(5, 9)$. Does it follow that the answer we are looking for is $5 \cdot 7!$? No, because $\alpha$ and $\beta$ can be interchanged, a move that doubles the number of possibilities. The total number of arrangements having the property that the two Greeks are separated by exactly three people is

$$2 \cdot 5 \cdot 7! = 50{,}400$$

(**d**) Here the nationalities are irrelevant, and no restrictions are being imposed: any arrangement of the nine individuals is admissible (see Figure 4.21). By Corollary 4.1 the number of such arrangements is

$$9! = 362{,}880$$

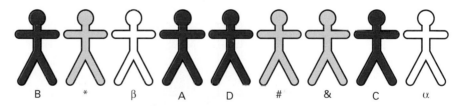

**FIG. 4.21**

---

*EXAMPLE 4.15 (Optional)*    In chess a rook can move vertically and horizontally. It can capture any unobstructed piece located anywhere in its own row or column (see Figure 4.22). In how many ways can eight rooks be placed on a chessboard (having eight rows and eight columns) so that no two can capture one another?

*Solution*    At first glance you might think that the answer is two, the two configurations occurring when the rooks are lined up along either diagonal (Figure 4.23). A careful analysis, though, reveals a much larger number. To begin, keep in mind the condition that every admissible arrangement needs to satisfy: no two rooks can be in the same row or the same column. Now, imagine

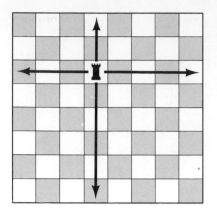

**FIG. 4.22**

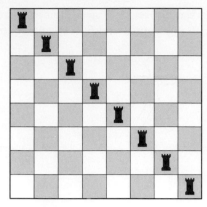

**FIG. 4.23**

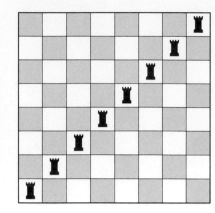

the rows of the board numbered 1 through 8. The rook in row 1 (and there *must* be a rook in row 1) can be positioned in any one of eight columns. Seven places are available for the rook in row 2, six for the rook in row 3, and so on.

Figure 4.24 shows the number of choices available at each row, together with one of the admissible arrangements. By Corollary 4.1 the number of noncapturing rook configurations is

$$_8P_8 = 40,320$$

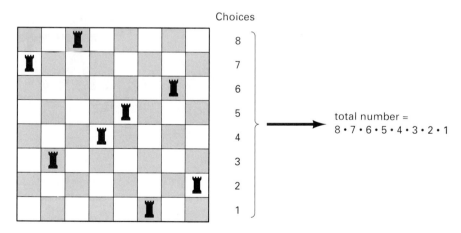

**FIG. 4.24**

---

**QUESTIONS**

4.15  The board of a large corporation has six members willing to be nominated as officers. How many different president–vice-president–treasurer slates can be submitted to the stockholders?

**4.16** Social security numbers are nine digits long. How many have the property that all nine digits are different?

**4.17** The top four contestants in a state beauty pageant are designated "winner," "first runner-up," "second runner-up," and "third runner-up." If 15 women enter the contest, in how many ways can those top four positions be assigned?

**4.18** In how many ways can a set of four tires be put on a car? How many ways are possible if two of the four are snow tires?

**4.19** In how many ways can a 12-member cheerleading squad (6 men and 6 women) be paired to form 6 male-female teams? (Hint: Imagine the six male members (A through F) lined up in a row. How many females can be paired with A?)

**4.20** In her sonnet with the famous first line, "How do I love thee? Let me count the ways," Elizabeth Barrett Browning listed eight. Suppose that Ms. Browning had gone into the greeting-card business and composed verses using all possible choices and orderings of four of those ways. For how many years could she have corresponded with her favorite beau on a daily basis and never sent the same card twice? Assume that no way is to be used more than once in a given verse.

**4.21** License plates in a certain state have the format

letter letter — number number number number

(a) How many plates can be stamped if no letter can appear more than once and no number can appear more than once?

(b) How many plates can be stamped if no letter can appear more than once but the numbers can be repeated?

(c) How many plates can be stamped if letters and numbers can be repeated?

**4.22** Remember the Mahler Maulers in Example 4.10? Suppose that six of the team can play any of the infield positions (including pitcher and catcher) and the other three can play any of the outfield positions, but none of the infielders can play the outfield and none of the outfielders can play the infield. How many different fielding lineups can the Mauler manager produce?

**4.23** In how many ways can the digits 1 through 9 be arranged so that

(a) all the even digits precede all the odd digits?

(b) all even digits remain adjacent?

**4.24** Suppose the scriptwriters working on *Friday the 13th, Part X* (recall Example 4.12) wish to alternate the male and female victims. How many scenarios are possible? (Hint: Think of one specific admissible scenario and use it to suggest how all the others can be generated.)

**4.25** Suppose a long symphony is recorded on all four sides of a two-record album. In how many ways can the four sides be played so that at least one is out of sequence? (Hint: In how many ways can the sides be played so that *none* of them is out of sequence?)

**4.26** A freshman needs to take three humanities courses sometime during his first four semesters. The three are to be selected from a list of six. In how many ways can he schedule the classes, assuming he never wants to take more than one of the courses in any given term? (Hint: How many ways can the courses be scheduled if they are to be taken in the first, second, and third semesters?)

**4.27** Uncle Harry and Aunt Minnie will both be attending your next family reunion. Unfortunately, they hate each other. Unless they are seated with at least two people between them, they are likely to start a food fight. The side of the table at which they will sit has five chairs. How many seating arrangements are possible for those five people if a safe distance is to be maintained between your aunt and your uncle?

# 4.4

## MORE PERMUTATIONS (Optional)

Having addressed the question of permuting *distinct* objects, we might naturally inquire about the solvability of the more general problem—counting permutations in situations where some of the available objects are alike. Imagine, for example, permuting the set $(A, A, B, C)$. If the two $A$'s were different (say an $A_1$ and an $A_2$), there would be $_4P_4 = 4!$, or 24, ordered arrangements. If the $A$'s are identical, though, a direct enumeration shows that that number of permutations drops to 12:

| | |
|---|---|
| $(A, A, B, C)$ | $(B, A, A, C)$ |
| $(A, B, A, C)$ | $(C, A, A, B)$ |
| $(A, B, C, A)$ | $(B, A, C, A)$ |
| $(A, A, C, B)$ | $(B, C, A, A)$ |
| $(A, C, A, B)$ | $(C, A, B, A)$ |
| $(A, C, B, A)$ | $(C, B, A, A)$ |

In general, we might have $r$ different types of elements, each represented $n_i$ times, $i = 1, 2, \ldots, r$. (Here, $r = 3$, $n_1 = $ number of $A$'s $= 2$, $n_2 = $ number of $B$'s $= 1$, and $n_3 = $ number of $C$'s $= 1$.) How many different permutations can such a set of $n_1 + n_2 + \cdots + n_r$ objects generate? The answer is not difficult to derive, and it greatly extends the range of problems we can solve.

---

**THEOREM 4.3** A Generalized Permutation Rule

The number of ways to arrange $n$ objects, $n_1$ being of one kind, $n_2$ of a second kind, $\ldots$, and $n_r$ of an $r$th kind, is

$$\frac{n!}{n_1! \, n_2! \cdots n_r!}$$

where $\sum_{i=1}^{r} n_i = n$.

---

*Proof*   Let $N$ denote the number of arrangements we are trying to count. Suppose that for any one of those $N$ arrangements the identical objects are treated as if they were distinct. The number of permutations that particular arrangement could then generate is $n_1! \, n_2! \cdots n_r!$ (Why?). But each of the $N$ arrangements could produce that same number of variations, so the total number of permutations of the $n$ objects is $N \cdot n_1! \, n_2! \cdots n_r!$. We know from Corollary 4.1, though, that the number of ways to permute $n$ distinct objects is $n!$. Therefore, $N \cdot n_1! \, n_2! \cdots n_r!$ must equal $n!$, which implies that

$$N = \frac{n!}{n_1! \, n_2! \cdots n_r!} \quad \blacksquare$$

---

EXAMPLE 4.16   A sandwich in a vending machine costs 85¢. In how many ways can a customer put in two quarters (Q, Q), three dimes (D, D, D), and one nickel (N)?

Solution   If we treat all coins of a given value as being identical, then a typical denomination sequence, say Q D D Q N D (Figure 4.25), can be thought of as a permutation of $n = 6$ objects of $r = 3$ types, where

$$n_1 = \text{number of nickels} = 1$$

$$n_2 = \text{number of dimes} = 3$$

$$n_3 = \text{number of quarters} = 2$$

1   2   3   4   5   6

Order in which coins are deposited

**FIG. 4.25**

By Theorem 4.3, there are 60 such sequences:

$$\frac{n!}{n_1! \, n_2! \, n_3!} = \frac{6!}{1! \, 3! \, 2!} = 60$$

If we had assumed that the coins were distinct (because they were minted at different places and at different times), the number of different entry sequences would be $6! = 720$.

*EXAMPLE 4.17*   Recall the four Nigerians, three Chinese, and two Greeks we left standing in line at the World's Fair (Example 4.14). Suppose a vacationing Martian strolls by and wants to photograph the nine for her scrapbook. A bit myopic, the Martian is quite capable of discerning gross differences in human anatomy but finds herself unable to distinguish one Nigerian (N) from another, one Chinese (C) from another, or one Greek (G) from another. Instead of perceiving the individuals in Figure 4.21 as

$$B * \beta \, A \, D \, \# \, \& \, C \, \alpha$$

she sees

$$N \, C \, G \, N \, N \, C \, C \, N \, G$$

From the Martian's perspective, in how many different ways can the three kinds of funny-looking Earthlings line themselves up?

*Solution*   The answer follows directly from Theorem 4.3: if $n$ is the number of Earthlings, $n_1$ the number of Nigerians, $n_2$ the number of Chinese, and $n_3$ the number of Greeks, then

$$\frac{n!}{n_1! \, n_2! \, n_3!} = \frac{9!}{4! \, 3! \, 2!} = 1260$$

---

*EXAMPLE 4.18*   Vanderbilt's football season is 11 games. In how many ways can they win two, lose seven, and tie twice?

*Solution*   Notice that each of the seasons we are trying to count is a permutation of length 11 where the objects being permuted are two W's, seven L's, and two T's. For example, winning the first and third games, tying the last two, and losing all the others is the ordered sequence of Figure 4.26. Let

$$n_1 = \text{number of W's} = 2$$

$$n_2 = \text{number of L's} = 7$$

$$n_3 = \text{number of T's} = 2$$

| W | L | W | L | L | L | L | L | L | T | T |
|---|---|---|---|---|---|---|---|---|----|----|
| 1 | 2 | 3 | 4 | 5 | 6 | 7 | 8 | 9 | 10 | 11 |

Games

**FIG. 4.26**

Then

$$n = n_1 + n_2 + n_3 = 2 + 7 + 2 = 11$$

From Theorem 4.3 the number of permutations of the type in Figure 4.26 is

$$\frac{n!}{n_1! \, n_2! \, n_3!} = \frac{11!}{2! \, 7! \, 2!} = 1980$$

Would all 1980 of these sequences be equally likely? No. Certain games during the year are particularly difficult to win. As a result, certain sequences will be more improbable than others. Traditionally, two of Vanderbilt's toughest opponents are Alabama and Georgia. If those games were the second and fifth, for example, any permutation of the form shown in Figure 4.27 would be rather unlikely. In general, the problem of assessing the relative likelihoods of ordered sequences is very important. We will take a first look at the question in Section 4.6 and then discuss it at length in Chapter 5.

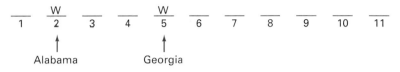

**FIG. 4.27**

---

*EXAMPLE 4.19*    Before the seventeenth century there were no scientific journals, which made it difficult for researchers to document discoveries. If a scientist sent a copy of his findings to a colleague, the colleague might claim the work as his own. The obvious alternative—wait to get enough material to publish a book—invariably resulted in lengthy delays and would not be appropriate in most cases anyway. As a compromise, scientists sometimes sent each other *anagrams*—letter puzzles that when properly unscrambled, summarized in a sentence or two what had been discovered. Because they were dated, anagrams helped establish who discovered what, and when.

When Christiaan Huygens (1629–1695) looked through his primitive telescope and saw a ring around Saturn, he composed the following anagram and sent it to a friend (171):

> *aaaaaaa, ccccc, d, eeeee, g, h, iiiiiii, lll, mm,*
>
> *nnnnnnnnn, oooo, pp, q, rr, s, ttttt, uuuuu*

In how many ways can the 61 letters in Huygens's message be arranged?

*Solution*    Let $n_1$ denote the number of $a$'s, $n_2$ the number of $c$'s, and so on. Substituting into Theorem 4.3 gives

$$\frac{61!}{7!\,5!\,1!\,5!\,1!\,1!\,7!\,3!\,2!\,9!\,4!\,2!\,1!\,2!\,1!\,5!\,5!} \doteq 3.6 \times 10^{60}$$

as the total number of different anagram arrangements. Huygens was not taking any chances; that figure is more than a million billion times as large as the total number of molecules in the earth's atmosphere! Having so many possible permutations precluded anyone from simply playing with the letters and finding the correct ordering by luck. (When appropriately rearranged, the anagram becomes "annulo cingitur tenui, plano, nusquam cohaerente, ad eclipticam inclinato," which translates to "surrounded by a thin ring, flat, suspended nowhere, inclined to the ecliptic.")

---

**4.28**   Linda is taking a five-course load her first semester: English, math, French, psychology, and history. In how many different ways can she earn three A's and two B's? Enumerate the entire set of possibilities. Use Theorem 4.3 to verify your answer.

**4.29**   In the game of pinochle, a player who is dealt two jacks of diamonds and two queens of spades has "300 pinochle." In how many orders can a player be dealt those four cards? Enumerate the possibilities. Does your total agree with Theorem 4.3?

**4.30**   Which state name can generate more permutations,

TENNESSEE   or   FLORIDA?

**4.31**   Of the 10 stock cars entered in a race, 3 are Fords, 2 are Plymouths, 4 are Dodges, and 1 is a Chevrolet. In how many ways can the 10 entrants finish, assuming we are interested only in the companies that built the cars?

**4.32**   An interior decorator is trying to arrange a shelf containing eight books, three with red covers, three with blue covers, and two with brown covers.

(a) Assuming the titles and the sizes of the books are irrelevant, in how many ways can she arrange the eight books?

(b) In how many ways could the books be arranged if they were all considered distinct?

(c) In how many ways could the books be arranged if the red books were considered indistinguishable, but the other five were considered distinct?

**4.33**   How many numbers greater than 4,000,000 can be formed from the digits 2, 3, 4, 4, 5, 5, 5?

**4.34**   Make an anagram out of the familiar expression STATISTICS IS FUN. In how many ways can the letters in the anagram be permuted?

**4.35**   Imagine six points in a plane, no three of which lie on a straight line. In how many ways can the six points be used as vertices to form two triangles? (Hint: Number the points 1 through 6. Call one of the triangles A and the other B. What does the permutation

A   A   B   B   A   B

1   2   3   4   5   6

represent?)

**4.36**   A palindrome is a phrase whose letters are in the same order whether they are read backward or forward, such as Napoleon's lament

Able was I ere I saw Elba

or the often cited

Madam, I'm Adam.

Words themselves can become the units in a palindrome, as in the sentence

Girl, bathing on Bikini, eyeing boy,
finds boy eyeing bikini on bathing girl.

Suppose the members of a set consisting of four objects of one type, six of a second type, and two of a third type are to be lined up in a row. How many of those permutations are palindromes?

**4.37** The freshman dorm is five blocks west and two blocks south of Bunky's first class.

(a) In how many ways can he drive from the dorm to his class without going out of his way? (Hint: How would you characterize the path shown below in terms of a permutation?)

(b) How many different round-trips can he make?

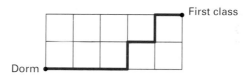

# 4.5

## COMBINATIONS

Order is not always a meaningful characteristic of a collection of objects, our efforts of the previous three sections notwithstanding. If a poker player is dealt a five-card hand, it makes no difference whether the 2 of hearts, 4 of clubs, 9 of clubs, jack of hearts, and ace of diamonds are dealt *in that order* or in any of the other $5! - 1$ permutations of those five cards—the hand is still the same.

We call a collection of $r$ *unordered* objects a **combination of size r**. For example, given a set of $n = 4$ distinct objects ($A$, $B$, $C$, and $D$) there are six ways to form combinations of size 2:

| | |
|---|---|
| $(A, B)$ | $(B, C)$ |
| $(A, C)$ | $(B, D)$ |
| $(A, D)$ | $(C, D)$ |

(Why isn't $(B, A)$ listed? Because with order not being a factor, $(A, B)$ and $(B, A)$ are equivalent.) A general formula for counting combinations can be derived quite easily from what we already know about counting permutations.

**combination**—An unordered collection of objects. The total number of combinations of size $r$ that can be formed from a set of $n$ distinct objects is

$$_nC_r \quad \text{or} \quad \binom{n}{r},$$

where $\quad _nC_r = \dfrac{n!}{r! \, (n - r)!}$

---

**THEOREM 4.4**   Combination Rule

The number of ways to form combinations of size $r$ from a set of $n$ distinct objects, repetitions not allowed, is $_nC_r$, where[1]

$$_nC_r = \frac{n!}{r!(n-r)!}$$

---

*Proof*   Our argument follows the same lines as the proof of Theorem 4.3. Let $_nC_r$ denote the number of combinations satisfying the conditions of the theorem. Each combination can be ordered in $r!$ ways (why?), so $r! \, _nC_r$ must equal the number of *permutations* of length $r$ that can be formed from $n$ distinct objects. But by Theorem 4.2, $n$ distinct objects can be formed into permutations of length $r$ in $n!/(n-r)!$ ways. Therefore,

$$r! \cdot \, _nC_r = \frac{n!}{(n-r)!}$$

which implies that

$$_nC_r = \frac{n!}{r! \, (n-r)!} \quad \blacksquare$$

---

*EXAMPLE 4.20*   The sisters of Alpha Beta Zeta intend to recruit five new members during fall rush. If 15 girls want to join the sorority, how many different pledge classes can be formed?

*Solution*   Think of the 15 potential initiates as chips—lettered A through O—in an urn (Figure 4.28). Drawing a sample of five of those chips is equivalent to choosing a particular pledge class (see Table 4.2). It follows that the total

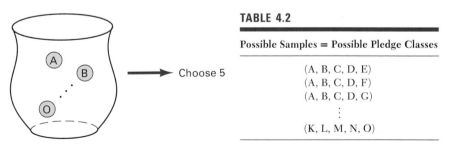

**FIG. 4.28**

**TABLE 4.2**

| Possible Samples = Possible Pledge Classes |
|---|
| (A, B, C, D, E) |
| (A, B, C, D, F) |
| (A, B, C, D, G) |
| ⋮ |
| (K, L, M, N, O) |

---

[1] The symbol $\binom{n}{r}$ is often used in place of $_nC_r$. The former is read "$n$ choose $r$."

number of pledge classes must be equal to the total number of samples of size $r = 5$ that can be drawn from a set of $n = 15$ distinct objects. By Theorem 4.4,

$$\text{Number of pledge classes} = {}_{15}C_5 = \frac{15!}{5!\ 10!} = 3003$$

---

*EXAMPLE 4.21*   Eight politicians meet at a fundraising dinner. How many greetings can be exchanged if each politician shakes hands with every other politician exactly once?

*Solution*   Once again the urn model can be invoked to reduce this question to one we have already answered. Imagine the politicians to be represented by eight chips in an urn, numbered 1 through 8 (see Figure 4.29). A handshake corresponds to an unordered sample of size 2 (we can think of the sample (2, 3), for instance, as corresponding to politicians 2 and 3 shaking hands). Theorem 4.4, therefore, gives not only the number of samples of size 2 but also, by analogy, the number of possible handshakes:

$$\text{Number of handshakes} = {}_8C_2 = \frac{8!}{2!\ 6!} = 28$$

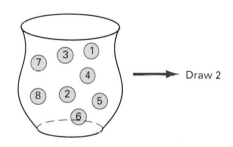

**FIG. 4.29**

---

*EXAMPLE 4.22*   The basketball recruiter for Swampwater Tech has scouted 16 former NBA starters that he thinks he can pass off as junior college transfers. Six are guards, seven are forwards, and three are centers. Unfortunately, his slush fund of illegal alumni donations is at an all-time low, and he can afford to buy new Trans-Ams for only nine of the players. If he wants to keep three guards, four forwards, and two centers, in how many ways can he parcel out the cars?

*Solution*   This is a combination problem that also requires an application of the multiplication rule. First, note there are ${}_6C_3$ *sets* of three guards that could be chosen to receive Trans-Ams (think of drawing a sample of three names out of an urn containing six names). Similarly, the forwards and centers can be bribed in $_7C_4$ and $_3C_2$ ways, respectively. It follows from the multiplication rule that the total number of ways to divvy up the cars is

$$_6C_3 \cdot {}_7C_4 \cdot {}_3C_2 = \frac{6!}{3!\,3!} \cdot \frac{7!}{4!\,3!} \cdot \frac{3!}{2!\,1!} = 2100$$

---

**EXAMPLE 4.23**    Diana is trying to put a little zing into her cabaret act by telling four jokes at the beginning of each show. Her current engagement is booked to run four months. If she gives one performance a night and never wants to repeat the same *set* of jokes on any two nights, what is the minimum number of jokes she must have in her repertoire?

**Solution**    Notice that this problem is the reverse of what we have seen up to this point. In Examples 4.20, 4.21, and 4.22, both $n$ and $r$ were given and our objective was to find $_nC_r$. Here we are trying to find the smallest $n$ that will generate at least four months worth of combinations of size $r = 4$. Assume four months translates into approximately 120 performances. Our intuition might suggest that $n$ would have to be moderately large to have the capability of generating 120 combinations. Surprisingly, that proves not to be so. Table 4.3 shows $_nC_4$ for $n$ values of 7, 8, and 9. By inspection we see that Diana can get by for four months knowing only *nine* jokes.

**TABLE 4.3**

| $n$ | $_nC_4$ | Greater than 120? |
|---|---|---|
| 7 | 35 | No |
| 8 | 70 | No |
| 9 | 126 | Yes |

---

**EXAMPLE 4.24 (Optional)**    The symbol $_nC_r$ has many interesting properties. One of the most famous involves *Pascal's triangle*, a numerical array where each entry is equal to the sum of the two entries appearing diagonally above it (see Figure 4.30). If we number the rows (beginning with 0 for the top of the triangle) and the columns (going from left to right), each entry in Figure 4.30 can be replaced by a corresponding $_nC_r$ (see Figure 4.31). From the latter array we

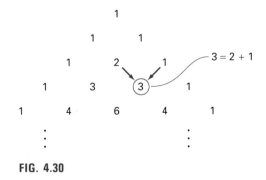

**FIG. 4.30**

$$_0C_0$$

$$_1C_0 \qquad _1C_1$$

$$_2C_0 \qquad _2C_1 \qquad _2C_2$$

$$_3C_0 \qquad _3C_1 \qquad _3C_2 \qquad _3C_3$$

$$_4C_0 \qquad _4C_1 \qquad _4C_2 \qquad _4C_3 \qquad _4C_4$$

$$\vdots \qquad\qquad\qquad \vdots$$

**FIG. 4.31**

can see that the formation of Pascal's triangle can be summarized by the equation

$$_{n+1}C_r = {}_nC_r + {}_nC_{r-1} \tag{4.1}$$

Equation 4.1 is called a *combinatorial identity*. It holds for any values of $n$ and $r$ for which the symbol $_nC_r$ is defined.

---

**QUESTIONS**

4.38 The final exam in History 101 consists of five essay questions that the professor chooses from a pool of seven that are given to the students a week in advance. For how many possible sets of questions does a student need to be prepared? In this situation does order matter?

4.39 Cathie is a popular young lady who has received five different invitations (from Andy, Bob, Chuck, Dave, and Ed) to go to three parties over the weekend. How many sets of three different boys can she go out with? Enumerate the possibilities and verify that your answer agrees with Theorem 4.4.

4.40 How many straight lines can be drawn connecting five points $(A, B, C, D, E)$, no three of which lie on the same straight line? Do the problem using an appropriate combinatorial computation and then verify your answer by drawing the entire set of possible lines.

4.41 How many different 5-card hands can be dealt from a standard 52-card poker deck?

4.42 Four identical cars pull into a parking lot that has eight empty spaces.

(a) In how many ways can the cars occupy four of those spots?

(b) How many arrangements are such that the leftmost and the rightmost spaces remain unoccupied?

4.43 Kumar needs to take three more math courses and two more basket weaving labs to complete the requirements for his math–basket weaving double major. The math electives are to be chosen from a list of eight approved courses, and the basket weaving labs from a list of four. In how many different ways can Kumar complete his requirements?

4.44   Nine students, five men and four women, interview for four summer internships sponsored by a city newspaper.

(a) In how many ways can the newspaper choose a set of four interns?

(b) In how many ways can the newspaper choose a set of four interns if it must include two men and two women in each set?

(c) How many sets of four can be picked such that not everyone in a set is of the same sex?

4.45   A policeman has spotted you and five other drivers drag racing through a school zone. He and his partner can pull over two of the cars, but the other four will get away.

(a) How many different sets of two cars can be ticketed?

(b) How many different sets of ticketed drivers include you?

(c) Intuitively, what are your chances of getting a ticket?

4.46   Evaluate $_5C_0$, $_5C_1$, $_5C_2$, $_5C_3$, $_5C_4$, and $_5C_5$ by determining the fifth row in Pascal's triangle (see Example 4.24).

4.47   Refer to Example 4.23. What is the smallest number of jokes Diana would need to know if she wanted to give 120 performances and never tell the same *ordered* set of four jokes twice?

4.48   Recall Question 4.39. Suppose Cathie cared who took her to which party.

(a) How many sets of three dates could she arrange such that her escort to each party was a different person?

(b) How many sets of three dates could she arrange if she allowed the possibility of going out with the same person more than once?

4.49   In Example 4.7 we used the multiplication rule to show that 256 different burgers can be made with the options available at Hamburger Heaven. Find that same number by treating the extras added to a burger as a sample of size $r$ chosen from a set of $n = 8$ possible toppings.

4.50   One algebraic application for $_nC_r$ occurs in the expansion of $(x + y)^n$. It can be shown that

$$(x + y)^n = \sum_{r=0}^{n} {}_nC_r x^r y^{n-r}$$

(a) Evaluate both sides of this equation for $x = y = 1$. This result justifies, *in general*, the equivalence of the two solutions in Question 4.49 (where $n = 8$).

(b) Use the right side of the equation to compute the number of possible letters in the Braille alphabet (see Example 4.6).

**4.51**  In how many ways can the word ABRACADABRA be formed in the array shown below? Assume that you must start at the top A and move diagonally downward to the bottom A.

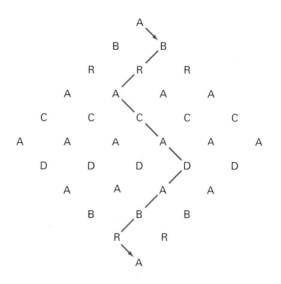

(Hint: Note that each path consists of 10 transitions, 5 moving to the right and 5 moving to the left. Furthermore, each path can be characterized by the positions of the transitions to the right. For example, if the transitions are numbered 1 through 10, the path shown in the diagram has moves to the right occurring in positions 1, 4, 5, 6, and 10.)

# 4.6

## COMBINATORIAL PROBABILITY

**probability**—The fraction of the time, *in the long run*, that an event will occur if an experiment is repeated again and again. If the sample space associated with an experiment contains *n* outcomes, equally likely, then the probability of an event can also be thought of as the ratio $m/n$, where $m$ of the experiment's possible outcomes belong to the given event. The $m/n$ ratio is called the *classical definition of probability*.

Up to now our efforts have focused on counting the ways an operation or sequence of operations can be performed. In this section we will combine those enumeration results with the notion of probability. Putting the two together makes a lot of sense. In many combinatorial problems an enumeration alone does not provide particularly relevant information. A poker player, for example, is not likely to be interested in knowing the total *number* of ways he can draw to an inside straight; he is interested, though, in his *probability* of drawing to an inside straight.

Making the transition from an enumeration to a probability is usually straightforward. If there are *n* ways, *all equally likely*, to perform a certain operation, and if *m* of those *n* satisfy some stated condition A, then the **probability of A** is defined to be $m/n$.

Historically, the $m/n$ idea is what motivated the early work of Cardano, Pascal, Fermat, and Huygens. Today we recognize that not all probabilities are so easily characterized. Nevertheless, the $m/n$ model is entirely appropriate for describing a wide variety of phenomena.

---

**EXAMPLE 4.25**   Imagine rolling a pair of fair dice, one red and one green. What is the probability that the sum of the faces showing is a 7?

**Solution**   Since each die can come up six different ways, the total number of rolls according to the multiplication rule is 36 ($=6 \cdot 6$). Figure 4.32 shows the entire set of possible outcomes as a $6 \times 6$ matrix. By inspection we see that *six* of those outcomes share the characteristic that their sum is a 7:

$$(1, 6), \quad (2, 5), \quad (3, 4), \quad (4, 3), \quad (5, 2), \quad (6, 1)$$

Therefore, with $n = 36$ and $m = 6$, the probability of rolling a seven is 6/36, or 1/6.

Green die

|  | 1 | 2 | 3 | 4 | 5 | 6 | sum = 7 |
|---|---|---|---|---|---|---|---|
| 1 | (1,1) | (1,2) | (1,3) | (1,4) | (1,5) | (1,6) | |
| 2 | (2,1) | (2,2) | (2,3) | (2,4) | (2,5) | (2,6) | |
| 3 | (3,1) | (3,2) | (3,3) | (3,4) | (3,5) | (3,6) | |
| 4 | (4,1) | (4,2) | (4,3) | (4,4) | (4,5) | (4,6) | |
| 5 | (5,1) | (5,2) | (5,3) | (5,4) | (5,5) | (5,6) | |
| 6 | (6,1) | (6,2) | (6,3) | (6,4) | (6,5) | (6,6) | |

Red die

**FIG. 4.32**

---

4.52   For the outcomes in Figure 4.31:

(a) What is the probability that the sum of the two dice is 9?

(b) What is the probability that the face showing on one die is twice the number showing on the other?

4.53   Suppose three fair dice, one red, one white, and one blue, are tossed.

(a) How many different outcomes are possible?

(b) What is the probability of rolling a sum equal to 5?

4.54   Suppose $n$ fair dice are rolled. What is the probability that all $n$ faces will be the same?

4.55   An urn contains four red chips, three white chips, and two blue chips. One chip is drawn at random.

(a) What is the probability that the chip drawn is white?

(b) What is the probability that the chip drawn is either blue or white?

(c) What is the probability that the chip drawn is not red?

**4.56**   Recall Example 4.12. Suppose the scriptwriters had worked on all possible killing-order scenarios. If they simply picked one scenario at random, what is the probability that it would be one in which all the males were eliminated before any of the females?

**4.57**   One card is dealt from a standard 52-card poker deck.

(a) What is the probability that the card is an ace?

(b) What is the probability that the card is a club?

(c) What is the probability that the card is an ace of clubs?

(d) What is the probability that the card is black?

### Interpreting the Probability of an Event

experiment—Any procedure capable of producing a well-defined set of possible outcomes.

sample space—The entire set of possible outcomes (finite or infinite) associated with an experiment.

event—A collection of outcomes in the sample space that share, or satisfy, some condition.

In the terminology of probability, the set of possible outcomes associated with an *experiment* is called the *sample space*; any collection of outcomes belonging to the sample space and satisfying a stated condition is an *event*. The *probability of an event* $A$ is denoted by $P(A)$. In Example 4.25 the experiment is the act of rolling two fair dice, and the sample space is the set of 36 outcomes listed in Figure 4.32. Let $A$ denote the event "sum is a 7." Then

$$P(A) = \frac{6}{36} = \frac{1}{6}$$

Suppose we toss a fair coin and define $A$ to be the event "head appears." The associated sample space contains $n = 2$ equally likely outcomes (heads or tails), $m = 1$ of which is favorable to the event $A$. Therefore, we can write $P(A) = 1/2$. But how do we *interpret* that probability? Does it mean that once out of every two tosses a head will appear? Of course not. We are all familiar enough with random events to know that that degree of regularity will not occur. What $P(A) = 1/2$ does mean is that *in the long run* if we were to toss a fair coin many times the ratio of heads to total tosses would converge to 1/2.

Table 4.4 shows the results of 35 computer-simulated flips of a fair coin. The accumulated proportion of heads, which serves as a running "estimate" of $P(A)$, is the quotient $x/y$. Notice that for small $y$ the variability in $x/y$ is considerable. But as $y$ gets larger, the ratio of the total number of heads to the total number of tosses becomes much more stable and does appear to be converging to 1/2. In the terminology of calculus,

$$P(A) = \lim_{y \to \infty} \frac{x}{y} = \frac{1}{2}$$

**TABLE 4.4**

| Toss, $y$ | Outcome | Accumulated Heads, $x$ | $x/y$ | Toss, $y$ | Outcome | Accumulated Heads, $x$ | $x/y$ |
|---|---|---|---|---|---|---|---|
| 1 | H | 1 | 1.00 | 19 | T | 11 | 0.58 |
| 2 | H | 2 | 1.00 | 20 | T | 11 | 0.55 |
| 3 | T | 2 | 0.67 | 21 | T | 11 | 0.52 |
| 4 | H | 3 | 0.75 | 22 | T | 11 | 0.50 |
| 5 | T | 3 | 0.60 | 23 | T | 11 | 0.48 |
| 6 | H | 4 | 0.67 | 24 | T | 11 | 0.46 |
| 7 | H | 5 | 0.71 | 25 | H | 12 | 0.48 |
| 8 | T | 5 | 0.62 | 26 | H | 13 | 0.50 |
| 9 | H | 6 | 0.67 | 27 | H | 14 | 0.52 |
| 10 | H | 7 | 0.70 | 28 | T | 14 | 0.50 |
| 11 | H | 8 | 0.73 | 29 | H | 15 | 0.52 |
| 12 | H | 9 | 0.75 | 30 | H | 16 | 0.53 |
| 13 | T | 9 | 0.69 | 31 | H | 17 | 0.55 |
| 14 | T | 9 | 0.64 | 32 | T | 17 | 0.53 |
| 15 | T | 9 | 0.60 | 33 | T | 17 | 0.52 |
| 16 | T | 9 | 0.56 | 34 | T | 17 | 0.50 |
| 17 | H | 10 | 0.59 | 35 | T | 17 | 0.49 |
| 18 | H | 11 | 0.61 | | | | |

**QUESTIONS**

4.58  (a) A fair coin is tossed three times. How many outcomes are in the sample space?

   (b) Let $A$ be the event "exactly two heads appear." Enumerate the outcomes in $A$ and compute $P(A)$.

4.59  Suppose two cards are dealt from a standard 52-card poker deck.

   (a) How many outcomes are in the sample space?

   (b) Let $A$ be the event "both cards drawn are aces." Find $P(A)$.

4.60  A typical U.S. coin is well-balanced with respect to flipping; that is, it will have a probability very close to 1/2 of coming up heads. However, when a coin is placed on its edge and spun, it often will not be so well-balanced. Spin a coin 30 times and keep a tally of $y$, the spin number, and $x$, the accumulated number of heads, as shown in Table 4.4. Compute

$$\frac{x}{y} \quad \text{for} \quad y = 1, 5, 10, 15, 20, 25, \text{ and } 30$$

What would you estimate is the probability that that coin, when spun, will come up heads?

*EXAMPLE 4.26*  Three chips are drawn from an urn containing 10 chips, numbered 1 through 10. What is the probability that the largest chip in the sample is a 6?

*Solution*  From Theorem 4.4 the number of different samples of size 3 that can be drawn is $_{10}C_3 = 120\ (=n)$. All 120 are equally likely. How many of those share the characteristic that the largest of the three numbers is a 6? Look at Figure 4.33. In order for the largest number to be a 6, we must (**a**) draw the 6 and (**b**) draw two other numbers less than 6. The former can be done in $_1C_1$ ways; the latter, in $_5C_2$ ways. Therefore, $m$, the number of admissible samples (ones having 6 as the largest entry) is

$$_1C_1 \cdot {}_5C_2 = 1 \cdot 10 = 10 \quad \text{(Why?)}$$

It follows that the probability of the largest number being a 6 is $10/120 = 0.08$.

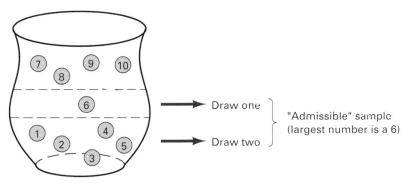

**FIG. 4.33**

---

**QUESTIONS**

**4.61**  List the 10 outcomes in Example 4.26 that comprise the event "largest number in sample is a 6."

**4.62**  An urn contains four chips, numbered 1, 2, 3, and 4. Two are chosen at random and added together. What is the probability that the sum of the numbers drawn is five? Answer the question by enumerating the sample space and computing the sum for each possible outcome.

**4.63**  Country $X$ launches a tactical nuclear strike at country $Y$. Fifteen missiles are fired, five of which carry nuclear warheads. Country $Y$ has a defense system that will enable it to shoot down six of the incoming missiles, but the system has no way of knowing which missiles have nuclear warheads and which do not. What is the probability that exactly two of the nuclear-armed missiles penetrate country $Y$'s defense?

---

**EXAMPLE 4.27**  In the Illinois state lottery six numbers are drawn from an urn containing 44 numbers. A $1 ticket buys two sets of six numbers. For a ticket to win, all six numbers drawn must match one of the two sets of six numbers showing on the ticket. What is the probability that a $1 investment will win you the Illinois lottery?

*Solution*   By Theorem 4.4, $_{44}C_6$ different sets of potential winning numbers can be drawn. Two of those sets are on a $1 ticket. Your chance of winning, then, is

$$\frac{2}{_{44}C_6} = \frac{2}{6,894,888}$$

or about one in $3\frac{1}{2}$ million—in other words, don't count on it!

---

QUESTIONS

**4.64**   What would your chances be of winning the Illinois state lottery with a $1 ticket if the *order* of the six numbers mattered?

---

*EXAMPLE 4.28*   One of the more important applications of probability is the measurement of "improbability." We use the latter as input for making certain kinds of decisions. In statistics, for example, a frequently encountered question is whether the extent to which a set of observations deviate from their "norm" is due to (**a**) chance or (**b**) an assignable cause. If the "chance" hypothesis can be shown to have a very low probability of producing the observed $y_i$'s, we *reject* it and conclude that the data are the result of a genuine, nonrandom effect. Chapter 8 will expand on that argument in considerable detail.

A similar, but simpler, problem arises in connection with experiments whose objective is to demonstrate that animals can develop primitive language skills. Suppose a chimpanzee, for example, is asked to respond to a question by forming a sequence of symbols. If it answers correctly—that is, if it produces the proper sequence—how can we effectively rule out the possibility that the chimp got the correct answer just by luck? *By showing how improbable that would be.*

In a frequently cited experiment by Premack (127), a chimpanzee was instructed to "read" a sentence and to formulate a response. At the chimp's disposal was a set of eight different symbols. Assuming the chimp knew that the correct answer required exactly four symbols, none repeated, what would be its probability of getting the right answer just by chance?

*Solution*   Figure 4.34 shows the experimental situation reduced to a combinatorial problem. Every possible answer is a permutation of length 4. (Notice that order is critical, meaning that permutations rather than combinations need to be considered.) Since $n = 8$, the number of permutations of length 4 is, by Theorem 4.2,

$$_8P_4 = 8 \cdot 7 \cdot 6 \cdot 5 = 1680$$

Only *one* of those sequences, though, is the correct answer. Therefore, the chimp's probability of responding appropriately *by chance alone* is

$$\frac{1}{1680} = 0.000595$$

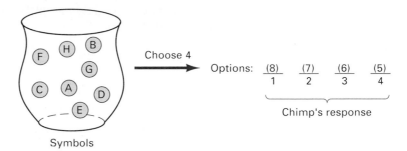

**FIG. 4.34**

The latter figure is so small that if the chimp did answer correctly we would be obliged to conclude that something other than chance was the cause. For all practical purposes, either (1) the chimp actually had acquired some language skills, (2) the chimp was picking up unintentional cues from the experimenters, or (3) the results were faked.

---

**EXAMPLE 4.29** Among the most intriguing examples of combinatorial probability is the *birthday problem*. Its statement is simple, its solution is straightforward, but its answer goes strongly contrary to our intuition. Suppose a group of $k$ randomly selected individuals is assembled. What is the probability that at least two of them will have the same birthday?

**Solution** Most people would guess that for $k$ relatively small, say less than 70, a match would be very unlikely. Not so. The odds of at least one match are better than 50–50 if as few as 23 people are present. And when the group does number 70, there is a 99.9% chance of at least one match!

The counting results we need to appeal to here are Theorems 4.1 and 4.2. Imagine a group of $k$ people lined up in a row. Omitting leap year, each of those $k$ people might have any one of 365 birthdays. By the multiplication rule the group as a whole generates a total of $365^k$ different sets of birthday sequences, the collection of which comprise the sample space $S$ (see Figure 4.35).

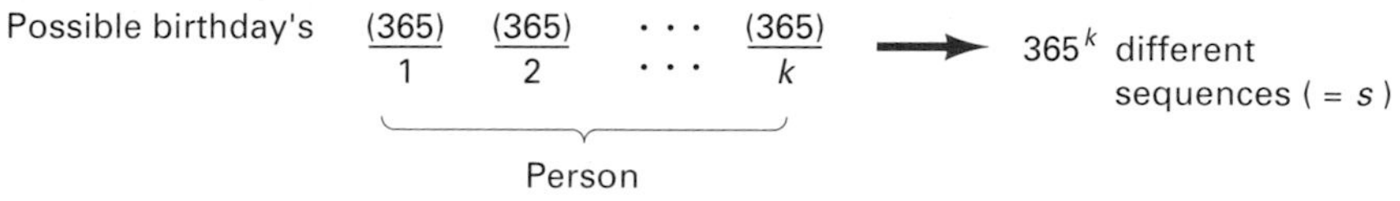

Possible birthday's $\quad \underline{(365)} \quad \underline{(365)} \quad \cdots \quad \underline{(365)} \quad \longrightarrow \quad 365^k$ different
$\qquad\qquad\qquad\qquad 1 \qquad\quad 2 \quad \cdots \quad\; k \qquad\qquad\qquad\quad$ sequences $(=s)$

$\underbrace{\qquad\qquad\qquad\qquad}_{\text{Person}}$

**FIG. 4.35**

Define the event $A$ to be "at least two people have the same birthday." Our objective is to express $P(A)$ as a function of $k$. Doing that will require that we evaluate the numerator and denominator of an appropriate $m/n$ ratio. Specifically, if each person is assumed to have an equal chance of being born on any given day, the $365^k$ sequences in the sample space $S$ will be equally likely, and

$$P(A) = \frac{\text{Number of sequences in } A}{365^k}$$

Notice that each birthday sequence in the sample space will belong to exactly one of two categories (see Figure 4.36):

1. At least two people will have the same birthday.
2. All $k$ people will have different birthdays.

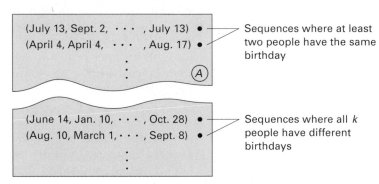

Sample space:  all birthday sequences of length k (contains $365^k$ outcomes)

**FIG. 4.36**

It follows that

Number of sequences in $A$                                                 (4.2)

$= 365^k -$ number of sequences where all $k$ people have
different birthdays

Equation 4.2 is a key step because we can easily find the number of birthday sequences having no matches by using Theorem 4.2. A simple subtraction will then give the number of outcomes in $A$. Counting the number of outcomes in $A$ *directly*, though, would be prohibitively difficult.

The number of ways to form birthday sequences for $k$ people subject to the restriction that all $k$ birthdays must be different is simply the number of ways to form permutations of length $k$ from a set of 365 distinct objects:

$$_{365}P_k = 365(364) \cdots (365 - k + 1)$$

Therefore,

$$P(A) = P \text{ (at least two people have the same birthday)}$$

$$= \frac{365^k - 365(364) \cdots (365 - k + 1)}{365^k}$$

**TABLE 4.5**

| $k$ | $P(A) = P$ (at least two have same birthday) |
|-----|-----|
| 15 | 0.253 |
| 22 | 0.475 |
| 23 | 0.507 |
| 40 | 0.891 |
| 50 | 0.970 |
| 70 | 0.999 |

Table 4.5 shows $P(A)$ for $k$ values of 15, 22, 23, 40, 50, and 70. Notice how the $P(A)$'s greatly exceed what our intuition would suggest.

Presidential biographies offer one opportunity to "confirm" the unexpectedly large values that Table 4.4 gives for $P(A)$. Among our $k = 40$ presidents, two did have the same birthday: Harding and Polk were both born on November 2. More surprising, though, are the death dates of the presidents: Adams, Jefferson, and Monroe all died on July 4, and Fillmore and Taft both died on March 8.

**QUESTIONS**

4.65 Suppose you have one brother and two sisters. What is the probability that at least two of you have the same birthday?

4.66 Suppose six fair dice are rolled.

(a) How many outcomes are in the sample space?

(b) What is the probability that each face appears exactly once?

(Hint: Think of the six dice lined up in a row. How many permutations have the property that all six faces are different?)

4.67 An apartment building has eight floors. If seven people get on the elevator on the first floor, what is the probability that they all want to get off on different floors? on the same floor? Assume that each person is equally likely to get off at any floor.

4.68 A three-digit number is formed at random from the digits 1 through 7, with no digit being used more than once. What is the probability that the number will be less than 289?

4.69 Bordering a large estate is a stand of 15 European white birches, all in a row. Three adjacent trees have leaves damaged by a parasite.

(a) What probability would you associate with that event?

(b) Does it seem reasonable to argue that the infestation is contagious? Explain.

4.70 The American League has seven teams each in its eastern and western divisions. The National League has six teams in each division.

(a) What is the probability that during a single day's slate of interdivisional games, the eastern division teams win every one?

(b) What probability would you associate with the event that, on a given day, all the victories in a league are recorded by teams in one division?

Assume that each team has a 50% chance of winning any game. (Hint: Think of the 13 games as an ordered sequence. If each game has two possible outcomes—eastern division team wins, or western division team wins—how many sequences make up the sample space?)

---

**EXAMPLE 4.30 (Optional)**    Two monkeys, Mickey and Marian, are strolling along a moonlit beach when Mickey sees an abandoned Scrabble set. Investigating, he notices that some of the letters are missing. A closer scrutiny reveals that the following 59 tiles are still in the box:

| A | B | C | D | E | F | G | H | I | J | K | L | M |
|---|---|---|---|---|---|---|---|---|---|---|---|---|
| 4 | 1 | 2 | 2 | 7 | 1 | 1 | 3 | 5 | 0 | 3 | 5 | 1 |

| N | O | P | Q | R | S | T | U | V | W | X | Y | Z |
|---|---|---|---|---|---|---|---|---|---|---|---|---|
| 3 | 2 | 0 | 0 | 2 | 8 | 4 | 2 | 0 | 1 | 0 | 2 | 0 |

Mickey, being of a romantic bent, would like to impress Marian, so he re-arranges the letters in hopes of spelling something clever. (The rearranging would be random because Mickey can't spell; fortunately, Marian can't read, so it really doesn't matter.) What are Mickey's chances of getting lucky and putting together

> She walks in beauty, like the night
>     Of cloudless climes and starry skies

**Solution**    As we might imagine, Mickey would have to get *very* lucky! The number of ways to permute 59 letters (four *A*'s, one *B*, two *C*'s, and so on) is a direct application of Theorem 4.3:

$$\frac{59!}{4!\,1!\,2!\cdots 2!\,0!}$$

But only one of those is the hoped for couplet. So, since he arranges the letters randomly, making all permutations equally likely, the probability of his spelling out Byron's lines is a discouraging $1.7 \times 10^{-61}$.

$$\frac{\text{Number of ``correct'' arrangements}}{\text{Total number of arrangements}} = \frac{1}{\dfrac{59!}{4!\,1!\,2!\cdots 2!\,0!}} = 1.7 \times 10^{-61}$$

Love may conquer all, but it won't beat those odds!

**EXAMPLE 4.31 (Optional)**  After partying a bit too much, an inebriated conventioneer finds himself in the embarrassing predicament of not knowing whether he is walking forward or backward. He has a 50–50 chance at each step of going in either direction. Suppose he hazards $2n$ steps under those conditions. What is the probability that he ends up right back where he started?

**Solution**  First, consider the sample space. Each step presents two options: the conventioneer can stagger forward (F) or backward (B). Since he intends to take $2n$ such steps, his total number of possible paths—that is, the number of (equally likely) outcomes in the sample space—is $2^{2n}$ (see Figure 4.37).

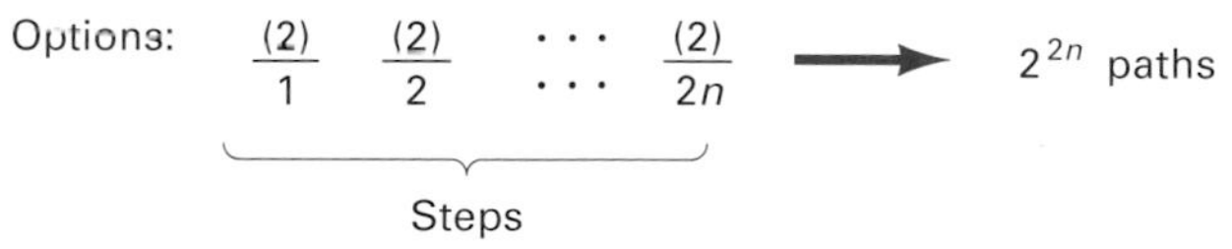

$$\text{Options:} \quad \underset{1}{\underline{(2)}} \quad \underset{2}{\underline{(2)}} \quad \overset{\cdots}{\cdots} \quad \underset{2n}{\underline{(2)}} \quad \longrightarrow \quad 2^{2n} \text{ paths}$$

$$\underbrace{\qquad\qquad\qquad\qquad}_{\text{Steps}}$$

**FIG. 4.37**

Let $A$ denote the event "net displacement is zero." Which of the $2^{2n}$ paths in the sample space belong to $A$? The ones having exactly $n$ steps forward and $n$ steps backward. But $n$ F's and $n$ B's can be arranged in $\dfrac{(2n)!}{n!\,n!}$ ways (by Theorem 4.3). Therefore,

$$P(A) = \frac{(2n)!/(n!\,n!)}{2^{2n}}$$

Table 4.6 lists $P(A)$ for various values of $2n$.

**TABLE 4.6**

| Total No. of Steps ($2n$) | $P$ (net displacement $= 0$) |
| --- | --- |
| 2 | 0.500 |
| 4 | 0.375 |
| 6 | 0.312 |
| 8 | 0.273 |
| 10 | 0.246 |

**QUESTIONS**

**4.71**  Suppose the 12 objects in Question 4.36 are arranged at random. What is the probability that the resulting permutation will be a palindrome?

**4.72**  Imagine that the test tube pictured below contains $2n$ grains of sand, $n$ white and $n$ black. Suppose the tube is vigorously shaken. What is the probability that the two colors of sand will completely separate; that is, all of one color fall to the bottom, and all of the other color lie on top? (Hint:

Consider the $2n$ grains to be aligned in a row. In how many ways can the $n$ white and the $n$ black grains be permuted?)

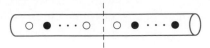

# 4.7

## SUMMARY

Learning how to solve combinatorial problems is basically an exercise in learning how to ask the right questions about the *structure* of an outcome. Typically, a combinatorial problem requires that we arrange a series of objects (or perform a sequence of operations) *subject to certain restrictions*. The number of ways we can do that depends ultimately on the answers to two questions:

**1.** Are the objects all distinct?

**2.** Does order matter?

The first step in approaching any combinatorial problem is to picture what a typical outcome meeting the stated restrictions looks like. Then we need to ask ourselves what changes can be made to a typical outcome and still keep it an admissible outcome. Most likely, it will be necessary to appeal to some or all of the four theorems in Table 4.7.

### TABLE 4.7  Basic Counting Techniques

**Theorem 4.1**   Multiplication Rule.   Given $n_i$ options at each of $k$ steps, $i = 1, 2, \ldots, k$, the number of different ordered sequences of the form (step 1, step 2, ..., step $k$) is

$$n_1 \cdot n_2 \cdots n_k$$

**Theorem 4.2**   Permutation Rule.   The number of permutations of length $r$ that can be formed from a set of $n$ distinct elements is

$$_nP_r = n(n-1)(n-2) \cdots (n-r+1) = \frac{n!}{(n-r)!}$$

**Theorem 4.3**   (Optional) Generalized Permutation Rule.   The number of ways to permute an entire set of $n$ elements *not all distinct* ($n_1$ of one type, $n_2$ of a second type, ..., $n_r$ of an $r$th type) is

$$\frac{n!}{n_1! \, n_2! \cdots n_r!}$$

**Theorem 4.4**   Combination Rule.   The number of combinations of size $r$ that can be formed from a set of $n$ distinct elements is

$$_nC_r = \binom{n}{r} = \frac{n!}{r! \, (n-r)!}$$

Combinatorics led us quite naturally in this chapter to a discussion of probability. Suppose $A$ represents a certain condition satisfied by a subset of the possible outcomes associated with a given experiment. If there are $m$ outcomes in $A$, a total of $n$ associated with the experiment, and if all possible outcomes are equally likely, we define the *probability of* $A$ to be the ratio $m/n$. This definition is fundamentally important. Not every probability problem can be formulated as a simple quotient, but the equally likely model has many applications, and we will see it frequently in the chapters ahead.

| REVIEW QUESTIONS | | |
|---|---|---|

**4.73** Residents of a condominium have an automatic garage door opener that has a row of eight buttons. Each garage door has been programmed to respond to a particular set of buttons being pushed. If the condominium houses 250 families, can residents be assured that no two garage doors will open on the same signal? If so, how many additional families can be added before the eight-button code becomes inadequate?

**4.74** In how many ways can the letters in the word

*MISCREANT*

be arranged so that the first and last letters are vowels?

**4.75** Suppose 12 basketball players suited up for a game include 3 centers, 5 guards, and 4 forwards.

(a) In how many orders can the 12 players be introduced if the centers have to be announced first, then the guards, then the forwards?

(b) Suppose the 12 players (A, B, . . . , L) are lined up at random for a team photo. What is the probability that the centers are in the three left-most positions?

(c) Suppose the coach picks three players at random to demonstrate a new drill. What is the probability that he picks 1 center, 1 guard, and 1 forward?

**4.76** An urn contains 10 chips, numbered 1 through 10. Two chips are drawn at random, without replacement. What is the probability that the sum of the chips drawn is 10?

**4.77** Four fair dice are tossed. What is the probability that the sum of the faces showing is 6?

**4.78** In how many different ways can the members of five families—each consisting of a mother, a father, and a child—be seated in a row of 15 chairs such that members of a family are sitting together?

**4.79** Ten marketing assistants for a major department store are candidates for promotion to associate buyer. Seven are men and three are women. If the company intends to promote four of the workers at random, what is the probability that exactly two of the four are women?

**4.80**   A committee of 50 politicians is to be chosen from among the 100 U.S. senators. If the selection is done at random, what is the probability that each state will be represented?

**4.81**   Dana is not the world's best poker player. She is dealt a 2 of diamonds, an 8 of diamonds, an ace of hearts, an ace of clubs, and an ace of spades. She discards the three aces. What are her chances of drawing a flush?

**4.82**   The victim of a mugging is asked to look at a police lineup consisting of eight people. All three of the persons responsible for the mugging are present in the lineup. If the victim points at three people at random, what is the probability that one of the three is innocent?

**4.83**   Factorials of large numbers can be approximated by using *Stirling's formula*, which states that

$$\log n! \doteq 0.3990899 + \left(n + \frac{1}{2}\right)\log n - 0.4342945n$$

Taking the antilog of $\log n!$ gives $n!$. (On a hand calculator, we compute the righthand-side of Stirling's formula and then hit INV and LOG.) Use Stirling's formula to approximate $30!$.

**4.84**   In how many ways can the letters in the word

$$E\,L\,E\,E\,M\,O\,S\,Y\,N\,A\,R\,Y$$

be arranged so that the S is always immediately followed by a Y?

**4.85**   A total of $2n$ students meet on the school baseball diamond to play a softball game. In how many ways can the group be divided into two teams, each having $n$ players? Check your answer by enumerating all the possibilities when $2n = 4$.

**4.86**   In the expansion

$$(a + b)^2 = (a + b)(a + b) = a^2 + 2ab + b^2$$

the coefficient 2 that multiplies $ab$ represents the number of ways to form the product $ab$ by choosing an $a$ from one factor and a $b$ from the other factor. Generalizing that idea, deduce the coefficient of $xy^3z$ in the expansion of $(x + y + z)^5$.

**4.87**   Consider the following sequence of figures formed by arranging circles in a triangular pattern:

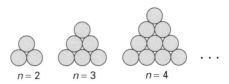

be arranged so that the S is always immediately followed by a Y?

Show that the number of circles in the $k$th triangle is $_{k+1}C_2$. (Hint: Look at Review Exercise 1.15.)

4.88    Listed in Question 2.15 are the mate selections of 39 female cockroaches. On 18 occasions it was the dominant male who won the lady's heart.

   (a)  In how many ways can exactly 18 yeses be positioned in the data table?

   (b)  Suppose the experiment is repeated and the new data again show 18 yeses. What is the probability that all 18 are adjacent?

4.89    Three Democrats and three Republicans are to be seated in six adjacent chairs along one side of a banquet table.

   (a)  If all six people are considered distinct, in how many different ways can they be arranged?

   (b)  In how many ways can they be arranged if the Democrats are considered all the same and the Republicans are considered all the same?

   (c)  Assume that all six people are distinct. Suppose that Dominique is one of the Republicans and Louie is one of the Democrats. If the seating arrangements are made at random, what is the probability that Dominique and Louie will be next to one another?

4.90    Suppose a pitcher faces a batter who never swings. For how many ball/strike sequences will the batter be called out on strikes on the fifth pitch? Enumerate the possibilities.

4.91    How many different numbers can be formed from the digits 1 through 9 if each digit is to be used exactly once? What does the expression

$$\sum_{i=1}^{9} \frac{9!}{(i-1)!}$$

represent in the context of forming numbers from the digits 1 through 9? (Hint: What does $9!/0! = 9!$ represent? What does $9!/1! = 9 \cdot 8 \cdot 7 \cdots 2$ represent? What does $9!/8! = 9$ represent?)

4.92    Show that the number of combinations of $n$ things taken $k$ at a time is equal to the number of combinations of $n$ things taken $n - k$ at a time.

4.93    Icosahedron dice have twenty sides, two sides are labeled 0, two sides are labeled 1, and so on. Each side is an equilateral triangle having the same area. If your intention is to maximize the probability of rolling a 0, is it better to label two adjacent faces 0 or to label the top face 0 and the bottom face 0?

4.94    Does a monkey have a better chance of rearranging

$$A\,C\,C\,L\,L\,U\,U\,S \quad \text{to spell} \quad C\,A\,L\,C\,U\,L\,U\,S$$

or

$$A\,A\,B\,E\,G\,L\,R \quad \text{to spell} \quad A\,L\,G\,E\,B\,R\,A?$$

4.95    Your statistics teacher has given you a 20-page reading assignment on Monday, that is to be finished by Thursday morning. You intend to read the first $x_1$ pages Monday, the next $x_2$ pages Tuesday, and the last $x_3$ pages

Wednesday, where $x_1 + x_2 + x_3 = 20$ and each $x_i \geq 1$. In how many ways can you choose a set of values for $x_1$, $x_2$, and $x_3$? (Hint: Consider the 19 spaces between a set of 20 lines. In the diagram below, positioning two arrows at spaces 3 and 11 would be equivalent to reading 3 pages Monday, 8 pages Tuesday, and 9 pages Wednesday.)

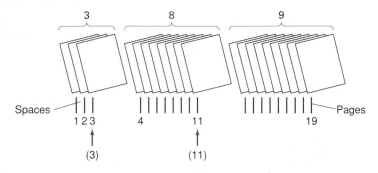

4.96   A poker player is dealt a 7 of diamonds, a queen of diamonds, a queen of hearts, a queen of clubs, and an ace of hearts. He discards the 7. What is his probability of drawing to either a full house or four-of-a-kind?

4.97   Thirteen tombstones in a country churchyard are arranged in three rows, 4 in the first row, 5 in the second, and 4 in the third. Suppose that two men are buried in each row. Assuming each arrangement to be equally likely, what is the probability that in each row the men occupy the two leftmost positions? Is it reasonable to assume that the arrangements are equally likely? Explain.

## Appendix   CARD PROBLEMS (Optional)

One of the most instructive and, to some, useful applications of combinatorics is the calculation of probabilities associated with poker hands. What are the chances of being able to "open" (being dealt a pair or better)? How much more unlikely is a full house than a flush? What is the probability of drawing to an inside straight?

From a pedagogical standpoint poker hands provide an excellent opportunity for practicing the problem-solving strategy suggested in Section 4.7. First, we get a clear idea of what a specific admissible hand looks like; then we decide what changes can be made that will generate similar hands having the same basic underlying structure. Poker hands are an exercise in the repeated use of two results: Theorems 4.1 and 4.4.

We assume that five cards are dealt from a poker deck and that no other cards are showing, although some may already have been dealt. Since the order of the cards in a hand is irrelevant, the sample space (the set of all possible five-card hands) has $_{52}C_5 = 2{,}598{,}960$ outcomes, all equally likely.

We will derive the probabilities of being dealt a full house, one pair, and a straight. Probabilities for other poker hands are obtained similarly.

"Seeing" a poker hand is an important first step in computing its probability. A good way, in general, to visualize card problems is to picture the deck as a grid having four rows representing the four suits:

diamonds, hearts, clubs, and spades

and 13 columns representing the 13 denominations:

2, 3, 4, 5, 6, 7, 8, 9, 10, jack, queen, king, ace

Specific cards can then be indicated as ×'s in the appropriate rows and columns.

### Full House

A full house consists of three cards of one denomination and two of another. Figure A4.1 shows the full house consisting of the 7 of hearts, 7 of clubs, 7 of spades, queen of hearts, and queen of spades. Notice that the formation of a full house is essentially a four-step operation: we need to choose (1) a *denomination* for the three-of-a-kind (2) *suits* for the three-of-a-kind, (3) a *denomination* for the pair, and (4) *suits* for the pair. We can easily see the number of ways to do each step by looking at the figure. How many denominations can be chosen for the three-of-a-kind? $_{13}C_1$. How many suits can be chosen for the three of a kind? $_4C_3$. How many denominations can be chosen for the pair (having already selected a denomination for the three of a kind)? $_{12}C_1$. How many suits can be chosen for the pair? $_4C_2$. By the multiplication rule, then, the number of full houses is

$$_{13}C_1 \cdot {_4}C_3 \cdot {_{12}}C_1 \cdot {_4}C_2$$

| | 2 | 3 | 4 | 5 | 6 | 7 | 8 | 9 | 10 | J | Q | K | A |
|---|---|---|---|---|---|---|---|---|---|---|---|---|---|
| D | | | | | | | | | | | | | |
| H | | | | | | × | | | | | × | | |
| C | | | | | | × | | | | | | | |
| S | | | | | | × | | | | | × | | |

**FIG. A4.1**

It follows that the *probability* of being dealt a full house is the number of full houses divided by the number of hands:

$$P \text{ (full house)} = \frac{_{13}C_1 \cdot {_4}C_3 \cdot {_{12}}C_1 \cdot {_4}C_2}{_{52}C_5}$$

$$= \frac{13 \cdot 4 \cdot 12 \cdot 6}{2{,}598{,}960} = 0.00144$$

### One Pair

To qualify as a one-pair hand, the five cards must include two of the same denomination and three single cards—cards whose denominations match neither the pair nor each other (see Figure A4.2). Tallying up all the possible one-pair hands requires a series of suit and denomination counts, much in the spirit of what was done to compute $P$ (full house). Specifically, the number of ways to choose a denomination for the pair is $_{13}C_1$; for each of those possible denominations the suits can be chosen in $_4C_2$ ways. How many choices do we have for the denominations of the three single cards? Since one denomination is already taken, the three that we need must be chosen from the available set of 12, which can be done in $_{12}C_3$ ways. Each single card can have any of $_4C_1$ suits. Therefore, appealing once again to the multiplication rule, we have

$$P \text{ (one-pair hand)} = \frac{_{13}C_1 \cdot {_4C_2} \cdot {_{12}C_3} \cdot {_4C_1} \cdot {_4C_1} \cdot {_4C_1}}{_{52}C_5}$$

$$= \frac{13 \cdot 6 \cdot 220 \cdot 4 \cdot 4 \cdot 4}{2{,}598{,}960} = 0.42$$

|   | 2 | 3 | 4 | 5 | 6 | 7 | 8 | 9 | 10 | J | Q | K | A |
|---|---|---|---|---|---|---|---|---|----|---|---|---|---|
| D |   |   | × |   |   |   |   |   |    |   |   |   | × |
| H |   |   |   |   |   | × |   | × |    |   |   |   |   |
| C |   |   |   |   |   | × |   |   |    |   |   |   |   |
| S |   |   |   |   |   |   |   |   |    |   |   |   |   |

**FIG. A4.2**

### Straight

A straight is five cards having consecutive denominations not all in the same suit; for example, a 4 of diamonds, 5 of hearts, 6 of hearts, 7 of clubs, and 8 of diamonds (see Figure A4.3). An ace may be counted "high" or "low," meaning that

10, jack, queen, king, ace    and    ace, 2, 3, 4, 5

|   | 2 | 3 | 4 | 5 | 6 | 7 | 8 | 9 | 10 | J | Q | K | A |
|---|---|---|---|---|---|---|---|---|----|---|---|---|---|
| D |   |   | × |   |   |   | × |   |    |   |   |   |   |
| H |   |   |   | × | × |   |   |   |    |   |   |   |   |
| C |   |   |   |   |   | × |   |   |    |   |   |   |   |
| S |   |   |   |   |   |   |   |   |    |   |   |   |   |

**FIG. A4.3**

are both straights. (If five consecutive cards are all in the same suit, the hand is called a ***straight flush***. The latter is considered a fundamentally different type of hand in the sense that any straight flush beats any straight.) To get the numerator for $P$ (straight), we first ignore the condition that all five cards not be in the same suit and simply count the number of hands having consecutive denominations. Note that there are 10 sets of consecutive denominations of length 5:

ace, 2, 3, 4, 5;   2, 3, 4, 5, 6;   . . . ;   and 10, jack, queen, king, ace

With no restrictions on the suits, each card can be a diamond, heart, club, or spade. It follows that the number of five-card hands having consecutive denominations is

$$10 \cdot {_4C_1} \cdot {_4C_1} \cdot {_4C_1} \cdot {_4C_1} \cdot {_4C_1}$$

But 40 ($= 10 \cdot {_4C_1}$) of those hands are straight flushes (why?) and need to be discounted (i.e., subtracted). Therefore,

$$P \text{ (straight)} = \frac{10 \cdot ({_4C_1})^5 - 40}{2,598,960} = 0.00392$$

**TABLE A4.1**

| Hand | Probability |
|---|---|
| One pair | 0.42 |
| Two pairs | 0.048 |
| Three of a kind | 0.021 |
| Straight | 0.0039 |
| Flush | 0.0020 |
| Full house | 0.0014 |
| Four of a kind | 0.00024 |
| Straight flush | 0.000014 |
| Royal flush | 0.0000015 |

Table A4.1 shows the probabilities associated with all the different poker hands. The less likely a hand is, the better it is, meaning that hand $i$ beats hand $j$ if $P(\text{hand } i) < P(\text{hand } j)$. According to Table A4.1, a full house beats a flush, a flush beats a straight, a straight beats three of a kind, and so on.

QUESTIONS

**4.98**   A *royal flush* in poker is a 10, jack, queen, king, and ace in the same suit. What is the probability of being dealt a royal flush?

**4.99**   A *flush* consists of five cards in the same suit, where not all five have consecutive denominations. The total number of possible flushes is ${_4C_1} \cdot {_{13}C_5} - 40$. Explain the significance of each term in that expression.

**4.100**   In bridge, players are dealt hands of 13 cards. If no card is an ace or higher than a 9, the cards are said to be a *coke hand*. What is the probability of being dealt such a hand?

**4.101**   A pinochle deck has 48 cards, two of each of six denominations (9, jack, queen, king, 10, ace) and the usual four suits. Among the many hands that count for *meld* is a *roundhouse*, which occurs when a player has a king and a queen of each suit. In a hand of 12 cards, what is the probability of getting *exactly* one king and queen of each suit?

*EXAMPLE A4.1*   The same techniques that allow us to compute the probability of being *dealt* a particular kind of hand can be used to determine the likelihood of *improving* a hand if we have the option of replacing certain cards by drawing others.

For example, consider a poker player who has been dealt a 3 of diamonds, an 8 of clubs, a 9 of clubs, a 10 of spades, and an ace of hearts (see Figure A4.4). In draw poker his options are to stand pat or to discard one, two, or three cards. Suppose he decides to discard the 3 and the ace. What is his probability of drawing to a straight?

| | 2 | 3 | 4 | 5 | 6 | 7 | 8 | 9 | 10 | J | Q | K | A |
|---|---|---|---|---|---|---|---|---|---|---|---|---|---|
| D | | × | | | | | | | | | | | |
| H | | | | | | | | | | | | | × |
| C | | | | | | | × | × | | | | | |
| S | | | | | | | | | × | | | | |

**FIG. A4.4**

*Solution*    Looking at Figure A4.4, we can see that there are three basic ways to complete a straight: (1) draw a 6 and a 7, (2) draw a jack and a queen, or (3) draw a 7 and a jack. Since there are four 6's and four 7's available, the number of ways to draw both a 6 and a 7 is

$$_4C_1 \cdot {}_4C_1 = 16$$

By the same argument, there are 16 ways to draw to each of the other two kinds of straights. Therefore,

$$P \text{ (player draws to straight)} = \frac{3 \cdot {}_4C_1 \cdot {}_4C_1}{{}_{47}C_2} = 0.044$$

QUESTIONS

4.102    Why is the denominator in the preceding expression $_{47}C_2$?

4.103    A poker player is dealt a 4 of clubs, a 6 of hearts, an 8 of hearts, a 9 of hearts, and a king of diamonds. She discards the 4 and the king. What is her probability of drawing to a straight flush? to a flush?

4.104    A poker player is dealt a 6 of diamonds, a 6 of clubs, a 9 of diamonds, a jack of hearts, and an ace of spades. He discards the 9 and the jack. What is his probability of drawing to a full house?

# 5

# *Probability*

*C*oincidences, in general, are
great stumbling blocks in the way of that
class of thinkers who have been educated to
know nothing of the theory of probabilities.

*Edgar Allan Poe*

## 5.1

### INTRODUCTION

Section 4.6 introduced one type of probability problem: if an experiment can lead to *n* possible outcomes, *all equally likely*, and if the event *A* comprises *m* of those *n* outcomes, then the probability of *A* is *m/n*. Situations where probabilities are defined in this fashion are not uncommon. Games of chance, for instance, such as dice, lotteries, and poker are familiar examples. Not all experiments, however, are simple enough to be described by the *m/n* model. For example, suppose dice are loaded or a lottery is rigged? What happens if the number of possible outcomes is infinite? Either of these conditions will invalidate the approach we have taken thus far.

It is also true that not all events are as easily characterized as the various winning throws and winning hands described in Section 4.6. In many situations the outcomes in *A* must be defined indirectly—in terms of their membership in some other set of events $A_1, A_2, \ldots, A_n$. So we need to rethink our definition of probability and to generalize the way we manipulate events. The theorems in the next several sections will expand enormously the range of problems we can address.

## 5.2

### ALGEBRA OF SETS

As a starting point, we will consider events derived in various ways from other events.

---

**DEFINITION 5.1**  Operations with Sets

Let *A* and *B* be any two events defined on a sample space S.

1. The **union** of *A* and *B*, written $A \cup B$, is the event whose outcomes belong to *A or B* or both.
2. The **intersection** of *A* and *B*, written $A \cap B$, is the event whose outcomes belong to *A and B*. If two events have no outcomes in common, the events are **mutually exclusive** and we write $A \cap B = \varnothing$. (The symbol $\varnothing$ denotes the **null event**, the event that has no outcomes.)
3. The **complement** of *A*, written $A^c$, is the event consisting of all outcomes in S that are not in *A*.

---

EXAMPLE 5.1   One card is dealt from a poker deck. Let $A$ be the event that a heart is drawn; let $B$ be the event that a jack is drawn. Which cards are in $A \cup B$? in $A \cap B$? in $A^c$?

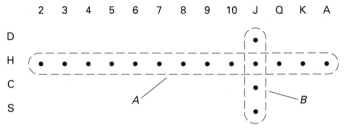

**FIG. 5.1**

Solution   Figure 5.1 shows the sample space represented as a grid having 4 rows (suits) and 13 columns (denominations). Each of the experiment's 52 possible outcomes corresponds to a point on the grid. We can see that $A \cup B$ contains 16 outcomes, $A \cap B$ contains 1, and $A^c$ contains 39:

$$A \cup B = \{2 \text{ of hearts}, 3 \text{ of hearts}, \ldots, \text{jack of hearts}, \ldots,$$
$$\text{ace of hearts, jack of diamonds, jack of clubs,}$$
$$\text{jack of spades}\}$$

$$A \cap B = \{\text{jack of hearts}\}$$

$$A^c = \{2 \text{ of diamonds}, \ldots, \text{ace of diamonds}, 2 \text{ of clubs}, \ldots,$$
$$\text{ace of clubs}, 2 \text{ of spades}, \ldots, \text{ace of spades}\}$$

COMMENT:   The concept of intersection figured prominently in some of the card problems in the appendix to Chapter 4. For example, if S is the set of all straights and if $F$ is the set of all flushes, then $S \cap F$ is the set of straight flushes. Out of the $_{52}C_5 = 2{,}598{,}960$ five-card hands in the sample space S, 40 belong to $S \cap F$.

EXAMPLE 5.2   Recall the birthday problem in Example 4.29. Recognizing an event's complement was a crucial step in computing the probability of getting at least one match. The sample space of $365^k$ birthday sequences was divided into the events $A = \{$at least two people have the same birthday$\}$ and $\{$all $k$ people had different birthdays$\}$. The second set would have been called $A^c$ had we known about Definition 5.1. The outcomes in $A^c$ are easy to count *directly*:

$$\text{Outcomes in } A^c = (365)(364) \cdots (365 - k + 1)$$

We determined the outcomes in $A$ *indirectly* by subtracting the outcomes in $A^c$ from $365^k$.

*EXAMPLE 5.3*    Two eyewitnesses to a hit-and-run accident have given information to the police. One witness is certain that the first letter of the license plate of the car that left the scene was $G$; the second witness said that the last digit was 6. If license plates have the form

letter    letter    number    number    number    number

how large is the pool of potential suspects?

*Solution*    Let $A$ and $B$ denote the sets of plates compatible with the information given by the witnesses. Only drivers whose plates agree with *both* eyewitness accounts qualify as suspects, so we want the *intersection* of $A$ and $B$. Plates in $A \cap B$ have the structure

G    ___    ___    ___    ___    6

letter    number    number    number

Since 26 choices are available for the second letter and 10 for each of the missing numbers, the maximum possible size of $A \cap B$ is

$$26 \cdot 10 \cdot 10 \cdot 10 = 26{,}000$$

---

*EXAMPLE 5.4*    A commuter encounters three traffic lights on her way to work. If at least two lights are red, she arrives late. If the first light is red, she gets angry. For which sequences of traffic lights will she arrive late *or* be angry? For which sequences will she arrive late *and* be angry?

*Solution*    Let $L$ denote the set of traffic light sequences for which the commuter arrives late, and let $M$ be the sequences that make her angry. Think of the lights encountered during a ride to work as an ordered sequence of length 3. Two outcomes are possible at each light: red ($r$) and green ($g$). For example, $(r, r, g)$ is the sequence where the first two lights are red and the last one is green.

Listed in the first column of Table 5.1 is the commuter's entire set of $8 (= 2 \cdot 2 \cdot 2)$ possible light sequences. The final four columns indicate which of those eight belong to $L$, $M$, $L \cup M$, and $L \cap M$.

**TABLE 5.1**

| Outcome | In $L$? | In $M$? | In $L \cup M$? | In $L \cap M$? |
|---------|---------|---------|----------------|----------------|
| $(r, r, r)$ | Yes | Yes | Yes | Yes |
| $(r, r, g)$ | Yes | Yes | Yes | Yes |
| $(r, g, r)$ | Yes | Yes | Yes | Yes |
| $(g, r, r)$ | Yes | No | Yes | No |
| $(r, g, g)$ | No | Yes | Yes | No |
| $(g, r, g)$ | No | No | No | No |
| $(g, g, r)$ | No | No | No | No |
| $(g, g, g)$ | No | No | No | No |

**QUESTIONS**

5.1 A card is drawn from a poker deck. Let $R$ denote the event that the card is a diamond or a heart. Let $F$ denote the event that the card is a jack, a queen, or a king. Which cards are in $R \cap F$?

5.2 Two coins are tossed. Let $(H, T)$ denote the outcome where the first coin is a head and the second is a tail.

(a) List all outcomes in the sample space $S$.

(b) Let $A$ denote the event where both coins show the same face. Which outcomes are in $A^c$?

5.3 Let $S$ be the set of integers from 1 to 10. Define the events $A$ and $B$ as follows:

$$A = \{1, 2, 3, 6, 8\} \quad \text{and} \quad B = \{2, 3, 7, 9, 10\}$$

Which outcomes are in

(a) $A^c$    (b) $A \cup B$    (c) $A \cap B$

(d) $A \cap B^c$    (e) $A^c \cup B^c$    (f) $(A \cap B)^c$

5.4 One number is drawn from the integers $1, 2, \ldots, 12$. Let $A$ be the set of numbers divisible by 2, and let $B$ be the set of numbers divisible by 3. Which numbers are in $A \cap B$? Which numbers are in $(A \cup B)^c$?

5.5 Let $A$ and $B$ denote the sets of points lying in the interiors of the squares. Draw diagrams for $A \cup B$ and $A \cap B$.

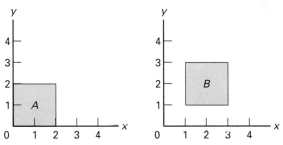

5.6 The Cleveland Indians and Boston Red Sox are playing a three-game series in Cleveland. List the possible outcomes—for example, let $(B, C, C)$ represent the event that Boston wins the first game and Cleveland wins the second and the third. Let $W$ denote the event that Cleveland wins the series—that is, they win at least two of the games. Let $T$ denote the event that Cleveland wins the third game. List the outcomes in

(a) $W$    (b) $T$    (c) $W \cup T$

(d) $W \cap T$    (e) $(W \cup T)^c$    (f) $W^c \cap T$

5.7 In craps two dice are thrown. A person wins on the first toss if the sum on the up faces of the dice is 7 or 11. Define two sets whose union is the set of outcomes for which the player wins on the first toss. How many outcomes are contained in that union? What is the probability that a person wins on the first throw? (Assume the dice are fair.)

5.8   Let $A$ be the set of $x$'s for which $x^2 + 2x = 8$; let $B$ be the set of $x$'s for which $x^2 + x = 6$. Find $A \cup B$ and $A \cap B$.

5.9   When asked which historical figures he would most like to invite to dinner, Art Buchwald came up with the following list (176):

1. Richard Nixon
2. John Wilkes Booth
3. Jack the Ripper
4. Adolf Hitler
5. Lizzie Borden
6. Nero
7. Joseph Stalin
8. Judas
9. Mao Tse-tung
10. Cain
11. Marquis de Sade

In defending his choices, Buchwald wrote, "I know I will have a seating problem, but I believe the conversation would be quite interesting and certainly worth a decent Chateau Mouton-Rothschild." Let $A$ denote the two guests with whom Lizzie Borden would feel the most uncomfortable. Let $B$ denote the guest you would not ask for help if a fire broke out in the kitchen. Let $C$ denote the two guests who would have the most fun carving the turkey.

(a) Are $A$ and $B$ mutually exclusive?

(b) Are $A$ and $C$ mutually exclusive?

5.10   A country club has 100 members. Fifty members are lawyers, and rumor has it that 30 are liars. If 10 members are both, how many are neither? Express the set of members who are neither lawyers nor liars in terms of $A$ and $B$. Let $A$ be the set of lawyers, and $B$ the set of liars.

### Venn Diagrams

**Venn diagram**—A graphical representation of events defined on a sample space.

A helpful device for displaying relationships among events is the **Venn diagram**. Colored areas (usually circles) inside the diagram represent outcomes of events. The entire area enclosing the Venn diagram (usually a rectangle) denotes the sample space. See Figure 5.2.

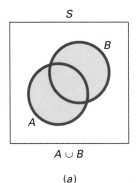

$A \cup B$

(a)

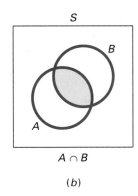

$A \cap B$

(b)

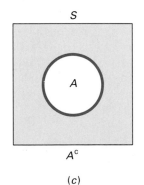

$A^c$

(c)

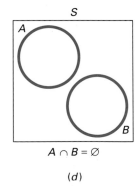

$A \cap B = \varnothing$

(d)

**FIG. 5.2**

*EXAMPLE 5.5*   Represented by a Venn diagram, the eight outcomes of the sample space in Example 5.4 would appear as eight points arranged arbitrarily inside a rectangle S as shown in Figure 5.3.

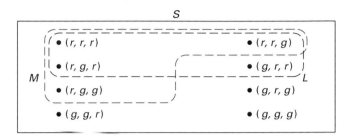

**FIG. 5.3**

*EXAMPLE 5.6*   When only one event is defined on a sample space, the potential probability questions that can be formulated are limited: we can ask only for the probability that the event *does* occur or the probability that the event *does not* occur. When two or more events are defined, however, other meaningful questions arise. Problems asking for the probability that *at least one occurs*, *at most one occurs*, or *exactly one occurs* appear often.

More specifically, let A and B be any two events defined on a sample space S. Draw the Venn diagrams and write the formulas for the following statements:

1. *At least one* of the events occurs.
2. *At most one* of the events occurs.
3. *Exactly one* of the events occurs.

*Solution*   Figure 5.4 shows the appropriate Venn diagrams. Convince yourself that the shaded regions are valid representations of the three verbal statements.

*Formulas* for the different shaded areas can be motivated by writing each of the desired events in a little more detail. For example, *at least one event occurs* means that A occurs or B occurs or both occur. Since these

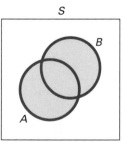
At least one

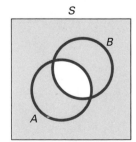
At most one

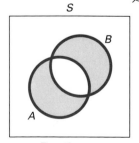
Exactly one

**FIG. 5.4**

conditions were used to define a union, we can write *at least one event occurs* $= A \cup B$. Similarly, *at most one event occurs* is satisfied by all the outcomes in the sample space except those belonging to the intersection of $A$ and $B$ (where two events occur). Therefore, *at most one event occurs* $= (A \cap B)^c$. The last statement is the most difficult of the three to translate into unions, complements, and intersections. *Exactly one event occurs* means that $A$ occurs and $B$ doesn't occur, *or* that $A$ doesn't occur and $B$ does occur.[1] Replacing the *and*'s with intersections and the *or*'s with unions gives

$$\text{Exactly one event occurs} = (A \cap B^c) \cup (A^c \cap B)$$

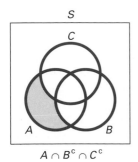

$A \cap B^c \cap C^c$

**FIG. 5.5**

### Multiple Events

Extending the notions of union and intersection to more than two events is easy. The event $A \cup B \cup C$ is the set of outcomes belonging to $A$ *or* $B$ *or* $C$, or to some combination of $A$, $B$, and $C$. The event $A \cap B \cap C$ is the set of outcomes belonging to $A$ *and* $B$ *and* $C$. Figure 5.5 shows the Venn diagrams $A \cup B \cup C$ and $A \cap B \cap C$ for three arbitrary events defined on a sample space S.

We can also include complements in multievent configurations. Figure 5.6 is the Venn diagram for $A \cap B^c \cap C^c$, the event where $A$ occurs but $B$ and $C$ do not occur.

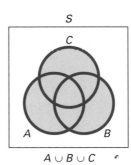

$A \cup B \cup C$

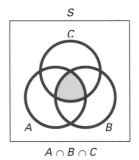

$A \cap B \cap C$

**FIG. 5.6**

QUESTIONS   **5.11**   Draw Venn diagrams to show the nature of the relationship between events $A$ and $B$ if

(a) $A \cap B = B$       (b) $A \cup B = B$

---

[1] This statement is *not* the same as "at most one event occurs," which allows the possibility that *no* event occurs. "Exactly one" means one event must occur.

5.12 Consider the Venn diagram shown below. What does $(B \cap A) \cup (B \cap A^c)$ represent?

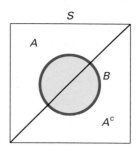

5.13 For the three events

$$A = \text{all females}$$
$$B = \text{all math professors}$$
$$C = \text{all professors with tenure}$$

describe in words:

(a) $A \cap B \cap C^c$    (b) $(A \cap B \cap C)^c$    (c) $A^c \cap B^c \cap C$

5.14 Suppose a single digit is chosen at random from the digits 0 through 9. Define the events

$$A - \{0, 1, 2, 3, 4\} \qquad B = \{5, 6, 7, 8, 9\} \qquad C = \{3, 4, 5, 6\}$$

Which outcomes are in:

(a) $A \cup B \cup C$    (b) $A \cap B \cap C$    (c) $A \cap B^c \cap C$

(d) $A^c \cap B^c \cap C^c$    (e) $(A^c \cap B \cap C)^c$    (f) $(A \cap B^c \cap C^c)^c$

5.15 The following information was collected on six students who volunteered to help with a psychology experiment.

| ID | Parents Divorced? | Left-Handed? | Number of Years to Finish High School |
|----|----|----|----|
| U | No | Yes | 3 |
| V | Yes | No | 4 |
| W | No | No | 3 |
| X | Yes | No | 7 |
| Y | No | Yes | 2 |
| Z | No | No | 4 |

Define the following events:

$$A = \text{student's parents are divorced}$$
$$B = \text{student is left-handed}$$
$$C = \text{student completed high school in 3 years or less}$$

Draw a Venn diagram showing $A$, $B$, and $C$ and each of the six outcomes listed in the table. Which individuals are in the following sets:

(a) $A \cup B$          (b) $A \cap B \cap C^c$

(c) $A^c \cap B^c \cap C^c$      (d) $A^c \cup C^c$

5.16    Muffy has a starring role in a new biker movie. She appears in 60% of the film's scenes, her stunt double appears in 20%, and both of them appear in 5% of the scenes. If the film has 100 scenes, in how many

     (a) do neither Muffy nor her stunt double appear?

     (b) does at most one of the two appear?

     (c) does at least one of the two appear?

     (d) does exactly one of the two appear?

5.17    The mathematics department at Swampwater Tech has 32 faculty members. Six are easy graders, 20 are good teachers, and 8 are neither.

     (a) How many professors are easy graders or good teachers?

     (b) How many professors are easy graders or good teachers but not both?

5.18    A fair coin is tossed three times. Partition the sample space of all possible head-tail sequences by defining a set of four $A_i$'s, where each $A_i$ represents one of the possible outcomes of the first two tosses. Draw a Venn diagram showing the four $A_i$'s and the outcomes each one contains.

## 5.3

### THE PROBABILITY FUNCTION

We are now ready to develop the principles and applications of probability more fully. What we derived in Section 4.6 on combinatorial probability is simply a special case of the more general treatment of the next few sections.

     A key step in using the algebraic structure of Section 5.2 is recognizing that probability is a function. A *function* is a formula that associates a num-

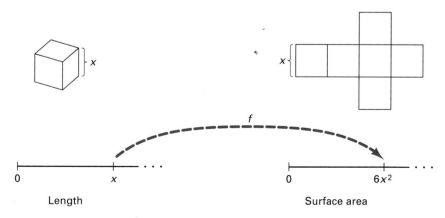

**FIG. 5.7**

ber $x$ with a possibly different number $f(x)$. Often $f(x)$ is physically significant. For example, let $x$ be the length of the side of a cube. Then $f(x) = 6x^2$ gives the cube's surface area (see Figure 5.7).

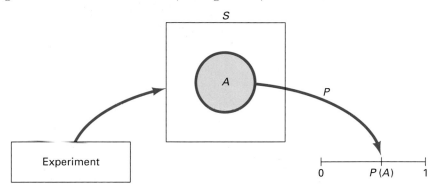

**FIG. 5.8**

We can also define a function between an event and a number. Figure 5.8 shows the general structure of such a relationship. Associated with an experiment is a sample space S. Defined on S is an event $A$. We want to associate with $A$ a number $P(A)$, which represents the proportion of times $A$ would be expected to occur if the experiment were repeated many, many times (recall Section 4.6). We call $P(A)$ the *probability of A*. More formally, $P$ is a function whose domain consists of sets defined on S and whose range is the interval from 0 to 1.

What distinguishes one problem from another is the way in which $P(A)$ is defined. The only probability formula we know at this point is $P(A) = m/n$, but there are many others. Chapters 6 and 7 show how to evaluate $P(A)$ in specific contexts. Here we are concerned with manipulating probabilities in general.

Defining $P(A)$ as a proportion of occurrences $m$ over attempts $n$ leads to several properties that the probability function must automatically satisfy, regardless of its explicit formulation. The first two should be apparent:

---

**Properties of Probability Functions**

  1. For any event $A$, $0 \le P(A) \le 1$.
  2. $P(S) = 1$; $P(\varnothing) = 0$.
  3. If events $A$ and $B$ are mutually exclusive, then
  $$P(A \cup B) = P(A) + P(B)$$
  4. $P(A^c) = 1 - P(A)$.
  5. For any two events $A$ and $B$ defined on S,
  $$P(A \cup B) = P(A) + P(B) - P(A \cap B)$$

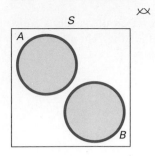

**FIG. 5.9**

Consider Property 3. Figure 5.9 is the Venn diagram of mutually exclusive events $A$ and $B$. Since $A$ and $B$ have no overlap, $P(A \cup B)$ must be the sum of the probabilities associated with $A$ and $B$.

Property 5 generalizes Property 3. If $A$ and $B$ are *any* two events defined on $S$, they may overlap (Figure 5.10). If we simply add $P(A)$ to $P(B)$ in this situation, we will count twice the probability associated with the overlapping portion, which is just $A \cap B$. Since the probability of every outcome in $A \cup B$ must be counted *only once* in determining $P(A \cup B)$, we must subtract $P(A \cap B)$ from $P(A) + P(B)$. If $A$ and $B$ are mutually exclusive, then $A \cap B = \varnothing, P(A \cap B) = 0$ (by Property 2), and Property 5 reduces to Property 3.

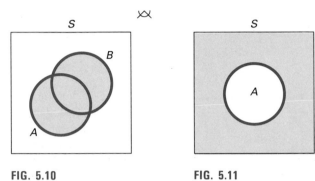

**FIG. 5.10**          **FIG. 5.11**

Property 4 is among the simplest and most useful results in probability. See Figure 5.11 for the Venn diagram. Since

$$S = A \cup A^c$$

$$P(S) = P(A) + P(A^c) \qquad \text{by Property 3}$$

$$1 = P(A) + P(A^c) \qquad \text{by Property 2}$$

$$P(A^c) = 1 - P(A)$$

---

*EXAMPLE 5.7*    Two fair dice are rolled. The shooter wins on the first roll if the sum showing on the dice is 7 or 11; the shooter loses on the first roll if the sum is 2, 3, or 12. What is the probability that the game is concluded on the first roll?

*Solution*    Let $A$ be the event "shooter wins on first roll," and $B$ the event "shooter loses on first roll." Figure 5.12 shows the corresponding Venn diagram. Clearly,

$$P \text{ (game is concluded on first roll)}$$

$$= P \text{ (shooter wins on first roll}$$
$$\text{or shooter loses on first roll)}$$

$$= P(A \cup B)$$

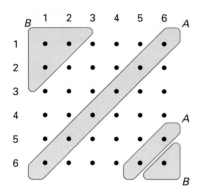

**FIG. 5.12**

Since $A$ and $B$ are mutually exclusive, Property 3 applies and

$$P(A \cup B) = P(A) + P(B) = P(7 \text{ or } 11) + P(2, 3, \text{ or } 12)$$

But

$$P(A) = P(\{(1, 6), (6, 1), (2, 5), (5, 2), (3, 4), (4, 3), (5, 6), (6, 5)\})$$

$$= \frac{8}{36} = \frac{2}{9}$$

and

$$P(B) = P(\{(1, 1), (1, 2), (2, 1), (6, 6)\}) = \frac{4}{36} = \frac{1}{9}$$

Therefore,

$$P(A \cup B) = \frac{2}{9} + \frac{1}{9} = \frac{1}{3}$$

---

**EXAMPLE 5.8**   Five passengers, each going to a different city, check their luggage at an airport terminal. The redcaps, in a playful mood, close their eyes and put the destination tags on the luggage *at random* (one to a bag). What is the probability that the luggage for at least one passenger gets sent to the wrong place?

**Solution**   Each outcome in the "experiment" being performed by the redcaps is a possible assignment of the five destination tags to the five pieces of luggage. In how many ways can that assignment be made? According to Corollary 4.1, there are $_5P_5 = 5! = 120$ ways, each of which occurs with probability $1/120$.

We can imagine two events, then, being defined on that sample space of 120 different luggage assignments:

$A = $ the event that all 5 travelers are properly
united with their luggage

$A^c = $ the event that *not* all 5 travelers are properly
united with their luggage

We are looking for $P(A^c)$, but $P(A)$ is much easier to find: since only one of the 120 possible ticket assignments satisfies the event $A$, $P(A) = \frac{1}{120}$. We can now easily compute $P(A^c)$ by applying Property 4:

$$P(A^c) = P(\text{at least one mixup})$$

$$= 1 - P(A) = 1 - \frac{1}{120} = \frac{119}{120}$$

Always keep Property 4 in mind when doing probability problems. Why struggle over a difficult $P(A)$ if computing $P(A^c)$ is trivial?

---

*EXAMPLE 5.9* Winthrop, a premed, has been summarily rejected by all 126 accredited U.S. medical schools. Desperate, he sends his transcripts and MCATs to the two least selective correspondence schools he can think of, Liability College and Malpractice Tech. Based on the success his friends have had there, he estimates that his probability of being accepted at Liability is 0.7 and at Malpractice 0.4. He also suspects there is a 25% chance that both schools will accept him. What is the probability he gets in somewhere?

*Solution* Let $A$ be the event "Liability College accepts him," and $B$ the event "Malpractice Tech accepts him." We are given that $P(A) = 0.7$, $P(B) = 0.4$, and $P(A \cap B) = 0.25$. From Example 5.6 we know that the event "getting in somewhere" is the union of $A$ and $B$. Therefore, by Property 5,

$$P \text{ (at least one school accepts him)} = P(A \cup B)$$
$$= P(A) + P(B) - P(A \cap B)$$
$$= 0.7 + 0.4 - 0.25$$
$$= 0.85$$

*EXAMPLE 5.10* Let $A$ and $B$ be any two events defined on a sample space S. Suppose that $P(A) = 0.4$, $P(B) = 0.5$, and $P(A \cap B) = 0.1$. Find the probability that $A$ or $B$, but not both, occurs.

*Solution* We want the probability that *exactly one* event occurs. The shaded region in Figure 5.13 is the set of outcomes whose probability we are trying to find (recall Figure 5.4). By inspection,

$$P \text{ (A or B but not both occur)} = P(A) + P(B) - 2P(A \cap B)$$
$$= 0.4 + 0.5 - 2(0.1)$$
$$= 0.7$$

(Why do we subtract $P(A \cap B)$ *twice*?)

**FIG. 5.13**

*EXAMPLE 5.11* Consolidated Industries has come under considerable pressure to eliminate its discriminatory hiring practices. Company officials have agreed that during the next five years, 60% of its new employees will be females and 30% of those female employees will be black. However, one out of four new employees will be white males. What percentage of black females is Consolidated committed to hiring?

*Solution*   Four basic events are relevant to this problem:

$$B = \text{employee is black} \qquad F = \text{employee is female}$$
$$W = \text{employee is white} \qquad M = \text{employee is male}$$

The three probabilities given, $P(F) = 0.60$, $P(B) = 0.30$, and $P(W \cap M) = 0.25$, can be pictured in a $2 \times 2$ table (Table 5.2). Entries in the table are intersection probabilities; the row and column totals represent single-event probabilities $P(F)$, $P(M)$, $P(B)$, and $P(W)$. We want to find $P(B \cap F)$.

Note first that if $P(B) = 0.30$, then $P(W) = 1 - P(B) = 0.70$. Also,

$$P(W) = P(W \cap F) + P(W \cap M) \quad \text{or} \quad 0.70 = P(W \cap F) + 0.25$$

so

$$P(W \cap F) = 0.45$$

(see the circled entries in Table 5.2). Thus

$$P(B \cap F) = P(F) - P(W \cap F) = 0.60 - 0.45 = 0.15$$

so 15% of all newly hired employees will be black females.

**TABLE 5.2**

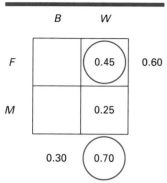

---

*EXAMPLE 5.12*   When Swampwater Tech's class of '55 held its thirtieth reunion, 100 graduates attended. Fifteen alumni are lawyers, and, according to rumor, 30 of the 100 graduates are psychopaths. If 10 alumni are both lawyers and psychopaths, what proportion of the class suffered from neither of these afflictions?

*Solution*   Let $L$ denote the set of alumni who are lawyers and $Y$ the set of alumni who are psychopaths. Figure 5.14 is a Venn diagram of the sets $L$ and $Y$ as subsets of the class of '55. We are looking for the probability that an alum does *not* belong to $L \cup Y$.

Since

$$P(L) = \frac{15}{100} = 0.15, \quad P(Y) = \frac{30}{100} = 0.30$$

and

$$P(L \cap Y) = \frac{10}{100} = 0.10$$

it follows from Property 5 that

$$P(L \cup Y) = P(L) + P(Y) - P(L \cap Y)$$
$$= 0.15 + 0.30 - 0.10 = 0.35$$

But if 35% of the alumni are *either* lawyers or psychopaths, then 65% are *neither*:

$$P((L \cup Y)^c) = P \text{ (alum is neither lawyer nor psychopath)}$$
$$= 1 - 0.35 = 0.65$$

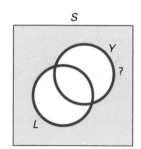

**FIG. 5.14**

QUESTIONS

**5.19**  An experiment has two posssible outcomes: the first occurs with probability $k$, and the second with probability $4k$. Find $k$.

**5.20**  One card is dealt from a deck of 52. Let $A$ be the event that the card is an ace. Let $B$ be the event that the card is a heart. What is the probability that the card is an ace or a heart?

**5.21**  A new "no-frills" airline runs a commuter shuttle service using a refurbished World War I bomber. The two-propeller plane will fly if either or both of its engines function. Suppose that $P$(port engine fails) = 0.10, $P$(starboard engine fails) = 0.15, and $P$(both engines fail) = 0.015. What is the probability that the plane will complete its next flight safely?

**5.22**  If $P(A) = \frac{1}{3}$, $P(B) = \frac{1}{2}$, and $P(A \cup B) = \frac{3}{4}$, find

(a) $P(A \cap B)$     (b) $P(A^c \cap B)$     (c) $P(A \cap B^c)$

(Hint: Draw a Venn diagram and show the probabilities associated with the different areas.)

**5.23**  In the game of odd man out each player tosses a fair coin. If all the coins turn up the same *except for one*, the player tossing the "minority" coin is declared the odd man out and is eliminated from the contest. Suppose three people play the game. What is the probability that someone will be eliminated on the first toss? (Hint: Consider the complement—that no one will be eliminated.)

**5.24**  A fair die is tossed once. Define the following events:

$$A = \text{number showing is divisible by 2}$$

$$B = \text{number showing is divisible by 3}$$

Draw a Venn diagram showing the six outcomes in S. Indicate on the diagram the events $A$ and $B$. Find $P(A \cup B)$.

**5.25**  Refer to Example 5.9.

(a) What is the probability that Winthrop gets accepted to neither school?

(b) What is the probability that he gets accepted to exactly one school?

(c) What is the probability that he gets accepted to Liability College but not to Malpractice Tech?

(d) What is the probability that he gets accepted to Malpractice Tech but not to Liability College?

Show the events being asked for as areas on a Venn diagram.

**5.26**  If Vanderbilt's football team has a 0.1 probability of defeating Saturday's opponent (event $A$), a 0.3 probability of defeating the following Saturday's opponent (event $B$), and a 0.65 probability of losing both games, what is the probability that they win exactly one?

**5.27**  Donna checks her campus mailbox once a day. She keeps a record of all the mail she receives and knows that the probability of getting junk mail on any day is 0.6, her probability of getting real mail is 0.1, and her probability of getting some of each is 0.03. What is the probability that when she goes to her mailbox tomorrow nothing will be there?

**5.28**   Refer to Example 5.12. In addition to sets $Y$ and $L$, define $H$ to be the set of alumni who are honest. Assume that $H$ and $L$ are mutually exclusive and that 8 of Swampwater's 60 honest alumni are psychopaths.

(a)  Draw a Venn diagram indicating the sets $L$, $Y$, and $H$.

(b)  Suppose an alum is picked at random. What is the probability that that person is an honest lawyer?

(c)  What proportion of lawyers are psychopaths?

(d)  What is the probability of $(L \cup Y \cup H)^c$?

# 5.4

## CONDITIONAL PROBABILITY

Probabilities sometimes have to be reevaluated in light of additional information. An event $A$, for example, defined on a sample space $S$ has an initial probability based on the outcomes it contains. Suppose we learn, however, that some other event on $S$, say $B$, has already occurred. It could well be that $A$'s probability, *now that B has occurred*, is different than it was initially. If $A$ and $B$ are mutually exclusive, for example, the revised probability of $A$ is zero, no matter how likely it was before the occurrence of $B$. This section and the next examine the effect that the occurrence of one event has on the probability of another.

Suppose we toss two fair dice. What is the probability of getting an odd sum? Recognizing that "odd sum" is an event consisting of the outcomes 3, 5, 7, 9, and 11, all mutually exclusive (see Figure 5.15), we can write

$$P \text{ (odd sum)} = P(3, 5, 7, 9, 11)$$

$$= P(3) + P(5) + P(7) + P(9) + P(11)$$

$$= \frac{2}{36} + \frac{4}{36} + \frac{6}{36} + \frac{4}{36} + \frac{2}{36} = \frac{1}{2}$$

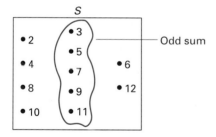

**FIG. 5.15**

This is the sort of computation we considered in Sections 4.6 and 5.3.

Now, suppose we roll the dice *blindfolded* so that we cannot see the outcome. However, a friend is watching, and, though he refuses to tell us exactly how the faces came up, he does give us a hint by revealing that the sum was at least as large as nine. Given that partial information, how should we revise our estimate of the probability that the sum is odd? Or, phrased in the language we will soon be using, what is the probability that the sum is odd *given* that the sum is nine or greater?

Figure 5.16 is the Venn diagram of the situation. The clue provided by our friend restricts the outcome to being 9, 10, 11, or 12. Of those four,

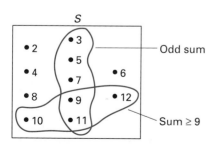

**FIG. 5.16**

two are odd. If the four outcomes were equally likely, it would seem reasonable to write the probability we are looking for as 2/4. But the outcomes are not equally likely, and the probability is not 2/4. Still, that rationale can be easily generalized. The probability of rolling a 9 or an 11 knowing that we threw a 9, 10, 11, or 12 should be the ratio of the probability associated with the event "9 or 11" divided by the probability associated with the event "sum ⩾ 9":

$P$ (sum is odd given that sum ⩾ 9)

$$= P \text{ (sum is 9 or 11 given that sum equals 9, 10, 11, or 12)}$$

$$= \frac{P \text{ (sum is 9 or 11)}}{P \text{ (sum equals 9, 10, 11, or 12)}}$$

$$= \frac{\frac{4}{36} + \frac{2}{36}}{\frac{4}{36} + \frac{3}{36} + \frac{2}{36} + \frac{1}{36}} = \frac{3}{5}$$

(Notice that the numerator is the intersection of the two events "sum is odd" and "sum is 9 or greater.")

These two computations show that the probability of an event $A$ (here, that the sum is odd) may have to be revised if we know that some other event $B$ (that the sum equals 9, 10, 11, or 12) has already occurred. Being able to calculate such *conditional probabilities* is extremely important, simply because we often encounter situations where partial information is available.

Mathematically, the knowledge that some event $B$ has already occurred has a direct effect on the sample space. Certain outcomes, those in $B^c$, no longer need to be considered. The original sample space $S$ is therefore replaced by a *new* sample space, the event $B$. And, as a consequence, the "new" probability of $A$ becomes the ratio of $P(A \cap B)$ to $P(B)$.

**conditional probability**—A probability computed on the presumption that some other event has already occurred.

---

**DEFINITION 5.2**   Conditional Probability

Let $A$ and $B$ be any two events defined on a sample space $S$ such that $P(B) > 0$. The *conditional probability of A given B*, written $P(A|B)$, is

$$P(A|B) = \frac{P(A \cap B)}{P(B)} \tag{5.1}$$

---

**COMMENT:**   We can rewrite Equation 5.1 to obtain an expression for the probability of an intersection:

$$P(A \cap B) = P(A|B)P(B) \tag{5.2}$$

---

*EXAMPLE 5.13*   A certain metropolitan area is serviced by 100 banks: some are family owned, and some are guilty of fraud. Table 5.3 shows a cross-tabulation of the city's banking establishment according to those two criteria. Suppose a

**TABLE 5.3**

|  |  | FAMILY OWNED? | |
|---|---|---|---|
|  |  | Yes | No |
| **GUILTY OF** | Yes | 6 | 4 |
| **BANK FRAUD?** | No | 34 | 56 |

Federal Reserve Board examiner randomly picks one of the banks to audit. What is the probability that the bank

(a) is guilty of fraud?

(b) is guilty of fraud *if it is family owned?*

*Solution*  (a) Let $A$ be the event "bank audited is guilty of fraud." The *probability* of $A$ is simply the proportion of the entire set of banks that are guilty of fraud, regardless of whether they are family owned. According to Table 5.3, 10 banks are guilty of fraud. Therefore,

$$P(A) = P \text{ (bank audited is guilty of fraud)} = \frac{10}{100} = 0.1$$

(b) Let $B$ be the event "bank audited is family owned." The *conditional probability* of $A$ given $B$ is the ratio of $P(A \cap B)$ to $P(B)$. From Table 5.3 a total of 6 banks comprise the event $A \cap B$; and 40 belong to $B$. It follows from Equation 5.1, then, that

$$P(A|B) = P \text{ (bank audited is guilty of fraud given that}$$
$$\text{bank audited is family-owned)}$$

$$= \frac{P(A \cap B)}{P(B)} = \frac{\frac{6}{100}}{\frac{40}{100}} = 0.15$$

The events $A$ and $B$ are probabilistically *dependent*. Our estimate of the likelihood of a bank's involvement in illegal activities changes if we learn that the bank is family owned. Specifically,

$$P(A|B) > P(A)$$

---

*EXAMPLE 5.14*  A card is drawn at random from a poker deck. What is the probability that the card is a club, given that the card is a king?

*Solution*  Intuitively, the answer is 1/4: the king is equally likely to be a heart, diamond, club, or spade. More formally, let $C$ be the event "card is a club"; let $K$ be the event "card is a king." But

$$P(K) = \tfrac{4}{52} \quad \text{and} \quad P(C \cap K) = P \text{ (card is a king of clubs)} = \frac{1}{52}$$

Therefore, by Equation 5.1,

$$P(C|K) = \frac{P(C \cap K)}{P(K)} = \frac{\frac{1}{52}}{\frac{4}{52}} = \frac{1}{4}$$

Note that, in contrast to Example 5.13, the conditional probability $P(C|K)$ is numerically equal to the unconditional probability $P(C)$. Thus our knowledge that $K$ has occurred tells us nothing about the likelihood of $C$ occurring. Two events having this property are said to be *independent*. We will come back to the notion of independence in Section 5.5.

---

**EXAMPLE 5.15**    Riesling grapes are grown in two wine-producing valleys, X and Y, in southern California. Vineyards in both areas are susceptible to a parasite capable of retarding growth. Let $A$ be the event that valley X is infested, and $B$ the event that valley Y is infested. Suppose that local viticulturists estimate that

$$P(A) = \frac{2}{5}, \qquad P(B) = \frac{3}{4}, \quad \text{and} \quad P(A \cup B) = \frac{4}{5}$$

If state inspectors find the parasite on a sample of grapes grown in valley Y, what is the probability that wineries in valley X are facing a similar problem?

*Solution*    We are looking for the conditional probability $P(A|B)$. We already know $P(B)$, but we must find $P(A \cap B)$. By Property 5 on page 231,

$$P(A \cup B) = P(A) + P(B) - P(A \cap B)$$

so

$$P(A \cap B) = P(A) + P(B) - P(A \cup B)$$

$$= \frac{2}{5} + \frac{3}{4} - \frac{4}{5} = \frac{7}{20}$$

Therefore    $P(A|B) = P$ (valley X is infested *given that* valley Y is infested)

$$= \frac{\frac{7}{20}}{\frac{3}{4}} = \frac{7}{15}$$

---

**EXAMPLE 5.16**    Some experts suggest importing liquefied natural gas (LNG) as a good way of coping with a future energy crunch. Other disagree. Critics point out the LNG is highly volatile and poses an enormous safety hazard. Any major spill near a U.S. port could precipitate a fire of catastrophic proportions. The *likelihood* of a spill, therefore, becomes critical input for future policy-makers. How can that probability be determined?

*Solution*    We need two numbers (1) the probability that a tanker will have an accident near a port, and (2) the probability that a major spill will develop *given* that an accident has happened. Although no significant spills of LNG have yet occurred anywhere in the world, these probabilities can be approximated from records kept on similar tankers transporting less dangerous cargo (46). Thus, the probability that an LNG tanker will have an accident on any one trip is estimated to be 8/50,000. And given that an accident *has* occurred, it is suspected that the probability of a major spill is 3/15,000. From Equation 5.2, then, the single-trip probability for a major LNG disaster is $3.2 \times 10^{-8}$:

$$P \text{ (accident occurs and spill develops)}$$

$$= P \text{ (spill develops}|\text{accident occurs)}P(\text{accident occurs})$$

$$= \left(\frac{3}{15,000}\right)\left(\frac{8}{50,000}\right) = 0.000000032$$

**CASE STUDY 5.1 (Optional)**

*Simulating Shakespeare*

There once was a brainy baboon
Who always breathed down a bassoon
　　For he said, "It appears
　　That in billions of years
I shall certainly hit on a tune."

Eddington

The image of a monkey sitting at a typewriter and pecking away at random until it gets lucky and types out a perfect copy of the complete works of William Shakespeare has long been a favorite model of statisticians and philosophers to illustrate the distinction between something that is theoretically possible but, for all practical purposes, impossible. But if that monkey and its typewriter are replaced by a high-technology computer and if we program in the right conditional probabilities, the prospects for generating *something* intelligible become less far-fetched, perhaps disturbingly less far-fetched (8).

Simulating nonnumerical English text requires that 28 characters be dealt with: the 26 letters, the space, and the apostrophe. The simplest approach is to assign each character a number from 1 to 28. Then a random number in that range is generated, and the character corresponding to that number is printed. A second random number is generated, a corresponding second character is printed, and so on.

Would that be a reasonable model? No, because E's, for example, are much more frequently used than X's, so their chances of being selected should be greater. At the very least, weights should be assigned to all the characters proportional to their relative probabilities. Table 5.4 shows the empirical distribution of the 26 letters, the space, and the apostrophe in

**TABLE 5.4**

| Character | Frequency | Probability | Random Number Range | Character | Frequency | Probability | Random Number Range |
|---|---|---|---|---|---|---|---|
| Space | 6934 | 0.1968 | 00001–06934 | Y | 783 | 0.0222 | 30329–31111 |
| E | 3277 | 0.0930 | 06935–10211 | W | 716 | 0.0203 | 31112–31827 |
| O | 2578 | 0.0732 | 10212–12789 | F | 629 | 0.0178 | 31828–32456 |
| T | 2557 | 0.0726 | 12790–15346 | C | 584 | 0.0166 | 32457–33040 |
| A | 2043 | 0.0580 | 15347–17389 | G | 478 | 0.0136 | 33041–33518 |
| S | 1856 | 0.0527 | 17390–19245 | P | 433 | 0.0123 | 33519–33951 |
| H | 1773 | 0.0503 | 19246–21018 | B | 410 | 0.0116 | 33952–34361 |
| N | 1741 | 0.0494 | 21019–22759 | V | 309 | 0.0088 | 34362–34670 |
| I | 1736 | 0.0493 | 22760–24495 | K | 255 | 0.0072 | 34671–34925 |
| R | 1593 | 0.0452 | 24496–26088 | ' | 203 | 0.0058 | 34926–35128 |
| L | 1238 | 0.0351 | 26089–27326 | J | 34 | 0.0010 | 35129–35162 |
| D | 1099 | 0.0312 | 27327–28425 | Q | 27 | 0.0008 | 35163–35189 |
| U | 1014 | 0.0288 | 28426–29439 | X | 21 | 0.0006 | 35190–35210 |
| M | 889 | 0.0252 | 29440–30328 | Z | 14 | 0.0004 | 35211–35224 |

the 35,224 characters making up Act III of *Hamlet*. Ranges of random numbers corresponding to each character's frequency are listed in the last column. If two random numbers were generated, say 27,351 and 11,616, the computer would print the characters *D* and *O*. Doing that, of course, is equivalent to printing a *D* with probability 0.0312

$$\left( = \frac{28{,}425 - 27{,}327}{35{,}244} = \frac{1099}{35{,}244} \right)$$

and an *O* with probability 0.0732

$$\left( = \frac{12{,}789 - 10{,}212}{35{,}244} = \frac{2576}{35{,}244} \right)$$

Extending this idea to *sequences* of letters requires an application of Equation 5.1. What is the probability, for example, that a *T* follows an *E*? Clearly,

$$P \,(T \text{ follows an } E) = P(T|E) = \frac{\text{Number of } ET\text{'s}}{\text{Number of } E\text{'s}}$$

Therefore, the analog of Table 5.4 is an array having 28 rows and 28 columns. The entry in the $i$th row and $j$th column is $P(i|j)$, the conditional probability that letter $i$ follows letter $j$. Similarly, we could estimate conditional probabilities for longer sequences. For example, the probability that an *A* follows the sequence *QU* is

$$P \,(A \text{ follows } QU) = P(A|QU) = \frac{\text{Number of } QUA\text{'s}}{\text{Number of } QU\text{'s}}$$

What does our monkey gain by having a typewriter programmed with probabilities of sequences? Quite a bit. Figure 5.17 shows three lines of computer text generated by a program knowing only single-letter frequencies (Table 5.4). No words are spelled correctly. Contrast that with Figure 5.18, showing computer text generated by a program that had been given estimates for conditional probabilities corresponding to all 614,656 ($=28^4$) *four*-letter sequences. What we get is still garbled, but the improvement is astounding: more than 80% of the letter combinations are actual words.

```
AOOAAORH ONNNDGELC TEFSISO VTALIDMA POESDHEMHLESWON
PJTOMJ FTL FIM TAOFERLMT O NORDEERH HMFIOMRETWOVRCA
OSRIE IEOBOTOGIM NUDSEEWU WHHS AWUA HIDNEVE NL SELTS
```

**FIG. 5.17**

```
A GO THIS BABE AND JUDGEMENT OF TIMEDIOUS RETCH AND NOT LORD
WHAL IF THE EASELVES AND DO AND MAKE AND BASE GATHEM I AY
BEATELLOUS WE PLAY MEANS HOLY FOOL MOUR WORK FROM INMOST
BED BE CONFOULD HAVE MANY JUDGEMENT WAS IT YOU MASSURE'S TO
LADY WOULD HAT PRIME THAT'S OUR THROWN AND DID WIFE FATHER'ST
LIVENGTH SLEEP TITH I AMBITION TO THIN HIM AND FORCE AND LAW'S
MAY BUT SMELL SO AND SPURSELY SIGNOR GENT MUCH CHIEF MIXTURN
```

**FIG. 5.18**

One can only wonder how intelligible computer-generated texts would be if conditional probabilities for, say, seven- or eight-letter sequences were available. Given the rate that computer technology is developing, they soon will be. Our monkey will probably still never come up with text as creative as Hamlet's soliloquy, but something ridiculously simpleminded, like a political speech (or a statistics lecture), might not be out of the question.

---

**QUESTIONS**

**5.29** Two fair dice are tossed. What is the probability that the sum is even, given that the sum is 4 or less?

**5.30** Home security experts estimate that an untrained housedog has a 70% probability of detecting an intruder and a 50% probability of scaring the intruder away if he or she has been detected (1). What is the probability that Fido will successfully thwart a burglar? (Hint: Let $A$ be the event "Fido scares away burglar," and let $B$ be the event "Fido detects burglar." Find $P(A \cap B)$.)

**5.31** (a) For $P(A) = 0.3$, $P(B) = 0.4$, and $P(A|B) = 0.5$, find $P(A \cup B)$.

(b) For $P(A \cap B) = 0.1$, $P(A|B) = 0.2$, and $P(B|A) = 0.25$, find $P(A \cup B)$.

**5.32** Let $A$ be the event that a person dies between the ages of 80 and 85. Let $B$ be the event that a person lives to be at least 70.

(a) If $P(A) = 0.03$ and $P(B) = 0.16$, find the probability that a person dies between the ages of 80 and 85 given that that person has lived to be at least 70.

(b) Which probability would you expect to be larger, $P(A)$ or $P(A|B)$?

5.33   Three fair coins are tossed. Let $A$ be the event that the first coin is a head. Let $B$ be the event that exactly two of the coins are heads. Find $P(A|B)$. (Hint: Draw a Venn diagram showing the eight possible outcomes and indicate which outcomes belong to the sets $A$ and $B$.)

5.34   Two pieces of information are recorded for each of the first 50 students attending the annual Vanderbilt-Tennessee football game: (1) whether they are UT fans or Vandy fans and (2) whether they are wearing shoes. The results are summarized below:

|  | UT Fans | Vanderbilt Fans |
|---|---|---|
| Wearing Shoes | 4 | 30 |
| Not Wearing Shoes | 16 | 0 |

Let $A$ be the event that the student is a Vandy fan. Let $B$ be the event that the student is wearing shoes. Find

(a) $P(A)$      (b) $P(B)$      (c) $P(A \cap B)$

(d) $P(A|B)$      (e) $P(B|A)$.

5.35   Suppose a family has two children. What is the probability that both are boys, given that at least one is a boy? Assume that the probability of any child being a boy is 1/2.

5.36   A chip is drawn from an urn containing one red chip and one white chip. If the chip drawn is red, it and two more red chips are put back into the urn. A second chip is then drawn. If the first chip selected is white, it is simply returned to the urn, and a second chip is drawn. What is the probability that both chips drawn are red? (Hint: Let $A$ be the event that a red chip is selected on the second draw. Let $B$ be the event that a red chip is selected on the first draw. Find $P(A \cap B)$.)

5.37   The table below shows the prediction record of a TV weather forecaster over the past several years. Entries in the table are intersection probabilities.

|  |  | FORECAST | | |
|---|---|---|---|---|
|  |  | Sunny | Cloudy | Rain |
| ACTUAL WEATHER | Sunny | 0.3 | 0.05 | 0.05 |
|  | Cloudy | 0.04 | 0.2 | 0.02 |
|  | Rain | 0.1 | 0.04 | 0.2 |

(a) What proportion of the days were sunny?

(b) How often was the forecaster wrong?

(c) What was the probability of rain on the days the forecast was sunny?

5.38   Misty, Tupper, and Jasmine audition for parts in a big-budget remake of *Bedtime for Bonzo*. On the basis of their past experience and the caliber of competition they face, Misty has a 40% chance of being hired, Tupper a 50% chance, and Jasmine a 30% chance. If exactly two of the three are cast, what is the probability that Misty was rejected? (Hint: Let $A$ be the event that Misty is rejected. Let $B$ be the event that exactly two are hired. Find $P(A|B)$.)

# 5.5

## TWO PROBABILITY THEOREMS

### Partitioning a Sample Space

In this section we examine two theorems that apply the concept of conditional probability to *partitioned* sample spaces. A set of $k$ events, $A_1, A_2, \ldots, A_k$, *partitions* the sample space S if every outcome in S belongs to one *and only one* $A_i$. Figure 5.19 shows the situation graphically: no portion of S is left "uncovered" (in other words, $A_1 \cup A_2 \cup \ldots \cup A_k = S$), and no two of the $A_i$'s overlap (that is, $A_i \cap A_j = \varnothing$ for all $i \neq j$).

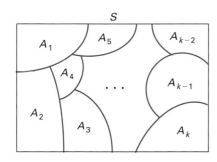

FIG. 5.19

Suppose three chips are drawn from an urn that has five chips numbered 1 through 5. A typical outcome would be the set (2, 3, 5). Consider the following three $A_i$'s:

$A_1$ = largest number in sample is 3

$A_2$ = largest number in sample is 4

$A_3$ = largest number in sample is 5

The number of possible samples is $_5C_3 = 10$, and each sample belongs to exactly one $A_i$ (Figure 5.20). Thus, by definition, these three $A_i$'s partition S.

Now, let $B$ denote any other event defined on S. Notice that part of $B$ overlaps $A_1$, another part overlaps $A_2$, and a third part overlaps $A_3$. The event $B$, in other words, is really the union of the intersections $A_1 \cap B$, $A_2 \cap B$, and $A_3 \cap B$. That is,

$$B = (A_1 \cap B) \cup (A_2 \cap B) \cup (A_3 \cap B)$$

In general, if S is partitioned by a set of $k$ events, $A_1, A_2, \ldots, A_k$, then

$$B = (A_1 \cap B) \cup (A_2 \cap B) \cup \cdots \cup (A_k \cap B)$$

Furthermore, all the intersections are necessarily mutually exclusive since the $A_i$'s themselves are mutually exclusive. Therefore,

$$P(B) = P(A_1 \cap B) + P(A_2 \cap B) + \cdots + P(A_k \cap B)$$

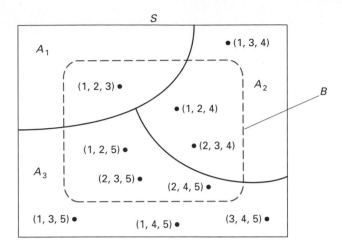

**FIG. 5.20**

or, applying Equation 5.2, to each of the $k$ intersections,

$$P(B) = P(B|A_1)P(A_1) + P(B|A_2)P(A_2) + \cdots + P(B|A_k)P(A_k)$$

$$= \sum_{i=1}^{k} P(B|A_i)P(A_i)$$

We have just proved Theorem 5.1, a highly important result because of its many real-world applications. The theorem tells us how to piece together a set of conditional probabilities

$$P(B|A_i), \qquad i = 1, 2, \ldots, k$$

to obtain an unconditional probability $P(B)$.

---

**THEOREM 5.1**

Let $A_1, A_2, \ldots, A_k$ be a set of events that partition a sample space S. Let $B$ be any event defined on S. Then

$$P(B) = \sum_{i=1}^{k} P(B|A_i)P(A_i) \tag{5.3}$$

---

*EXAMPLE 5.17*    A toy manufacturer buys ball bearings from three different suppliers: 50% of his total order comes from supplier 1, 30% from supplier 2, and 20% from supplier 3. Past experience has shown that the quality control standards of the three suppliers are not equal. Of the ball bearings produced by supplier 1, 2% are defective; suppliers 2 and 3 send defective bearings 3% and 4% of the time, respectively. What proportion of the ball bearings in the manufacturer's inventory is defective?

*Solution*  Let $A_i$ be the event "bearing came from supplier $i$," $i = 1, 2, 3$. Let $B$ be the event "bearing in manufacturer's inventory is defective." We know that

$$P(A_1) = 0.5, \qquad P(A_2) - 0.3, \qquad P(A_3) = 0.2$$

and that

$$P(B|A_1) = 0.02, \qquad P(B|A_2) = 0.03, \qquad P(B|A_3) = 0.04$$

By Equation 5.3,

$$P(B) = P(B|A_1)P(A_1) + P(B|A_2)P(A_2) + P(B|A_3)P(A_3)$$

$$= (0.02)(0.5) + (0.03)(0.3) + (0.04)(0.2)$$

$$= 0.027$$

Thus the manufacturer can expect 2.7% of his ball-bearing stock to be defective.

---

**EXAMPLE 5.18**  The crew of the Starship *Enterprise* is considering launching a surprise attack against the Klingons in a neutral quadrant. Possible interference by the Romulans, though, is causing Captain Kirk and Mr. Spock to reassess their strategy. According to Spock's calculations, the probability of the Romulans joining forces with the Klingons is 0.2384. Captain Kirk feels that the probability of a successful attack is 0.8 if the *Enterprise* can catch the Klingons alone, but only 0.3 if they have to engage both adversaries. Spock claims that the attack would be a tactical misadventure if its probability of success were not at least 0.7306. Should they attack?

*Solution*  Let $A_1$ and $A_2$ denote the partitioning events "Romulans join in" and "Romulans do not join in," respectively; let $B$ be the event "attack is a success." Substituting directly into Equation 5.3, we get

$$P(B) = P(B|A_1)P(A_1) + P(B|A_2)P(A_2)$$

$$= (0.3)(0.2384) + (0.80)(0.7616)$$

$$= 0.6808$$

so that the proposed mission's chance of ending favorably is only 68%, a probability not sufficiently promising. Therefore it would behoove Kirk and Spock to direct their xenophobic hostilities elsewhere.

---

**EXAMPLE 5.19**  A card is drawn off the top of a poker deck and dealt face down. Then the next card is turned over. What is the probability that that second card is an ace?

*Solution*  Set up the $A_i$'s to represent what happens on the *first* draw. Specifically, let $A_1$ denote the event that an ace is drawn, and $A_2$ that a non-ace is drawn.

Let $B$ be the event that the *second* card is an ace. Our objective is to find $P(B)$.

Note that

$$P(B|A_1) = \frac{3}{51}, \quad P(B|A_2) = \frac{4}{51}, \quad P(A_1) = \frac{4}{52}, \quad P(A_2) = \frac{48}{52}$$

According to Equation 5.3,

$$P(B) = P(B|A_1)P(A_1) + P(B|A_2)P(A_2)$$

$$= \left(\frac{3}{51}\right)\left(\frac{4}{52}\right) + \left(\frac{4}{51}\right)\left(\frac{48}{52}\right) = \frac{4}{52}$$

which is *the same as the probability that the first card is an ace*. Thus our answer implies that the probability of a card being, say, an ace is the same *regardless of where in the deck it is located*. Recall the poker hand probabilities we computed in the appendix to Chapter 4. In doing those problems, we imagined that the five cards being dealt came from the very top of the deck. The answers that were derived, though, are really much more general than that by what we have just shown. The probability of getting a full house, for example, with the *top* five cards is no greater or smaller than the probability of getting a full house with *any* five cards. Where a player sits relative to the dealer, therefore, has absolutely no effect on his probability of receiving any particular hand.

---

*EXAMPLE 5.20*    At the university computing center, 30% of all the programs submitted are written in COBOL, 60% are in PASCAL, and 10% are in FORTRAN. Past experience has shown that 10% of the COBOL programs, 25% of the PASCAL programs, and 20% of the FORTRAN programs compile on their first run. Suppose a program is selected at random from the backlog waiting to be loaded. What is the probability that that program will compile on its first run?

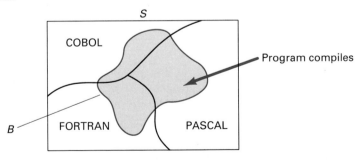

**FIG. 5.21**

*Solution*    Figure 5.21 shows the appropriate Venn diagram. The partitioning events are the languages COBOL, PASCAL, and FORTRAN; $B$ is the event that the program compiles on its first run. We are given three conditional probabilities and three unconditional probabilities:

$$P \text{ (program compiles} \mid \text{program is in COBOL)} = 0.10$$

$$P \text{ (program compiles} \mid \text{program is in PASCAL)} = 0.25$$

$$P \text{ (program compiles} \mid \text{program is in FORTRAN)} = 0.20$$

$$P \text{ (program is in COBOL)} = 0.30$$

$$P \text{ (program is in PASCAL)} = 0.60$$

$$P \text{ (program is in FORTRAN)} = 0.10$$

Therefore,

$$P(B) = P \text{ (program compiles)} = (0.10)(0.30) + (0.25)(0.60) + (0.20)(0.10)$$

$$= 0.20$$

---

**EXAMPLE 5.21 (Optional)**

Opinion polls, although widely used, are treacherous to interpret: nuances in the way a question is worded or the context in which it is asked can profoundly influence the way a respondent is likely to answer. This situation, not good under the best of conditions, gets even worse when the question is sufficiently sensitive that respondents feel pressured to answer in a certain way. Surveys addressing topics such as drug use or sexual preferences certainly fall into this category. In this example we show that by using a *randomized response survey* (23), it is possible to ask sensitive questions and still convince the respondents that their anonymity is completely guaranteed (so that they have nothing to fear by being truthful).

Suppose a sociologist is trying to estimate the percentage of college students who have smoked marihuana. We assume that 100 students have been randomly selected to make up the sample. Taking a direct approach and asking each person, "Have you ever smoked marihuana?" would be foolish, and the answers would be worthless. A much better strategy is an indirect approach.

Each student is asked to pick a random number between 00 and 99 (without revealing its identity). Students with numbers between, say, 00 and 69 are asked to answer the marihuana question. The others are instructed to answer a trivial, nonsensitive question such as, "Does your student ID end with an odd digit?" The students write their answers on slips of paper without indicating the question to which they are responding.

Suppose 44% of *all* the responses are "yes." What proportion of yes answers can we infer are coming from students answering the marihuana question?

**Solution**

The answer comes from a rather unusual "reverse" application of Theorem 5.1. Letting the two original questions be the set of partitioning events $A_1$ and $A_2$ and letting a "yes" response correspond to the event $B$, we can write

$$P \text{ (yes)} = P \text{ (yes}|\text{sensitive question)}P \text{ (sensitive question)}$$

$$+ P \text{ (yes}|\text{nonsensitive question)}P \text{ (nonsensitive question)}$$

Substituting for the four probabilities we know gives

$$0.44 = P \text{ (yes}|\text{sensitive question)}(0.7) + (0.5)(0.3)$$

Therefore,

$$P \text{ (yes}|\text{sensitive question)} = P \text{ (student has smoked marihuana)}$$

$$= \frac{0.44 - (0.5)(0.3)}{0.7} = 0.41$$

Strange as it may sound, what we have done here is analyze a set of answers without knowing the questions!

---

QUESTIONS

**5.39** The governor of a certain state has decided to come out strongly for prison reform and is preparing a new early-release program. Its guidelines are simple: if a prisoner is related to a member of the governor's staff, he has a 90% chance of being released early; if he is not a relative, his probability of early release is 0.01. Suppose that 40% of all inmates are related to someone on the governor's staff. What is the probability that a prisoner selected at random will be released early?

**5.40** If men constitute 47% of the population and tell the truth 78% of the time, whereas women tell the truth 63% of the time, what is the probability that a person selected at random will answer a question truthfully?

**5.41** One urn contains three white chips and four red chips: a second urn contains seven white chips and six red chips. One of the urns is picked at random, and from it a chip is selected. What is the probability that that chip is red? (Hint: Let $A_1$ and $A_2$ be the events that urn 1 and urn 2 are chosen, respectively. Let $B$ be the event that the chip selected is red.)

**5.42** A study claims that the probability of a female college graduate getting married is 0.97 if she was a math major and 0.80 if she majored in something else. Suppose that 5% of all female college graduates are math majors. What is the probability that a female college graduate, chosen at random, will get married?

**5.43** Bunky's grade on an upcoming statistics exam will be adversely affected if he has to take a geology exam just before he takes the math exam. He estimates that his probability of getting a B or better on the statistics exam is 0.75 if he comes in fresh, but it drops to 0.35 if he has to take the two exams back-to-back. If the probability of both instructors' scheduling exams for the same day is 0.60, what are Bunky's chances of getting a B or better on the statistics exam?

5.44  Recall the two urns in Question 5.41. Between them, they contain 10 red chips and 10 white chips.

(a) How should we allocate the 20 chips to the two urns to maximize the probability of drawing a red chip?

(b) What is the probability of drawing a red chip if the optimal allocation is used?

5.45  In a local congressional race, the incumbent Republican ($R$) is running against a field of three Democrats ($D_1$, $D_2$, and $D_3$), each seeking the nomination. Suppose political pundits estimate that the probabilities of $D_1$, $D_2$, and $D_3$ winning the primary are 0.35, 0.40, and 0.25, respectively. Furthermore, results from polls suggest that $R$ has a 40% chance of defeating $D_1$ in the general election, a 35% chance of defeating $D_2$, and a 60% chance of defeating $D_3$. If these estimates are accurate, what are the chances that the Republican will retain his seat?

5.46  An urn contains 40 red chips and 60 white chips. Six chips are drawn out and discarded, and a seventh chip is drawn. What is the probability that the seventh chip is red? (Hint: Remember Example 5.19.)

5.47  Urn I contains two red chips and four white chips; urn II contains three red chips and one white chip. A chip is drawn at random from urn I and transferred to urn II. Then a chip is drawn from urn II. What is the probability that the chip drawn from urn II is red? (Hint: Let the partitioning events $A_1$ and $A_2$ correspond to the color of the chip being transferred from urn I to urn II. Let $B$ be the event that a red chip is drawn from urn II.)

5.48  A telephone solicitor is responsible for canvassing three suburbs. In the past, 60% of her calls to Belle Meade resulted in contributions, compared with 55% for Oak Hill and 35% for Antioch. The caller's list of telephone numbers includes 1000 households from Belle Meade, 1000 from Oak Hill, and 2000 from Antioch. Suppose she picks a number at random from the list of 4000 and places the call. What is the probability that she gets a donation?

5.49  Does Equation 5.3 allow for the possibility that $B$ and certain $A_i$'s might be mutually exclusive? Explain.

**Bayes' rule**—A formula that "reverses" conditional probabilities. If events $A_1, A_2, \ldots, A_k$ partition a sample space $S$ and if $B$ is any event defined on $S$ such that $P(B) > 0$, then Bayes' rule gives $P(A_j | B)$, for any $j$, in terms of the conditional probabilities $P(B|A_i)$, $i = 1, 2, \ldots, k$.

## Bayes' Rule (optional)

We have seen how repeated applications of Equation 5.1 led to Equation 5.3. Further manipulation of Equation 5.1 leads to another important probability result, **Bayes' rule**.

We again assume that an experiment's sample space $S$ is partitioned by a set of events $A_1, A_2, \ldots, A_k$. Let $B$ be any event defined on $S$ such that $P(B) > 0$. By Equation 5.1 we can write

$$P(A_j|B) = \frac{P(A_j \cap B)}{P(B)}, \qquad j = 1, \ldots, k$$

which implies that $P(A_j \cap B)$ can be written as either

$$P(A_j \cap B) = P(A_j|B)P(B) \quad \text{or} \quad P(A_j \cap B) = P(B|A_j)P(A_j)$$

If we choose the latter, then

$$P(A_j|B) = \frac{P(B|A_j)P(A_j)}{P(B)}$$

Now, using Equation 5.3 to rewrite the denominator, we have Bayes' rule:

$$P(A_j|B) = \frac{P(B|A_j)P(A_j)}{\sum_{i=1}^{k} P(B|A_i)P(A_i)} \qquad (5.4)$$

Equation 5.4 is a curious result. If we know $P(B|A_i)$ for all $i$, we can compute conditional probabilities *in the other direction*—that is, we can use the set of $P(B|A_i)$'s to compute $P(A_j|B)$.

---

### THEOREM 5.2    Bayes' Rule

Let $A_1, A_2, \ldots, A_k$ be a set of events that partition S. Let $B$ be any event defined on S such that $P(B) > 0$. Then

$$P(A_j|B) = \frac{P(B|A_j)P(A_j)}{\sum_{i=1}^{k} P(B|A_i)P(A_i)} \qquad (5.4)$$

for any $1 \leqslant j \leqslant k$.

---

*EXAMPLE 5.22*    Urn I contains two white chips ($w_1, w_2$) and one red chip ($r_1$); urn II has one white chip ($w_3$) and two red chips ($r_2, r_3$). One chip is drawn at random from urn I and transferred to urn II. Then one chip is drawn from urn II (see Figure 5.22). Suppose a red chip is selected from urn II. What is the probability that the chip *transferred* was white?

*Solution*    When setting up a Bayes' problem, start by looking for whatever "happened" and call that the event $B$. Here $B$ is the event "red chip is drawn from urn II." Then identify the partitioning set $A_1, \ldots, A_k$. Let $A_1$ and $A_2$ denote the events "red chip is transferred from urn I" and "white chip is transferred from urn I," respectively. The probability being asked for is $P(A_2|B)$.

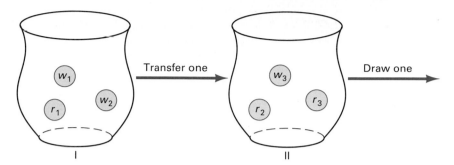

**FIG. 5.22**

We will find it instructive to examine the problem at its most basic level by enumerating the sample space. If $(r_i, w_j)$ denotes the outcome of transferring the $i$th red chip from urn I and then drawing the $j$th white chip from urn II, the sample space reduces to a set of 12 such points, all equally likely. Figure 5.23 shows S together with the events $A_1$, $A_2$, and $B$. By inspection, $P(A_2|B) = \frac{4}{7}$ (why?).

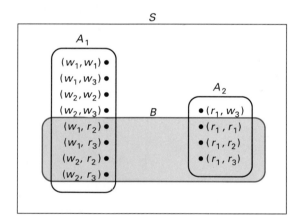

**FIG. 5.23**

We can get the same answer by Bayes' rule without enumerating the sample space. Note that $P(A_1) = \frac{4}{12}$ and $P(A_2) = \frac{8}{12}$. Furthermore, *if* a red chip is transferred, then urn II will contain three reds and one white, and the probability of drawing a red from that particular configuration is $\frac{3}{4}$. That is, $P(B|A_1) = \frac{3}{4}$. By a similar argument, $P(B|A_2) = \frac{2}{4}$. From Equation 5.4,

$$P(A_2|B) = \frac{P(B|A_2)P(A_2)}{P(B|A_1)P(A_1) + P(B|A_2)P(A_2)}$$

$$= \frac{(\frac{2}{4})(\frac{8}{12})}{(\frac{3}{4})(\frac{4}{12}) + (\frac{2}{4})(\frac{8}{12})} = \frac{4}{7}$$

*EXAMPLE 5.23*   A recent innovation in medicine is the use of computers to help physicians make diagnoses. Bayes' rule is the key. A patient has syndrome $B$ (e.g., loss of appetite, muscular pains, and dizziness). The possible causes are a set of diseases $A_1, A_2, \ldots, A_k$. From past experience the likelihoods of persons with diseases $A_1, \ldots, A_k$ suffering from syndrome $B$ are known. That is, the computer has been given

$$P(B|A_1), \quad P(B|A_2), \quad \ldots, \quad \text{and} \quad P(B|A_k)$$

The unconditional prevalences of the diseases $P(A_1), P(A_2), \ldots,$ and $P(A_k)$ are also well known. The computer then calculates $P(A_j|B)$, from Equation 5.4, for all $j$, and its diagnosis is the disease giving the highest conditional probability. The situation described in this example is a simplified application of the fundamental principle involved in the computer diagnostic process. It shows mathematically how symptoms affect the probability associated with a diagnosis.

Suppose that 0.5% of all students seeking treatment at a school infirmary are eventually diagnosed as having mononucleosis. Of those who do have mono, 90% complain of a sore throat. But 30% of those not having mono also have sore throats. If a student comes to the infirmary and says that he has a sore throat, what is the probability he has mono?

*Solution*   Let $B$ be the event that the student complains of a sore throat. Let $A_1$ and $A_2$ denote the events that the student *does* and *does not* have mono, respectively. We know that

$$P(A_1) = 0.005 \qquad P(A_2) = 0.995$$

$$P(B|A_1) = 0.90 \qquad P(B|A_2) = 0.30$$

We are looking for $P(A_1|B)$.

Substituting directly into Bayes' rule gives

$$P(A_1|B) = \frac{(0.90)(0.005)}{(0.90)(0.005) + (0.30)(0.995)} = 0.015$$

Since $P(A_1)$ was 0.005, the effect of the symptom (the sore throat) is to *triple* the student's probability of having mono.

---

*EXAMPLE 5.24*   State College is playing Backwater A & M for the conference football championship. If Backwater's first-string quarterback is healthy, A & M has a 75% chance of winning. If they have to start their backup quarterback, their chances of winning drop to 40%. The team physician says there is a 70% chance that the first-string quarterback will play.

(a) What is the probability that Backwater will win the game?

(b) Suppose you miss the game but read in the headlines of Sunday's paper that Backwater won. What is the probability that the second-string quarterback started?

*Solution*    (a) Let $B$ be the event "Backwater wins," and let $A_1$ and $A_2$ be the events "first-string quarterback starts" and "second-string quarterback starts," respectively. Then, from Equation 5.3,

$$P(B) = P(B|A_1)P(A_1) + P(B|A_2)P(A_2)$$

$$= (0.75)(0.70) + (0.40)(0.30) = 0.645$$

(b) Here we need Bayes' rule. Using the notation from Part (a),

$$P(A_2|B) = \frac{P(B|A_2)P(A_2)}{P(B|A_1)P(A_1) + P(B|A_2)P(A_2)}$$

$$= \frac{(0.40)(0.30)}{0.645} = 0.186$$

That is, given that Backwater won, there is approximately a 19% chance that the backup quarterback started.

*EXAMPLE 5.25*    Your next-door neighbor has a rather old and temperamental burglar alarm. If someone breaks into his house, the probability of the alarm sounding is 0.95. In the last two years, though, it has gone off on five different nights, each time for no apparent reason. Police records show that the chances of a home being burglarized in your community on any given night are 2 in 10,000. If your neighbor's alarm goes off tomorrow night, what is the probability that a crime is being committed?

*Solution*    Often, as here, the partitioning set has only two members: an event and its complement. What are the two options in this situation? Either your neighbor is being burglarized ($A_1$), or he is *not* being burglarized ($A_2 = A_1^c$). We already know that

$$P(B|A_1) = 0.95$$

$$P(B|A_2) = \frac{5}{730} = 0.0068$$

$$P(A_1) = \frac{2}{10,000} = 0.0002$$

$$P(A_2) = 1 - P(A_1) = 1 - 0.0002 = 0.9998$$

What we are trying to find $P(A_1|B)$. But

$$P(A_1|B) = \frac{P(B|A_1)P(A_1)}{P(B|A_1)P(A_1) + P(B|A_2)P(A_2)}$$

$$= \frac{(0.95)(0.0002)}{(0.95)(0.0002) + (0.0068)(0.9998)} = 0.027$$

(Notice the magnitude of $P(A_1|B)$. Is it numerically in line with what your intuition would have predicted?)

Random drug and alcohol testing is being used by more and more institutions to help curb drug addiction. Guidelines for these plans are simple: workers to be tested are chosen at random and without prior notice. Whether their on-the-job behavior suggests they have a drug problem is not considered.

Screening programs of this sort raise difficult medical and legal questions. But they also raise a serious statistical question. Using Bayes' rule, we can show that screening procedures will typically produce an astonishingly high percentage of false alarms. That is, the probability of someone actually having a drug problem, *given that the screening test says he does*, is very small. The next example typifies the numbers and probabilities associated with screening procedures.

*EXAMPLE 5.26*  A government task force is considering the feasibility of setting up a national screening program to detect child abuse. Consultants for the group estimate that

1. One child in 90 is abused.
2. A physician can detect an abused child 90% of the time.
3. A screening program would incorrectly label 3% of all nonabused children as abused.

What is the probability that a child is actually abused given that the screening program diagnoses him as such?

*Solution*  Let $B$ be the event "screening program diagnoses child as abused." Define $A_1$ and $A_2$ to be the events "child is abused" and "child is not abused," respectively. According to the consultants' estimates,

$$P(A_1) = \frac{1}{90} \qquad\qquad P(A_2) = \frac{89}{90}$$

$$P(B|A_1) = 0.90 \qquad\qquad P(B|A_2) = 0.03$$

Our objective is to find $P(A_1|B)$. At first glance, the screening test seems very precise: $P(B|A_1)$ is quite large, and $P(B|A_2)$ is quite small. We would probably suspect, intuitively, that $P(A_1|B)$ is also large, as it should be. But it isn't. From Theorem 5.5.2,

$$P(A_1|B) = \frac{(0.90)(\frac{1}{90})}{(0.90)(\frac{1}{90}) + (0.03)(\frac{89}{90})} = 0.25$$

That is, only 25% of the children identified as "victims" would actually have been abused. Or, viewed from the other perspective, 75% of the parents, relatives, and friends suspected of being child abusers would, in fact, be innocent.

QUESTIONS

**5.50** A dashboard warning light is supposed to flash red if a car's oil pressure is too low. On a certain model the probability of the light flashing when it should is 0.95; 2% of the time, though, it flashes for no apparent reason. If there is a 10% chance that the oil pressure really is low, what is the probability that a driver needs to be concerned if the warning light goes on? (Hint: Let $A_1$ be the event "oil pressure is low." Let $A_2$ be the event "oil pressure is not low." Let $B$ be the event "warning light flashes." Find $P(A_1|B)$.)

**5.51** From what she has heard about the two instructors who will teach a freshman statistics course, Jasmine estimates that her chances of passing the course are 0.85 if she gets Professor X, but only 0.60 if she gets Professor Y. The registrar determines the section into which she is put. Suppose her chances of being assigned to Professor X are 4 out of 10. Fifteen weeks later we learn that Jasmine did, indeed, pass the course. What is the probability that she was enrolled in Professor X's section? (Hint: Let $A_1$ be the event "Jasmine was in Professor X's section." Let $A_2$ be the event "Jasmine was in Professor Y's section." Let $B$ be the event "Jasmine passed course.")

**5.52** A liquor store owner is willing to cash personal checks for amounts up to $50, but she has become wary of customers who wear sunglasses. Of checks written by persons wearing sunglasses, 50% have bounced. In contrast, 98% of the checks written by persons not wearing sunglasses clear the bank. She estimates that 10% of her customers wear sunglasses. If the bank returns a check marked "insufficient funds," what is the probability it was written by someone wearing sunglasses? (Hint: Let $B$ be the event "check bounces.")

**5.53** A biased coin, twice as likely to come up heads as tails, is tossed once. If it shows heads, a chip is drawn from urn I, which contains three white chips and four red chips; if it shows tails, a chip is drawn from urn II, which contains six white chips and three red chips. Given that a white chip was drawn, what is the probability that the coin came up tails? (Hint: Let the $A_i$'s represent the possible urns being sampled.)

**5.54** During a power blackout 100 persons are arrested on suspicion of looting and are given polygraph tests. From past experience it is known that the polygraph is 90% reliable when administered to a guilty suspect and 98% reliable when given to an innocent person. Suppose that only 12 of the 100 persons taken into custody were actually involved in any wrongdoing. What is the probability that a suspect is innocent given that the polygraph says he is guilty?

**5.55** A family has two dogs (Rex and Rover) and a little boy (Russ). None of them is fond of the mailman. Given that they are outside, Rex and Rover have a 30% and a 40% chance, respectively, of biting the mailman. Russ, if he is outside, has a 15% chance of doing the same thing. Suppose that only one of the three is outside when the mailman comes.

(a) If Rex is outside 50% of the time, Rover 20% of the time, and Russ 30% of the time, what is the probability the mailman will be bitten?

(b) *If* the mailman is bitten, what are the chances that Russ did it?

5.56   Urn I has four chips, three red and one white; urn II also has four chips, two red and two white. A chip is drawn at random from urn I and transferred to urn II. Then a chip is drawn from urn II. Suppose the chip drawn from urn II is white. What is the probability that the chip transferred was red?

5.57   Brett and Margo have each thought about murdering their rich Uncle Basil in hopes of claiming their inheritance a bit early. Hoping to take advantage of Basil's predilection for immoderate desserts, Brett has put rat poison in the cherries flambe; Margo, unaware of Brett's activities, has laced the chocolate mousse with cyanide. Given the amounts likely to be eaten, the probability of the rat poison being fatal is 0.60; the cyanide, 0.90. Based on other dinners where Basil was presented with the same dessert options, we can assume that he has a 50% chance of asking for the cherries flambe, a 40% chance of ordering the chocolate mousse, and a 10% chance of skipping dessert altogether. No sooner are the dishes cleared away when Basil drops dead. In the absence of any other evidence, who should be considered the prime suspect?

5.58   (a) Recompute $P(A_1|B)$ in Example 5.26, assuming the prevalence of abused children is 1 in 900 instead of 1 in 90.

(b) What would you conclude in general about the relationship between the prevalence of the condition or disease being investigated and the magnitude of the associated misclassification probabilities?

5.59   A sample space S is partitioned by events $A_1$ and $A_2$. Let B be any other event defined on S such that $P(B) > 0$. If $P(A_1|B) = p$, must $P(A_2|B)$ equal $1 - p$? (Hint: Draw a Venn diagram showing $A_1$, $A_2$, and B.)

5.60   Consider the problem of screening for tuberculosis. Let T be the event that a person has TB. Let + be the event that the person's x ray suggests that he has TB. Assume that $P(T) = 0.0001$, $P(+|T) = 0.90$ (the test correctly identifies 90% of all the people who have the disease), and $P(+|T^c) = 0.001$ (the test gives one false positive, on the average, out of every 1000 subjects). Find $P(T|+)$, the probability a person actually has TB given that the x ray says he does.

# 5.6

## INDEPENDENCE

In many situations a knowledge that an event B has occurred gives *no* additional insight about the likelihood of an event A occurring. For these problems $P(A|B)$ and $P(A)$ are numerically the same. Recall Example 5.14: being told that a card is a king was of no help in predicting whether that card is a club; that is, $P(C|K) = P(C)$. In this section we want to examine the consequences of two events being independent and to look at the kinds of problems that make use of that property.

The definition of independence can be stated in several ways. From what we have just noted, events $A$ and $B$ are *independent* if $P(A|B) = P(A)$. Definition 5.3 gives a second characterization, one that will prove especially useful in solving problems.

---

DEFINITION 5.3    Independent Events

Two events $A$ and $B$ are **independent** if

$$P(A|B) = P(A)$$

or, equivalently, if

$$P(A \cap B) = P(A)P(B)$$

---

EXAMPLE 5.27    Spike is not a terribly bright student. His chances of passing chemistry and his chances of passing mathematics are 0.35 and 0.40, respectively. His probability of succeeding at *both* is 0.12. Are the events "Spike passes chemistry" and "Spike passes mathematics" independent?

Solution    Let $A$ be the event "Spike passes chemistry," and let $B$ be the event "Spike passes mathematics." We know that

$$P(A) = 0.35, \qquad P(B) = 0.40, \quad \text{and} \quad P(A \cap B) = 0.12$$

For $A$ and $B$ to be independent it must be true that

$$P(A \cap B) = P(A)P(B)$$

But $$0.12 \neq (0.35)(0.4) = 0.14$$

Therefore the events are not independent.

---

EXAMPLE 5.28    Let $A$ and $B$ denote any two events, each with probability greater than 0. If $A$ and $B$ are mutually exclusive, are they also independent?

Solution    From the sound of the words, we might be tempted to say "yes," but the correct answer is "no," and it follows from the first statement in Definition 5.3. If $A$ and $B$ are mutually exclusive, $P(A|B) = 0$. By assumption, $P(A) > 0$, so $P(A|B)$ cannot equal $P(A)$. It must necessarily be true, then, that $A$ and $B$ are *dependent*.

**5.61** Let $A$ and $B$ be two independent events such that $P(A) = 0.2$ and $P(B) = 0.4$. Compute $P(A \cup B)$.

**5.62** If $P(A) = 0.5$, $P(B) = 0.3$, and $P(A \cup B) = 0.7$, are $A$ and $B$ independent?

**5.63** Suppose two fair coins are tossed. Let $A$ be the event that the first coin shows heads. Let $B$ be the event that the second coin shows tails. Are $A$ and $B$ independent?

**5.64** Suppose event $B$ is a subset of event $A$ such that $P(A^c) > 0$. Are $A$ and $B$ independent? (Hint: Draw the corresponding Venn diagram.)

**5.65** Suppose two fair dice, one red and one white, are thrown. Let $A$ be the event that a 3, 4, or 5 shows on the red die. Let $B$ be the event that the dice sum is 4, 11, or 12. Are $A$ and $B$ independent?

**5.66** Events can be dependent to different extents. Suppose an urn contains $n$ red chips and $n$ white chips. Draw out one chip. Then, without replacing that first one, draw out a second chip. Let $A$ be the event that the first chip is red, and let $B$ be the event that the second chip is red. Compare $P(B|A)$ to $P(B)$ for $n = 1$ and $n = 1000$.

**5.67** A large company is responding to an affirmative-action commitment by setting hiring quotas by race and sex for office personnel. So far they have agreed to employ the 120 people indicated in the table below. How many black women must they include if they want to be able to claim that the race and sex of the people on their staff are independent?

|        | White | Black |
|--------|-------|-------|
| Male   | 50    | 30    |
| Female | 40    |       |

### Independence of More Than Two Events

Establishing that two events are independent (or not independent) is not the typical application to which Definition 5.3 is put. There are many situations where we can argue on *physical* grounds that $A$ and $B$ "must" be independent. In those cases we can use Definition 5.3 as a very useful formula for the probability of an intersection:

$$P(A \cap B) = P(A)P(B)$$

For example, suppose a fair coin is tossed twice, and we let $A$ be the event that a head appears on the first toss, and $B$ the event that a head appears on the second toss. Since the outcome of one flip has no influence on the outcome of another, $A$ and $B$ must be independent. Therefore, the probability that *both* occur is simply the product of the probabilities of each occurring:

$$P(A \cap B) = P(H, H) = P(A)P(B) = P(H)P(H) = \frac{1}{2} \cdot \frac{1}{2} = \frac{1}{4}$$

The details will be omitted, but the second statement in Definition 5.3 can be extended to more than two events. That is, if events $A_1, A_2, \ldots, A_k$ are all independent, then

$$P(A_1 \cap A_2 \cap \cdots \cap A_k) = P(A_1)P(A_2) \cdots P(A_k) \tag{5.5}$$

---

**EXAMPLE 5.29**    Diane and Lew are physicists employed by the Department of Public Health. One of their duties is to be on call during nonworking hours to handle any nuclear-related incidents that might endanger the public safety. Each carries a pager that can be activated by personnel at Civil Defense. Lew is a conscientious worker and is within earshot of his pager 80% of the time. Not nearly as reliable, Diane can respond to a pager alert only 40% of the time. If Diane and Lew report into Civil Defense independently, what is the probability that at least one of them could be contacted in the event of a nuclear emergency?

*Solution*    Let $D$ and $L$ denote the events "Diane responds to alert" and "Lew responds to alert," respectively. The response capability we are trying to find is the probability of the union $P(D \cup L)$. Recall that

$$P(D \cup L) = P(D) + P(L) - P(D \cap L)$$

We are given that $P(D) = 0.40$ and $P(L) = 0.80$. Also, since $D$ and $L$ are independent,

$$P(D \cap L) = P(D)P(L) = (0.40)(0.80) = 0.32$$

Therefore, there is an 88% chance that *someone* will respond to an alert:

$$P(D \cup L) = 0.40 + 0.80 - 0.32 = 0.88$$

---

**EXAMPLE 5.30**    An insurance company has three clients (one in Alaska, one in Missouri, and one in Vermont) whose estimated chances of living to the year 2000 are 0.7, 0.9, and 0.3, respectively. What is the probability that by the end of 1999 the company will have had to pay death benefits to exactly one of the three?

*Solution*    Let the events $A$, $M$, and $V$ denote the survival of a client:

$A$ = Alaska client survives through 1999

$M$ = Missouri client survives through 1999

$V$ = Vermont client survives through 1999

Let $E$ be the event "exactly one of the three dies before the year 2000." Notice that $E$ can occur in three different ways, each way being an intersection of three events:

$$E \text{ occurs if } \begin{cases} & A \cap M \cap V^c \\ \text{or} & A \cap M^c \cap V \\ \text{or} & A^c \cap M \cap V \end{cases}$$

More formally, we would write

$$E = (A \cap M \cap V^c) \cup (A \cap M^c \cap V) \cup (A^c \cap M \cap V)$$

Since the three sets of intersections in $E$ are mutually exclusive,

$$P(E) = P(A \cap M \cap V^c) + P(A \cap M^c \cap V) + P(A^c \cap M \cap V) \quad (5.6)$$

Furthermore, the geographical distances between the three clients suggest that for all practical purposes their fates are independent. Each intersection probability in Equation 5.6, then, can be replaced by a product. Substituting the values given for $P(A)$, $P(M)$, and $P(V)$, we find that the probability is 0.543 that the company will be paying death benefits to exactly one of the three by the year 2000:

$$P(E) = P(A)P(M)P(V^c) + P(A)P(M^c)P(V) + P(A^c)P(M)P(V)$$

$$= (0.70)(0.90)(1 - 0.30) + (0.70)(1 - 0.90)(0.30)$$

$$+ (1 - 0.70)(0.90)(0.30)$$

$$= 0.543$$

---

**EXAMPLE 5.31**    On her way to work, a commuter encounters four traffic signals. We will assume that the distance between each of the four is sufficiently great that her probability of getting a green light at any intersection is independent of what happened at any previous intersection. If each light is green for 40 seconds of every minute, what is the probability that she has to stop at least three times?

*Solution*    Think of each trip to work as an ordered sequence of four traffic lights. Each light can be red ($R$) or green ($G$) with probabilities

$$P(R) = \frac{20}{60} = \frac{1}{3} \quad \text{and} \quad P(G) = \frac{40}{60} = \frac{2}{3}$$

Let $E$ be the event that the driver encounters at least three red lights. How many ways can event $E$ happen? *Five*, as shown below:

$$(R \ R \ R \ R)$$
$$(R \ R \ R \ G)$$
$$(R \ R \ G \ R)$$
$$(R \ G \ R \ R)$$
$$(G \ R \ R \ R)$$

Recall the analysis in Example 5.30. Since the five ordered sequences in $E$ are all mutually exclusive and each is an intersection of independent events, we can write

$$P(E) = P(R)P(R)P(R)P(R) + P(R)P(R)P(R)P(G) + P(R)P(R)P(G)P(R)$$

$$+ P(R)P(G)P(R)P(R) + P(G)P(R)P(R)P(R)$$

$$= \left(\frac{1}{3}\right)^4 + 4\left(\frac{1}{3}\right)^3\left(\frac{2}{3}\right)^1$$

$$= \frac{1}{9}$$

---

*EXAMPLE 5.32 (Optional)*   During the 1978 baseball season, Pete Rose of the Cincinnati Reds set a National League record by hitting safely in 44 consecutive games. Assume that Rose is a .300 hitter who comes to bat four times each game. Consider each at-bat to be an independent event. What probability might reasonably be associated with Rose hitting safely in a given set of 44 consecutive games?

*Solution*   For this problem we need to invoke Equation 5.5 *twice*—once for the four at-bats making up a game, and once for the 44 games making up the streak. Let $A_i$ denote the event "Rose hits safely in $i$th game," $i = 1, 2, \ldots, 44$. Then

$P$ (Rose hits safely in 44 consecutive games)

$$= P(A_1 \cap A_2 \cap \cdots \cap A_{44}) = P(A_1)P(A_2) \cdots P(A_{44}) = (P(A_1))^{44}$$

the last equality holding because each $A_i$ has the same probability.

The last piece in the puzzle—evaluating $P(A_1)$—is best accomplished by considering the complement of $A_1$:

$$P(A_1) = 1 - P(A_1^c)$$

$$= 1 - P \text{ (Rose does } not \text{ hit safely in game 1)}$$

$$= 1 - P \text{ (Rose makes four outs)}$$

$$= 1 - (0.700)^4 \quad \text{(why?)}$$

$$= 0.76$$

The probability, then, of a .300 hitter putting together a 44-game streak is

$$(0.76)^{44} = 0.0000057$$

---

*EXAMPLE 5.33 (Optional)*   Scientists have calculated that the number of molecules in the earth's atmosphere is on the order of $10^{44}$. Furthermore, that number is fairly stable; the $10^{44}$ that are floating around now are pretty much the same $10^{44}$ that were here thousands of years ago (21). Experiments have also shown that a typical breath consists of approximately $10^{24}$ molecules. Those two facts raise an interesting bit of speculation. When Julius Caesar cried out "Et tu, Brute" in his dying breath, he presumably exhaled $10^{24}$ molecules. More than 2000 years have elapsed since the occurrence of that event—plenty of time for those $10^{24}$ molecules to have been completely "mixed" throughout the atmosphere. The question is this: What are the chances that in your very *next* breath there will be at least one molecule from Caesar's *last* breath?

*Solution*    Intuitively, the answer seems obvious: zero. The prospects of inhaling a Caesar molecule would seem to be even more dismal than finding the proverbial needle in a haystack. As we have seen before, though, intuitions can be fooled.

Mathematically, the answer to our question reduces to an application of Equation 5.5 where $k = 10^{24}$. Think of the earth's atmosphere as a giant urn containing $10^{44}$ molecules. Included in that number are the $10^{24}$ molecules that made up Caesar's last breath. Your *next* breath is simply a sample of size $10^{24}$ chosen from the original $10^{44}$ (see Figure 5.24).

Let $A_i$, $i = 1, 2, \ldots, 10^{24}$, denote the event that the $i$th molecule in your next breath is *not* a Caesar molecule. Then

$P$ (at least one Caesar molecule is in your next breath)

$= 1 - P$ (*no* Caesar molecules are in your next breath)

$= 1 - P(A_1 \cap A_2 \cap \cdots \cap A_{10^{24}})$

To get a numerical answer for the probability of the intersection, we need to make a little approximation. From Figure 5.24 it seems clear that

$$P(A_1) = \frac{10^{44} - 10^{24}}{10^{44}}$$

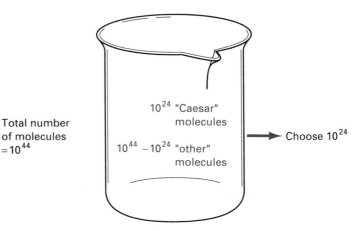

Total number
of molecules
$= 10^{44}$

$10^{24}$ "Caesar"
molecules

$10^{44} - 10^{24}$ "other"
molecules

Choose $10^{24}$

**FIG. 5.24**

Furthermore, if $A_1$ does occur, then

$$P(A_2) = \frac{10^{44} - 10^{24} - 1}{10^{44} - 1}$$

which is different from $P(A_1)$. The numbers in the numerator and denominator of $P(A_2)$, though, are so large that subtracting 1 from both has virtually no effect on the ratio. That is,

$$\frac{10^{44} - 10^{24}}{10^{44}} \doteq \frac{10^{44} - 10^{24} - 1}{10^{44} - 1}$$

By the same argument *all* the $P(A_i)$'s have essentially the same probability, in which case

$$P(A_1 \cap A_2 \cap \cdots \cap A_{10^{24}}) \doteq (P(A_1))^{10^{24}}$$

Therefore,

$P$(at least one Caesar molecule is in your next breath)

$$= 1 - \left(\frac{10^{44} - 10^{24}}{10^{44}}\right)^{10^{24}} = 1 - (1 - 10^{-20})^{10^{24}} > 0.9999999999999 \qquad (5.7)$$

Inequality 5.7 is a shocker, to say the least. It implies that your next breath is almost certain to contain at least one molecule from Caesar's last breath!

---

**QUESTIONS**

5.68 A fair coin is tossed eight times. Which sequence of outcomes is more likely,

$$(H\,H\,H\,H\,H\,H\,H\,H) \quad \text{or} \quad (H\,T\,T\,H\,T\,T\,T\,H)?$$

5.69 Suppose the probability that a coin comes up heads is 1/3. What is the probability that the first head occurs on the fourth toss?

5.70 Four spiders are living in four student post office boxes. One box belongs to a freshman, one to a sophomore, one to a junior, and one to a senior. If undisturbed by any incoming mail, each spider will spin a web in its box each day. Suppose the probabilities of getting mail on any given day are 0.7 for the freshman, 0.5 for the sophomore, 0.4 for the junior and 0.3 for the senior. Assume the mail delivered to each student is independent of what is delivered to any other student. What is the probability that on any given day exactly three of the spiders will have a chance to spin a web?

5.71 Urn I has three red, two black, and five white chips; urn II has two red, four black, and three white chips. One chip is drawn at random from each urn. What is the probability that both chips are the same color? (Hint: Write the event "both chips are the same color" as a union of three mutually exclusive intersections.)

**5.72**   Trip, Brandon, and Dante enter a round-robin tennis tournament where each of them plays the other two exactly once. If

$$P \text{ (Trip beats Brandon)} = 0.6$$

$$P \text{ (Trip beats Dante)} = 0.8$$

$$P \text{ (Brandon beats Dante)} = 0.9$$

who is more likely to win the tournament, Trip or Brandon?

**5.73**   Two fair dice are rolled. What is the probability that the number showing on one of the two dice will be twice the number showing on the other (Hint: Let $A$ be the event "number on one die is twice the number on the other die." Which outcomes are in $A$?)

**5.74**   In a certain third world nation, statistics show that only 2000 out of 10,000 children born in the early 1960s reached the age of 21. Assume the same mortality rate holds for the next generation. If a family has six children, what is the probability that at least one survives to adulthood? Assume that the survival of each child is an independent event. (Hint: Consider the complement of "at least one.")

**5.75**   In general, the union of three events $A$, $B$, and $C$ is given by the formula

$$P(A \cup B \cup C) = P(A) + P(B) + P(C) - P(A \cap B)$$
$$- P(A \cap C) - P(B \cap C) + P(A \cap B \cap C)$$

Look again at Example 5.29. Suppose a third employee, Mike, is added to the team. If Mike's probability of answering a pager alert is 0.60, how much does his presence improve the group's response capability?

**5.76**   A 50-lb load ($L$) is hung on the support structure pictured below. Compression studies indicate that a load of that magnitude will cause arms $a$, $b$, and $c$ to crack with probabilities 0.05, 0.04, and 0.03, respectively. What is the probability that the support will hold? Assume that (1) the arms fail independently and that (2) if at least one arm cracks the entire structure will collapse. (Hint: What must happen if the support is to hold?)

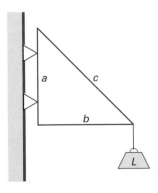

**5.77**   Suppose the distribution of blood types in the general population for both men and women can be characterized by the following figures. What is the probability that two people getting married will have different blood types? What assumption are you making? Is it reasonable?

| Blood Type | Probability |
|------------|-------------|
| A | 0.40 |
| B | 0.10 |
| AB | 0.05 |
| O | 0.45 |

**5.78** A string of eight Christmas tree lights is wired in series. (Thus, if at least one bulb fails, the entire string will not light.) If the probability of any bulb failing sometime during the holiday season is 0.05 and if the failures are independent events, what is the probability that the string will not remain lit? (Hint: Let $A_i$ denote the event that the $i$th bulb remains lit, $i = 1, 2, \ldots, 8$. The string remains lit only if $A_1 \cap A_2 \cap \cdots \cap A_8$ occurs.)

**5.79** Two myopic deer hunters fire rifles simultaneously and independently at a nearby rooster. The probability of hunter A's shot killing the rooster is 0.2; hunter B's probability is 0.3. Suppose the rooster is hit and killed by only one bullet. What is the probability that hunter B fired the fatal shot?

# 5.7

## SUMMARY

The key to understanding Chapter 5 is to realize that probability is a *set function*. That is, it associates each set (or event) in a sample space S with a number between 0 and 1, where the number represents the likelihood of that event occurring. For example, let S be the sample space defined for the experiment of tossing two fair coins, and let A be the event "at least one head appears." Each of the four outcomes in S is equally likely, and three of them belong to A (see Figure 5.25). The probability function P associates A with the number 3/4, and we write $P(A) = 3/4$.

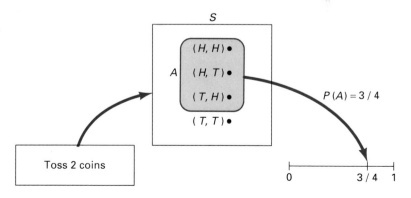

**FIG. 5.25**

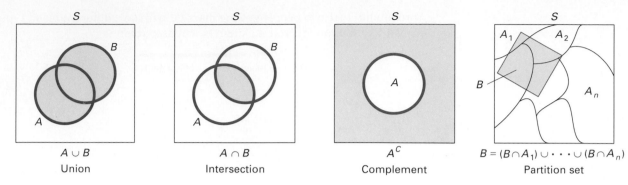

| $A \cup B$ | $A \cap B$ | $A^C$ | $B = (B \cap A_1) \cup \cdots \cup (B \cap A_n)$ |
| Union | Intersection | Complement | Partition set |

**FIG. 5.26**

Given that we intended to make extensive use of the general structure in Figure 5.25, we began the chapter by examining some of the algebra of sets. The basic set definitions are *union*, *intersection*, and *complement*. These rules govern how sets are combined and manipulated to generate other sets. Knowing these operations is important because the events in which we are usually interested are not "simple" sets (like $A$) but a composite formed of unions, intersections, or complements. Venn diagrams are often used to clarify the relationships that exist among events (see Figure 5.26).

With the definitions of Section 5.2 serving as a backdrop, we deduced a number of properties that any probability function $P$ must satisfy (Table 5.5).

*Conditional probability*, a probability revised to take into account additional information, appeared in Section 5.4 and figured prominently in the rest of the chapter. If $A$ and $B$ are any two events defined on a sample space $S$, and if $P(B) > 0$, then the conditional probability of $A$ given $B$, denoted $P(A|B)$, is $P(A \cap B)/P(B)$. Rewritten, the definition of conditional probability also leads to a useful formula for the probability of an intersection: since $P(A|B) = P(A \cap B)/P(B)$, then

$$P(A \cap B) = P(A|B)P(B)$$

In Section 5.5, we applied the notion of conditional probability to sample spaces partitioned by a set of events $A_1, A_2, \ldots, A_k$ and derived two important theorems, one for $P(B)$ and one for $P(A_j|B)$. Table 5.6 summarizes the results that involve conditional probability.

Sometimes the occurrence of an event $B$ has no impact on the likelihood of an event $A$. That is, $P(A|B) = P(A)$. Such events are said to be *independent*. An obvious example of independent events occurs in the familiar context of tossing coins. If a coin is flipped twice and we let $A$ be the event that a head appears on the first toss and $B$, that a head appears on the second toss, then $A$ and $B$ are clearly independent. To argue otherwise is to believe that the coin, when tossed the second time, "remembered" what happened the first time. Problems involving independence are taken up in Section 5.6. Many require that the probability of an intersection of $k$

**TABLE 5.5  Properties of Probability Functions**

1. For any event $A$, $0 \le P(A) \le 1$.
2. $P(S) = 1$; $P(\phi) = 0$.
3. If events $A$ and $B$ are mutually exclusive, then

$$P(A \cup B) = P(A) + P(B)$$

4. $P(A^c) = 1 - P(A)$.
5. For any two events $A$ and $B$ defined on $S$,

$$P(A \cup B) = P(A) + P(B) - P(A \cap B)$$

**TABLE 5.6**

| Formula | Equation | |
|---|---|---|
| $P(A\mid B) = \dfrac{P(A \cap B)}{P(B)}$ | 5.1 | |
| $P(A \cap B) = P(A\mid B)P(B)$ | 5.2 | |
| $P(B) = \sum\limits_{i=1}^{k} P(B\mid A_i)P(A_i)$ | 5.3 | |
| $P(A_j\mid B) = \dfrac{P(B\mid A_j)P(A_j)}{\sum\limits_{i=1}^{k} P(B\mid A_i)P(A_i)}$ | 5.4 | (Bayes' rule) |

events be computed. If the $k$ $A_i$'s are all independent, then

$$P(A_1 \cap A_2 \cap \cdots \cap A_k) = P(A_1)P(A_2) \cdots P(A_k).$$

Few results in probability are more useful.

**REVIEW QUESTIONS**

**5.80**  Suppose you have a 40% probability of getting a date for Friday night, a 60% probability of getting a date for Saturday night, and a 20% probability of getting a date for Sunday night. What is the probability that you get at least one date over the weekend? Assume that the dates are independent events.

**5.81**  Let $A$ be the set of integers from 1 through 12. Let $B$ be the set of integers in that range that are divisible by 3. Let $C$ be the set of integers in that range that are divisible by 2. What numbers are in the set $(A \cap B \cap C)^c$?

**5.82**  A weather forecaster says there is a 40% probability that a hurricane will hit the Florida coast next week. If it does, there is a 10% probability that the damage will be severe. What is the probability that the storm hits *and* the damage is severe?

**5.83**  Let $A$ be the set of letters in $T R O U B L E$ that are in the first half of the alphabet. Let $B$ be the set of vowels in $T R O U B L E$. Suppose a letter is chosen from the word at random. Compute $P(A\mid B)$.

**5.84**  Jasmine's parents have told her that they *may* buy her a new BMW for Christmas. Whether they do depends to some extent on the grades she earns the Fall semester. If Jasmine's GPA is 3.5 or better, the probability is 0.8 that her parents will buy the car; if her GPA is between 2.5 and 3.49, her probability of getting the car is 0.6; if her GPA is less than 2.5, her chances of getting the car drop to 0.10. Based on her grades going into finals, Jasmine estimates that her likelihood of getting a GPA in the 3.5+ range is 0.05; in the 2.5–3.49 range, 0.75; and in the "less than 2.5" range, 0.20. What is the probability that Jasmine gets the BMW?

**5.85**  Two events $A$ and $B$ are defined on a sample space such that

$$P(A) = 0.3, \quad P(B) = 0.2, \quad \text{and} \quad P(A \cap B) = 0.1$$

What is the probability that exactly one of the two events occurs?

**5.86**  One hundred voters were asked how they felt about the two candidates, $A$ and $B$, running for mayor. Sixty said they liked $A$, fifty-five said they liked $B$, and twenty-five said they liked both.

(a) What is the probability that someone likes neither?

(b) What is the probability that someone likes only one of the two?

(c) What is the probability that someone doesn't like $A$?

(d) What is the probability that someone likes $A$ given that he likes $B$?

(e) Are the events "voter likes $A$" and "voter likes $B$" independent? Are they mutually exclusive?

**5.87**  Dana and Cathie are roommates. Of the telephone calls they receive, 40% are for Dana and 60% are for Cathie. If their phone rings on four different occasions tomorrow night, what is the probability that the fourth call is Dana's third? Assume the calls are independent events.

**5.88**  A fast-food chain is running a new promotion. For each purchase, a customer is given a game card that may win $10. The company claims that the probability of a person winning at least once in five tries is 0.32. What is the probability that a customer wins $10 on his first purchase?

**5.89**  Urn I contains 7 chips, 2 white, 2 red, and 3 black; urn II contains 9 chips, 3 white, 2 red, and 4 black. One chip is drawn from urn I and transferred to urn II. Then one chip is drawn from urn II. What is the probability that the chip drawn from urn II is not white?

**5.90**  Assume that events $A_1$ and $A_2$ partition a sample space S, such that

$$P(A_1) = p_1 \quad \text{and} \quad P(A_2) = p_2.$$

If $2p_1 - p_2 = \frac{1}{4}$, find $p_2$.

**5.91**  Recall Example 5.20. Suppose the next program submitted to the computer center compiles on its first run. What is the probability that the program was written in COBOL?

**5.92**  Two hundred adults were surveyed as part of a study on child abuse. Thirty said they were abused as children. Among those thirty, 25 admitted to being abusive parents. Of the 170 respondents who were not abused as children, 145 were not, themselves, abusive parents. Are the events "child is abused" and "parent is abusive" independent? (Hint: Summarize the data in a table having two rows and two columns.)

**5.93**  Let $A_i$ be the set of numbers from 0 to $1/i$, for $i = 1, 2, \ldots, k$. Which numbers are in the sets

$$A_1 \cup A_2 \cup \cdots \cup A_k \quad \text{and} \quad A_1 \cap A_2 \cap \cdots \cap A_k$$

**5.94**  Suppose a baseball player has a 0.70 probability of getting at least one hit in any particular game. If five games are played, what is the probability that the player hits safely in at least three consecutive games? What does your analysis imply about the 44-game-streak probability calculated in Example 5.32?

**5.95** A box contains one two-headed coin and eight fair coins. One coin is selected at random and tossed 8 times. Suppose all the tosses come up heads. What is the probability that the coin being tossed has two heads? (Hint: Let $B$ be the event that all 8 tosses come up heads; let $A_1$ be the event that the coin tossed has two heads; and let $A_2$ be the event that the coin tossed is fair.)

**5.96** Fifteen criminals are being transported to a maximum security institution when their bus breaks down and one of them escapes. Six of the fifteen are homicidal, five are maniacs, and six are neither. Assuming they are all equally devious, what is the probability that the one who escaped is a homicidal maniac?

**5.97** Suppose a dart lands at random inside the $2' \times 2'$ target shown below. Intuitively, what is the probability that the dart lands inside the shaded square?

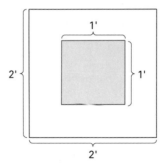

**5.98** Records are kept daily on the number of nuclear reactor workers whose on-the-job radiation exposure is high enough to trigger a safety alarm. The table below is a frequency distribution of the number of such incidents reported during the last 200 days.

| Number of Incidents/Day, $x$ | Number of Days |
|---|---|
| 0 | 130 |
| 1 | 50 |
| 2 | 17 |
| 3 | 1 |
| 4 | 2 |
| | 200 |

(a) What is the probability that a single incident will occur tomorrow?

(b) If $A$ is the event that $x \geqslant 2$ and $B$ is the event that $1 \leqslant x \leqslant 3$, what is $P(A|B)$?

5.99    Let $A$, $B$, and $C$ be the events that Bubba fails art, Bubba fails biology, and Bubba fails computer science, respectively. Assume that

$$P(A) = 0.6, \quad P(B) = 0.3, \quad P(C) = 0.8$$

and that events $A$, $B$, and $C$ are independent. Find the probability that Bubba fails at least one course. Note: For any three events,

$$P(A \cup B \cup C) = P(A) + P(B) + P(C) - P(A \cap B) - P(A \cap C)$$
$$- P(B \cap C) + P(A \cap B \cap C)$$

5.100    Let $A$, $B$, and $C$ be any three events defined on a sample space S. In general, is it true that $(A \cup B \cup C)^c = (A \cap B \cap C)^c$? Answer the question by drawing Venn diagrams.

5.101    Two fair dice are rolled. What is the probability that at least one shows a 5 given that their sum is even? (Hint: Let $A$ be the event that at least one shows a 5 and let $B$ be the event that their sum is even. Which outcomes are in $A \cap B$?)

5.102    Five helicopters are sent to look for a small schooner that may have run aground. Based on past performance, each helicopter crew has a 30% chance of locating a lost boat. If the crews search independently, what is the probability the schooner is found?

5.103    Shown below is the summary of a city's weather last year. Assume that

| Conditions | Number of Days |
|---|---|
| Sunny | 260 |
| Cloudy | 60 |
| Rainy | 45 |
| | 365 |

the same overall pattern will prevail this year. What, if anything, is wrong with the following computation:

$P$ (next Monday is rainy $\cap$ next Tuesday is sunny)

$= P$ (next Monday is rainy) $\cdot P$ (next Tuesday is sunny)

$$= \frac{45}{365} \cdot \frac{260}{365} = 0.088$$

5.104    Foreign policy experts estimate that the probability is 0.65 that war will break out next year between two Middle East countries if either side significantly escalates its terrorist activities; otherwise, the likelihood of war is estimated to be 0.05. Based on what has happened this year, the chances of terrorism reaching a critical level in the next twelve months are thought to be 3 in 10. What is the probability that the two countries will not go to war?

# 6 Random Variables

*Some people hate the very name of statistics,
but I find them full of beauty and interest.
Whenever they are not brutalized, but
delicately handled by the higher methods, and are
warily interpreted, their power of dealing
with complicated phenomena is extraordinary.
They are the only tools by which an opening
can be cut through the formidable thicket of
difficulties that bars the path of those who
pursue the Science of man.*

**Sir Francis Galton**

## 6.1

### INTRODUCTION

Chapter 5 introduced the idea that probability should be viewed as a function $P$ that assigns a number (between 0 and 1) to each possible event defined on a sample space S. Simple examples of such functions should by now be comfortably familiar. Let $A$ be the event that the outcome of the toss of a fair die is 4 or 5. Intuitively, $A$ will occur 2/6 of the time, and we write $P(A) = 2/6$ (see Figure 6.1).

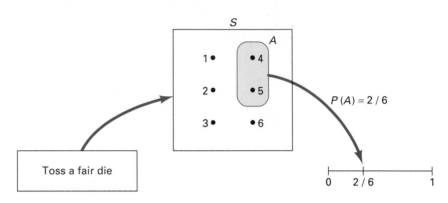

**FIG. 6.1**

In general, the *nature* of the function $P$ was not a prime concern in Chapter 5. The value for $P(A)$ was either *given* (recall, for instance, Examples 5.9, 5.10, and 5.11), or else the possible outcomes were all equally likely, in which case we found $P(A)$ by simply computing the ratio $m/n$ (Examples 5.7, 5.8, and 5.12). For many real-world applications, though, finding $P(A)$ is not always so easy.

This chapter examines in more detail the way in which probabilities are assigned to sample outcomes. We will also look closely at the sample space itself. This material is important. What we learn here will provide the conceptual background and notation for much of the remainder of the book.

## 6.2

### RANDOM VARIABLES

#### A Probability Function

The function $P$ is applied to *events* belonging to a sample space S (Figure 6.1). It will be helpful to define a second probability function $p$ on individual *outcomes* in S. To simplify matters, we will initially assume that the number of outcomes in S is finite.

---

> ### DEFINITION 6.1
>
> Let $p$ be a function that assigns a number to every element $s$ in a finite sample space S. The function $p$ is called a **probability function** if
>
> 1. $0 \leqslant p(s) \leqslant 1$    for each $s$ in S.
> 2. $\sum\limits_{\text{all } s \text{ in } S} p(s) = 1.$

---

**EXAMPLE 6.1**    A chip is drawn at random from an urn containing five chips numbered 1 through 5. What are the values of the corresponding probability function, and for which outcomes is it defined?

*Solution*    The sample space has five outcomes: $S = \{1, 2, 3, 4, 5\}$. Since each outcome has the same chance of being drawn, the values assigned by the probability function $p$ are

$$p(1) = \frac{1}{5}, \qquad p(2) = \frac{1}{5}, \quad \ldots, \quad p(5) = \frac{1}{5}$$

We are simply applying new notation to the equally likely model that appeared so many times in Chapters 4 and 5.

The relationship between the $p$ function of Definition 6.1 and the $P$ function in Chapter 5 is straightforward. If $A$, for example, is the event that the number drawn is odd, then the probability of $A$ is simply the sum of the probabilities of the individual outcomes that $A$ comprises:

$$P(A) = P \text{ (number is odd)} = P(1, 3, \text{ or } 5)$$

$$= p(1) + p(3) + p(5) = \frac{1}{5} + \frac{1}{5} + \frac{1}{5} = \frac{3}{5}$$

---

**EXAMPLE 6.2**    Consider a dart board divided into six regions of equal area. Three of the regions are red ($r$), two are white ($w$), and one is blue ($b$). A dart is thrown at random at the board and the color of the region in which it lands is recorded. How should $p$ be defined?

*Solution*    First, consider S. The sample space has three outcomes, $S = \{r, w, b\}$, but the three are not equally likely. The geometry of the dart board indicates the way in which probabilities should be assigned to $r$, $w$, and $b$:

$$p(r) = \frac{3}{6}, \qquad p(w) = \frac{2}{6}, \qquad p(b) = \frac{1}{6}$$

Suppose $A$ is the event that the dart does not land in a white region. Then $A$ occurs if the outcome is $r$ or $b$. The *probability* of $A$, $P(A)$, is therefore the sum of the $p$ values assigned to $r$ and $b$:

$$P(A) = P(r, b) = p(r) + p(b)$$

$$= \frac{3}{6} + \frac{1}{6} = \frac{2}{3}$$

---

**DEFINITION 6.2**

Let $p$ be a probability function satisfying the conditions of Definition 6.1. Let $A$ be any event defined on S. If $P(A)$ denotes the probability that event $A$ occurs, then

$$P(A) = \sum_{\text{all } s \text{ in } A} p(s)$$

---

**EXAMPLE 6.3**

Dishonest gamblers load a die by placing a small plug of lead behind each spot. Suppose the weights are such that the probability of a face coming up is proportional to its number of spots. What values should the probability function $p$ assign to each possible outcome?

**Solution**

If probabilities are to be proportional to outcomes, there must be a constant $r$ such that

$$p(1) = r \qquad p(4) = 4r$$
$$p(2) = 2r \qquad p(5) = 5r$$
$$p(3) = 3r \qquad p(6) = 6r$$

How do we find $r$? We use Property 2 of Definition 6.1. Since the total probability associated with any sample space is 1, it follows that

$$\sum_{\text{all } s} p(s) = r + 2r + 3r + 4r + 5r + 6r = 21r = 1$$

Thus, $r = \frac{1}{21}$, which makes

$$p(1) = \frac{1}{21}, \quad p(2) = \frac{2}{21}, \quad p(3) = \frac{3}{21}, \quad p(4) = \frac{4}{21}, \quad p(5) = \frac{5}{21}, \quad \text{and} \quad p(6) = \frac{6}{21}.$$

---

**EXAMPLE 6.4**

Making and breaking codes has challenged our ingenuity for thousands of years. Understandably, solving codes requires many mathematical tricks and techniques. Among the most prominent tools of the cryptographer's trade, though, is the probability function $p(s)$.

*Solution* In general, to encipher a message we need to replace each letter in the original text by another letter, according to some prescribed plan. If the substitution scheme, for example, called for the new letters to be shifted one place to the left of the old (called a Caesar cipher), then the word DOG would be encoded CNF. Breaking this type of cipher is not difficult. In the English language, *e* is by far the most common letter. If a message has been encoded by simply shifting each letter a certain number of places to the right or to the left, then the "new" letter that appears most often should probably be decoded as *e*. Once the *e* is found, all the other letters can easily be recovered.

Modern codes are much more sophisticated than a Caesar cipher, but their analysis is still often based on letter probabilities. Empirical estimates for the latter are credited to Dewey (37), who made a tally of 438,023 letters of text and found the relative frequencies listed in Table 6.1.

**TABLE 6.1**

| Letter | Relative Frequency | Letter | Relative Frequency |
|--------|--------------------|--------|--------------------|
| E | 0.1268 | F | 0.0256 |
| T | 0.0978 | M | 0.0244 |
| A | 0.0788 | W | 0.0214 |
| O | 0.0776 | Y | 0.0202 |
| I | 0.0707 | G | 0.0187 |
| N | 0.0706 | P | 0.0186 |
| S | 0.0634 | B | 0.0156 |
| R | 0.0594 | V | 0.0102 |
| H | 0.0573 | K | 0.0060 |
| L | 0.0394 | X | 0.0016 |
| D | 0.0389 | J | 0.0010 |
| U | 0.0280 | Q | 0.0009 |
| C | 0.0268 | Z | 0.0006 |

The 26 numbers shown in columns 2 and 4 define the probability function $p$ associated with the experiment of choosing a letter at random from a line of English text. The probability $P$ of an event, such as "vowel is chosen," would be

$$P \text{ (vowel is chosen)} = p(A) + p(E) + p(I) + p(O) + p(U) + p(Y)$$

$$= 0.0788 + 0.1268 + 0.0709 + 0.0776$$

$$+ 0.0280 + 0.0202$$

$$= 0.4021$$

Specific cryptology applications of the letter frequency probability function given in Table 6.1 are too far afield for us to consider here. An excellent reference is *The Codebreakers: The Story of Secret Writing* by David Kahn (85).

**6.1** For the sample space $S = \{0, 1, 2, 3, 4, 5\}$ define a function $p$ on $S$ by the table

| $s$ | 0 | 1 | 2 | 3 | 4 | 5 |
|---|---|---|---|---|---|---|
| $p(s)$ | 0.03 | 0.16 | 0.31 | 0.31 | 0.16 | 0.03 |

Use Definition 6.1 to verify that $p$ is a probability function. If $A$ is the event that the sample point is odd, find $P(A)$.

**6.2** Let the sample space $S = \{e, t, a, o, i, n, s\}$. The following table defines a function $p$:

| $s$ | $e$ | $t$ | $a$ | $o$ | $i$ | $n$ | $s$ |
|---|---|---|---|---|---|---|---|
| $p(s)$ | 0.22 | 0.17 | 0.13 | 0.13 | 0.12 | 0.12 | 0.11 |

Use Definition 6.1 to verify that $p$ is a probability function. Let $A$ be the event that the sample point is a consonant. Find $P(A)$.

**6.3** For the sample space $S = \{0, 1, 2, 3, 4\}$ consider the following assignment of numbers to outcomes in $S$:

| $s$ | 0 | 1 | 2 | 3 | 4 |
|---|---|---|---|---|---|
| $p(s)$ | 0.06 | 0.25 | ? | 0.25 | 0.06 |

What will the missing value equal if $p$ is to be a probability function?

**6.4** Which of the following $p(s)$'s qualify as probability functions?

(a) $p(s) = \dfrac{(s - 1)}{10}$     for   $s = 2, 3, 4, 5$

(b) $p(s) = \dfrac{(s - 2)}{4}$     for   $s = 1, 2, 3, 4$

(c) $p(s) = \dfrac{(s + 1)}{10}$     for   $s = 1, 2, 3, 4$

**6.5** Verify that the function $P$ given in Definition 6.2 is a probability function as defined in Section 5.3.

**6.6** Consider a die loaded so that the probability of each odd face is twice that of each even face.

(a) Find the corresponding probability function.

(b) What is the probability of an odd face?

**6.7** Suppose a single letter is chosen at random from a line of English text (see Example 6.4). What is the probability that it will be a $K$, $L$, $M$, $N$, or $O$?

**6.8** Suppose an urn contains two chips numbered 1, three chips numbered 2, one chip numbered 3, and one chip numbered 4. A single chip is drawn at random. Let the sample space be $S = \{1, 2, 3, 4\}$.

(a) What is the appropriate probability function $p$?

(b) Let $A$ be the event that the chip drawn has an odd number. Find $P(A)$.

6.9 From the urn of Question 6.8, draw one chip, note its number, replace it, and draw a second chip (this procedure is called drawing a sample of size 2 *with replacement*).

(a) Find the sample space S and the corresponding probability function $p$.

(b) Let $A$ be the event that the sum of the chips drawn is an even number. Find $P(A)$.

### Defining the Sample Space

We need to look a little more closely at the notion of a sample space. For the examples described thus far, the outcomes and the probability structure of S have seemed clear-cut. But that simplicity is a bit misleading. Recall the dart game in Example 6.2. Suppose the six regions on the board were numbered 1 through 6 in addition to being red, white, or blue. Should we define S to be a set of numbers or a set of colors? The answer is that we define the outcomes in S according to whichever criterion best suits our objectives.

---

**EXAMPLE 6.5** Suppose two dice are rolled. How should we define the sample space?

*Solution* Perhaps the most obvious choice for S would be the set of all possible "pairs" of the form $(x, y)$, where $x$ is the face showing on the first die, and $y$ the face showing on the second die (recall Example 4.25). Are the 36 $(x, y)$'s in Figure 4.31 the sample space a gambler is likely to have in mind? No. For most dice games the *sum* of the two faces showing determines whether a player wins or loses. A $(3, 2)$, in other words, is equivalent to a $(2, 3)$, a $(4, 1)$, or a $(1, 4)$. A better choice for S, then, might be the set of 11 possible sums,

$$S = \{2, 3, 4, 5, 6, 7, 8, 9, 10, 11, 12\}$$

Conceptually, we have "collapsed" the original sample space of 36 outcomes to a more relevant sample space of 11 outcomes. Notice that the probability structure of the two sample spaces is different. If the two dice are fair, $p$ assigns the value $\frac{1}{36}$ to each outcome in the first sample space:

$$p((1, 1)) = p((1, 2)) = p((2, 1)) = \cdots = p((6, 6)) = \frac{1}{36}$$

Table 6.2 shows the $p$ function appropriate for the second sample space. (Why is $p(8)$, for example, equal to $\frac{5}{36}$?)

**TABLE 6.2**

| Sum, $s$ | $p(s)$ | Sum, $s$ | $p(s)$ |
|----------|--------|----------|--------|
| 2 | 1/36 | 8 | 5/36 |
| 3 | 2/36 | 9 | 4/36 |
| 4 | 3/36 | 10 | 3/36 |
| 5 | 4/36 | 11 | 2/36 |
| 6 | 5/36 | 12 | 1/36 |
| 7 | 6/36 | | |

*EXAMPLE 6.6*    Using the probability structure in one sample space, say $p((x, y)) = \frac{1}{36}$, to determine the $p$ function in another (Table 6.2) is the most important concept in this chapter. Examples are not hard to find. Suppose four coins are tossed and our interest is in the number of heads that occur. The most relevant sample space has five possible outcomes: $S = \{0, 1, 2, 3, 4\}$. We find the *probabilities* of those outcomes by considering the more detailed sample space of sequences of four tosses.

*Solution*    The first and third columns in Table 6.3 show the 16 possible sequences of length 4. The second and fourth columns show the number of heads in each of those sequences. One sequence, for example, gives rise to 0 heads, four give rise to 1 head, and so on. If the four coins were fair, each sequence has probability 1/16. Therefore, we determine the probability structure in $S$, by multiplying the number of times an entry appears in the second or fourth columns by 1/16. Table 6.4 summarizes those computations.

**TABLE 6.3**

| Sequence | Number of Heads | Sequence | Number of Heads |
|----------|-----------------|----------|-----------------|
| (T, T, T, T) | 0 | (T, H, H, T) | 2 |
| (H, T, T, T) | 1 | (T, H, T, H) | 2 |
| (T, H, T, T) | 1 | (T, T, H, H) | 2 |
| (T, T, H, T) | 1 | (H, H, H, T) | 3 |
| (T, T, T, H) | 1 | (H, H, T, H) | 3 |
| (H, H, T, T) | 2 | (H, T, H, H) | 3 |
| (H, T, H, T) | 2 | (T, H, H, H) | 3 |
| (H, T, T, H) | 2 | (H, H, H, H) | 4 |

**TABLE 6.4**

| Number of Heads, $s$ | $p(s)$ |
|----------------------|--------|
| 0 | 1/16 |
| 1 | 4/16 |
| 2 | 6/16 |
| 3 | 4/16 |
| 4 | 1/16 |

**QUESTIONS**

6.10    Suppose an urn contains five chips numbered 1 through 5. Two chips are drawn simultaneously.

(a) List the $_5C_2$ outcomes in the sample space of all possible unordered pairs of chips. What probability is associated with each outcome?

(b) Suppose the outcome $s$ that we record for a sample of two chips is the sum of the numbers drawn. Use your answer to part (a) to tabulate the probability function associated with the sample space of all possible sums.

6.11    Suppose the outcome $s$ recorded for each possible sample of the two chips in Question 6.10 is the *product* of the numbers drawn. Construct a table showing the values of the function $p(s)$.

### Classes of Random Variables

The motivating idea behind both the dice and the coin examples we just described was the desire to relabel the outcomes of the "original" sample space and to formulate a more relevant "new" sample space. A similar technique, but in a different context, runs through much of high school mathematics. Relabeling an outcome in a sample space is analogous in algebra to evaluating a function $f$ at some point $x$. If $x$, for instance, is the radius of a circle, but our interest is in the *area* of that circle, we would define the function $f$ to compute the latter. Specifically, we would let $f(x) = \pi x^2$.

In probability, functions that relabel sample spaces are denoted not by $f$ but by capital letters, usually $X$ or $Y$. Recall Example 6.5. Suppose the ordered pair $(2, 4)$ denotes the event of rolling a 2 on the first die and a 4 on the second. If $X$ is the function associating each original outcome with the sum of the two faces showing, we would write

$$X((2, 4)) = 6$$

Similarly, $X((1, 1)) = 2$, $X((4, 2)) = 6$, and so on.

---

DEFINITION 6.3

A function that assigns a numerical value to each outcome in a sample space is called a *random variable*. The latter will be denoted by capital letters, often $X$ or $Y$. Particular values of a random variable will be designated by the corresponding lowercase letters.

---

discrete random variable—A random variable whose set of possible values is either finite or countably infinite.

continuous random variable—A random variable that can take on any value in a certain interval—that is, its possible values are uncountably infinite.

There are two broad classes of random variables, with the distinction being based on the nature of the sample space. If S has a finite or countably infinite[1] number of outcomes, then any random variable defined on that sample space is said to be *discrete*. If the number of outcomes in S is *uncountably infinite*, then any random variable defined on that sample space is said to be *continuous*.

It will be helpful to have some formal notation for describing the probability function associated with the values of a random variable. As the two examples in this section emphasized, the probability structure of a relabeled sample space (that is, the probability structure associated with the values taken on by a random variable) is derived from the probability structure of the original sample space. Don't be intimidated by Definition 6.4. All it does is state in general terms exactly what we did in Examples 6.5 and 6.6.

---

[1] Countably infinite means that the $y$ values associated with the random variable can be paired off one-to-one with the set of positive integers.

---

### DEFINITION 6.4

Suppose $Y$ is a discrete random variable. Let $y$ be any value of $Y$; that is, $y = Y(s)$ for some $s$ in the sample space. Define the event "$Y = y$" to be the set of *all* outcomes in S whose $Y$ value is $y$:

$$\{Y = y\} = \{s \in S: Y(s) = y\}$$

Let $f_Y(y)$ denote the probability of the event $Y = y$. Then

$$f_Y(y) = P(Y = y) = P(\{s \in S: Y(s) = y\})$$

The function $f_Y(y)$ is called a ***probability density function*** (pdf). A pdf is also called a ***distribution***. (If $y$ is *not* a value taken on by $Y$, we write $f_Y(y) = 0$.)

---

**EXAMPLE 6.7**    Suppose two fair coins are tossed. Let the random variable $Y$ denote the number of heads that appear. Find $f_Y(y)$.

*Solution*    Figure 6.2 shows the four possible sequences of heads and tails. Clearly, the random variable $Y$ can only be 0, 1, or 2. According to Definition 6.4, the probability that $Y = 0$ is the probability of the set of outcomes in S whose $Y$ value is 0. By inspection, only one $s$ in S, namely, $(T, T)$, has a $Y$ value of 0. Therefore,

$$f_Y(0) = P(Y = 0) = P(\{s \in S: Y(s) = 0\}) = P((T, T)) = \frac{1}{4}$$

Notice that *two* outcomes in S, $(H, T)$ and $(T, H)$, have $Y$ values of 1. The pdf of $Y$ evaluated at 1, then, is the sum of the probabilities associated with those two sequences:

$$f_Y(1) = P(Y = 1) = P(\{s \in S: Y(s) = 1\})$$

$$= P(\{(H, T), (T, H)\}) = p(H, T) + p(T, H) = \frac{1}{4} + \frac{1}{4} = \frac{1}{2}$$

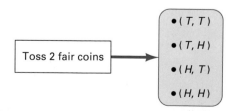

**FIG. 6.2**

**TABLE 6.5**

| y = Number of Heads | $f_Y(y)$ |
|:---:|:---:|
| 0 | 1/4 |
| 1 | 1/2 |
| 2 | 1/4 |

The third value for $Y$—like the first—occurs for only one outcome in S:

$$f_Y(2) = P(Y = 2) = P(\{s \in S: Y(s) = 2\}) = p(H, H) = \frac{1}{4}$$

Table 6.5 shows the entire "distribution" of the random variable $Y$. (If $y$ is any number other than 0, 1, or 2, then $f_Y(y) = 0$).

QUESTIONS

6.12    Suppose an urn contains three chips numbered 1, 2, and 3. Draw two with replacement. Let $Y$ be the larger number drawn. Find $f_Y(y)$. (Hint: First list the nine different ordered pairs of chips that can be drawn.)

6.13    Suppose an urn contains five chips, three red and two white. Draw two simultaneously. Let $Y$ denote the number of red chips drawn. Find $f_Y(y)$.

6.14    Three chips are drawn simultaneously from an urn containing 10 chips numbered 1 through 10. Let $Y$ be the largest number of the three drawn. Find $f_Y(y)$. (Hint: See Example 4.22)

6.15    Ace-six flats are a type of crooked dice where the cube is shortened in the one-six direction. Suppose an ace-six die has

$$p(1) = p(6) = \frac{1}{4} \quad \text{and} \quad p(2) = p(3) = p(4) = p(5) = \frac{1}{8}$$

Imagine two such dice being rolled, with $Y$ denoting the sum of the faces showing.

Find $f_Y(7)$.

Is it more likely that a 7 will be rolled with two ace-six flats or two fair dice?

6.16    Suppose a pair of dice are rolled that have been loaded in the way described in Example 6.3. Let $Y$ denote the sum of the faces showing. Find $f_Y(2), f_Y(3)$, and $f_Y(4)$.

6.17    Suppose five cards are drawn from a standard poker deck. Let $Y$ denote the number of aces among the five cards. Find a formula for $f_Y(y)$. (Hint: Think of the deck as an urn containing four objects of one kind (the aces) and 48 objects of a second kind (the non-aces). What must be true of a hand if its $Y$ value is, say, 3?)

# 6.3

## THE BINOMIAL DISTRIBUTION

The number of heads in a sequence of coin flips is the prototype of one of the most important random variables in statistics. We want to find a general formula here for the pdf of that random variable.

Imagine an experiment that has exactly two outcomes, such as heads or tails, yes or no, male or female, acceptable or unacceptable. In general, any two such outcomes are referred to as "success" and "failure." Let $p$ denote the probability that the outcome is a success. In our earlier coin-tossing problems, for example, $p$ is the probability of a coin coming up heads. (If the coin is assumed to be fair, $p = 1/2$.) Furthermore, if $P$ (success) $= p$, then $P$ (failure) $= 1 - p$. For convenience we occasionally replace $1 - p$ by $q$.

In practice, such an experiment is repeated $n$ times, with each performance being done under presumably identical conditions. We call each of those $n$ repetitions a *trial*. We also assume that the outcome of any trial has no effect on the outcome of any other trials. In other words, the trials are *independent*.

The structure we have just described is known formally as **$n$ repeated independent Bernoulli trials**. (The Bernoulli referred to is Jakob, one of a Swiss family that produced eight distinguished scientists. Jakob's book *Ars Conjectandi*, published in 1713, was the first major treatise on mathematical statistics.) The sample space S corresponding to a series of $n$ Bernoulli trials is the set of all possible ordered sequences of length $n$. Since each trial ends in one of two possible outcomes, there are $2^n$ such sequences (recall Theorem 4.1).

Assigning probabilities to the $2^n$ sequences in S is an exercise in the use of Equation 5.5. Figure 6.3 shows a Bernoulli sequence of length 4 where the first trial ends in success and the last three end in failure. Suppose $p = P$ (success). Since the sequence is an intersection of four independent events, its probability is the product of the four probabilities associated with the individual trials:

$$P((S, F, F, F)) = P(S)P(F)P(F)P(F) = pqqq = pq^3$$

where $q = 1 - p$. More generally, if a sequence of length $n$ consists of $k$ successes and $n - k$ failures, its probability is the product of $k$ $p$'s and $n - k$ $q$'s: $p^k q^{n-k}$.

Table 6.6 shows the sample space and the probability structure associated with the experiment of tossing three biased coins, where $p = P$ (heads). Notice that the *order* of the heads and tails has no effect on the probability of a sequence; all that matters is the *number* of heads and the *number* of tails. The outcome $(H, T, H)$, for example, has the same probability as the outcome $(T, H, H)$. If $p = \frac{1}{2}$, even the number of heads and the number of tails are of no consequence, and each outcome has probability $(\frac{1}{2})^3 = \frac{1}{8}$.

Our primary interest in a series of $n$ Bernoulli trials is not with the individual sequences themselves. For reasons that will soon become apparent, we need to focus on the *number* of successes that occur in the $n$ trials. More specifically, we want to find the probability density function for that number.

Look again at Table 6.6. Let the random variable $X$ denote the number of heads that occur in the three tosses. By inspection, $X$ can be 0, 1, 2, or 3, and it has the probability structure summarized in Table 6.7 (recall Definition 6.4).

**Bernoulli trial**—An experiment with only two possible outcomes: success and failure.

| S | F | F | F |
|---|---|---|---|
| 1 | 2 | 3 | 4 |

Trial number

**FIG. 6.3**

**TABLE 6.6**

| Outcome | Probability |
|---|---|
| $(H, H, H)$ | $ppp = p^3$ |
| $(H, H, T)$ | $ppq = p^2q$ |
| $(H, T, H)$ | $pqp = p^2q$ |
| $(T, H, H)$ | $qpp = p^2q$ |
| $(H, T, T)$ | $pqq = pq^2$ |
| $(T, H, T)$ | $qpq = pq^2$ |
| $(T, T, H)$ | $qqp = pq^2$ |
| $(T, T, T)$ | $qqq = q^3$ |

**TABLE 6.7**

| Number of Heads, $k$ | $\{s \in S: X(s) = k\}$ | $P(\{s \in S: X(s) = k\})$ $= P(X = k)$ |
|---|---|---|
| 0 | $\{(T, T, T)\}$ | $q^3$ |
| 1 | $\{(H, T, T), (T, H, T), (T, T, H)\}$ | $3pq^2$ |
| 2 | $\{(H, H, T), (H, T, H), (T, H, H)\}$ | $3p^2q$ |
| 3 | $\{(H, H, H)\}$ | $p^3$ |

The last column of Table 6.7. suggests an approach for finding a formula for the pdf of $X$. Consider, for example, the probability that $X$ equals 1. According to the table.

$$P(X = 1) = f_X(1) = 3pq^2$$

The 3 in the equation represents the number of ways to permute one head and two tails. The $pq^2$ is the probability of any sequence having one head and two tails.

Now, think of the general case: let $X$ be the number of successes in $n$ Bernoulli trials, where $p = P$ (success). What is the formula for $f_X(k) = P(X = k)$?. By analogy,

$$f_X(k) = P(X = k)$$

$= $ (number of ways to arrange $k$ successes and $n - k$ failures)
$\times$ (probability of any particular sequence of $k$ successes and $n - k$ failures)

The first factor, from Theorem 4.3, is

$$\frac{n!}{k! \, (n - k)!}$$

or $_nC_k$. The second factor, from Equation 5.5, is

$$p^k q^{n-k}$$

Therefore,

$$f_X(k) = P(X = k) = {_nC_k} p^k q^{n-k}$$

for $k = 0, 1, 2, \ldots, n$.

---

**DEFINITION 6.5  The Binomial Distribution**

---

Let the random variable $X$ denote the number of successes in $n$ independent Bernoulli trials, where the probability of success at any trial is $p$ (and the probability of failure is $q = 1 - p$). Then $X$ is said to have a **binomial pdf** or **binomial distribution** and

$$f_X(k) = P(X = k) = {_nC_k} p^k q^{n-k}, \qquad k = 0, 1, \ldots, n \qquad (6.1)$$

COMMENT:   A well-known result in algebra gives a formula for expanding the binomial expression $(p + q)^n$:

$$(p + q)^n = \sum_{k=0}^{n} {}_nC_k p^k q^{n-k} \tag{6.2}$$

The terms being added, here, on the right-hand-side of Equation 6.2 are identical to the set of values for $P(X = k)$ as given in Equation 6.1. The name *binomial distribution* was coined in deference to that similarity.

---

*EXAMPLE 6.8*   An investment analyst has tracked a certain blue-chip stock for the past six months and found that on any given day it either goes up a point or down a point. Furthermore, it went up on 25% of the days and down on 75%. What is the probability that at the close of trading four days from now the price of the stock will be the same as it is today? Assume that the daily fluctuations are independent events.

*Solution*   The days in this example constitute a series of $n = 4$ independent Bernoulli trials. Let $X$ denote the number of days for which the price of the stock increases by a point. Then

$$p = P \text{ (success)} = P \text{ (stock goes up)} = \frac{1}{4}$$

and

$$q = P \text{ (failure)} = P \text{ (stock goes down)} = \frac{3}{4}$$

For the price four days from now to be the same as it is today, the stock must go up exactly twice and down exactly twice. The number of successes, in other words, must be 2. Therefore,

$P$ (stock returns to original price after four days)

$\quad = P$ (stock goes up twice and down twice)

$$= P(X = 2) = {}_4C_2 \left(\frac{1}{4}\right)^2 \left(\frac{3}{4}\right)^{4-2} = \frac{4!}{2! \, 2!} \cdot \frac{1}{16} \cdot \frac{9}{16} = 0.21$$

---

*EXAMPLE 6.9*   In a nuclear reactor, the fission process is controlled by inserting special rods into the radioactive core to absorb neutrons and to slow down the nuclear chain reaction. When functioning properly, these rods serve as a first-line defense against a core meltdown.

Suppose a reactor has 10 control rods (actually, there would be more than 100), each operating independently and each having a 0.80 probability of being inserted properly in the event of an incident. Furthermore, suppose that a meltdown will be prevented if at least half the rods perform satisfactorily. What is the probability that, when put to the test, the system will fail?

*Solution*  If $X$ denotes the number of control rods that function as they should, a system failure occurs if $X \leqslant 4$. In the notation of Definition 6.5, $n$ = number of rods = 10 and $p = P$ (success) = $P$ (rod is inserted properly) = 0.80. The probability of a system failure is the sum of the probabilities associated with $X$ values of 0, 1, 2, 3, and 4:

$$P \text{ (system will fail)} = P(X \leqslant 4)$$

$$= \sum_{k=0}^{4} {}_{10}C_k(0.80)^k(0.20)^{10-k}$$

$$= {}_{10}C_0(0.80)^0(0.20)^{10} + \cdots + {}_{10}C_4(0.80)^4(0.20)^6$$

$$= 0.000 + \cdots + 0.006 = 0.007$$

---

*EXAMPLE 6.10*  Suppose the impossible happens and the next World Series is a matchup between the Cleveland Indians of the American League East and the Chicago Cubs of the National League West. If the Indians are a slightly better team and have a 55% chance of prevailing on any given day, what is the probability that the Tribe wins the series in six games? (Assume the games are independent events.)

*Solution*  Mathematically, the solution we seek involves a binomial computation embedded in a slightly more general problem. Figure 6.4 shows the sort of scenario that must transpire in order for Cleveland to win in six. Let the random variable $X$ denote the number of Cleveland victories among the first five games. If the contests are independent events, then $X$ is a binomial random variable with

$$n = 5 \quad \text{and} \quad p = P \text{ (Cleveland wins individual game)} = 0.55$$

Cleveland's winning the series in six games is really an intersection of two events: (1) they need to win three of the first five games, *and* (2) they need to win the sixth game. Since both those events are independent, the

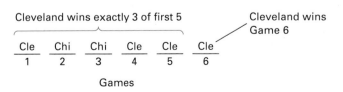

FIG. 6.4

probability of their intersection is the product of the individual probabilities:

$$P \text{ (Cleveland wins World Series in 6 games)}$$
$$= P \text{ (Cleveland wins 3 of first 5} \cap \text{Cleveland wins sixth)}$$
$$= P \text{ (Cleveland wins 3 of first 5)} P \text{ (Cleveland wins sixth)}$$
$$= P(X = 3) P \text{ (Cleveland wins sixth)}$$
$$= {}_5C_3 (0.55)^3 (0.45)^{5-3} (0.55)$$
$$= 0.18$$

QUESTIONS

**6.18** Let $X$ be a binomial random variable. Find $f_X(k)$ for

(a) $n = 4$,  $p = 0.5$,  $k = 2$    (b) $n = 6$,  $p = 0.5$,  $k = 4$

(c) $n = 4$,  $p = 0.7$,  $k = 2$    (d) $n = 6$,  $p = 0.7$,  $k = 4$

(e) $n = 9$,  $p = 0.47$,  $k = 6$    (f) $n = 10$,  $p = 0.32$,  $k = 7$

**6.19** For a binomial random variable $X$, find the indicated probabilities:

(a) $P(X > 3)$   when   $n = 4, p = 0.5$

(b) $P(X < 2)$   when   $n = 4, p = 0.3$

(c) $P(X \geqslant 5)$   when   $n = 7, p = 0.4$

(d) $P(X \leqslant 2)$   when   $n = 7, p = 0.4$

**6.20** Recall Example 6.10. What is the probability that Chicago wins the series in seven games?

**6.21** The captain of a Navy gunboat orders a volley of $n$ missiles to be fired at random along a 500-ft stretch of shoreline that he hopes to establish as a beachhead. Dug into the ground is a 30-ft long bunker housing the enemy. If $n = 25$, what is the probability that exactly three shells will hit the bunker?

**6.22** Suppose a couple intend to have four children. If the probability of a child being a boy is 1/2, is it more likely they will have two boys and two girls or three of one sex and one of the other? (Hint: Let the random variable $X$ denote the number of boys among the four children.)

**6.23** A new drug is being tested that researchers believe has a 60% chance of reducing a person's cholesterol level. If the drug is administered to 10 subjects, what is the probability that

(a) exactly five show a reduced cholesterol level?

(b) at least eight show a reduced cholesterol level?

**6.24** Carol spent all weekend partying and knows absolutely nothing about the 10 true/false questions on a Monday morning psychology quiz. What is the probability that she gets at least 8 right by simply guessing?

6.25   Suppose that, since the early 1950s some 10,000 independent UFO sightings have been reported to civil authorities. If the probability that any sighting is genuine is on the order of 1 in 100,000, what is the probability that at least 1 of the 10,000 was genuine? (Hint: Let the random variable $X$ denote the number of genuine sightings (out of 10,000). What is the complement of "at least one sighting was genuine?")

6.26   In a large metropolitan area 70% of the registered voters are thought to favor a new zoning ordinance. Suppose six voters chosen at random are interviewed. What is the probability that fewer than half support the ordinance?

6.27   The great English diarist Samuel Pepys asked his friend Sir Isaac Newton the following question: Is it more likely to get at least one 6 when 6 dice are rolled, at least two 6's when 12 dice are rolled, or at least three 6's when 18 dice are rolled? After considerable correspondence (see (148)), Newton convinced the skeptical Pepys that the first probability is the most likely. Compute the three probabilities.

6.28   The gunner on a small assault boat fires six missiles at an attacking plane. Each has a 20% chance of being on target. If two or more of the shells find their mark, the plane will crash. At the same time, the pilot of the plane fires 10 air-to-surface rockets, each of which has a 0.05 chance of critically disabling the boat. Would you rather be on the plane or the boat?

6.29   What is the probability that a baseball player who is batting .330 goes "4-for-5" in tomorrow's game (i.e., gets four hits in five at-bats)? What assumption are you making?

6.30   In his 1889 publication *Natural Inheritance*, the renowned British scientist Sir Francis Galton described a pinball-type board that he called a *quincunx*. As pictured, the quincunx has five rows of pegs, the pegs in each row being the same distance apart. At the bottom of the board are five cells, numbered 0 through 4. A ball pushed between the pegs in the first row will hit the middle peg in the second row, veer to the right or to the left, strike a peg in the third row, veer to the right or to the left, and so on. If the ball has a 50–50 chance of going in either direction each time it hits a peg, what is the probability that it ends up in cell 3?

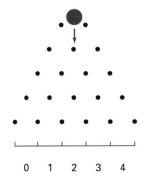

# 6.4

## THE HYPERGEOMETRIC DISTRIBUTION

The experiment of drawing chips from an urn provides a useful model for a variety of real-world applications. Two important variations are frequently encountered. The distinction, which is the starting point of this section, depends on whether the chips are drawn *with* replacement or *without* replacement.

Figure 6.5 shows an urn with 10 chips, of which 7 are blue and 3 are white. We draw a chip, note its color, return it to the urn, and repeat the process, say, five times. What can we say about the probabilistic structure of this experiment?

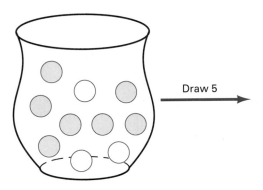

Draw 5

**FIG. 6.5**

First, the replacement of the chips guarantees that each successive drawing is independent of its predecessors. Second, the probability $p$ of choosing, say, a red chip is the same from drawing to drawing because the composition of the urn is not changing. Both of those conditions should sound familiar. Drawing chips *with replacement* is simply a series of Bernoulli trials, here with

$$n = 5 \quad \text{and} \quad p = P \text{ (success)} = P \text{ (red chip is drawn)} = 0.7$$

If we let the random variable $X$ denote the number of red chips in the sample, then, by Equation 6.1,

$$P(X = k) = f_X(k) = {}_{10}C_k(0.7)^k(0.3)^{10-k}, \qquad k = 0, 1, \ldots, 10$$

An alternative scheme is sampling *without replacement* (i.e., the chips are not returned to the urn after they are drawn). What is the corresponding sample space? It depends. Sometimes the outcomes of interest are all possible *ordered* samples. For example, if five chips are drawn from the urn, two such outcomes are $(R, R, W, R, R)$ and $(R, W, R, R, R)$. Usually, however, the outcomes of interest are *unordered*. The experimenter cares only about the *number* of red chips and the *number* of white chips that are chosen; their sequence is irrelevant. Mathematically, forming an unordered sample of five chips is analogous to reaching into the urn and physically choosing that handful of five simultaneously.

Finding the probability density function for $X$, the number of red chips in a sample drawn without replacement, is an application of two ideas from Chapter 4: the multiplication rule (Theorem 4.1) and the enumeration of combinations (Theorem 4.4). Figure 6.5 shows an urn containing seven red chips and three white chips. Let $X$ be the number of red chips in an unordered sample of size 5. Our objective—a formula for $f_X(k) = P(X = k)$. Remember that a sample of size 5 contains exactly $k$ red chips only if it also contains exactly $5 - k$ white chips. By Theorem 4.4 the number of ways to choose $k$ red chips is $_7C_k$, and the number of ways to choose $5 - k$ white chips is $_3C_{5-k}$. By the multiplication rule the number of ways to form samples of size 5 consisting of $k$ red chips *and* $5 - k$ white chips is the product, $_7C_k \cdot _3C_{5-k}$. The *total* number of unordered samples of size 5 is $_{10}C_5$. Since the latter are all equally likely, the probability that $X = k$ follows from the $m/n$ rule of Section 4.6:

$$P(X = k) = f_X(k) = \frac{_7C_k \cdot _3C_{5-k}}{_{10}C_5} \tag{6.3}$$

**TABLE 6.8**

| $k$ | $P(X = k)$ |
|---|---|
| 2 | $21/252 = 0.083$ |
| 3 | $105/252 = 0.417$ |
| 4 | $105/252 = 0.417$ |
| 5 | $21/252 = 0.083$ |

Table 6.8 shows the actual numerical values computed from Equation 6.3. Notice that $k$ ranges only from 2 to 5, inclusive (why?).

We can generalize Equation 6.3 by not specifying the precise numbers of red chips and white chips or the size of the sample.

---

**DEFINITION 6.6 The Hypergeometric Distribution**

Suppose an urn contains $r$ red chips and $w$ white chips. Let $N = r + w$. A sample of size $n$ is drawn without replacement. Let $X$ denote the number of red chips in the sample. Then $X$ has a **hypergeometric distribution** and

$$f_X(k) = P(X = k) = \frac{_rC_k \cdot _wC_{n-k}}{_NC_n} \tag{6.4}$$

---

*EXAMPLE 6.11*   Nevada keno is among the most popular games in Las Vegas, despite the fact that its odds are overwhelmingly in favor of the house. (Betting on keno is only slightly less foolish than playing a slot machine!) A keno card has 80 numbers, 1 through 80, from which the player selects a sample of size $n$ ($n = 1$ to 15). The caller then announces 20 winning numbers, chosen at random from the 80. How much the player wins depends on how many of his numbers match the 20 identified by the caller. A popular choice among keno players is to bet on $n = 10$ numbers. What are the chances that such a bet "catches" five numbers? That is, what is the probability that exactly 5 of the 10 numbers picked by the gambler are among the 20 winners announced by the caller?

*Solution*　Imagine an urn containing 80 chips, of which 20 are winners and 60 are losers. By betting on a 10-spot ticket, the player, in effect, is drawing a sample of size 10 without replacement from that urn. Let $X$ denote the number of winners among the player's 10 selections. We are trying to find $P(X = 5)$.

　　In the terminology of Definition 6.6 the random variable $X$ has a hypergeometric distribution with $r = 20$, $w = 60$, $n = 10$, $N = r + w = 80$, and $k = 5$. Substituting into Equation 6.4 gives

$$f_X(5) = P(X = 5) = \frac{{}_{20}C_5 \cdot {}_{60}C_{10-5}}{{}_{80}C_{10}} = 0.05$$

---

*EXAMPLE 6.12*　Muffy is preparing for the final in her history of France course. The exam will consist of 5 essay questions selected at random from a list of 10 the professor has handed out in advance. Not exactly a Napoleon buff, Muffy would like to avoid researching all 10 questions but still be reasonably assured of getting a fairly good grade. In particular, she wants to have at least an 85% chance of getting at least four of the five questions right. Will studying 8 of the 10 questions be sufficient preparation?

*Solution*　Think of the questions as two kinds of chips in an urn: the eight whose answers Muffy will know, and the two for which she will be unprepared (see Figure 6.6). In making out the test, the professor is drawing a random sample of size 5. Let $X$ denote the number of questions in the sample (i.e., test) coming from the set of eight for which Muffy will have prepared. The

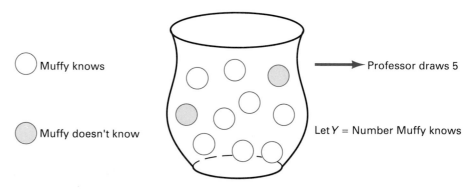

○ Muffy knows

◉ Muffy doesn't know

→ Professor draws 5

Let $Y$ = Number Muffy knows

**FIG. 6.6**

probability at issue is $P(X \geqslant 4)$. Unfortunately,

$$P(X \geqslant 4) = P(X = 4) + P(X = 5) = \frac{{}_8C_4 \cdot {}_2C_1}{{}_{10}C_5} + \frac{{}_8C_5 \cdot {}_2C_0}{{}_{10}C_5} = 0.78$$

so it's back to the books! By preparing for eight questions, Muffy has only a 78% chance (rather than 85%) of getting at least four correct.

---

**EXAMPLE 6.13**  The hypergeometric distribution figures prominently in an important area of statistics known as *acceptance sampling*. Consider the problem faced by a manufacturer who orders a supply of parts from an outside contractor. When the shipment arrives, the manufacturer would like to have some assurance that the parts meet her specifications. She could, of course, inspect every item but that would be costly, time-consuming, and maybe unreasonable (what if the part were a flashbulb?). A better approach is to take a random sample of size $n$, inspect all $n$, and accept the shipment as being of sufficiently high quality only if the sample is of sufficiently high quality.

Suppose a shipment contains 100 items, out of which she selects a sample of size $n = 2$. Let the random variable $X$ denote the number of defectives in the sample. As a *decision rule*, she decides to reject the shipment if $X \geqslant 1$. What is the probability that she will accept a shipment that is 10% defective?

**Solution**  If she intends to reject the shipment when $X \geqslant 1$, it follows that she will accept the shipment when $X = 0$. Furthermore, $X$ is clearly a hypergeometric random variable (see Figure 6.7). From Equation 6.4

$$P \text{ (she accepts shipment)} = P(X = 0) = \frac{{}_{10}C_0 \cdot {}_{90}C_2}{{}_{100}C_2} = 0.81$$

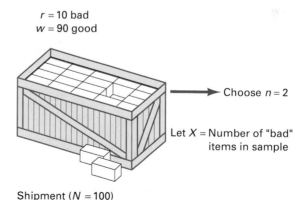

*r* = 10 bad
*w* = 90 good

Choose *n* = 2

Let *X* = Number of "bad" items in sample

Shipment (*N* = 100)

**FIG. 6.7**

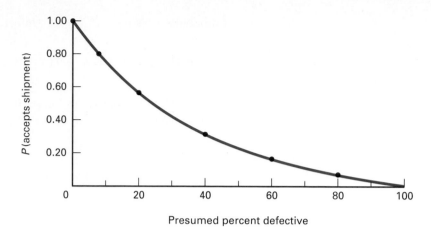

**FIG. 6.8**

Similar acceptance probabilities can be computed for other presumed levels of incoming quality. If the shipment were 20% defective, for instance, it would be accepted (by the same decision rule) 64% of the time:

$$P \text{ (she accepts shipment)} = P(X = 0)$$

$$= \frac{_{20}C_0 \cdot {}_{80}C_2}{_{100}C_2} = 0.64$$

Figure 6.8 is a graph summarizing these kinds of computations. It plots the manufacturer's acceptance probability (in this case, $P(X = 0)$) as a function of the shipment percent defective. Graphs giving this kind of information are called *operating characteristic curves*: they show the ability of sampling plans to detect lapses in product quality.

From the manufacturer's standpoint the information given in Figure 6.8 is a bit disturbing. It clearly shows that this sampling plan is not very good at protecting her interests. If the shipment were 60% defective, for example—certainly an intolerable state of affairs—the plan would still recommend 15% of the time that the order be accepted. (Intuitively, how should the sampling plan be modified?)

---

**QUESTIONS**

**6.31**  For a hypergeometric random variable $Y$ with $r = 4$, $w = 5$, and $n = 3$, calculate

(a) $P(Y = 2)$      (b) $P(Y = 4)$      (c) $P(Y > 1)$

**6.32**  For a hypergeometric random variable $Y$ with $r = 7$, $w = 4$, and $n = 5$, find

(a) $P(Y = 3)$      (b) $P(Y = 6)$      (c) $P(Y = 0)$      (d) $P(Y \leqslant 2)$

**6.33**  Let $Y$ be a hypergeometric random variable with $r = 5$, $w = 3$, and $n = 4$. Find $f_Y(k)$ (specify the values of $k$ for which $f_Y(k)$ is nonzero).

6.34   Recall Example 6.11. Suppose a player chooses a 12-spot ticket. What is his probability of getting exactly five winners?

6.35   A Scrabble set consists of 54 consonants and 44 vowels. What is the probability that your initial draw (of seven letters) will be all consonants? six consonants and one vowel? five consonants and two vowels?

6.36   Six terminals, numbered 1 through 6, are on-line to a DEC-10 computer; all are ready to execute their programs. You and a friend are working on terminals 2 and 5. At random the computer selects three terminals and advances them in the access priority queue. What is the probability that both your terminal and your friend's terminal are among the three selected to be advanced?

6.37   Urn I contains four red chips, three white chips, and two blue chips; urn II, three red, four white, and five blue. Two chips are drawn at random and without replacement from each urn. What is the probability that all four chips are the same color? (Hint: Write the event "all four chips are the same color" as the union of three intersections.)

6.38   X rays show that 6 of Missy's 32 teeth have cavities. Unfortunately, her dentist is a bit myopic and drills six of her teeth *at random*. What is the probability that fewer than half of the cavities are properly drilled?

6.39   An IRS auditor has 10 income tax forms on her desk. She has time to examine only 6 of the 10. Suppose that 5 of the 10 contain serious errors. What is the probability that at least four of the five incorrect returns are included in the six she chooses to audit?

6.40   Recall Example 6.12. Suppose the exam is to consist of 6 questions chosen at random from a set of 12. If Muffy studies eight of the questions, what is her probability of getting at least five correct?

6.41   Recall the acceptance sampling discussion in Example 6.13. Suppose that 5 items are to be sampled from the shipment (of 100), with acceptance requiring that $Y \leqslant 1$, where $Y$ is the number of defectives. Construct the corresponding operating characteristic curve and compare it to Figure 6.8.

6.42   A camera manufacturer receives a shipment of 100 semiautomatic lens housings. For his sampling plan he decides to select 10 of the housings at random and accept the shipment if no more than one is defective. Construct the corresponding operating characteristic curve. For approximately what incoming quality will he accept the shipment 50% of the time?

# 6.5

## INFINITE DISCRETE SAMPLE SPACES (optional)

### The Geometric Distribution

All the sample spaces we have considered so far have been *finite*. It is easy to imagine an experiment, though, where the number of possible outcomes is *infinite*. One example is a coin flipped until a head comes up for the first time. Although in practice that first head is likely to appear fairly soon, theoretically we could go on flipping forever. The number of outcomes in S, then, is infinite. At the same time the "structure" of those outcomes is easy to characterize: each is a sequence of a different number of tails followed by a head:

$$S = \{H, TH, TTH, TTTH, TTTTH, \ldots\}$$

Assigning a probability to each of the infinitely many outcomes in S is straightforward. Since the individual tosses are independent events, the probability of the first head coming on, say, the kth toss; that is, the probability of

$$\underbrace{TT \cdots TH}_{k-1 \text{ tails}}$$

is a direct application of Equation 5.5:

$$P(\underbrace{TT \cdots TH}_{k-1 \text{ tails}}) = P \text{ (first head comes on kth toss)}$$

$$= \underbrace{P(T)P(T) \cdots P(T)P(H)}_{k-1 \text{ factors}}$$

The first two columns of Table 6.9 show the first six outcomes and their probabilities if the chances of getting a head on any given toss are 4 in 5; that is, if $P(H) = 4/5$ and $P(T) = 1/5$. The last column sums the probabilities associated with the outcomes listed in the first column. The total probability associated with *all* the outcomes in S must be 1. The set of sequence longer than $TTTTTH$, therefore, has a *combined* probability of only 0.0001.

**TABLE 6.9**

| Outcome | Probability | Cumulative Probability |
|---------|-------------|------------------------|
| H | $\left(\frac{1}{5}\right)^0 \left(\frac{4}{5}\right) = 0.8000$ | 0.8000 |
| TH | $\left(\frac{1}{5}\right)^1 \left(\frac{4}{5}\right) = 0.1600$ | 0.9600 |
| TTH | $\left(\frac{1}{5}\right)^2 \left(\frac{4}{5}\right) = 0.0320$ | 0.9920 |
| TTTH | $\left(\frac{1}{5}\right)^3 \left(\frac{4}{5}\right) = 0.0064$ | 0.9984 |
| TTTTH | $\left(\frac{1}{5}\right)^4 \left(\frac{4}{5}\right) = 0.0013$ | 0.9997 |
| TTTTTH | $\left(\frac{1}{5}\right)^5 \left(\frac{4}{5}\right) = 0.0002$ | 0.9999 |

In general, suppose an experiment consists of repeating a series of independent Bernoulli trials until the first success occurs. Let $s$ and $f$ denote the occurrence of a success and a failure, respectively, at any given trial. Then outcomes of the experiment will have the form $s$, $fs$, $ffs$, $fffs$, $ffffs$, and so on. Furthermore, if $p = P(s)$ and $q = 1 - p = P(f)$, then

$$P(s) = p$$

$$P(fs) = qp$$

$$P(ffs) = q^2 p$$

$$P(fffs) = q^3 p$$

$$\vdots$$

It can be shown that

$$p + qp + q^2 p + q^3 p + q^4 p + \cdots = \sum_{k=1}^{\infty} q^{k-1} p = 1$$

Now, suppose we define the random variable $Y$ to be the number of trials necessary to get the *first* success. The corresponding probability density function for $Y$ should be apparent:

$$P(Y = k) = P \text{ (first success occurs on } k\text{th trial)}$$

$$= P(\underbrace{ff \cdots fs}_{k-1 \text{ failures}}) = q^{k-1} p$$

The possible values of $k$ are the (infinite) set of positive integers.

---

**DEFINITION 6.7 The Geometric Distribution**

Suppose a sequence of independent Bernoulli trials, where $p = P$ (success) and $q = 1 - p = P$ (failure), is continued until the *first* success occurs. Let $Y$ be the number of trials required. Then $Y$ is said to be a **geometric random variable**, and

$$f_Y(k) = P(Y = k) = q^{k-1} p, \qquad k = 1, 2, \ldots \qquad (6.5)$$

---

*EXAMPLE 6.14*   Suppose transistors have a 0.98 probability of working and a 0.02 probability of being defective. Each transistor is tested before it is wired into a circuit. What is the probability that the *third* transistor tested is the first one that works?

*Solution*   Let the random variable $Y$ denote the "number" of the first transistor that works. We are looking for $P(Y = 3)$. Notice that the structure of $Y$ makes it a geometric random variable: each transistor tested is an independent

Bernoulli trial with

$$p = P \text{ (transistor works)} = 0.98 \quad \text{and}$$

$$q = 1 - p = P \text{ (transistor is defective)} = 0.02.$$

From Equation 6.5,

$$P(Y = 3) = f_Y(3) = (0.02)^{3-1}(0.98) = 0.000392$$

Questions can also be formulated that require $f_Y(k)$ to be summed over different values of $k$. What is the probability, for instance, that *more than three* transistors need to be tested—that is, $P(Y \geq 4)$? Approached directly, $P(Y \geq 4)$ is not just a simple sum, but the sum of an *infinite* number of terms:

$$P(Y \geq 4) = P(Y = 4) + P(Y = 5) + P(Y = 6) + \cdots = \sum_{k=4}^{\infty} P(Y = k)$$

Fortunately, we can avoid any computational unpleasantness here by working with the *complement* of "$Y \geq 4$," an event that contains only three terms ($Y = 1, 2,$ and $3$). Subtracting the probability of the latter from 1 will give $P(Y \geq 4)$:

$$P(Y \geq 4) = 1 - P(Y \leq 3) \qquad \text{(why?)}$$

$$= 1 - (P(Y = 1) + P(Y = 2) + P(Y = 3))$$

$$= 1 - (0.98 + (0.02)(0.98) + (0.02)^2(0.98)) = 0.000008$$

---

*EXAMPLE 6.15*   Your local supermarket is sponsoring a sales promotion where each customer making a purchase is given a star that can be one of six different colors. Anyone collecting a set of all six colors receives $10 worth of groceries free. According to contest rules, the store has an equal number of stars of each color, and the cashiers give the different colors away at random. Furthermore, the number of stars of each color is so large that, for all practical purposes, the probability of getting any color on any day remains constant. Suppose you already have five of the six. If you buy something at the store every weeknight, what is the probability that sometime during the next five days you will qualify for the $10 prize?

*Solution*   Let $Y$ denote the number of purchases necessary to get the sixth star. In order for $Y$ to equal some number $k$, you would need to (1) receive colors you already have for the next $k - 1$ days and (2) receive the sixth color on the $k$th day. Since you are starting with five of the six colors, your probability of receiving one of those five again on any day is 5/6, and the probability of getting the one you *don't* have is 1/6. From Equation 6.5,

$$P(Y = k) = f_Y(k) = \left(\frac{5}{6}\right)^{k-1}\left(\frac{1}{6}\right), \qquad k = 1, 2, 3, \ldots$$

Your chances of success sometime during the week would be the sum of $f_Y(k)$ for $k = 1, 2, 3, 4,$ and 5:

$P$ (qualify for prize sometime during next 5 days)

$$= P(1 \leqslant Y \leqslant 5)$$

$$= P(Y = 1) + P(Y = 2) + P(Y = 3) + P(Y = 4) + P(Y = 5)$$

$$= \left(\frac{1}{6}\right) + \left(\frac{5}{6}\right)^1 \left(\frac{1}{6}\right) + \left(\frac{5}{6}\right)^2 \left(\frac{1}{6}\right) + \left(\frac{5}{6}\right)^3 \left(\frac{1}{6}\right) + \left(\frac{5}{6}\right)^4 \left(\frac{1}{6}\right) = 0.60$$

---

**QUESTIONS**

**6.43**  Let $Y$ be a geometric random variable with $p = 1/3$. Find

(a) $P(Y = 0)$  (b) $P(Y = 2)$  (c) $P(Y \leqslant 2)$  (d) $P(Y > 4)$

**6.44**  Let $Y$ be a geometric random variable with $p = 0.7$. Compute

(a) $P(Y = 2)$  (b) $P(Y \leqslant 2)$  (c) $P(Y > 3)$

**6.45**  A fair die is tossed. What is the probability that the first 5 occurs on the fourth roll?

**6.46**  A basketball player has a 70% chance of making a free throw. Let $Y$ denote the attempt on which the first basket is made.

(a) Write a formula for $f_Y(k)$.

(b) What is the probability that the first basket is made on the third toss? Assume each throw is an independent event.

**6.47**  A somewhat uncoordinated woman is attempting to get a driver's license. Having successfully completed the written exam, she needs only to pass the road test. Her abilities in that area, though, are less than exceptional: an unbiased observer would give her no more than a 10% chance of passing on any given try. What is the probability that she will have to take the test at least three times in order to get a license? Assume that her driving skills remain at the same level, regardless of how many times she fails the test.

**6.48**  A young couple plan to continue having children until they have their first girl. Suppose the probability that a child is a girl is 1/2 and the outcome of each birth is an independent event. Can they be at least 90% certain that achieving their objective will result in a family of three children or less?

**6.49**  (a) Suppose a fair die is rolled until the *second* 5 occurs. What is the probability of that happening on the fourth roll?

(b) More generally, let the random variable $Y$ denote the roll on which the second 5 occurs. Find the probability density function for $Y$. (Hint: If the second 5 occurs on the $k$th roll, what must happen during the first $k - 1$ rolls?)

### The Poisson Approximation

Another random variable leading to a sample space with an infinite number of outcomes grew out of the work of the eminent French mathematician and physicist, Siméon-Denis Poisson (1791–1840). The catalyst for the latter's efforts was the binomial distribution. Definition 6.5 was well known in Poisson's time, and the binomial was already recognized as an important probability model. Actually computing $P(Y = k)$ was difficult, though, when $n$ was large. What Poisson derived was an easy-to-use approximation to the binomial.

---

#### THEOREM 6.1

Let $X$ be a binomial random variable defined on $n$ independent trials, where $p = P$ (success). If $n$ is large and $p$ is small,

$$P(X = k) = {}_nC_k p^k (1 - p)^{n-k} \doteq \frac{e^{-np}(np)^k}{k!}, \qquad k = 0, 1, \ldots, n \quad (6.6)$$

where $e \doteq 2.71828$ is the base of the natural logarithms.

---

**EXAMPLE 6.16**    Theorem 6.1 claims that $e^{-np}(np)^k/k!$ approximates $P(X = k)$ when $n$ is "large" and $p$ is "small," but doesn't state precisely what those terms mean. Does 30 qualify as a large $n$? Is 0.05 a small $p$? There are no easy answers to those questions. Ultimately, $n$ is large enough and $p$ is small enough if, in our opinion, the approximation is good enough.

*Solution*    Tables 6.10 and 6.11 show a numerical comparison between

$$ {}_nC_k p^k (1 - p)^{n-k} \quad \text{and} \quad \frac{e^{-np}(np)^k}{k!}$$

for two sets of values for $n$ and $p$. The entries in Table 6.10 give binomial probabilities and their Poisson approximations when $n = 5$ and $p = 1/5$. A row-by-row comparison of the second and third columns shows that the approximation is not very good. The exact binomial probability, for example, that $X = 0$ is 0.328, but the Poisson approximation is a considerably larger 0.368. Either an $n$ of 5 is not sufficiently large, or a $p$ of 1/5 is not sufficiently small, or both.

In contrast, Table 6.11 shows similar computations when $n = 100$ and $p = 0.01$. The row-by-row agreement in this case is quite good. Any binomial problem, therefore, where $n > 100$ and $p < 0.01$ is certainly a suitable candidate for a Poisson approximation.

**TABLE 6.10**

| $k$ | $_5C_k\left(\dfrac{1}{5}\right)^k\left(\dfrac{4}{5}\right)^{5-k}$ | $\dfrac{e^{-5(1/5)}(5(1/5))^k}{k!}$ |
|---|---|---|
| 0 | 0.328 | 0.368 |
| 1 | 0.410 | 0.368 |
| 2 | 0.205 | 0.184 |
| 3 | 0.051 | 0.061 |
| 4 | 0.006 | 0.015 |
| 5 | 0.000 | 0.003 |
| 6+ | 0 | 0.001 |
|  | 1.000 | 1.000 |

**TABLE 6.11**

| $k$ | $_{100}C_k(0.01)^k(0.99)^{100-k}$ | $\dfrac{e^{-100(0.01)}(100(0.01))^k}{k!}$ |
|---|---|---|
| 0 | 0.366032 | 0.367879 |
| 1 | 0.369730 | 0.367879 |
| 2 | 0.184865 | 0.183940 |
| 3 | 0.060999 | 0.061313 |
| 4 | 0.014942 | 0.015328 |
| 5 | 0.002898 | 0.003066 |
| 6 | 0.000463 | 0.000511 |
| 7 | 0.000063 | 0.000073 |
| 8 | 0.000007 | 0.000009 |
| 9 | 0.000001 | 0.000001 |
| 10 | 0.000000 | 0.000000 |
|  | 1.000000 | 0.999999 |

**EXAMPLE 6.17**  In a certain city, data show that among children aged 0 through 14, leukemia occurs at an annual rate of 3.5 cases per 100,000. If that city has 160,000 children, what is the probability that fewer than four new leukemia cases will be diagnosed during the next 12 months?

**Solution**  Fundamentally, this is a binomial problem. Each of the $n = 160,000$ children acts as a Bernoulli trial: a child either (1) gets leukemia or (2) does *not* get leukemia. Also,

$$p = P\,(\text{child gets leukemia}) = \frac{3.5}{100,000} = 0.000035$$

Let the random variable $X$ denote the number of new leukemia cases among the city's 160,000 children. From Definition 6.5,

$$P(X < 4) = P(X \leqslant 3)$$

$$= \sum_{k=0}^{3} {}_{160,000}C_k(0.000035)^k(0.999965)^{160,000-k}$$

Evaluating $P(X \leqslant 3)$ by the exact formula is not easy (try it!). Getting a Poisson approximation, though, is fairly routine. First, write

$$np = (160,000)(0.000035) = 5.6.$$

Then

$$P(X \leqslant 3) \doteq \sum_{k=0}^{3} \frac{e^{-5.6}(5.6)^k}{k!}$$

$$= e^{-5.6}\left\{\frac{(5.6)^0}{0!} + \frac{(5.6)^1}{1!} + \frac{(5.6)^2}{2!} + \frac{(5.6)^3}{3!}\right\} = 0.19$$

Would 0.19 be a good approximation to the exact binomial probability? Definitely. Recall from Table 6.11 how closely the Poisson agrees with the binomial when $n$ is 100 and $p$ is 0.01. Here, $n$ is *much* larger (160,000) and $p$ is *much* smaller (0.000035). Under these conditions the approximation, for all practical purposes, is an equality.

---

**6.50**  Use the Poisson approximation to estimate the following binomial probabilities when $n = 200$ and $p = 0.02$:

(a) $P(X = 1)$      (b) $P(1 \leqslant X \leqslant 3)$      (c) $P(X \geqslant 2)$

**6.51**  Suppose each of 1000 independent Bernoulli trials has probability 0.005 of ending in success. Let the random variable $X$ denote the total number of successes occurring in the 1000 trials. Use the Poisson approximation to estimate

(a) $P(4 \leqslant X \leqslant 6)$      (b) $P(X \geqslant 3)$

**6.52**  A chromosome mutation believed to be linked with color blindness is known to occur, on the average, once in every 10,000 births. If 20,000 babies are born this year in a certain city, approximate the probability that exact three of the children have the mutation. (Hint: Think of how the problem would be set up as a binomial. What values would be assigned to $n$, $p$, and $X$?)

**6.53**  Use the Poisson approximation to estimate the probability that at most 1 person in 500 will have a birthday on Christmas. Assume there are 365 days in a year and that each is equally likely to be someone's birthday.

**6.54**  You are behind 10 people in a supermarket express line. Assume that (1) each person in front of you is buying five items and (2) the probability of any item requiring a price check, and thereby causing a delay, is 0.01. Use the Poisson approximation to estimate your chances of being delayed. (Hint: Let $X$ denote the number of items requiring a price check. For what values of $X$ will you be delayed?)

### The Poisson Distribution

For more than 50 years after its discovery, Theorem 6.1 was used in precisely the way Poisson intended, as a numerical approximation to the binomial when $n$ is large and $p$ is small. Beginning in the late nineteenth century, however, scientists from a variety of disciplines began using Poisson's formula as a probability model in its own right. Situations where it was first applied typically involved phenomena for which the measurement recorded was the number of occurrences of a rare event over a prescribed time. Telephone calls coming into a switchboard per minute, accidents occurring in a factory per week, and $\alpha$ particles detected by a Geiger counter per second are all examples of "rare" phenomena. All that fundamentally distinguishes one from another is the rate $\lambda$ at which the events occur.

It can be shown that the kind of situation just described is similar mathematically to counting the number of successes occurring in a series of Bernoulli trials where $n$ is large and $p$ is small. We should not be sur-

prised to find, then, that the pdf describing the number of occurrences of
a rare event is nothing other than Poisson's formula.

---

**DEFINITION 6.8**   Poisson Distribution

If the pdf for a random variable $X$ is given by

$$f_X(k) = P(X = k) = \frac{e^{-\lambda}\lambda^k}{k!}, \qquad k = 0, 1, 2, \dots \qquad (6.7)$$

where $\lambda > 0$, then $X$ is said to have a **Poisson distribution**.

---

COMMENT:   As the discussion on pp. 302 suggests, an important application
of the Poisson pdf arises in connection with rare events. Specifically, if $\lambda$
denotes the *rate* at which a rare event occurs and the random variable $X$
denotes the *number* that occur, then

$$P(X = k) \doteq \frac{e^{-\lambda}\lambda^k}{k!}$$

---

**EXAMPLE 6.18**   A certain young lady is quite popular with her male classmates. She receives, on the average, six phone calls a night. What is the probability that
tomorrow night she receives exactly eight calls?

*Solution*   Units are the key in setting up Poisson problems: $X$ and $\lambda$ must be defined
comparably. Here, we should let the random variable $X$ denote the number
of calls she receives tomorrow night. By assumption, the *rate* of calls received per night ($\lambda$) is 6. If $X$, then, can be described by a Poisson pdf,

$$P(X = 8) = f_X(8) \doteq \frac{e^{-6}6^8}{8!} = \frac{(0.002479)(1{,}679{,}616)}{40{,}320} = 0.10$$

Is the assumption that $X$ is Poisson reasonable? Yes. Suppose that by
"night" we really mean the 4-hr interval from 7:00 P.M. to 11:00 P.M.
Imagine partitioning that interval into a large number of very small subintervals, so small that the chances of more than one call arriving during
any given subinterval are essentially zero. (See Figure 6.9, where the 4 hr
have been divided into 240 min.) Each subinterval is then a Bernoulli trial:
either it ends in "success" (a call is received) or it ends in "failure" (a call
is *not* received). If, on the average, six calls arrive every 4 hr, then $6/240 =$
$0.025$ calls arrive, on the average, every minute. Now, let $X$ denote the
number of calls arriving during the 240 min. Clearly, the conditions set out in
Theorem 6.1 are met: $X$ is binomial, $n$ is large, and $p$ is small. Therefore,

$$P(X = 8) \doteq \frac{e^{-240(6/240)}(240(6/240))^8}{8!} = \frac{e^{-6}6^8}{8!} = 0.10$$

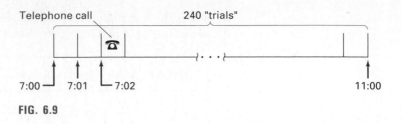

**FIG. 6.9**

*Detecting α Particles*　Among the early research projects investigating the nature of radiation was a 1910 study of α emission by Ernest Rutherford and Hans Geiger (143). For each of 2608 eighth-min intervals, the two physicists recorded the number of α particles emitted from a polonium source and subsequently detected by a Geiger counter. The numbers of times $k$ particles ($k = 0, 1, 2, 3, \ldots$) were detected in a given eighth-min are summarized in the first three columns of Table 6.12. Two α particles, for example, were detected in each of 383 eighth-min intervals. The corresponding *relative frequency*, or *empirical probability*, is the quotient shown in column 3: $383/2608 = 0.15$.

**TABLE 6.12**

| Number of Particles Detected, $k$ | Frequency | Relative Frequency | $P(X = k) = \dfrac{e^{-3.87}(3.87)^k}{k!}$ |
|:---:|:---:|:---:|:---:|
| 0 | 57 | 0.02 | 0.02 |
| 1 | 203 | 0.08 | 0.08 |
| 2 | 383 | 0.15 | 0.16 |
| 3 | 525 | 0.20 | 0.20 |
| 4 | 532 | 0.20 | 0.20 |
| 5 | 408 | 0.16 | 0.15 |
| 6 | 273 | 0.10 | 0.10 |
| 7 | 139 | 0.05 | 0.05 |
| 8 | 45 | 0.02 | 0.03 |
| 9 | 27 | 0.01 | 0.01 |
| 10 | 10 | 0.00 | 0.00 |
| 11+ | 6 | 0.00 | 0.00 |
| | | 1.0 | 1.0 |

If 0 particles are observed in each of 57 intervals, 1 particle is 203 intervals, 2 particles is 383 intervals, and so on, the *total* number of particles can be calculated as a sum of products:

$$\begin{pmatrix}\text{Total number} \\ \text{of particles}\end{pmatrix} = 0 \cdot 57 + 1 \cdot 203 + 2 \cdot 383 + \cdots + 11 \cdot 6$$

$$= 10{,}097$$

It then follows that the *rate* at which the particles are detected (per eighth-min) is the previous total divided by the total number of eighth-min:

$$\text{Detection rate} = \lambda = \frac{\text{Total number of particles detected}}{\text{Total number of intervals}}$$

$$= \frac{10{,}097}{2{,}608} = 3.87$$

Let $X$ denote the number of particles detected in a given eighth-min. *If $X$ is a Poisson random variable, we would expect the entries in column 3 to be closely approximated by the expression*

$$\frac{e^{-3.87}(3.87)^k}{k!}$$

Column 4 shows the latter. Quite clearly, the row-by-row agreement between columns 3 and 4 is excellent. Our conclusion? That radiation, as a phenomenon, can be modeled every effectively by the Poisson distribution.

---

**QUESTIONS**

**6.55**  Suppose a 520-page book contains 390 typographical errors. What is the probability that a randomly selected page will be free of errors?

**6.56**  A radioactive source is metered for 2 hr, during which time 482 α particles are counted. What is the probability that

(a) Exactly three particles will be counted during the next minute? (Hint: What is the detection rate *per minute*?)

(b) No more than three particles will be counted?

**6.57**  Historically, one of the first applications of Definition 6.8 addressed the phenomenon of Prussian cavalry soldiers being kicked to death by their horses (11). Ten cavalry corps were monitored for 20 years. Recorded for each year and each corps was $X$, the number of fatalities due to kicks. Shown below is the entire distribution of the 200 observations.

(a) Following the procedure described in Case Study 6.1, "fit" these data with a Poisson model.

(b) Is it reasonable to conclude that $X$ can be thought of as a Poisson random variable?

| Number of Deaths, $k$ | Number of Corps-Years in Which $k$ Deaths Occurred |
|:---:|:---:|
| 0 | 109 |
| 1 | 65 |
| 2 | 22 |
| 3 | 3 |
| 4 | 1 |
| | 200 |

**6.58** Flaws in one type of metal sheeting occur at an average rate of one per 10 ft². What is the probability of two or more flaws in a 5-by-8-ft sheet?

**6.59** The Brown's Ferry incident of 1975 focussed national attention on the ever-present danger of fires breaking out in nuclear power plants. The Nuclear Regulatory Commission has estimated that with present technology there will be, on the average, one fire for every 10 nuclear-reactor years. Suppose a company puts two reactors on line in 1990. Assuming the incidence of fires can be described by a Poisson distribution, find the probability that by 1995 at least two fires will have occurred.

**6.60** In the 432 years from 1500 to 1931, war broke out somewhere in the world 299 times (136). (By definition a military action was a war if it was legally declared, involved over 50,000 troops, or resulted in significant boundary realignments. To achieve greater uniformity from war to war, major confrontations were split into smaller "subwars": World War I, for example, was treated as five separate wars.) Let $X$ denote the number of wars beginning in a given year (see the table). Can $X$ be considered a Poisson random variable?

| Number of Wars ($k$) Beginning in a Given Year | Observed Frequency |
|:---:|:---:|
| 0 | 223 |
| 1 | 142 |
| 2 | 48 |
| 3 | 15 |
| 4+ | 4 |
| | 432 |

# 6.6

## CENTRAL TENDENCY AND DISPERSION

We have seen that pdfs provide a global overview of a random variable's behavior; that is, they give the probability associated with each of its possible values. Detail that explicit, though, is not always necessary or even helpful. Sometimes a better strategy is to focus the information in $f_Y(y)$ by summarizing some of its features with special numbers. The search for those numbers, and how they should be applied and interpreted, is the primary topic of this section.

The first feature of a pdf that we will examine is *central tendency*, or *location*. The most frequently used measured of central tendency is the *expected value*. We will see that the expected value of a random variable (or, equivalently, of its pdf) is a generalized version of the familiar concept of an arithmetic average.

**expected value**—A number that measures the "center" of a pdf. $E(Y)$ does for a probability density function what $\bar{y}$ does for a sample.

A particularly useful way to think about expected values (which also helps explain their name) is in the context of gambling. If a random variable's values represent the amounts won (or lost) on the different possible outcomes in a game of chance, then the expected value gives the amount won (or lost), *on the average*, each time the game is played.

The casino game of roulette provides a good illustration of the notion of expected value. A roulette wheel contains 38 slots, numbered

$$00, \quad 0, \quad 1, \quad 2, \quad \ldots, \quad 36$$

The 00 and 0 slots are green, 18 slots are red, and 18 are black. The wheel is spun, and a metal ball is thrown around its rim in the opposite direction. A few seconds later the ball comes to rest in one of the slots. The number and color of that slot are declared winners (e.g., 32, red). Although many a disgruntled gambler believes otherwise, roulette wheels are carefully balanced, and each of the 38 possible outcomes is equally likely.

Suppose we bet $10 on red. If the random variable $Y$ denotes our winnings, then $Y$ takes the value $+10$ if the ball comes to rest in a red slot, and $-10$ (that is, we lose $10) if it lands in either a black slot or a green slot. What is the pdf for $Y$? By the equally likely assumption,

$$f_Y(+10) = P(Y = +10) = \frac{18}{38} = \frac{9}{19} \tag{6.8}$$

and
$$f_Y(-10) = P(Y = -10) = \frac{20}{38} = \frac{10}{19} \tag{6.9}$$

Equations 6.8 and 6.9 imply that 9/19 of the time we will win $10 and 10/19 of the time we will lose $10. Intuitively, then, if we persist in this foolishness, we stand to lose, on the average, 53¢ each time we play the game:

$$\text{Expected winnings} = \$10\left(\frac{9}{19}\right) + (-\$10)\left(\frac{10}{19}\right)$$

$$= -\$0.53$$

The quantity $-0.53$ is the expected value of $Y$. (For the casino the expected value of a customer's $10 bet is *plus* $0.53. That's why casinos stay in business and gamblers go broke!)

Physically, an expected value can be thought of as the "center of gravity" for $f_Y(y)$. Figure 6.10 shows two bars of "weight" 10/19 and 9/19 positioned along an axis at the points $-10$ and $+10$, respectively. If a fulcrum were placed at $-0.53$ the weights would be in balance. Actually the expected value of every random variable has that same property. It thus makes sense to interpret the expected value, in general, as a measure of the center of a pdf.

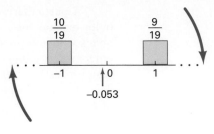

**FIG. 6.10**

---

**6.61** In roulette a player can also bet on a single number rather than on a color. If that number comes up, a $10 wager wins $360. Otherwise, the $10 is lost. What is the corresponding expected value?

**6.62** European casinos use roulette wheels that have a 0 but no 00.

(a) Calculate the expected value of a $10 bet on red in a European casino.

(b) If a trip to Europe costs $1500, how much would a player have to bet in order to justify gambling in Europe rather than Las Vegas?

### A Formula for Expected Value

Suppose $Y$ is a random variable taking the values 1, 3, 5, 7, and 9, each with probability 1/5. What is the expected value of $Y$? Simply the arithmetic average of all its possible values:

$$\text{Expected value of } Y = 1\left(\frac{1}{5}\right) + 3\left(\frac{1}{5}\right) + 5\left(\frac{1}{5}\right) + 7\left(\frac{1}{5}\right) + 9\left(\frac{1}{5}\right)$$

$$= 4.8$$

More generally, if we had denoted the values of $Y$ by $y_1 = 1$, $y_2 = 3$, $y_3 = 5$, $y_4 = 7$, and $y_5 = 9$, we would have written the expected value as

$$\text{Expected value of } Y = y_1 f_Y(y_1) + y_2 f_Y(y_2) + y_3 f_Y(y_3) + y_4 f_Y(y_4) + y_5 f_Y(y_5)$$

$$= \sum_{i=1}^{5} y_i f_Y(y_i) \tag{6.10}$$

Equation 6.10 suggests that the expected value is a "weighted" average, each value of $Y$ being weighted by the probability of its occurrence. Definition 6.9 states that result formally. It applies to any random variable whose sample space is finite (like the binomial) or countably infinite (like the Poisson).

---

**DEFINITION 6.9**

Let $Y$ be a random variable with pdf $f_Y(y)$. The **expected value** (or **mean**) **of** $Y$ is denoted $E(Y)$, or $\mu$, where

$$E(Y) = \mu = \sum_{\text{all } y} y f_Y(y) \qquad (6.11)$$

---

*EXAMPLE 6.19*   Let $Y$ be the number of successes in $n = 4$ Bernoulli trials, with

$$p = P\,(\text{success}) = \frac{1}{3}$$

Find $E(Y)$.

*Solution*   Letting $n = 4$ and $p = 1/3$ in Equation 6.1 gives the values in columns 1 and 2 of Table 6.13. Multiplying the entries in those two columns row by row (see column 3) generates the terms whose sum is $E(Y)$. Therefore,

$$E(Y) = \sum_{\text{all } y} y f_Y(y) = 0 + \frac{32}{81} + \frac{48}{81} + \frac{24}{81} + \frac{4}{81} = \frac{4}{3}$$

**TABLE 6.13**

| $y$ | $f_y(y) = {}_4C_y\left(\dfrac{1}{3}\right)^y\left(\dfrac{2}{3}\right)^{4-y}$ | $y f_Y(y)$ |
|:---:|:---:|:---:|
| 0 | $\dfrac{16}{81}$ | $0 \cdot \dfrac{16}{81} = 0$ |
| 1 | $\dfrac{32}{81}$ | $1 \cdot \dfrac{32}{81} = \dfrac{32}{81}$ |
| 2 | $\dfrac{24}{81}$ | $2 \cdot \dfrac{24}{81} = \dfrac{48}{81}$ |
| 3 | $\dfrac{8}{81}$ | $3 \cdot \dfrac{8}{81} = \dfrac{24}{81}$ |
| 4 | $\dfrac{1}{81}$ | $4 \cdot \dfrac{1}{81} = \dfrac{4}{81}$ |
| | | $\dfrac{108}{81}$ |

Notice in this example that $E(Y)$ is numerically equal to $n$ times $p$: $E(Y) = 4(\frac{1}{3}) = \frac{4}{3}$. This equality is no coincidence—it holds for any binomial random variable. It also saves us an enormous amount of work. Think of how difficult it would be to find $E(Y)$ using Equation 6.11 if $n$ was 100.

---

### THEOREM 6.2

If $Y$ is the number of successes in $n$ independent Bernoulli trials, where $p = P$ (success), then

$$E(Y) = np. \qquad (6.12)$$

---

*EXAMPLE 6.20*   In the stock fluctuation problem of Example 6.8, let $Y$ denote the number of days (out of four) that the stock goes up. Find $E(Y)$.

*Solution*   We could find $E(Y)$ by using Equation 6.11, but Equation 6.12 is much quicker. Note that $Y$ is a binomial random variable—it represents the number of successes (stock price increases) in $n = 4$ independent Bernoulli trials. Also,

$$p = P \text{ (success)} = P \text{ (stock price goes up)} = \frac{1}{4}$$

The expected number of increases, then, is 1:

$$E(Y) = np = 4 \cdot \frac{1}{4} = 1$$

---

*EXAMPLE 6.21*   Suppose an urn contains 10 chips, 4 red and 6 white. Three are drawn at random without replacement. On the average, how many red chips will be in the sample?

*Solution*   Let $Y$ denote the number of red chips in the sample. Recalling Section 6.4, we should recognize that $Y$ is a hypergeometric random variable with $r = 4$, $w = 6$, $N = r + w = 10$, and $n = 3$. Its pdf, therefore, is given by Equation 6.4:

$$P(Y = k) = f_Y(k) = \frac{{}_4C_k \cdot {}_6C_{3-k}}{{}_{10}C_3}, \qquad k = 0, 1, 2, 3$$

Table 6.14 shows the computations necessary to find $E(Y)$.

### TABLE 6.14

| $k$ | $f_Y(k) = \dfrac{{}_4C_k \cdot {}_6C_{3-k}}{{}_{10}C_3}$ | $kf_Y(k)$ |
|---|---|---|
| 0 | 20/120 | 0 |
| 1 | 60/120 | 60/120 |
| 2 | 36/120 | 72/120 |
| 3 | 4/120 | 12/120 |

Adding the entries in the last column gives

$$E(Y) = \sum_{k=0}^{3} k f_Y(k) = 0 + \frac{60}{120} + \frac{72}{120} + \frac{12}{120} = \frac{6}{5}$$

We saw in Example 6.20 that a simple formula can be found for the expected value of a binomial random variable. Is there a similar shortcut when the random variable is hypergeometric? Yes.

---

**THEOREM 6.3**

---

Suppose an urn contains $r$ red chips and $w$ white chips. Let $n$ chips be drawn without replacement. If the random variable $Y$ denotes the number of red chips in the sample, then

$$E(Y) = n\left(\frac{r}{r+w}\right) \tag{6.13}$$

---

**EXAMPLE 6.22**  Nevada keno is a game whose odds are based on the hypergeometric distribution (recall Example 6.11). In general, the probability of catching $k$ winners is

$$P(Y = k) = f_Y(k) = \frac{{}_{20}C_k \cdot {}_{60}C_{10-k}}{{}_{80}C_{10}}, \qquad k = 0, 1, \ldots, 10$$

Suppose a gambler bets $1 on a game of Keno. How much can he expect to win?

**Solution**  According to the rules of the game, the amount $W$ a gambler wins depends on $k$. If fewer than 5 of the gambler's 10 picks are among the 20 chosen by the caller, he loses the amount bet ($1). If exactly five match, he wins $2; if six match, he wins $18. Columns 1 and 2 of Table 6.15 show all the possible payoffs for a $1 bet.

We calculate the probabilities in column 3 associated with each payoff by using Equation 6.4. A payoff of $2 occurs, for example, when $Y = 5$. Therefore, $f_W(2) = f_Y(5) = 0.0514$. Similarly, $f_W(18) = f_Y(6) = 0.0115$, and so on. *Our objective is to find $E(W)$.* Adding the entries in column 4 of Table 6.15 gives

$$E(W) = \sum_{\text{all } w} w f_W(w) = -0.935 + 0.103 + \cdots + 0.001 = -0.144$$

On the average, then, the gambler loses (and the house wins) almost 15¢ out of every dollar wagered.

**TABLE 6.15**

| Number of Matches, $k$ | Payoff, $w$ | $f_W(w)$ | $wf_W(w)$ |
|:---:|:---:|:---:|:---:|
| <5 | −1 | 0.935 | −0.935 |
| 5 | 2 | 0.0514 | 0.103 |
| 6 | 18 | 0.0115 | 0.207 |
| 7 | 180 | 0.0016 | 0.288 |
| 8 | 1,300 | $1.35 \times 10^{-4}$ | 0.176 |
| 9 | 2,600 | $6.12 \times 10^{-6}$ | 0.016 |
| 10 | 10,000 | $1.12 \times 10^{-7}$ | 0.001 |
| | | | −0.144 |

*EXAMPLE 6.23*   Suppose that 50 people are to be given a blood test to see who might have a certain disease. The obvious laboratory procedure would be to examine each person's blood, meaning that 50 tests would eventually be run. Is that the most efficient approach? Probably not. It certainly sounds reasonable, but if the disease being looked for is rare, testing all 50 samples individually is foolish. To see why is an exercise in expected values.

*Solution*   Consider the following alternative. Divide each of the 50 samples into two parts, $A$ and $B$. Put all 50 $A$'s into a common container and test the pooled sample. If that combined sample is negative for the disease, then all 50 individuals must be negative. One test in that event has done the work of 50. If the pooled sample is positive for the disease, we need to go back and retest all the "$B$'s" individually, meaning that 51 tests will be required.

The key in comparing individual testing with pooled testing is the expected number of tests required for the latter. Let $p$ denote the probability of a person having the disease, and let $Y$ denote the number of tests required if the samples are pooled. From what we have just described, $Y$ is a random variable that takes only two possible values, 1 or 51: ·

$$P(Y = 1) = f_Y(1) = P \text{ (pooled sample is negative)}$$

$$= P \text{ (all 50 subjects are negative)}$$

$$= P \text{ (1st subject is negative } and \text{ 2nd subject is negative } and \ldots \text{ 50th subject is negative)}$$

$$= P \text{ (1st subject is negative)} P \text{ (2nd subject is negative)} \ldots P \text{ (50th subject is negative)} \quad (\text{why?})$$

$$= (1 - p)^{50}$$

and
$$P(Y = 51) = f_Y(51) = P \text{ (pooled sample is positive)}$$

$$= 1 - P \text{ (pooled sampled is negative)}$$

$$= 1 - (1 - p)^{50}$$

The number of tests required by the pooling strategy, on the average, is an application of Equation 6.11:

$$E(Y) = \text{Expected number of tests} = \sum_{\text{all } y} y f_Y(y)$$

$$= 1 f_Y(1) + 51 f_Y(51)$$

$$= 1(1-p)^{50} + 51(1-(1-p)^{50})$$

$$= 51 - 50(1-p)^{50}$$

**TABLE 6.16**

| $p$ | $E(Y)$ |
|---|---|
| 0.5 | 51.0 |
| 0.1 | 50.7 |
| 0.01 | 20.8 |
| 0.001 | 3.4 |
| 0.0001 | 1.2 |

Table 6.16 shows $E(Y)$ as a function of $p$. For example, when each subject has a probability of 0.001 of having the disease, pooling requires, on the average, only 3.4 tests to reach a final answer, as compared with 50 if the samples are analyzed individually.

**QUESTIONS**

**6.63** Find $E(Y)$ for a random variable having the following pdf:

| $y$ | $f_Y(y)$ |
|---|---|
| −2 | 2/14 |
| −1 | 3/14 |
| 0 | 4/14 |
| 1 | 3/14 |
| 2 | 2/14 |

**6.64** Given the following $f_Y(y)$, find $E(Y)$:

| $y$ | $f_Y(y)$ |
|---|---|
| −1 | 0.1 |
| 0.25 | 0.2 |
| 0.50 | 0.3 |
| 1 | 0.4 |

**6.65** Swampwater Tech has not had a particularly good football team for the past 10 years. Five times the Muskrats won only two games, twice they won three games, and in three seasons they went winless. How many games did they win on the average? (Hint: If $Y$ denotes the number of games won in a given season, what is $f_Y(0)$? $f_Y(2)$? $f_Y(3)$?)

**6.66** A couch potato has been keeping records of the number of commercials aired during one of his favorite late-night reruns. He has seen 20 different episodes of the program and endured the following distribution of commercials:

| Number of Commercials | Number of Programs |
|:---------------------:|:------------------:|
| 2 | 3 |
| 3 | 5 |
| 4 | 10 |
| 5 | 1 |
| 6 | 1 |

Let $Y$ denote the number of commercials aired per show. Find $E(Y)$. (Hint: Use column 2 to find $f_Y(y)$.)

6.67　Suppose $Y$ is a binomial random variable with $n = 4$ and $p = 0.25$. Find $E(Y)$, using (a) Definition 6.9 and (b) Theorem 6.2.

6.68　Suppose $Y$ is a hypergeometric random variable with $r = 3$, $w = 4$, and $n = 5$. Find $E(Y)$, using (a) Definition 6.9 and (b) Theorem 6.3.

6.69　Refer to Case Study 2.3. Assume that each bee landing on the test table can be thought of as an independent Bernoulli trial. Let $p$ denote the probability that a bee lands on a flower with lines.

(a) If $Y$ represents the *number* of bees landing on flowers with lines, what is $E(Y)$?

(b) What is $E(Y)$ if the bees are not influenced by nectar guides?

6.70　If a baseball player has a .280 batting average, how many hits will he get, on the average, in a game where he has five official at-bats?

6.71　In your pocket are three quarters and four nickels. You take out two coins at random. Let $Y$ denote the total value of the two coins selected. Find $E(Y)$.

6.72　Three chips are drawn from an urn containing six chips numbered 1 through 6. Let the random variable $Y$ denote the largest number in the sample. Find $E(Y)$. (Hint: See Example 4.26.)

6.73　Country $X$ launches a tactical nuclear strike at Country $Y$. Six missiles are fired, four of which carry nuclear warheads. Country $Y$ has a defense system that enables it to shoot down three incoming missiles at random. On the average, how many missiles with nuclear warheads will hit country $Y$?

6.74　Recall Example 5.30. How many red lights will the commuter encounter on a typical ride to work?

6.75　If a game is fair, its expected payoff is zero. Look again at Table 6.15. If all the other payoffs stayed the same, how much would the gambler have to win when $k = 5$ in order for keno to have $E(W) = 0$?

### Variance and Standard Deviation

Expected values serve well as measures of central tendency, but $E(Y)$ alone does not summarize everything we need to know about a random variable's behavior. We at least need more information about the variable's dispersion (how spread out its distribution is). A football squad, for example, may have an interior line consisting of five players weighing 200-lb, or it may team four 150-lb men with a 400-lb behemoth at left tackle. In both cases the *average* player weight is 200 lb, but the *variation* in weights is quite different, and would certainly not go unnoticed by the opposing coach.

How should we measure dispersion? We could average, in the generalized sense, the deviations of the random variable $Y$ from their "center" $E(Y)$. In other words, we might define dispersion as $E(Y - \mu)$, where $\mu = E(Y)$. In its favor, $E(Y - \mu)$ has the beauty of simplicity; unfortunately, it also has the fatal drawback of being useless, because $E(Y - \mu) = 0$ for *any* random variable. Why? Because the positive deviations from $\mu$ equal the negative deviations, so their "average" is zero. Another strategy that *does* work is to measure dispersion in terms of *squared* deviations.

---

**DEFINITION 6.10**

The **variance** of a random variable $Y$, denoted Var $(Y)$ or $\sigma^2$, is the expected value of its squared deviation from $\mu = E(Y)$:

$$\text{Var}\,(Y) = E[(Y - \mu)^2] = \sum_{\text{all } y} (y - \mu)^2 f_Y(y) \qquad (6.14)$$

---

An unfortunate consequence of Definition 6.10 is that the variance and the random variable have different units. If $Y$, for example, is a linear measurement, say feet, then Var $(Y)$ is in square feet, which makes it a measurement of area. It becomes difficult, then, to relate the value of the variance to the original values of $Y$. For that reason, Var $(Y)$ is often replaced as a measure of variability by its square root.

---

**DEFINITION 6.11**

The **standard deviation** $\sigma$ of a random variable $Y$ is the square root of its variance:

$$\sigma = \text{standard deviation of } Y = \sqrt{\text{Var}\,(Y)}$$
$$= \sqrt{\sum_{\text{all } y} (y - \mu)^2 f_Y(y)} \qquad (6.15)$$

---

**EXAMPLE 6.24**    Let $Y$ be a binomial random variable with $n = 4$ and $p = 1/3$ (see Example 6.19). Find the variance and the standard deviation of $Y$.

*Solution*    By Equation 6.12, $E(Y) = \mu = np = 4(\frac{1}{3}) = \frac{4}{3}$. Table 6.17 shows the intermediate steps needed to evaluate Equation 6.14. Adding the entries in the last column gives Var $(Y)$:

$$\text{Var }(Y) = \sum_{\text{all } y} (y - \mu)^2 f_Y(y) = \frac{256}{729} + \frac{32}{729} + \cdots + \frac{64}{729} = 0.89$$

**TABLE 6.17**

| $y$ | $f_Y(y)$ | $y - \mu$ | $(y - \mu)^2$ | $(y - \mu)^2 f_Y(y)$ |
|---|---|---|---|---|
| 0 | $\dfrac{16}{81}$ | $-\dfrac{4}{3}$ | $\dfrac{16}{9}$ | $\dfrac{256}{729}$ |
| 1 | $\dfrac{32}{81}$ | $-\dfrac{1}{3}$ | $\dfrac{1}{9}$ | $\dfrac{32}{729}$ |
| 2 | $\dfrac{24}{81}$ | $\dfrac{2}{3}$ | $\dfrac{4}{9}$ | $\dfrac{96}{729}$ |
| 3 | $\dfrac{8}{81}$ | $\dfrac{5}{3}$ | $\dfrac{25}{9}$ | $\dfrac{200}{729}$ |
| 4 | $\dfrac{1}{81}$ | $\dfrac{8}{3}$ | $\dfrac{64}{9}$ | $\dfrac{64}{729}$ |

Therefore, by Equation 6.15

$$\sigma = \sqrt{0.89} = 0.94$$

---

**QUESTIONS**    6.76    Suppose the random variable $Y$ has the pdf shown below. Find Var $(Y)$.

| $y$ | $f_Y(y)$ |
|---|---|
| $-2$ | $\dfrac{2}{14}$ |
| $-1$ | $\dfrac{3}{14}$ |
| $0$ | $\dfrac{4}{14}$ |
| $1$ | $\dfrac{3}{14}$ |
| $2$ | $\dfrac{2}{14}$ |

6.77 Find the variance and the standard deviation of a random variable $Y$ with the following pdf:

| $y$ | $f_Y(y)$ |
|------|---------|
| $-1$ | 0.1 |
| 0.25 | 0.2 |
| $-0.50$ | 0.3 |
| $-1$ | 0.4 |

6.78 For the Muskrat data in Question 6.65, let $Y$ denote the number of victories won in a given season. Find Var $(Y)$.

6.79 An urn contains four chips numbered 1 through 4. Two are drawn without replacement. Let $Y$ be the larger of the two numbers drawn. Find Var $(Y)$. (Hint: Start by listing the $_4C_2$ possible samples. What probability is associated with each one?)

6.80 In Question 6.66 compute the standard deviation of the number of commercials broadcast during a late-night television program.

### Another Variance Formula

Equation 6.14 is not always the most convenient way to compute a variance. For some random variables Equation 6.16 is easier to use. Algebraically, both equations are equivalent.

---

**THEOREM 6.4**

For any random variable $Y$,

$$\text{Var}(Y) = E(Y^2) - \mu^2 = \sum_{\text{all } y} y^2 f_Y(y) - \mu^2 \qquad (6.16)$$

---

*EXAMPLE 6.25* An urn contains five chips, two red and three white. Two are drawn at random. Let $Y$ denote the number of red chips in the sample. Verify that Equations 6.14 and 6.16 give the same value for Var $(Y)$.

*Solution* We need the pdf of $Y$. But $Y$ is hypergeometric, so

$$f_Y(y) = P(Y = y) = \frac{_2C_y \cdot {}_3C_{2-y}}{_5C_2}, \qquad y = 0, 1, 2$$

**TABLE 6.18**

| $y$ | $f_Y(y)$ | $yf_Y(y)$ | $y^2f_Y(y)$ | $(y-\mu)^2$ | $(y-\mu)^2f_Y(y)$ |
|---|---|---|---|---|---|
| 0 | $\dfrac{3}{10}$ | 0 | 0 | $\dfrac{64}{100}$ | $\dfrac{192}{1000}$ |
| 1 | $\dfrac{6}{10}$ | $\dfrac{6}{10}$ | $\dfrac{6}{10}$ | $\dfrac{4}{100}$ | $\dfrac{24}{1000}$ |
| 2 | $\dfrac{1}{10}$ | $\dfrac{2}{10}$ | $\dfrac{4}{10}$ | $\dfrac{144}{100}$ | $\dfrac{144}{1000}$ |
| | | $\dfrac{8}{10}$ | $\dfrac{10}{10}$ | | $\dfrac{360}{1000}$ |
| | | $\dfrac{10}{10}$ | | | |

Table 6.18 shows all the terms appearing on the right sides of $E(Y)$, $E(Y^2)$, and the two formulas for Var$(Y)$. Adding the entries in column 3 gives $E(Y)$:

$$E(Y) = \sum_{\text{all } y} yf_Y(y) = 0 + \frac{6}{10} + \frac{2}{10} = \frac{8}{10}$$

The sum of the entries in column 4 is $E(Y^2)$:

$$E(Y^2) = \sum_{\text{all } y} y^2f_Y(y) = 0 + \frac{6}{10} + \frac{4}{10} = 1$$

By Equation 16.16

$$\text{Var}(Y) = E(Y^2) - \mu^2 = 1 - \left(\frac{8}{10}\right)^2 = \frac{36}{100}$$

To find Var$(Y)$ using Equation 6.14, we need to add the entries in column 6:

$$\text{Var}(Y) = \sum_{\text{all } y} (y-\mu)^2f_Y(y) = \frac{192}{1000} + \frac{24}{1000} + \frac{144}{1000} = \frac{36}{100}$$

As claimed, both answers for Var$(Y)$ are the same.

---

**QUESTIONS**

6.81    For a random variable $Y$ with pdf

| $y$ | $f_Y(y)$ |
|---|---|
| 0 | 0.2 |
| 1 | 0.7 |
| 4 | 0.1 |

use Equation 6.16 to find Var$(Y)$.

6.82    Use Equation 6.16 to find the variance for the pdf in Question 6.76.

6.83    A statistically minded cashier has kept records on the number of items purchased by customers in the express lane at a supermarket. Her data are summarized below. Find Var$(Y)$ two different ways.

| Number of items, $y$ | $f_Y(y)$ |
|:---:|:---:|
| 1 | 0.1 |
| 2 | 0.1 |
| 3 | 0.2 |
| 4 | 0.4 |
| 5 | 0.2 |

### Additional Variance Formulas

Earlier we pointed out that the binomial and hypergeometric random variables have special formulas for $E(Y)$. The variances of those two variables also have special formulas.

---

**THEOREM 6.5**   Variance of a Binomial Random Variable

Suppose $Y$ is the number of successes in $n$ independent Bernoulli trials, where $p = P$ (success). Then

$$\text{Var}(Y) = np(1 - p) = npq \qquad (6.17)$$

---

**THEOREM 6.6**   Variance of a Hypergeometric Random Variable

Suppose an urn contains $r$ red chips and $w$ white chips. A sample of size $n$ is drawn. Let $Y$ denote the number of red chips in the sample. Then

$$\text{Var}(Y) = \frac{nrw(N - n)}{N^2(N - 1)} \qquad (6.18)$$

where $N = r + w$.

---

QUESTIONS

**6.84**   Verify Equation 6.17 for the binomial random variable described in Example 6.66.

**6.85**   Suppose a bowler has a 30% chance of throwing a strike to lead off any given frame. Let $Y$ denote the number of strikes rolled in a 10-frame game. Find Var $(Y)$

**6.86**   Refer to Question 6.70. What is the variance of the number of hits that a player with a .280 average makes in a game in which he comes to bat five times?

**6.87**   Verify Equation 6.18 for the urn problem described in Example 6.25.

**6.88**   Six men and four women, all equally qualified, apply for three lab technician positions. Lacking any rational basis for ranking the candidates from best to worst, the personnel director decides to select three at random. Let $Y$ represent the number of men hired. Find the standard deviation of $Y$.

# 6.7

## CONTINUOUS DISTRIBUTIONS

The random variables considered so far have all been *discrete*; in other words, they have had a finite or a countably infinite number of possible values. In the sense that experimenters are limited by the accuracy of their measuring instruments, all random variables are finite. (We can't determine a person's weight, for example, to an infinite number of decimal places.) Nevertheless, we can effectively model many phenomena by using random variables that we assume are capable of taking on *any* number in a certain range. Such random variables are said to be *continuous* and the sets of possible $y$ values they can equal are said to be *uncountably infinite*.

Other areas in mathematics and science find it useful to construct similar models. In geometry, for example, a line is considered to be an uncountable infinity of dimensionless points. In physics, time is treated as a continuous flow of instantaneous moments. For our situation, "dimensionless points" and "instantaneous moments" are analogous to saying that a continuous random variable has the property that $P(Y = y) = 0$ for any number $y$.

The motivation for introducing continuous random variables becomes even clearer if we put the problem in its historical perspective. In general, the notion of continuous quantities has pervaded science and mathematics since the classical era, so being precise about the origins of the special case of continuous random variables is difficult. Even so, we can safely say that solving the problem of approximating binomial probabilities had a major impact on the development of the notion of continuous random variables.

We have already seen that binomial random variables provide useful models for many real-world phenomena. Most of the examples we considered, though, involved a rather small $n$, the result being that the actual computation of binomial probabilities was not a significant problem. But suppose $Y$ is a binomial random variable with $n = 1600$, $p = 0.80$, and we need a numerical value for the probability that $Y$ lies between 1240 and 1325, inclusive. The appropriate formula is

$$P(1240 \leqslant Y \leqslant 1325) = \sum_{k=1240}^{1325} {}_{1600}C_k (0.80)^k (0.20)^{1600-k} \qquad (6.19)$$

but evaluating that formula is not easy. Even using a computer does not guarantee success: a typical programmer would most likely be unable to calculate the sum indicated on the right-side of Equation 6.19. Then how could a scientist of the 1700s—someone without access to an Apple or an IBM—handle such a problem? By making use of some very sophisticated approximations. Indeed, the discovery of an enormous amount of mathematics was driven by the need to obtain numerical values for complicated formulas.

### The De Moivre-Laplace Theorem

The first step in dealing with the numerical evaluation of intimidating expressions like Equation 6.19 is to realize that exact solutions are probably not possible. Instead, we seek good approximations. For binomial problems, the first useful approximation was derived in 1718 by the English mathematician, Abraham De Moivre. However, his result did not initially attract the attention it deserved. Almost 100 years later, though, it reappeared, somewhat generalized and without credit, in Laplace's *Theorie Analytique des Probabilities*.

---

**THEOREM 6.7** De Moivre-Laplace

Let $X$ be the number of successes in $n$ Bernoulli trials, each having probability $\rho$ of success. For $n$ sufficiently large[1] and for any numbers $c$ and $d$, where $c < d$,

$$P\left(c \leqslant \frac{X - np}{\sqrt{np(1 - p)}} \leqslant d\right)$$

is approximately equal to the area under the curve

$$y = \left(\frac{1}{\sqrt{2\pi}}\right) e^{-z^2/2} \qquad \text{from} \quad z = c \text{ to } z = d$$

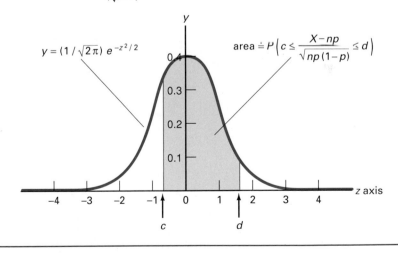

---

[1] The phrase "$n$ sufficiently large" may sound unnecessarily vague, but to make it precise would take us too far afield. A workable rule of thumb is to assume that $n$ is large enough (that is, the approximation can be used) if $np \geqslant 5$ and $n(1 - p) \geqslant 5$.

COMMENT:   The "$e$" on the right-hand-side of

$$y = \left(\frac{1}{\sqrt{2\pi}}\right)e^{-z^2/2}$$

is the same constant that appeared in the Poisson pdf described in Section 6.5. Although it has no exact decimal representation, its numerical value is approximately 2.718. Theoretically, $e$ is the sum of an infinite number of reciprocals of factorials:

$$e = 1 + 1 + \frac{1}{2!} + \frac{1}{3!} + \frac{1}{4!} + \cdots \doteq 2.718.$$

It is convenient to have special notation for the curve pictured in Theorem 6.7. We let $\phi(z)$ denote the height of the curve at the point $z$. That is,

$$y = \phi(z) = \left(\frac{1}{\sqrt{2\pi}}\right)e^{-z^2/2}$$

for any number $z$. Also, we use $\Phi(c)$ to represent the *area* under the curve to the left of $c$ (see Figure 6.11). (The *total* area under the curve, $\Phi(\infty)$, is 1.)

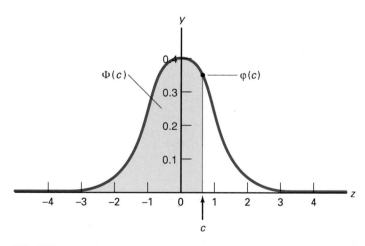

**FIG. 6.11**

Finding areas under the curve $y = \phi(z)$ will be necessary for solving a number of different kinds of problems. Fortunately, we will not have to work directly with the function. Areas under $\phi(z)$ are tabulated in Appendix A.1.

### Finding Areas under $\Phi(z)$

*Area to the Left of a Point c.* This area is precisely $\Phi(c)$. For example, suppose we want to find the area under the curve to the left of $c = 1.79$. Look in the far left column of Appendix A.1 for the value 1.7. Move across that row to the column headed 9. The entry 0.9633 is the area to the left of 1.79. That is,

$$\Phi(1.79) = 0.9633$$

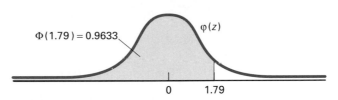
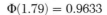

*Area to the Right of a Point c.* Since the total area under $\Phi(z)$ is 1, the area to the right of $c$ is $1 - \Phi(c)$. For example, the area to the right of the point 1.64 is

$$1 - \Phi(1.64) = 1 - 0.9495 = 0.0505.$$

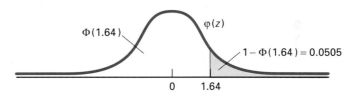

*Area between Two Points c and d.* The area we want is the total area lying to the left of $d$ minus the area lying to the left of $c$ (i.e., $\Phi(d) - \Phi(c)$). For instance, the area under $\phi(z)$ between $c = -0.65$ and $d = 1.32$ is

$$\Phi(1.32) - \Phi(-0.65) = 0.9066 - 0.2578 = 0.6488.$$

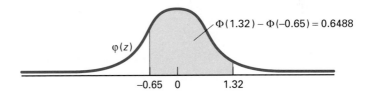

**QUESTIONS**

6.89  Use Appendix A.1 to find the following areas under the curve $y = \phi(z)$. In each case draw a picture of the area in question.

(a) the area to the left of $c = -0.67$

(b) the area to the left of $c = 0.00$

(c) the area to the right of $c = 1.23$

(d) the area to the right of $c = -0.94$

(e) the area between $c = 0.52$ and $d = 2.68$

(f) the area between $c = -2.68$ and $d = -0.52$

**6.90** Draw a diagram of the following areas under $y = \varphi(z)$ and use Appendix A.1 to evaluate each one numerically.

(a) the area to the left of $c = 1.23$

(b) the area to the right of $c = -2.16$

(c) the area between $c = -0.07$ and $d = 1.47$

(d) the area to the left of $c = -6.72$

### Approximating Binomial Probabilities

We now show how the De Moivre-Laplace theorem is used to approximate binomial probabilities. Notice in each case the way the original interval of $X$'s is transformed into an equivalent interval of $\dfrac{X - np}{\sqrt{np(1 - p)}}$'s.

---

*EXAMPLE 6.26*  Suppose that $X$ is a binomial random variable with $n = 25$ and $p = 0.4$. Approximate $P(8 \leqslant X \leqslant 12)$.

*Solution*  Theorem 6.7 does not apply to $X$ directly; instead, we must "standardize" the values of $X$. Specifically, we must subtract $np = 25(0.4) = 10$ from each term of the original inequality and then divide each difference by

$$\sqrt{np(1 - p)} = \sqrt{25(0.4)(0.6)} = \sqrt{6} = 2.449$$

Therefore,

$$P(8 \leqslant X \leqslant 12) = P\left( \frac{8 - 10}{2.449} \leqslant \frac{X - 10}{2.449} \leqslant \frac{12 - 10}{2.449} \right)$$

$$= P\left( -0.82 \leqslant \frac{X - 10}{2.449} \leqslant 0.82 \right)$$

Recognizing that the latter expression has the format of Theorem 6.7, we can write

$$P(8 \leqslant X \leqslant 12) \doteq \Phi(0.82) - \Phi(-0.82) = 0.7939 - 0.2061 = 0.5878$$

---

*EXAMPLE 6.27*  A radio manufacturer produces electronic components domestically but subcontracts cabinets to a foreign supplier. The foreign supplier is inexpensive but has poor quality control: on the average, only 80% of the shipment will be usable. Suppose the manufacturer purchases 1600 cabinets and needs at least 1240 usable ones to fill orders already on hand. Suppose, also, that there is storage space for only 1325 cabinets. What is the probability that, of the 1600 cabinets delivered, between 1240 and 1325 will be usable?

*Solution*    Let $X$ denote the number of usable cabinets (out of 1600). Our objective is to find $P(1240 \leqslant X \leqslant 1325)$. Equation 6.19 is a formula for the *exact* probability that $X$ lies in that particular range. Fortunately, $n$ in this case is so large that the approximation in Theorem 6.7 is more than adequate.

Note that

$$np = 1600(0.80) = 1280 \quad \text{and} \quad \sqrt{np(1-p)} = \sqrt{1600(0.80)(0.20)} = 16$$

Therefore, from the De Moivre-Laplace approximation,

$$P(1240 \leqslant X \leqslant 1325) = P\left(\frac{1240 - 1280}{16} \leqslant \frac{X - 1280}{16} \leqslant \frac{1325 - 1280}{16}\right)$$

$$= P\left(-2.50 \leqslant \frac{X - 1280}{16} \leqslant 2.81\right)$$

$$\doteq \Phi(2.81) - \Phi(-2.50) = 0.9975 - 0.0062 = 0.9913$$

Based on the latter figure, the manufacturer should feel reassured: the probability of his needs being met is very high.

---

## CASE STUDY 6.2

*Extrasensory Sense?*    Research in extrasensory perception has ranged from the slightly unconventional to the downright bizarre. Toward the latter part of the nineteenth century and even well into the twentieth century, much of what was done involved spiritualists and mediums. But around 1910, experimenters moved out of the seance parlors and into the laboratory, where they began setting up controlled studies that could be analyzed statistically. In 1938 Pratt and Woodruff, working out of Duke University, did an experiment that became a prototype for an entire generation of ESP research (67).

The investigator and a subject sat at opposite ends of a table. Between them was a screen with a large gap at the bottom. Five blank cards, visible to both participants, were placed side by side on the table beneath the screen. On the subject's side of the screen, one of the standard ESP symbols (see figure below) was hung over each of the blank cards.

The experimenter shuffled a deck of ESP cards, looked at the top one, and concentrated. The subject tried to guess the card's identity. If he thought it was a circle, he would point to the blank card on the table that was beneath the circle card hanging on his side of the screen. The procedure was then repeated. Altogether, 32 subjects, all students, took part in the experiment. They made 60,000 guesses and were correct 12,489 times.

How many correct guesses "should" they have made? With five answers possible, the probability of a subject making a correct identification just by chance is 1/5. Since the binomial model applies, the expected number of correct guesses ($np$) is

$$60{,}000\left(\frac{1}{5}\right) = 12{,}000$$

or 489 fewer than their actual score. Statistically, then, the question is clear: how "near" to 12,000 is 12,489? Should we write off the observed excess of 489 as nothing more than luck, or can we conclude that ESP has been demonstrated?

To effect a resolution here between the conflicting luck and ESP hypotheses, we need to compute the probability, for $p = 1/5$, that the students would get 12,489 *or more* correct answers. Only if the latter probability is very small can 12,489 be construed as evidence in support of ESP. (We will examine the basis for that rationale in Chapter 8.)

Let the random variable $X$ denote the number of correct responses in 60,000 tries. Then

$$P(X \geqslant 12{,}489) = \sum_{x=12{,}489}^{60{,}000} {}_{60{,}000}C_x \left(\frac{1}{5}\right)^x \left(\frac{4}{5}\right)^{60{,}000-x}$$

Rather than compute the 47,512 binomial probabilities indicated in the equation (or the 12,489 making up its complement), we appeal to Theorem 6.7:

$P(X \geqslant 12{,}489)$

$$= P\left(\frac{X - 60{,}000(\frac{1}{5})}{\sqrt{60{,}000(\frac{1}{5})(\frac{4}{5})}} \geqslant \frac{12{,}489 - 60{,}000(\frac{1}{5})}{\sqrt{60{,}000(\frac{1}{5})(\frac{4}{5})}}\right) \doteq 1 - \Phi(4.99)$$

$$= 0.0000003$$

(This last value was obtained from a more extensive version of Appendix A.1.)

The fact that $P(X \geqslant 12{,}489)$ is so extremely small makes the luck hypothesis ($p = 1/5$) untenable. Something other than chance seems to be responsible for the occurrence of so many correct guesses. But it does not follow the ESP has been demonstrated. Flaws in the experimental setup and/or errors in reporting the scores could have inadvertently produced what appears to be a statistically significant result. Suffice it to say that a great many scientists remain highly skeptical of ESP research in general and of the Pratt-Woodruff experiment in particular. [For a more thorough critique of the data we have just described, see (48).]

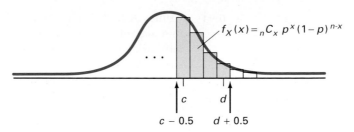

**FIG. 6.12**

COMMENT:   When normal curves are used to estimate binomial probabilities, the quality of the approximation can be improved by using a *continuity correction*. Suppose the bars shown in Figure 6.12 are part of the histogram for a binomial pdf. Notice that the $c$ bar really starts at $c - 0.5$ and the $d$ bar ends at $d + 0.5$. The area under the curve, then, that corresponds to the event $c \leqslant X \leqslant d$ is really the area that extends from $c - 0.5$ to $d + 0.5$. It follows that we should write $P(c \leqslant X \leqslant d)$ as $P(c - 0.5 \leqslant X \leqslant d + 0.5)$ before applying the transformation of Theorem 6.7.

In practice, whether the continuity correction is necessary depends on the value of $n$ and the range of $X$. In Case Study 6.2, for example, the answer is essentially the same whether we compute $P(X \geqslant 12{,}489)$ or $P(X \geqslant 12{,}488.5)$. In Example 6.26, on the other hand,

$$P(8 \leqslant X \leqslant 12) = 0.5878$$

but

$$P(7.5 \leqslant X \leqslant 12.5) = \Phi(1.02) - \Phi(-1.02) = 0.6922$$

QUESTIONS

**6.91**   If $X$ is a binomial random variable with $n = 25$ and $p = 0.4$, use Theorem 6.7 to approximate the probability that $0 \leqslant X \leqslant 12$. Use the continuity correction.

**6.92**   If $X$ is a binomial random variable with $n = 100$ and $p = 0.4$, use Theorem 6.7 to approximate the probability that $38 \leqslant X \leqslant 43$. Use the continuity correction.

**6.93**   Some parapsychologists embrace a theory that a person's ESP ability can be enhanced by hypnosis. To test that hypothesis (24), experimenters hypnotized 15 students and asked them to guess the identity of an ESP card on which another hypnotized person was concentrating. Each student made 100 guesses. The total number correct (out of 1500) was 326. Write the binomial expression giving the probability of the event $X \geqslant 326$ and then use the De Moivre-Laplace theorem to approximate the probability numerically.

**6.94**  Airlines A and B offer identical service on two flights leaving at the same time (meaning the probability of a passenger choosing either is 1/2). Suppose both airlines are competing for the same pool of 400 potential passengers. Airline A sells tickets to everyone who requests one, and the capacity of its plane is 230. Approximate the probability that airline A overbooks.

**6.95**  A new variety of pumpkin seed has a probability of 0.8 of germinating. If a package has 100 seeds, what is the probability that at least 75 will germinate?

**6.96**  Suppose a basketball team has a 72% foul-shooting average. Estimate the probability that they make fewer than 40 of their next 60 attempts. Use the continuity correction.

**6.97**  If 45% of the voters in a metropolitan area favor a wheel tax, approximate the probability that a random sample of 100 voters would show a majority in favor of the proposition. Use the continuity correction.

**6.98**  A random sample of 747 obituaries published in Salt Lake City newspapers revealed that 46% of the decedents died in the three-month period following their birthdays (113). Assess the statistical significance of that finding by estimating the probability that 46% or more would die in that interval if deaths occurred randomly throughout the year. What would you conclude on the basis of your answer?

### Properties of Normal Curves

The argument that areas under $\varphi(z)$ can be used to approximate binomial probabilities suggests that similar areas might be used as theoretical probabilities for continuous phenomena. For example, the data in Table 6.19 summarize the chest measurements of 5738 soldiers, as reported in an 1846 book of letters to the Duke of Saxe-Cobourg and Gotha (128). Figure 6.13 is a graphical representation of the table's first and third columns. Each bar of the histogram has a base 1 in long and is centered on the inch. The bar corresponding to soldiers with a 42-in chest circumference, for example, has its base beginning at 41.5 in and extending to 42.5 in. Its height is $658/5738 = 0.115$ (why?).

Look at the curve superimposed over the histogram. It is not $\varphi(z)$ (which peaks at 0, not 40), but it has the same general bell shape that characterized $\varphi(z)$. Both $\varphi(z)$ and the curve in Figure 6.16 are members of the same family of pdfs. Most of the remainder of this section is devoted to properties and applications of that family.

Before dealing specifically with bell-shaped curves, we need to resolve some questions about continuous random variables in general. For our purposes, $Y$ will be called a continuous random variable if $P(Y = y) = 0$ for every number $y$. *Intervals* of numbers, however, can have nonzero probabilities, and these are assigned through a probability density function (pdf). We introduced probability density functions earlier in connection with discrete random variables, but discrete pdfs and continuous pdfs are quite different.

**TABLE 6.19**

| Measures de la Poitrine | Nombre d'Hommes | Nombre Proportionnel | Probabilité d'après l'Observation | Rang dans la Table | Rang d'après le Calcul. | Probabilité d'après la Table | Nombre d'Observations Calculé |
|---|---|---|---|---|---|---|---|
| Pouces |  |  |  |  |  |  |  |
| 33 | 3 | 5 | 0.5000 |  |  | 0.5000 | 7 |
| 34 | 18 | 31 | 0.4995 | 52 | 50 | 0.4993 | 29 |
| 35 | 81 | 141 | 0.4964 | 42.5 | 42.5 | 0.4964 | 110 |
| 36 | 185 | 322 | 0.4823 | 33.5 | 34.5 | 0.4854 | 323 |
| 37 | 420 | 732 | 0.4501 | 26.0 | 26.5 | 0.4531 | 732 |
| 38 | 749 | 1305 | 0.3769 | 18.0 | 18.5 | 0.3799 | 1333 |
| 39 | 1073 | 1867 | 0.2464 | 10.5 | 10.5 | 0.2466 | 1838 |
|  |  |  | 0.0597 | 2.5 | 2.5 | 0.0628 |  |
| 40 | 1079 | 1882 | 0.1285 | 5.5 | 5.5 | 0.1359 | 1987 |
| 41 | 934 | 1628 | 0.2913 | 13 | 13.5 | 0.3034 | 1675 |
| 42 | 658 | 1148 | 0.4061 | 21 | 21.5 | 0.4130 | 1096 |
| 43 | 370 | 645 | 0.4706 | 30 | 29.5 | 0.4690 | 560 |
| 44 | 92 | 160 | 0.4866 | 35 | 37.5 | 0.4911 | 221 |
| 45 | 50 | 87 | 0.4953 | 41 | 45.5 | 0.4980 | 69 |
| 46 | 21 | 38 | 0.4991 | 49.5 | 53.5 | 0.4996 | 16 |
| 47 | 4 | 7 | 0.4998 | 56 | 61.8 | 0.4999 | 3 |
| 48 | 1 | 2 | 0.5000 |  |  | 0.5000 | 1 |
|  | 5738 | 1,0000 |  |  |  |  | 1,0000 |

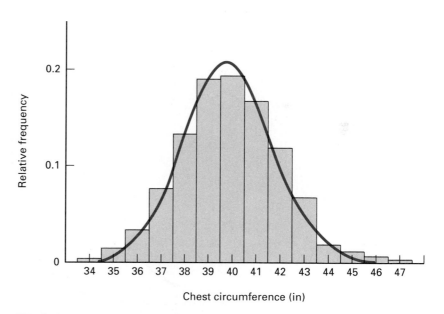

**FIG. 6.13**

---

### DEFINITION 6.12

---

The **probability density function** (**pdf**) for a *continuous* random variable $Y$ is a function $f_Y$ such that

(a) For each value $y$, $f_Y(y) \geqslant 0$.

(b) The area under the graph of $f_Y$ and above the horizontal axis is 1.

(c) For each pair of numbers $c$ and $d$, where $c < d$, $P(c \leqslant Y \leqslant d)$ is the area under the graph of $f_Y$ above the interval from $c$ to $d$.

---

**COMMENTS:**

1. Since $P(Y = y) = 0$ for every point $y$, it is immaterial whether we use "$<$" or "$\leqslant$" in probability statements involving a continuous random variable:

$$P(c \leqslant Y \leqslant d) = P(c \leqslant Y < d) = P(c < Y \leqslant d) = P(c < Y < d)$$

2. Property c relates probabilities of continuous random variables to one of the most basic concepts in calculus, the definite integral:

$$P(c \leqslant Y \leqslant d) = \int_c^d f_Y(y)\, dy$$

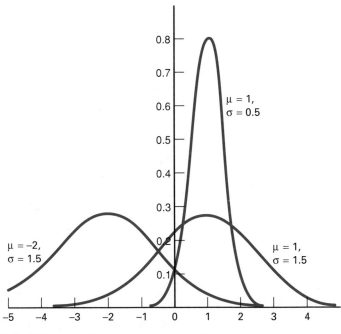

**FIG. 6.14**

Of all continuous pdfs, by far the most important is the family of bell-shaped, or *normal*, curves represented by $\varphi(z)$ and by the graph in Figure 6.16. Members of that family differ only in terms of two parameters: one controlling location ($\mu$), the other controlling dispersion ($\sigma$) (see Figure 6.14).

---

**DEFINITION 6.13**

1. A continuous random variable $Y$ has a ***normal distribution with mean $\mu$ and standard deviation $\sigma$*** if

$$f_Y(y) = \frac{1}{\sqrt{2\pi}\,\sigma}\, e^{-\frac{1}{2}[(y-\mu)/\sigma]^2}, \qquad -\infty < y < \infty \qquad (6.20)$$

2. A continuous random variable $Z$ has a ***standard normal distribution*** if

$$f_Z(z) = \varphi(z) = \frac{1}{\sqrt{2\pi}}\, e^{-z^2/2}, \qquad -\infty < z < \infty \qquad (6.21)$$

---

The standard normal pdf is a special case of the normal pdf: if $\mu = 0$ and $\sigma = 1$, then $f_Y(y) = f_Z(z)$. That relationship is important. We have already seen that areas (i.e., probabilities) associated with standard normal pdfs can be looked up in Appendix A.1. What about areas under "arbitrary" normal curves? How would we compute $P(Y \leqslant c)$? Although there are no tables of areas under $f_Y(y)$, we can find any probability involving $Y$ by phrasing an equivalent question in terms of $Z$. The transformation in Theorem 6.8 is the key.

---

**THEOREM 6.8** *Z* Values

Let $Y$ be a normal random variable with mean $\mu$ and standard deviation $\sigma$. Then

$$Z = \frac{Y - \mu}{\sigma} \qquad (6.22)$$

is a standard normal random variable.

---

*EXAMPLE 6.28*  Suppose $Y$ has a normal distribution with $\mu = 2$ and $\sigma = 3$. Find $P(1.45 \leqslant Y \leqslant 2.71)$. (That is, find the area under $f_Y(y)$ that lies between $Y = 1.45$ and $Y = 2.71$.)

*Solution*   To use Theorem 6.8, we must transform the original question, expressed in terms of $Y$, into an interval of $Z$ values. The necessary steps are analogous to those we followed earlier when applying the De Moivre-Laplace theorem:

$$P(1.45 \leqslant Y \leqslant 2.71) = P(1.45 - 2 \leqslant Y - 2 \leqslant 2.71 - 2)$$

$$= P\left(\frac{1.45 - 2}{3} \leqslant \frac{Y - 2}{3} \leqslant \frac{2.71 - 2}{3}\right)$$

$$= P(-0.18 \leqslant Z \leqslant 0.24)$$

By Appendix A.1

$$P(1.45 \leqslant Y \leqslant 2.71) = P(-0.18 \leqslant Z \leqslant 0.24)$$

$$= \Phi(0.24) - \Phi(-0.18) = 0.5948 - 0.4286 = 0.1662$$

Figure 6.15 shows that the $Y$ area and the $Z$ area are equal.

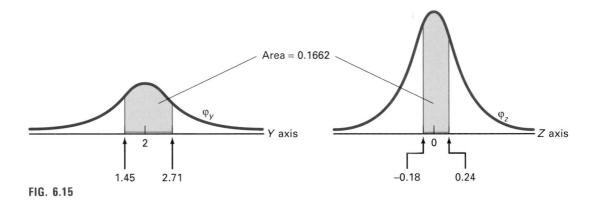

**FIG. 6.15**

*EXAMPLE 6.29*   Look again at Table 6.19. Column 1 lists, in inches, the range of possible chest measurements. Column 2 gives the corresponding observed frequencies, which are then divided by 5738 and reappear in column 3 as *relative frequencies* or, as they are sometimes called, *empirical probabilities*. Decimal points have been omitted. The last column lists "expected" probabilities, all computed under the assumption that chest measurements follow a normal distribution. Specifically, $\mu$ was estimated to be 39.8 in, and $\sigma$ to be 2.05 in, so these final entries were based on the model

$$f_Y(y) = \left(\frac{1}{\sqrt{2\pi}\,(2.05)}\right) e^{-[(y - 39.8)/2.05]^2/2}$$

Verify that the predicted proportion of soldiers with 42-in chests is, as stated, 0.1096.

*Solution*  Remember that the set of individuals assigned to the 42-in class includes anyone with a chest measurement from 41.5 to 42.5 in. Therefore,

$$P(Y \text{ "equals" } 42) = P(41.5 \leqslant Y \leqslant 42.5)$$

$$= P\left(\frac{41.5 - 39.8}{2.05} \leqslant \frac{Y - 39.8}{2.05} \leqslant \frac{42.5 - 39.8}{2.05}\right)$$

$$= P(0.83 \leqslant Z \leqslant 1.32) = \Phi(1.32) - \Phi(0.83) = 0.1096$$

Notice how closely the calculated probabilities in the last column agree with the empirical probabilities in column 3. The normal distribution is clearly a very appropriate model for chest measurements.

---

*EXAMPLE 6.30*  In most states a driver is legally drunk if his blood alcohol concentration, $Y$, is at least 0.10%. When a suspected DUI offender is pulled over, police often request that he submit to a breath analyzer test. Any such test, though, exhibits a certain amount of measurement error, so the possibility exists that a driver whose reading is over 0.10% is *not* intoxicated. Experience has shown that the variability in $Y$ can be described by a normal distribution with $\mu$ equal to the person's *true* blood alcohol concentration and $\sigma$ equal to 0.004%. What is the probability that a driver with a true blood alcohol concentration of 0.095% will be incorrectly booked on a DUI charge?

*Solution*  Since a DUI booking occurs when $Y \geqslant 0.10\%$, we are looking for $P(Y \geqslant 0.10)$ when $\mu = 0.095$. But

$$P(Y \geqslant 0.10) = P\left(\frac{Y - 0.095}{0.004} \geqslant \frac{0.10 - 0.095}{0.004}\right) = P(Z \geqslant 1.25)$$

$$= 1 - \Phi(1.25) = 1 - 0.8944 = 0.1056$$

A person with a blood alcohol concentration of 0.095, in other words, has an almost 11% chance of being declared legally drunk, the mistake being a result of measurement errors inherent in the breath analyzer.

## CASE STUDY 6.7

*Pregnant*
*Implications*  Applications of statistical reasoning are not always confined to laboratory research or field surveys, nor do they always appear in technical journals. The following letter showed up in a Dear Abby column (164):

> Dear Abby: You wrote in your column that a woman is pregnant for 266 days. Who said so? I carried my baby for ten months and five days, and there is no doubt about it because I know the exact date

my baby was conceived. My husband is in the Navy and it couldn't have possibly been conceived any other time because I saw him only once for an hour, and I didn't see him again until the day before the baby was born.

I don't drink or run around, and there is no way this baby isn't his, so please print a retraction about the 266-day carrying time because otherwise I am in a lot of trouble.

*San Diego Reader*

Although the full implications of San Diego Reader's plight are not entirely mathematical, it is possible for us to assess quantitatively the statistical likelihood of a pregnancy lasting 310 days (10 months and 5 days). By the same reasoning used in Case Study 6.2, the latter is done by computing the probability that by chance alone the length of a pregnancy will be 310 days *or longer*. If $Y$, in other words, denotes the duration of an arbitrary pregnancy, we need to find $P(Y \geqslant 310)$. The smaller that probability is, the less credibility San Diego Reader has.

According to well-documented norms, the mean of $Y$ is 266 days. It is also known that the standard deviation of pregnancy durations is 16 days. Assuming the distribution of $Y$ is normal, we can write

$$P(Y \geqslant 310) = P\left(\frac{Y - 266}{16} \geqslant \frac{310 - 266}{16}\right)$$

$$= P(Z \geqslant 2.75) = 1 - \Phi(2.75) = 1 - 0.9970 = 0.0030$$

Figure 6.16 shows the two areas involved: the original one under the normal curve with $\mu = 266$ and $\sigma = 16$, and the transformed one under $\varphi(z)$.

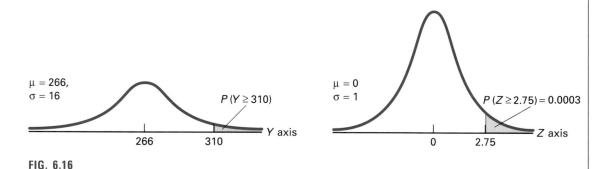

$\mu = 266,$
$\sigma = 16$

$P\,(Y \geq 310)$

Y axis

266    310

$\mu = 0$
$\sigma = 1$

$P\,(Z \geq 2.75) = 0.0003$

Z axis

0    2.75

**FIG. 6.16**

Insurance companies sometimes use probability computations of this type in deciding whether to reimburse married couples for maternity costs. Certain policies contain a clause to the effect that if $Y$ is too much *smaller* than 266, as measured from the couple's date of marriage, the company can refuse to pay.

**QUESTIONS**

**6.99** Suppose $Y$ has a normal distribution with $\mu = -7$ and $\sigma = 4$. Find

(a) $P(Y \leqslant -7)$     (b) $P(Y \geqslant -6)$     (c) $P(-5.5 < Y < -3.2)$

**6.100** Let $Y$ be a normal random variable with $\mu = 3$ and $\sigma = 5$. Find

(a) $P(Y < -2)$     (b) $P(Y > 8)$

(c) $P(-1 \leqslant Y \leqslant 7)$     (d) $P(-7 \leqslant Y < 13)$

**6.101** Assume that the number of miles a driver gets on a set of radial tires is normally distributed with a mean of 30,000 miles and a standard deviation of 5000 miles. Would the manufacturer of these tires be justified in claiming that 90% of all drivers will get at least 25,000 miles?

**6.102** The Stanford-Binet IQ test is scaled to give a mean score of 100 and a standard deviation of 16.

(a) Assuming the distribution of scores is normal, what is the probability that a child has an IQ less than 80? greater than 145?

(b) Suppose children having IQs less than 80 or greater than 145 are deemed to need special attention. Given a population of 2000 children, what will be the expected demand for these additional services?

**6.103** The diameter of the connecting rod in the steering mechanism of a certain foreign sports car must be between 1.480 and 1.500 cm to be usable. The distribution of connecting-rod diameters produced by the manufacturing process is normal with $\mu = 1.495$ cm and $\sigma = 0.005$ cm. What percentage of rods will have to be scrapped?

**6.104** The systolic blood pressure of 18-year-old women is normally distributed with $\mu = 120$ mmHg and $\sigma = 12$ mmHg. What is the probability that the blood pressure of a randomly selected 18-year-old woman will be greater than 150? less than 115? between 110 and 130?

**6.105** Recall Example 6.30. What is the probability that a driver whose true blood alcohol concentration is 0.11% gets a blood analyzer reading that incorrectly shows him *not* to be intoxicated?

**6.106** A criminologist has developed a questionnaire for predicting whether a teenager will become a delinquent. Scores on the questionnaire can range from 0 to 100, with higher values reflecting a presumably greater criminal tendency. As a rule of thumb, the criminologist decides to classify a teenager as a potential delinquent if his or her score exceeds 75. The questionnaire has already been tested on a large sample of teenagers, both delinquent and nondelinquent. Among those considered nondelinquent, scores were normally distributed with a mean of 60 and a standard deviation of 10. Among those considered delinquent, scores were normally distributed with a mean of 80 and a standard deviation of 5.

(a) What proportion of the time will the criminologist misclassify a non-delinquent as a delinquent?

(b) What proportion of the time will the criminologist misclassify a delinquent as a nondelinquent?

6.107   At State University, the average score of first-year students on the verbal part of the SAT is 565, with a standard deviation of 75. If the distribution of scores is normal, what proportion of that school's first-year students have verbal SATs over 650? under 500?

6.108   Dental structure provides an effective criterion for classifying certain fossils. Not long ago a baboon skull of unknown origin was discovered in a cave in Angola (106); the length of its third molar was 9.0 mm. Speculation arose that the baboon in question might be a "missing link" and belong to the genus *Papio*. Members of that genus have third molars measuring, on the average, 8.18 mm long, with a standard deviation of 0.47 mm.

(a)   Quantify the significance of the 9.0-mm molar.

(b)   What do you infer about the baboon's lineage?

6.109   Listed below are a recent year's traffic death rates (fatalities per 100 million motor-vehicle miles) for each of the 50 states (103):

| | | | | | |
|------|-----|--------|-----|-------|-----|
| Ala. | 6.4 | La. | 7.1 | Ohio | 4.5 |
| Alaska | 8.8 | Maine | 4.6 | Okla. | 5.0 |
| Ariz. | 6.2 | Mass. | 3.5 | Oreg. | 5.3 |
| Ark. | 5.6 | Md. | 3.9 | Pa. | 4.1 |
| Calif. | 4.4 | Mich. | 4.2 | R.I. | 3.0 |
| Colo. | 5.3 | Minn. | 4.6 | S.C. | 6.5 |
| Conn. | 2.8 | Miss. | 5.6 | S.Dak. | 5.4 |
| Del. | 5.2 | Mo. | 5.6 | Tenn. | 7.1 |
| Fla. | 5.5 | Mont. | 7.0 | Tex. | 5.2 |
| Ga. | 6.1 | N.C. | 6.2 | Utah | 5.5 |
| Hawaii | 4.7 | N.Dak. | 4.8 | Va. | 4.5 |
| Idaho | 7.1 | Nebr. | 4.4 | Vt. | 4.7 |
| Ill. | 4.3 | Nev. | 8.0 | W.Va. | 6.2 |
| Ind. | 5.1 | N.H. | 4.6 | Wash. | 4.3 |
| Iowa | 5.9 | N.J. | 3.2 | Wis. | 4.7 |
| Kans. | 5.0 | N.Mex. | 8.0 | Wyo. | 6.5 |
| Ky. | 5.6 | N.Y. | 4.7 | | |

(a)   Make a relative frequency histogram for these data. Define the classes as 2.0–2.9, 3.0–3.9, and so on.

(b)   Compute the expected probability for the 4.0–4.9 class, assuming the rates are normally distributed with $\mu = 5.3$ and $\sigma = 1.3$. (Hint: Think of the 4.0–4.9 class as a set of $Y$ values from 3.95 to 4.95.)

## Computing Percentiles

Most probability problems involving normal curves are solved by following the procedure illustrated in the last few examples:

1.   Transform the desired interval of $Y$ values into an equivalent interval of $Z$ values.

2.   Use Appendix A.1 to find an appropriate area under $\varphi(z)$.

Occasionally, a situation arises where Appendix A.1 needs to be used "backwards"; that is, an area is given, and the problem is to find the corresponding interval of $Y$ values (or $Z$ values).

---

**DEFINITION 6.14**

Let $Z$ be a standard normal random variable. Then $z_\alpha$ is the value along the horizontal axis such that the area under $\varphi(z)$ to the right of $z_\alpha$ is $\alpha$. That is,

$$P(Z \geqslant z_\alpha) = \alpha$$

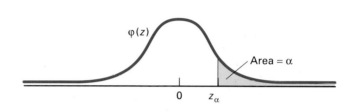

---

*EXAMPLE 6.31*   Find $z_\alpha$ for $\alpha = 0.10$ (see Figure 6.17).

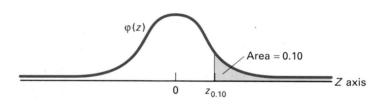

**FIG. 6.17**

*Solution*   Keep in mind that if the area to the right of $z_{0.10}$ is 0.10, then the area to its left is $1 - 0.10 = 0.90$. That is, $\Phi(z_{0.10}) = 0.90$. Now look in Appendix A.1 for the entry closest to 0.90. What we find is the number 0.8997, which appears at the intersection of the 1.2 row and the 8 column. Therefore,

$$P(Z \leqslant 1.28) = 0.8997 \doteq 0.90$$

or, equivalently,

$$P(Z \geqslant 1.28) = 1 - 0.8997 = 0.1003 \doteq 0.10$$

which means that $z_{0.10} = 1.28$. (In the terminology of Section 3.2, 1.28 is the *90th percentile* of the standard normal distribution.)

*EXAMPLE 6.32*    Mensa is an international society whose membership is limited to people with IQs above the 98th percentile of the general population (*mens* is the Latin word for mind). If IQ scores are normally distributed with a mean of 100 and a standard deviation of 16, what is the *lowest* IQ that will qualify a person for membership in Mensa?

*Solution*    Let $Y$ denote the IQ of a randomly chosen person. According to Mensan policy, the minimum score to qualify for membership, call it $y^*$, must satisfy the equation

$$P(Y \geqslant y^*) = 0.02 \qquad \text{(Why?)}$$

Transforming $Y \geqslant y^*$ to a statement about $Z$ gives

$$P(Y \geqslant y^*) = P\left(\frac{Y - 100}{16} \geqslant \frac{y^* - 100}{16}\right)$$

$$= P\left(Z \geqslant \frac{y^* - 100}{16}\right) = 0.02$$

or, in terms of an "area to the left,"

$$P\left(Z \leqslant \frac{y^* - 100}{16}\right) = 0.98$$

According to this equation $\dfrac{y^* - 100}{16}$ is the 98th percentile of $\varphi(z)$. From Appendix A.1, though,

$$P(Z \leqslant 2.05) = 0.9798 \doteq 0.98$$

so 2.05 is *also* the 98th percentile of $\varphi(z)$. Equating the two expressions for the 98th percentile gives

$$\frac{y^* - 100}{16} = 2.05$$

from which it follows that

$$y^* = 16(2.05) + 100 = 133$$

According to the stated criterion, therefore, a person must have an IQ of at least 133 to be a card-carrying Mensan.

---

QUESTIONS

6.110    If $Z$ is a standard normal random variable, find

(a) $z_{0.01}$      (b) $z_{0.99}$      (c) $z_{0.50}$      (d) $z_{0.25}$

6.111    Let $Z$ be a standard normal random variable.

(a) What value of $z$ satisfies the equation $P(Z \geqslant z) = 0.17$?

(b) What percentile does $z$ represent?

6.112   Find $z_\alpha$ if

(a) $\alpha = 0.50$      (b) $\alpha = 0.85$      (c) $\alpha = 0.001$

6.113   Suppose $Y$ has a normal distribution with $\mu = 50$ and $\sigma = 10$. Find $y^*$ such that $P(Y \geqslant y^*) = 0.20$.

6.114   Find the 60th percentile of the verbal SAT scores in Question 6.107.

6.115   A college professor teaches Chemistry 101 each fall to a large class of freshmen. She uses standardized exams that she knows from past experience produce bell-shaped grade distributions with $\mu = 70$ and $\sigma = 12$. Her philosophy of grading is to impose standards that will yield, in the long run, 14% A's, 20% B's, 32% C's, 20% D's, and 14% F's. Where should the cutoff be between

(a) the A's and the B's?

(b) the B's and the C's?

# 6.8

## COMPUTER NOTES

One convenience of using computers for statistical problems is that we don't need to use probability tables. Using numerical techniques or tables stored in memory, software packages provide probabilities for most of the standard pdf's.

Figure 6.18 demonstrates a Minitab session for obtaining the probability table for a binomial random variable with $n = 5$ and $p = 1/5$. The same session produced a table for a Poisson random variable with parameter 1.

```
BINOMIAL 5  .2

   BINOMIAL PROBABILITIES FOR N =   5  AND P =   .200000

        K            P(X = K)          P(X LESS OR = K)
        0             .3277               .3277
        1             .4096               .7373
        2             .2048               .9421
        3             .0512               .9933
        4             .0064               .9997
        5             .0003              1.0000


POISSON 1

   POISSON PROBABILITIES FOR MEAN =    1.000

        K            P( X =  K)          P(X LESS OR = K)
        0             .3679               .3679
        1             .3679               .7358
        2             .1839               .9197
        3             .0613               .9810
        4             .0153               .9963
        5             .0031               .9994
        6             .0005               .9999
        7             .0001              1.0000
```

**FIG. 6.18**

```
DATA ONE;
   INPUT Z;
   CARDS;
      -2
      -1
       0
       1
       2
DATA;
   SET ONE;
   PHI = PROBNORM(Z);
PROC PRINT;

OBS    Z        PHI

 1    -2     0.022750
 2    -1     0.158655
 3     0     0.500000
 4     1     0.841345
 5     2     0.977250
```

**FIG. 6.19**

```
DATA TWO;
   INPUT PHI;
   CARDS;
      1
      2
      3
DATA;
   SET TWO;
   Z = PROBIT(PHI);
PROC PRINT;

OBS    PHI       Z

 1    0.050   -1.6449
 2    0.900    1.2816
 3    0.975    1.9600
```

**FIG. 6.20**

```
DO I=1 TO 10;
   Y = 3*NORMAL(0) + 7;
   OUTPUT;
   END;
PROC PRINT;
   VAR Y;

OBS              Y

 1           3.4722
 2          11.8449
 3           5.7152
 4           3.1016
 5          10.1896
 6           8.8128
 7           5.2034
 8           6.7421
 9           7.7079
10           8.0282
```

**FIG. 6.21**

Note that the computer ceases to generate Poisson values when they are less than 0.0001. In both tables the second column contains the values of the pdf. The third column sums these values, so it gives the cumulative distribution function (cdf). Compare these tables with Table 6.10.

Most statistical software packages have a variety of programs dealing with the normal distribution. The SAS function PROBNORM gives values of the normal cdf $\Phi$. The program of Figure 6.19 gives $\Phi(z)$ for $z = -2$, $-1$, 0, 1, and 2. Compare the results with Appendix A.1.

The normal *inverse probability function*, known as PROBIT in SAS, is also useful. Its values correspond to the $z_\alpha$'s. In Figure 6.20 the Z values presented are $-z_{0.05}$, $z_{0.10}$, and $z_{0.025}$, respectively.

Simulation, in which some process is imitated by a computer program, is being used more often. At the heart of many such programs is the generation of a set of independent observations from some random variable. Also, being able to apply a statistical procedure to a data set of known origin is necessary in some practical settings and useful for pedagogical purposes. SAS has a complete library of pdfs from which observations can be generated. As an example, Figure 6.21 is a sample of 10 independent observations of a normal random variable with mean 7 and standard deviation 3.

# 6.9

## SUMMARY

*Random variables* are functions defined on the set of possible outcomes associated with an experiment. Their objective is to summarize those outcomes in the way that is most relevant to the experimenter. Bernoulli trials are a good example. Initially, the sample space associated with a series of $n$ independent Bernoulli trials is the set of all $2^n$ possible sequences of successes and failures. Frequently, we are interested only in the *number* of successes that occur—where they actually appear in a given sequence is irrelevant. If that should be the case, it makes sense to focus on that number by defining the binomial random variable $Y$.

The immediate consequence of defining a random variable is that a new probability structure emerges. With Bernoulli trials, for example, the probability associated with a particular sequence of $k$ successes and $n - k$ failures in the original sample space is $p^k(1 - p)^{n-k}$. The probability, though, associated with a total of $k$ successes, *regardless of order*, is

$$P(Y = k) = {_nC_k}p^k(1 - p)^{n-k}$$

Formulas that describe the probabilistic behavior of random variables are called *probability density functions* (pdf's).

Certain experimental structures reappear in various settings. The Bernoulli trial model, for instance, applies to phenomena as disparate as coin-tossing, stock fluctuations, consumer surveys, pharmaceutical trials, war games, and baseball. Each is analyzed in exactly the same way—by defining a binomial random variable. For obvious reasons, any random variable that can be applied to a broad range of phenomena is intrinsically important, and we accord it considerable attention. Five such random variables were discussed: the binomial, the hypergeometric, the Poisson, the geometric, and the normal (see Table 6.20 on page 342).

Perhaps the most important concept in Chapter 6 is the distinction between discrete and continuous random variables. A random variable is *discrete* if its range of possible values is either finite or countably infinite; the binomial, hypergeometric, geometric, and Poisson all fall into that category. If a random variable can theoretically take on *any* value in a certain interval, we say that that variable is *continuous*; the most familiar example being the normal.

Just as random variables summarize outcomes in sample spaces, we sometimes find it convenient to summarize random variables. We defined two numbers for that purpose: the mean $E(Y)$ and the standard deviation $\sigma$. The first quantifies the *location* of $f_Y(y)$; the second measures its *dispersion*. The real significance of $E(Y)$ and $\sigma$ will become clearer when we study statistical inference.

**REVIEW QUESTIONS**

6.116  The amount of money collected daily by a television ministry is normally distributed with a mean of $2000 and a standard deviation of $500. How likely is it that tomorrow's donations will exceed $3000?

6.117  The prosecuting and defense attorneys working on a murder case have reduced the pool of mutually acceptable jurors to 10 men and 15 women. If the final selection of 12 jury members is done at random, what is the probability that the number of men seated is either 5 or 6?

**TABLE 6.20**

| Name | Type | pdf |
|------|------|-----|
| Binomial | Discrete | $f_Y(k) = P(Y = k) = {}_nC_k p^k (1 - p)^{n-k}$ |

$$E(Y) = np$$

$$\text{Var}\,(Y) = np(1 - p)$$

$Y$ = number of successes in $n$ independent trials, where each trial has only two possible outcomes (success or failure) and $p = P$ (success) remains constant from trial to trial.

| | | |
|------|------|-----|
| Hypergeometric | Discrete | $f_Y(k) = P(Y = k) = \dfrac{{}_rC_k \cdot {}_wC_{n-k}}{{}_NC_n}$ |

$$E(Y) = \frac{nr}{N}$$

$$\text{Var}\,(Y) = \frac{nrw(N - n)}{N^2(N - 1)}$$

$Y$ = number of red chips in a sample of size $n$ drawn from an urn containing $r$ red chips and $w$ white chips ($r + w = N$).

| | | |
|------|------|-----|
| Geometric | Discrete | $f_Y(k) = P(Y = k) = pq^{k-1}$ |

$$E(Y) = \frac{1}{p}$$

$$\text{Var}\,(Y) = \frac{q}{p^2}$$

$Y$ = trial number at which first success occurs (each trial has two possible outcomes—success or failure—and $p = P$ (success) remains constant from trial to trial.)

| | | |
|------|------|-----|
| Poisson | Discrete | $f_Y(k) = P(Y = k) = \dfrac{e^{-\lambda}\lambda^k}{k!}$ |

$$E(Y) = \lambda$$

$$\text{Var}\,(Y) = \lambda$$

$Y$ = number of occurrences of a rare event; $\lambda$ represents the occurrence *rate* and is assumed to be constant.

| | | |
|------|------|-----|
| Normal | Continuous | $f_Y(y) = \left(\dfrac{1}{\sqrt{2\pi}\,\sigma}\right) e^{-\frac{1}{2}[(y - \mu)/\sigma]^2}$ |

$$E(Y) = \mu$$

$$\text{Var}\,(Y) = \sigma^2$$

The normal pdf is the familiar bell-shaped curve. It is unquestionably the single most important pdf in statistics.

6.118   A *run* in a sequence of Bernoulli trials is a series of similar consecutive outcomes. For example, the sequence

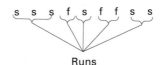

Runs

has 5 runs. Suppose an experiment consists of three Bernoulli trials, where $p = P$ (success). Let the random variable $Y$ denote the number of runs in an outcome. Find $f_Y(y)$ when (a) $p = \frac{1}{2}$ and (b) $p = \frac{2}{3}$.

6.119   The cross-sectional area of plastic tubing manufactured for use in pulmonary resuscitators is normally distributed with $\mu = 12.5$ mm$^2$ and $\sigma = 0.2$ mm$^2$. If that area is less than 12.0 mm$^2$ or greater than 13.0 mm$^2$, the tube will not fit properly. Assume that the tubes are shipped in boxes of 1000. How many tubes per box can doctors expect not to fit?

6.120   Lucky Horseshoe Casino is promoting a new card game that pays off $Y$ dollars on a \$5 bet according to the probabilities summarized below:

| $Y$ = Amount Player Wins | $f_Y(y)$ |
|---|---|
| $-\$3$ | 7/16 |
| $-\$1$ | 2/16 |
| 2 | 3/16 |
| 3 | 1/16 |
| 4 | 1/16 |
| 5 | 1/16 |
| 6 | 1/16 |
| | 1 |

Is this a game that you would want to play? Explain.

6.121   According to the Registrar, enrollment in a child psychology course averages 35.2 students a semester. Let the Poisson random variable $X$ denote next semester's enrollment. Approximate the probability that $X$ will be between 32 and 40, respectively. (Hint: If $X$ is a Poisson random variable with parameter $\lambda > 5$,

$$P(c \leqslant X \leqslant d) \doteq P\left(\frac{c - 0.5 - \lambda}{\sqrt{\lambda}} < Z < \frac{d + 0.5 - \lambda}{\sqrt{\lambda}}\right))$$

6.122   What is the difference between the sample space for a binomial random variable and the sample space for a geometric random variable? Between the sample space for a Poisson random variable and the sample space for a normal random variable?

6.123　An urn contains three red chips numbered 1, 2, and 3, and two white chips numbered 0 and 4. Two chips are drawn at random, without replacement. Define two different random variables on the original sample space of $_5C_2$ outcomes, and find the pdf for each.

6.124　Suppose the driver referred to in Example 6.30 takes the breath analyzer test five times. What is the probability that on at least one occasion he is found to be legally drunk?

6.125　Let $k > 0$. Assume that $X$ is normally distributed with mean $\mu$ and $\sigma = 1$ and that $Y$ is normally distributed with mean $\mu$ and $\sigma = 10$. Is $P(\mu - k < X < \mu + k)$ less than, greater than, or equal to $P(\mu - k < Y < \mu + k)$? (Hint: Draw a diagram of the two pdfs. How are they the same? How do they differ?)

6.126　An appliance dealer has agreed to stock a new dishwasher on a trial basis. The number of units that he sells over a 15-week-period are summarized below. Define the "obvious" random variable and compute $f_Y(y)$, $E(Y)$, and Var $(Y)$.

| Number of Dishwashers Sold per Week | Number of Weeks |
| --- | --- |
| 0 | 5 |
| 1 | 3 |
| 2 | 3 |
| 3 | 2 |
| 4 | 2 |
| | 15 |

6.127　The probability that a telephone solicitor is able to give away free dancing lessons on any given call is 0.08. She intends to quit for the night when she gets her first acceptance. What is the probability that she will not have to dial a fourth call?

6.128　Refer to Example 6.8. Let $X$ denote the number of days the stock goes up. Draw the histogram of $f_X(x)$. Shade the region that corresponds to $P(X \geqslant 3)$.

6.129　A nursery plants 200 azaleas, each of which has an 80% chance of reaching maturity. Is it reasonable to say that the standard deviation of the number of azaleas reaching maturity is

$$\sqrt{200(0.80)(0.20)} = 5.6?$$

Explain.

6.130　Three freshmen and four sophomores apply for positions on the same committee. If the four committee members are chosen at random, what is the probability that exactly two will be freshmen?

**6.131**   Let $\varphi(z)$ denote the pdf of a standard normal random variable. Which is larger, $\varphi(-1.20)$ or $\varphi(1.20)$?

**6.132**   Spot welds on an automobile chassis are graded as either acceptable or unacceptable by a line foreman. Assume that each weld has an 80% probability of being rated acceptable. Use the normal curve to approximate the probability that the number of acceptable welds in the next 400 attempts will be between 310 and 325 inclusive. Use the continuity correction.

**6.133**   Records show that an average of 200 telephone calls are received every four hours at a central switchboard. Estimate the probability that no more than one call will come in during the next minute. Note: $e^{-0.83} = 0.43$.

**6.134**   An actuarial consulting firm gives prospective employees a test that measures their potential for a managerial career. Scores on the test are known to be normally distributed with $\mu = 120$ and $\sigma = 10$. Anyone getting above 142 is considered a prime candidate for a supervisory position. What proportion of their applicants score in that range?

**6.135**   Use Table 6.1 to translate the following Caesar cipher:

> HVOCZHVODXDVIN VMZ GDFZ AMZIXCHZI;
>
> RCVOZQZM TJP NVT OJ OCZH,
>
> OCZT OMVINGVOZ DIOJ OCZDM JRI GVIBPVBZ
>
> VIY AJMOCRDOC DO DN NJHZOCDIB ZIODMZGT YDAAZMZIO.
>
> BJZOCZ

**6.136**   On the average, 520 highway deaths occur during a certain three-day holiday weekend. Is it unlikely that next year's toll will exceed 600? Justify your answer quantitatively. (Hint: See Question 6.121.)

**6.137**   Suppose a continuous random variable has a pdf whose graph is a horizontal line extending from 2 to 4, as shown below. What is the height of that line?

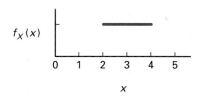

**6.138**   If $f_Y(y) = (1/\lambda)e^{-y/\lambda}$, $y > 0$, the (continuous) random variable $Y$ is said to have an *exponential distribution with parameter* $\lambda$ (where $\lambda = E(Y)$ is a constant greater than 0). For any two points $c$ and $d$, where $0 < c < d$, it can be shown using calculus that

$$P(c \leqslant Y \leqslant d) = e^{-c/\lambda} - e^{-d/\lambda}$$

(see the graph below). Suppose reliability studies suggest that lifetimes of 60-watt bulbs are described by an exponential pdf with $\lambda = 500$ hr. What proportion of bulbs will last between 400 and 700 hr?

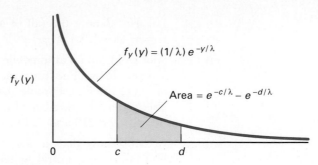

6.139    Recall the cockroach data in Question 2.15. Let the random variable $Y$ denote the number of times that females chose dominant males for mates.

(a) Write a formula for $P(Y = 18)$ under the assumption that dominance is not a factor in the female's selection of a mate.

(b) If dominance is not a factor, how many yeses would we expect to find among the 39 trials?

(c) Use the DeMoivre-Laplace theorem to approximate $P(Y \geqslant 17.5)$. What role does $P(Y \geqslant 17.5)$ play in the interpretation of the results?

6.140    In the kinetic theory of gases, the distance $Y$ that a molecule travels before colliding with another molecule is described by the exponential pdf in Question 6.138. The parameter $\lambda$ is called the mean free path, and represents the average distance that a molecule travels between collisions. Find $P(Y < \lambda/2)$.

# 7 Sampling Distributions

*I know of scarcely anything so apt to impress the imagination as the wonderful form of cosmic order expressed by the "Law of Frequency of Error" (the normal distribution). The law would have been personified by the Greeks and deified, if they had known of it. It reigns with serenity and in complete self effacement amidst the wildest confusion. The huger the mob, and the greater the anarchy, the more perfect is its sway. It is the supreme law of Unreason.*

**Sir Francis Galton**

## 7.1

### INTRODUCTION

This chapter is a conceptual "boundary." It ends our development of probability and begins our study of statistics. Are the subjects so different? Yes. Probability deals with problems where the distribution of a random variable is known, and the objective is to calculate the likelihood of various events defined in terms of that random variable. Statistics does just the opposite. Based on the occurrences of certain events, statisticians try to make useful inferences about the pdf of the random variable that generated those events.

A probabilist, for example, looks at coin-tossing problems on the assumption that the bias, or lack of bias, in the coin is already known. If $p = P(\text{head})$ is assumed to be $1/3$, for instance, what is the probability of the sequence $HHHT$? Since the tosses are independent events,

$$P(HHHT) = P(H)P(H)P(H)P(T) = \left(\frac{1}{3}\right)^3\left(\frac{2}{3}\right) = \frac{2}{81}$$

A statistician, on the other hand, looks at the data (that is, the sequence $HHHT$) and asks, "What is $P(\text{head})$?"

But probability and statistics do have much in common. Mathematically, they are two peas in the same pod. Four important ideas from Chapter 6 are also prominent in Chapter 7: (1) continuous random variables, (2) expected value, (3) standard deviation, and (4) normal curves.

This chapter is not difficult. It deals more with concepts and definitions than it does with problems. Much of the material is background for specific applications that will come later. The central idea in the chapter, though— the notion of a sampling distribution—is extremely important. Learn it well.

## 7.2

### SAMPLING DISTRIBUTIONS

An example is the best way to illustrate the interplay between the probability ideas we have already seen and the statistical questions that lay ahead. Suppose we are interested in the weight reduction potential of the latest fad diet. Let the random variable $Y$ denote the weight loss of a person who has followed the diet faithfully for a month. Assume that previous studies have shown $Y$ to be a normal random variable with $\sigma = 5$ lb. The *mean* weight loss, though, is unknown. Our objective is to estimate $\mu$, or at least be able to assert whether or not $\mu > 0$. The question is, how do we do it?

A single observation $Y$ would be totally inadequate for satisfying either objective. We need to examine the weight losses of several people, the more the better. Furthermore, the $Y_i$'s that we eventually measure should be independent. Having all measurements from a single family, for example, where a large weight loss for one member suggests a likelihood of the same for others, would tell us less about $\mu$ than a set of observations coming from a group of subjects who are unrelated.

We assume, then, that $n$ unrelated subjects are found, they agree to participate, and their resulting weight losses are $y_1, y_2, \ldots, y_n$. Then the $y_i$'s are usually combined to produce a number that summarizes the information contained in the sample. We might compute, for instance, the sample mean of the $y_i$'s:

$$\bar{y} = \frac{y_1 + y_2 + \cdots + y_n}{n}$$

Figure 7.1 shows the general "process" of taking and summarizing a set of $n$ observations. Let $Y_i$ dénote the measurement recorded on the $i$th subject. Then the *set* of random variables, $Y_1, Y_2, \ldots, Y_n$, has two important properties; (1) the $Y_i$'s are *independent*, which means that knowing the numerical value of $Y_i$ tells us nothing about the numerical value of $Y_j$, and (2) each $Y_i$ has the same pdf $f_Y$. Any set of random variables having these two properties is called a ***random sample of size n*** (from $f_Y$).

**random sample of size** $n$—A set of $n$ independent observations taken from the same pdf under presumably identical conditions.

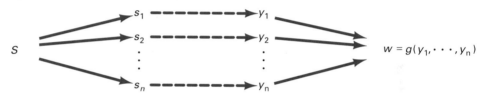

| Sample space | Elements chosen independently from $S$ | Values of the random variable, $Y$ | Defining a function $g$ that summarizes the $y_i$'s |

**FIG. 7.1**

To focus whatever information in the random variables is pertinent to the experimenter's objectives, we combine the $Y_i$'s according to some function $g$. The result is a new random variable, $W$:

$$W = g(Y_1, Y_2, \ldots, Y_n)$$

**statistic**—A function computed from a random sample.

Any such function is called a ***statistic***. Two of the most frequently-used statistics should be familiar from Chapter 3: the *sample mean*,

$$\bar{Y} = g(Y_1, Y_2, \ldots, Y_n) = \frac{Y_1 + Y_2 + \cdots + Y_n}{n}$$

and the *sample standard deviation*,

$$S = g(Y_1, Y_2, \ldots, Y_n) = \sqrt{\left(\frac{1}{n-1}\right) \sum_{i=1}^{n} (Y_i - \bar{Y})^2}$$

**sampling distribution**—The pdf describing the probabilistic behavior of a statistic.

In general, statistics serve as the principle tool for making inferences about unknown parameters. The pdf of a statistic is called a ***sampling distribution***.

***EXAMPLE 7.1***    Suppose $Y$ is a random variable with the following pdf:

| $y$ | $f_Y$ |
|-----|-------|
| 0 | 1/6 |
| 1 | 2/6 |
| 2 | 3/6 |

Imagine drawing a random sample of size $n = 2$ from $f_Y$ and computing the corresponding sample mean: $\bar{Y} = (Y_1 + Y_2)/2$. Find $f_{\bar{Y}}$, the probability density function of $\bar{Y}$.

*Solution*    Conceptually, finding an $f_{\bar{Y}}$ is no different from finding an $f_Y$. We simply list all the values that $\bar{Y}$ can equal and compute the corresponding probabilities. Table 7.1 shows all possible samples of size 2. The probability of each sample pair is listed in the third column. Since the individual observations are independent,

$$P(Y_1 = y_1 \quad \text{and} \quad Y_2 = y_2) = P(Y_1 = y_1)P(Y_2 = y_2)$$

For example,

$$P(Y_1 = 0 \quad \text{and} \quad Y_2 = 1) = P(Y_1 = 0)P(Y_2 = 1)$$

$$= \left(\frac{1}{6}\right)\left(\frac{2}{6}\right) = \frac{2}{36}$$

**TABLE 7.1**

| $y_1$ | $y_2$ | $P(Y_1 = y_1 \text{ and } Y_2 = y_2)$ | $\bar{y} = (y_1 + y_2)/2$ |
|-------|-------|----------------------------------------|----------------------------|
| 0 | 0 | 1/36 | 0 |
| 0 | 1 | 2/36 | 1/2 |
| 0 | 2 | 3/36 | 1 |
| 1 | 0 | 2/36 | 1/2 |
| 1 | 1 | 4/36 | 1 |
| 1 | 2 | 6/36 | 3/2 |
| 2 | 0 | 3/36 | 1 |
| 2 | 1 | 6/36 | 3/2 |
| 2 | 2 | 9/36 | 2 |

**TABLE 7.2**

| $\bar{y}$ | $f_Y$ |
|-----------|-------|
| 0 | 1/36 |
| 1/2 | 4/36 |
| 1 | 10/36 |
| 3/2 | 12/36 |
| 2 | 9/36 |

Column 4 gives the sample means associated with each pair of values for $y_1$ and $y_2$. Adding the probabilities corresponding to a fixed value of $(y_1 + y_2)/2$ defines the pdf $f_{\bar{Y}}$ for $\bar{Y}$. For instance,

$$f_{\bar{Y}}\left(\frac{1}{2}\right) = \frac{2}{36} + \frac{2}{36} = \frac{4}{36}$$

Table 7.2 shows the entire probability distribution (that is, *sampling distribution*) of the statistic $\bar{Y}$.

Notice that the $Y$ distribution and the $\bar{Y}$ distribution have the same mean:

$$E(Y) = \sum_{\text{all } y} y f_Y(y) = 0\left(\frac{1}{6}\right) + 1\left(\frac{2}{6}\right) + 2\left(\frac{3}{6}\right)$$

$$= \frac{4}{3}$$

and

$$E(\bar{Y}) = \sum_{\text{all } \bar{y}} \bar{y} f_{\bar{Y}}(\bar{y})$$

$$= 0\left(\frac{1}{36}\right) + \left(\frac{1}{2}\right)\left(\frac{4}{36}\right) + 1\left(\frac{10}{36}\right)$$

$$+ \left(\frac{3}{2}\right)\left(\frac{12}{36}\right) + 2\left(\frac{9}{36}\right) = \frac{4}{3}$$

It is also true—although we will omit the computations—that the variance of $\bar{Y}$ is equal to the variance of $Y$ divided by the sample size 2:

$$\text{Var}(\bar{Y}) = E(\bar{Y}^2) - (E(\bar{Y}))^2 = \frac{5}{18}$$

$$= \frac{\frac{5}{9}}{2}$$

$$= \frac{E(Y^2) - (E(Y))^2}{2} = \frac{\text{Var}(Y)}{2}$$

The relationships between $E(Y)$ and $E(\bar{Y})$ and between $\text{Var}(Y)$ and $\text{Var}(\bar{Y})$ *always* hold, no matter what $f_Y$ might be.

---

### THEOREM 7.1

Suppose $Y_1, Y_2, \ldots, Y_n$ is a random sample of size $n$ from any pdf with mean $\mu$ and variance $\sigma^2$. Let

$$\bar{Y} = \frac{Y_1 + Y_2 + \cdots + Y_n}{n}$$

Then

(a) $E(\bar{Y}) = \mu = E(Y)$

(b) $\text{Var}(\bar{Y}) = \frac{\sigma^2}{n} = \frac{\text{Var}(Y)}{n}$

Equivalently,

(a') $E(Y_1 + Y_2 + \cdots + Y_n) = E(Y_1) + \cdots + E(Y_n) = n\mu$

(b') $\text{Var}(Y_1 + Y_2 + \cdots + Y_n) = \text{Var}(Y_1) + \cdots + \text{Var}(Y_n) = n\sigma^2$

Notice that Theorem 7.1 addresses the relationship between the means of $Y$ and $\bar{Y}$ and the variances of $Y$ and $\bar{Y}$, but says nothing about the pdf of $\bar{Y}$. Why? Because there is no easily stated relationship, in general, between $f_Y$ and $f_{\bar{Y}}$. But there are exceptions, and one is especially important: if the $Y_i$'s are *normal* random variables, then $\bar{Y}$ is also a *normal* random variable.

---

### THEOREM 7.2

Suppose $Y_1, Y_2, \ldots, Y_n$ is a random sample of size $n$ from a normal pdf with mean $\mu$ and variance $\sigma^2$. Let

$$\bar{Y} = \frac{Y_1 + Y_2 + \cdots + Y_n}{n}$$

Then $f_{\bar{Y}}$ is *normal* with mean $\mu$ and variance $\sigma^2/n$. Also, the ratio

$$\frac{\bar{Y} - \mu}{\sigma/\sqrt{n}} = \frac{\sqrt{n}(\bar{Y} - \mu)}{\sigma}$$

is a *standard normal* Z.

---

Figure 7.2 illustrates Theorem 7.2 for a particular case. The bottom curve is the original pdf, a normal with $\mu = 1$ and $\sigma = 2$. The middle curve is the pdf of $\bar{Y}$ when the sample size is $n = 5$. Note that $f_{\bar{Y}}$ (1) is normal, (2) has the same mean as $f_Y$, and (3) has a standard deviation smaller than the original $\sigma$ ($\sigma_{\bar{Y}} = \sigma/\sqrt{5}$). The top curve shows the sampling distribution of $\bar{Y}$

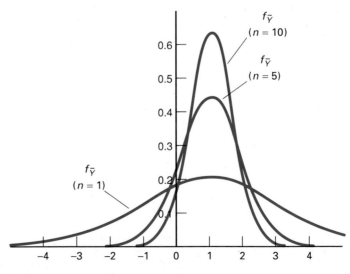

**FIG. 7.2**

when $n = 10$. The latter is even more concentrated around the true mean than is its predecessor. In effect, $\bar{Y}$ is becoming a more "reliable" estimator for $\mu$ as the sample size increases (and the standard deviation of $\bar{Y}$ decreases).

---

**EXAMPLE 7.2**    An elevator in the athletic dorm at Swampwater Tech has a maximum capacity of 2400 lb. Suppose that 10 football players get on at the twentieth floor. If the weights of Tech's players are normally distributed with a mean of 220 lb and a standard deviation of 20 lb, what is the probability that there will be 10 fewer Muskrats at tomorrow's practice?

*Solution*    Let the random variables $Y_1, Y_2, \ldots, Y_{10}$ denote the weights of the 10 players. We are looking for the probability that the *sum* of those weights exceeds the elevator's capacity; that is, $P\left( \sum_{i=1}^{10} Y_i > 2400 \right)$. But

$$P\left( \sum_{i=1}^{10} Y_i > 2400 \right) = P\left( \frac{1}{10} \sum_{i=1}^{10} Y_i > \frac{1}{10} \cdot 2400 \right) = P(\bar{Y} > 240.0)$$

$$= P\left( \frac{\bar{Y} - 220}{20/\sqrt{10}} > \frac{240.0 - 220}{20/\sqrt{10}} \right) = P(Z > 3.16)$$

$$= 0.0008$$

so it's unlikely that anyone will have to scrape Muskrats off the bottom of the elevator shaft. (What if *11* players got on?)

---

QUESTIONS    **7.1**    Suppose $Y$ has the pdf given below.

(a) Find the pdf of $\bar{Y}$ based on samples of size $n = 2$.

(b) Compute $E(Y)$, $E(\bar{Y})$, Var $(Y)$, and Var $(\bar{Y})$.

(c) Verify the relationships claimed in the first part of Theorem 7.1.

| $y$ | $f_Y$ |
|---|---|
| 1 | 2/8 |
| 2 | 5/8 |
| 3 | 1/8 |

**7.2**    (a) For the random variable $Y$ given in Example 7.1, find the pdf of the sample mean $\bar{Y}$ for $n = 3$.

(b) Calculate $E(Y)$, $E(\bar{Y})$, Var $(Y)$, and Var $(\bar{Y})$.

(c) Verify the relationships given in the first part of Theorem 7.1.

7.3 Suppose $Y_1$, $Y_2$, $Y_3$, $Y_4$, $Y_5$ is a random sample from a normal pdf with mean 3 and standard deviation 7.

(a) What are the mean and standard deviation of $Y_1 + \cdots + Y_5$?

(b) What are the mean and standard deviation of $\bar{Y}$?

7.4 Suppose $Y_1$, $Y_2$, ..., $Y_6$ are independent and each is a binomial random variable with $n = 3$ and $p = 0.25$.

(a) What are the mean and standard deviation of $Y_1 + Y_2 + \cdots + Y_6$?

(b) What are the mean and standard deviation of $\bar{Y}$?

7.5 Suppose $\bar{Y}$ is based on a random sample of size 10 from a normal pdf with $\mu = 15$ and $\sigma = 4$.

(a) What is the probability that $\bar{Y} > 15.6$?

(b) What is the probability that $\bar{Y}$ lies between 15.3 and 16.0?

7.6 The IQ's of nine randomly selected persons are recorded. Let $\bar{Y}$ denote their average.

(a) Assuming the distribution from which the IQ's are drawn is normal with $\mu = 100$ and $\sigma = 16$, find $P(\bar{Y} > 103)$?

(b) What is the probability that an individual $Y_i > 103$?

7.7 The mileage of a certain new car on an indoor test track is normally distributed with $\mu = 32.4$ mi/gal and $\sigma = 1.6$ mi/gal. Suppose six runs are made. What is the probability that the average mileage exceeds 33.5 mi/gal?

7.8 See Example 6.30. Suppose a driver with a true blood alcohol concentration of 0.095% is pulled over by police.

(a) Is it to the driver's advantage to request that *two* readings be taken and that he be charged with DUI only if $\bar{y}$ exceeds 0.10%? Compute the relevant probability.

(b) Should a driver with a true blood alcohol concentration of 0.11% request that his booking be based on the average of two readings?

7.9 Suppose that measurement errors associated with a radioactive dating technique are normally distributed with $\mu = 0$ and $\sigma = 25$ million years. If 10 mineral samples of the same age are dated, what is the probability that the resulting average is in error by more than 1 million years?

7.10 Suppose $Y_1$, $Y_2$, ..., $Y_n$ is a random sample from a normal pdf with mean 2 and standard deviation 4. How large must $n$ be so that

$$P(1.9 \leqslant \bar{Y} \leqslant 2.1) \geqslant 0.99?$$

# 7.3

## THE STUDENT t DISTRIBUTION

High on the list of important problems in statistics is the need to draw inferences about the mean of a normal distribution on the basis of a random sample of size $n$. Doing so requires a statistic that 1) involves both $\bar{Y}$ and $\mu$

and 2) has a known pdf. Have we already seen such a statistic? Yes and no. From Theorem 7.2 we know that

$$Z = \frac{\bar{Y} - \mu}{\sigma/\sqrt{n}}$$

has a standard normal distribution, but this equation is of potential use only if a numerical value is known for $\sigma$. However, in situations, where $\mu$ is unknown, which is why the experimenter is collecting data in the first place, it is unlikely that $\sigma$ *is* known. More often, *both* parameters are unknown.

Turn-of-the-century statisticians faced the problem of unknown $\sigma$'s by ignoring it; they simply replaced $\sigma$ with the sample standard deviation S. They assumed that making that substitution had no mathematical repercussions. Specifically, they felt that the two statistics

$$Z = \frac{\bar{Y} - \mu}{\sigma/\sqrt{n}} \qquad \text{and} \qquad T = \frac{\bar{Y} - \mu}{S/\sqrt{n}}$$

had the same sampling distribution.

Not everyone agreed. Challenging the conventional wisdom of the day was a young man working out of a most unlikely environment—the Messrs. Arthur Guinness Son & Co., a brewing firm in Dublin that is still in operation. In 1899 William Sealy Gosset, a graduate of New College, Oxford, where he studied chemistry and mathematics, took a job at Guinness. Brewers had already begun to use scientific methods to study the relationships among raw materials, conditions of production, and outgoing quality. For Gosset that was the good news. The bad news was that the brewery management felt that small samples were sufficient for providing whatever information was needed.

It soon became apparent to Gosset, however, that statistical techniques developed for large samples did not apply well to the small $n$'s with which he was forced to work. When the number of observations is less than 10, for example, it matters a great deal whether S is defined as

$$\sqrt{\frac{1}{n} \sum_{i=1}^{n} (Y_i - \bar{Y})^2} \qquad \text{or} \qquad \sqrt{\frac{1}{n-1} \sum_{i=1}^{n} (Y_i - \bar{Y})^2}$$

although, at the time, both formulas were used interchangeably. Yet when Gosset asked Karl Pearson, the leading statistician of the day, which to use, Pearson replied that it really didn't matter.

In spite of Pearson's indifference, Gosset pursued the question. From his *empirical* discovery that the random variable

$$T = \frac{\bar{Y} - \mu}{S/\sqrt{n}}$$

does *not* have a normal pdf, he progressed to a more mathematical analysis of the problem. In 1908 his results were published in a landmark paper, "The Probable Error of a Mean" (160). Since the Guinness Company had

**chi-square distribution**—The pdf that describes the behavior of the statistic $(n-1)S^2/\sigma^2$, where $S^2$ is computed from a random sample of size $n$ taken from a normal distribution with variance $\sigma^2$.

**Student $t$ distribution**—The pdf that describes the behavior of the statistic

$$\frac{\bar{Y}-\mu}{S/\sqrt{n}}$$

where $\bar{Y}$ and $S$ are calculated from a random sample of size $n$ drawn from a normal distribution with mean $\mu$.

**degrees of freedom**—Constants related to sample size that appear in the pdfs for Student $t$, $\chi^2$, and $F$ random variables.

a policy of discouraging its employees from publishing, Gosset's work appeared under a pseudonym. From that day on he was known as *Student*.

Several important results came from Gosset's paper, and two are the nucleus for almost everything in this chapter: He found the sampling distributions for

1. $\dfrac{(n-1)S^2}{\sigma^2} = \dfrac{1}{\sigma^2} \displaystyle\sum_{i=1}^{n} (Y_i - \bar{Y})^2$

2. $T = \dfrac{\bar{Y}-\mu}{S/\sqrt{n}}$

The first distribution is known as the ***chi-square distribution with $n-1$ degrees of freedom***; the second is called, in Gosset's honor, the ***Student $t$ distribution with $n-1$ degrees of freedom***. Virtually every inference procedure involving parameters of the normal distribution uses either or both of these results.

We will look at the $t$ distribution first. Like many of the other pdfs we have seen, the formula for the Student $t$ involves a parameter. For reasons not obvious and unimportant to our objectives, that parameter is called ***degrees of freedom*** (df). It can be any positive integer. Figure 7.3 shows several Student $t$ curves (with different degrees of freedom) superimposed on a graph of the standard normal.

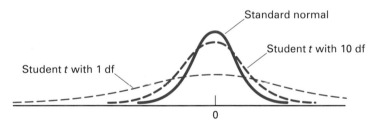

**FIG. 7.3**

In general, $t$ curves are symmetric, bell-shaped, and have a mean of zero. They are slightly flatter than the standard normal. However, as the number of degrees of freedom increases, the corresponding $t$ curve looks more and more like the standard normal. For $n-1 > 100$, $t$ curves and the $Z$ curve are essentially the same.

As we did for normal random variables, areas under the pdf for $T$ must be found in a table; they cannot be calculated from a simple formula. The format of that table is different from what we encountered with the standard normal. Recall that all normal curves can be reduced to a *standard* normal via the $Z$ transformation. That being the case, only one normal curve, $Z$, needs to be tabulated, and it can be treated in considerable detail. Areas under the pdf for $T$, however, cannot be expressed in terms of a "standard" Student $t$. Any $t$ table, therefore, must treat the $t$ curve with 1 df as a distribution entirely different from the $t$ curve with 2 df or 3 df, and so on.

To keep such tables from being prohibitively long, we must limit the amount of information they provide for each specific curve.

*EXAMPLE 7.3*   Table 7.3 shows the top portion of the Student $t$ table as it appears in Appendix A.2. Each row corresponds to a different Student $t$ curve. Numbers in the body of the table cut off *to their right* the areas (labeled $\alpha$) appearing in the column headings.

**TABLE 7.3**

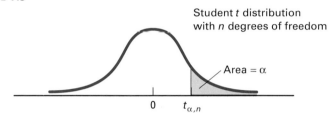

Student $t$ distribution
with $n$ degrees of freedom

Area = $\alpha$

0   $t_{\alpha,n}$

|   |   |   |   | $\alpha$ |   |   |   |
|---|---|---|---|---|---|---|---|
| *df* | **0.20** | **0.15** | **0.10** | **0.05** | **0.025** | **0.01** | **0.005** |
| 1 | 1.376 | 1.963 | 3.078 | 6.3138 | 12.706 | 31.821 | 63.657 |
| 2 | 1.061 | 1.386 | 1.886 | 2.9200 | 4.3027 | 6.965 | 9.9248 |
| 3 | 0.978 | 1.250 | 1.638 | 2.3534 | 3.1825 | 4.541 | 5.8409 |
| 4 | 0.941 | 1.190 | 1.533 | 2.1318 | 2.7764 | 3.747 | 4.6041 |
| 5 | 0.920 | 1.156 | 1.476 | 2.0150 | 2.5706 | 3.365 | 4.0321 |
| 6 | 0.906 | 1.134 | 1.440 | 1.9432 | 2.4469 | 3.143 | 3.7074 |
| 7 | 0.896 | 1.119 | 1.415 | 1.8946 | 2.3646 | 2.998 | 3.4995 |
| 8 | 0.889 | 1.108 | 1.397 | 1.8595 | 2.3060 | 2.896 | 3.3554 |
| 9 | 0.883 | 1.100 | 1.383 | 1.8331 | 2.2622 | 2.821 | 3.2498 |

Consider the Student $t$ random variable

$$T = \frac{\bar{Y} - \mu}{S/\sqrt{4}}$$

where $n = 4$ and df $= 4 - 1 = 3$. What is the number $b$ for which $P(T > b) = 0.05$ (see Figure 7.4)?

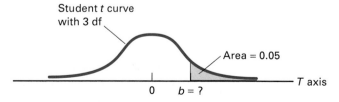

Student $t$ curve
with 3 df

Area = 0.05

0   $b = ?$   $T$ axis

**FIG. 7.4**

*Solution*　Go down the df column to 3. Move across that row to the column headed 0.05. The entry there is the $b$ we are looking for, 2.3534. That is,

$$P(T > 2.3534) = 0.05$$

In general, if $T$ has $k$ degrees of freedom, Appendix A.2 gives $b$ such that $P(T > b) = \alpha$ for $\alpha = 0.20, 0.15, 0.10, 0.05, 0.025, 0.01, 0.005$, and for $k = 1, 2, \ldots, 100$. We denote $b$ by $t_{\alpha, k}$. The number 2.3534, for example, is referred to as $t_{0.05, 3}$.

Like the standard normal, Student $t$ curves are symmetric around 0. Therefore, if a certain number cuts off an area of $\alpha$ to its right, then the negative of that number cuts off an equal area to its left. We saw, for instance, that if $T$ has 3 degrees of freedom,

$$P(T > 2.3534) = 0.05$$

It will necessarily be true, then, that

$$P(T < -2.3534) = 0.05$$

(see Figure 7.5).

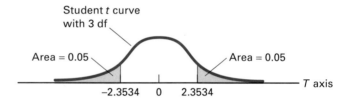

**FIG. 7.5**

---

**DEFINITION 7.1**

For a Student $t$ random variable with $k$ degrees of freedom, $t_{\alpha, k}$ is the number along the horizontal axis such that

$$P(T > t_{\alpha, k}) = \alpha \qquad \text{and} \qquad P(T < -t_{\alpha, k}) = \alpha$$

---

QUESTIONS

**7.11**　Use Appendix A.2 to find the values of $t_{\alpha, 4}$ for $\alpha = 0.10, 0.05, 0.025$, and $0.01$.

**7.12**　Use Appendix A.2 to find the values of $t_{0.05, k}$ for $k = 10, 20, 40, 75$, and 100. Compare those numbers to $z_{0.05}$.

**7.13**　For $t_{0.10, k} = 1.333$, find $k$.

**7.14**   For $t_{\alpha, 8} = 2.3060$, find $\alpha$.

**7.15**   Find $t_{0.90, 7}$.

**7.16**   Suppose $T$ has 10 degrees of freedom. Determine:

(a) $P(T > 3.1693)$

(b) $P(T < -1.093)$

(c) $P(T < 1.372)$

(d) $P(-3.1693 < T < 3.1693)$

**7.17**   Find the number $c$ so that $P(T > c) = 0.025$, where $T$ has 6 degrees of freedom.

**7.18**   Find the number $b$ for which $P(T < b) = 0.025$, where $T$ has 6 degrees of freedom.

**7.19**   Find the number $d$ for which $P(T < d) = 0.95$, where $T$ has 9 degrees of freedom.

**7.20**   Let $Y_1, Y_2, \ldots, Y_6$ be a random sample from a standard normal random variable. Define $T = \sqrt{6}\,\bar{Y}/S$. Find

(a) $P(T > 3.365)$

(b) $P(T < 0.920)$

(c) $P(-2.5706 < T < 2.5706)$

# 7.4

## APPROXIMATE SAMPLING DISTRIBUTIONS: THE CENTRAL LIMIT THEOREM

It was pointed out in Section 7.2 that the average of a set of normal random variables is again normal (Theorem 7.2). It would be too much to ask that an average of *nonnormal* variables also be normal. Even so, under very general conditions such an average can be *approximately* normal. The significance of that fact cannot be overstated. Without question, the approximate normality of $f_{\bar{Y}}$ is one of the most spectacular results in all of mathematics, and certainly the single most important theorem in statistics. Its elegance is exceeded only by its practicality.

We are not encountering at this point the notion of a "normal approximation" for the first time: the De Moivre-Laplace theorem, which dates back to the early eighteenth century, had already shown that the sum (or average) of a set of $n$ Bernoulli random variables has a pdf that can be approximated by the standard normal (Theorem 6.7). What we are about to see is a sweeping generalization of that idea.

**central limit theorem**—A powerful result showing that sample means are approximately normal, regardless of the pdf from which the individual $Y_i$'s are drawn.

Doing much of the work on this problem was the late nineteenth-century and early twentieth-century school of Russian probabilists. In 1901 they produced a powerful theorem credited to A. M. Lyapunov. The version of his result given in Theorem 7.3 is often called the ***central limit theorem***, a name coined by the eminent mathematician, George Polya, in 1920.

> **THEOREM 7.3**   **Central Limit Theorem**
>
> Suppose $Y_1, Y_2, \ldots, Y_n$ is a sequence of independent random variables, each with the same distribution. Let $\mu$ and $\sigma$ be the mean and the standard deviation, respectively, of each of the $Y_i$'s. For large $n$, the pdf of
>
> $$\frac{Y_1 + \cdots + Y_n - n\mu}{\sqrt{n}\,\sigma} = \frac{\bar{Y} - \mu}{\sigma/\sqrt{n}}$$
>
> is approximately the same as the standard normal.

COMMENT:   Meaningful statements about how large $n$ must be in order for the normal approximation to be adequate are difficult to formulate. Not the least of the problems is defining what "adequate" means. Nevertheless, two general statements can be made: (1) If the pdf of the $Y_i$'s is roughly bell-shaped to begin with, convergence to normality will be more rapid than if the variables are decidedly nonnormal (2) Except for "pathological" situations, sample sizes of 10 to 15 will produce $\bar{Y}$'s that are well-approximated by normal curves.

---

*EXAMPLE 7.4*   Suppose 100 fair dice are rolled. Let $Y_i$ be the number showing on the $i$th die. Estimate the probability that the sum of the faces equals or exceeds 360; that is, approximate

$$P(Y_1 + \cdots + Y_{100} \geqslant 360)$$

*Solution*   Each $Y_i$ is a discrete random variable that equals any of the integers 1 through 6 with probability 1/6:

| $y_i$ | $f_{Y_i}(y_i) = P(Y_i = y_i)$ |
|-------|-------------------------------|
| 1 | 1/6 |
| 2 | 1/6 |
| 3 | 1/6 |
| 4 | 1/6 |
| 5 | 1/6 |
| 6 | 1/6 |

Therefore,

$$E(Y_i) = \mu = \sum_{\text{all } y_i} y_i f_{Y_i}(y_i)$$

$$= 1\left(\frac{1}{6}\right) + 2\left(\frac{1}{6}\right) + \cdots + 6\left(\frac{1}{6}\right) = \frac{7}{2}$$

and
$$E(Y_i^2) = \sum_{\text{all } y_i} y_i^2 f_{Y_i}(y_i)$$

$$= 1^2\left(\frac{1}{6}\right) + 2^2\left(\frac{1}{6}\right) + \cdots + 6^2\left(\frac{1}{6}\right) = \frac{91}{6}$$

From Equation 6.16

$$\sigma = \sqrt{\text{Var}(Y_i)} = \sqrt{E(Y_i^2) - \mu^2} = \sqrt{\frac{91}{6} - \left(\frac{7}{2}\right)^2} = \sqrt{\frac{35}{12}}$$

We can now apply the Central Limit Theorem:

$$P(Y_1 + \cdots + Y_{100} \geqslant 360)$$

$$= P\left(\frac{Y_1 + \cdots + Y_{100} - 100(\frac{7}{2})}{\sqrt{100}\sqrt{\frac{35}{12}}} \geqslant \frac{360 - 100(\frac{7}{2})}{\sqrt{100}\sqrt{\frac{35}{12}}}\right)$$

$$\doteq P(Z \geqslant 0.59) = 1 - \Phi(0.59) = 1 - 0.7224 = 0.2776$$

In other words, the probability is almost 0.28 that the sum of the 100 dice will be 360 or greater.

---

**EXAMPLE 7.5**  It is incumbent on businesses such as banks and groceries to minimize the time that patrons spend in line. The widespread introduction of optical scanners in supermarkets is one method. Verifying the effectiveness of new technology is not always easy, however, and can lead to a statistical problem whose solution hinges on the Central Limit Theorem.

Suppose a supermarket has installed a new scanner and cash register system. Let the random variable $Y$ denote the time elapsed from the arrival of the shopper in the checkout line until he or she completes payment. The company marketing the new system advertises that $E(Y) = 6.7$ min and $\sigma = 1.8$ min. Skeptical, the manager of the supermarket watches a random sample of 30 customers as they move through the line: he finds that their average waiting time $(\bar{y})$ is 7.4 min. Is the vendor's claim that $E(Y) = 6.7$ min credible?

*Solution*  We need to look at $P(\bar{Y} \geqslant 7.4)$. If the latter is very small, our confidence in the vendor's integrity would surely diminish (why?). Computing $P(\bar{Y} \geqslant 7.4)$ is similar to problems we did in Section 7.2, where the solution involved a $Z$ transformation (recall Theorem 7.2). That earlier approach, though, was contingent on the $Y_i$'s being a random sample of normal random variables. We may have no reason to make that assumption here. But even if a histogram of the $y_i$'s shows clearly that $Y$ is *not* a normal random variable, we can still approximate $P(\bar{Y} \geqslant 7.4)$ by invoking the central limit theorem.

If the vendor is telling the truth, $\bar{Y}$ should have mean 6.7 and standard deviation $1.8/\sqrt{30}$. Furthermore, Theorem 7.3 tells us that $\bar{Y}$ is approximately normal. Therefore,

$$P(Y \geqslant 7.4) = P\left(\frac{\bar{Y} - 6.7}{1.8/\sqrt{30}} \geqslant \frac{7.4 - 6.7}{1.8/\sqrt{30}}\right)$$

$$\doteq P(Z \geqslant 2.12) = 1 - \Phi(2.12) = 1 - 0.9830 = 0.017$$

The 0.017 implies that the vendor's figures for $\mu$ and/or $\sigma$ are probably incorrect. If the true mean *is* 6.7 min and the true standard deviation *is* 1.8 min, we would expect to get, as the computation shows, a *sample* mean as large or larger than 7.4 only 1.7% of the time. It's possible, of course, that this particular sample of 30 shoppers waited an inordinately long time. More likely, the vendor has painted an overly rosy picture of the capability of his system!

---

**QUESTIONS**

**7.21**　A random sample $Y_1, Y_2, \ldots, Y_{80}$ is taken from a pdf $f_Y$. It is known that $Y$ has a mean of 50 and a standard deviation of 12. Let

$$S_{80} = Y_1 + Y_2 + \cdots + Y_{80}$$

Approximate

(a) $P(S_{80} \leqslant 4800)$　　(b) $P(S_{80} \geqslant 5200)$　　(c) $P(4700 \leqslant S_{80} \leqslant 5000)$

**7.22**　A random sample of size 80 is chosen from a pdf $f_Y$. It is known that $Y$ has $\mu = 50$ and $\sigma = 12$. Approximate

(a) $P(\bar{Y} \leqslant 48)$　　(b) $P(\bar{Y} \geqslant 52)$　　(c) $P(47 \leqslant \bar{Y} \leqslant 50)$

**7.23**　A random sample of size 150 is taken from $f_Y$. Past experience with $f_Y$ indicates that $\mu = 25$ and $\sigma = 10$. Approximate

(a) $P(\bar{Y} \leqslant 24)$　　(b) $P(\bar{Y} \geqslant 27)$　　(c) $P(24.5 \leqslant \bar{Y} \leqslant 24.9)$

**7.24**　For Example 7.4 let $W$ denote the sum of the 100 numbers showing. Approximate $P(330 \leqslant W \leqslant 380)$.

**7.25**　For the $W$ defined in Question 7.24, approximate $P(2 \leqslant W/100 \leqslant 3)$.

**7.26**　An urn contains 10 chips numbered 1 through 10. Suppose 100 chips are drawn with replacement. Let $W$ be the sum of the numbers drawn. Approximate $P(W \geqslant 75)$.

**7.27**　For the $W$ defined in Question 7.26, approximate $P(W/100 \geqslant 6)$.

**7.28**　Suppose a fair coin is tossed 200 times. Let $X_i = 1$ if the $i$th toss is a head, and $X_i = 0$ otherwise, $i = 1, 2, \ldots, 200$. Let $X$ be the sum of the $X_i$'s. Use the central limit theorem to approximate the probability that

(a) $X < 90$　　(b) $X$ is within 5 of $E(X)$

**7.29**　An urn contains four chips, one of them red. Suppose 50 chips are drawn with replacement. Approximate the probability that the number of red chips drawn is between 10 and 15.

**7.30**   An agency of the Commerce Department in a certain state wishes to check the accuracy of weights in supermarkets. They decide to weigh 15 packages of ground meat, each labeled as 1 lb. They will investigate any supermarket where the average weight of the packages examined is less than 15.5 oz. Approximate the probability that they will investigate an honest market. Assume the standard deviation of package weights is 0.6 oz.

# 7.5

## THE CHI SQUARE DISTRIBUTION [1]

As mentioned earlier, Gosset's research efforts centered around two statistics. Both assumed that the underlying $Y_i$'s were a random sample of size $n$ from a *normal* pdf. The first was the $T$ ratio described in Section 7.3:

$$T = \frac{\bar{Y} - \mu}{S/\sqrt{n}}$$

As we will see in Chapter 9, the latter is used to make inferences about $\mu$. His other statistic focused on the sample variance, and led to his discovery of a second important sampling distribution, the *chi square pdf*.

---

### THEOREM 7.4

Suppose $Y_1, Y_2, \ldots, Y_n$ is a random sample of size $n$ from a normal pdf with mean $\mu$ and variance $\sigma^2$. Let

$$S^2 = \frac{1}{n-1} \sum_{i=1}^{n} (Y_i - \bar{Y})^2$$

Then $(n-1)S^2/\sigma^2$ has a **chi-square distribution with $n-1$ degrees of freedom.**

---

Figure 7.6 shows how the shape of a chi-square pdf varies with $n - 1$, the number of degrees of freedom. Because chi-square pdfs are not symmetric, the notation for denoting numbers that cut off specified tail areas is slightly different than the $z_\alpha$ and $t_{\alpha,k}$ we encountered for the Z and the Student $t$.

---

[1] This material is not needed until the end of Chapter 9. We include it here because of its conceptual similarity to the Student $t$ distribution. It can be omitted for now with no loss of continuity.

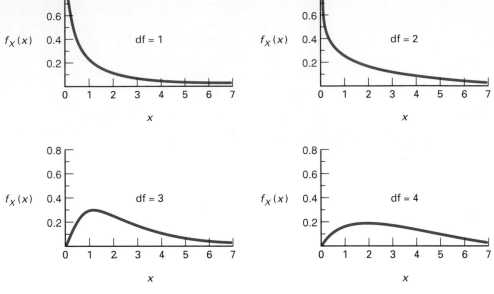

**FIG. 7.6**

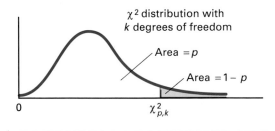

**DEFINITION 7.2**

Suppose $X$ is a chi-square variable with $k$ degrees of freedom. Let $\chi^2_{p,k}$ denote that value for which

$$P(X \leqslant \chi^2_{p,k}) = p, \qquad 0 \leqslant p \leqslant 1$$

That is, $\chi^2_{p,k}$ is the distribution's $p$th percentile. Equivalently, $\chi^2_{p,k}$ must satisfy the equation

$$P(X > \chi^2_{p,k}) = 1 - p$$

As was true for Student $t$ curves, tabulating chi-square probabilities fully would require a separate table for each value of $k$. That being prohibitive, chi-square tables restrict their attention to a limited number of tail

**TABLE 7.4**

| | $p$ | | | | | | | |
|---|---|---|---|---|---|---|---|---|
| *df* | **0.010** | **0.025** | **0.050** | **0.10** | **0.90** | **0.95** | **0.975** | **0.99** |
| 1 | 0.000157 | 0.000982 | 0.00393 | 0.0158 | 2.706 | 3.841 | 5.024 | 6.635 |
| 2 | 0.0201 | 0.0506 | 0.103 | 0.211 | 4.605 | 5.991 | 7.378 | 9.210 |
| 3 | 0.115 | 0.216 | 0.352 | 0.584 | 6.251 | 7.815 | 9.348 | 11.345 |
| 4 | 0.297 | 0.484 | 0.711 | 1.064 | 7.779 | 9.488 | 11.143 | 13.277 |
| 5 | 0.554 | 0.831 | 1.145 | 1.610 | 9.236 | 11.070 | 12.832 | 15.086 |
| 6 | 0.872 | 1.237 | 1.635 | 2.204 | 10.645 | 12.592 | 14.449 | 16.812 |
| 7 | 1.239 | 1.690 | 2.167 | 2.833 | 12.017 | 14.067 | 16.013 | 18.475 |
| 8 | 1.646 | 2.180 | 2.733 | 3.490 | 13.362 | 15.507 | 17.535 | 20.090 |
| 9 | 2.088 | 2.700 | 3.325 | 4.168 | 14.684 | 16.919 | 19.023 | 21.666 |
| 10 | 2.558 | 3.247 | 3.940 | 4.865 | 15.987 | 18.307 | 20.483 | 23.209 |
| 11 | 3.053 | 3.816 | 4.575 | 5.578 | 17.275 | 19.675 | 21.920 | 24.725 |
| 12 | 3.571 | 4.404 | 5.226 | 6.304 | 18.549 | 21.026 | 23.336 | 26.217 |

areas. Table 7.4 shows the top portion of the chi-square table as it appears in Appendix A.3. Entries in the body of the table are values of $\chi^2_{p,k}$ for $k$ between 1 and 50 and for $p = 0.01$, 0.025, 0.05, 0.10, 0.90, 0.95, 0.975, and 0.99. Values listed under $p = 0.01$, 0.025, 0.050, and 0.10 are *left-tail* cutoffs; numbers shown in the last four columns are *right-tail* cutoffs.

Chi-square tables differ from Student $t$ tables in two important ways.

1. The $\alpha$ in $t$ tables is the size of the area *to the right* (of $t_{\alpha,k}$); the $p$ in chi-square tables is the size of the area *to the left* (of $\chi^2_{p,k}$). Thus $t_{0.05,k}$ is in a right-hand tail, but $\chi^2_{0.05,k}$ is in a left-hand tail.
2. In chi-square tables, left-tail cutoffs are *not* the negatives of right-tail cutoffs, as they are for $t$ tables.

***EXAMPLE 7.6*** Find $\chi^2_{0.95,4}$.

***Solution*** Look across the df row labeled 4 to the column headed 0.95. The corresponding entry, our answer, is 9.488. If $X$ denotes a chi-square random variable with 4 df, we could write

$$P(X \leqslant 9.488) = 0.95$$

***EXAMPLE 7.7*** Suppose $X$ is a chi-square variable with 7 df. For what value $x$ is it true that $P(X \leqslant x) = 0.05$?

***Solution*** From Definition 7.2, $x$ is just another name for $\chi^2_{0.05,7}$, and we find the latter in Appendix A.3 (or Table 7.4) to be 2.167.

**EXAMPLE 7.8**    Let $X$ be a chi-square variable with 6 df. Find the number $x$ such that $P(X > x) = 0.05$.

*Solution*    First, rewrite the equation as $P(X \leqslant x) = 0.95$. From Appendix A.3,

$$x = 12.592 \, (= \chi^2_{0.95, 6})$$

---

**EXAMPLE 7.9**    Suppose $X$ is a chi-square variable such that $P(X \leqslant 14.067) = 0.95$. How many degrees of freedom are associated with $X$?

*Solution*    Seven. We find the degrees of freedom by looking down the 0.95 column and finding the row in which 14.067 is located.

---

**EXAMPLE 7.10**    Theorem 7.4 and Appendix A.3 allow us to make certain probabilistic statements about $S^2$. Suppose a random sample of size $n = 10$ is taken from a normal distribution with $\sigma^2 = 4$. Find the values for $x$ such that
(a) $P(9S^2/4 \leqslant x) = 0.01$ and (b) $P(S^2 \leqslant x) = 0.90$

*Solution*    (a) Note that $9S^2/4$ is really $(n-1)S^2/\sigma^2$, so by Theorem 7.4 it has a chi-square distribution with 9 df. It follows that $x$ must be $\chi^2_{0.01, 9} = 2.088$.

(b) We first need to rewrite the left-hand side of the inequality in the form $(n-1)S^2/\sigma^2$. To do that requires that *both* sides of the inequality be multiplied by 9/4:

$$P(S^2 \leqslant x) = 0.90 = P\left(\frac{9S^2}{4} \leqslant \frac{9x}{4}\right) = 0.90$$

But the latter equation implies that

$$\frac{9x}{4} = \chi^2_{0.90, 9} = 14.684, \quad \text{so} \quad x = \left(\frac{4}{9}\right)(14.684) = 6.526$$

---

**QUESTIONS**

7.31    Find $\chi^2_{0.975, 8}$.

7.32    Suppose $X$ is a chi-square variable with 11 df. For what $x$ does $P(X \leqslant x) = 0.025$?

7.33    Let $X$ be a chi-square variable with 5 df. Find $x$ such that $P(X > x) = 0.10$.

7.34    Suppose $X$ is a chi-square variable such that $P(X > 8.547) = 0.90$. How many degrees of freedom are associated with $X$?

7.35    Suppose $X$ is a chi-square variable such that $P(X \leqslant 12.017) = 0.90$. How many degrees of freedom are associated with $X$?

7.36   Suppose $S^2$ is derived from a normal random sample of size 17, where the members of the random sample all have variance 5. Find $x$ such that $P(16S^2/5 \leqslant x) = 0.025$.

7.37   For Question 7.36,

(a) find $x$ such that $P(S^2 \leqslant x) = 0.025$.

(b) for which $x$ does $P(S^2 > x) = 0.025$?

## 7.6

## THE F DISTRIBUTION[2]

**F distribution**—The pdf that describes the behavior of the quotient of two independent chi-square random variables, each divided by its degrees of freedom.

The two-sample, *k*-sample, and randomized block data structures cited in Chapter 2 sometimes require statistical techniques that involve the comparison of two separate measures of variability. We usually make these comparisons by examining *quotients* of random variables, an operation that leads to still another sampling pdf, the **F distribution**.

The British statistician Sir Ronald A. Fisher pioneered the use of this procedure, and the distribution is named in his honor.

---

### DEFINITION 7.3

Suppose *U* and *V* are chi-square random variables with *m* and *n* degrees of freedom, respectively. Let *U* and *V* be independent. Then the pdf of the random variable.

$$\frac{U/m}{V/n}$$

is called an **F distribution with m and n degrees of freedom**.

---

Like the other distributions considered in this chapter, only the most useful tail areas of *F* distributions are tabulated. Table 7.5 shows a portion of the *F* table taken from Appendix A.4. In the Student *t* and chi-square tables, each row represented a degree of freedom and tail areas were listed across the top. Since the *F* distribution has *two* parameters, both the row and the column headings must be used for indicating degrees of freedom. Hence, we need a separate table for each desired tail area. Entries in Table 7.5, for example, cut off areas to their right of 0.05.

---

[2] The material in this section is not used until Chapter 10. It can be omitted for now and returned to later.

**TABLE 7.5 Upper Percentiles of F Distribution**
(tail area = 0.05)

| Denominator df | Numerator df | | | | | | |
|---|---|---|---|---|---|---|---|
| | 1 | 2 | 3 | 4 | 5 | 6 | 7 |
| 1 | 161.4 | 199.5 | 215.7 | 224.6 | 230.2 | 234.0 | 236.8 |
| 2 | 18.51 | 19.00 | 19.16 | 19.25 | 19.30 | 19.33 | 19.36 |
| 3 | 10.13 | 9.55 | 9.28 | 9.12 | 9.01 | 8.94 | 8.88 |
| 4 | 7.71 | 6.94 | 6.59 | 6.39 | 6.26 | 6.16 | 6.09 |
| 5 | 6.61 | 5.79 | 5.41 | 5.19 | 5.05 | 4.95 | 4.88 |

*EXAMPLE 7.11*  Suppose $W$ is an $F$ random variable with 6 and 4 degrees of freedom. For which $b$ does $P(W \geqslant b) = 0.05$?

*Solution*  The number we want is precisely the kind of entry being tabulated. The first parameter cited refers to the numerator degrees of freedom, which appears as a column heading. The second parameter is the denominator degrees of freedom, which appears as a row heading. Go down column 6 and across row 4. The number 6.16 is the desired cutoff (see Figure 7.7).

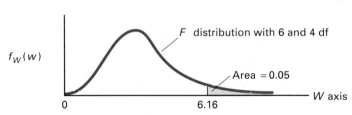

**FIG. 7.7**

*EXAMPLE 7.12*  We are sometimes interested in statistically *small* values of an $F$ variable, situations for which we need *left-tail* cutoffs. Fortunately, these do not require separate tables. With a little work, we can calculate lower percentiles from right-tail cutoffs. Suppose, for instance, $W$ is an $F$ random variable with 7 and 5 degrees of freedom. Find the $c$ for which $P(W \leqslant c) = 0.05$.

*Solution*  Appendix A.4 does not directly apply here, but it can still be used if we first rewrite the question. If

$$P(W \leqslant c) = 0.05$$

then
$$P\left(\frac{1}{W} \geqslant \frac{1}{c}\right) = 0.05 \quad \text{(why?)}$$

Furthermore, from what was given,

$$W = \frac{U/7}{V/5}$$

where $U$ is a chi-square variable with 7 df and $V$ is a chi-square variable with 5 df. Therefore,

$$\frac{1}{W} = \frac{V/5}{U/7}$$

which implies that $1/W$ is a chi-square variable with 5 and 7 degrees of freedom.

Now, look again at the equation

$$P\left(\frac{1}{W} \geqslant \frac{1}{c}\right) = 0.05$$

The cutoff $1/c$ must be the entry in Appendix A.4 appearing at the intersection of column 5 and row 7, namely, 3.97. But if $1/c = 3.97$, then $c = 1/3.97 = 0.25$. In terms of the original $W$, therefore,

$$P(W \leqslant 0.25) = 0.05$$

---

**QUESTIONS**

**7.38**   Suppose $W$ is an $F$ random variable with 12 and 10 df. Find $b$ such that $P(W \geqslant b) - 0.025$.

**7.39**   Suppose $W$ is an $F$ random variable with 10 and 12 df. Find $b$ for which $P(W \geqslant b) = 0.025$.

**7.40**   Let $W$ be an $F$ random variable with 20 and 20 df. Find $b$ so that $P(W \geqslant b) = 0.01$.

**7.41**   Suppose $U$ and $V$ are chi-square random variables with 7 and 9 df, respectively. Find $b$ so that

$$P\left(\frac{U/7}{V/9} \geqslant b\right) = 0.05$$

**7.42**   If $U$ and $V$ are chi-square variables with 8 and 12 df, respectively, find $b$ for which $P(1.2U/V \geqslant b) = 0.05$.

**7.43**   Suppose $U$ and $V$ are independent chi-square random variables, each with 10 degrees of freedom. Find the $b$ for which $P(U/V \geqslant b) = 0.05$.

**7.44**   Let $U$ and $V$ be chi-square random variables with 10 and 8 df, respectively. Find $b$ such that $P(U/V \geqslant b) = 0.05$.

**7.45**   Let $W$ have an $F$ distribution with 10 and 10 df. Find the value of $c$ that makes $P(W \leqslant c) = 0.01$.

**7.46**   Suppose $W$ has an $F$ distribution with 15 and 9 df. Find the value of $c$ such that $P(W \leqslant c) = 0.025$.

7.47    Suppose $W$ is an $F$ random variable with 9 and 15 df. Find $c$ such that $P(W \leqslant c) = 0.025$.

7.48    Let $U$ and $V$ be chi-square random variables with 7 and 9 df, respectively. Find the number $c$ for which

$$P\left(\frac{U/7}{V/9} \leqslant c\right) = 0.05$$

### Applications of the F Distribution

Applications of the $F$ distribution typically arise when Definition 7.3 is used in conjunction with Definition 7.2. Suppose a random sample of size $n$, $X_1, X_2, \ldots, X_n$, is taken from a normal distribution with variance $\sigma_X^2$, and a second (independent) random sample of size $m$, $Y_1, Y_2, \ldots, Y_m$ is taken from a normal distribution with variance $\sigma_Y^2$. Let $S_X^2$ and $S_Y^2$ be the two *sample* variances. Then

1.  $U = \dfrac{(n-1)S_X^2}{\sigma_X^2}$ has a $\chi^2$ distribution with $n-1$ df (by Definition 7.2).

2.  $V = \dfrac{(m-1)S_Y^2}{\sigma_Y^2}$ has a $\chi^2$ distribution with $m-1$ df (by Definition 7.2).

3.  $\dfrac{U/(n-1)}{V/(m-1)} = \dfrac{S_X^2/\sigma_X^2}{S_Y^2/\sigma_Y^2}$ has an $F$ distribution with $n-1$ and $m-1$ df (by Definition 7.3).

Furthermore, if $\sigma_X^2$ is assumed to equal $\sigma_Y^2$ (which will often be the case), the ratio

$$\frac{U/(n-1)}{V/(m-1)}$$

reduces to $S_X^2/S_Y^2$. The value of the latter is what we need to answer certain kinds of questions.

---

*EXAMPLE 7.13*    Apex Cereal has a machine that puts 16 oz of corn flakes, *on the average*, into boxes. But some boxes are overfilled and others are underfilled. The company's research division is experimenting with some new equipment that they claim is more precise. Management decides to compare 25 boxes filled by the current machine with 20 filled by the new one. They agree to pursue the research division's claim if the 25 boxes filled by the old machine have weights that are significantly more variable than the 20 boxes filled by the new machine. If $S_X^2$, in other words, is the sample variance of the weights filled by the current machine and $S_Y^2$ is the sample variance of the weights filled by the new machine, then management will give credence to the research group's contention if $S_X^2/S_Y^2 \geqslant b$. What numerical value should we assign to $b$?

*Solution* Intuitively, two conflicting objectives need to be reconciled in our selection of $b$. If $b$ is assigned a value too close to 1, the probability of $S_X^2/S_Y^2$ exceeding that value, *even though* $\sigma_X^2 = \sigma_Y^2$, will be unacceptably high. The company in those cases will be wasting money by investing in a machine that is no better than what they already have. On the other hand, if $b$ is assigned a value too large, it will almost never be true that $S_X^2/S_Y^2 \geqslant b$, *even when* $\sigma_X^2 > \sigma_Y^2$. If that happens, the company will lose money by not developing better equipment. Among the suitable compromises (for reasons we will explain more fully in Chapter 8) is to choose $b$ so that, when $\sigma_X^2 = \sigma_Y^2$,

$$P\left(\frac{S_X^2}{S_Y^2} \geqslant b\right) = 0.01 \tag{7.1}$$

We said earlier that when the *true* variances are equal, the *sample* variances have a ratio whose behavior is described by an $F$ distribution. Therefore, the $b$ in Equation 7.1 is simply the 99th percentile of the $F$ distribution with 24 $(=n-1)$ and 19 $(=m-1)$ degrees of freedom. From Appendix A.4, $b = 2.92$.

## QUESTIONS

**7.49** Suppose Apex Cereal performs the experiment described in Example 7.13. If the values of the sample variances are $s_X^2 = 1.76$ and $s_Y^2 = 0.85$, what decision will be made about the future of the new machine?

**7.50** In Example 7.13, suppose management decides that they are imposing too harsh a standard for judging the new machine. They propose instead to commit research funds if

$$P\left(\frac{S_X^2}{S_Y^2} \geqslant c\right) = 0.05$$

Find $c$.

# 7.7

## SUMMARY

We should by now be comfortably familiar with the notion of a probability function $f_Y(y)$ describing the behavior of a random variable $Y$. Chapter 6 examined that particular idea and explored in detail a number of different applications, ranging from the performance of control rods in a nuclear reactor to the incidence of childhood leukemia in a large city to the problem of testing for ESP. Among the pdf's considered, especially important were the *normal*,

$$f_Y(y) = \frac{1}{\sqrt{2\pi}\,\sigma} e^{-[(y-\mu)/\sigma]^2/2}, \quad -\infty < y < \infty$$

and the *binomial*,

$$f_Y(k) = {}_nC_k p^k q^{n-k}, \quad k = 0, 1, \ldots, n$$

In a very similar fashion we can derive probability functions that describe the behavior of quantities computed from *sets* of random variables. Suppose we observe $Y_1, Y_2, \ldots, Y_n$ and compute the sample mean

$$\bar{Y} = \frac{1}{n} \sum_{i=1}^{n} Y_i$$

The variability of $\bar{Y}$ from sample to sample can be described by a function $f_{\bar{Y}}$ in the same sense that the variability from observation to observation is described by a function $f_Y$.

Functions of observations like $\bar{Y}$, are called *statistics*, and the pdfs that describe their behavior are called *sampling distributions*. Certain functions are particularly important because of the role they play in statistical inference. Three are singled out in Chapter 7:

$$\frac{\bar{Y} - \mu}{S/\sqrt{n}}, \quad \frac{(n-1)S^2}{\sigma^2}, \quad \text{and} \quad \frac{S_X^2}{S_Y^2}.$$

The pdfs describing those functions are known, respectively, as the *Student t distribution*, the *chi-square distribution*, and the *F distribution*.

Another key concept in this chapter is the central limit theorem, which states that under very general conditions the statistic

$$\frac{\bar{Y} - \mu}{\sigma/n}$$

has approximately a standard normal distribution, regardless of the form of the pdf $f_Y(y)$ from which the $Y_i$'s are drawn. The consequences of the central limit theorem, both practical as well as theoretical, are profound. It is without question the single most important result in statistics.

---

**REVIEW QUESTIONS**

**7.51**  To be eligible for a secretarial job, Mindy must average at least 80 words per minute (wpm) on two typing tests. In the past, her scores on individual tests have been normally distributed with $\mu = 78$ wpm and $\sigma = 5$ wpm. What is the probability that she qualifies for the job?

**7.52**  If the daily profits of a small business are normally distributed with a mean of \$1500 and a standard deviation of \$200, what is the probability that the sum of its earnings for the next five work days falls below \$7000? Answer the question two ways: (1) by applying a Z transformation to $\sum_{i=1}^{5} Y_i$ and (2) by applying a Z transformation to $\bar{Y}$.

**7.53**  A gambler plays 10 hands in a card game. For each hand he wins $5 with probability 0.45 and loses $5 with probability 0.65. On the average, how much money will he have lost when the 10 hands are concluded?

**7.54**  It can be proved that if $Z_1, Z_2, \ldots, Z_n$ is a random sample of size $n$ from a standard normal pdf, then $Z_1^2 + Z_2^2 + \cdots + Z_n^2$ has a $\chi^2$ distribution with $n$ df. For what value $w$ does

$$P(Z_1^2 + Z_2^2 + Z_3^2 > w) = 0.95?$$

**7.55**  At State Tech, GPA's are normally distributed with a mean of 2.85 (out of 4.00) and a standard deviation of 0.50. Suppose that next semester you intend to share a suite with five other students. What is the probability that the average GPA of your suitemates exceeds 3.00?

**7.56**  Find the following values:

(a) $t_{0.10, 17}$    (b) $t_{0.80, 4}$    (c) $t_{0.50, 175}$    (d) $t_{0.27, \infty}$

**7.57**  A loaded die having the probability structure shown below is tossed 200 times. What is the probability that the sum of the faces showing exceeds 850?

| Face Showing | Probability |
|:---:|:---:|
| 1 | 1/21 |
| 2 | 2/21 |
| 3 | 3/21 |
| 4 | 4/21 |
| 5 | 5/21 |
| 6 | 6/21 |

**7.58**  An electric circuit has six resistors wired in series. The resistance $Y_i$ of the $i$th resistor is normally distributed with a mean of 5 ohms and a standard deviation of 0.1 ohm. The circuit resistance $R$ is the sum of the six individual resistances. Problems will develop if the circuit resistance falls below 28.5. Is that likely to happen?

**7.59**  A random sample of size 2 is drawn from the pdf $f_Y(y)$, where $Y = 1$ with probability $p$ and $Y = 0$ with probability $1 - p$. Show that $E(Y) = E(\bar{Y})$.

**7.60**  The heights of Christmas trees grown by a nursery are normally distributed with $\mu = 8.0$ ft and $\sigma = 0.6$ ft. One hundred trees are cut down and sent to a local charity. What is the probability that the average height of the trees cut by the nursery is between 7.90 and 8.06 ft?

**7.61**  Let $\bar{Z}$ denote the mean of a random sample of size $n$ taken from a standard normal pdf. If the 90th percentile of $f_{\bar{Z}}$ is 0.32, what is $n$?

7.62   Let $Y_1, Y_2, \ldots, Y_{11}$ be a random sample of size 11 from a normal distribution with mean $\mu$ and standard deviation $\sigma$. Find the numbers $c$ and $d$ that satisfy each of the following equations:

(a) $P\left(c \leqslant \dfrac{\bar{Y} - \mu}{S/\sqrt{11}} \leqslant d\right) = 0.95$      (b) $P\left(c \leqslant \dfrac{10S^2}{\sigma^2} \leqslant d\right) = 0.95$

7.63   Intuitively, why do the 95th and 99th percentiles of the $F$ distribution with $n$ and $m$ df approach 1 as $n$ and $m$ go to infinity? (Hint: Consider the ratio $\dfrac{S_X^2/\sigma_X^2}{S_Y^2/\sigma_Y^2}$. What happens, for example, to $S_X^2/\sigma_X^2$ as $n$ gets large?)

7.64   Plans call for the base of a patio retaining wall to contain a row of 50 10-in bricks, each separated by 1/2-in-thick mortar. The bricks used are randomly chosen from a population of bricks whose mean length is 10 in and whose standard deviation is 1/32 inch. The mason will make the mortar 1/2 in thick, on the average, but the actual dimension will vary, the standard deviation of the thicknesses being 1/16 in. What is the standard deviation of $L$, the length of the first row of the wall? (Hint: Write $L$ as the sum of 99 random variables and use $b'$ of Theorem 7.1.)

7.65   A random sample of size 4 is chosen from a normal distribution with $\mu = 30$ and $\sigma = 5$. Find the 60th percentile of the $\bar{Y}$ distribution.

7.66   Recall the card game whose probability structure was summarized in Question 6.120. Suppose you play the game 20 times. Approximate the probability that you end up losing money.

7.67   Other than the fact that one refers to a Student $t$ distribution and one refers to a $\chi^2$ distribution, what is the difference between $t_{0.05,5}$ and $\chi^2_{0.05,5}$?

7.68   According to Definition 7.1, $t_{0.60,k}$ is the 40th percentile of the student $t$ distribution with $k$ df. As $k$ increases, does $t_{0.60,k}$ move toward zero or away from zero?

7.69   Let $V$ be a chi-square random variable with $n$ df. It can be shown that the mean and variance of $V$ are $n$ and $2n$, respectively, and that $(V - n)/\sqrt{2n}$ has approximately a standard normal pdf. Use the latter to estimate the 70th percentile of the $\chi^2_{200}$ pdf.

7.70   Suppose an experimenter wishes to estimate the mean ($\mu$) of a normal distribution. If cost and effort were not factors, why is it better to take 20 observations rather than 10 observations? Is it possible that a sample of size one will give a more accurate estimate of $\mu$ than a sample of size 1000?

7.71   Let $Y_1, Y_2, \ldots, Y_5$ be a random sample from a normal distribution with mean $\mu$ and standard deviation $\sigma$. Is it true that

$$P\left(\bar{Y} - 2.7764 \cdot \frac{S}{\sqrt{n}} \leqslant \mu \leqslant \bar{Y} + 2.7764 \cdot \frac{S}{\sqrt{n}}\right) = 0.95$$

(Hint: For what value $b$ does $P\left(-b \leqslant \dfrac{\bar{Y} - \mu}{S/\sqrt{5}} \leqslant b\right) = 0.95$?)

7.72 In general, $z_{\alpha/2}^2 = \chi_{1-\alpha,1}^2$. Verify that statement by comparing appropriate values from Appendix A.1 and A.3 for $\alpha = 0.10$, $\alpha = 0.05$, and $\alpha = 0.01$.

7.73 Having nothing better to do during math class, Bunky spends his time counting the number of words on the first 100 pages of his statistics book. He makes a histogram of the totals and concludes that the number of words per page is normally distributed with $\mu = 236$ and $\sigma = 55$. Suppose he selects 10 pages at random from the remainder of the book. What is the probability that the average number of words on those 10 pages exceeds 250?

7.74 A baseball player with a 300 batting average plays a four-game weekend series and gets five at-bats in each game. If $Y_i$ denotes the number of hits that he gets in the $i$th game, what is

(a) the expected number of hits that he gets during the series

(b) the variance of the number of hits that he gets during the series?

(Hint: If $Y$ denotes the total number of hits that he makes during the series, write $Y = Y_1 + Y_2 + Y_3 + Y_4$.)

7.75 Suppose $\bar{Y}$ and $S$ are to be computed for 10 observations taken from a normal distribution with $\mu = 100$ and $\sigma = 20$. Let $k > 0$. Which is larger,

$$P\left(\frac{\bar{Y} - 100}{20/\sqrt{10}} > k\right) \quad \text{or} \quad P\left(\frac{\bar{Y} - 100}{S/\sqrt{10}} > k\right)?$$

# 8 Principles of Hypothesis Testing

"*The time has come*," *the Walrus said*
"*To talk of many things:*
*Of shoes—and ships—and sealing wax*
*Of cabbages—and kings.*"

*Carroll*

### INTRODUCTION

Making decisions is a common, and sometimes painful, part of life. To a statistician, making decisions is a way of life. Experimenters formulate options, collect data, and then look to statistics as a way of sorting out, in some mathematical way, the good choices from the bad.

The role of statistics in these deliberations varies considerably, depending on the situation. Nevertheless, one type of question occurs again and again—the need to make a choice between two contradictory hypotheses: a quality control engineer decides whether a product does or does not live up to its advertisements; a sociologist follows a prison population over a period of years and concludes that a new counseling program has or has not significantly reduced the recidivism rate; an archaeologist looks at two samples of pottery shards and makes a case that they were the handiwork of the same community or different communities.

In general, statistical methods for making decisions are based on a random sample of size $n$ drawn from a pdf $f_Y(y)$. The two hypotheses to be tested are framed in terms of assumptions made about the parameters of that pdf. Then probability is used to assess the relative likelihoods of the two competing hypotheses. The first "theory" being tested is called the *null hypothesis* (written $H_0$); the second is called the *alternative hypothesis* ( written $H_1$). The two are treated somewhat differently. We concentrate on the null hypothesis in Section 8.2; Section 8.3 focuses on $H_1$.

Hypothesis testing is only one form of *statistical inference*, the science of making generalizations from samples to populations. A second major problem addressed by statistical inference is the estimation of unknown parameters, a procedure that seeks to bring respectability to the art of guessing. Section 8.4 takes a first look at the construction of *confidence intervals*, one of the principle techniques used in statistical estimation.

The concepts presented in this chapter are fundamental. Much of the rest of the book deals with special cases building on the data models introduced in Chapter 2 and the statistical principles laid down here.

**null hypothesis ($H_0$)**—A statement about the value of a parameter that reflects either the status quo or the *absence* of any special effect associated with the treatment being investigated.

**alternative hypothesis ($H_1$)**—A statement about the value of a parameter that reflects the *presence* of an effect.

**statistical inference**—the science of making generalizations from samples to populations.

**confidence interval**—A random interval having the property that it "covers" an unknown parameter with a certain probability, often 0.95 or 0.99.

## 8.2

### HYPOTHESIS TESTING

#### The Decision Rule

We will consider the basic concepts of hypothesis testing in the context of a fictitious, but not unrealistic, example. The Visamaster Credit Card Company must evaluate applications for credit cards. The company would like as many customers as possible, but it doesn't want to issue cards to bad credit risks. Visamaster has been contacted by the Computer Credit Evaluation Group (CCEG), a company claiming to have a superior method for identifying potential slackers. What Visamaster needs to determine is whether CCEG can do what it says it can do.

To test CCEG's ability, Visamaster executives prepare 18 folders, each containing an application from a known credit risk and one from a valued customer. They challenge CCEG to decide which is which. Two weeks later CCEG returns with its decisions, and it turns out that they arrived at the correct answer for *12* of the folders. Has CCEG demonstrated any genuine ability?

At first glance the answer may seem to be "yes." After all, if CCEG were only guessing, we would expect them to get only *nine* right (if the binomial random variable $Y$ denotes the number they identify correctly by chance, then $E(Y) = np = 18(1/2) = 9$). However, the additional three "successess" does not necessarily mean that Visamaster should be convinced of CCEG's ability. Why? Because the three extra correct choices *might* be evidence that CCEG's system works, or they might be the result of nothing more than dumb luck. Our problem is to devise a strategy that allows us to make an intelligent choice between those two conflicting interpretations of the observed data.

The first proposition we consider—"CCEG is just guessing"—is the *null hypothesis*, denoted $H_0$. In general, $H_0$ says no special ability is present, hence the name "null." The competing proposition—"CCEG has some genuine ability"—is the *alternative hypothesis*, denoted $H_1$.

A statistician chooses between $H_0$ and $H_1$ analogous to the way a jury deliberates in a court trial. The null hypothesis is the "defendant" and is presumed true until the data argue overwhelmingly to the contrary. That strategy certainly makes good sense: Visamaster would not want to acknowledge CCEG's claim unless the latter presented some rather convincing evidence.

How can we apply a courtroom philosophy to Visamaster's problem? We have agreed to give $H_0$ the benefit of the doubt. But $H_0$ implies that CCEG should get, on the average, 9 right. Therefore, if the number of correct identifications is 9 *or close to 9*, we should conclude that CCEG has *not* demonstrated its superiority. Ultimately, then, what we decide hinges on what "close" means. In normal conversation, "close" is measured by distance. In statistics, "close" is measured by probability. For us to do the latter requires that the data recorded be a random variable with a known distribution. Fortunately, that is precisely the situation we have.

The experiment performed by CCEG is a sequence of 18 independent trials, each of which can result in a "success"—CCEG finds the poor credit risk—or a "failure." Suppose the probability $p$ of a success is the same for each trial. If $Y$ represents the number of successes in the 18 trials, then $Y$ has a binomial distribution, and the two hypotheses reduce to statement about $p$:

$$H_0: \quad p = \frac{1}{2} \quad \text{(CCEG is just guessing)}$$

vs.

$$H_1: \quad p > \frac{1}{2} \quad \text{(CCEG has some genuine ability)}$$

The possible values of $Y$ (0, 1, ... , 18) can be viewed as a credibility scale for $H_0$. Values for $Y$ of 9 or less are certainly grounds for *accepting* the null hypothesis, and so are values slightly larger than 9 (remember we are committed to giving $H_0$ the benefit of the doubt). On the other hand, values of $Y$ close to 18 should be considered strong evidence against the null hypothesis, leading us to *reject $H_0$*.

It follows that somewhere between 9 and 18 there is a point $y^*$ where, for all practical purposes, the credibility of $H_0$ ends. Using courtroom terminology, we would say that $Y \geqslant y^*$ implies that $H_0$ is false *beyond all reasonable doubt*. The point $y^*$, separating $Y$ values where we accept $H_0$ from those where we reject $H_0$, is called a ***critical value***.

In practice, finding the critical value for a binomial hypothesis test is straightforward, but the rationale is a bit circuitous. The difficulty comes from having to translate "beyond all reasonable doubt" into something numerical. The next few paragraphs show the probabilistic consequences associated with various arbitrary choices for $y^*$. Properly interpreted, these calculations suggest a specific answer to the credit rating problem while leading us to a general method for dealing with binomial hypotheses.

Suppose we let $y^* = 11$ (Figure 8.1). If $H_0$ is true,

$$P(Y \geqslant 11) = P(Y = 11) + P(Y = 12) + \cdots + P(Y = 18)$$

$$= \sum_{k=11}^{18} {}_{18}C_k \left(\frac{1}{2}\right)^k \left(1 - \frac{1}{2}\right)^{18-k} = 0.24$$

(Figure 8.2). Thus if 11 were chosen as the critical value, we would incorrectly reject $H_0$ 24% of the time. For most people this would not be a

**critical value**—The number that marks the beginning of the critical region. If the test statistic is either equal to the critical value or more extreme than the critical value, we reject $H_0$. Typically, critical values will be obtained by using either a $Z$ table, a $t$ table, a $\chi^2$ table, or an $F$ table.

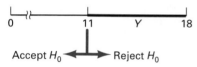

**FIG. 8.1**

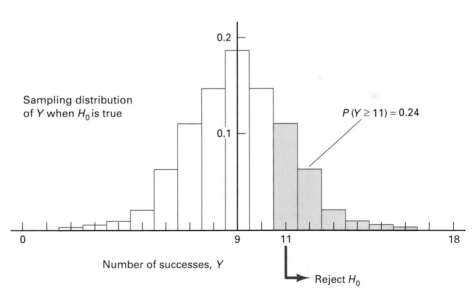

**FIG. 8.2**

satisfactory definition of reasonable doubt. (No jury, for example, would convict a defendant knowing it had a 25% chance of being wrong.)

Clearly, we must look for a critical value closer to 18. Suppose $y^* = 15$. The probability of $Y$ exceeding this new choice—again assuming $H_0$ to be true—is 0.004:

$$P(Y \geqslant 15) = P(Y = 15) + P(Y = 16) + P(Y = 17) + P(Y = 18)$$

$$= \sum_{k=15}^{18} {}_{18}C_k \left(\frac{1}{2}\right)^k \left(1 - \frac{1}{2}\right)^{18-k} = 0.004$$

(Figure 8.3). Now we may have gone too far to the other extreme. Requiring so many correct responses is comparable to a jury not convicting a defendant unless the prosecution could produce a host of eyewitnesses, an obvious motive, and a signed confession!

The consequences of these two choices for $y^*$ suggest that a "reasonable" critical value lies between 11 and 15. Equivalently, the probability of rejecting $H_0$ when $H_0$ is true should be less than 0.24 but greater than 0.004. Of course, we can't argue that this rejection probability *ought to be* 0.10 or 0.18 or 0.009 or anything else. But over the years users of statistics have agreed that, *in many situations the beginning of reasonable doubt is taken as the critical value that is equaled or exceeded only 5% of the time (when $H_0$ is true)*. That is, we should choose $y^*$ so that $P(Y \geqslant y^*) = 0.05$. Specifically,

$$P(Y \geqslant y^*) = P(Y = y^*) + P(Y = y^* + 1) + \cdots + P(Y = 18)$$

$$= \sum_{k=y^*}^{18} {}_{18}C_k \left(\frac{1}{2}\right)^k \left(1 - \frac{1}{2}\right)^{18-k} = 0.05 \qquad (8.1)$$

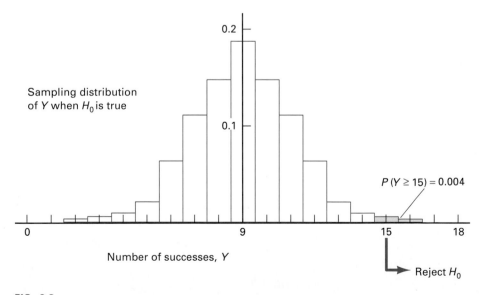

FIG. 8.3

Trial and error shows that the solution to Equation 8.1 is $y^* = 13$. Thus, Visamaster should reject $H_0$ if and only if CCEG correctly classifies 13 or more folders. Since only 12 were identified properly, the appropriate decision is to *accept* $H_0$. In other words, based on our 5% criterion, CCEG has failed to demonstrate any genuine ability.

**level of significance ($\alpha$)**—The probability that the test statistic falls in the critical region even though $H_0$ is true. A test's level of significance is a measure of how much evidence is being required to overturn the initial presumption that $H_0$ is true. The smaller the level, the more difficult it is to reject $H_0$.

**COMMENT:** In general, the probability of rejecting $H_0$ when $H_0$ is true is called the **level of significance**, and is denoted $\alpha$. For the CCEG analysis, $\alpha = 0.05$. We will examine the consequences of $\alpha$ more fully in Section 8.3.

**QUESTIONS**

**8.1** Find $y^*$ for the problem just discussed if a 10% decision rule is used; that is, if $y^*$ is chosen so that the probability of $Y \geqslant y^*$ is as close as possible to 0.10 without exceeding 0.10.

**8.2** (a) Suppose $n = 12$. Find the 0.05 decision rule for choosing between $H_0$: $p = \frac{3}{4}$ and $H_1$: $p > \frac{3}{4}$.

(b) Suppose $n = 14$. Find the 0.05 decision rule for choosing between $H_0$: $p = \frac{3}{5}$ and $H_1$: $p < \frac{3}{5}$. (Hint: What values for $Y$ would be interpreted as strong evidence against $H_0$ and in favor of $H_1$?)

(c) Suppose $n = 20$. Find the 0.01 decision rule for choosing between $H_0$: $p = \frac{1}{3}$ and $H_1$: $p > \frac{1}{3}$.

**8.3** A graphologist is given a set of 10 folders. Each folder contains handwriting samples of a "normal" person and a schizophrenic. The graphologist wants to identify which writings are the work of schizophrenics.

(a) Define the appropriate null and alternative hypotheses and find the corresponding 0.05 decision rule.

(b) When the experiment was actually performed (118), the graphologist made six correct identifications. Did she demonstrate a statistically significant ability to distinguish the writing of a schizophrenic from that of a normal person?

### Large-Sample Binomial

Finding the critical value $y^*$ in the credit rating example was not entirely straightforward. Fortunately, special cases of that type of problem have easier solutions, particularly when $n$, the number of trials, is large and we can use a normal approximation.

Suppose we want to test $H_0$: $p = p_0$ against $H_1$: $p > p_0$, where $p$ is the probability of a success. Let $Y$ denote the number of successes in the $n$ trials. We need to find $y^*$ so that

$$P(Y \geqslant y^* | H_0 \text{ is true}) = 0.05$$

Now, for any $p$, the inequality $Y \geqslant y^*$ is algebraically equivalent to the inequality

$$\frac{Y - np}{\sqrt{npq}} \geqslant \frac{y^* - np}{\sqrt{npq}} \qquad \text{(why?)}$$

By the De Moivre-Laplace limit theorem, the latter is *approximately* equivalent to

$$Z \geqslant \frac{y^* - np}{\sqrt{npq}}$$

where $Z$ is a standard normal. Therefore,

$$P(Y \geqslant y^* \mid H_0 \text{ is true}) = 0.05$$

$$\doteq P\left( Z \geqslant \frac{y^* - np}{\sqrt{npq}} \,\middle|\, H_0 \text{ is true} \right)$$

$$= P\left( Z \geqslant \frac{y^* - np_0}{\sqrt{np_0 q_0}} \right) = 0.05$$

At the next case study shows, finding $y^*$ under these *large-sample* conditions is much easier than solving expressions like Equation 8.1.

## CASE STUDY 8.1

***Point Spreads***   Betting on football games is illegal in almost every state, yet a Lou Harris poll found that one out of every four fans bets regularly. Wagers are placed with the help of a colorful class of entrepeneurs affectionately known as bookies. Mathematically, betting on football games or other sporting events is quite different from gambling on craps or roulette. For casino games probability laws dictate the odds and ultimately determine the payoffs, but bookies have to set their own odds.

Recall from Section 6.6 the roulette example. Red and black each occur with probability 18/38. Let the random variable $Y$ denote the amount a player wins. If a player bets \$1 on red and red comes up, then he "should" win $Y = 20/18$ dollars. Why? Because under those conditions the game would be fair—that is, the expected value of the amount won would be 0:

$$E(Y) = \text{expected winnings for a \$1 bet}$$

$$= \left( \frac{\$20}{18} \right) P\,(\text{red appears}) + (-\$1) P\,(\text{red does not appear})$$

$$= \left( \frac{\$20}{18} \right)\left( \frac{18}{38} \right) + (-\$1)\left( \frac{20}{38} \right) = 0$$

But the actual payoff is \$1, not \$20/18, so a gambler's expected winnings are *less than* zero:

$$E(Y) = (\$1)\left( \frac{18}{38} \right) + (-\$1)\left( \frac{20}{38} \right) = -\$0.0526$$

To say that $E(Y) = -0.05$ (for a $1 bet) is equivalent to saying that the casino wins, over the long run, 5% of the total amount wagered.

The bookie's profit arises in a different way. If a sporting event is a toss-up, meaning that each participant is equally likely to win, then the odds on the event are set at 10–11; that is, the gambler wagers $11 to win $10. Under those conditions,

$$E(Y) = \text{(amount won)}P \text{ (gambler wins)} + \text{(amount bet)}P \text{ (gambler loses)}$$

$$= (\$10)\left(\frac{1}{2}\right) + (-\$11)\left(\frac{1}{2}\right) = -\$0.50$$

showing that the bookie's profits, like the casino's, are about 5% of the total amount wagered.

---

Actually, many events are not toss-ups. When the Cleveland Browns play the Atlanta Falcons, not even the most rabid Falcon fan thinks that his team has a 50–50 chance of winning. But what is the probability that the Falcons will prevail? How should the odds be set? Rather than grapple with that problem head-on, the gambling community has settled on the point spread as a device for making outcomes "equally" likely. The spread is the number of points an underdog is allowed to add to its score. For example, if the point spread is "Falcons (+7)" and the final score is Browns 20, Falcons 14, then Atlanta wins 21–20, as far as gamblers are concerned.

How the point spread is determined and nationally disseminated requires more knowlege of the gambling community than is necessary or healthy. One influential Las Vegas oddsmaker (104) uses a melange of records, ratings, black magic, and intutition to set his "line." His intent is to establish a differential that makes a bettor equally disposed to backing either team. "I ask myself whether I would bet a game at a number. I keep moving the number until I feel I wouldn't bet either side. That's the number."

If point spreads are "correct," we would expect the favorites to beat the spread roughly half the time. Do they?

Fifteen years ago it would have been difficult to test any sort of hypothesis concerning point spreads; unless a person patronized a (probably illegal) gambling establishment, there was no way to find out what the spreads were. Today, to the great distress of the moguls of the NFL, the betting line is reported in most metropolitan newspapers. Table 8.1 gives the point spreads for the fourth week of the 1984 NFL season as reported in the *Daily Oklahoman*.

Ignoring the St. Louis–New Orleans game, where the teams were rated even, and the Monday night game, which was not given, we see that the favorite beat the line exactly half the time. That "score" certainly speaks well for the oddsmakers on that particular week, but we really need to look at their performance over a longer period of time.

**TABLE 8.1**

| Favorite | Underdog | Line | Score | Winner |
|---|---|---|---|---|
| Cincinnati | *Los Angeles | 6 | 24–14 | Los Angeles |
| NY Jets | *Buffalo | 3 | 28–26 | NY |
| Detroit | *Minnesota | 6 | 29–28 | Minnesota |
| *San Francisco | Philadelphia | $4\frac{1}{2}$ | 21–9 | San Francisco |
| *Atlanta | Houston | 9 | 42–10 | Atlanta |
| *Pittsburg | *Cleveland | $2\frac{1}{2}$ | 20–10 | Cleveland |
| *Washington | New England | 1 | 26–10 | Washington |
| St. Louis | New Orleans | 0 | 34–24 | New Orleans |
| Kansas City | *Denver | 1 | 21–0 | Denver |
| *Miami | Indianapolis | 11 | 44–7 | Miami |
| NY Giants | *Tampa Bay | 6 | 17–14 | NY |
| *Dallas | Green Bay | 6 | 20–6 | Dallas |

* The asterisk denotes the team winning according to the spread.

In 124 NFL games played during weeks 4 through 13 of the 1984 season, the favorite beat the spread 67 times (and failed to beat the spread 57 times). What do those data tell us about the "accuracy" of the odds-makers? Let $p = P$(favorite beats spread). The question reduces to a choice between two hypotheses about $p$:

$$H_0: \quad p = \frac{1}{2}$$

vs.

$$H_1: \quad p \neq \frac{1}{2}$$

We should reject the null hypothesis if the number of successes—that is, the number of favorites beating the spread—is too large *or* too small. Let 0.05 be the level of significance. Then *two* critical values, $y_1^*$ and $y_2^*$, must be defined, one in either tail of the $Y$ distribution. Written formally,

$$P(Y \leqslant y_1^* \,|\, H_0 \text{ is true}) = 0.025$$

$$\text{and} \quad P(Y \geqslant y_2^* \,|\, H_0 \text{ is true}) = 0.025$$

Note that *half* the predetermined probability for reasonable doubt is put into each tail (Figure 8.4). The decision rule calls for the rejection of $H_0$ if either $y \leqslant y_1^*$ or $y \geqslant y_2^*$. Our problem is to find numerical values for $y_1^*$ and $y_2^*$.

Let $Y$ denote the number of times in the 124 games that the favorite beats the line. The critical value $y_2^*$ must satisfy the equation

$$P(Y \geqslant y_2^*) = P(Y = y_2^*) + P(Y = y_2^* + 1) + \cdots + P(Y = 124)$$

$$= \sum_{k=y_2^*}^{124} {}_{124}C_k \left(\frac{1}{2}\right)^k \left(1 - \frac{1}{2}\right)^{124-k} = 0.025 \qquad (8.2)$$

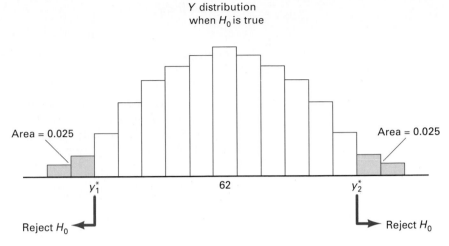

**FIG. 8.4**

Solving Equation 8.2 exactly is difficult, but we can get a good estimate for $y_2^*$ by setting up a normal approximation. Note that

$$P(Y \geqslant y_2^* \mid H_0 \text{ is true}) = P\left(\frac{Y - 124\left(\frac{1}{2}\right)}{\sqrt{124\left(\frac{1}{2}\right)\left(\frac{1}{2}\right)}} \geqslant \frac{y_2^* - 124\left(\frac{1}{2}\right)}{\sqrt{124\left(\frac{1}{2}\right)\left(\frac{1}{2}\right)}}\right)$$

$$= 0.025$$

We know from the De Moivre-Laplace theorem that the distribution of

$$\frac{Y - np}{\sqrt{npq}} = \frac{Y - 124\left(\frac{1}{2}\right)}{\sqrt{124\left(\frac{1}{2}\right)\left(\frac{1}{2}\right)}}$$

is approximately normal (when $H_0$ is true). Therefore,

$$P(Y \geqslant y_2^* \mid H_0 \text{ is true}) \doteq P\left(Z \geqslant \frac{y_2^* - 124\left(\frac{1}{2}\right)}{\sqrt{124\left(\frac{1}{2}\right)\left(\frac{1}{2}\right)}}\right)$$

$$= 0.025$$

From Appendix A.1, though,

$$P(Z \geqslant 1.96) = 0.025$$

Since the right-hand-sides of the last two inequalities are representations of the same point, they can be set equal:

$$1.96 = \frac{y_2^* - 124\left(\frac{1}{2}\right)}{\sqrt{124\left(\frac{1}{2}\right)\left(\frac{1}{2}\right)}} = \frac{y_2^* - 62}{\sqrt{31}}$$

Solving for $y_2^*$ gives

$$y_2^* = 62 + (1.96)\sqrt{31} = 72.9$$

Applying the same sequence of steps to $y_1^*$ gives

$$y_1^* = 62 + (-1.96)\sqrt{31} = 51.1$$

Now we know that $H_0$ should be rejected if $y \leqslant 51.1$ or $y \geqslant 72.9$. In our case the observed $y$ (the test statistic) is 67, which is *not* in the rejection region. Hence, we conclude that oddsmakers are doing a good job, and point spreads show no significant bias toward the favorite or the underdog.

COMMENT: A little algebra shows that rejecting $H_0$ when $y \leqslant 51.1$ or $y \geqslant 72.9$ is equivalent to rejecting $H_0$ when

$$\frac{y - 62}{\sqrt{31}} \leqslant -1.96 \qquad \text{or} \qquad \frac{y - 62}{\sqrt{31}} \geqslant +1.96$$

Depending on the situation, we will use both formats to express decision rules.

COMMENT: For the credit rating example, the alternative hypothesis was written $H_1: p > \frac{1}{2}$. This is called a ***one-sided alternative***. The nature of that problem suggested that values of $p$ less than $\frac{1}{2}$ were unrealistic and could be ignored. In Case Study 8.1, however, values of the parameter on *either* side of the $H_0$ value make sense. For that reason the alternative hypothesis is then said to be ***two-sided*** (and we write $H_1: p \neq \frac{1}{2}$). The "sidedness" of $H_1$ dictates the way we define the critical region. If $H_1$ is two-sided, the critical region has two parts, one interval in either tail of the test statistic's pdf. One-sided alternatives have only a single interval making up their critical region.

**one-side alternative**—An $H_1$ where parameter deviations in only one direction from $p_0$ (or $\mu_0$) are expected.

**two-sided alternative**—An $H_1$ that allows for the possibility that the parameter being tested may be larger or smaller than the value specified in $H_0$.

---

### THEOREM 8.1

Suppose $y$ successes are observed in a series of $n$ independent Bernoulli trials, where $n$ is sufficiently large (see the footnote to Theorem 6.7). Let $p$ be the probability of success for any trial, and let $\alpha$ denote the predetermined level of significance. Let

$$z = \frac{y - np_0}{\sqrt{np_0(1 - p_0)}}$$

1. To test $H_0: p = p_0$ versus $H_1: p \neq p_0$, reject $H_0$ if $z \leqslant -z_{\alpha/2}$ or $z \geqslant z_{\alpha/2}$.
2. To test $H_0: p = p_0$ versus $H_1: p > p_0$, reject $H_0$ if $z \geqslant z_\alpha$.
3. To test $H_0: p = p_0$ versus $H_1: p < p_0$, reject $H_0$ if $z \leqslant -z_\alpha$.

**EXAMPLE 8.1**  Commercial fishermen working certain parts of the Atlantic Ocean some-times have problems with whales scaring away fish. Often the interference is short-lived: sonar operators have confirmed that 40% of all whales leave fishing areas of their own accord, probably to get away from the noise of the boat. Efforts have been made to increase that figure by transmitting underwater the sounds of a killer whale. Suppose the technique has been tried on 52 whales, of which 24, or 46%, immediately left the area. Can we conclude at the $\alpha = 0.05$ level of significance that the observed 46% departure rate is significantly larger than the 40% we would have expected had the additional sounds not been transmitted?

*Solution*  Let $p$ denote the probability that a whale leaves an area after hearing the transmitted sound of a killer whale. The hypotheses to be tested are

$$H_0: \quad p = 0.40$$

vs.

$$H_1: \quad p > 0.40$$

(Notice that the appropriate alternative here is one-sided: there is no reason to believe that transmitting the sounds of a killer whale would *lower* $p$.) From statement 2 of Theorem 8.1 we should reject $H_0$ if

$$z = \frac{y - np_0}{\sqrt{np_0(1 - p_0)}}$$

is greater than or equal to

$$z_\alpha = z_{0.05} = 1.64$$

(Figure 8.5).

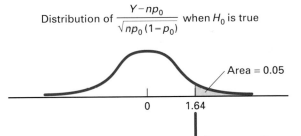

Distribution of $\dfrac{Y - np_0}{\sqrt{np_0(1 - p_0)}}$ when $H_0$ is true

Area = 0.05

0    1.64

Reject $H_0$

**FIG. 8.5**

Evaluating the test statistic gives

$$z = \frac{y - np_0}{\sqrt{np_0(1 - p_0)}} = \frac{24 - 52(0.40)}{\sqrt{52(0.40)(1 - 0.40)}} = 0.90$$

Since $0.90 < 1.64$, we accept $H_0$. The 46% *observed* departure rate (after the transmission of the killer whale sounds) does not constitute enough evidence to overturn the initial presumption that the *true* departure rate has remained at 40%.

**QUESTIONS**

**8.4**   Suppose $n = 100$. Describe the test statistic and the critical value(s) for testing the binomial parameter at the $\alpha = 0.05$ level for

(a) $H_0$: $p = 3/8$   versus   $H_1$: $p \neq 3/8$

(b) $H_0$: $p = 3/8$   versus   $H_1$: $p < 3/8$

(c) $H_0$: $p = 3/8$   versus   $H_1$: $p > 3/8$

**8.5**   Answer Question 8.4, using $\alpha = 0.01$.

**8.6**   Answer Question 8.4, using $n = 200$.

**8.7**   Perform the following hypothesis tests:

(a) $H_0$: $p = 1/4$   versus   $H_1$: $p \neq 1/4$; $n = 30$, $\alpha = 0.05$, $y = 12$

(b) $H_0$: $p = 1/4$   versus   $H_1$: $p \neq 1/4$; $n = 30$, $\alpha = 0.01$, $y = 12$

(c) $H_0$: $p = 1/4$   versus   $H_1$: $p > 1/4$; $n = 30$, $\alpha = 0.01$, $y = 12$

(d) $H_0$: $p = 1/4$   versus   $H_1$: $p < 1/4$; $n = 30$, $\alpha = 0.01$, $y = 3$

**8.8**   Refer to Question 2.13. Let $p$ be the probability that next year's Super Bowl is won by an NFC team. On the basis of the outcomes for the first 20 Super Bowls, test $H_0$: $p = \frac{1}{2}$ against $H_1$: $p \neq \frac{1}{2}$. Let $\alpha = 0.05$.

**8.9**   Set up and test an appropriate $H_0$ and $H_1$ for the cockroach data described in Question 2.15. Let $\alpha = 0.01$ and make the alternative two-sided.

**8.10**   If we test $H_0$: $p = 0.4$ versus $H_1$: $p > 0.4$ at the 0.05 level with $n = 100$, what is the smallest value of $y$ that will cause us to reject $H_0$?

**8.11**   If we test $H_0$: $p = 0.7$ versus $H_1$: $p < 0.7$ at the 0.05 level with $n = 100$, what is the largest value of $y$ that will cause us to reject $H_0$?

**8.12**   There is a theory (122) that the anticipation of a birthday can prolong a person's life. While examining that notion statistically, experimenters found that only 60 of 747 people whose obituaries were published in Salt Lake City newspapers died in the three-month period preceding their birthdays (113). Test the appropriate hypotheses at the 0.01 level.

**8.13**   A manufacturer of razor blades controls 40% of the market. The firm decides to put on a major advertising campaign to increase sales. After two months they survey 500 potential customers and find that 220 bought their product. Can they conclude at the 0.05 level that the advertising campaign was successful?

### A Test for μ

The next case study shows the construction of a hypothesis test for normally distributed data. The rationale is much the same as we used for the large-sample binomial.

---

**CASE STUDY 8.2**

*Pregnant Durations*

No amount of preparation can remove all the uncertainties connected with childbirth. One principle unknown is when the baby will arrive. Although much is known about their distribution, individual pregnancy durations are quite variable. Let the length of a pregnancy be denoted by the random variable $Y$. For women who receive adequate medical attention, records show that (1) $Y$ is normally distributed, (2) $E(Y) = 266$ days, and (3) $\sigma = 16$ days.

Women not receiving adequate prenatal care, however, tend to have pregnancy durations shorter than 266 days, and short pregnancies, in general, can lead to serious health problems for the newborn. Therefore, identifying potential mothers whose lack of medical care puts them at risk becomes an important community health objective.

Table 8.2 lists 70 pregnancy durations reported from an urban hospital in Nashville, Tennessee (100). Many of the patients are from lower socio-economic groups. Can we conclude from this sample of 70 that the community, at large, needs to have more effective prenatal health care?

**TABLE 8.2 Pregnancy Durations (days)**

| | | | | | | | | | |
|---|---|---|---|---|---|---|---|---|---|
| 251 | 264 | 234 | 283 | 226 | 244 | 269 | 241 | 276 | 274 |
| 263 | 243 | 254 | 276 | 241 | 232 | 260 | 248 | 284 | 253 |
| 265 | 235 | 259 | 279 | 256 | 256 | 254 | 256 | 250 | 269 |
| 240 | 261 | 263 | 262 | 259 | 230 | 268 | 284 | 259 | 261 |
| 268 | 268 | 264 | 271 | 263 | 259 | 294 | 259 | 263 | 278 |
| 267 | 293 | 247 | 244 | 250 | 266 | 286 | 263 | 274 | 253 |
| 281 | 286 | 266 | 249 | 255 | 233 | 245 | 266 | 265 | 264 |

Let $\alpha = 0.05$, and let $\mu$ denote the true average pregnancy duration characteristic of women living in this area. The concerns being raised can be reduced to a choice between the hypotheses,

$$H_0: \quad \mu = 266$$

vs.

$$H_1: \quad \mu < 266$$

Intuitively, a sensible statistic for testing the *true mean* $\mu$ would be the sample mean $\bar{Y}$. Recall that if a set of $n$ normal $Y_i$'s has mean $\mu$ and standard deviation $\sigma$, then $\bar{Y}$ is normal with mean $\mu$ and standard deviation $\sigma/\sqrt{n}$ (Theorem 7.2). Under $H_0$, then,

$$E(\bar{Y}) = 266 \quad \text{and} \quad \sigma_{\bar{Y}} = \frac{16}{\sqrt{70}} = 1.912$$

When should we reject $H_0$? When $\bar{Y}$ is too much *smaller* than 266. There is some value $\bar{y}^*$ ($<266$) that marks the point where we can no longer support the null hypothesis. By analogy with our approach for setting up binomial hypothesis tests, we should choose $\bar{y}^*$ to be that value for which $P(\bar{Y} \leqslant \bar{y}^* | H_0 \text{ is true}) = 0.05$. But

$$P(\bar{Y} \leqslant \bar{y}^* | H_0 \text{ is true}) = 0.05$$

$$= P\left(\frac{\bar{Y} - 266}{1.912} \leqslant \frac{\bar{y}^* - 266}{1.912}\right) = 0.05$$

$$= P\left(Z \leqslant \frac{\bar{y}^* - 266}{1.912}\right) = 0.05$$

From Appendix A.1, though,

$$P(Z \leqslant -1.64) = 0.05$$

Since the right-hand-sides of the two previous inequalities both represent the 5th percentile of a standard normal pdf, the two numbers must be the same. That is,

$$-1.64 = \frac{\bar{y}^* - 266}{1.912}$$

which implies that

$$\bar{y}^* = 266 - 1.64(1.912) = 262.9$$

Is $\bar{y} < \bar{y}^*$? Yes. For the 70 $y_i$'s in Table 8.2, $\bar{y} = 260.3$. Therefore, we should reject $H_0$: the typical pregnancy duration for these women *is* significantly shorter than the 266-day norm.

---

**COMMENT:** As in the case of the large-sample binomial hypothesis test, we can express the decision rule for $\mu$ in two equivalent ways. If, for example, the hypotheses being tested are $H_0$: $\mu = \mu_0$ versus $H_1$: $\mu < \mu_0$, we can write the decision rule as either

1. Reject $H_0$ if $\quad \bar{y} \leqslant \mu_0 - z_\alpha \dfrac{\sigma}{\sqrt{n}} \qquad (=\bar{y}^*)$

or

2. Reject $H_0$ if $\quad \dfrac{\bar{y} - \mu_0}{\sigma/\sqrt{n}} \leqslant -z_\alpha.$

For some problems, the first form is more convenient, but more often we will use the second.

Theorem 8.2 summarizes the three variations that arise (because of the nature of $H_1$) in testing $H_0: \mu = \mu_0$. Notice how the critical region in each case reflects the alternative hypothesis.

---

**THEOREM 8.2**

Let $Y_1, Y_2, \ldots, Y_n$ be a random sample from a normal distribution where $\sigma^2$ is known, and let

$$z = \frac{\bar{y} - \mu_0}{\sigma/\sqrt{n}}$$

1. To test

$$H_0: \quad \mu = \mu_0$$

vs.

$$H_1: \quad \mu \neq \mu_0$$

at the $\alpha$ level of significance, reject $H_0$ if either (a) $z \leqslant -z_{\alpha/2}$ or (b) $z \geqslant z_{\alpha/2}$.

2. To test

$$H_0: \quad \mu = \mu_0$$

vs.

$$H_1: \quad \mu > \mu_0$$

at the $\alpha$ level of significance, reject $H_0$ if $z \geqslant z_\alpha$.

3. To test

$$H_0: \quad \mu = \mu_0$$

vs.

$$H_1: \quad \mu < \mu_0$$

at the $\alpha$ level of significance, reject $H_0$ if $z \leqslant -z_\alpha$.

---

*EXAMPLE 8.2*  Among the most widely used predictors of college success is a battery of tests developed by American College Testing (ACT). Each student taking these examinations is given an overall verbal and quantitative aptitude score, called the ACT composite. For the national 1985 high school graduating class, ACT composites were normally distributed with a mean ($\mu$) of 18.6 and a standard deviation ($\sigma$) of 6. Suppose that Backwater High is concerned that their academic programs may not be up to par. They poll their graduates and find that their 45 seniors had an average ACT of 17.4. At the 0.05 level can they conclude that their senior class did significantly worse than the nation as a whole?

*Solution*  If $\mu$ denotes the true average composite ACT indicative of Backwater's students, the hypotheses to be tested are

$$H_0: \quad \mu = 18.6$$

vs.

$$H_1: \quad \mu < 18.6$$

According to Statement 3 in Theorem 8.2, $H_0$ should be rejected if

$$z = \frac{\bar{y} - \mu_0}{\sigma/\sqrt{n}} \leqslant -z_\alpha = -z_{0.05} = -1.64$$

But

$$z = \frac{17.4 - 18.6}{6/\sqrt{45}} = -1.34$$

Since $-1.34 \nleqslant -1.64$, our conclusion is to accept $H_0$—even though Backwater's average ACT is less than the national average, it is not *significantly* less.

---

QUESTIONS

**8.14**  Perform the hypothesis tests indicated below. Assume that $\sigma = 2.5$, $n = 30$, and $\bar{y} = 4.85$.

  (a) $H_0: \mu = 4$  versus  $H_1: \mu \neq 4$;  $\alpha = 0.05$

  (b) $H_0: \mu = 4$  versus  $H_1: \mu > 4$;  $\alpha = 0.05$

  (c) $H_0: \mu = 4$  versus  $H_1: \mu \neq 4$;  $\alpha = 0.10$

  (d) $H_0: \mu = 4$  versus  $H_1: \mu > 4$;  $\alpha = 0.10$

**8.15**  Do the following hypothesis tests. Let $\sigma = 6.7$ and $n = 25$. Assume that $\bar{y} = 11.7$.

  (a) $H_0: \mu = 15$  versus  $H_1: \mu \neq 15$;  $\alpha = 0.01$

  (b) $H_0: \mu = 15$  versus  $H_1: \mu < 15$;  $\alpha = 0.01$

  (c) $H_0: \mu = 15$  versus  $H_1: \mu \neq 15$;  $\alpha = 0.05$

  (d) $H_0: \mu = 15$  versus  $H_1: \mu < 15$;  $\alpha = 0.05$

**8.16**  For the data

| | | | | |
|---|---|---|---|---|
| 14.49 | 7.57 | 10.40 | 7.84 | 6.16 |
| 10.03 | 7.72 | 10.20 | −0.51 | 5.45 |

test $H_0: \mu = 7.5$ versus $H_1: \mu \neq 7.5$ at the 0.05 level of significance. Assume that the observations are a random sample from a normal population with $\sigma = 3.2$.

**8.17**  Let $\bar{Y}$ be the mean of a sample of size 25 from a normal distribution with $\sigma = 2.5$. Suppose $H_0: \mu = 10$ is tested against $H_1: \mu \neq 10$ by rejecting the null hypothesis if $\bar{y} \leqslant 8.915$ or $\bar{y} \geqslant 11.085$. Find $\alpha$.

**8.18**  Sixteen observations are taken from a normal pdf having $\sigma = 5$. It is decided to test $H_0: \mu = 12$ versus $H_1: \mu > 12$ by rejecting the null hypothesis if $\bar{y} \geqslant 14.35$. What is the corresponding level of significance?

**8.19**   The systolic blood pressure of 18-year-old women is known to be normally distributed with $\mu = 120$ mmHg and $\sigma = 12$ mmHg. Is it reasonable to assume that the following 10 observations are a random sample from that distribution? Test the appropriate hypothesis at the 0.05 level of significance.

|     |     |     |     |     |
|-----|-----|-----|-----|-----|
| 119 | 130 | 133 | 128 | 127 |
| 134 | 126 | 129 | 122 | 129 |

**8.20**   A standard test for measuring IQ is scored in such a way that $\mu = 100$ and $\sigma = 16$. A sample of 75 subjects given a new version of the test has scored an average of $\bar{y} = 102.7$. Is the latter compatible with the assumption that the true mean has remained at 100? Do the appropriate hypothesis test at the 0.05 level of significance. Use a two-sided alternative.

**8.21**   A manufacturer wants to advertise that its 60-watt light bulbs have an average life of 1000 hr. To examine the truthfulness of such a claim, their quality control department puts a sample of 10 bulbs "on test" and records the number of hours that each one lasts. Suppose the 10 observed lifetimes are 1034, 1029, 980, 932, 889, 930, 859, 862, 973, and 981. Assuming that the lifetimes are normally distributed and $\sigma = 75$ hr, test at the 0.01 level of significance the null hypothesis that the true mean is 1000. (Note: The company is not concerned if the true mean is *greater* than 1000.)

**8.22**   A biologist is worried about the possible temperature rise in the water downstream from a nuclear reactor. If an increase of more than 3°F occurs, the ecosystem may suffer irreparable damage. She decides to test $H_0: \mu = 3$°F versus $H_1: \mu > 3$°F, where $\mu$ is the true average temperature increase in the water. Sixteen water samples will be collected, with the choice between $H_0$ and $H_1$ being made on the basis of the sample average temperature increase $\bar{y}$. Similar studies suggest that any such temperature changes will be normally distributed with a standard deviation of 1°F.

(a) What is the appropriate decision rule if $\alpha = 0.05$?

(b) Suppose the biologist decides to reject $H_0$ if $\bar{y} \geqslant 3.5$°F. What is the corresponding $\alpha$?

**8.23**   Restate Theorem 8.2, using $\bar{y}$ rather than $z$ as the test statistic.

# 8.3

## TYPE I AND TYPE II ERRORS

Errors are an inevitable by-product of hypothesis testing (like fleas on a dog . . .). No matter what mathematical decision-making procedure is used, there is no way to avoid the possibility of drawing an incorrect inference. One kind of error—rejecting $H_0$ when $H_0$ is true—was the motivating concept in Section 8.2. We argued that critical regions should be defined to keep the probability of making such errors small, say 0.05 or 0.01. Another kind of error, with probabilities often much larger than 0.05 or 0.01, also figures prominently in the decision-making process. Computing and interpreting the latter is our objective in this section.

**Type I error**—Rejecting $H_0$ when $H_0$ is true. The *probability* of committing a Type I error is denoted $\alpha$ and referred to as the *level of significance*.

**Type II error**—Accepting $H_0$ when $H_1$ is true. The *probability* of committing a Type II error is denoted by $\beta$.

In any hypothesis test two kinds of errors can be committed: (1) we can reject $H_0$ when $H_0$ is true, or (2) we can accept $H_0$ when $H_0$ is false. These are called **Type I errors** and **Type II errors**, respectively. Similarly, two kinds of correct decisions can be made: (1) we can accept a true $H_0$, or (2) we can reject a false $H_0$. Table 8.3 shows these four possible "decision/state of nature" combinations.

**TABLE 8.3**

|  |  | TRUE STATE OF NATURE | |
|---|---|---|---|
|  |  | $H_0$ is True | $H_1$ is True |
| | Accept $H_0$ | Correct decision | Type II error |
| OUR DECISION | Reject $H_0$ | Type I error | Correct decision |

Once an inference is made, there is no way to know if an error was committed, because we never know what the answer should have been. However, we can compute the *probability* of making an error. Recall the credit rating example of Section 8.2: given $n = 18$ observations on a Bernoulli random variable, we wanted to test $H_0$: $p = \frac{1}{2}$ versus $H_1$: $p > \frac{1}{2}$. The decision rule stated that $H_0$ should be rejected if $y$, the number of correct identifications, equals or exceeds 13. It follows then that what we are calling a Type I error—rejecting $H_0$ when $H_0$ is true—will occur 5% of the time:

$$P \text{ (Type I error)} = P \left( H_0 \text{ is rejected} \mid H_0 \text{ is true} \right)$$

$$= P \left( Y \geqslant 13 \middle| p = \frac{1}{2} \right)$$

$$= P(Y = 13) + P(Y = 14) + \cdots + P(Y = 18)$$

$$= \sum_{y=13}^{18} {}_{18}C_y \left( \frac{1}{2} \right)^y \left( 1 - \frac{1}{2} \right)^{18-y}$$

$$= 0.05$$

The 0.05 should come as no surprise. We specifically set the critical value equal to 13 so that the probability associated with the critical region *would* be 0.05. In other words, the probability of committing a Type I error is what we referred to earlier as $\alpha$, the level of significance.

Whatever we call it, the probability associated with rejecting $H_0$ when $H_0$ is true is important because it defines what we mean by "beyond all reasonable doubt." Indeed, when reporting the results of any statistical test, we must always include two pieces of information: (1) whether $H_0$ was accepted or rejected and (2) the level of significance. Knowing only the former tells us very little. What the level of significance adds is a single-number summary of the "rules" by which the decision was made. In short, $\alpha$ reflects the amount of evidence that the experimenter is demanding to see before abandoning the null hypothesis. (An alternative way of summarizing hypothesis tests involves the use of P *values*. We will discuss these later, after we become more comfortable with the notions of Type I and Type II errors.)

---

**QUESTIONS**

**8.24** Suppose $H_0$: $p = p_0$ is tested against $H_1$: $p > p_0$.

(a) If the null hypothesis is rejected at the $\alpha = 0.05$ level of significance, will it necessarily be rejected at the $\alpha = 0.01$ level of significance?

(b) If $H_0$ is rejected at the $\alpha = 0.01$ level, will it necessarily be rejected at the $\alpha = 0.05$ level?

So far, we have stressed the importance of the Type I error. In the credit rating example, the emphasis was on trying to avoid the mistake of adopting the services of CCEG when it could not, as claimed, distinguish bad credit risks. But suppose that CCEG does have superior abilities, but Visamaster, in its extreme caution, demands such overwhelming evidence that it dismisses the genuine talent of CCEG. Not only would Visamaster lose those useful services, but its competitor, National Express, might be less cautious and employ CCEG. Clearly, the designer of hypothesis tests cannot ignore the Type II error: failing to recognize a "true" $H_1$ may result in a valuable new insight going unnoticed or a promising new technique remaining undeveloped.

Computing the probability of a Type II error is quite different from finding the probability of a Type I error. The latter is the level of significance and is simply set by the experimenter, so no computations are required. The Type II error, on the other hand, is not specified explicitly by the experimenter. Furthermore, for every value of the parameter in $H_1$, there is a different value for the probability of committing a Type II error.

As an example, suppose we want to find the probability of committing a Type II error in the Visamaster example if the true $p$ (the probability that CCEG finds the bad credit risk) is 0.7. By definition, we are looking for the probability that $Y$, the number of correct identifications, is in the *acceptance* region when $p = 0.7$. That is,

$$P \text{ (Type II error} \mid p = 0.7) = P \text{ (we accept } H_0 \mid p = 0.7)$$

$$= P(Y \leqslant 12 \mid p = 0.7)$$

$$= P(Y = 0) + \cdots + P(Y = 12)$$

$$= \sum_{y=0}^{12} {}_{18}C_y (0.7)^y (1 - 0.7)^{18-y}$$

$$= 0.47$$

Thus, if $p$ were actually 0.7, CCEG's ability would go "undetected" (that is, we would wrongly accept $H_0$) 47% of the time.

The probability of a Type II error is denoted by $\beta$. Figure 8.6 shows the sampling distributions of $Y$ when $p = \frac{1}{2}$ and $p = 0.7$; shaded are the areas corresponding to $\alpha$ and $\beta$. Notice that $\alpha$ is under the "$H_0$ distribution" and $\beta$ is under the "$H_1$ distribution."

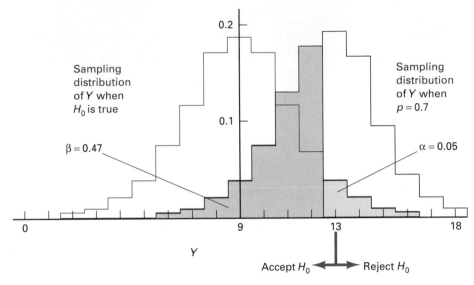

**FIG. 8.6**

The magnitude of $\beta$ clearly depends on the presumed value for $p$. If, for example, CCEG is correct 90% of the time (that is, $p = 0.9$), then the probability of Visamaster making a Type II error is only 0.006:

$$\beta = P \,(\text{Type II error} \,|\, p = 0.9) = P \,(\text{we accept } H_0 \,|\, p = 0.9)$$

$$= P(Y \leqslant 12 \,|\, p = 0.9)$$

$$= P(Y = 0) + \cdots + P(Y = 12)$$

$$= \sum_{y=0}^{12} {}_{18}C_y (0.9)^y (1 - 0.9)^{18 - y}$$

$$= 0.006$$

(see Figure 8.7).

The notation for the probability of committing a Type II error is well established, which is unfortunate because it tends to be a bit misleading. The use of $\beta$ suggests a single number, but, as we have seen, $\beta$ is actually a *function* (of the parameter being tested) and functional notation would be more appropriate. In the previous example, for instance, it would have made sense to write $\beta(0.7) = 0.47$ and $\beta(0.9) = 0.006$. We will sometimes use functional notation in connection with Type II errors when $\beta$'s dependence on the parameter needs to be emphasized.

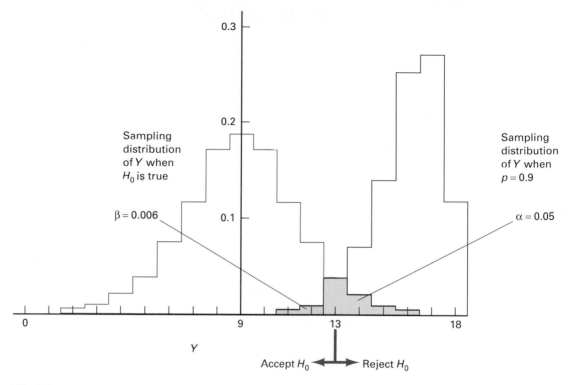

**FIG. 8.7**

COMMENT:  Usually a change in the critical region that produces a *decrease* in $\beta$ causes an *increase* in $\alpha$, and vice versa. A simultaneous decrease in both $\alpha$ and $\beta$ generally requires a larger sample size.

COMMENT:  The basic notions of Type I and Type II errors first arose in the context of quality control. The pioneering work was done at the Bell Telephone Laboratories: there the terms *producer's risk* and *consumer's risk* were introduced for what we are calling $\alpha$ and $\beta$. Those first attempts to quantify the decision-making process were generalized by two statisticians, Jerzy Neyman and E. S. Pearson, in the 1930s, and soon evolved into the theory of hypothesis testing as we know it today.

---

*EXAMPLE 8.3*   For the point spread data in Case Study 8.1, we derived a two-sided 0.05 decision rule: reject $H_0$: $p = \frac{1}{2}$ if $y \leqslant 51.1$ or $y \geqslant 72.9$, where $y$ is the number of times (out of 124) that the favorite beat the spread. Suppose the true probability of the favorite beating the spread is $p = 0.55$. If so, how often will a Type II error be committed?

*Solution*   By definition,

$$\beta = P \text{ (Type II error} \,|\, p = 0.55) = P \text{ (we accept } H_0 \,|\, p = 0.55)$$

$$= P(51.1 \leqslant Y \leqslant 72.9 \,|\, p = 0.55)$$

Using the De Moivre-Laplace approximation, we can write

$$\beta = P\left(\frac{51.1 - 124(0.55)}{\sqrt{124(0.55)(0.45)}} \leq \frac{Y - 124(0.55)}{\sqrt{124(0.55)(0.45)}} \leq \frac{72.9 - 124(0.55)}{\sqrt{124(0.55)(0.45)}}\right)$$

$$\doteq P(-3.09 \leq Z \leq 0.85) = 0.8023$$

The magnitude of $\beta$, here, is very revealing. It shows that taking 124 observations does not give us enough "power" to detect consistently a shift away from $p_0$ as small as 0.05 $(= 0.55 - 0.50)$.

---

**EXAMPLE 8.4**   In Case Study 8.2 we tested the null hypothesis that the mean pregnancy duration for a certain group of women is 266 days (against the one-sided alternative that $\mu < 266$). The 0.05 decision rule called for $H_0$ to be rejected if $\bar{y} \leq 262.9$ (based on a random sample size of 70). Suppose the true value for the mean is 261. What is our probability of committing a Type II error?

*Solution*   Since we accept $H_0$ when $\bar{y} > 262.9$, $\beta$ is the likelihood of that happening when $\mu = 261$:

$$\beta = P\,(\text{Type II error}\,|\,\mu = 261) = P\,(\text{we accept } H_0\,|\,\mu = 261)$$

$$= P(\bar{Y} > 262.9\,|\,\mu = 261) = P\left(\frac{\bar{Y} - 261}{16/\sqrt{70}} > \frac{262.9 - 261}{16/\sqrt{70}}\right)$$

$$= P(Z > 0.99) = 1 - \Phi(0.99) = 1 - 0.8389 = 0.1611$$

Slightly more than 16% of the time, then, the *sample* mean (when $\mu = 261$) will take on a value that will "fool" us into thinking that the *true* mean is still 266.

---

**QUESTIONS**

**8.25**   In Example 8.3 find $\beta$ when $p = 0.45$.

**8.26**   An experiment consists of 12 Bernoulli trials. Let $H_0\colon p = \frac{1}{3}$ be tested against $H_1\colon p > \frac{1}{3}$ by using the decision rule "reject $H_0$ if $y$, the number of successes, equals or exceeds 8."

(a) Find $\alpha$.

(b) Find $\beta$ for $p = \frac{1}{2}$ and $p = \frac{2}{3}$.

**8.27**   An urn contains 10 chips. An unknown number of the chips are red; the others are white. We wish to test

$$H_0\colon \quad \text{exactly half the chips are red}$$

vs.

$$H_1\colon \quad \text{more than half the chips are red}$$

Three chips will be drawn, without replacement, and $H_0$ will be rejected if two or more are red.

(a) Find $\alpha$.

(b) Find $\beta$ when the urn is 60% red.

(c) Find $\beta$ when the urn is 70% red.

8.28   In Example 8.4 find $\beta$ when $\mu = 258$ and when $\mu = 269$.

8.29   Suppose a sample of size 40 is chosen from a normal population whose mean ($\mu$) is unknown but whose standard deviation ($\sigma$) is 7.1.

(a) Find the decision rule (in terms of $\bar{y}$) for testing $H_0: \mu = 50$ versus $H_1: \mu < 50$ when $\alpha = 0.10$, 0.05, and 0.01.

(b) Find $\beta$ if $\alpha = 0.05$ and $\mu = 48$.

8.30   Suppose $\bar{y}$ is the mean of a normal random sample of size 25, where $\sigma = 10$. Test $H_0: \mu = 75$ versus $H_1: \mu < 75$ by rejecting the null hypothesis when $\bar{y} \leqslant 71.08$.

(a) Find $\alpha$.

(b) Find $\beta$ for $\mu = 71$.

(c) Find $\beta$ for $\mu = 73$.

8.31   Carla is a pizza inspector for the city health department. She has received numerous complaints about a certain pizzeria for failing to comply with its advertisements. The pizzeria claims that, on the average, each of its large pizzas is topped with 2 oz of pepperoni. The dissatisfied customers feel that the actual amount of pepperoni used is considerably less. To settle the matter, Carla decides to do a hypothesis test. She assumes that the distribution of pepperoni weights on a pizza is normal with mean $\mu$ and standard deviation $\sigma = 0.5$. Her objective is to test $H_0: \mu = 2$ versus $H_1: \mu < 2$. For data she takes one large pizza and weighs the pepperoni. If the weight is less than or equal to 1.3 oz, she will reject $H_0$.

(a) Compute $\alpha$.

(b) Compute $\beta(1.8)$.

(c) What decision rule would make $\alpha = 0.01$?

(d) Compute $\beta(1.8)$ for the decision rule in part (c).

8.32   Polygraphs used in criminal investigations typically measure five bodily functions: (1) thoracic respiration (2) abdominal respiration, (3) blood pressure and pulse rate, (4) muscular movement and pressure and (5) galvanic skin response. The magnitudes of these responses presumably indicate whether the subject is telling the truth. But the test is not infallible (73). Seven experienced polygraph examiners were given a set of 40 records, 20 from innocent subjects and 20 from guilty ones. Each subject had been asked seven questions. On the basis of each question, the examiner was to make a summary judgment: "innocent" or "guilty." The results are shown below:

|  |  | ACTUAL STATUS | |
|  |  | Innocent | Guilty |
|---|---|---|---|
| EXAMINER'S DECISION | Innocent | 131 | 15 |
| | Guilty | 9 | 125 |

(a) What are the numerical values of $\alpha$ and $\beta$ in this context?

(b) In a judicial setting should Type I and Type II errors carry equal weight?

### Power of a Test

**power**—The probability that $H_0$ is rejected when $H_1$ is true. Numerically, it equals $1 - \beta$, where $\beta$ is the probability of committing a Type II error. A graph of $1 - \beta$ versus the possible values of the parameter being tested is called a *power curve*.

If $\beta$ is the probability that we *accept* $H_0$ when $H_1$ is true, then $1 - \beta$ is necessarily the probability that we *reject* $H_0$ when $H_1$ is true. We call $1 - \beta$ the *power* of a test: it represents the probability of correctly recognizing that $H_0$ is false. Recall the CCEG example. When

$$p = P(\text{CCEG identifies credit risk}) = 0.7$$

$\beta$ was shown to equal 0.47. The power of the test, then, against a $p$ of 0.7 is 0.53:

$$1 - \beta(0.7) = 1 - 0.47 = 0.53$$

Similarly, if $p = 0.9$, $\beta(0.9) = 0.006$, so the probability of the test rejecting $H_0$ (as it should) is

$$1 - \beta(0.9) = 1 - 0.006 = 0.994$$

In general, the dependence of $1 - \beta$ on the parameter being tested can be pictured by drawing a *power curve*, which is simply a graph showing $1 - \beta$ on the vertical axis and values of the parameter on the horizontal axis. Figure 8..8 shows the power curve for testing the CCEG hypotheses

$$H_0: \quad p = \frac{1}{2}$$

vs.

$$H_1: \quad p > \frac{1}{2}$$

where $p$ is the probability of success in 18 Bernoulli trials and the decision rule is "reject $H_0$ if $y \geqslant 13$." The two ×'s on the curve show the $(p, 1 - \beta)$ coordinates we just described: (0.7, 0.53) and (0.9, 0.994).

To a statistician or an experimenter, power curves are quite useful. They show at a glance the overall performance capability of a hypothesis

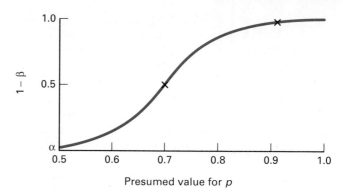

**FIG. 8.8**

test. Specifically, they give the probability that the test will be able to "spot" a certain shift away from $H_0$. If, for a given parameter value, $1 - \beta$ is not as large as the experimenter would like, then the hypothesis test needs to be redesigned. The usual way of increasing $1 - \beta$ for a given value of $p$ (or $\mu$) is to increase the sample size.

QUESTIONS

8.33 Consider the power curve for the decision rule developed in Case Study 8.1. Use the De Moivre-Laplace theorem to approximate $1 - \beta$ for $p = 0.60$, 0.65, 0.68, and 0.70. Use the curve to estimate the probability of committing a Type II error if $p = 0.64$.

8.34 Sketch the power curve for the hypothesis test described in Question 8.26.

8.35 Sketch the power curve for Carla's pepperoni hypothesis test in Question 8.31.

8.36 In general, what is the power of a test when the parameter is equal to the value specified in $H_0$?

# 8.4

## INTERVAL ESTIMATION

Hypothesis testing provides a very workable framework for a great many questions that experimenters typically want answered. In forthcoming chapters we will see that posing problems with "accept $H_0$" or "reject $H_0$" as possible solutions can be as effective in biology and engineering as in sociology and economics. Nevertheless, in some situations forcing a "yes/no" conclusion is not a good idea and may not even be possible.

Imagine a businessman assessing the marketability of a new product. What he would like to have, ideally, is a numerical value for $p$, the proportion of consumers who would buy the product. But $p$ is a population parameter, so there is no way its value can be known exactly. What we offer

**interval estimation**—Using random intervals to estimate unknown parameters. Typically the intervals are constructed so that a high proportion, often 95% or 99%, contain the unknown parameter as an interior point.

in this section is a compromise for finding a range of values for that proportion (say 0.15 to 0.26) where the range has a high probability of including the true $p$. We will take a first look at this notion of **interval estimation** in the next several pages. It will appear in many forms in later chapters.

Perhaps the best way to understand both the mechanics as well as the theory of interval estimation is to consider an example. Suppose a company is testing automobile tires manufactured by a new process that promises to increase tire life. Let the random variable $Y$ denote the number of miles a driver gets on a set of these new tires. On the basis of previous experience with similar tires, we assume that the distribution of $Y$ is normal and $\sigma = 4000$ miles. How should we compute an estimate for the true average tire life $\mu$?

One answer should be obvious. Let a sample of $n$ drivers test the tires and record the number of miles they get, $y_1, y_2, \ldots, y_n$, before the tires have to be replaced. Then, in the spirit of Section 3.2, compute the sample average tire life $\bar{y}$, where

$$\bar{y} = \frac{1}{n} \sum_{i=1}^{n} y_i$$

**point estimate**—A single number that represents our best guess for the value of an unknown parameter.

In the terminology of statistics the sample mean $\bar{y}$ is called a **point estimate** for the true mean $\mu$. If our objective is to "guess" the center of the $Y$ distribution by producing one numerical value, then $\bar{y}$ is the best possible choice. But $\bar{y}$ has a serious deficiency: it tells us nothing about the "precision" of the estimate. If we find that $\bar{y} = 35,200$ miles, for example, how likely is it that the true mean is within 500 miles of that number? or 5000 miles? Clearly, any procedure that would allow us to associate a probability with the distance between $\bar{y}$ and $\mu$ would be very enlightening. It would give an experimenter a way of distinguishing a precise estimate from one that was not. The most widely used technique for assigning probabilities to point estimates is the construction of *confidence intervals*.

If $Y_1, Y_2, \ldots, Y_n$ represents a sample of (normally distributed) tire mileages, then the ratio

$$\frac{\bar{Y} - \mu}{\sigma/\sqrt{n}}$$

is a standard normal random variable $Z$ (recall Theorem 7.2). Suppose, for reasons soon to be apparent, we want to find two numbers along the horizontal axis of a $Z$ distribution, equidistant from 0, such that the area under $\varphi_Z(z)$ between those two numbers is 0.95. From Appendix A.1,

$$P(-1.96 \leqslant Z \leqslant 1.96) = 0.95$$

Equivalently,

$$P\left(-1.96 \leqslant \frac{\bar{Y} - \mu}{\sigma/\sqrt{n}} \leqslant 1.96\right) = 0.95 \tag{8.3}$$

(see Figure 8.9).

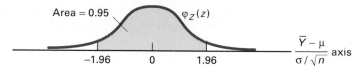

**FIG. 8.9**

By doing a little algebra, we can reexpress Equation 8.3 in a very useful way by isolating $\mu$ in the center of the two inequalities:

$$P\left(-1.96 \leqslant \frac{\bar{Y} - \mu}{\sigma/\sqrt{n}} \leqslant 1.96\right) = 0.95$$

$$= P\left(-1.96\,\frac{\sigma}{\sqrt{n}} \leqslant \bar{Y} - \mu \leqslant 1.96\,\frac{\sigma}{\sqrt{n}}\right)$$

$$= P\left(-\bar{Y} - 1.96\,\frac{\sigma}{\sqrt{n}} \leqslant -\mu \leqslant -\bar{Y} + 1.96\,\frac{\sigma}{\sqrt{n}}\right)$$

$$= P\left(\bar{Y} - 1.96\,\frac{\sigma}{\sqrt{n}} \leqslant \mu \leqslant \bar{Y} + 1.96\,\frac{\sigma}{\sqrt{n}}\right) = 0.95 \qquad (8.4)$$

Notice what has happened. We started out in Equation 8.3 with a probability statement about a random variable, $\dfrac{\bar{Y} - \mu}{\sigma/\sqrt{n}}$, lying between two constants, $-1.96$ and $+1.96$. After the algebraic manipulations, we end up in Equation 8.4 with a probability statement about a constant $\mu$ lying between two random variables,

$$\bar{Y} - 1.96\,\frac{\sigma}{\sqrt{n}} \quad \text{and} \quad \bar{Y} + 1.96\,\frac{\sigma}{\sqrt{n}}$$

In effect, we turned Equation 8.3 inside out!

What is important about the transformation from Equation 8.3 to Equation 8.4 is the interpretation that can be assigned to the latter. First, we can say that the random interval extending from

$$\bar{Y} - 1.96\,\frac{\sigma}{\sqrt{n}} \quad \text{to} \quad \bar{Y} + 1.96\,\frac{\sigma}{\sqrt{n}}$$

has a 95% probability of containing the true $\mu$ as one of its interior points. Second, we can view the *length* of that interval,

$$\bar{Y} + 1.96\,\frac{\sigma}{\sqrt{n}} - \left(\bar{Y} - 1.96\,\frac{\sigma}{\sqrt{n}}\right) = 3.92\,\frac{\sigma}{\sqrt{n}}$$

as a useful measure of the precision of $\bar{Y}$: the shorter the interval, the more reassured we can feel that the sample estimate is close to the population mean.

As an example of how these ideas are put into practice, suppose the experiment being hypothesized is actually carried out by a team of $n = 25$ drivers. When the study is over, we find that the sample average tire life was 43,126 miles ($= \bar{y}$). As a result, the random interval we just described is now reduced to a *specific* set of values:

$$\left( 43{,}126 - 1.96\, \frac{4000}{\sqrt{25}},\ 43{,}126 + 1.96\, \frac{4000}{\sqrt{25}} \right)$$

$$= (43{,}126 - 1568,\ 43{,}126 + 1568)$$

$$= (41{,}558 \text{ miles},\ 44{,}694 \text{ miles})$$

(remember that we assumed $\sigma = 4000$ miles). The set of mileages from 41,558 to 44,694 is called a 95% *confidence interval for $\mu$*. We "hope" that one of the mileages in that range is the true average tire life.

Confidence intervals are often misinterpreted. The 95% refers to the procedure, not to the actual interval (41,558, 44,694). Before any data are collected, the random interval

$$\left( \bar{Y} - 1.96\, \frac{\sigma}{\sqrt{n}},\ \bar{Y} + 1.96\, \frac{\sigma}{\sqrt{n}} \right)$$

has a 95% chance of "covering" the true $\mu$. In other words, if we repeatedly took samples of size 25 and calculated the corresponding 95% confidence intervals, it would be true that 95% of those intervals would contain $\mu$ as an interior point and 5% would not (see Figure 8.10). After the data are collected, any resulting interval, such as (41,588, 44,694), is no longer random: it contains $\mu$ either 100% of the time or 0% of the time.

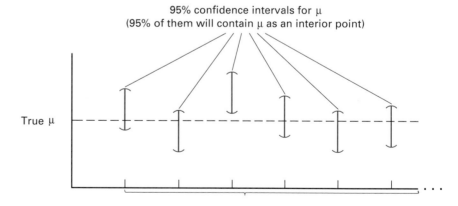

95% confidence intervals for $\mu$
(95% of them will contain $\mu$ as an interior point)

True $\mu$

Possible samples of size $n$

**FIG. 8.10**

We do have the reassurance, though, of having used a procedure that produces "correct" intervals, in the long run, 19 times out of 20. It is in the latter sense that we can hope that our particular interval is one of the

95% that are correct and not one of the 5% that misses the parameter altogether.

Conceptually, constructing a 95% confidence interval to estimate $\mu$ is like playing a rather perverse dart game where the board is thrown at the dart, instead of the other way around. Our formula,

$$\left( \bar{Y} - 1.96 \frac{\sigma}{\sqrt{n}}, \bar{Y} + 1.96 \frac{\sigma}{\sqrt{n}} \right)$$

gives us the ability to throw the board with sufficient skill that it hits the unseen dart, in the long run, 95% of the time. Once the board is thrown on any particular occasion, though, it either hits the dart or it misses, and we have no way of knowing what actually happened.

Confidence intervals do not have to be 95% confidence intervals (any more than a hypothesis test has to have $\alpha = 0.05$). A 90% confidence interval, for example, would be a range of values having a 90% probability of containing the true mean. Since

$$P(-1.64 \leqslant Z \leqslant 1.64) = 0.90$$

it follows that we would define such an interval by the formula

$$\left( \bar{Y} - 1.64 \frac{\sigma}{\sqrt{n}}, \bar{Y} + 1.64 \frac{\sigma}{\sqrt{n}} \right)$$

(recall the steps leading from Equation 8.3 to Equation 8.4). These intervals are shorter than 95% confidence intervals because they need not cover the true $\mu$ as often.

Mathematically, we can adjust the confidence "coefficient" of a random interval to any level desired simply by changing the multiple of $\sigma/\sqrt{n}$. For a 95% interval we use $\pm 1.96$; for a 90% interval the multiples are $\pm 1.64$, and so on. However, only 50%, 90%, 95%, and 99% intervals are widely used in practice.

A general formula for a confidence interval for $\mu$ is given in Theorem 8.3. Don't be confused by the notation: the expression "$100(1 - \alpha)\%$" is nothing more than a way of denoting an *arbitrary* confidence coefficient. If $\alpha = 0.05$, for instance, then $100(1 - \alpha)\%$ reduces to $100(1 - 0.05)\% = 95\%$.

---

### THEOREM 8.3

Let $y_1, y_2, \ldots, y_n$ be a random sample of size $n$ from a normal distribution with known variance $\sigma^2$ and unknown mean $\mu$. A $100(1 - \alpha)\%$ confidence interval for $\mu$ is the range of numbers,

$$\left( \bar{y} - z_{\alpha/2} \frac{\sigma}{\sqrt{n}}, \bar{y} + z_{\alpha/2} \frac{\sigma}{\sqrt{n}} \right)$$

where $P(Z \geqslant z_{\alpha/2}) = \alpha/2$.

*EXAMPLE 8.5*   Consider again the medical problem described in Case Study 8.2. If the average pregnancy duration $\mu$ of women living in a certain area indicates the general health status of those women, then having a good estimate for that parameter is important. A reasonable statistical objective for public health officials in that case would be the construction of, say, a 99% confidence interval for $\mu$. Find that interval, using the data of Table 8.2.

*Solution*   Since pregnancy durations are normally distributed and their standard deviation is known ($\sigma = 16$ days), Theorem 8.3 applies. From Appendix A.1,

$$P(-2.58 \leqslant Z \leqslant 2.58) = 0.99$$

so $z_{\alpha/2} = z_{0.01/2} = z_{0.005} = 2.58$. Also, for the data referred to, $n = 70$ and $\bar{y} = 260.3$ days. The 99% confidence interval for the true average pregnancy duration, therefore, is

$$\left( \bar{y} - 2.58 \frac{\sigma}{\sqrt{n}}, \bar{y} + 2.58 \frac{\sigma}{\sqrt{n}} \right)$$

$$= \left( 260.3 - 2.58 \frac{16}{\sqrt{70}}, 260.3 + 2.58 \frac{16}{\sqrt{70}} \right)$$

$$= (260.3 - 4.9, 260.3 + 4.9)$$

$$= (255.4 \text{ days}, 265.2 \text{ days})$$

QUESTIONS

**8.37**   Suppose a random sample of size 100 is drawn from a normal population with mean $\mu$ and standard deviation $\sigma = 5$. Find the 95% confidence interval for $\mu$ if the sample mean is 46.7. How would the length of the confidence interval change if the sample size were doubled?

**8.38**   Construct a 50% confidence interval for the pregnancy data given in Table 8.2. What is the statistical interpretation of that interval?

**8.39**   Refer to the data in Question 2.7. Let $\mu$ denote the true average radiation level characteristic of TV display areas in department stores. Suppose that other studies have shown that the standard deviation of radiation levels in these areas is approximately 0.20 mr/hr. Construct a 90% confidence interval for $\mu$.

**8.40**   Four factors appear in the formula for a confidence interval: $\bar{y}$, $z_{\alpha/2}$, $\sigma$, and $n$. Which factors affect the length of the interval?

**8.41**   Construct the confidence interval for $\mu$ appropriate for each of the following situations:

(a) $n = 49$,   $100(1 - \alpha) = 80$,   $\sigma = 10.4$,   $\bar{y} = 14.3$

(b) $n = 36$,   $100(1 - \alpha) = 95$,   $\sigma = 4.1$,   $\bar{y} = 35.7$

(c) $n = 40$, $100(1 - \alpha) = 50$, $\sigma = 6.3$, $\bar{y} = 13.9$

(d) $n = 4$, $100(1 - \alpha) = 99$, $\sigma = 2.0$, $\bar{y} = 15.0$

8.42 Suppose a 95% confidence interval is to be constructed for the mean of a normal distribution where $\sigma = 10$. If the experimenter wants the length of the interval to be no longer than 8, what is the smallest sample size that can be taken?

8.43 What confidence coefficient is associated with the following random interval:

$$\left( \bar{Y} - 1.28 \frac{\sigma}{\sqrt{n}}, \bar{Y} + 1.64 \frac{\sigma}{\sqrt{n}} \right)?$$

8.44 Recall the discussion in Example 8.2 of ACT composites. Suppose school officials take a random sample of scores earned by eight of their graduating seniors and get the following results:

| | |
|---|---|
| 17 | 29 |
| 19 | 17 |
| 21 | 18 |
| 14 | 26 |

Define the relevant parameter and estimate it with a 95% confidence interval.

# 8.5

## SUMMARY

Chapter 8 has provided us with a first glimpse of the principles of statistical inference. We saw that many questions in science can be meaningfully reduced to a choice between two competing hypotheses, where each of those hypotheses is a statement about the value of a parameter. Although the number of parameters that might be of interest is quite large, we will usually focus on either the mean $\mu$ of a normal distribution or the success probability $p$ of a Bernoulli trial. The procedure by which we analyze a set of data and decide to either "accept $H_0$" or "reject $H_0$" is called a *hypothesis test*.

Specific formulas for carrying out a hypothesis test depend on the parameter being investigated and the underlying nature of the data (two-sample, regression-correlation, and so on), but the *structure* of a test—how we proceed conceptually—is always the same.

1. We begin by presuming that $H_0$ is true. Inevitably, that "predicts" to some extent what the data should look like.

2. We then compare, by using a *test statistic*, what *did* happen with what the null hypothesis *predicted* would happen.

3. If the data and the predictions are sufficiently close, we accept $H_0$; otherwise, we reject $H_0$.

An important key to understanding hypothesis testing is being aware of the errors to which our decision may lead. Specifically, we may be "deceived" (by the test statistic) into rejecting $H_0$ when, in fact, $H_0$ is true. A mistake of that sort is called a *Type I error* and the probability of its occurrence, $\alpha$, is fixed at the outset by the experimenter. Decision rules are typically set up in such a way that the likelihood of committing that particular error—what we refer to as a test's *level of significance* (or $\alpha$ *level*), is 0.10, 0.05, 0.01, or 0.001.

Unfortunately, a Type I error is not the only kind of decision-making blunder to which we can fall victim. If $H_1$ is actually true but the test statistic takes on a value that tells us to accept $H_0$, we commit a *Type II error*. The probability of incorrectly accepting $H_0$ is denoted by $\beta$: its magnitude is a good indicator of the precision of a test. Ideally, we would like $\beta$ to be as small as possible.

The second theme of Chapter 8 takes up where the first leaves off: it addresses situations where hypothesis testing is either inappropriate, irrelevant, or impossible. It will sometimes be the case, for instance, that there is no obvious choice for either $p_0$ or $\mu_0$. When that happens, no testing is possible because there is no way of defining either $H_0$ or $H_1$. There will also be occasions where an "accept $H_0$" or "reject $H_0$" response oversimplifies a problem, and what an experimenter really wants is a more numerically oriented inference.

*Confidence intervals* are "plan B." When hypothesis testing—for whatever reason—does not work, interval estimation will. Section 8.4 describes the construction of confidence intervals for the mean of a normal distribution when the standard deviation is known. The idea is to find a random interval that has a high probability (usually 0.95 or 0.99) of "covering" the true $\mu$. Like the material on hypothesis testing, the notion of a confidence interval will reappear in many other contexts in the chapters ahead.

| To Test | Versus | Reject $H_0$ | Formula |
|---------|--------|--------------|---------|
| $H_0$:  $p = p_0$ | $H_1$:  $p \neq p_0$ | $z \leqslant -z_{\alpha/2}$, $z \geqslant z_{\alpha/2}$ | $z = \dfrac{y - np_0}{\sqrt{np_0(1 - p_0)}}$ |
|  | $H_1$:  $p > p_0$ | $z \geqslant z_\alpha$ |  |
|  | $H_1$:  $p < p_0$ | $z \leqslant -z_\alpha$ |  |
| $H_0$:  $\mu = \mu_0$ | $H_1$:  $\mu \neq \mu_0$ | $z \leqslant -z_{\alpha/2}$, $z \geqslant z_{\alpha/2}$ | $z = \dfrac{\bar{y} - \mu_0}{\sigma/\sqrt{n}}$ |
|  | $H_1$:  $\mu > \mu_0$ | $z \geqslant z_\alpha$ |  |
|  | $H_1$:  $\mu < \mu_0$ | $z \leqslant -z_\alpha$ |  |

$100(1 - \alpha)\%$ confidence interval for $\mu$:

$$\left( \bar{y} - z_{\alpha/2} \frac{\sigma}{\sqrt{n}}, \bar{y} + z_{\alpha/2} \frac{\sigma}{\sqrt{n}} \right)$$

**REVIEW QUESTIONS**

**8.45** A marketing research study has shown that only 25% of all customers browsing in Ye Olde Book Shoppe buy anything. Hoping to increase that figure, the store's manager has begun promoting heavily discounted paperbacks as daily specials. If 80 of the next 200 browsers buy something, can the manager conclude that her efforts have had an effect? Set up and carry out the appropriate hypothesis test. Let $\alpha = 0.01$.

**8.46** Listed below are the pay raises (expressed as per cent increases) for eight typists selected at random from a large secretarial pool. For the entire office staff, raises are normally distributed with $\sigma = 1.3$.

$$
\begin{array}{cccc}
3.6 & 2.4 & 5.2 & 3.4 \\
7.1 & 4.6 & 4.3 & 6.1
\end{array}
$$

Construct a 90% confidence interval for $\mu$, the true average secretarial pay raise.

**8.47** Last year 8% of a city's automobiles were in violation of at least one federally regulated emission standard. This year 10% of a sample of 400 cars were found to be out of compliance. At the 0.01 level, is the difference between 8% and 10% statistically significant?

**8.48** Which statement is stronger: "reject $H_0$ at the $\alpha = 0.01$ level of significance" or "reject $H_0$ at the $\alpha = 0.05$ level of significance"?

**8.49** Skip believes he is a better tennis player than Palmer and challenges his friend to an 8-game match. They agree that Skip will have demonstrated his superiority if he wins at least six of the games. Let $p = P$ (Skip wins game). Define the appropriate $H_0$ and $H_1$. What level of significance is being used?

**8.50** A random sample $Y_1, Y_2, \ldots, Y_n$ is taken from a normal distribution with standard deviation $\sigma$. The pdf on the left is the sampling distribution of $\bar{Y}$ when $H_0$ is true; the pdf on the right is the sampling distribution of $\bar{Y}$ when $\mu = 50$.

(a) State $H_0$ and $H_1$.

(b) What does $P(\bar{Y} > 47 | \mu = 50)$ represent?

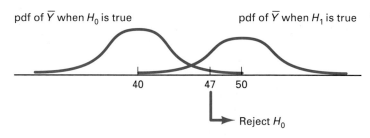

pdf of $\bar{Y}$ when $H_0$ is true          pdf of $\bar{Y}$ when $H_1$ is true

40          47   50

Reject $H_0$

**8.51** Six rain gauges, all located within an $\frac{1}{8}$-mile radius, gave the following precipitation readings (in inches) during a thunderstorm:

$$
\begin{array}{ccc}
1.02 & 1.31 & 0.98 \\
1.14 & 1.20 & 1.05
\end{array}
$$

The manufacturer claims that the standard deviation associated with the gauges is 0.075 inch. Construct a 90% confidence interval for $\mu$, the true average rainfall. Assume that the rain gauge readings are normally distributed.

8.52    Suppose the random variable $Y$ has a geometric pdf,

$$f_Y(y) = q^{y-1}p, \qquad y = 1, 2, \ldots$$

where $p$ is unknown. To test $H_0$: $p = 1/20$ versus $H_1$: $p > 1/20$, an experimenter decides to reject $H_0$ if $y \leqslant 2$. (a) Find $\alpha$. (b) Compute the power of the test if $p = 1/8$.

8.53    The time it takes (in minutes) for a low rpm mixer to homogenize industrial waste is known to be normally distributed with $\sigma^2 = 12.0$. On the average, a batch should be processed in 56.0 minutes, but the foreman suspects the machine is partially clogged and not working as efficiently as it should. The last five runs have yielded mixing times of 63.0, 55.0, 60.0, 66.0, and 57.0. Set up and test the relevant hypotheses. Let $\alpha = 0.10$.

8.54    An experimenter is finishing the statistical write-up on four unrelated projects. Each analysis requires the construction of a 95% confidence interval. What is the probability that exactly two of the intervals contain the unknown parameters being estimated?

8.55    State the decision rule for testing

$$H_0: \quad p = 0.55$$
vs.
$$H_1: \quad p \neq 0.55$$

at the $\alpha = 0.37$ level of significance. If $n = 60$ and $y = 40$, should $H_0$ be rejected?

8.56    Twelve elderly patients suffering from a nervous disorder are given a new drug that may improve the 30% cure rate reported for the medication currently available.

(a) What $H_0$ and $H_1$ need to be tested?

(b) State a decision rule for which $\alpha$ will be approximately 0.05.

8.57    In terms of what they represent, what is the difference between the intervals $\bar{y} + s$ and $\bar{y} \pm s/\sqrt{n}$?

8.58    Suppose that (6.7, 14.9) is a 95% confidence interval for $\mu$ based on a random sample of size 9 from a normal distribution. Find $\sigma$.

8.59    Shown below are two power curves for tests of $H_0$: $p = \frac{1}{2}$ versus $H_1$: $p > \frac{1}{2}$. Both decision rules have the same level of significance and both use the test statistic described in Theorem 8.1. Test $A$ is based on $n$ observations, and Test $B$ on $m$ observations. Which is larger, $n$ or $m$?

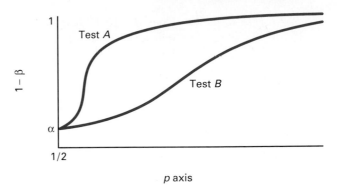

8.60   Nationwide, 17% of all inexpensive VCR's required factory repairs within two years of purchase. In contrast, a survey showed that only 10% of 2000 moderately priced VCR's sold two years ago have required factory servicing. At the 0.01 level of significance, can the manufacturers of the more expensive VCR's claim that their products are more reliable?

8.61   Let $Y_1, Y_2, \ldots, Y_n$ be a random sample from a normal distribution. Which would be shorter, a 90% confidence interval for $\mu$ when $n = 16$ or a 99% confidence interval for $\mu$ when $n = 25$?

8.62   Otto is given the problem of constructing a decision rule for testing whether a coin is fair. For reasons known only to himself, he decides to flip the coin 100 times and reject $H_0: p = 1/2$ if $49 \leqslant y \leqslant 51$, where the random variable $Y$ denotes the number of heads that appear. Find $\alpha$ (use the continuity correction described on p. 327). Intuitively, what is wrong with the critical region that Otto has defined?

8.63   In general, the power of a test is the probability that $H_0$ is rejected given that $H_1$ is true. To what value does $1 - \beta$ approach as the presumed parameter value (in $H_1$) gets closer to the value of the parameter specified in $H_0$?

8.64   Suppose that Will made 30 of his last 50 free throws. Is it reasonable to argue that his "true" shooting percentage is better than 45%? Support your position quantitatively.

8.65   Six observations are taken on a binomial random variable $Y$ for the purpose of testing $H_0: p = 0.7$ versus $H_1: p > 0.7$. We intend to reject $H_0$ if $y = 6$. What is the power of the test if $p = 0.8$?

8.66   Customer satisfaction surveys show that the owners of a certain model compact car get gas mileages that are normally distributed with $\mu = 33.4$ and $\sigma = 2.6$. Another model made by the same company is virtually identical except for a different fuel injection system. Ten owners of the second model reported the gas mileages shown below:

|    |    |    |    |    |
|----|----|----|----|----|
| 36 | 35 | 41 | 36 | 33 |
| 33 | 39 | 32 | 38 | 31 |

Does the fuel injection system appear to have an effect on the car's gas mileages? Formulate and carry out the relevant hypothesis test. Let 0.01 be the level of significance.

**8.67**    Suppose it can be assumed that the standard deviation of the pdf from which the following observations were drawn is known. Would it be correct to construct a confidence interval for $\mu$ using Theorem 8.3? Explain.

| | | | | |
|------|------|------|------|------|
| 0.68 | 0.96 | 0.71 | 0.99 | 0.09 |
| 0.91 | 0.49 | 0.95 | 0.60 | 0.92 |
| 0.38 | 0.87 | 0.82 | 0.96 | 0.93 |
| 0.74 | 0.77 | 0.23 | 0.87 | 0.69 |
| 0.96 | 0.73 | 0.44 | 0.64 | 0.42 |
| 0.24 | 0.63 | 0.97 | 0.99 | 0.72 |
| 0.87 | 0.99 | 0.68 | 0.78 | 0.42 |
| 0.64 | 0.79 | 0.33 | 0.79 | 0.60 |

**8.68**    The *P*-value of a test is the probability, under $H_0$, of getting a test statistic as extreme or more extreme than the one actually observed. For example, if $H_0$: $p = p_0$ is being tested against $H_1$: $p > p_0$, the *P* value is the probability that a standard normal random variable Z exceeds

$$\frac{y - np_0}{\sqrt{np_0(1 - p_0)}}$$

Find the *P* value for the hypothesis test in Example 8.1.

**8.69**    Refer to Question 8.68. What relationship exists between the *P* value of a test and whether we accept or reject $H_0$? (Hint: If we test $H_0$: $p = p_0$ versus $H_1$: $p > p_0$ at the $\alpha$ level of significance and we find that the *P* value is *less than* $\alpha$, where is $z$ relative to $z_\alpha$?)

# 9 *The One-Sample Problem*

*There are three kinds of lies—lies, damned lies, and statistics.*

*Disraeli*

### INTRODUCTION

This chapter begins the systematic study of statistical methods for the seven data models identified in Chapter 2. Having studied probability in Chapters 4 through 7 and the principles of inference in Chapter 8, we are now fully prepared to *analyze* what we learned to *recognize* in Chapter 2.

The data from any experiment with a one-sample structure consist of a single random sample of size $n$ $(y_1, y_2, \ldots, y_n)$. The examples that we will consider belong to one of two categories: (1) the $Y_i$'s have a *normal* distribution; (2) the $Y_i$'s represent the outcomes of $n$ Bernoulli trials, and $Y = \sum_{i=1}^{n} Y_i$ has a *binomial* distribution. For the former, our usual objective is to draw an inference about $\mu$. Whether the format of that inference is a hypothesis test or a confidence interval, the *mechanism* is the same—we make use of the result from Chapter 7 that $\dfrac{\bar{Y} - \mu}{S/\sqrt{n}}$ has a Student $t$ distribution with $n - 1$ degrees of freedom.

If, on the other hand, the $Y_i$'s are observations on a Bernoulli random variable, we are interested in $p$, the probability that any trial ends in success. For problems lending themselves to a "yes/no" resolution, we choose between $H_0$ and $H_1$ on the basis of the test statistic,

$$\frac{Y - np_0}{\sqrt{np_0(1 - p_0)}}$$

which has approximately a standard normal distribution when $H_0$ is true (recall Theorem 8.1). Confidence intervals for $p$ derive from a similar ratio,

$$\frac{Y/n - p}{\sqrt{\dfrac{(Y/n)(1 - Y/n)}{n}}}$$

whose distribution is also approximated by a standard normal.

Occasionally, the $Y_i$'s are normally distributed, but our interest focuses on $\sigma^2$ rather than on $\mu$. Inference procedures involving $\sigma^2$ are based on a second result from Chapter 7—that $\dfrac{(n - 1)S^2}{\sigma^2}$ has a $\chi^2$ distribution with $n - 1$ degrees of freedom.

An important experimental design question emerges in Section 9.4: how many observations need to be taken in order to estimate $\mu$ (or $p$) with a specified precision? In practice, choosing an appropriate sample size is one of the very first decisions an experimenter is forced to make, no matter what data structure is involved. For the one-sample model, we can identify a minimum $n$ by defining "precision" in terms of the radius of a confidence interval. Theorems 9.4 and 9.5 give formulas for $n$ that cover two different situations where the question can be expected to arise: (1) estimating $\mu$ when $\sigma$ is known, and (2) estimating $p$. The latter formula for binomial data is especially important.

Read the case studies carefully. Imagine yourself as the experimenter; see if the statistical question being formulated is the one you would want answered.

## 9.2
## THE ONE-SAMPLE t TEST

If $Y_1, Y_2, \ldots, Y_n$ is a random sample from a normal distribution with unknown mean $\mu$ and known standard deviation $\sigma$, then

$$\frac{\bar{Y} - \mu}{\sigma/\sqrt{n}}$$

has the same pdf as a standard normal random variable Z (recall Theorem 7.2). For our first look at hypothesis testing, that was a key result: it enabled us to set up a decision rule for testing $H_0$: $\mu = \mu_0$. Theorem 8.2 gave the details.

From an experimenter's standpoint, however, Theorem 8.2 is not particularly useful. Why? Because if we do not know $\mu$, which is presumably the reason for testing $H_0$ in the first place, it is highly unlikely that we *do* know $\sigma$. More typically, an experimenter knows nothing about either parameter. Unable to base a test, then, on the statistic $\dfrac{\bar{Y} - \mu}{\sigma/\sqrt{n}}$, which has a standard normal distribution, we make the obvious modification—replace $\sigma$ with S and redefine the test using the ratio $\dfrac{\bar{Y} - \mu}{S/\sqrt{n}}$, which has a Student $t$ distribution with $n - 1$ degrees of freedom (recall Section 7.3).

---

**THEOREM 9.1**

Let $Y_1, Y_2, \ldots, Y_n$ be a random sample from a normal distribution where $\mu$ and $\sigma^2$ are unknown. Let

$$t = \frac{\bar{y} - \mu_0}{s/\sqrt{n}}$$

1. To test

$$H_0: \quad \mu = \mu_0$$

vs.

$$H_1: \quad \mu \neq \mu_0$$

at the $\alpha$ level of significance, reject $H_0$ if either

(a) $t \leqslant -t_{\alpha/2,\,n-1}$ or (b) $t \geqslant t_{\alpha/2,\,n-1}$.

2. To test

$$H_0: \quad \mu = \mu_0$$

vs.

$$H_1: \quad \mu > \mu_0$$

at the $\alpha$ level of significance, reject $H_0$ if $t \geqslant t_{\alpha,\,n-1}$.

3. To test

$$H_0: \quad \mu = \mu_0$$

vs.

$$H_1: \quad \mu < \mu_0$$

at the $\alpha$ level of significance, reject $H_0$ if $t \leqslant -t_{\alpha,\,n-1}$.

## CASE STUDY 9.1

*Were Etruscans Really Italians?*

In the eighth century B.C. the Etruscan civilization was the most advanced in all of Italy. Its art forms and political innovations were destined to leave indelible marks on the entire Western world. Originally located in the region now known as Tuscany, it spread rapidly across the Apennines and eventually overran much of Italy. But as quickly as it came, it faded. Militarily it was no match for the burgeoning Roman legions, and by the dawn of Christianity it was all but gone.

No chronicles of the Etruscan empire have ever been found, and to this day its origins remain shrouded in mystery. Were the Etruscans native Italians or were they immigrants? And if they were immigrants, where did they come from? Much of our knowledge of the Etruscans derives from archaeological investigations and anthropometric studies. The data presented here are a typical example of the latter: body measurements are being used to determine racial characteristics and, ultimately, ethnic origins.

For modern Italian males the maximum head breadth averages 132.4 mm. Table 9.1 lists the skull widths of the fossil remains of 84 male Etruscans uncovered in archaeological digs throughout Italy (6).

Ignoring any possible shifts due to evolution, we would expect the 84 widths in Table 9.1 to be comparable to the modern Italian average—132.4 mm—*if the Etruscans were native Italians*. A $\bar{y}$ close to 132.44 mm, in other words, suggests a common origin. On the other hand, if the Etruscan skull

**TABLE 9.1**

Maximum Head Breadths (mm),
Etruscan Males

| | | | | | | |
|---|---|---|---|---|---|---|
| 141 | 148 | 132 | 138 | 154 | 142 | 150 |
| 146 | 155 | 158 | 150 | 140 | 147 | 148 |
| 144 | 150 | 149 | 145 | 149 | 158 | 143 |
| 141 | 144 | 144 | 126 | 140 | 144 | 142 |
| 141 | 140 | 145 | 135 | 147 | 146 | 141 |
| 136 | 140 | 146 | 142 | 137 | 148 | 154 |
| 137 | 139 | 143 | 140 | 131 | 143 | 141 |
| 149 | 148 | 135 | 148 | 152 | 143 | 144 |
| 141 | 143 | 147 | 146 | 150 | 132 | 142 |
| 142 | 143 | 153 | 149 | 146 | 149 | 138 |
| 142 | 149 | 142 | 137 | 134 | 144 | 146 |
| 147 | 140 | 142 | 140 | 137 | 152 | 145 |

widths are markedly different from what we know to be characteristic of Italians, the conclusion is the reverse: that one group is *not* the direct descendant of the other.

Let $\mu$ denote the true average maximum head breadth for Etruscans. Couched in the terminology of hypothesis testing, the preceding comments suggest that we can infer the relationship between Etruscans and Italians by testing

$$H_0: \quad \mu = 132.4$$

vs.

$$H_1: \quad \mu \neq 132.4$$

Rejecting $H_0$ is equivalent to saying that the Etruscans were not native Italians. Take $\alpha = 0.05$.

First, we need to compute $\bar{y}$ and $s$. From Table 9.1,

$$\sum_{i=1}^{84} y_i = 12,077 \quad \text{and} \quad \sum_{i=1}^{84} y_i^2 = 1,739,315$$

Therefore,

$$\bar{y} = \frac{12,077}{84} = 143.8$$

and

$$s = \sqrt{\frac{84(1,739,315) - (12,077)^2}{84(83)}} = 6.0$$

Since $n = 84$, critical values for the test statistic

$$\frac{\bar{y} - \mu_0}{s/\sqrt{84}}$$

come from a Student $t$ distribution with 83 $(= n - 1)$ degrees of freedom. Specifically, the numbers that cut off tail areas of $\alpha/2 = 0.025$ are $\pm 1.9890$ (see Figure 9.1).

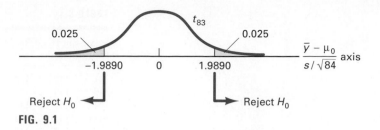

**FIG. 9.1**

According to part 1 of Theorem 9.1, we should reject $H_0$ if the "observed $t$" is either (1) $\leqslant -1.9890$ or (2) $\geqslant 1.9890$. In this case, the decision is clear-cut:

$$t = \frac{143.8 - 132.4}{6.0/\sqrt{84}} = 17.4$$

so we should reject $H_0$. On the basis of skull width, it would appear that home for the Etruscans was somewhere other than Italy.

## CASE STUDY 9.2

***Do Isomers Prolong Sleep?***

Certain chemical compounds have more than one structure. Each of those structures is called an *isomer*. Two isomers of a compound have the same molecular formula, but their atoms are not arranged the same way. Variations in a compound's structure can make its isomers respond differently to certain physical phenomena, such as light. *Optical isomers*, for example, cause polarized light to be rotated in different directions. An isomer causing rotation to the right is given the prefix *dextro*, or $d$; if the rotation is to the left, the prefix is *levo* or $l$.

One of Gosset's early "$t$ test" applications was his analysis of a 1904 experiment on optical isomers conducted by A. R. Cushny and A. R. Peebles (34). The original study focused on whether the optical isomers of the drug hyoscyamine hydrobromide could prolong sleep. Table 9.2 shows the additional hours of sleep reported by each of 10 patients given $d$-hyoscyamine hydrobromide.

Let $\mu$ denote the true average increase in sleep attributable to $d$-hyoscyamine hydrobromide. If the drug has no effect, $\mu$ should be 0; if it does have an effect, $\mu$ should be greater than zero. The sample average increase, $\bar{y}$, is 0.75 hours:

$$\bar{y} = \frac{1}{10} \sum_{i=1}^{10} y_i = \frac{1}{10}(7.5) = 0.75 \text{ hr}$$

The question to be answered, then, is whether the difference between 0.75 hr and 0 hr is statistically significant.

**TABLE 9.2**

| Patient | Additional Hours of Sleep |
|---------|---------------------------|
| 1 | 0.7 |
| 2 | −1.6 |
| 3 | −0.2 |
| 4 | −1.2 |
| 5 | −0.1 |
| 6 | 3.4 |
| 7 | 3.7 |
| 8 | 0.8 |
| 9 | 0.0 |
| 10 | 2.0 |

Let $\alpha = 0.05$. What we need to test is

$$H_0: \quad \mu = 0$$

vs.

$$H_1: \quad \mu > 0$$

Since the alternative hypothesis is one-sided, the appropriate critical value from the $T$ distribution with 9 ($= 10 - 1$) df is $t_{0.05, 9} = 1.8331$ (see Figure 9.2). According to part 2 of Theorem 9.1, we should reject $H_0$ if $(\bar{y} - 0)/(s/\sqrt{10}) \geqslant 1.8331$.

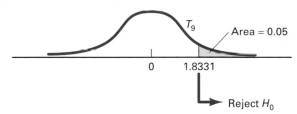

**FIG. 9.2**

From Table 9.2,

$$\sum_{i=1}^{10} y_i^2 = 34.43$$

Substituting the latter into the computing formula for $s$ gives

$$s = \sqrt{\frac{10(34.43) - (7.5)^2}{10(9)}} = 1.79$$

The resulting $t$ ratio is 1.325:

$$t = \frac{0.75 - 0}{1.79/\sqrt{10}} = 1.325$$

Since 1.325 is not to the right of $t_{0.05, 9}$, we accept $H_0$—the data do not argue convincingly that the $d$ isomer prolongs sleep.

Table 9.3 shows a similar set of data for the $l$ isomer. For these $y_i$'s, $\bar{y} = 2.33$, $s = 2.00$, and

$$t = \frac{2.33 - 0}{2.00/\sqrt{10}} = 3.684$$

But $3.684 > t_{0.05, 9} = 1.8331$, so now we reject $H_0$—the $l$ isomer does seem to prolong sleep.

**TABLE 9.3**

| Patient | Additional Hours of Sleep |
|---------|---------------------------|
| 1 | 1.9 |
| 2 | 0.8 |
| 3 | 1.1 |
| 4 | 0.1 |
| 5 | −0.1 |
| 6 | 4.4 |
| 7 | 5.5 |
| 8 | 1.6 |
| 9 | 4.6 |
| 10 | 3.4 |

COMMENT: Although Gosset's methods of data analysis were quickly accepted in the brewing industry, they were mostly ignored elsewhere. A wider audience began to be reached in 1923 with R. A. Fisher's publication of a mathematically rigorous treatment of Gosset's "ratio."

COMMENT: The validity of the $t$ statistic is contingent on the data being normal, but when the sample size is small there is no good way of verifying that particular assumption. Gosset believed that nonnormality was not a major problem. When announcing his results, he said: "This assumption [of normality] is accordingly made in the present paper, so its conclusions are not strictly applicable to populations not known to be normally distributed; yet it appears probable that deviations from normality must be very extreme to lead to serious error."

Studies have confirmed Gosset's remarkable intuition: any nonnormality in the population being sampled tends to have only minimal effects on the probabilistic behavior of the $t$ ratio. Thus, the $t$ test remains "valid," even though the data do not come from a normal distribution. In describing this property, we say that the $t$ test is **robust** *with respect to departures from normality*.

**robustness**—The ability of the sampling distribution of a test statistic to be relatively unaffected by violations of its underlying assumptions.

## CASE STUDY 9.3

*Golden Rectangles*

Not all rectangles are created equal. Since antiquity, societies have expressed aesthetic preferences for rectangles having certain width ($w$) to length ($l$) ratios. Plato, for example, wrote that rectangles whose sides are in a $1 : \sqrt{3}$ ratio are especially pleasing (these are the rectangles formed from the two halves of an equilateral triangle). Another "standard," one that is much better known, calls for the width-to-length ratio to be equal to the ratio of the length to the sum of the width and the length. That is,

$$\frac{w}{l} = \frac{l}{w + l} \tag{9.1}$$

Equation 9.1, if satisfied, implies that $w \doteq 0.618l$ (see Question 9.9). The Greeks called that particular configuration a *golden rectangle* and used it often in their architecture (see Figure 9.3). Many other cultures were similarly inclined. The Egyptians, for example, built their pyramids with stones whose faces were golden rectangles. In our society the golden rectangle remains an architectural and artistic standard; even items such as drivers'

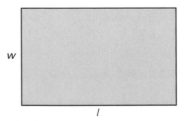

$w$

$l$

**FIG. 9.3**

**TABLE 9.4**

| | | |
|---|---|---|
| 0.693 | 0.628 | 0.576 |
| 0.662 | 0.609 | 0.670 |
| 0.690 | 0.844 | 0.606 |
| 0.606 | 0.654 | 0.611 |
| 0.570 | 0.615 | 0.553 |
| 0.749 | 0.668 | 0.933 |
| 0.672 | 0.601 | |

licenses, business cards, and picture frames often have $w/l$ ratios close to 0.618.

The study described here is an example from a field known as experimental aesthetics. The data in Table 9.4 are $w/l$ ratios of beaded rectangles used by Shoshoni Indians to decorate their leather goods (39). The question is whether the golden rectangle can be considered an aesthetic standard for the Shoshonis, just as it was for the Greeks and the Egyptians. If it can, it would follow that the sample mean of the 20 $y_i$'s in Table 9.4 should be close to 0.618. More formally, if the true average $w/l$ ratio for the Shoshonis is $\mu$, we want to test

$$H_0: \quad \mu = 0.618$$

vs.

$$H_1: \quad \mu \neq 0.618.$$

Let $\alpha = 0.05$. The critical values for the $t$ ratio are the numbers that cut off areas of 0.025 in either tail of the $T$ distribution with 19 df. According to Appendix A.2, those numbers are $\pm 2.0930$ (see Figure 9.4).

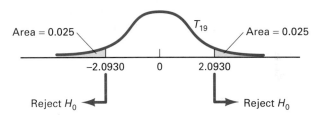

**FIG. 9.4**

From Table 9.4, $\sum_{i=1}^{20} y_i = 13.210$, and $\sum_{i=1}^{20} y_i^2 = 8.8878$. Therefore,

$$\bar{y} = \frac{13.210}{20} = 0.661$$

and

$$s = \sqrt{\frac{20(8.8878) - (13.210)^2}{20(19)}} = 0.093$$

Thus

$$t = \frac{0.661 - 0.618}{0.093/\sqrt{20}} = 2.05$$

Our conclusion is to accept $H_0$, but the proximity of the observed $t$ to the upper cutoff value ($= 2.0930$) should serve as a warning that the evidence *against* $H_0$ was not inconsiderable. In fact, had $\alpha$ been chosen to be 0.10, the critical values would have been $\pm 1.7291$ and $H_0$ would have been rejected.

**QUESTIONS**

**9.1** An experimenter has taken 25 observations from a normal distribution for the purpose of testing $H_0: \mu = 100$. Find the critical value(s) appropriate for

(a) $H_1: \mu > 100, \quad \alpha = 0.05$

(b) $H_1: \mu > 100, \quad \alpha = 0.01$

(c) $H_1: \mu < 100, \quad \alpha = 0.025$

(d) $H_1: \mu \neq 100, \quad \alpha = 0.05$

(e) $H_1: \mu \neq 100, \quad \alpha = 0.10$

**9.2** A random sample of size 20 from a normal distribution gives $\bar{y} = 11.7$ and $s = 2.54$. Test $H_0: \mu = 12$ versus $H_1: \mu < 12$ at the 0.05 level of significance.

**9.3** Suppose $H_1$ in Question 9.2 was two-sided. If everything else stays the same, what conclusion would you reach?

**9.4** Test $H_0: \mu = 15$ versus $H_1: \mu \neq 15$ if a random sample of size 20 gives $\bar{y} = 9.6$ and $s = 4.7$. Let $\alpha = 0.05$.

**9.5** Recall the toxic tort data described in Case Study 2.1. Let $\mu$ denote the true average $FEV_1/VC$ ratio characteristic of workers exposed to the *Bacillus subtilis* enzyme.

(a) At the $\alpha = 0.05$ level of significance, test $H_0: \mu = 0.80$ versus $H_1: \mu < 0.80$.

(b) Why does $H_0$ single out the value 0.80?

**9.6** Lead content measurements taken at eight Australian wineries were listed in Table 3.1. The recommended maximum level of lead set by the Australian government is 0.20 mg/liter. Can we conclude that wineries, on the average, exceed the government limit? Let $\alpha = 0.05$.

**9.7** Refer to the mosquito bite duration data given in Table 3.12. Are those five observations compatible with the hypothesis that the true average bite durations is 3 min? Set up and carry out the appropriate hypothesis test with $\alpha = 0.05$.

**9.8** Are the six observations in Table 3.14 consistent with the hypothesis that nine-banded armadillos sleep, on the average, 19 hr a day? Test the appropriate $H_0$ and $H_1$ at the $\alpha = 0.10$ level of significance.

**9.9** Solve Equation 9.1 for $w/l$. (Hint: Let $q = w/l$, take the reciprocal of both sides of the equation, and use the quadratic formula.)

**9.10** A psychologist wants to know if persons with high IQs (over 130) are significantly faster at working a certain puzzle than are persons with average IQs. Based on past experience, it is known that persons with average IQs complete the puzzle in a mean time of 4.6 min. When the psychologist presents the puzzle to four persons with IQs over 130, she records completion times (in minutes) of 4.0, 4.2, 4.6, and 4.2. State the appropriate $H_0$ and $H_1$ and carry out the test at the $\alpha = 0.05$ level of significance.

9.11   A contractor buys cement from a local manufacturer. Each bag is supposed to weigh 95 lb. The contractor decides to test 10 bags to see if he is getting his money's worth. The weights he records are 94.1, 94.4, 93.8, 94.4, 96.4, 94.5, 95.0, 94.8, 93.9, and 95.2 lb. Do the appropriate hypothesis test at the $\alpha = 0.05$ level of significance.

9.12   Given the aesthetic appeal of the golden rectangle, it is not unreasonable to ask whether food cans are dimensioned so they look like golden rectangles when stacked on a shelf. Listed below are the diameters and heights for a sample of 11 cans.

| Can Number | Diameter (cm) | Height (cm) |
|:----------:|:-------------:|:-----------:|
| 1 | 8.0 | 11.1 |
| 2 | 10.2 | 11.8 |
| 3 | 10.2 | 13.8 |
| 4 | 8.0 | 11.1 |
| 5 | 7.9 | 11.0 |
| 6 | 6.8 | 10.0 |
| 7 | 7.5 | 11.2 |
| 8 | 6.8 | 12.3 |
| 9 | 6.7 | 10.1 |
| 10 | 8.6 | 11.6 |
| 11 | 8.6 | 11.0 |

Test at the 0.01 level whether the diameter-to-height ratios can be assumed to have come from a population with $\mu = 0.618$.

# 9.3

## CONFIDENCE INTERVALS

### Estimating $\mu$

We pointed out in Section 8.4 that the "yes/no" framework of hypothesis testing is sometimes inappropriate. Situations exist for which there is no obvious numerical choice for $\mu_0$ or $p_0$, in which case there is no way to define either $H_0$ or $H_1$. What we do in those cases is forget about hypothesis testing altogether and turn to the confidence interval format instead.

   Section 8.4 emphasized why we use a confidence interval: we want to estimate a parameter by constructing a random interval having a high probability of containing that parameter as an interior point. Theorem 8.3 gave the $100(1 - \alpha)\%$ confidence interval for the mean of a normal distribution when the standard deviation is known. The discussion that led to that theorem touched on all the key issues associated with confidence intervals. Applications of that result, though, are infrequent because of the assumption it makes about $\sigma$. In the first part of this section we will develop a more useful version of Theorem 8.3; specifically, the construction of a confidence interval for $\mu$ when $\sigma$ is not known.

Let $Y_1, Y_2, \ldots, Y_n$ be a random sample from a normal distribution where $\mu$ and $\sigma$ are unknown. From Section 7.3 we know that

$$T = \frac{\bar{Y} - \mu}{S/\sqrt{n}}$$

has a Student $t$ distribution with $n - 1$ degrees of freedom. Recall that $\pm t_{\alpha/2, n-1}$ are the numbers that cut off areas of $\alpha/2$ in either tail of that curve. Therefore,

$$P\left(-t_{\alpha/2, n-1} \leqslant \frac{\bar{Y} - \mu}{S/\sqrt{n}} \leqslant t_{\alpha/2, n-1}\right) = 1 - \alpha \qquad (9.2)$$

(see Figure 9.5). (In practice, we usually choose $\alpha$ so that $1 - \alpha$ is 0.50, 0.90, 0.95, or 0.99.)

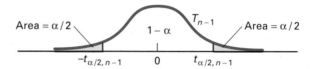

**FIG. 9.5**

To get a formula for a confidence interval we must isolate $\mu$ in the center of the two inequalities in Equation 9.2. Following the same sequence of steps used on Equation 8.3, we can transform Equation 9.2 to

$$P\left(\bar{Y} - t_{\alpha/2, n-1} \frac{S}{\sqrt{n}} \leqslant \mu \leqslant \bar{Y} + t_{\alpha/2, n-1} \frac{S}{\sqrt{n}}\right) = 1 - \alpha \qquad (9.3)$$

We interpret Equation 9.3 by saying that the random interval

$$\left(\bar{Y} - t_{\alpha/2, n-1} \frac{s}{\sqrt{n}}, \bar{Y} + t_{\alpha/2, n-1} \frac{s}{\sqrt{n}}\right)$$

has a $100(1 - \alpha)\%$ probability of containing the true $\mu$ as an interior point. After the data are collected, we replace $\bar{Y}$ by $\bar{y}$ and S by $s$. The resulting range of numbers is called the $100(1 - \alpha)\%$ *confidence interval for $\mu$*. In the long run, $100(1 - \alpha)\%$ of all such intervals will contain $\mu$ as an interior point (and $100\alpha\%$ will not).

---

**THEOREM 9.2**

Let $y_1, y_2, \ldots, y_n$ be a random sample of size $n$ from a normal distribution with unknown mean $\mu$ and unknown standard deviation $\sigma$. A $100(1 - \alpha)\%$ confidence interval for $\mu$ is the range of values

$$\left(\bar{y} - t_{\alpha/2, n-1} \frac{s}{\sqrt{n}}, \bar{y} + t_{\alpha/2, n-1} \frac{s}{\sqrt{n}}\right)$$

COMMENT: In Section 8.4 the confidence intervals were centered around $\bar{y}$, so their location varied from sample to sample (see Figure 8.10). The width of each interval, though, remained the same. In contrast, intervals constructed using Theorem 9.2 have the property that both location and width will vary from sample to sample. The latter happens because the width is equal to

$$2 \cdot t_{\alpha/2, n-1} \frac{s}{\sqrt{n}}$$

and $s$, like $\bar{y}$, varies from sample to sample.

## CASE STUDY 9.4

**TABLE 9.5**

| Catch Number | Detection Distance, $y_i$ (cm) |
|:---:|:---:|
| 1 | 62 |
| 2 | 52 |
| 3 | 68 |
| 4 | 23 |
| 5 | 34 |
| 6 | 45 |
| 7 | 27 |
| 8 | 42 |
| 9 | 83 |
| 10 | 56 |
| 11 | 40 |
| | 532 |

*Bat Snacks* The bat, nature's only flying mammal, suffers from a less than sterling public image. To most people, bats are ugly, live in caves with other crawly things, and spend most of their waking hours getting tangled up in people's hair. Only Dracula seems to show them any affection. The "truth," of course, is something entirely different: not only are most bats perfectly harmless, they provide a valuable service by devouring large numbers of insects. They are also the objects of considerable scientific research, the study described in the next several paragraphs being a case in point (59).

To hunt flying insects, bats emit high-frequency sounds and then listen for their echoes. Until an insect is located, the spacing between pulses is about 50 to 100 milliseconds. When an insect is detected, the bat immediately reduces the pulse-to-pulse interval, sometimes to as short as 10 milliseconds. The higher transmission frequency enables the bat to pinpoint the position of its prey. All of this raises an interesting question: how far apart are the bat and the insect when the bat first senses that the insect is there? Or, put another way, what is the effective range of the bat's echolocation system?

Scientists put a hungry bat and an ample supply of fruit flies into an 11-by-16-ft room. Two synchronized 16 mm sound-on-film cameras recorded the drosophila carnage. By studying the two sets of pictures frame by frame, the scientists could follow the bat's flight pattern and simultaneously, monitor its pulse frequency. For each insect caught, it was then possible to calculate the distance between the bat and the insect at the precise moment the bat's pulse-to-pulse interval decreased. Table 9.5 lists the 11 detection distances.

Let $\mu$ denote the true average detection distance. We assume that this is the first time such an experiment has been done. That being the case, there is no "standard" value for $\mu_0$ around which to formulate a hypothesis test. Our objective, instead, will be to estimate $\mu$ by constructing a 95% confidence interval.

From Table 9.5,

$$\sum_{i=1}^{11} y_i = 532 \quad \text{and} \quad \sum_{i=1}^{11} y_i^2 = 29{,}000$$

Therefore,

$$\bar{y} = \frac{532}{11} = 48.4 \text{ cm}$$

and

$$s = \sqrt{\frac{11(29{,}000) - (593)^2}{11(10)}} = 18.1 \text{ cm}$$

Here,

$$\pm t_{\alpha/2, n-1} = \pm t_{0.05/2, 11-1} = \pm t_{0.025, 10} = \pm 2.2281$$

By Theorem 9.2,

$$\left( 48.4 - 2.2282 \frac{18.1}{\sqrt{11}}, \ 48.4 + 2.2281 \frac{18.1}{\sqrt{11}} \right)$$

or

$$(36.2 \text{ cm}, 60.6 \text{ cm})$$

is the 95% confidence interval for $\mu$.

---

QUESTIONS

**9.13** Suppose a sample of size 20 from a normal distribution gives $\bar{y} = 5.78$ and $s = 0.93$. Find

(a) a 90% confidence interval for $\mu$

(b) a 95% confidence interval for $\mu$.

**9.14** Fifteen observations taken from a normal distribution have $\bar{y} = 105.1$ and $s = 17.4$. Find 95% and 99% confidence intervals for $\mu$.

**9.15** Staffs from 15 different hospitals participated in a surveillance program to monitor the number of patients experiencing adverse reactions to prescribed medication. The percentages for the 15 hospitals were 5.8, 5.3, 4.5, 3.9, 4.6, 5.4, 7.9, 8.2, 6.9, 5.7, 4.6, 6.3, 8.4, 4.6, and 7.3. Let $\mu$ denote the true average percentage of adverse reactions. Construct

a 60% confidence interval for $\mu$

a 95% confidence interval for $\mu$.

**9.16** Using the data in Question 1.1, construct a 90% confidence interval for $\mu$, the true average number of years between the conception and the realization of an invention.

**9.17** Case Study 2.2 listed the ages of 12 scientists when they made their most famous discoveries. Construct a 90% confidence interval for the corresponding parameter.

**9.18** Construct a 70% confidence interval for $\mu$, using the radiation data in Question 2.7.

What does $\mu$ represent in this context?

**9.19** Maximum heart rates for 11 parachutists were listed in Question 2.10. Find the length of the 95% confidence interval for the true average maximum heart rate. Note:

$$\sum_{i=1}^{11} y_i = 1817 \quad \text{and} \quad \sum_{i=1}^{11} y_i^2 = 301{,}553$$

**9.20** For the data given in Question 3.8, let $\mu$ denote the true average cockpit noise level in commercial aircraft. Construct a 90% confidence interval for $\mu$.

**9.21** What "confidence" would be associated with each of the following intervals:

(a) $\left( \bar{y} - 2.0930 \, \dfrac{s}{\sqrt{20}}, \, \bar{y} + 2.0930 \, \dfrac{s}{\sqrt{20}} \right)$

(b) $\left( \bar{y} - 1.345 \, \dfrac{s}{\sqrt{15}}, \, \bar{y} + 1.345 \, \dfrac{s}{\sqrt{15}} \right)$

(c) $\left( \bar{y} - 1.7056 \, \dfrac{s}{\sqrt{27}}, \, \bar{y} + 2.7787 \, \dfrac{s}{\sqrt{27}} \right)$

(d) $\left( -\infty, \, \bar{y} + 1.7247 \, \dfrac{s}{\sqrt{21}} \right)$

**9.22** Two samples, each of size $n$, are taken from a normal distribution with unknown mean $\mu$ and unknown standard deviation $\sigma$. A 90% confidence interval for $\mu$ is constructed with the first sample, and a 95% confidence interval with the second. Will the 95% interval necessarily be longer than the 90% interval? Explain.

### Estimating $p$

Confidence intervals can also be applied to problems where the data come from a binomial distribution and the parameter to be estimated is $p$. Finding an *exact* confidence interval for $p$ is quite difficult, but we can derive an approximate $100(1 - \alpha)\%$ interval, by using the De Moivre-Laplace version of the Central Limit Theorem. Recall from Theorem 6.7 that for $n$ sufficiently large, the random variable

$$\frac{Y - np}{\sqrt{np(1 - p)}} = \frac{Y/n - p}{\sqrt{p(1 - p)/n}}$$

has approximately a standard normal distribution. It is also true that the ratio

$$\frac{Y/n - p}{\sqrt{\dfrac{(Y/n)(1 - Y/n)}{n}}}$$

has a pdf similar to $\varphi(z)$. Therefore,

$$P\left(-z_{\alpha/2} \leqslant \frac{Y/n - p}{\sqrt{\dfrac{(Y/n)(1 - Y/n)}{n}}} \leqslant z_{\alpha/2}\right) \doteq 1 - \alpha$$

or, equivalently,

$$P\left(\frac{Y}{n} - z_{\alpha/2}\sqrt{\frac{(Y/n)(1 - Y/n)}{n}} \leqslant p \leqslant \frac{Y}{n} + z_{\alpha/2}\sqrt{\frac{(Y/n)(1 - Y/n)}{n}}\right) \doteq 1 - \alpha$$

$$(9.43)$$

(recall the steps that followed Equations 8.3 and 9.2). The end points surrounding $p$ in Equation 9.4 define the interval we seek.

---

### THEOREM 9.3

Let $y$ be the total number of successes in $n$ independent Bernoulli trials, where $p$ is the probability that a success occurs at any trial. An approximate $100(1 - \alpha)\%$ confidence interval for $p$ is the set of values ranging from

$$\frac{y}{n} - z_{\alpha/2}\sqrt{\frac{(y/n)(1 - y/n)}{n}} \qquad \text{to} \qquad \frac{y}{n} + z_{\alpha/2}\sqrt{\frac{(y/n)(1 - y/n)}{n}}$$

Note: In order for the approximation to be sufficiently precise, it should be true that

$$0 \leqslant \frac{y}{n} - 2\sqrt{\frac{(y/n)(1 - y/n)}{n}} \qquad \text{and} \qquad \frac{y}{n} + 2\sqrt{\frac{(y/n)(1 - y/n)}{n}} \leqslant n$$

---

### CASE STUDY 9.5

*Tasters' Choice*  Surveys to determine consumer preferences are a routine part of the development and marketing of many products. Most such surveys, though, are conducted with a great deal less visibility than the live television taste tests sponsored by the Joseph Schlitz Brewing Company in the winter of 1980–1981. Those particular commercials were undertaken to counter Schlitz' rapidly decreasing share of the premium beer market. In 1974, an ill-advised change in its brewing procedure coupled with a slight price increase caused the company to fall from second in sales to a distant third by 1977. Even though management responded by hiring a master brewer as president, customers stayed away. In desperation, the firm decided to take on its leading competitors, mug-to-mug, using television as a coast-to-coast battleground.

A typical commercial in the Schlitz campaign was broadcast on December 28, 1980, during halftime of an NFL playoff game. One hundred imbibers, all of whom signed affidavits that they drank at least two six-packs of Budweiser a week, were each given two identical unlabeled beer mugs—one containing Budweiser; the other Schlitz. Veteran NFL referee Tommy Bell, wearing his "zebra" shirt, presided over the test. Each drinker sampled both beers off-camera (because of broadcasting regulations) and then appeared on-camera and pressed a button showing his preference. The vote was 46 for Schlitz and 54 for Budweiser.

Statistically, each subject was acting as a Bernoulli trial, where the relevant parameter, from Schlitz's point of view, was the probability $p$ that a drinker prefers Schlitz to Bud. A reasonable objective would be to estimate $p$ by constructing, say, a 95% confidence interval.

The data consist of $n = 100$ observations $(y_1, y_2, \ldots, y_{100})$ where

$$y_i = \begin{cases} 1 & \text{if } i\text{th subject prefers Schlitz} \\ 0 & \text{otherwize} \end{cases}$$

Let $y = \sum_{i=1}^{100} y_i$. Here $y = 46$. If the interval is to have a confidence coefficient of 95%, its two $z$ multiples must be

$$\pm z_{0.05/2} = \pm z_{0.025} = \pm 1.96 \quad (\text{why?})$$

From Theorem 9.3 the 95% confidence interval for $p$ is

$$\left( 0.46 - 1.96 \sqrt{\frac{(0.46)(0.54)}{100}}, \; 0.46 + 1.96 \sqrt{\frac{(0.46)(0.54)}{100}} \right) = (0.36, 0.56)$$

The fact that $p = 0.50$ is an interior point of the confidence interval should be a comfort to Schlitz. Why? Because even though the sample showed an overall preference for Bud, it is not inconceivable that in the entire population both beers are rated equal (that is, the true $p$ may still be 0.50). Only if the sample data had produced an interval that lay entirely to the left of 0.50 would Schlitz have had cause for serious concern. Any 95% confidence interval positioned like the one shown in Figure 9.6 is equivalent to rejecting $H_0$: $p = 0.50$ in favor of $H_1$: $p < 0.50$ (at the $\alpha = 0.05$ level of significance).

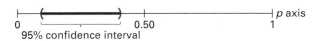

FIG. 9.6

9.23 Suppose $n = 100$ Bernoulli trials are observed, where $p$ is unknown. The sample consists of 43 successes and 57 failures. Find (a) a 95% confidence interval for $p$ and (b) a 90% confidence interval for $p$.

9.24 A series of 75 Bernoulli trials results in 28 successes. Find a 99% confidence interval for $p$, the true probability of success at any given trial.

9.25 A food processing company is considering marketing a new spice mix for Creole and Cajun cooking. They interview 250 consumers and find that 53 would purchase the product. Find a 90% confidence interval for $p$, the true proportion of potential buyers.

9.26 A baseball player has a .315 batting average based on 300 official at-bats. Let $p$ denote his true probability of hitting safely on a typical trip to the plate. Construct an 80% confidence interval for $p$.

9.27 Food-poisoning outbreaks are often caused by contaminated salads. In 1967 the New York City Department of Health examined 220 tuna salads marketed and found that 179 had high bacterial levels (152). Find a 95% confidence interval for $p$, the true proportion of contaminated tuna salads in New York City.

9.28 See Case Study 2.3. Let $p$ denote the true probability that a bee lands on a flower with lines. Construct a 50% confidence interval for $p$.

9.29 In most professional sports, players cannot play for whomever they wish. Instead, the league grants a team the right to draft and sign a given player, usually through the mechanism of a draft. Later in their careers, though, players can become "free agents," which means they can offer their services to the highest bidder. But are they worth their high salaries? The Kansas City Royals pointed out that of 98 free agents signed by teams over the last three years, 61 experienced "significantly subpar seasons" the first year after their new contracts (28). Construct a 95% confidence interval for $p$, the true probability that a player will have a subpar season the year after becoming a free agent.

9.30 Intelligent life on other planets is a cinematic theme that continues to be box office magic. Theater-goers seem equally enthralled by intergalactic brethren portrayed as hostile, as in H. G. Wells's *War of the Worlds*, or benign, as in Stephen Spielberg's *E. T.* What is not so clear is the extent to which people actually believe that such creatures exist. In a close encounter of the statistical kind, a 1985 Media General–Associated Press poll found that, of 1517 respondents, 713 said they believed there was intelligent life on other worlds. Find a 95% confidence interval for the true proportion of Americans who believe we are not alone.

# 9.4

## SAMPLE SIZE DETERMINATION

### Estimating μ

Suppose a would-be politician decides to explore her attractiveness as a candidate for local office. She commissions a poll and finds that 52 of the 100 voters interviewed would support her campaign. Having recently read Section 9.3, she realizes that data of that sort can best be analyzed by constructing a confidence interval. Letting $\alpha = 0.05$, she uses Theorem 9.2 to calculate that

$$\left( 0.52 - 1.96 \sqrt{\frac{(0.52)(0.48)}{100}}, \, 0.52 + 1.96 \sqrt{\frac{(0.52)(0.48)}{100}} \right)$$

or                                        $(0.42, 0.62)$

is the 95% confidence interval for $p$, the true proportion of voters who would endorse her candidacy.

What does the interval $(0.42, 0.62)$ tell her? Not very much. It predicts that she might be headed for anything ranging from a colossal defeat (if $p$ is actually close to 0.42) to a landslide win (if $p$ is close to 0.62)! All in all, not the *quality* of input a decision maker likes to have.

Consider a second scenario. Instead of commissioning a poll of 100 voters, she raises the sample size to 2400. And of those 2400, suppose that 1248 promise her their votes. Just as before, the proportion in the sample professing their support is 0.52 $(= 1248/2400)$. The 95% confidence interval, though, changes dramatically. With $n = 2400$, Theorem 9.2 gives

$$\left( 0.52 - 1.96 \sqrt{\frac{(0.52)(0.48)}{2400}}, \, 0.52 + 1.96 \sqrt{\frac{(0.52)(0.48)}{2400}} \right) = (0.50, 0.54)$$

This interval tells her a good deal. The range of possible $p$ values is much smaller (indicating that the parameter is being estimated more precisely), and its location is very encouraging. If this interval, for example, is one of the 19 out of 20 $(= 95\%)$ that is "correct" (that is, if it actually covers the true $p$), then she will win.

Since a larger sample size shortens a confidence interval and makes it easier to interpret, why not routinely take $n$ to be in the thousands? Because every experiment comes down to a trade-off between the precision a researcher wants to attain and the amount of time and money that can be spent collecting data. Ideally, we would like to take the smallest possible sample size that will achieve a level of precision sufficient for the experimenter's objectives.

The "precision" of a confidence interval is usually stated in terms of its *radius*, the distance from the center of the interval to either end point: the

**radius (of a confidence interval)**— The distance from the center of a confidence interval to either end point. The radius is a measure of the precision with which a parameter is being estimated.

431

smaller the radius the more precise the confidence interval. The examples in this section illustrate how the length of the radius is used to determine the number of observations an experimenter needs to take. Specifically, we want the smallest sample size $n$ that "guarantees" (with probability $1 - \alpha$) that the radius of a confidence interval will be less than some predetermined value.

---

**EXAMPLE 9.1**   Recall Case Study 8.2. The point of contention was that the accepted value of 266 days for an average pregnancy duration might not be valid for all populations (and deviations from 266 might have diagnostic significance). Suppose a researcher wants to estimate with a 90% confidence interval the true average pregnancy duration for a group of women. She believes that the estimate will not be useful unless the confidence interval has a radius of no more than 3 days. How many women should be included in the sample?

**Solution**   As before, we assume that pregnancy durations are normally distributed with $\sigma = 16$ days. Let $\mu$ denote the true average pregnancy duration for the group being studied. From Theorem 8.3 the radius of a 90% confidence interval for $\mu$ (based on a sample of size $n$) is

$$z_{0.10/2} \frac{\sigma}{\sqrt{n}} = z_{0.05} \frac{\sigma}{\sqrt{n}} = 1.64 \frac{16}{\sqrt{n}}$$

Requiring that the radius be no longer than 3 is equivalent to saying that $n$ must satisfy the inequality

$$1.64 \frac{16}{\sqrt{n}} \leqslant 3$$

Therefore,
$$n \geqslant \left[ \frac{1.64(16)}{3} \right]^2 = 76.5$$

so the researcher needs a sample of at least 77 women to be able to estimate $\mu$ with the desired precision.

Extending what we have done to other confidence coefficients is straightforward. If, for example, an experimenter wishes to construct a 95% confidence interval having a radius no longer than 3, then

$$n \geqslant \frac{(1.96)^2(16)^2}{3^2} = 109.3$$

because $P(-1.96 \leqslant Z \leqslant 1.96) = 0.95$. Theorem 9.4 gives a general formula for the minimum sample size in terms of the interval's confidence coefficient, its radius, and the standard deviation $\sigma$.

---

---

**THEOREM 9.4**

Let $Y_1, Y_2, \ldots, Y_n$ be a random sample of size $n$ from a normal distribution where $\sigma$ is known. The minimum sample size that guarantees that a $100(1 - \alpha)\%$ confidence interval for $\mu$ has a radius less than $d$ is the smallest value of $n$ satisfying the inequality

$$n \geqslant \left( \frac{z_{\alpha/2}\sigma}{d} \right)^2$$

---

QUESTIONS

9.31  In Example 9.1 suppose that the confidence coefficient is raised to 99%. What is the minimum sample size that needs to be taken?

9.32  Find the smallest value of $n$ assuring that a 95% confidence interval for $\mu$ has a radius no longer than 0.2 when

(a) $\sigma = 5$     (b) $\sigma = 10$.

9.33  For each of the following $\alpha$ levels, find the smallest value of $n$ guaranteeing that a $100(1 - \alpha)\%$ confidence interval for $\mu$ has a radius no longer than 0.35: (a) 0.05, (b) 0.10, (c) 0.01. Assume $\sigma = 7$.

9.34  What is the smallest value of $n$ that will make a 95% confidence interval for $\mu$ have radius $d$ no greater than (a) 0.1, (b) 0.2. Assume $\sigma = 12$.

9.35  The cylinders of a certain internal combustion engine are manufactured by a process that produces a normal distribution of cylinder diameters having $\sigma = 0.4$ cm. We want to estimate $\mu$ of this distribution with a 95% confidence interval. How many cylinders should we measure if we insist that the estimation be sufficiently precise to make the radius of the resulting interval no greater than 0.08?

9.36  Suppose we want a 95% confidence interval to have a radius less than $\sigma/5$, regardless of the value of $\sigma$. What sample size is required? Generalize your answer to a $100(1 - \alpha)\%$ confidence interval.

9.37  See Example 8.2. Suppose we can assume that the ACT scores of next year's graduates will have $\sigma = 6$, but a possibly different mean. How large a sample must we take to estimate $\mu$ with an 80% confidence interval if we want the entire length of the interval to be no greater than 0.7?

*Estimating $\rho$*

Finding a minimum sample size to estimate a binomial parameter $p$ is more complicated than what we just derived for $\mu$, but the basic principle is the same. Suppose $Y$ denotes the number of successes in $n$ Bernoulli trials. Then

$$P\left(-z_{\alpha/2} \leqslant \frac{Y/n - p}{\sqrt{p(1-p)/n}} \leqslant z_{\alpha/2}\right) \doteq 1 - \alpha$$

or, equivalently,

$$P\left(-z_{\alpha/2}\sqrt{\frac{p(1-p)}{n}} \leqslant \frac{Y}{n} - p \leqslant z_{\alpha/2}\sqrt{\frac{p(1-p)}{n}}\right) \doteq 1 - \alpha \qquad (9.5)$$

(recall Theorem 6.7).

It follows from Equation 9.5 that $Y/n$, our estimate of the true proportion, will be within a distance $d$ of $p$ if

$$z_{\alpha/2}\sqrt{\frac{p(1-p)}{n}} \leqslant d$$

Solving for $n$ gives

$$n \geqslant \frac{z_{\alpha/2}^2 p(1-p)}{d^2}$$

But this inequality is not in a form that immediately answers our question, because it gives $n$ *as a function of* $p$, and $p$ is what we are trying to estimate. Fortunately, we can still salvage an approximation.

Since $p$ is a probability, it must be a number between 0 and 1. Furthermore, for any number in that range, $p(1-p) \leqslant \frac{1}{4}$ (Figure 9.7). It follows that if we replace $p(1-p)$ with $\frac{1}{4}$ in the last inequality, the resulting $n$ always gives a confidence interval with a radius no greater than $d$.

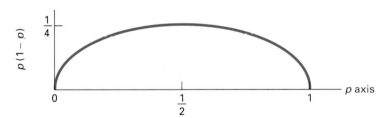

**FIG. 9.7**

---

**THEOREM 9.5**

Let $Y$ denote the number of successes in $n$ Bernoulli trials, where $p = P$ (success occurs on any given trial). To estimate $p$ with a $100(1 - \alpha)\%$ confidence interval having a radius less than or equal to $d$, it is necessary that the sample size $n$ satisfies the inequality

$$n \geqslant \frac{z_{\alpha/2}^2}{4d^2} \qquad (9.6)$$

*EXAMPLE 9.2*    Recall Case Study 9.5. Let $p$ denote the true proportion of drinkers who prefer Schlitz. In setting up the taste test, suppose Schlitz wants to choose enough subjects so that the probability is 0.95 that $Y/n$, the proportion in the sample who prefer Schlitz, is within 0.05 of the true $p$. What is the smallest value of $n$ that will produce that level of precision?

*Solution*    From Theorem 9.5 the minimum sample size we should consider is the smallest integer greater than or equal to

$$\frac{z_{\alpha/2}^2}{4d^2} = \frac{(1.96)^2}{4(0.05)^2} = 384.12$$

Therefore, take $n = 385$; a sample size that large will necessarily produce a 95% confidence interval for $p$ whose radius is less than or equal to 0.05.

QUESTIONS

9.38    Find the smallest value of $n$ so that a 95% confidence interval for $p$ has radius less than or equal to $d$ if

(a) $d = 0.02$    (b) $d = 0.04$.

9.39    Find the smallest value of $n$ guaranteeing that a $100(1 - \alpha)$% confidence interval for $p$ has radius no greater than 0.03 if

(a) $\alpha = 0.05$    (b) $\alpha = 0.01$    (c) $\alpha = 0.50$.

9.40    An attorney for a civil liberties group is interviewing female nurses in a large hospital to see if they believe they have been victims of sexism. Let $p$ denote the true proportion of nurses who would consider themselves the victims of such discrimination. How many nurses should the attorney interview in order to be 60% certain that the sample proportion of nurses who say, "Yes, I have been discriminated against," is within 0.05 of $p$?

9.41    A public health survey is to be done in an inner-city area to estimate the proportion of children, ages 0 to 14, having adequate polio immunization. The final estimate should be within 0.05 of the true proportion with probability 0.98. What is the minimum sample size required?

9.42    Harold owns a prosperous chain of stores selling clothes for college men. He is contemplating the expansion of his business to include women's wear and would like to know $p$, the proportion of coeds who would shop at his establishment. How many female students must he survey to be 90% certain that his estimate for $p$ $(= Y/n)$ will be within 0.075 of the true $p$?

## 9.5

# INFERENCES ABOUT THE VARIANCE

### Confidence Intervals

Most of the inferences we have considered thus far have dealt with means or proportions. For good reason. In practice, the parameters in one-sample problems that are most likely to be of interest to an experimenter are, indeed,

$\mu$ if the data are normal and $p$ if the data are binomial. Nevertheless, there are situations where the precision of a measurement is more important than its location. For those kinds of problems, any inference procedures—whether confidence intervals or hypothesis tests—should focus on the variance rather than on the mean.

Not surprisingly, the test statistic appropriate for drawing conclusions about the *population variance* $\sigma^2$ is the *sample variance* $S^2$, where

$$S^2 = \frac{1}{n-1} \sum_{i=1}^{n} (Y_i - \bar{Y})^2$$

It was established in Chapter 7 that $(n-1)S^2/\sigma^2$ has a chi-square distribution with $n-1$ degrees of freedom. That result is the key for everything we do in this section. Inverted, it leads to a confidence interval for $\sigma^2$. And if $\sigma^2$ is replaced by $\sigma_0^2$, it becomes the test statistic for choosing between $H_0 \colon \sigma^2 = \sigma_0^2$ and $H_1 \colon \sigma^2 \neq \sigma_0^2$.

We begin with the problem of finding a confidence interval for $\sigma^2$. Since $(n-1)S^2/\sigma^2$ has a $\chi^2$ distribution with $n-1$ degrees of freedom,

$$P\left( \chi^2_{\alpha/2,\,n-1} \leqslant \frac{(n-1)S^2}{\sigma^2} \leqslant \chi^2_{1-\alpha/2,\,n-1} \right) = 1 - \alpha$$

Therefore,

$$P\left( \frac{1}{\chi^2_{1-\alpha/2,\,n-1}} \leqslant \frac{\sigma^2}{(n-1)S^2} \leqslant \frac{1}{\chi^2_{\alpha/2,\,n-1}} \right) = 1 - \alpha \quad \text{(Why?)}$$

Multiplying by $(n-1)S^2$ will isolate $\sigma^2$ in the center of the inequalities:

$$P\left( \frac{(n-1)S^2}{\chi^2_{1-\alpha/2,\,n-1}} \leqslant \sigma^2 \leqslant \frac{(n-1)S^2}{\chi^2_{\alpha/2,\,n-1}} \right) = 1 - \alpha \qquad (9.7)$$

The two end points in Equation 9.7 define the $100(1-\alpha)\%$ confidence interval for $\sigma^2$.

---

### THEOREM 9.6

Let $Y_1, Y_2, \ldots, Y_n$ be a random sample of size $n$ from a normal distribution where $\mu$ and $\sigma^2$ are unknown. A $100(1-\alpha)\%$ confidence interval for $\sigma^2$ is the range of values

$$\left( \frac{(n-1)s^2}{\chi^2_{1-\alpha/2,\,n-1}}, \frac{(n-1)s^2}{\chi^2_{\alpha/2,\,n-1}} \right)$$

Also, a $100(1-\alpha)\%$ confidence interval for the standard deviation $\sigma$ is the range

$$\left( \sqrt{\frac{(n-1)s^2}{\chi^2_{1-\alpha/2,\,n-1}}}, \sqrt{\frac{(n-1)s^2}{\chi^2_{\alpha/2,\,n-1}}} \right)$$

## CASE STUDY 9.6

*Radioactive Dating*  The chain of events that we call the geological evolution of the earth began hundreds of millions of years ago. Fossils have played a key role in documenting the *relative* times those events occurred, but to establish an *absolute* chronology, scientists rely primarily on radioactive decay. In this study we look at the variability associated with a dating technique based on a mineral's potassium-argon ratio.

Almost all minerals contain potassium (K) as well as certain of its isotopes, including $^{40}$K. But $^{40}$K is unstable and decays into isotopes of argon and calcium, $^{40}$Ar and $^{40}$Ca, respectively. By knowing the rates at which those daughter products are formed, and by measuring the amounts of $^{40}$Ar and $^{40}$K that a specimen contains, geologists can deduce the rock's age.

Table 9.6 gives the ages (in millions of years) estimated for 19 mineral samples collected from Germany's Black Forest (102). Each data point was based on a rock's potassium-argon ratio. Since all the specimens have the same *true* age, any variation in their *estimated* ages is reflecting the inherent precision of the potassium-argon dating procedure. The best way to quantify that precision is to construct, say, a 95% confidence interval for $\sigma$.

For the 19 ages listed

**TABLE 9.6**

| Specimen | Estimated Age (millions of years) | |
|----------|-----|------|
| 1 | $y_1 =$ | 249 |
| 2 | $y_2 =$ | 254 |
| 3 | $y_3 =$ | 243 |
| 4 | $y_4 =$ | 268 |
| 5 | $y_5 =$ | 253 |
| 6 | $y_6 =$ | 269 |
| 7 | $y_7 =$ | 287 |
| 8 | $y_8 =$ | 241 |
| 9 | $y_9 =$ | 273 |
| 10 | $y_{10} =$ | 306 |
| 11 | $y_{11} =$ | 303 |
| 12 | $y_{12} =$ | 280 |
| 13 | $y_{13} =$ | 260 |
| 14 | $y_{14} =$ | 256 |
| 15 | $y_{15} =$ | 278 |
| 16 | $y_{16} =$ | 344 |
| 17 | $y_{17} =$ | 304 |
| 18 | $y_{18} =$ | 283 |
| 19 | $y_{19} =$ | 310 |
| | | 5261 |

$$\sum_{i=1}^{19} y_i = 5261 \quad \text{and} \quad \sum_{i=1}^{19} y_i^2 = 1{,}469{,}945$$

so the sample variance is 733.4:

$$s^2 = \frac{19(1{,}469{,}945) - (5261)^2}{19(18)} = 733.4$$

Since $n = 19$ and $\alpha = 0.05$, the $\chi^2$ multiples appearing in Theorem 9.6 are the 2.5th and 97.5th percentiles of a $\chi^2$ distribution with $18 (= n - 1)$ degrees of freedom. From Appendix A.3,

$$\chi^2_{0.025,\,18} = 8.23 \quad \text{and} \quad \chi^2_{0.975,\,18} = 31.53$$

Therefore, a 95% confidence interval for $\sigma$ is the range of values

$$\left( \sqrt{\frac{(19-1)(733.4)}{31.53}}, \sqrt{\frac{(19-1)(733.4)}{8.23}} \right)$$

which reduces to

(20.5 million years, 40.0 million years)

**QUESTIONS**

**9.43**   Suppose that a random sample of size $n$ yielded the values of $s$ shown below. For each situation construct a 95% confidence interval for $\sigma^2$.

(a) $n = 10$   and   $s = 14.5$      (b) $n = 20$   and   $s = 8.5$

**9.44**   The following random sample of size 10 is drawn from a distribution assumed to be normal: 19.5, 29.2, 26.8, 21.6, 20.8, 22.6, 22.7, 19.4, 24.0, and 21.1.

(a) Construct a 90% confidence interval for $\sigma^2$.

(b) Construct a 90% confidence interval for $\sigma$.

(c) Suppose the data were measured in inches. What are the units of your answers in parts (a) and (b)?

**9.45**   Assume that the $w/l$ ratios of the Shoshoni rectangles in Case Study 9.3 are observations coming from a normal distribution. Construct a 95% confidence interval for $\sigma^2$.

**9.46**   Using the data in Question 1.1, construct a 90% confidence interval for the standard deviation of the number of years between an invention's conception and its realization.

**9.47**   Which is likely to be more useful to an applied statistician, a confidence interval for $\sigma$ or a confidence interval for $\sigma^2$? Explain.

**9.48**   Besides involving a different test statistic and a different sampling distribution, how else are confidence intervals for $\sigma^2$ different from what we constructed for $\mu$ or $p$?

**9.49**   For the 19 $FEV_1/VC$ ratios in Table 2.2, construct two different *one-sided* 95% confidence intervals for $\sigma^2$.

**9.50**   Construct a 95% confidence interval for $\sigma$, using the mosquito bite data in Table 3.12.

**9.51**   What confidence coefficient is associated with the following interval based on a set of measurements that gave $s^2 = 32.7$:

$$\left( \frac{(11-1)(32.7)}{20.483}, \frac{(11-1)(32.7)}{3.940} \right)?$$

### Hypothesis Tests

For problems where the precision of a measurement is the characteristic of primary concern, and where a "standard" value for that precision ($\sigma_0^2$) is readily available, any inferences we make will generally be in the form of hypothesis tests rather than confidence intervals. If $H_0: \sigma^2 = \sigma_0^2$ is true, then $(n-1)S^2/\sigma_0^2$ will have a $\chi^2$ distribution with $n-1$ degrees of freedom. Depending on the nature of $H_1$, values of $(n-1)s^2/\sigma_0^2$ in either or both tails of that distribution will lead us to reject $H_0$.

---

### THEOREM 9.7

Let $Y_1, Y_2, \ldots, Y_n$ be a random sample from a normal distribution where $\mu$ and $\sigma^2$ are unknown. Let

$$\chi^2 = \frac{(n-1)s^2}{\sigma_0^2}$$

1. To test

$$H_0: \quad \sigma^2 = \sigma_0^2$$

vs.

$$H_1: \quad \sigma^2 \neq \sigma_0^2$$

at the $\alpha$ level of significance, reject $H_0$ if either (a) $\chi^2 \leqslant \chi^2_{\alpha/2, n-1}$ or (b) $\chi^2 \geqslant \chi^2_{1-\alpha/2, n-1}$.

2. To test

$$H_0: \quad \sigma^2 = \sigma_0^2$$

vs.

$$H_1: \quad \sigma^2 > \sigma_0^2$$

at the $\alpha$ level of significance, reject $H_0$ if $\chi^2 \geqslant \chi^2_{1-\alpha, n-1}$.

$$\frac{(n-1)s^2}{\sigma_0^2} \geqslant \chi^2_{1-\alpha, n-1}$$

3. To test

$$H_0: \quad \sigma^2 = \sigma_0^2$$

vs.

$$H_1: \quad \sigma^2 < \sigma_0^2$$

at the $\alpha$ level of significance, reject $H_0$ if $\chi^2 \leqslant \chi^2_{\alpha, n-1}$.

---

*EXAMPLE 9.3*   Another procedure for dating rocks radioactively was widely used prior to the discovery of the potassium-argon method described in Case Study 9.6. Based on a mineral's lead content, this earlier method could yield estimates having a standard deviation of 30.4 million years. Recall that the sample standard deviation for the 19 Black Forest rocks (i.e., for the potassium-argon method) was *smaller*—specifically, it was $\sqrt{733.4}$, or 27.1 million years. Can the latter be taken as "proof" that the potassium-argon procedure is significantly more precise than its predecessor? That is, if $\sigma^2$ denotes the true variance for the potassium-argon method, can we reject $H_0: \sigma^2 = (30.4)^2$ in favor of $H_1: \sigma^2 < (30.4)^2$?

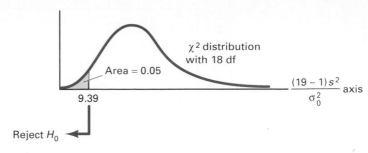

**FIG. 9.8**

*Solution*   Let $\alpha = 0.05$. For a $\chi^2$ random variable with 18 degrees of freedom,

$$P(\chi^2_{18} \leqslant 9.39) = 0.05$$

(see Figure 9.8). From part 3 of Theorem 9.7, $H_0$ should be rejected if

$$\chi^2 = \frac{(n-1)s^2}{\sigma_0^2} \leqslant 9.39$$

But
$$\frac{(n-1)s^2}{\sigma_0^2} = \frac{(19-1)(733.4)}{(30.4)^2} = 14.3$$

implying that the difference between 30.4 million years and 27.1 million years is *not* statistically significant. We have not demonstrated, in other words, that the potassium-argon method is more precise than its predecessor.

---

**QUESTIONS**

**9.52**   Find the critical value(s) for testing $H_0$: $\sigma^2 = 10$ if

(a) $H_1$: $\sigma^2 > 10$, $\alpha = 0.05$, $n = 8$

(b) $H_1$: $\sigma^2 < 10$, $\alpha = 0.05$, $n = 8$

(c) $H_1$: $\sigma^2 \neq 10$, $\alpha = 0.10$, $n = 8$

(d) $H_1$: $\sigma^2 \neq 10$, $\alpha = 0.05$, $n = 8$

(e) $H_1$: $\sigma^2 > 10$, $\alpha = 0.05$, $n = 12$

**9.53**   For the data

113.3   93.7   118.5   106.3   108.1   113.1   112.0

test $H_0$: $\sigma^2 = 15$ versus $H_1$: $\sigma^2 > 15$. Let $\alpha = 0.01$.

**9.54**   In Question 8.19 it was claimed that systolic blood pressures of 18-year-old women are normally distributed and that $\sigma = 12$ mmHg. Do the data in that exercise (119, 130, 133, 128, 127, 134, 126, 129, 122, 129) support the second claim. Test the appropriate $H_0$ and $H_1$, using $\alpha = 0.05$.

**9.55**   Use the data in Question 8.21 to test whether the standard deviation of 60-W light bulbs is 75 hr, as claimed. Make the alternative one-sided ($H_1$: $\sigma < 75$). Let $\alpha = 0.05$.

9.56 The A above middle C is the note given to an orchestra, usually by the oboe, for tuning purposes. Its pitch is defined to be the sound of a tuning fork vibrating at 440 cycles per second (cps). No tuning fork, of course, vibrates at *exactly* 440 cps; rather, its pitch $Y$ is a random variable. Assume that $Y$ is normally distributed with $\mu = 440$ and variance $\sigma^2$. The quality of a tuning fork is obviously related to $\sigma^2$: the smaller the variance, the higher the quality. Suppose that tuning forks manufactured a "standard" way have $\sigma^2 = 15$. Proponents of a new process, though, are claiming that their procedure will produce tuning forks having a smaller variance. As a test, six tuning forks are manufactured by the new technology. The resulting vibration frequencies are 424.0, 447.3, 440.6, 411.1, 442.6, and 419.6 cps. Set up and choose between the appropriate $H_0$ and $H_1$. Let $\alpha = 0.05$.

9.57 A sheet metal firm intends to run quality control tests on the shear strength of spot welds. As a guideline to choosing a sample size, they decide to take the smallest number of observations for which there is at least a 95% chance that the ratio of the sample variance to the true variance is less than 2; that is, they want

$$P\left(\frac{S^2}{\sigma^2} < 2\right) \geqslant 0.95$$

What is the smallest sample size they can use?

# 9.6

## COMPUTER NOTES

Performing one-sample inference with statistical software is not difficult. Just remember that hypothesis testing is usually done with $P$ values. The $P$ value is the probability under $H_0$ that the test statistic is greater than the observed value of the test statistic. Most textbooks perform hypothesis tests by choosing the level of significance $\alpha$, finding the critical value(s) from the appropriate table, and comparing the test statistic with the critical values. When we test with $P$ values, we reject the null hypothesis if the $P$ value is *less than* the chosen $\alpha$.

Let us compare the two methods of conducting a test for Case Study 9.3. There $\alpha = 0.05$ and $t = 2.05$. Using Appendix A.2, we reject $H_0$ if $t > 2.0930$ or $t < -2.0930$ (Theorem 9.1). Since $t = 2.05$, we accept $H_0$. We do the same test with $P$ values by finding the probability $P(Y < -2.05$ or $Y > 2.05)$ where $Y$ is a Student $t$ random variable with 19 df. We reject $H_0$ if this probability is less than 0.05.

Most readily available statistical materials do not contain tables of $P$ values, but they are provided by the computer routines. For example, Figure 9.9 reproduces a portion of an SAS printout for the data of Case Study 9.3. The test statistic value appears as the entry $T: \text{MEAN} = 0 \quad 2.05452$. The entry $\text{PROB} > |T|$ gives the $P$ value 0.0539413 for a $t$ variable with 19 df, assuming that $H_0: \mu = 0$ is true. Since the $P$ value $> 0.05$, we accept

```
DATA;
  INPUT WLRATIO;
  Y = WLRATIO - .618;
  CARDS;
```

(data from table 9.4, one observation per line)

```
PROC UNIVARIATE;

VARIABLE=Y
```

### MOMENTS

| | | | |
|---|---|---|---|
| N | 20 | SUM WGTS | 20 |
| MEAN | 0.0425 | SUM | 0.85 |
| STD DEV | 0.0925109 | VARIANCE | 0.00855826 |
| SKEWNESS | 1.74772 | KURTOSIS | 3.37592 |
| USS | 0.198732 | CSS | 0.162607 |
| CV | 27.673 | STD MEAN | 0.0206861 |
| T:MEAN=0 | 2.05452 | PROB> |T| | 0.0539413 |
| SGN RANK | 46 | PROB> |S| | 0.0893598 |
| NUM ¬= 0 | 20 | | |

**FIG. 9.9**

$H_0$. By this process we have performed a one-sample $t$ test with a two-sided alternative hypothesis.

Producing confidence intervals by computer is straightforward. In Minitab we merely enter the data and ask for a confidence interval for the specified confidence level. Figure 9.10 demonstrates this process for Case Study 9.4. The right-hand confidence interval differs slightly from the one given previously. Such discrepancies are common when we use computers or calculators, due to different computational techniques or methods of rounding.

```
SET C1
  62   52   68   23   34   45   27   42   83   56   40
TINTERVAL 95 CONFIDENCE FOR C1
```

| | N | MEAN | STDEV | SE MEAN | 95.0 PERCENT C.I. |
|---|---|---|---|---|---|
| C1 | 11 | 48.4 | 18.1 | 5.5 | ( 36.2, 60.5) |

**FIG. 9.10**

# 9.7

## SUMMARY

Conceptually and experimentally, one-sample problems are the simplest of the seven data models identified in Chapter 2. Nevertheless, how we treat one-sample data varies considerably from problem to problem. In general, the direction the analysis takes depends on the answers to three questions: (1) Are the data normal or binomial? (2) Which parameter is the focus of attention? (3) How should the inference be phrased, as a hypothesis test or a confidence interval?

Three sampling distributions figure prominently in this chapter—the Student $t$, $\chi^2$, and Z. If the $Y_i$'s are normal, $\dfrac{\bar{Y} - \mu}{S/\sqrt{n}}$ has a Student $t$ distribution with $n - 1$ df; also, $(n - 1)S^2/\sigma^2$ has a $\chi^2$ distribution with $n - 1$ df. We use the first ratio to make inferences about $\mu$; the second, to make inferences about $\sigma^2$. If the $Y_i$'s are Bernoulli random variables and $n$ is sufficiently large, then

$$\frac{Y/n - p}{\sqrt{\dfrac{(Y/n)(1 - Y/n)}{n}}}$$

behaves essentially the same as a standard normal, where $Y = \sum\limits_{i=1}^{n} Y_i$. Inverting the approximate Z gives confidence intervals for $p$.

Related to the construction of a confidence interval is a frequently encountered sample size problem—how many observations need to be taken in order for the sample estimate of $\mu$ (or $p$) to have a sufficiently high probability of being within a distance $d$ of the parameter's true value? Answers to such questions are important. For different reasons, sample sizes too small and sample sizes too large both waste money. Theorems 9.4 and 9.5 offer some helpful guidelines in choosing an appropriate $n$.

| Test | Versus | Reject $H_0$ | Formulas |
|------|--------|--------------|----------|
| $H_0$:  $\mu = \mu_0$ | $H_1$:  $\mu \neq \mu_0$ | $t \leqslant -t_{\alpha/2, n-1}$,   $t \geqslant t_{\alpha/2, n-1}$ | $t = \dfrac{\bar{y} - \mu_0}{s/\sqrt{n}}$ |
|  | $H_1$:  $\mu > \mu_0$ | $t \geqslant t_{\alpha/n - 1}$ | |
|  | $H_1$:  $\mu < \mu_0$ | $t \leqslant -t_{\alpha/n - 1}$ | |
| $H_0$:  $\sigma^2 = \sigma_0^2$ | $H_1$:  $\sigma^2 \neq \sigma_0^2$ | $\chi^2 \leqslant \chi^2_{\alpha/2, n-1}$,   $\chi^2 \geqslant \chi^2_{1 - \alpha/2, n-1}$ | $\chi^2 = \dfrac{(n - 1)s^2}{\sigma_0^2}$ |
|  | $H_1$:  $\sigma^2 > \sigma_0^2$ | $\chi^2 \geqslant \chi^2_{1 - \alpha, n-1}$ | |
|  | $H_1$:  $\sigma^2 < \sigma_0^2$ | $\chi^2 \leqslant \chi^2_{\alpha, n-1}$ | |

| Distribution | $100(1 - \alpha)\%$ Confidence Interval | For |
|--------------|------------------------------------------|-----|
| Normal | $\left( \bar{y} - t_{\alpha/2, n-1} \dfrac{s}{\sqrt{n}}, \;\; \bar{y} + t_{\alpha/2, n-1} \dfrac{s}{\sqrt{n}} \right)$ | $\mu$ |
| Normal | $\left( \dfrac{(n - 1)s^2}{\chi^2_{1 - \alpha/2, n-1}}, \;\; \dfrac{(n - 1)s^2}{\chi^2_{\alpha/2, n-1}} \right)$ | $\sigma^2$ |
| Binomial | $\left( \dfrac{y}{n} - z_{\alpha/2} \sqrt{\dfrac{(y/n)(1 - y/n)}{n}}, \;\; \dfrac{y}{n} + z_{\alpha/2} \sqrt{\dfrac{(y/n)(1 - y/n)}{n}} \right)$ | $p$ |

**REVIEW QUESTIONS**

**9.58**   As part of a project to determine the extent to which an aquifer has been contaminated by a chemical plant's improper waste disposal, engineers have measured the heavy metal concentration in eight water samples. The results (in ppb) are listed below.

| | |
|---|---|
| 2.7 | 3.0 |
| 3.1 | 2.9 |
| 2.6 | 2.4 |
| 2.9 | 3.6 |

Construct a 95% confidence interval for the true average heavy-metal concentration.

**9.59**   Three hundred elderly adults are given a new influenza vaccine. Two weeks later, 180 show a six-fold serum titer rise, indicating that the vaccine has produced the desired antibody response. Use a 90% confidence interval to estimate the true proportion of elderly adults on whom the vaccine would have a similar effect.

**9.60**   In the past, the amount of sick leave taken by maintenance workers in a nuclear plant averaged 2.2 days per worker per month. Management has tried to reduce that figure by including sick leave as an item on a worker's performance review. For the last six months, sick leave averages were 1.2, 1.8, 0.9, 1.9, 2.1, and 2.4. Can management claim that their efforts have been successful? Set up and carry out the appropriate hypothesis test. Let $\alpha = 0.05$.

**9.61**   Let $p$ denote the true proportion of college students who support the movement to colorize classic films. Let the random variable $Y$ denote the number of students (out of $n$) who prefer colorized versions to black and white. What is the smallest sample size for which the probability is 90% that the difference between $Y/n$ and $p$ is less than 0.02?

**9.62**   Brightness is one of the characteristics of stationery that the quality control department of a printing company monitors carefully. If given the same paper sample repeatedly, the machine that measures brightness is supposed to produce readings with a standard deviation of 0.6. Listed below are five such measurements. Test $H_0$: $\sigma = 0.6$ versus $H_1$: $\sigma \neq 0.6$ at the $\alpha = 0.10$ level of significance.

$$6.1 \quad 5.8 \quad 7.8 \quad 4.3 \quad 8.9$$

**9.63**   A random sample of 43 restaurants that opened in the early 1980's revealed that only 27 were still in business three years later. Define an appropriate $p$ and construct the corresponding 90% confidence interval.

**9.64**   A former politician recently released from jail wants to conduct a poll to see whether he remains a viable candidate for future office. Suppose $p$ denotes the proportion of voters who would absolutely refuse to support his candidacy. How large a sample is needed to guarantee that 99% of the time the sample estimate $(Y/n)$ will be within 0.03 of $p$?

**9.65**   The turn-around time for programs run on the university mainframe computer averages 12.5 hr. In the past, the standard deviation of those times was 3.0 hr, but users are becoming concerned that $\sigma$ may be increasing. Listed below are the turn-around times for 10 programs chosen at random.

| | |
|---|---|
| 12.0 | 22.5 |
| 18.5 | 10.0 |
| 13.0 | 2.5 |
| 14.5 | 3.0 |
| 6.0 | 12.0 |

Set up and test a relevant $H_0$ and $H_1$. Let $\alpha = 0.05$.

**9.66**   Suppose the 90% confidence interval for the mean of a normal distribution based on a sample of size 22 is (210.6, 231.8). Find $s$.

**9.67**   What is the difference between the sampling distributions of

$$\frac{\bar{Y} - \mu_0}{S/\sqrt{n}} \quad \text{and} \quad \frac{\bar{Y} - \mu_0}{\sigma/\sqrt{n}}?$$

How is the difference between the two pdf's related to $n$? What determines which statistic is used in testing $H_0$: $\mu = \mu_0$?

**9.68**   The sample mean and sample standard deviation for the following 20 numbers are 2.6 and 3.6, respectively. Is it correct to say that a 95% confidence interval for $\mu$ is

$$\left(2.6 - 2.0930\,\frac{3.6}{\sqrt{20}},\ 2.6 + 2.0930\,\frac{3.6}{\sqrt{20}}\right) = (0.9,\ 3.3)?$$

Explain.

| | | | |
|---|---|---|---|
| 2.5 | 0.1 | 0.2 | 1.3 |
| 3.2 | 0.1 | 0.1 | 1.4 |
| 0.5 | 0.2 | 0.4 | 11.2 |
| 0.4 | 7.4 | 1.8 | 2.1 |
| 0.3 | 8.6 | 0.3 | 10.1 |

**9.69**   The mortgage rates charged by a random sample of six local banks are 10.2, 10.8, 9.9, 10.0, 11.1, and 10.7, respectively. At the $\alpha = 0.10$ level, can we conclude that the true average mortgage rate of the population of banks represented by the sample is not equal to 10.9?

**9.70**   What sample size gives (14.53, 60.68) as a 95% confidence interval for $\sigma^2$ if the sample variance is 26.2?

**9.71**   If the single number that best approximates the mean of a normal distribution is $\bar{y}$, why do we complicate the estimation process by constructing confidence intervals?

**9.72**   Of the 312 students who took Statistics 101 in the last few years, 210 passed. Assume that Bubba's ability in statistics is typical. Is it likely that his chances of passing are less than 50%? Answer the question by constructing a 95% confidence interval.

9.73   What is the difference between the intervals

$$\left( \bar{Y} - t_{\alpha/2, n-1} \frac{S}{\sqrt{n}}, \, \bar{Y} + t_{\alpha/2, n-1} \frac{S}{\sqrt{n}} \right)$$

and

$$\left( \bar{y} - t_{\alpha/2, n-1} \frac{s}{\sqrt{n}}, \, \bar{y} + t_{\alpha/2, n-1} \frac{s}{\sqrt{n}} \right)?$$

9.74   As a way of monitoring the conditioning of his pitchers, a baseball manager uses a radar gun to time the fastballs each one throws to the first couple batters he faces. The 12 readings below (in mph) are speeds recorded over the past several games for the team's bullpen ace. Last year his average speed was 98.4 mph. Is he losing something off his fastball? Answer the question by doing a hypothesis test with $\alpha = 0.10$.

| | | |
|---|---|---|
| 98.2 | 100.1 | 90.6 |
| 94.6 | 94.2 | 92.1 |
| 95.1 | 93.6 | 93.4 |
| 93.3 | 95.1 | 88.6 |

Note: $\sum_{i=1}^{12} y_i = 1128.9$   and   $\sum_{i=1}^{12} y_i^2 = 106{,}304.21$.

9.75   How does an experimenter know whether to test $H_0: \mu = \mu_0$ or construct a confidence interval for $\mu$? Is it ever possible to do both?

9.76   Is it likely that the following sample of size 9 came from a normal distribution with $\mu = 50.0$? Set up and test the appropriate $H_0$ versus $H_1$. Let $\alpha = 0.10$.

| | |
|---|---|
| 28.0 | 59.2 |
| 30.6 | 36.2 |
| 45.5 | 43.1 |
| 41.8 | 40.6 |
| 53.4 | |

Note: $\sum_{i=1}^{9} y_i = 378.4$   and   $\sum_{i=1}^{9} y_i^2 = 16{,}710.46$.

9.77   Recall the experiment proposed in Question 9.61. What would be the appropriate sample size if it can be assumed that no more than 35% of the students favor colorization?

9.78   Construct a 90% confidence interval for $\sigma$ using the data in Question 9.62. Does the interval predict the outcome of testing $H_0: \sigma = 0.6$ versus $H_1: \sigma \neq 0.6$? Explain.

9.79   Listed below are the amounts of money (in thousands) spent by seven middle-aged vacationers on a three-day Caribbean cruise. Are the data consistent with the travel agency's belief that $3500 is the average amount spent by members of that age group? Do an appropriate hypothesis test at the $\alpha = 0.10$ level.

| | |
|---|---|
| 0.8 | 2.1 |
| 3.2 | 3.4 |
| 1.6 | 6.1 |
| 0.4 | |

9.80   Which of the following two intervals has the greater probability of containing the binomial parameter $p$:

$$\left(\frac{Y}{n} - 0.67\sqrt{\frac{(Y/n)(1 - Y/n)}{n}}, \frac{Y}{n} + 0.67\sqrt{\frac{(Y/n)(1 - Y/n)}{n}}\right) \quad \text{or} \quad \left(\frac{Y}{n}, \infty\right)?$$

9.81   Suppose $Y_1, Y_2, \ldots, Y_{20}$ come from a normal distribution with $\sigma = 3$. For what value of $d$ does

$$P(-d < \bar{Y} - \mu < d) = 0.95?$$

9.82   According to the Metro Transit schedule, the bus leaving York Road at 6:30 A.M. is supposed to arrive at West 25th Street at 7.15 A.M. Skeptical that the bus completes the trip in only 45 minutes, a commuter records route times for the next ten days. The results (in minutes) are listed below. Analyze the data using the hypothesis testing format. Let $\alpha = 0.05$.

| | |
|----|----|
| 42 | 50 |
| 50 | 49 |
| 53 | 48 |
| 52 | 46 |
| 48 | 49 |

# 10 The Two-Sample Problem

*Certainty is no substitute for hope*

*Moltmann*

# 10.1

## INTRODUCTION

Variations on a theme. To most of us, the phrase conjures up an image of a composer restructuring a melody—adding a phrase here, deleting a passage there. What eventually reappears is something definitely new, yet sounding vaguely familiar. As it happens, Beethoven and his musical compatriots are not the only ones who play that game—so do statisticians. Everything we cover in the rest of this book is little more than a series of variations applied to the themes in Chapters 8 and 9. *Keeping that fact uppermost in your mind is the best way to learn and organize what lies ahead.* Look at each new procedure for what it is—a minor modification of a problem already solved. What distinguishes these procedures, one from another, is critically important, but no more important than what they have in common.

We begin by stating explicitly the themes we should be looking for. Common denominators other than the five listed below can be found, but these will prove to be the most important for our purposes.

1. Inferences are typically directed at the unknown parameters of the distribution that the data presumably represent ($\mu$ and/or $\sigma$ if the data are normal; $p$ if the data are binomial).
2. Procedures for analyzing normal data are different from those for analyzing binomial data.
3. Inferences about parameters take two different forms: hypothesis tests and confidence intervals. The nature of the problem as well as the experimenter's objectives dictate which should be used.
4. Associated with every hypothesis test and every confidence interval is a function that involves both the data and the unknown parameter. Knowing the probability distributions of these functions (for example, knowing that

$$\frac{\bar{Y} - \mu}{S/\sqrt{n}}$$

has a Student $t$ distribution with $n - 1$ degrees of freedom) is the key to doing any inference procedure.
5. We never know whether the result of a statistical procedure is correct. However, the *probabilities* of making errors can be computed.

The variations on these five ideas that we need to address derive from the seven data models profiled in Chapter 2. The formulas, for example, necessary for making inferences in the one-sample problem—what we explored in Chapter 9—are slightly different from those appropriate for the two-sample problem. And those, in turn, need to be modified to accommodate the paired-data problem, and so on.

In this chapter we focus on the *two-sample problem*. Two independent sets of data are observed:

$$x_1, x_2, \ldots x_n \qquad \text{and} \qquad y_1, y_2, \ldots y_m$$

We assume the $x$'s are a random sample from population $X$, and the $y$'s a random sample from population $Y$. Our objective is to make some sort of comparison between the parameters in the two populations. Three distinct cases will be considered. We will compare

1. The means of two normal distributions.
2. The standard deviations of two normal distributions.
3. The "success" probabilities of two binomial distributions.

Cases 1 and 3 are especially important.

## 10.2

### THE TWO-SAMPLE FORMAT

Figure 10.1 reviews the two types of data structures that qualify as two-sample problems (recall Section 2.3): either two different treatments ($X$ and $Y$) are applied to two sets of similar subjects (Figure 10.1a), or the same treatment is applied to two sets of dissimilar subjects (Figure 10.1b). In Figure 10.1a we want to compare the treatments; in Figure 10.1b we want to compare the subjects. In both cases the $x_i$'s and $y_i$'s are assumed to be *independent* (knowing the value of $x_i$ does not help to predict the value of

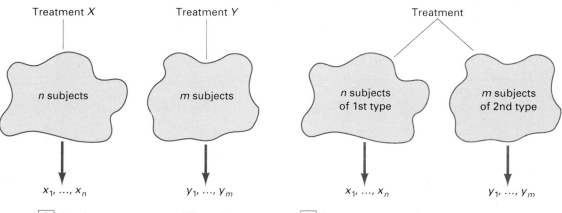

**FIG. 10.1**

$y_i$). An example of the first type would be a football trainer recording rehabilitation times associated with two different therapy regimens ($X$ and $Y$) designed for athletes recovering from arthroscopic surgery. An example of the second type would be a psychiatrist studying whether male secretaries experience the same levels of job-related stress as female secretaries do.

The distinction between types "a" and "b" has no bearing on how the data are analyzed. What *does* matter, though, are the distribution assumptions we put on the $x$'s and the $y$'s and on which parameters of those distributions we choose to focus.

In general, the $x_i$'s and the $y_i$'s might represent random samples from *any* probability functions $f_X(x)$ and $f_Y(y)$, respectively. For example, $f_X(x)$ and $f_Y(y)$ might both be geometric distributions (recall Section 6.5). However, the vast majority of two-sample problems likely to be encountered (and all those that we will be concerned with) will be such that the $X$'s and $Y$'s are either normal random variables or Bernoulli random variables. Sections 10.3 and 10.4 treat the "normal" case; inference procedures appropriate for Bernoulli data are covered in Section 10.5.

# 10.3

## TESTING $H_0$: $\mu_X = \mu_Y$—THE TWO-SAMPLE $t$ TEST

Let $x_1, x_2, \ldots, x_n$ be a random sample of size $n$ from a normal distribution with mean $\mu_X$ and standard deviation $\sigma$. Let $y_1, y_2, \ldots, y_m$ be a random sample of size $m$ from a normal distribution with mean $\mu_Y$ and standard deviation $\sigma$ (see Figure 10.2). The $x_i$'s are assumed to be independent of the $y_i$'s, and the parameters $\mu_X$, $\mu_Y$, and $\sigma$ are assumed to be unknown. Our objective is to test $H_0$: $\mu_X = \mu_Y$. In practice, testing $H_0$: $\mu_X = \mu_Y$ is the most common of all two-sample problems because the true average

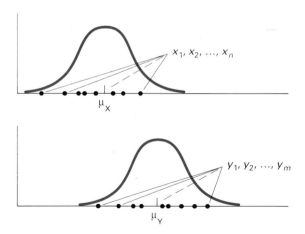

**FIG. 10.2**

response associated with a treatment (i.e., $\mu_X$ or $\mu_Y$) is generally perceived to be the best single-number descriptor of the effectiveness of that treatment (recall our discussion of expected values in Section 6.6). That being the case, it should not be surprising that a comparison of two treatments often comes down to a comparison of two means.

The statistic for testing $\mu_X = \mu_Y$ is similar to what we used for $H_0: \mu = \mu_0$ in Section 9.2. Note, first, that if $H_0: \mu_X = \mu_Y$ is true, then $\mu_X - \mu_Y = 0$, and we would expect $\bar{x} - \bar{y}$ to be "close" to zero (since $\bar{x}$ estimates $\mu_X$ and $\bar{y}$ estimates $\mu_Y$). How far away from zero must $\bar{x} - \bar{y}$ be before we feel obligated to reject $H_0$? For the one-sample problem we answered a similar question by forming a (one-sample) $t$ ratio

$$\frac{\bar{y} - \mu_0}{s/\sqrt{n}}$$

Here we set up a **two-sample** $t$ ratio.

**two-sample $t$ test**—A procedure for testing $H_0: \mu_X = \mu_Y$. Data for a two-sample $t$ test should consist of independent random samples from two normal distributions, each having the same variance.

---

### THEOREM 10.1

Let $x_1, x_2, \ldots, x_n$ and $y_1, y_2, \ldots, y_m$ be independent random samples from two normal distributions with means $\mu_X$ and $\mu_Y$, respectively, and with the same standard deviation $\sigma$. Let

$$s_p = \sqrt{\frac{\sum\limits_{i=1}^{n} X_i^2 - \dfrac{\left(\sum\limits_{i=1}^{n} X_i\right)^2}{n} + \sum\limits_{i=1}^{m} Y_i^2 - \dfrac{\left(\sum\limits_{i=1}^{m} Y_i\right)^2}{m}}{n + m - 2}}$$

Then the random variable

$$\frac{\bar{X} - \bar{Y} - (\mu_X - \mu_Y)}{s_p \sqrt{\dfrac{1}{n} + \dfrac{1}{m}}}$$

has a Student $t$ distribution with $n + m - 2$ degrees of freedom. Let

$$t = \frac{\bar{x} - \bar{y}}{s_p \sqrt{\dfrac{1}{n} + \dfrac{1}{m}}}$$

1. To test

$$H_0: \quad \mu_X = \mu_Y$$

vs.

$$H_1: \quad \mu_X \neq \mu_Y$$

at the $\alpha$ level of significance, reject $H_0$ if either **(a)** $t \leqslant -t_{\alpha/2, n+m-2}$ or **(b)** $t \geqslant t_{\alpha/2, n+m-2}$.

2. To test

$$H_0: \quad \mu_X = \mu_Y$$

vs.

$$H_1: \quad \mu_X > \mu_Y$$

at the $\alpha$ level of significance, reject $H_0$ if $t \geqslant t_{\alpha, n+m-2}$.

3. To test

$$H_0: \quad \mu_X = \mu_Y$$

vs.

$$H_1: \quad \mu_X < \mu_Y$$

at the $\alpha$ level of significance, reject $H_0$ if $t \leqslant -t_{\alpha, n+m-2}$.

---

**pooled standard deviation ($s_p$)**—The square root of a weighted average of the sample variances of the $x$'s and $y$'s. It is an estimate of the unknown $\sigma$ that is presumed to be the same for the $X$ and $Y$ distributions.

COMMENT:   The $s_p$ that appears in the formula for the two-sample $t$ ratio is often called the ***pooled standard deviation***. It estimates the unknown $\sigma$ by combining (or "pooling") the dispersion information contained in the $x_i$'s with the dispersion information contained in the $y_i$'s. More specifically, $s_p$ is a weighted average of the two sample variances $s_X^2$ and $s_Y^2$:

$$s_p = \sqrt{\frac{(n-1)s_X^2 + (m-1)s_Y^2}{n+m-2}}$$

---

## CASE STUDY 10.1

*Mark Twain Revisited*

Disputed authorship was a topic introduced in Case Study 3.5. The focus there was whether Mark Twain and Quintus Curtius Snodgrass were different people or whether Snodgrass was simply a pen name, a creation of Twain's fertile imagination.

Our initial attempt at dealing with that question was necessarily graphical. It was pointed out that authors have distinctive patterns with which they use words of various lengths (see Figure 3.6). If Twain and Snodgrass were different people, we would expect their word-length relative frequency polygons to be different.

Conceptually, making a visual comparison of two sets of frequency polygons seems like a reasonable approach to take; analytically, it leaves a lot to be desired. How far apart must the polygons be for us to reject the null hypothesis that Twain and Snodgrass are one and the same? Answering that is not easy if we insist on considering the entire word-length pattern. A strategy that lends itself much better to a statistical analysis is to focus on words of one particular length.

Table 10.1 shows the proportions of three-letter words found in 8 Twain essays and in the 10 Snodgrass essays. (Each Twain essay was written at approximately the same time the Snodgrass essays appeared.) Let

$$x_1 = 0.225, \quad x_2 = 0.262, \quad \ldots, \quad x_8 = 0.217$$

**TABLE 10.1 Proportion of Three-Letter Words**

| Twain Letters | Proportion | QCS | Proportion |
|---|---|---|---|
| Sergeant Fathom | 0.225 | I | 0.209 |
| Madame Caprell | 0.262 | II | 0.205 |
| Letters in *Territorial Enterprise* | | III | 0.196 |
| First | 0.217 | IV | 0.210 |
| Second | 0.240 | V | 0.202 |
| Third | 0.230 | VI | 0.207 |
| Fourth | 0.229 | VII | 0.224 |
| First *Innocents Abroad* | | VIII | 0.223 |
| First half | 0.235 | IX | 0.220 |
| Second half | 0.217 | X | 0.201 |

We will think of these numbers as a random sample of size $n = 8$ from the distribution of all three-letter-word proportions associated with Twain's essays. We will further suppose that this distribution is normal. Call its (unknown) mean $\mu_X$. Similarly, write

$$y_1 = 0.209, \quad y_2 = 0.205, \quad \ldots, \quad y_{10} = 0.201$$

The unknown mean of the presumed normal distribution that the $y_i$'s represent will be denoted $\mu_Y$.

If Twain and Snodgrass are the same person, $\mu_X$ should equal $\mu_Y$. Therefore, to get a statistical solution to our disputed authorship problem, we should test

$$H_0: \quad \mu_X = \mu_Y$$

vs.

$$H_1: \quad \mu_X \neq \mu_Y$$

Notice that the alternative hypothesis here is two-sided because deviations in either "direction" from $\mu_X = \mu_Y$ suggest that Quintus Curtius Snodgrass was not a nom de plume of Mark Twain.

From Table 10.1,

$$\sum_{i=1}^{8} x_i = 1.855 \qquad \sum_{i=1}^{8} x_i^2 = 0.4316$$

and

$$\sum_{i=1}^{10} y_i = 2.097 \qquad \sum_{i=1}^{10} y_i^2 = 0.4406$$

A little further computation shows that the two sample means are 0.232 and 0.210 and the pooled standard deviation is 0.012:

$$\bar{x} = \frac{1.855}{8} = 0.232 \qquad \bar{y} = \frac{2.097}{10} = 0.210$$

$$s_p = \sqrt{\frac{0.4316 - \dfrac{(1.855)^2}{8} + 0.4406 - \dfrac{(2.097)^2}{10}}{8 + 10 - 2}} = 0.012$$

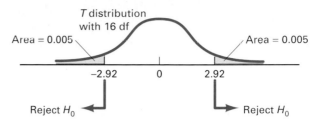

**FIG. 10.3**

Notice that $\bar{x} - \bar{y} = 0.022$. How should we interpret that figure? Is it close enough to zero to be compatible with $H_0$: $\mu_X = \mu_Y$, or is it so far away from zero that we need to reject $H_0$? A two-sample $t$ test will make the decision for us.

Let $\alpha = 0.01$. Looking at the Student $t$ distribution with 16 ($= 8 + 10 - 2$) degrees of freedom, we see that the two critical values are $\pm t_{0.005,\,16} = \pm 2.92$ (see Figure 10.3). The observed $t$ ratio is

$$t = \frac{\bar{x} - \bar{y}}{s_p \sqrt{\dfrac{1}{n} + \dfrac{1}{m}}} = \frac{0.232 - 0.210}{0.012 \sqrt{\dfrac{1}{8} + \dfrac{1}{10}}} = 3.86$$

implying that we should reject $H_0$—it would appear that Mark Twain and Quintus Curtius Snodgrass were not the same person.

---

## CASE STUDY 10.2

*Poverty Point Society*

Poverty Point is the name given to a number of widely scattered archaeological sites in Louisiana, Mississippi, and Arkansas. These sites are the remains of a society thought to have flourished from 1700 to 500 B.C. Among their characteristic artifacts were ornaments fashioned out of clay and then baked. Because of a property of clay known as thermoluminescence, anthropologists can use these ornaments to learn something about the comparative history of the Poverty Point settlements (77).

When certain substances are heated, they emit light in proportion to the amount of radiation to which they have been exposed. This is called *thermoluminescence*. When these same substances are heated to a high enough temperature (that is, *annealed*), they "lose" whatever exposure had been previously accumulated. Annealing is what happened to the Poverty Point ornaments when they were first baked over 2000 years ago. Each of them now provides a record of the total cosmic and background radiation it received since the time of its baking. By calibrating samples of clay to determine how much thermoluminescence is produced for a given amount of incident radiation, scientists can estimate the age of the artifacts.

**TABLE 10.2**

THERMOLUMINESCENT DATES (years B.C.)

| Terral Lewis Estimates, $x_i$ | Jaketown Estimates, $y_i$ |
|---|---|
| 1492 | 1346 |
| 1169 | 942 |
| 883 | 908 |
| 988 | 858 |

Table 10.2 gives the estimated ages of eight clay ornaments: four from the Terral Lewis settlement and four from Jaketown. The question to be answered is whether the technologies at these two places evolved at similar rates, as measured by when they were capable of making kiln-fired ornaments. That is, we want to use the property of thermoluminescence to date these two sites.

Let the true means of the distributions represented by the $x$'s and $y$'s in Table 10.2 be denoted by $\mu_X$ and $\mu_Y$, respectively. If Terral Lewis and Jaketown experienced simultaneously identical growth patterns, then $\mu_X = \mu_Y$. What we should test, therefore, is

$$H_0: \quad \mu_X = \mu_Y$$

vs.

$$H_1: \quad \mu_X \neq \mu_Y$$

(Why is the alternative hypothesis two-sided?) We will choose $\alpha = 0.05$ as the level of significance.

Note, first, that

$$\sum_{i=1}^{4} x_i = 4532 \qquad \sum_{i=1}^{4} x_i^2 = 5{,}348{,}458$$

$$\sum_{i=1}^{4} y_i = 4054 \qquad \sum_{i=1}^{4} y_i^2 = 4{,}259{,}708$$

so that

$$\bar{x} = \frac{4532}{4} = 1133.0 \qquad \bar{y} = \frac{4054}{4} = 1013.5$$

and

$$s_p = \sqrt{\frac{5{,}348{,}458 - \dfrac{(4532)^2}{4} + 4{,}259{,}708 - \dfrac{(4054)^2}{4}}{4 + 4 - 2}} = 246.5$$

The test statistic is then

$$t = \frac{\bar{x} - \bar{y}}{s_p \sqrt{1/n + 1/m}} = \frac{1133.0 - 1013.5}{246.5\sqrt{\frac{1}{4} + \frac{1}{4}}} = 0.68$$

According to Theorem 10.1, we reject $H_0$ if

$$t \leqslant -t_{\alpha/2,\,n+m-2} = -t_{0.025,6} = -2.45$$

or

$$t \geqslant t_{\alpha/2,\,n+m-2} = t_{0.025,6} = 2.45$$

Since 0.68 lies between $-2.45$ and $2.45$, we accept the null hypothesis. The $t$ test is telling us that the difference between the two estimated "ages" (1133.0 B.C. and 1013.5 B.C.) is not statistically significant.

## CASE STUDY 10.3

*Mercury Poisoning*     Mercury pollution is a serious ecological problem. Much of the mercury released into the environment originates as a by-product of coal burning and other industrial processes. It does not become really dangerous until it falls into large bodies of water, where microorganisms change it into methylmercury ($CH_3^{203}$), a particularly toxic organic form. Fish then become contaminated, and anyone eating those fish is at risk.

The severity of the medical problems that arise when an individual ingests too much methylmercury has prompted researchers to investigate mercury poisoning quite extensively. Among the topics on their agenda has been methylmercury ($CH_3^{203}$) metabolism, and whether it proceeds at a different rate for women than for men. Table 10.3 summarizes a study in which six females and nine males were each given an oral administration of protein-bound $CH_3^{203}$. The numbers listed are the resulting biological half-lives of the $CH_3^{203}$ in their systems (105). Is there evidence here that women metabolize $CH_3^{203}$ at a different rate than men do?

**TABLE 10.3**

METHYLMERCURY ($CH_3^{203}$) HALF-LIVES (IN DAYS)

| Females, $x_i$ | Males, $y_i$ |
| --- | --- |
| 52 | 72 |
| 69 | 88 |
| 73 | 87 |
| 88 | 74 |
| 87 | 78 |
| 56 | 70 |
| | 78 |
| | 93 |
| | 74 |

Let $\mu_X$ and $\mu_Y$ represent the true average $CH_3^{203}$ metabolism rates for women and for men, respectively. Our objective is to test

$$H_0: \quad \mu_X = \mu_Y$$

vs.

$$H_1: \quad \mu_X \neq \mu_Y$$

Let $\alpha = 0.05$.

Since

$$\sum_{i=1}^{6} x_i = 425 \qquad \sum_{i=1}^{6} x_i^2 = 31{,}243$$

and

$$\sum_{i=1}^{9} y_i = 714 \qquad \sum_{i=1}^{9} y_i^2 = 57{,}166$$

it follows that

$$\bar{x} = \frac{425}{6} = 70.8 \qquad \bar{y} = \frac{714}{9} = 79.3$$

and

$$s_p = \sqrt{\dfrac{31{,}243 - \dfrac{(425)^2}{6} + 57{,}166 - \dfrac{(714)^2}{9}}{6 + 9 - 2}} = 11.3$$

With $13 \ (= 6 + 9 - 2)$ degrees of freedom the critical values are $\pm 2.160$ (see Figure 10.4).

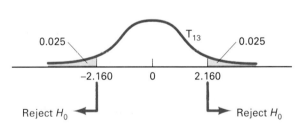

**FIG. 10.4**

Is the difference between the two sample means (70.8 and 79.3) statistically significant? No, because

$$t = \frac{70.8 - 79.3}{11.3\sqrt{\frac{1}{6} + \frac{1}{9}}} = -1.43$$

which falls between the two critical values. Our conclusion, therefore, is that the magnitude of the observed difference between the *sample* means is not incompatible with the hypothesis that the two *true* means are equal. That is, we should accept $H_0$.

QUESTIONS

**10.1**    Suppose two independent random samples of sizes $n$ and $m$, drawn from two normal distributions, give the results summarized below. Set up and carry out a two-sample $t$ test at the indicated level of significance. In each case use a two-sided alternative.

(a) $\bar{x} = 16.0$,    $\bar{y} = 14.6$,    $s_p = 1.2$,    $n = 5$,    $m = 6$,    $\alpha = 0.10$

(b) $\bar{x} = 42.9$,    $\bar{y} = 68.1$,    $s_p = 34.6$,    $n = 10$,    $m = 8$,    $\alpha = 0.05$

(c) $\bar{x} = 106.3$,    $\bar{y} = 104.8$,    $s_p = 0.9$,    $n = 21$,    $m = 21$,    $\alpha = 0.01$

**10.2**    Suppose the pooled standard deviation computed from independent random samples of sizes 6 and 10, respectively, is expected to be 4.9. What is the smallest difference between $\bar{x}$ and $\bar{y}$ that would allow $H_0: \mu_X = \mu_Y$ to be rejected in favor of $H_1: \mu_X > \mu_Y$ at the $\alpha = 0.05$ level of significance? (Hint: Write out the usual decision rule and "solve" for $\bar{x} - \bar{y}$.)

**10.3**    See Case Study 2.4. Let $\mu_X$ and $\mu_Y$ denote the true average silver contents in "early" coins and "late" coins, respectively. Test $H_0: \mu_X = \mu_Y$ versus $H_1: \mu_X \neq \mu_Y$ at the $\alpha = 0.05$ level of significance. Note:

$$\sum_{i=1}^{9} x_i = 60.7 \qquad \sum_{i=1}^{9} x_i^2 = 411.75$$

$$\sum_{i=1}^{7} y_i = 39.3 \qquad \sum_{i=1}^{7} y_i^2 = 221.43$$

**10.4**    For the first walking times in Table 2.5, test $H_0: \mu_X = \mu_Y$ versus $H_1: \mu_X < \mu_Y$ at the $\alpha = 0.05$ level of significance. Note:

$$\sum_{i=1}^{6} x_i = 60.75 \qquad \sum_{i=1}^{6} x_i^2 = 625.56$$

$$\sum_{i=1}^{6} y_i = 68.25 \qquad \sum_{i=1}^{6} y_i^2 = 794.31$$

Why is the alternative hypothesis one-sided?

**10.5**    Set up and carry out a two-sample $t$ test comparing the caffeine contents of spray-dried and freeze-dried coffee in Question 2.19. Let 0.01 be the level of significance.

**10.6**    For the "spectropenetration gradients" in Question 2.20,

$$\sum_{i=1}^{8} x_i = 43.4 \qquad \sum_{i=1}^{8} x_i^2 = 239.32$$

$$\sum_{i=1}^{8} y_i = 36.1 \qquad \sum_{i=1}^{8} y_i^2 = 166.95$$

Test $H_0: \mu_X = \mu_Y$ against a two-sided alternative. Let 0.05 be the level of significance.

**10.7**  See Case Study 3.11. Let $\mu_X$ and $\mu_Y$, respectively, denote the true average numbers of stories showing a positive parent-child relationship for mothers of normal children and mothers of schizophrenic children. At the 0.05 level test that those two means are equal. Use a two-sided alternative.

**10.8**  Using the values of $\bar{x}$, $\bar{y}$, $s_X$, and $s_Y$ given in Case Study 3.12, test whether the difference in life spans for short presidents and tall presidents is statistically significant. Let $\alpha = 0.05$ and use an appropriate one-sided alternative. Justify your choice of $H_1$.

**10.9**  If $\bar{x}$ and $\bar{y}$ are the sample means of $n$ and $m$ observations, respectively, drawn from normal distributions with *known* variances $\sigma_X^2$ and $\sigma_Y^2$, then the appropriate statistic for testing $H_0: \mu_X = \mu_Y$ is

$$\frac{\bar{x} - \bar{y}}{\sqrt{\dfrac{\sigma_X^2}{n} + \dfrac{\sigma_Y^2}{m}}}$$

which has a Z distribution. For the ACT discussion in Example 8.2, suppose that the standard deviation of scores for one ethnic group ($\sigma_X$) is 6.0 and that for another group ($\sigma_Y$) is 6.5. Six students from the first group and eight from the second have average ACT composites of 17.5 and 18.2, respectively. At the 0.01 level is the difference between those two averages statistically significant?

**10.10**  If random samples $X_1, X_2, \ldots, X_n$ and $Y_1, Y_2, \ldots, Y_m$ come from two normal distributions with means $\mu_X$ and $\mu_Y$ and the same standard deviation $\sigma$, then

$$\frac{\bar{X} - \bar{Y} - (\mu_X - \mu_Y)}{s_p \sqrt{\dfrac{1}{n} + \dfrac{1}{m}}}$$

has a Student $t$ distribution with $n + m - 2$ degrees of freedom. Derive a formula for a $100(1 - \alpha)\%$ confidence interval for $\mu_X - \mu_Y$. (Hint: Follow the sequence of steps that led to Theorem 9.2.)

**10.11**  Based on the answer to Question 10.10, construct a 95% confidence interval for $\mu_X - \mu_Y$, using the data in Question 3.52. Let $\mu_X$ and $\mu_Y$ denote the true average nod-swimming frequencies in American teals and European teals, respectively.

**10.12**  Case Study 9.5 pointed out that confidence intervals are related to hypothesis tests. Suppose a 95% confidence interval is constructed for $\mu_X - \mu_Y$.

(a)  How could that interval be used to test $H_0: \mu_X = \mu_Y$ versus $H_1: \mu_X \neq \mu_Y$?

(b)  What would be the level of significance?

## 10.4

### TESTING $H_0: \sigma_X^2 = \sigma_Y^2$—THE $F$ TEST (Optional)

**$F$ test**—A procedure for testing $H_0: \sigma_X^2 = \sigma_Y^2$. If both populations sampled are normal and have the same variance, then $S_Y^2/S_X^2$ has an $F$ distribution (with $m - 1$ and $n - 1$ degrees of freedom). Values of $s_Y^2/s_X^2$ too far from 1 are evidence that $H_0: \sigma_X^2 = \sigma_Y^2$ is not true.

For two-sample problems where the data are presumed to be random samples from normal distributions, the most common inference procedure—by far—is the two-sample $t$ test described in Section 10.3. Still, there are situations where the *variances* of the two distributions, rather than their means, are the parameters of primary interest. Suppose the foreman of an assembly line has to recommend the purchase of one of two different machines ($X$ and $Y$) for cutting metal strips into pieces 10 cm long. To get some idea of the output quality that might be expected from each machine, he measures the actual lengths, $x_1, \ldots, x_n$ and $y_1, \ldots, y_m$, of a sample of pieces cut by $X$ and by $Y$, respectively. How should he use that information to make his recommendation?

The true average length associated with each machine ($\mu_X$ and $\mu_Y$) would certainly be important parameters, and he would be well-advised to begin his analysis by testing $H_0: \mu_X = 10$ and $H_0: \mu_Y = 10$ (using the one-sample $t$ test of Section 9.2). But the means here are not the whole story. Figure 10.5 pictures a pair of distributions for which $\mu_X = \mu_Y$ but where $\sigma_Y$ is considerably larger than $\sigma_X$, the result being that Machine $Y$ produces a much greater proportion of pieces that are either too short or too long (and, therefore, unusable). Clearly, it makes sense here to test $H_0: \sigma_X^2 = \sigma_Y^2$.

Mathematically, testing variances proceeds along substantially different lines than does testing means. (This is true even for one-sample problems—recall Sections 9.2 and 9.5.) We have seen that the keys in testing $H_0: \mu_X = \mu_Y$ are the sample means—specifically, the *difference* in the sample means, $\bar{x} - \bar{y}$. In contrast, for testing $H_0: \sigma_X^2 = \sigma_Y^2$ we focus on the *ratio* of the sample variances, $s_Y^2/s_X^2$.

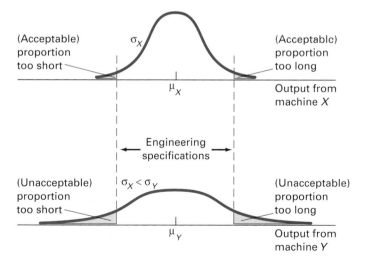

**FIG. 10.5**

---

**DEFINITION 10.1**

Let $X_1, X_2, \ldots, X_n$ be a random sample from a normal distribution with mean $\mu_X$ and variance $\sigma_X^2$. Let $Y_1, Y_2, \ldots, Y_m$ be an independent random sample from a normal distribution with mean $\mu_Y$ and variance $\sigma_Y^2$. Let $S_X^2$ and $S_Y^2$ be the two sample variances:

$$S_X^2 = \frac{1}{n-1} \sum_{i=1}^{n} (X_i - \bar{X})^2 = \frac{n \sum_{i=1}^{n} X_i^2 - \left(\sum_{i=1}^{n} X_i\right)^2}{n(n-1)}.$$

$$S_Y^2 = \frac{1}{m-1} \sum_{i=1}^{m} (Y_i - \bar{Y})^2 = \frac{m \sum_{i=1}^{m} Y_i^2 - \left(\sum_{i=1}^{m} Y_i\right)^2}{m(m-1)}$$

Then

$$F = \frac{S_Y^2/\sigma_Y^2}{S_X^2/\sigma_X^2}$$

has an *F distribution with $m-1$ and $n-1$ degrees of freedom.* The symbol for any such random variable is $F_{m-1, n-1}$.

---

In general, $F$ curves have shapes similar to $\chi^2$ distributions: they can never be negative, they extend to $+\infty$, and they are skewed to the right (see Figure 10.6). Appendix A.4 gives the numbers that cut off areas, *to their right*, of 0.01, 0.025, and 0.05 under $F$ distributions having various degrees of freedom.

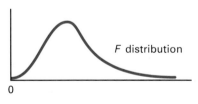

**FIG. 10.6**

---

*EXAMPLE 10.1*    Suppose an $X$ distribution and a $Y$ distribution, both presumed normal, are represented by samples of sizes $n = 11$ and $m = 9$, respectively. Let $S_X^2$, $\sigma_X^2$, $S_Y^2$, and $\sigma_Y^2$ be the corresponding sample variances and (unknown) true variances. Let

$$F = \frac{S_Y^2/\sigma_Y^2}{S_X^2/\sigma_X^2}$$

What is the 95th percentile of $F$?

*Solution*    According to Definition 10.1,

$$F = \frac{S_Y^2/\sigma_Y^2}{S_X^2/\sigma_X^2}$$

should have an $F$ distribution with 8 and 10 degrees of freedom. In Appendix A.4a, the intersection of the column labeled 8 and the row labeled 10, 3.07, is the value that cuts off an area of 0.05 to its right. Equivalently, 3.07 is the 95th percentile of the $F_{8,10}$ distribution (see Figure 10.7).

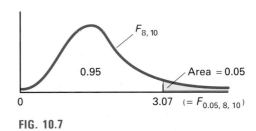

**FIG. 10.7**

The symbol $F_{\alpha, m-1, n-1}$ denotes the number cutting off an area of $\alpha$ *to its right* under an $F$ distribution with $m-1$ and $n-1$ degrees of freedom. Here, $F_{0.05, 8, 10} = 3.07$.

COMMENT:   Additional examples in the use of Appendix A.4 are worked in Section 7.6. Note that Definition 10.1 is a special case of Definition 7.3. The quotient

$$\frac{(m-1)S_Y^2/\sigma_Y^2}{m-1} = \frac{S_Y^2}{\sigma_Y^2}$$

for example, is equivalent to $U/(m-1)$. (Why?)

Once we know the sampling distribution that describes the behavior of $(S_Y^2/\sigma_Y^2)/(S_X^2/\sigma_X^2)$, it remains a simple matter to set up a procedure for testing $H_0$: $\sigma_X^2 = \sigma_Y^2$. The rationale is clear. If $H_0$: $\sigma_X^2 = \sigma_Y^2$ is true, then

$$\frac{S_Y^2/\sigma_Y^2}{S_X^2/\sigma_X^2}$$

reduces to $S_Y^2/S_X^2$ and we would expect the latter to be close to 1. If the ratio of the sample variances is *not* close to 1—as measured by the $F_{m-1, n-1}$ distribution—it makes sense to reject the null hypothesis that the true variances are equal. Theorem 10.2 gives the details.

---

### THEOREM 10.2

Let $x_1, x_2, \ldots, x_n$ be a random sample from a normal distribution with mean $\mu_X$ and variance $\sigma_X^2$. Let $y_1, y_2, \ldots, y_m$ be an independent random sample from a normal distribution with mean $\mu_Y$ and variance $\sigma_Y^2$. Define

$$F = \frac{s_Y^2}{s_X^2}$$

1. To test

$$H_0: \quad \sigma_X^2 = \sigma_Y^2$$
vs.
$$H_1: \quad \sigma_X^2 \neq \sigma_Y^2$$

reject $H_0$ at the $\alpha$ level of significance if either (a) $F \leqslant 1/F_{\alpha/2, n-1, m-1}$ or (b) $F \geqslant F_{\alpha/2, m-1, n-1}$.

2. To test

$$H_0: \quad \sigma_X^2 = \sigma_Y^2$$
vs.
$$H_1: \quad \sigma_X^2 > \sigma_Y^2$$

reject $H_0$ at the $\alpha$ level of significance if $F \leqslant 1/F_{\alpha, n-1, m-1}$.

3. To test

$$H_0: \quad \sigma_X^2 = \sigma_Y^2$$
vs.
$$H_1: \quad \sigma_X^2 < \sigma_Y^2$$

reject $H_0$ at the $\alpha$ level of significance if $F \geqslant F_{\alpha, m-1, n-1}$.

---

### CASE STUDY 10.4

***Brain Waves***    Electroencephalograms are records showing fluctuations of electrical activity in the brain. Among the several different kinds of brain "waves" produced, the dominant ones are usually $\alpha$ waves. These have a characteristic frequency of anywhere from 8 to 13 cycles per second.

A group of inmates in a Canadian prison volunteered for a study designed to see whether sensory deprivation has any effect on $\alpha$-wave frequencies. The 20 subjects were randomly split into two groups of equal size. Members of one group were placed in solitary confinement; those in the other group were allowed to remain in their own cells. A week later $\alpha$-wave frequencies were measured for all 20 subjects. Table 10.4 summarizes the results (55).

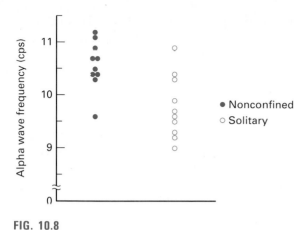

**FIG. 10.8**

**TABLE 10.4** $\alpha$-Wave Frequencies (cps)

| Nonconfined, $x_i$ | Solitary Confinement, $y_i$ |
|---|---|
| 10.7 | 9.6 |
| 10.7 | 10.4 |
| 10.4 | 9.7 |
| 10.9 | 10.3 |
| 10.5 | 9.2 |
| 10.3 | 9.3 |
| 9.6 | 9.9 |
| 11.1 | 9.5 |
| 11.2 | 9.0 |
| 10.4 | 10.9 |
| $\bar{x}$:   10.6 | $\bar{y}$:   9.8 |

In addition to the sample average frequencies being slightly different (10.6 versus 9.8), a graph of the data (Figure 10.8) suggests that the variances of the two groups might not be the same. To examine that possibility formally requires that we do an $F$ test.

Let $\sigma_X^2$ and $\sigma_Y^2$ denote the true variances of $\alpha$-wave frequencies for nonconfined and solitary confined prisoners, respectively. The hypotheses to be tested are

$$H_0: \quad \sigma_X^2 = \sigma_Y^2$$

vs.

$$H_1: \quad \sigma_X^2 \neq \sigma_Y^2$$

Let $\alpha = 0.05$. Given that

$$\sum_{i=1}^{10} x_i = 105.8 \qquad \sum_{i=1}^{10} x_i^2 = 1121.26$$

and

$$\sum_{i=1}^{10} y_i = 97.8 \qquad \sum_{i=1}^{10} y_i^2 = 959.70$$

the sample variances are

$$s_X^2 = \frac{10(1121.26) - (105.8)^2}{10(9)} = 0.21$$

and

$$s_Y^2 = \frac{10(959.70) - (97.8)^2}{10(9)} = 0.36$$

Dividing the sample variances gives an observed $F$ ratio of 1.71:

$$F = \frac{s_Y^2}{s_X^2} = \frac{0.36}{0.21} = 1.71$$

Since $n = \dot{m} = 10$, we would expect $S_Y^2/S_X^2$ to behave like an $F$ random variable with 9 and 9 degrees of freedom (assuming $H_0: \sigma_X^2 = \sigma_Y^2$ is true). From Appendix A.4b, $F_{0.025, 9, 9} = 4.03$. Also,

$$\frac{1}{F_{0.025, 9, 9}} = \frac{1}{4.03} = 0.248$$

According to part 1 of Theorem 10.2, we should reject $H_0$ if $s_Y^2/s_X^2 \leqslant 0.248$ or $s_Y^2/s_X^2 \geqslant 4.03$. Since $F = 1.71$, our conclusion is to accept $H_0$: sensory deprivation has no apparent effect on the variance of $\alpha$-wave frequencies.

COMMENT:   Keep in mind that $1/F_{\alpha/2, n-1, m-1}$, the lower critical value specified in Theorem 10.2, is simply the $(\alpha/2)$nd percentile of the $F_{m-1, n-1}$ distribution. Figure 10.9 shows the two critical values and their relation to the sampling distribution of $S_Y^2/S_X^2$ when $H_0$ is true.

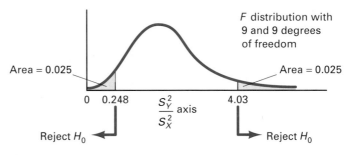

FIG. 10.9

Situations like Case Study 10.4 are not the only occasions where $H_0: \sigma_X^2 = \sigma_Y^2$ is an appropriate hypothesis—testing the equality of two variances is sometimes a necessary prelude to testing the equality of two means. Recall that the two-sample $t$ test of Section 10.3 requires that the $x$'s and $y$'s come from normal distributions with the same variance $\sigma^2$. If either or both of those assumptions are not true, the validity of the $t$ test is suspect. (Specifically, the probability associated with the critical region may not be the 0.01 or 0.05 that we "think" is the value of $\alpha$.) It follows that "testing" these two assumptions (i.e., verifying their reasonableness) should be done *before* doing a $t$ test on $\mu_X$ and $\mu_Y$.

We will learn a procedure for investigating the normality assumption in Chapter 13. The second assumption—that the $x$'s and $y$'s come from (normal) distributions *with the same variance*—can be examined by doing an $F$ test on $H_0: \sigma_X^2 = \sigma_Y^2$ versus $H_1: \sigma_X^2 \neq \sigma_Y^2$.

**EXAMPLE 10.2**   For the methylmercury half-lives in Table 10.3. The two sample variances for those data are considerably different:

for the females

$$s_X^2 = \frac{6(31{,}243) - (425)^2}{6(5)} = 227.77$$

for the males

$$s_Y^2 = \frac{9(57{,}166) - (714)^2}{9(8)} = 65.25$$

Does a sample variance ratio of $65.25/227.77 = 0.29$ invalidate the assumption that the two distributions represented by the $x$'s and $y$'s have the same variance?

**Solution**   To answer that question, we must do an $F$ test. Let $\sigma_X^2$ and $\sigma_Y^2$ denote the true variances associated with the half-life distributions for females and males, respectively. We need to test

$$H_0: \quad \sigma_X^2 = \sigma_Y^2$$

vs.

$$H_1: \quad \sigma_X^2 \neq \sigma_Y^2$$

Take $\alpha = 0.05$.

Since the $X$ sample contains six observations and the $Y$ sample contains nine, the probabilistic behavior of the ratio $S_Y^2/S_X^2$ will be described by an $F$ distribution with 8 and 5 degrees of freedom. Figure 10.10 shows the two critical values (recall part 1 of Theorem 10.2). The right-tail cutoff comes directly from Appendix A.4b:

$$6.76 = F_{0.025,\,8,\,5}$$

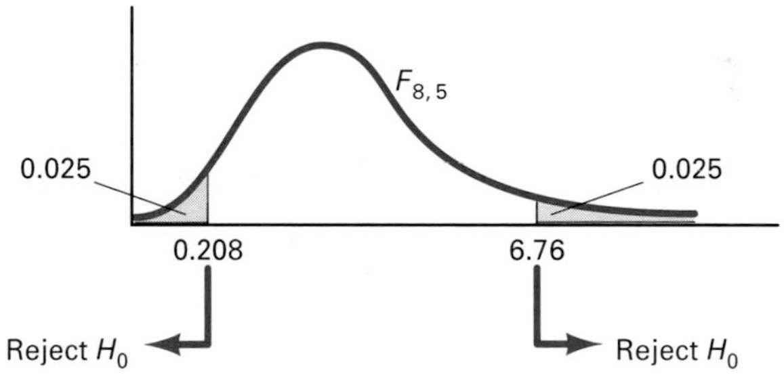

**FIG. 10.10**

The lower critical value is the reciprocal of $F_{0.025,\,5,\,8}$:

$$0.208 = \frac{1}{F_{0.025,\,5,\,8}} = \frac{1}{4.82}$$

Since the observed $F$ ratio ($=0.29$) falls between the two critical values, our conclusion is to accept $H_0$. Using a two-sample $t$ test for $H_0$: $\mu_X = \mu_Y$, therefore, would be perfectly valid (relative, at least, to the variance assumption).

---

QUESTIONS

10.13   (a) Find the first and 99th percentiles of an $F$ distribution with 6 and 9 degrees of freedom.

(b) Find the 2.5th and 97.5th percentiles of an $F$ distribution with 6 and 9 degrees of freedom.

(c) Find the 2.5th and 97.5th percentiles of an $F$ distribution with 9 and 17 degrees of freedom.

10.14   If independent random samples of sizes $n = 6$ and $m = 16$ are taken from two different normal populations with the same variance, what is

(a) the 95th percentile of the distribution of $S_Y^2/S_X^2$?

(b) the first percentile?

10.15   Suppose 6 observations are taken from population $X$, and 13 from population $Y$. If both populations are normal and have the same variance, complete the following equations:

(a) $P\left(? \leqslant \dfrac{S_Y^2}{S_X^2} \leqslant ?\right) = 0.95$     (b) $P\left(\dfrac{S_Y^2}{S_X^2} \geqslant ?\right) = 0.05$

(c) $P\left(\dfrac{S_Y^2}{S_X^2} \leqslant ?\right) = 0.01$

10.16   Suppose independent random samples of sizes $n$ and $m$ give sample variances of $s_X^2$ and $s_Y^2$, respectively. Assume that both distributions being represented are normal. We wish to test $H_0$: $\sigma_X^2 = \sigma_Y^2$. Find the critical value(s) and state your conclusion for each of the following situations:

(a) $H_1$: $\sigma_X^2 \neq \sigma_Y^2$,   $\alpha = 0.05$,   $n = 12$,   $m = 8$,   $s_X^2 = 26.4$,   $s_Y^2 = 10.7$

(b) $H_1$: $\sigma_X^2 > \sigma_Y^2$,   $\alpha = 0.05$,   $n = 6$,   $m = 4$,   $s_X^2 = 14.0$,   $s_Y^2 = 2.7$

(c) $H_1$: $\sigma_X^2 < \sigma_Y^2$,   $\alpha = 0.01$,   $n = 14$,   $m = 13$,   $s_X^2 = 42.8$,   $s_Y^2 = 94.9$

10.17   Two sets of survival times are listed in Question 2.18. The parameters of primary interest would be $\mu_X$ and $\mu_Y$, but the data suggest that a two-sample $t$ test might be inappropriate because the dispersions of the two samples appear to be quite different.

(a) Test $H_0$: $\sigma_X^2 = \sigma_Y^2$ versus $H_1$: $\sigma_X^2 \neq \sigma_Y^2$. Let $\alpha = 0.05$. Note:

$$\sum_{i=1}^{6} x_i = 51 \qquad \sum_{i=1}^{6} x_i^2 = 443$$

$$\sum_{i=1}^{6} y_i = 142 \qquad \sum_{i=1}^{6} y_i^2 = 3672$$

(b) Relative to the variance assumption, would a $t$ test of $H_0: \mu_X = \mu_Y$ be valid?

**10.18**  Test whether the variances of the two distributions of waving times represented by the $x$'s and $y$'s in Table 3.19 are equal. Note that $s_X = 15.7$ and $s_Y = 6.6$. Use a two-sided alternative and let $\alpha = 0.05$.

**10.19**  Refer to the presidential life span data in Case Study 3.12. Is it valid to use a two-sample $t$ test on $H_0: \mu_X = \mu_Y$? Support your answer as quantitatively as possible.

**10.20**  Is it reasonable to assume that the variances of the two populations represented by the end diastolic volumes in Question 3.51 are equal? Test $H_0: \sigma_X^2 = \sigma_Y^2$ against a two-sided alternative. Let $\alpha = 0.02$.

**10.21**  Let $S_X^2$ be the sample variance of $n$ observations from population $X$, and $S_Y^2$ the sample variance of $m$ observations from population $Y$. If the true variances of the two distributions are $\sigma_X^2$ and $\sigma_Y^2$, respectively, then

$$\frac{S_Y^2/\sigma_Y^2}{S_X^2/\sigma_X^2}$$

has an $F$ distribution with $m - 1$ and $n - 1$ degrees of freedom. Use the latter result to derive a formula for a $100(1 - \alpha)\%$ confidence interval for $\sigma_Y^2/\sigma_X^2$.

**10.22**  Suppose the range of values $(a, b)$ is a $100(1 - \alpha)\%$ confidence interval for $\sigma_Y^2/\sigma_X^2$.

(a) How can that interval be used to test $H_0: \sigma_X^2 = \sigma_Y^2$ against $H_1: \sigma_X^2 \neq \sigma_Y^2$?

(b) What would be the level of significance?

**10.23**  Use the answer to Question 10.21 on the thematic apperception data in Case Study 3.11 to construct a 95% confidence interval for $\sigma_Y^2/\sigma_X^2$.

# 10.5

## BINOMIAL DATA—TESTING $H_0: p_X = p_Y$

The third and final type of two-sample problems to be considered involve *binomial* data. These problems arise when the $x_i$'s and $y_i$'s each have the Bernoulli structure described in Section 2.2. That is, each observation equals 1 if the $i$th trial ends in success, and 0 otherwise. Using notation consistent with two-sample problems done earlier in this chapter, we can write

$$X_i = \begin{cases} 1 & \text{with probability } p_X \\ 0 & \text{with probability } 1 - p_X, \end{cases} \quad i = 1, 2, \ldots, n$$

and

$$Y_i = \begin{cases} 1 & \text{with probability } p_Y \\ 0 & \text{with probability } 1 - p_Y, \end{cases} \quad i = 1, 2, \ldots, m$$

In practice, the $X_i$'s represent responses to one treatment (or condition), and the $Y_i$'s responses to a second.

At issue are the unknown parameters $p_X$ and $p_Y$. Apart from representing the probabilities that a given outcome ends in success, $p_X$ and $p_Y$ "characterize" the two treatments in the same sense that $\mu_X$ and $\mu_Y$ characterized treatments in Section 10.3. If, for example, two treatments are equivalent, then $p_X = p_Y$. Therefore, what usually needs to be tested is the null hypothesis $H_0 \colon p_X = p_Y$.

Computing the *sum* of the responses associated with each treatment is the place to begin any two-sample binomial problem. Let

$$x = \text{Total number of successes with first treatment}$$

$$= x_1 + x_2 + \cdots + x_n$$

and

$$y = \text{Total number of successes with second treatment}$$

$$= y_1 + y_2 + \cdots + y_m$$

Then

$$\frac{x}{n} = \text{Sample proportion of trials ending in success}$$
$$\text{with first treatment}$$

and

$$\frac{y}{m} = \text{Sample proportion of trials ending in success}$$
$$\text{with second treatment}$$

If $p_X = p_Y$, then $x/n$ (which estimates $p_X$) should be numerically close to $y/m$ (which estimates $p_Y$). Equivalently, $x/n - y/m$ should be close to zero. With the difference between the sample proportions as the key, we can set up an approximate Z ratio for testing $H_0 \colon p_X = p_Y$.

---

### THEOREM 10.3

Let $X_1, X_2, \ldots, X_n$ be the outcomes of $n$ Bernoulli trials, each with parameter $p_X$. Let $X = X_1 + X_2 + \cdots + X_n$. Similarly, let $Y_1, Y_2, \ldots, Y_m$ be the outcomes of $m$ Bernoulli trials, each with parameter $p_Y$. Let $Y = Y_1 + Y_2 + \cdots + Y_m$. Assume that the $X_i$'s and $Y_i$'s are independent. If $H_0 \colon p_X = p_Y$ is true, then

$$Z = \frac{\left( \dfrac{X}{n} - \dfrac{Y}{m} \right) \sqrt{nm}}{\sqrt{\left( \dfrac{X+Y}{n+m} \right) \left( 1 - \dfrac{X+Y}{n+m} \right)(n+m)}}$$

has approximately a standard normal distribution.

1. To test

$$H_0: \quad p_X = p_Y$$
vs.
$$H_1: \quad p_X \neq p_Y$$

at the $\alpha$ level of significance, reject $H_0$ if either (a) $z \leqslant -z_{\alpha/2}$ or (b) $z \geqslant z_{\alpha/2}$.

2. To test

$$H_0: \quad p_X = p_Y$$
vs.
$$H_1: \quad p_X > p_Y$$

at the $\alpha$ level of significance, reject $H_0$ if $z \geqslant z_\alpha$.

3. To test

$$H_0: \quad p_X = p_Y$$
vs.
$$H_1: \quad p_X < p_Y$$

at the $\alpha$ level of significance, reject $H_0$ if $z \leqslant -z_\alpha$.

---

## CASE STUDY 10.5

*Kittiwake Mating*    The kittiwake is a seagull whose mating behavior is basically monogamous. Normally, the birds separate for several months after the completion of one breeding season and reunite at the beginning of the next. Whether the birds actually do reunite, though, may be affected by the success of their relationship" the season before.

A total of 769 kittiwake pair-bonds were studied (31) over the course of two breeding seasons; 609 successfully bred during the first season, while the remaining 160 were unsuccessful. The following season, 175 of the previously successful pair-bonds "divorced," as did 100 of the 160 whose previous relationship left something to be desired (see Table 10.5). Can we

**TABLE 10.5**

|  | BREEDING IN PREVIOUS YEAR | |
|---|---|---|
|  | Successful | Unsuccessful |
| Number Divorced | 175 | 100 |
| Number Not Divorced | 434 | 60 |
| Total | 609 | 160 |
| Percent Divorced | 29 | 63 |

conclude that the two divorce rates (29% versus 63%) are significantly different?

Note that the figures given here are *sums* of responses rather than individual $x_i$'s and $y_i$'s. In the terminology of Theorem 10.3, the first random sample represents kittiwakes whose breeding the previous year was successful. Each of the $n = 609$ $x_i$'s in that sample was either a 0 or a 1; it was assigned a 1 if that pair of birds divorced the next year, and a 0 if they stayed together. According to the table, a total of $x = 175$ of the $x_i$'s were ones and 434 were zeros. Similarly, the $m = 160$ $y_i$'s consisted of $y = 100$ ones and 60 zeros. The percent divorced row is simply $100(x/n)$ and $100(y/m)$.

Let $p_X$ denote the true proportion of kittiwakes who divorce the second year after having bred *successfully* the season before. Equivalently, $p_X = P(X_i = 1)$. The parameter $p_Y$ is defined analogously as the true proportion of kittiwakes who divorce the second year after having bred *unsuccessfully* the season before. Our objective is to test

$$H_0: \quad p_X = p_Y$$
vs.
$$H_1: \quad p_X \neq p_Y$$

If we accept $H_0$, we will be accepting the premise that the probability of a pair of kittiwakes getting divorced this year is unrelated to whether they bred successfully last year.

Let $\alpha = 0.05$. The two critical values are $\pm z_{0.025} = \pm 1.96$. From the data in Table 10.5, the test statistic equals $-7.93$:

$$z = \frac{\left(\dfrac{175}{609} - \dfrac{100}{160}\right)\sqrt{(609)(160)}}{\sqrt{\left(\dfrac{175 + 100}{609 + 160}\right)\left(1 - \dfrac{175 + 100}{609 + 160}\right)(609 + 160)}} = -7.93$$

Our conclusion is clear—reject $H_0$ (see Figure 10.11).

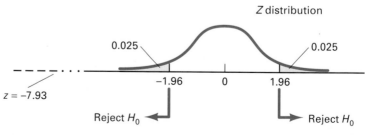

FIG. 10.11

**CASE STUDY 10.6**

*Nightmares*    Over the years numerous studies have sought to characterize the nightmare sufferer. Out of these has emerged the stereotype of someone with high anxiety, low ego strength, feelings of inadequacy, and poorer-than-average physical health. What is not so well known, though, is whether men fall into this pattern with the same frequency as women. Addressing that question, Hersen (72) looked at nightmare frequencies for a sample of 160 men and 192 women. Each subject was asked whether he or she experienced nightmares "often" (at least once a month) or "seldom" (less than once a month). The findings are summarized in Table 10.6.

**TABLE 10.6**

|                   | Men | Women |
|-------------------|-----|-------|
| Nightmares Often  | 55  | 60    |
| Nightmares Seldom | 105 | 132   |
| Totals            | 160 | 192   |

Let $p_M$ and $p_W$ denote the true proportions of men having nightmares often and women having nightmares often, respectively. The hypotheses to be tested are

$$H_0: \quad p_M = p_W$$

vs.

$$H_1: \quad p_M \neq p_W$$

If $\alpha$ is set at 0.01, the critical values become $\pm z_{0.005} = \pm 2.58$. Let men be the $X$ sample, and women the $Y$ sample. Define $X_i$ or $Y_i$ to be 1 if the $i$th subject experienced nightmares often, and 0 otherwise. In the terminology of Theorem 10.3, $x = 55$, $n = 160$, $y = 60$, and $m = 192$.

Substituting into the formula for the test statistic, we find

$$z = \frac{\left(\dfrac{55}{160} - \dfrac{60}{192}\right)\sqrt{(160)(192)}}{\sqrt{\left(\dfrac{55 + 60}{160 + 192}\right)\left(1 - \dfrac{55 + 60}{160 + 192}\right)(160 + 192)}} = 0.64$$

Since $z = 0.64$ falls between $-2.58$ and $+2.58$, we should accept $H_0$. These data provide no convincing evidence that frequencies of nightmares are different for men than for women.

**CASE STUDY 10.7**

*Law and Order*   Law and order was a key issue for the presidential contenders in 1968, particularly George Wallace. He and his supporters were the sharpest critics of the courts and the strongest advocates of a renewed commitment to vigorous law enforcement. Whenever one segment of society moralizes to another, though, there is a natural tendency for the accused to question whether the accusers are, themselves, above reproach. "Wallaceites" talked a good game of law and order, but did they practice what they preached? In Nashville, Tennessee, a team of sociologists carried out a rather unorthodox survey that seemed to indicate that maybe the Wallaceites didn't (186).

Before the general election of 1968, the government of Nashville-Davidson County had passed a law requiring any locally operated vehicle to have a "Metro sticker" (costing $15) displayed on its windshield. It was far from clear, though, how strictly the law would be enforced. Many motorists felt they could ignore the ordinance and not get caught.

For several days following the sticker deadline, investigators made spot checks of parking lots in and around Nashville to see if supporters of the candidates differed significantly in their compliance with the law. Table 10.7 shows the results for Wallace and the most liberal of his competitors, Hubert Humphrey.

**TABLE 10.7**

| In Support of[a] | Number of Cars | Number with Stickers |
|---|---|---|
| Humphrey | $n = 178$ | $x = 154$ |
| Wallace | $m = 361$ | $y = 270$ |

[a] A car was assumed to be owned by a Humphrey supporter, for example, if it displayed a Humphrey bumper sticker.

The sample proportions of Humphrey and Wallace supporters obeying the ordinance were $154/178 = 0.865$ and $270/361 = 0.748$, respectively. Presumably, those two figures are reasonable estimates of $p_X$ and $p_Y$, the *true* proportions of Humphrey and Wallace cars bearing a Metro sticker. Can we conclude that $p_X$ and $p_Y$ are not equal, as the two sample proportions would suggest, or is the observed difference of 0.117 ($= 0.865 - 0.748$) well within the normal bounds of sampling variability? Answering that requires that we test

$$H_0: \quad p_X = p_Y.$$

vs.

$$H_1: \quad p_X \neq p_Y$$

Let $\alpha = 0.01$.

Note, first, that the overall proportion of cars with Metro stickers is

$$\frac{154 + 178}{270 + 361} = \frac{424}{539} = 0.787$$

From Theorem 10.3 the observed Z ratio is 3.12

$$z = \frac{(0.865 - 0.748)\sqrt{(178)(361)}}{\sqrt{(0.787)(1 - 0.787)(539)}} = 3.12$$

The appropriate critical values are $\pm z_{0.005} = \pm 2.58$. Since $3.12 > 2.58$, our conclusion is to reject the null hypothesis: it would appear that the proportion of Wallace supporters breaking the law by not having a Metro sticker was significantly higher than the corresponding proportion of Humphrey supporters.

---

QUESTIONS

**10.24** Two independent sets of Bernoulli trials of sizes $n$ and $m$ yield $x$ and $y$ successes, respectively. We wish to test $H_0: p_X = p_Y$. Find the critical value(s) and carry out the hypothesis test for the following situations:

(a) $H_1: p_X \neq p_Y$, $\alpha = 0.05$, $x = 52$, $n = 100$, $y = 46$, $m = 115$

(b) $H_1: p_X > p_Y$, $\alpha = 0.01$, $x = 32$, $n = 76$, $y = 61$, $m = 200$

(c) $H_1: p_X < p_Y$, $\alpha = 0.10$, $x = 50$, $n = 114$, $y = 42$, $m = 71$

**10.25** The phenomenon of handedness has been studied extensively in humans. The percentages of adults who are right-handed, left-handed, and ambidextrous are well-documented. What is not so well known is that a similar phenomenon is present in lower animals. Dogs, for example, can be right-pawed or left-pawed. Suppose that in a random sample of 200 beagles it is found that 55 are left-pawed and that in a random sample of 200 collies 40 are left-pawed. Can we conclude that the true proportion of collies that are left-pawed is different from the true proportion of beagles that are left-pawed? Let $\alpha = 0.05$.

**10.26** A utility infielder for a National League baseball team batted .260 last season in 300 trips to the plate. This year he hit .250 in 200 at-bats. The owners are trying to cut his pay for next year on the grounds that his output has deteriorated. The player argues that his performances the last two seasons have not been significantly different, so his salary should not be reduced. Who is right? Defend your position quantitatively.

**10.27** An experiment investigating the possible effect of a special diet in retarding the process of arteriosclerosis was described in Case Study 2.6. Let $p_X$ denote the true proportion of people on the special diet whose ultimate cause of death would be an infarction. Define $p_Y$ analogously for the control group. Using the data in Table 2.6, test $H_0: p_X = p_Y$ against an appropriate one-sided alternative. Let $\alpha = 0.05$.

**10.28** Let $p_M$ and $p_W$ denote the true proportions of working men and working women, respectively, who claim to be highly satisfied with their jobs. Use the data in Question 2.22 to test $H_0: p_M = p_W$ versus $H_1: p_M \neq p_W$. Let $\alpha = 0.10$.

10.29   An experiment in water witching was described in Question 2.23. Let $p_W$ and $p_N$ denote the true proportions of witched wells and nonwitched wells, respectively, that prove to be successful. At the 0.01 level of significance, test $H_0: p_W = p_N$ against a two-sided alternative.

10.30   Set up and carry out an appropriate hypothesis test on the flying saucer data of Question 2.24. Define carefully the parameters you are testing. Let $\alpha = 0.05$.

10.31   As defined in Question 8.68, the $P$ value of a hypothesis test is the probability of getting a test statistic as extreme or more extreme than what actually occurred, assuming that $H_0$ is true. If $H_0: p_X = p_Y$ is being tested against $H_1: p_X \neq p_Y$, and the test statistic equals $z^*$, then the $P$ value is equal to $P(Z \geqslant z^*) + P(Z \leqslant -z^*)$. Compute the $P$ value for the nightmare data of Case Study 10.6.

10.32   Draw a diagram showing the areas corresponding to the $P$ value asked for in Question 10.31. In general, how is the $P$ value related to the level of significance? (Hint: For what values of $\alpha$ could $H_0$ in Case Study 10.6 be rejected?)

10.33   If $p_X$ and $p_Y$ denote the true success probabilities associated with two independent sets of Bernoulli trials, then

$$\frac{\dfrac{X}{n} - \dfrac{Y}{m} - (p_X - p_Y)}{\sqrt{\dfrac{(X/n)(1 - X/n)}{n} + \dfrac{(Y/m)(1 - Y/m)}{m}}}$$

has approximately a standard normal distribution. Using that result, derive a formula for a $100(1 - \alpha)\%$ confidence interval for $p_X - p_Y$.

10.34   Let $p_M$ and $p_W$ denote the true suicide rates for male and female members, respectively, of the American Chemical Society. Use the answer from Question 10.33 and the data in Question 2.25 to construct a 95% confidence interval for $p_M - p_W$.

# 10.6

## COMPUTER NOTES

The two-sample $t$ test for the equality of means is performed by a specific procedure in SAS. The output provides the $P$ value for the test described in Section 10.3, which assumes that the two samples come from normal random variables with the same variance. The printout also gives a $P$ value for a test of equality of means that does not make the assumption that $\sigma_X^2 = \sigma_Y^2$.

The printout of Figure 10.12 is for the test of Case Study 10.3. The $P$ value 0.1772 is on the line with EQUAL in the PROB > |T| column. Since $0.1772 > \alpha = 0.05$, we accept the null hypothesis that $\mu_X = \mu_Y$.

```
DATA;
  INPUT SEX $ HL @@;
  CARDS;

      (data from Table 10.3 here, in pairs, with sex first followed by
        half-life)

PROC TTEST;
  CLASS SEX;
  VAR HL;


VARIABLE: HL

SEX    N          MEAN           STD DEV        STD ERROR

M      9     79.33333333      8.07774721       2.69258240
F      6     70.83333333     15.09194045       6.16125889


    MINIMUM        MAXIMUM      VARIANCES           T      DF    PROB > |T|
70.00000000    93.00000000    EQUAL         -1.4269    13.0      0.1772
52.00000000    88.00000000    UNEQUAL       -1.2641     6.9      0.2471

FOR H0: VARIANCES ARE EQUAL,   F'=          .29 WITH 8 AND 5 DF
PROB > F' = 0.1143
```

**FIG. 10.12**

As a bonus we also get the test for equality of variances, as was done in Example 10.2. Since the $P$ value of 0.1143 exceeds 0.05, the null hypothesis is accepted. There is statistical support for our assumption of equal variances.

Minitab also has a routine to do two-sample inferences using a $t$ statistic. Figure 10.13 shows the result of the Minitab procedure for Case Study 10.1. The last line gives a $P$ value of 0.0013. The next-to-last line gives a 95% confidence interval (0.0101, 0.0343) for $\mu_X - \mu_Y$.

```
        SET TWAIN INTO C1
         .225  .262  .217  .240  .230  .229  .235  .217
        SET QCS INTO C2
         .209  .205  .196  .210  .202  .207  .224  .223  .220  .201
TWOSAMPLE T C1 C2;
  POOLED.

TWOSAMPLE T FOR C1 VS C2
       N      MEAN      STDEV     SE MEAN
C1     8     0.2319    0.0146     0.0051
C2    10     0.20970   0.00966    0.0031

95 PCT CI FOR MU C1 - MU C2: (0.0101, 0.0343)
TTEST MU C1 = MU C2 (VS NE): T=3.88 P=0.0013 DF=16.0
```

**FIG. 10.13**

# 10.7

## SUMMARY

Using independent samples for comparing two treatments, or comparing the responses of two different kinds of subjects to the same treatment, are objectives likely to be faced at one time or another by virtually all researchers, whatever their field. Is drug A more effective than drug B? Does tire $X$ provide better traction than tire $Y$? Are northerners and southerners equally supportive of presidential candidate C? Data models of this sort are referred to as *two-sample* problems. On a list of most frequently encountered experimental situations, two-sample problems would rank near the top.

Statistically, data for two-sample problems consist of two independent random samples, $x_1, x_2, \ldots, x_n$ and $y_1, y_2, \ldots, y_m$. The usual objective is a hypothesis test. The confidence interval format is not as widely used in the context of two-sample data as it was in Chapters 8 and 9 for one-sample data.

Three special cases of the two-sample problem have been the primary focus of Chapter 10. For $x$'s and $y$'s both coming from normal distributions, we developed a *two-sample t test* in Section 10.3 for assessing the likelihood of $H_0 \colon \mu_X = \mu_Y$. In Section 10.4 an $F$ *test* was set up to examine whether the variances of two normal distributions are the same ($H_0 \colon \sigma_X^2 = \sigma_Y^2$). Finally, binomial data were investigated in Section 10.5, where an approximate $Z$ *test* was formulated to handle $H_0 \colon p_X = p_Y$.

| To Test | Versus | Reject $H_0$ | Formula |
|---|---|---|---|
| $H_0 \colon \ \mu_X = \mu_Y$ | $H_1 \colon \ \mu_X \neq \mu_Y$ | $t \leqslant -t_{\alpha/2,\,n+m-2}, \quad t \geqslant t_{\alpha/2,\,n+m-2}$ | $t = \dfrac{\bar{x} - \bar{y}}{s_p \sqrt{\dfrac{1}{n} + \dfrac{1}{m}}}$ |
| | $H_1 \colon \ \mu_X > \mu_Y$ | $t \geqslant t_{\alpha,\,n+m-2}$ | |
| | $H_1 \colon \ \mu_X < \mu_Y$ | $t \leqslant -t_{\alpha,\,n+m-2}$ | $s_p = \sqrt{\dfrac{\sum\limits_{i=1}^{n} x_i^2 - \left(\sum\limits_{i=1}^{n} x_i\right)^2 / n + \sum\limits_{i=1}^{m} y_i^2 - \left(\sum\limits_{i=1}^{m} y_i\right)^2 / m}{n + m + 2}}$ |
| $H_0 \colon \ \sigma_X^2 = \sigma_Y^2$ | $H_1 \colon \ \sigma_X^2 \neq \sigma_Y^2$ | $F \leqslant \dfrac{1}{F_{\alpha/2,\,n-1,\,m-1}}, \quad F > F_{\alpha/2,\,m-1,\,n-1}$ | $F = \dfrac{s_Y^2}{s_X^2}$ |
| | $H_1 \colon \ \sigma_X^2 > \sigma_Y^2$ | $F \leqslant \dfrac{1}{F_{\alpha,\,n-1,\,m-1}}$ | |
| | $H_1 \colon \ \sigma_X^2 < \sigma_Y^2$ | $F \geqslant F_{\alpha,\,m-1,\,n-1}$ | |
| $H_0 \colon \ p_X = p_Y$ | $H_1 \colon \ p_X \neq p_Y$ | $z \leqslant -z_{\alpha/2}, \quad z \geq z_{\alpha/2}$ | $z = \dfrac{(x/n - y/m)\sqrt{nm}}{\sqrt{\left(\dfrac{x+y}{n+m}\right)\left(1 - \dfrac{x+y}{n+m}\right)(n+m)}}$ |
| | $H_1 \colon \ p_X > p_Y$ | $z \geqslant z_{\alpha}$ | |
| | $H_1 \colon \ p_X < p_Y$ | $z \leqslant -z_{\alpha}$ | |

**REVIEW QUESTIONS**

**10.35** The same tax information is given to 10 professional consultants, 5 of whom work for Company A and 5 for Company B. Summarized below are the differences between the amount of taxes each one said the client owed and the "true" tax owed (as determined by the IRS).

| Company A Estimates | Company B Estimates |
|---|---|
| $300 | −$55 |
| $50 | 0 |
| −$65 | −$15 |
| 0 | $125 |
| $45 | 0 |

Test $H_0: \sigma_A^2 = \sigma_B^2$ versus $H_1: \sigma_A^2 \neq \sigma_B^2$. Let $\alpha = 0.10$.

**10.36** Because of remarks he made that were thought to be chauvinistic, a congressman seeking reelection is concerned that he may be facing a gender gap. The latest poll shows that his support is stronger among men than it is among women:

|  | Men | Women |
|---|---|---|
| Support candidate | 118 | 96 |
| Do not support candidate | 92 | 94 |
|  | 210 | 190 |
| % in support | 56.2 | 50.5 |

Is the difference between 56.2% and 50.5% statistically significant? Test $H_0: p_M = p_W$ against a one-sided alternative. Take $\alpha = 0.05$.

**10.37** Two paint companies each claim to have the fastest drying enamel on the market. A random sample of six cans of each brand are applied to similar surfaces under identical environmental conditions. The drying times (in minutes) are shown below.

| Brand A: | 69 | 72 | 65 | 75 | 79 | 69 |
| Brand B: | 72 | 68 | 65 | 60 | 62 | 66 |

Can it be concluded at the 0.05 level of significance that the two brands do not dry in the same average amount of time? Assume that both populations of drying times are normal and have the same variance.

**10.38** To which of the two data types in Figure 10.1 do Case Studies 10.1, 10.2, and 10.3 belong?

10.39    Shown below are representative starting salaries (in thousands of dollars) for high school English teachers as reported by six counties in Tennessee and four counties in North Carolina. State precisely what $\mu_X$ and $\mu_Y$ represent and test $H_0: \mu_X = \mu_Y$ versus $H_1: \mu_X \neq \mu_Y$. Let $\alpha = 0.20$.

| Salaries in Tennessee | Salaries in North Carolina |
|---|---|
| 18.6 | 24.2 |
| 20.1 | 23.1 |
| 19.8 | 20.5 |
| 22.3 | 26.0 |
| 24.1 | |
| 23.6 | |

10.40    What is the smallest value for $\bar{x} - \bar{y}$ that will lead us to reject $H_0: \mu_X = \mu_Y$ in favor of $H_1: \mu_X > \mu_Y$ if $\alpha = 0.05$, $s_p = 18.6$, and $n = m = 15$?

10.41    A survey of 220 single-family dwellings found that 60% have inadequate fire alarm systems. Sixty-two per cent of 150 nearby apartments were similarly deficient. At the 0.05 level, is the difference between 60% and 62% statistically significant?

10.42    Court records confirm that average sentence lengths for the same crime vary considerably from judge to judge. Do standard deviations of sentence lengths show a similar inconsistency? Maybe. Shown below are eight sentences (in months) imposed by Judge Haynes and eight imposed by Judge Goddard, all for the same drug-related crime. Test $H_0: \sigma_H = \sigma_G$. Let $\alpha = 0.05$.

| Judge Haynes: | 30 | 24 | 48 | 40 | 35 | 32 | 36 | 60 |
|---|---|---|---|---|---|---|---|---|
| Judge Goddard: | 36 | 32 | 44 | 40 | 38 | 42 | 34 | 40 |

10.43    If $H_0: \sigma_X^2 = \sigma_Y^2$ is true for a set of two-sample data, then $(n + m - 2)S_p^2/\sigma^2$ has a $\chi^2$ distribution with $n + m - 2$ df. Use that result to construct a 95% confidence interval for $\sigma^2$ using the thermoluminesence data in Case Study 10.2.

10.44    Ten years ago a survey showed that 65 of 110 adults holding blue collar jobs were satisfied with their work. A recent questionnaire targeting the same population found 40 satisfied workers among a sample of 76. At the $\alpha = 0.10$ level of significance, can we conclude that job satisfaction among blue collar workers is declining?

10.45    Is the value of the pooled standard deviation dependent on whether $\mu_X = \mu_Y$? Why or why not?

10.46    Jason has purchased a number of video tapes from two mail order houses. Both companies charge the same prices and offer the same quality but the promptness of their service may be different. Orders placed with *Video Vintage* have returned on the average in 28.4 days, while turnaround times for orders placed with *TV Memories* have averaged 35.4 days. Is the difference between 28.4 and 35.4 statistically significant? Let $\alpha = 0.05$.

| Shipping Times for Video Vintage | Shipping Times for TV Memories |
|---|---|
| 36 | 28 |
| 28 | 34 |
| 26 | 36 |
| 22 | 41 |
| 30 | 38 |

10.47 Is it possible to test the significance of the difference between two proportions, $x/n$ and $y/m$, if the sample sizes $n$ and $m$ are not known?

10.48 Last week 2024 students ate lunch at the school cafeteria; 215 developed one or more symptoms of food poisoning. This week 1946 ate lunch but only 196 got sick. Can it be argued that the probability of a student getting sick the second week was less than it was the first week? Set up and carry out an appropriate hypothesis test at the $\alpha = 0.10$ level.

10.49 What *P*-value is associated with the hypothesis test asked for in Question 10.48?

10.50 Each of nine high schools, five public and four private, was asked to report the percentage of women on its faculty. Do the data below support the contention that the percentage of female faculty members tends to be higher at private schools? Test an appropriate hypothesis at the $\alpha = 0.10$ level of significance.

| Public High Schools | | Private High Schools | |
|---|---|---|---|
| 46 | 48 | 51 | 64 |
| 51 | 53 | 62 | 55 |
| 39 | | | |

10.51 Intuitively, why would $\sqrt{(s_X^2 + s_Y^2)/2}$ not be a reasonable way to define a pooled standard deviation?

10.52 Test $H_0: \sigma_A^2 = \sigma_B^2$ for the drying times in Question 10.37. Use $\alpha = 0.05$. What does your conclusion imply about the validity of using a two-sample *t* test on $H_0: \mu_A = \mu_B$?

10.53 Suppose that samples of size 100 are taken from two normal distributions. How might an experimenter use the data to test $H_0: p_X = p_Y$ rather than $H_0: \mu_X = \mu_Y$?

10.54    Suppose two independent samples of sizes $n = 50$ and $m = 40$ are collected on Treatments $X$ and $Y$ for the purpose of testing $H_0: \mu_X = \mu_Y$. Describe the steps you would take to verify the assumptions made by the two-sample $t$ test.

10.55    How do we know whether the null hypothesis to be tested should be $H_0: \mu = \mu_0$ or $H_0: \mu_X = \mu_Y$? Is it possible that both might be appropriate for the same set of data?

10.56    Refer to Question 10.35. In choosing a company to prepare your income tax, do you necessarily want the one whose variance is closest to zero? Do you necessarily want the company whose mean is closest to zero? Explain.

10.57    Why are confidence intervals not as important in analyzing two-sample data as they were in analyzing one-sample data?

10.58    What would the sampling distribution of $S_Y^2/S_X^2$ look like if $H_0: \sigma_X^2 = \sigma_Y^2$ is true and $n = m = 10{,}000$?

10.59    Suppose an experimenter plans to take a total of 40 observations on two treatment groups for the purpose of testing $H_0: \mu_X = \mu_Y$. Is it better to take $n = 20$ and $m = 20$ or $n = 10$ and $m = 30$?

# 11 *The Paired-Data Problem*

**W**e do not what we ought,
What we ought not, we do
And lean upon the thought
That Chance will bring us through.

*Arnold*

## 11.1

### INTRODUCTION

Comparing two distributions, one against the other, was the objective behind everything we did in Chapter 10. Depending on the situation and type of data, we compared either two means ($H_0$: $\mu_X = \mu_Y$), two variances ($H_0$: $\sigma_X^2 = \sigma_Y^2$), or two proportions ($H_0$: $p_X = p_Y$). A key factor in setting up each procedure was the underlying structure of the data—specifically, that the $x_i$'s and $y_i$'s were *independent*. In this chapter we also want to test $\mu_X = \mu_Y$. But now the $x_i$'s and $y_i$'s will be *dependent*.

For reasons we will defer until Section 11.4, the distinction between independent $x_i$'s and $y_i$'s and dependent $x_i$'s and $y_i$'s is crucial. Suffice it to say that the prudent use of dependent observations can often be an experimenter's best strategy for reducing the probability of committing a Type II error (accepting $H_0$ when $H_1$ is true). Without question, the development of a methodology for analyzing dependent data has been one of the most useful accomplishments of twentieth-century statistics. In this chapter we will learn the prototype of these techniques, the ***paired t test.***

The formulas defining a paired $t$ test should look familiar. Except for a minor revision in notation to reflect the change in the kind of data being considered, the statistic to be computed is exactly the same as the one-sample $t$ ratio of Section 9.2. Pay close attention to the types of dependencies between the $x_i$'s and $y_i$'s in the four case studies in Section 11.3. You will learn more from the *design* of these experiments than from their analysis.

## 11.2

### THE PAIRED-DATA FORMAT

Figure 11.1 shows the basic structure characteristic of paired data. Two treatments ($X$ and $Y$) are to be compared. Responses to treatment $X$ are denoted $x_1, x_2, \ldots, x_n$; responses to treatment $Y$ by $y_1, y_2, \ldots, y_n$. Furthermore, $x_i$ and $y_i$, $i = 1, 2, \ldots, n$, are *dependent* (row by row) in the sense that each $x_i$ and $y_i$ share certain relevant common traits. They may, for example, be measurements taken on the same person, taken at the same time, or taken in the same place.

| Pair | Treatment $X$ | Treatment $Y$ | Difference, $d_i$ |
|------|---------------|---------------|-------------------|
| 1 | $x_1$ | $y_1$ | $d_1 = y_1 - x_1$ |
| 2 | $x_2$ | $y_2$ | $d_2 = y_2 - x_2$ |
| $\vdots$ | $\vdots$ | $\vdots$ | $\vdots$ |
| $n$ | $x_n$ | $y_n$ | $d_n = y_n - x_n$ |

**FIG. 11.1**

The key to analyzing paired data is the set of $n$ *within-pair response differences*,

$$d_1 = y_1 - x_1, \quad d_2 = y_2 - x_2, \quad \ldots, \quad d_n = y_n - x_n$$

To these differences, and not to the original $x_i$'s and $y_i$'s, will the test statistic be applied. Notice that each $d_i$ is "clear" of whatever bias the conditions in pair $i$ may have introduced into both $x_i$ and $y_i$. Suppose, for example, the nature of pair $i$ is such that it inflates its associated responses by an amount $\Delta_i$. Whatever $\Delta_i$ is, it does not influence the within-pair response *difference* because the formula $d_i = y_i - x_i$ adds $\Delta_i$ in as part of $y_i$ but then subtracts it out as part of $x_i$.

Historically, it was agricultural research that provided some of the first applications for paired data. Imagine being assigned the task of comparing the yields of two varieties of corn ($X$ and $Y$). From what we learned in Chapter 10, the appropriate experimental design may seem obvious: grow variety $X$ at $n$ locations and variety $Y$ at $m$ locations. Ideally, we could determine the better variety by comparing the resulting yields,

$$x_1, x_2, \ldots, x_n \quad \text{and} \quad y_1, y_2, \ldots, y_m$$

with the two-sample $t$ test. Would this be a legitimate approach? Yes. Would it be a *good* approach? No.

While there is nothing that mathematically invalidates collecting—and then analyzing—yields from varieties $X$ and $Y$ according to a two-sample format, there are certain practical considerations that make that a decidedly poor strategy. In particular, what complicates the situation is that many factors *other than the two varieties* can significantly affect the responses. Variations in soil fertility, drainage, amount of sunlight, and so on all have the potential to bias the $x_i$'s and $y_i$'s.

All measurments, of course, in any experiment are subject to many influences beyond the control of the researcher. In agricultural work, though, the magnitude of the collective effect produced by these extraneous factors may be so great as to totally obscure the phenomenon being investigated. That being the case, *how* we collect the data becomes critically important.

One of the most important contributions that Sir Ronald Fisher made to applied statistics was to convince agricultural researchers that the obfuscating effect of uncontrollable environmental biases can be lessened by taking dependent observations. Fisher's strategy was to divide the total area to be planted into **n plots**, where the plots were chosen so that all the various growth factors within a plot were as uniform as possible. (Making the plots small would usually ensure the necessary homogeneity.) Each plot was then further divided into two *subplots*, with variety $X$ being randomly assigned to one subplot and variety $Y$ to the other (see Figure 11.2).

By collecting the data in this fashion, we make each $x_i$ and $y_i$ dependent because they both reflect whatever growth conditions are present in plot $i$.

**plot**—Another name for a pair. Historically, many of the early applications of the paired $t$ test involved agricultural research. The treatments being compared were applied within small, geographically contiguous pieces of land. The latter were called *plots*. Conditions within a plot relative to the response being measured were kept as homogeneous as possible.

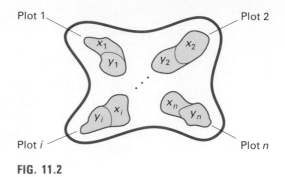

**FIG. 11.2**

At the same time the within-plot differences,

$$d_i = y_i - x_i, \quad i = 1, 2, \ldots, n$$

are *not* biased by any heterogeneity that might exist from plot to plot, for the reasons cited earlier. The $d_i$'s, therefore, are very good estimates of what we are trying to investigate, the difference between the yield potentials of variety $X$ and variety $Y$.

It did not take long for scientists working in other areas to recognize that the concept behind Fisher's plots (what we now call pairs) could be extended to their own disciplines. Some of those other applications are featured in the case studies in Section 11.3. The variety is really amazing. In practice, the definition of a pair is limited only by the imagination and ingenuity of the experimenter.

# 11.3

## THE PAIRED t TEST

Let $(x_1, y_1)$, $(x_2, y_2)$, . . . , $(x_n, y_n)$ be a set of $n$ paired observations. Let $d_i = y_i - x_i$, $i = 1, 2, \ldots, n$ be the corresponding within-pair response differences. We assume that the $x_i$'s and $y_i$'s come from distributions with means $\mu_X$ and $\mu_Y$, respectively, and that the $d_i$'s represent a random sample from a normal distribution with mean $\mu_D$ and standard deviation $\sigma_D$, both parameters being unknown. Our objective is to test $H_0: \mu_X = \mu_Y$.

Intuitively, if $d_i = y_i - x_i$, it should be true that $\mu_D = \mu_Y - \mu_X$. Therefore, testing $H_0: \mu_X = \mu_Y$ is equivalent to testing $H_0: \mu_D = 0$. For the latter we simply need to do a one-sample $t$ test on the $d_i$'s: if the *sample* average difference $\bar{d}$ is too far away from zero, we reject the null hypothesis that the *true* average difference $\mu_D$ is zero. In deference to the kind of data being analyzed, this procedure is called a **paired $t$ test**, even though the formulas involved are fundamentally the same as those used for the one-sample $t$ test.

**THEOREM 11.1**

Let $D_1 = Y_1 - X_1, D_2 = Y_2 - X_2, \ldots, D_n = Y_n - X_n$ be a set of within-pair response differences. Assume the $D_i$'s represent a random sample from a normal distribution with mean $\mu_D$ and standard deviation $\sigma_D$. Let

$$\bar{D} = \frac{1}{n} \sum_{i=1}^{n} D_i$$

and

$$S_D = \sqrt{\frac{1}{n-1} \sum_{i=1}^{n} (D_i - \bar{D})^2} = \sqrt{\frac{n \sum_{i=1}^{n} D_i^2 - \left(\sum_{i=1}^{n} D_i\right)^2}{n(n-1)}}$$

be the sample mean and sample standard deviation (of the differences), respectively. Then

$$T = \frac{\bar{D} - \mu_D}{S_D/\sqrt{n}}$$

has a Student $t$ distribution with $n - 1$ degrees of freedom.

Let $t = \bar{d}/(s_D/\sqrt{n})$.

1. To test

$$H_0: \quad \mu_D = 0$$
$$\text{vs.}$$
$$H_1: \quad \mu_D \neq 0$$

at the $\alpha$ level of significance, reject $H_0$ if $t \leqslant -t_{\alpha/2, n-1}$ or $t \geqslant t_{\alpha/2, n-1}$.

2. To test

$$H_0: \quad \mu_D = 0$$
$$\text{vs.}$$
$$H_1: \quad \mu_D < 0$$

at the $\alpha$ level of significance, reject $H_0$ if $t \leqslant -t_{\alpha, n-1}$.

3. To test

$$H_0: \quad \mu_D = 0$$
$$\text{vs.}$$
$$H_1: \quad \mu_D > 0$$

at the $\alpha$ level of significance, reject $H_0$ if $t \geqslant t_{\alpha, n-1}$.

*Pecking Orders*   Chickens in a confined area quickly establish a highly structured pecking order. But to what extent is that pecking order fixed? Can a chicken's behavior be modified in such a way that its pecking order changes?

Eight pens, each containing 14 chickens, were set up. The behavior of the chickens in each pen was closely observed, and the pecking orders were determined; the most dominant chicken in each pen was assigned a rank of 1, the least dominant a rank of 14. One chicken from each pen was then removed and thrown into a small cage with a big, mean-tempered rooster. The rooster was not overly thrilled at the prospect of having company and proceeded to thoroughly trash its new roommate. Suffice it to say that feathers flew, and it was not the rooster's plumage that suffered.

Later, after having been subjected to a goodly amount of the rooster's psychological and physical abuse, the chicken was removed and returned to its original pen, where it then had to reestablish its place in the pecking order. Would its new pecking order be affected by its unpleasant experience with the rooster? If so, how?

Table 11.1 lists the pecking order *changes:*

$$\text{new pecking order } (y_i) - \text{old pecking order } (x_i)$$

for the eight chickens who "volunteered" for the study. Values for the original $x$'s and $y$'s have been deleted (131). A positive change implies that a chicken has *fallen* in the pecking order and become less dominant.

**TABLE 11.1**

PECKING ORDERS ("old" and "new")

| Pen | Old $(x_i)$ | New $(y_i)$ | $d_i = y_i - x_i$ |
|-----|-------------|-------------|-------------------|
| 1 | $x_1$ | $y_1$ | 2 |
| 2 | $x_2$ | $y_2$ | 2 |
| 3 | $x_3$ | $y_3$ | 3 |
| 4 | $x_4$ | $y_4$ | $-2$ |
| 5 | $x_5$ | $y_5$ | 1 |
| 6 | $x_6$ | $y_6$ | 3 |
| 7 | $x_7$ | $y_7$ | 7 |
| 8 | $x_8$ | $y_8$ | 2 |

We assume that the $d_i$'s represent a random sample of size $n = 8$ from a normal distribution with mean $\mu_D$. If pecking order is unaffected by the prior conditioning endured by these chickens, then $\mu_D$ should be zero. What the experimenter wants to learn, therefore, can be phrased appropriately as a hypothesis test:

$$H_0: \quad \mu_D = 0$$

vs.

$$H_1: \quad \mu_D \neq 0$$

(Why is the alternative two-sided?)

Let $\alpha = 0.05$. From column 4 of Table 11.1 we compute

$$\sum_{i=1}^{8} d_i = 18 \quad \text{and} \quad \sum_{i=1}^{8} d_i^2 = 84$$

so that

$$\bar{d} = \frac{18}{8} = 2.25 \quad \text{and} \quad s_D = \sqrt{\frac{8(84) - (18)^2}{8(7)}} = 2.49$$

Since $n = 8$ and $H_1$ is two-sided, the appropriate critical values are the numbers that cut off areas of 0.025 in either tail of a Student $t$ curve with 7 df: $\pm t_{0.025, 7} = \pm 2.36$. The observed $t$ ratio from Theorem 11.1 is

$$t = \frac{\bar{d}}{s_D/\sqrt{n}} = \frac{2.25}{2.49/\sqrt{8}} = 2.56$$

which is located slightly above the upper critical value (Figure 11.3). Therefore, we reject $H_0$. (Would we have rejected $H_0$ if $\alpha$ were 0.01?)

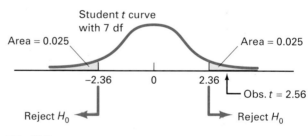

**FIG. 11.3**

***Blood-Clotting Times***   Letting each subject serve as its own pair—that is, measuring each one *twice*—may very well be the most common form of paired-data experiment. The appeal is obvious: what better way to eliminate extraneous influences from the $x_i$'s and $y_i$'s than to use the same subject for both measurements? Sometimes the $x_i$'s and $y_i$'s will represent responses to two different treatments, and sometimes there will be only a single treatment and the $x_i$'s and $y_i$'s are "before" and "after" measurements. What we are about to describe typifies the latter.

Blood coagulates only after a complex chain of chemical reactions has taken place. The final step is the transformation of fibrinogen, a protein found in the plasma, into fibrin, which forms the clot. The fibrinogen–fibrin reaction is triggered by another protein, thrombin, which itself is formed under the influence of still other proteins, including one called prothrombin.

A person's blood-clotting ability is often expressed in terms of a "prothrombin time," which is defined to be the interval between the initiation of the prothrombin-thrombin reaction and the formation of the final clot. Factors affecting the length of a person's prothrombin time clearly have profound medical significance. Research on the subject has been extensive. One study looked at the effect that aspirin might have (189). Twelve adult males participated. Their prothrombin times were measured before and 3 hr after each was given two aspirin tablets (650 mg). Table 11.2 shows the results.

**TABLE 11.2  Prothrombin Times (s)**

| Subject | Before Aspirin, $x_i$ | After Aspirin, $y_i$ | $d_i = y_i - x_i$ |
|---------|------------------------|-----------------------|--------------------|
| 1  | 12.3 | 12.0 | $-0.3$ |
| 2  | 12.0 | 12.3 | $+0.3$ |
| 3  | 12.0 | 12.5 | $+0.5$ |
| 4  | 13.0 | 12.0 | $-1.0$ |
| 5  | 13.0 | 13.0 | $0$ |
| 6  | 12.5 | 12.5 | $0$ |
| 7  | 11.3 | 10.3 | $-1.0$ |
| 8  | 11.8 | 11.3 | $-0.5$ |
| 9  | 11.5 | 11.5 | $0$ |
| 10 | 11.0 | 11.5 | $+0.5$ |
| 11 | 11.0 | 11.0 | $0$ |
| 12 | 11.3 | 11.5 | $+0.2$ |
|    |      |      | $-1.3$ |

Let $\mu_X$ and $\mu_Y$ denote a person's true average prothrombin time before taking aspirin and after taking aspirin, respectively. If $\mu_D = \mu_Y - \mu_X$, the hypotheses to be tested are

$$H_0: \quad \mu_D = 0$$

vs.

$$H_1: \quad \mu_D \neq 0$$

Let 0.05 be the level of significance.

From Table 11.2,

$$\sum_{i=1}^{12} d_i = -1.3 \quad \text{and} \quad \sum_{i=1}^{12} d_i^2 = 2.97$$

Therefore,
$$\bar{d} = (1/12)(-1.3) = -0.108$$

and
$$s_D = \sqrt{\frac{12(2.97) - (-1.3)^2}{12(11)}} = 0.507$$

Since $n = 12$, the critical values for the test statistic will be the 2.5th and 97.5th percentiles of the Student $t$ distribution with 11 df: $\pm 2.20$. The appropriate decision rule from Theorem 11.1 is to reject $H_0: \mu_D = 0$ if

$$\frac{\bar{d}}{s_D/\sqrt{12}} \leqslant -2.20 \qquad \text{or} \qquad \frac{\bar{d}}{s_D/\sqrt{12}} \geqslant 2.20$$

But the observed $t$ ratio is

$$t = \frac{-0.108}{0.507/\sqrt{12}} = -0.74$$

so our conclusion is to accept $H_0$. According to these data, aspirin has no demonstrable effect on prothrombin time.

For good reason, experimenters like to set up studies using the format of Case Study 11.2. Unfortunately, that structure does not always work. It may be impractical or physically impossible to apply both treatments to the same subject. The next best alternative, and one frequently invoked, is to use twins (or, for animal studies, littermates).

## CASE STUDY 11.3

***Are Rural Areas Healthier Than Urban Areas?***

Scientists can measure tracheobronchial clearance by having a subject inhale a radioactive aerosol and then later metering the radiation level in his lungs. In one experiment (22) seven pairs of monozygotic twins inhaled aerosols of radioactive Teflon particles. One member of each twin pair lived in a rural area; the other was a city dweller. The objective of the study was to see whether the data would support the popular contention that rural environments are more conducive to respiratory health than are urban environments. Table 11.3 gives the percentages of radioactivity retained in the lungs of each subject 1 hr after the initial inhalation.

Let $\mu_D$ denote the true average difference in retention percentages (urban − rural). If environment has no effect on tracheobronchial clearance,

**TABLE 11.3 Radioactivity Retention (%)**

| Twin Pair | Rural, $x_i$ | Urban, $y_i$ | $d_i = y_i - x_i$ |
|---|---|---|---|
| 1 | 10.1 | 28.1 | 18.0 |
| 2 | 51.8 | 36.2 | −15.6 |
| 3 | 33.5 | 40.7 | 7.2 |
| 4 | 32.8 | 38.8 | 6.0 |
| 5 | 69.0 | 71.0 | 2.0 |
| 6 | 38.9 | 47.0 | 8.1 |
| 7 | 54.6 | 57.0 | 2.4 |

$\mu_D$ should be zero; if, on the other hand, a rural environment is healthier, then $\mu_D > 0$. (Why?) What needs to be tested, therefore, is

$$H_0: \quad \mu_D = 0$$
$$\text{vs.}$$
$$H_1: \quad \mu_D > 0$$

Let $\alpha = 0.05$.

With $n = 7$ the appropriate critical value is $t_{0.05, 6} = 1.94$; that is, we should reject $H_0$ if $\bar{d}/(s_D/\sqrt{7}) \geqslant 1.94$. From Table 11.3,

$$\sum_{i=1}^{7} d_i = 28.1 \quad \text{and} \quad \sum_{i=1}^{7} d_i^2 = 730.57$$

Therefore, 
$$\bar{d} = \frac{28.1}{7} = 4.01$$

$$s_D = \sqrt{\frac{7(730.57) - (28.1)^2}{7(6)}} = 10.15$$

The observed $t$ ratio is

$$t = \frac{4.01}{10.15/\sqrt{7}} = 1.04$$

which lies to the left of $t_{0.05, 6}$. Therefore, we accept the null hypothesis—the observed $d_i$'s are not inconsistent with the assumption that environment has no effect on tracheobronchial clearance.

## CASE STUDY 11.4

*Bee Stings*    Many factors (other than pure orneriness) predispose a bee to sting. A person wearing dark clothing, for example, is more likely to be victimized than someone wearing light clothing. And someone whose movements are quick and jerky runs a higher risk than a person who moves slowly. Still another factor, particularly important to beekeepers, is whether the person has just been stung by another bee.

The purpose of this experiment was to simulate that last factor. (For obvious reasons, real beekeepers were not used!) Eight cotton balls wrapped in muslin were dangled up and down in front of the entrance to a beehive. Four of the balls had just been exposed to a swarm of angry bees and were filled with stingers; the other four were fresh. After a specified time the number of new stingers in each of the balls was counted. Eventually, the entire procedure was replicated eight times. Table 11.4 shows the results (52).

**TABLE 11.4  Number of Times Stung**

| Trial | Cotton Balls with Stings Already Present | Fresh Cotton Balls |
|-------|------------------------------------------|--------------------|
| 1 | 27 | 33 |
| 2 | 9 | 9 |
| 3 | 33 | 21 |
| 4 | 33 | 15 |
| 5 | 4 | 6 |
| 6 | 21 | 16 |
| 7 | 20 | 19 |
| 8 | 33 | 15 |
| 9 | 70 | 10 |

Let $d_1 = 33 - 27$, $d_2 = 9 - 9$, ..., $d_9 = 10 - 70$. Then

$$\sum_{i=1}^{9} d_i = -106 \quad \text{and} \quad \sum_{i=1}^{9} d_i^2 = 4458$$

If $\mu_D$ denotes the true average difference in the number of new stings (number in fresh cotton balls − number in previously stung cotton balls), the hypotheses that need to be tested are

$$H_0: \quad \mu_D = 0$$

vs.

$$H_1: \quad \mu_D < 0$$

Let $\alpha = 0.10$.

Figure 11.4 shows the appropriate critical region. If $\bar{d}/(s_D/\sqrt{9}) \leqslant -t_{0,10.8} = -1.40$, we should reject $H_0$. But

$$\bar{d} = \frac{-106}{9} = -11.8$$

and

$$s_D = \sqrt{\frac{9(4458) - (-106)^2}{9(8)}} = 20.0$$

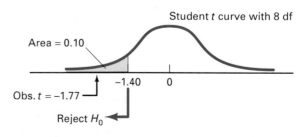

Student *t* curve with 8 df

Area = 0.10

−1.40    0

Obs. *t* = −1.77

Reject $H_0$

**FIG. 11.4**

so the observed $t$ ratio is

$$t = \frac{\bar{d}}{s_D/\sqrt{n}} = \frac{-11.8}{20.0/\sqrt{9}} = -1.77$$

Therefore, at the 0.10 level of significance, we should reject $H_0$: these data suggest that bees do show a predisposition for attacking objects that have already been stung.

---

QUESTIONS

11.1   A set of $n$ paired observations gives the values of $\bar{d}$ and $s_D$ listed below. Assume the distribution represented by the $d_i$'s is normal and has mean $\mu_D$. Set up and carry out the hypothesis test of $H_0$: $\mu_D = 0$.

(a) $H_1$: $\mu_D \neq 0$,    $\alpha = 0.05$,    $n = 14$,    $\bar{d} = 2.7$,    $s_D = 6.1$

(b) $H_1$: $\mu_D > 0$,    $\alpha = 0.01$,    $n = 25$,    $\bar{d} = 0.13$,    $s_D = 0.04$

(c) $H_1$: $\mu_D < 0$,    $\alpha = 0.10$,    $n = 6$,    $\bar{d} = -5.6$,    $s_D = 14.5$

11.2   Suppose an experimenter has a set of $n$ paired observations and wishes to test

$$H_0: \quad \mu_D = \mu_{D_0}$$

vs.

$$H_1: \quad \mu_D \neq \mu_{D_0}$$

at the $\alpha$ level of significance. State the decision rule.

11.3   (a) How would the aspirin/prothrombin time experiment of Case Study 11.2 be done as a two-sample problem?

(b) If you were the researcher, which approach (paired-data or two-sample) would you have taken? Why?

11.4   Two therapies for treating acrophobia were discussed in Case Study 2.8.

(a) Compare the two, using a paired $t$ test. Make the alternative two-sided, and let $\alpha = 0.05$.

(b) How might the method of pairing used in this example be adapted to experiments comparing two diets, where the variable to be measured is a person's weight loss?

11.5   For the mileage estimates in Question 2.27, test $H_0$: $\mu_D = 0$ versus $H_1$: $\mu_D \neq 0$. Let $\alpha = 0.01$. Note:

$$\sum_{i=1}^{20} d_i = -36.0 \quad \text{and} \quad \sum_{i=1}^{20} d_i^2 = 96.0$$

11.6   See Question 2.28. Let $\mu_X$ denote the true average percentage of daylight hours that quail living in condition A spend dustbathing. Define $\mu_Y$ analogously for condition B. Compare $\mu_X$ and $\mu_Y$ by testing $H_0$: $\mu_D = 0$ versus $H_1$: $\mu_D \neq 0$, where $\mu_D = \mu_Y - \mu_X$. Let $\alpha = 0.05$. To which of the case studies in Section 11.3 are the data in Question 2.28 most similar in terms of the way in which the pairing is defined?

11.7 Analyze the triticale data in Question 2.29 with an appropriate paired $t$ test. Define the parameter being tested and justify your choice of $H_1$. Let $\alpha = 0.05$.

11.8 It was speculated in Case Study 3.13 that the drug cyclandelate may have the ability to stimulate a person's cerebral circulation and help delay the onset of senility. Test that claim, using the data in Table 3.22. Let 0.05 be the level of significance. State $H_0$ and $H_1$, give the decision rule, and draw a conclusion.

11.9 Use the data of Question 3.57 to test whether type-token ratios have any potential as a diagnostic tool for identifying schizophrenics. Define the parameter being tested and justify your choice of $H_1$. Let $\alpha = 0.10$.

11.10 Hypnosis has been suggested as a mechanism that might increase a person's ESP ability. Do the data in Case Study 3.14 support that contention? Perform the necessary hypothesis test at the 0.01 level of significance.

11.11 Analyze the depth perception data in Question 3.59 with a paired $t$ test. Choose between $H_0: \mu_D = 0$ and $H_1: \mu_D > 0$, where $\mu_D = \mu_Y - \mu_X$. Let $\alpha = 0.05$.

# 11.4

## COMPUTER NOTES

Since paired-data inferences use one-sample techniques, any software with one-sample procedures will handle the paired-data problem. The computer just calculates the differences and applies the one-sample routines. As an example Figure 11.5 is a Minitab treatment of Case Study 11.3. Since the $P$ value of 0.17 exceeds $\alpha = 0.05$, we accept the null hypothesis.

```
SET C1
  10.1  51.8  33.5  32.8  69.0  38.9  54.6
SET C2
  28.1  36.2  40.7  38.8  71.0  47.0  57.0

LET C3 = C2 - C1
PRINT C3
  C3   18.0000  -15.6000  7.2000  6.0000  2.0000  8.1000  2.4000

TTEST C3;
  ALTERNATIVE = +1.

  TEST OF MU = 0 VS MU G.T.  0

                N      MEAN      STDEV    SE MEAN      T    P VALUE
  C3            7       4.0      10.1        3.7    1.05     0.17
```

**FIG. 11-5**

## 11.5

### SUMMARY

We have pointed out in general terms how the paired-data format represents a refinement over the two-sample format in the sense that the former's $d_i$'s are less likely to be affected by extraneous environmental influences than are like latter's $x_i$'s and $y_i$'s. It might be helpful as a way of summarizing this chapter to expand on that idea in just a little more detail. What we need to examine more closely are the two different test satistics, the paired $t$

$$\left( = \frac{\bar{d}}{(s_D/\sqrt{n})} \right)$$

and the two-sample $t$

$$\left( = \frac{\bar{x} - \bar{y}}{s_p\sqrt{(1/n) + (1/m)}} \right)$$

The factors on which we need to key are the standard deviations $s_D$ and $s_p$.

Consider again the bee experiment described in Case Study 11.4. We have all had enough experiences (painful and otherwise) with our buzzing friends to realize that their tolerance of humans is highly unpredictable. Whether or not they attack in a given situation is determined by a host of factors, most of which we would be unable even to identify, much less control. Still, among all those many factors, its seems reasonable to suppose that time and place are two of the most important. Therefore, though the $d_i$'s in Table 11.4 are influenced by various environmental factors, they are *not* affected by two of the most significant sources of variation (because of the nature of the pairing). Thus we can expect $s_D$ to be fairly small (compared to what it would have been had time and place not been taken into account).

In contrast, think of how to do this study as a two-sample problem. The experimenter would choose a set of, say, nine times and nine locations and dangle the fresh cotton balls in front of those hives at those times. In the same fashion, a second battery of times and places (different from the first) would be set aside for collecting the bees' responses to cotton balls already stung. The $x_i$'s and $y_i$'s measured in this way would be independent but highly variable since none of the environmental factors would have been screened out.

Where does the difference in how the data are collected show up? In the denominators of the test statistics. For the reasons just cited, we would expect

$$s_D/\sqrt{9} < s_p\sqrt{\tfrac{1}{9} + \tfrac{1}{9}}$$

Therefore, the paired $t$ ratio itself will likely be *larger* than the two-sample $t$ ratio. But a larger test statistic has a higher probability of being in the rejection region. Thus our chances of rejecting $H_0$: $\mu_X = \mu_Y$ (or $H_0$: $\mu_D = 0$) are likely to be greater if we use the paired-data format. Using the terminology of Section 8.3, we would say that the paired $t$ test relative to the two-sample $t$ test often has a smaller probability of committing a Type II error (accepting $H_0$ when $H_1$ is true).

In all fairness to the two-sample format, there is another factor to be considered when comparing these two designs—degrees of freedom. If $n$ paired measurements are recorded, the number of degrees of freedom associated with the corresponding paired $t$ test is $n - 1$; if the same number of independent observations are recorded, the number of degrees of freedom associated with the corresponding two-sample $t$ test in $2n - 2 = 2(n - 1)$. In general, if two statistical procedures are equivalent in all respects except degrees of freedom, the one with the larger df will have the smaller probability of committing a Type II error (because its critical values will be closer to zero).

What we have here, then, is a trade-off. The advantage of the paired-data format is that it gives a smaller value for the denominator of its test statistic (and, therefore, a larger value for its ratio); the advantage of the two-sample format is that its critical values are closer to zero because it has more degrees of freedom. Which advantage outweighs the other depends on the effectiveness of the pairing. If the pairs can be chosen so as to eliminate a major source of variation, the paired-data format is superior (in terms of having a smaller $\beta$). If the pairing is ineffective, the corresponding loss in degrees of freedom may result in the two-sample format being better. The noteworthiness of the paired-data format is that in many, many situations pairing *can* be defined that will be dramatically effective and will more than offset the "cost" of losing degrees of freedom. Of all the hypothesis tests we have seen thus far, the paired $t$ test is easily the most important.

| To Test | Versus | Reject $H_0$ | Formula |
|---|---|---|---|
| $H_0$: $\mu_D = 0$ | $H_1$: $\mu_D \neq 0$ | $t \leqslant -t_{\alpha/2, n-1}$, $\quad t \geqslant t_{\alpha/2, n-1}$ | $t = \dfrac{\bar{d}}{s_D/\sqrt{n}}$ |
| | $H_1$: $\mu_D < 0$ | $t \leqslant -t_{\alpha, n-1}$ | $\bar{d} = \dfrac{1}{n} \sum\limits_{i=1}^{n} d_i$ |
| | $H_1$: $\mu_D > 0$ | $t \geqslant t_{\alpha, n-1}$ | $s_D = \sqrt{\dfrac{n \sum\limits_{i=1}^{n} d_i^2 - \left( \sum\limits_{i=1}^{n} d_i \right)^2}{n(n-1)}}$ |

**REVIEW QUESTIONS**

**11.12** Five departments in a large corporation claim to have men and women performing the same duties. Hearing that, the company's affirmative action officer files a grievance, claiming that men are receiving significantly larger raises. At the $\alpha = 0.10$ level, do the following data support her contention?

| | LAST YEAR'S RAISES (in %) | |
|---|---|---|
| Department | Men | Women |
| Accounting | 8.6 | 7.3 |
| Legal services | 9.5 | 8.9 |
| Personnel | 5.4 | 5.4 |
| Quality control | 8.6 | 6.1 |
| Development | 7.9 | 8.1 |

11.13   A chemist is studying the effect of a nitroglycerin additive on the melting temperature of certain polymers. Do the temperatures below (in °F) suggest that nitroglycerin *raises* the melting temperature? Answer the question by doing an appropriate hypothesis test at the 0.05 level of significance.

| Polymers | Control | Nitroglycerin Additive |
|---|---|---|
| A260 | 576 | 584 |
| B12c | 515 | 563 |
| B244 | 562 | 555 |
| A006 | 540 | 561 |
| C66d | 582 | 586 |

11.14   Eight high school instructors have volunteered for an experiment designed to compare two methods for teaching computer literacy. None of the eight has had any previous experience with word processors. The sample is to be divided into four pairs, with one member of a pair being taught Method A and the other member, Method B. Based on the biographical information that follows, how would you define the pairs?

| Name | Years As a Teacher | Subject Taught |
|---|---|---|
| TK | 22 | Chemistry |
| JB | 4 | English |
| AD | 10 | Social Studies |
| HW | 3 | Mathematics |
| HM | 2 | Mathematics |
| LL | 5 | English |
| DM | 17 | Mathematics |
| EP | 13 | History |

11.15   The two finalists in a city's annual "Favorite Restaurant" competition are Maurice's and Crawfisher. Six local celebrities make up the panel that selects the winner. Do the scores below (on a scale of 1 to 100) allow us to conclude that the difference in ratings is statistically significant? Do an appropriate test at the $\alpha = 0.10$ level.

| Judge | Maurice's | Crawfisher |
|-------|-----------|------------|
| LT | 72 | 85 |
| TU | 87 | 95 |
| HA | 85 | 80 |
| MM | 92 | 76 |
| SM | 66 | 74 |
| SS | 71 | 82 |

11.16 Twelve ranch-style houses were divided into six pairs on the basis of square footage and exposure to the sun. Awnings were installed on one house in each pair. Summarized below are the average electric bills for each house during the three hottest months of the summer. Test $H_0: \mu_D = 0$ against the alternative that awnings conserve electricity. Let $\alpha = 0.10$.

| House Pair | Without Awnings | With Awnings |
|------------|-----------------|--------------|
| A | $110 | $98 |
| B | 86 | 90 |
| C | 125 | 114 |
| D | 105 | 109 |
| E | 78 | 74 |
| F | 145 | 141 |

11.17 If $D_1, D_2, \ldots, D_n$ is a random sample of differences from a normal distribution with mean $\mu_D$, then $\dfrac{\bar{D} - \mu_D}{S_D/\sqrt{n}}$ has a Student $t$ distribution with $n - 1$ degrees of freedom. Using that result, derive a formula for a $100(1 - \alpha)\%$ confidence interval for $\mu_D$.

11.18 Recall the tensile strength data of Question 2.30. Let $y_i$ denote the $i$th tube's tensile strength after having been annealed at a high temperature; let $x_i$ denote the analogous measurement when the annealing was done at a moderate temperature. Define $d_i = y_i - x_i$. Use the answer to Question 11.17 to construct a 95% confidence interval for $\mu_D$. Note:

$$\sum_{i=1}^{15} d_i = 1.1 \quad \text{and} \quad \sum_{i=1}^{15} d_i^2 = 1.71$$

11.19 What does the confidence interval constructed in Question 11.18 imply about testing $H_0: \mu_D = 0$ versus $H_1: \mu_D \neq 0$ at the $\alpha = 0.05$ level of significance? Explain.

11.20 Shown below are the number of words per minute that each of five secretaries typed on two different word processors (A and B). Is there enough evidence to conclude that the two machines are not entirely comparable? Do a test using $\alpha = 0.05$.

| Secretary | Wpm on Machine A, $x_i$ | Wpm on Machine B, $y_i$ |
|---|---|---|
| Sheridan | 85 | 80 |
| Cathie | 66 | 64 |
| Dana | 91 | 76 |
| Cindy | 78 | 79 |
| Linda | 92 | 84 |

11.21 An experimenter collects a set of 50 paired observations and computes the $d_i$'s shown below. Would it be appropriate to test $H_0: \mu_D = 0$ by using the statistic in Theorem 11.1? Why or why not?

| | | | | | | | | | |
|---|---|---|---|---|---|---|---|---|---|
| 14 | 22 | 12 | 16 | −7 | 25 | 25 | 8 | 14 | 32 |
| 3 | 12 | −18 | −4 | 2 | 7 | 12 | 4 | −11 | −6 |
| 3 | 0 | −21 | 19 | 6 | 9 | −17 | 28 | −27 | −7 |
| −14 | −7 | −9 | 38 | 6 | 7 | −3 | 5 | 7 | 4 |
| −17 | −5 | 16 | 24 | 19 | −9 | −3 | 11 | 2 | 3 |

11.22 Two cable TV companies (Newday and Viatron) have submitted bids to service six cities. The figures below are their minimum monthly subscription rates. Local factors such as population density and ease of access to telephone poles affect the cost from city to city. Assume that these rates are typical of what the companies will charge in other cities. Let

$$\mu_D = \text{true average Viatron rate} - \text{true average Newday rate}$$

Construct an 80% confidence interval for $\mu_D$.

| City | Newday | Viatron |
|---|---|---|
| Nashville | $18.50 | $16.00 |
| Charlotte | 16.00 | 15.25 |
| Knoxville | 17.40 | 17.10 |
| Memphis | 19.80 | 18.20 |
| Little Rock | 16.25 | 16.50 |
| Lexington | 19.60 | 16.80 |

11.23 A study is being planned that will use certain pulmonary indicators to assess the therapeutic effect of aspirin on cardiac patients. The experiment is to be done using the paired data structure, with one member of a pair receiving 350 mg of aspirin daily. How would you group the following eight subjects into four pairs?

| Subject | Age | Sex |
|---------|-----|-----|
| BG | 45 | F |
| WW | 40 | M |
| JS | 70 | M |
| LL | 75 | F |
| SG | 68 | M |
| BP | 38 | F |
| JM | 72 | F |
| NY | 42 | M |

11.24   Does it matter whether $d_i$ is defined to be $y_i - x_i$ or $x_i - y_i$? Would it ever make sense to define $d_i$ as $x_i + y_i$? Explain.

11.25   (a) How does a table of paired data look different from a table of two-sample data?

(b) How does a graph of paired data look different from a graph of two-sample data?

11.26   Two machines can be used for assembling the steering mechanism of a riding lawn mower. A management consultant has timed seven workers on both machines. On the basis of the data below, is there reason to believe that one machine may be easier to use than the other. Let $\alpha = 0.10$.

| Worker | ASSEMBLY TIMES (min) | |
|--------|-----------|-----------|
|        | Machine 1 | Machine 2 |
| LF | 18 | 12 |
| MS | 23 | 21 |
| LB | 16 | 14 |
| ML | 18 | 13 |
| JT | 17 | 19 |
| BB | 18 | 23 |
| JG | 25 | 21 |

11.27   Graph the data in Question 11.12.

11.28   Does the difference in degrees of freedom between the paired data statistic and the two-sample statistic ($n - 1$ versus $n + m - 2$) become more or less of a factor as $n$ increases? (Hint: Look at the variation in $t_{\alpha,n}$ values from $t_{\alpha, 1}$ to $t_{\alpha, \infty}$).

11.29   Suppose that for a set of paired data $s_X^2$ and $s_Y^2$ are numerically much larger than $s_D^2$. Is that good or bad? Explain.

11.30   Listed below are last year's home batting averages for a team's starting eight players. Construct a 99% confidence interval for $\mu_D$, the difference between a player's batting average during day games and his batting average during night games.

| Player | | Nighttime BA, $x_i$ | Daytime BA, $y_i$ |
|--------|------|------|------|
| HA, | rf | .310 | .320 |
| EM, | 3b | .286 | .290 |
| WC, | lf | .302 | .298 |
| JA, | 1b | .280 | .287 |
| DC, | c | .214 | .226 |
| RS, | 2b | .302 | .300 |
| JL, | ss | .276 | .290 |
| BB, | cf | .285 | .292 |

11.31   Suppose a set of $(x_i, y_i)$'s has been collected according to the paired data structure. But after looking at the data we see immediately that the variation among the $x_i$'s is minimal and so is the variation among the $y_i$'s. Can we legitimately ignore the pairing and analyze the data by using a two-sample $t$ test?

11.32   Which (if either) of the following sets of paired data suggest that treatments $X$ and $Y$ might more effectively be compared by using a two-sample format? Explain.

| | EXPERIMENT #1 | | | | EXPERIMENT #2 | | |
|------|------|------|------|------|------|------|------|
| Pair | $x_i$ | $y_i$ | $d_i$ | Pair | $x_i$ | $y_i$ | $d_i$ |
| 1 | 26 | 28 | 2 | 1 | 46 | 48 | 2 |
| 2 | 7 | 4 | $-3$ | 2 | 47 | 44 | $-3$ |
| 3 | 15 | 16 | 1 | 3 | 47 | 48 | 1 |
| 4 | 47 | 51 | 4 | 4 | 46 | 50 | 4 |
| 5 | 8 | 10 | 2 | 5 | 47 | 49 | 2 |

11.33   It was pointed out in Section 11.4 that $n$ pairs of observations collected under the paired data format will often lead to fewer Type II errors than will $n$ pairs of observations collected under the two-sample format. What else can an experimenter do to reduce the probability of committing a Type II error?

11.34   Suppose each of 20 subjects is to be taught two memory-improving techniques. Does the order in which the techniques are presented matter? Explain.

11.35   Earlier studies showed that a biofeedback therapy was capable of lowering a hypertensive's blood pressure by 15 mm Hg, on the average. Now researchers are touting a new technique, claiming it has the ability to lower blood pressures even more. Before and after measurements on six patients are listed below. Test $H_0: \mu_D = -15$ versus $H_1: \mu_D < -15$ at the $\alpha = 0.10$ level of significance (Hint: See Question 11.2.)

| | Blood Pressure | |
| Patient | Before, $x_i$ | After, $y_i$ |
|---|---|---|
| AW | 120 | 100 |
| PR | 146 | 113 |
| LB | 141 | 126 |
| NS | 163 | 132 |
| JG | 132 | 110 |
| JL | 129 | 112 |

11.36   Construct a one-sided 90% confidence interval for the $\mu_D$ in Question 11.35. Does your interval agree with the result of the hypothesis test? Explain.

# 12 Regression–Correlation Data

*Chance is nothing; there is no such thing as chance. What we call by that name is the effect which we see of a cause which we do not see.*

*Voltaire*

# 12.1

## INTRODUCTION

*Comparison* was the watchword for essentially everything we did in Chapters 7 through 11. Is the mean $\mu$ associated with a certain population equal to some previously established value $\mu_0$? Are success probabilities $p_X$ and $p_Y$, characteristic of two series of Bernoulli trials, the same? Does the variance of treatment $X$ equal the variance of treatment $Y$? Making comparisons in these situations is possible because the units of the variables being measured are the same. In Case Study 2.4 for example, both $x$ and $y$ are silver percentages; in Case Study 2.5 the two variables are recorded in months, and both represent first walking times.

For experiments, though, where units are *not* comparable, we need to rethink our objectives. Recall Case Study 2.10. The subjects were 11 different galactic clusters: determined for each was $x$, its distance from the earth (in millions of light-years), and $y$, its velocity away from the earth (in thousands of miles/s). Does it make sense to compare the mean of the $x$'s to the mean of the $y$'s? No. Does it make sense to look at the *relationship* between the $x$'s and the $y$'s? Yes.

Section 2.5 and 3.5 have given us a good start in learning how to deal with data where $x$ is measured in units different from $y$. Functions are the key. We try to find an equation that approximates the nature of the dependency between the two variables. Typically, that means plotting the $x_i$'s versus the $y_i$'s and hoping the appearance of the resulting *scatter diagram* suggests what the relationship might reasonably be. In Case Study 2.9, for example, $x$ represents a county's radiation exposure, and $y$ its cancer mortality rate. From Figure 2.10 it seems clear that those two variables have a *linear* relationship, meaning that $y$ tends to vary with $x$ according to a function of the form $y = a + bx$. An entirely different function is appropriate for the data in Figure 2.13. There, the $x$ variable has a *curvilinear* relationship with $y$: $y = ax^b$.

The nature of the functional relationship between two traits is not the only factor that distinguishes one regression–correlation problem from another. What we assume about the measurements *individually* also matters. Are the $y_i$'s, for example, simply constants (that is, numbers not associated with any underlying probability function), or should they be thought of as measurements made on a random variable $Y$? Are the $x_i$'s being "set" by the experimenter (in which case they are not random), or do they vary, unpredictably, from sample to sample? In practice, three sets of assumptions are frequently encountered:

1. Both $x_i$ and $y_i$ are constants; neither represents a distribution nor has a pdf.
2. The $x_i$'s are constants, but the $y_i$'s are random variables. Such data will be written $(x_i, Y_i)$, $i = 1, 2, \ldots, n$, in keeping with our earlier conventions for denoting random variables.

3. Both sets of measurements are random variables. The data are then written as $(X_i, Y_i)$, $i = 1, 2, \ldots, n$.

Not surprisingly, the assumptions associated with a set of regression–correlation data go a long way toward determining which particular statistical analysis the data lend themselves to and which analyses are inappropriate. What we did in Section 3.5, finding least squares lines, is the approach to take for the first set of assumptions. In this chapter we direct our efforts to the latter two scenarios, situations where at least one of the measurements is considered a random variable.

The difference between the range of questions we can pose in this chapter and what we were limited to asking in Chapter 3 is enormous. *Inference*, in a word, is the reason that our options here are so greatly expanded. If the $Y_i$'s are random variables, then so is any quantity computed from those $Y_i$'s, such as the estimate of the regression line's slope or the estimate of its $y$ intercept. Instead of finding $y = \hat{a} + \hat{b}x$, therefore, where $\hat{a}$ and $\hat{b}$ are constants(recall Section 3.5) we now look for random variables $A$ and $B$, which are estimators of $a$ and $b$. Since $A$ and $B$ are random variables, they have pdf's. And since they have pdfs, we can set up hypothesis tests and construct confidence intervals for the *true* slope and the *true* $y$ intercept. Making such inferences, as this chapter will show, is often precisely what an experimenter needs to do most.

# 12.2

## THE LINEAR MODEL

### MLE's

One of the dominant themes of each nonregression data structure we have seen is the notion that probability functions can be very useful in modeling the behavior of random phenomena. When we say that SAT scores, for example, are normally distributed, we mean that test scores, though individually unpredictable, show *in the aggregate* a discernible pattern (in this case, a pattern that can be approximated by a bell-shaped curve). Associating a probability function with a phenomenon pays enormous dividends. Born of that marriage has been every one of our inference procedures—hypothesis tests as well as confidence intervals.

It often makes sense to introduce a similar probabilistic component into regression analysis. The most common formulation is a set of assumptions known as the *linear model*. In this section we want to motivate the statistical foundations for the model and preview several inference questions to which it can be applied.

Imagine a situation where the $x$ variable is not a variable but a condition "set" by the experimenter. Consider a clinical trial where the objective is to study the relationship between $x$, the dosage level of a new drug, and $y$, a subject's blood pressure. Would $x$ be allowed to vary at random? Not likely. Chances are the experimenter would preselect a set of dosage levels and administer each of those levels to several subjects. If so, then $x$, in the terminology of Chapter 6, would *not* be a random variable.

The nature of $y$, on the other hand, is entirely different. Would all subjects given the same dosage react in the same way? Probably not. Common sense tells us that a host of other factors (age, weight, cholesterol level, and so on) are likely to influence a subject's blood pressure. For each value of $x$, therefore, we might reasonably expect to observe an entire *distribution* of $y$ values, which is equivalent to saying that "blood pressure" is a random variable. What we are measuring, therefore, is not $(x_i, y_i)$ but $(x_i, Y_i)$. Different $x$ values, of course, might very well generate different $Y$ distributions, ones having different shapes, different means, or different standard deviations (or some combination of all three).

Figure 12.1 illustrates what we have just described: pictured is a sample of six blood pressures measured on subjects receiving three different dosages ($x_1$, $x_2$, and $x_3$). The dashed lines represent the blood pressure distributions that might theoretically be generated if each dosage were applied to a large number of subjects.

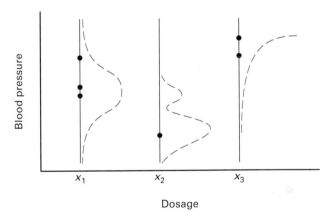

**FIG. 12.1**

linear model—A set of assumptions frequently invoked when doing a regression analysis. For each given value of $x$, the corresponding distributions of $Y$ are assumed to be normal with the same variance $\sigma^2$ and with means all lying on the same straight line $y = a + bx$.

Associating each value of $x$ with distributions as arbitrary as the three shown in Figure 12.1 would quickly lead to prohibitively difficult mathematical complications. Simplifying assumptions need to be made, and three are commonly invoked. Taken together, they define the ***linear model***:

1. For each value of $x$, $Y$ has a *normal* distribution.

2. Each $Y$ distribution has the same standard deviation $\sigma$.

3. The means of the $Y$ distributions all lie on the same straight line, $y = a + bx$.

A set of $n$ independent observations $(x_1, Y_1), \ldots, (x_n, Y_n)$, where each $Y_i$ satisfies the three conditions above, is a *random sample from the linear model*. Particular values of the sample $(x_1, y_1), \ldots, (x_n, y_n)$ will be called *regression data*.

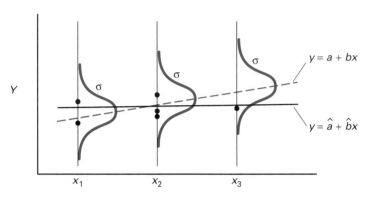

**FIG. 12.2**

**true regression line**—The line that connects the means of the $Y$ distributions in a linear model.

**estimated regression line**—A function that approximates the "true" relationship between two measurements.

Figure 12.2 summarizes the linear model graphically. The equation $y = a + bx$ is called the ***true regression line***. Its $y$ intercept $a$ and slope $b$ are unknown parameters (like $\mu$ in a one-sample $t$ test). The equation $y = \hat{a} + \hat{b}x$ is the ***estimated regression line***. We compute it from the actual data, using formulas we will see shortly. Ultimately, one of the primary objectives in this type of problem is to draw inferences about $a$ and $b$ using $\hat{a}$ and $\hat{b}$.

Before any kind of hypothesis test or confidence interval can be set up, all three parameters in the model ($a$, $b$, and $\sigma^2$) must be estimated. Recall in Section 3.5 that the first two were derived—in the non-random-variable-case—by appealing to a least squares criterion: that is, we choose $\hat{a}$ and $\hat{b}$ such that

$$\sum_{i=1}^{n} [y_i - (\hat{a} + \hat{b}x_i)]^2$$

was as small as possible. Here, the *method* of estimation is different, but numerically the answer is the same.

Let $(x_1, Y_1), (x_2, Y_2), \ldots, (x_n, Y_n)$ follow the assumptions defining the linear model. We estimate $a$ and $b$, using $A$ and $B$ respectively, where the "random" line $y = A + Bx$ is the "most probable" of all possible straight lines going through the data. A precise explanation of "most probable" will have to be omitted. In formal terminology, $A$ and $B$ are called *maximum likelihood estimators* (MLE's). Numerical values of the estimators $A$ and $B$ are referred to as *estimates* and denoted $\hat{a}$ and $\hat{b}$, respectively.

---

**THEOREM 12.1    Estimated Regression Line**

Let $(x_1, Y_1), (x_2, Y_2), \ldots, (x_n, Y_n)$ be a random sample from the linear model. The best estimators for $a$ and $b$ are

$$B = \frac{n \sum_{i=1}^{n} x_i Y_i - \left( \sum_{i=1}^{n} x_i \right) \left( \sum_{i=1}^{n} Y_i \right)}{n \sum_{i=1}^{n} x_i^2 - \left( \sum_{i=1}^{n} x_i \right)^2}$$

and

$$A = \frac{\sum_{i=1}^{n} Y_i - B \sum_{i=1}^{n} x_i}{n}$$

Furthermore, $B$ is a normal random variable with mean and variance given by $E(B)$ and Var $(B)$, respectively, where

$$E(B) = b \quad \text{and} \quad \text{Var } (B) = \frac{\sigma^2}{\sum_{i=1}^{n} (x_i - \bar{x})^2}$$

Similarly, $A$ is a normal random variable with

$$E(A) = a \quad \text{and} \quad \text{Var } (A) = \frac{\sigma^2 \sum_{i=1}^{n} x_i^2}{n \sum_{i=1}^{n} (x_i - \bar{x})^2}$$

---

What Theorem 12.1 states about the distributions of $A$ and $B$ is worth expanding. Suppose the true relationship between the means of the $Y$ distributions and the values of $x$ for some phenomenon is the straight line $y = a + bx$. Imagine taking $n$ observations $(x_1, y_1), (x_2, y_2), \ldots, (x_n, y_n)$ under those conditions and computing the estimated regression line $y = \hat{a} + \hat{b}x$. Now, imagine taking a second set of $n$ observations and computing a second estimated regression line. Would the $\hat{a}$ for the first sample be numerically the same as the $\hat{a}$ for the second sample? Would the two $\hat{b}$'s be the same? No to both questions, because the $(x_i, y_i)$'s for the two samples would be different.

If we continued taking samples of size $n$ and computing $\hat{a}$ and $\hat{b}$ on each occasion, we would generate an entire distribution of $\hat{a}$ values and a second distribution of $\hat{b}$ values. Figure 12.3 shows what Theorem 12.1 claims will be true of the $\hat{b}$ distribution, the more important of the two: it will have a bell shape, be centered at $b$, and have a variance

$$\frac{\sigma^2}{\sum_{i=1}^{n} (x_i - \bar{x})^2}$$

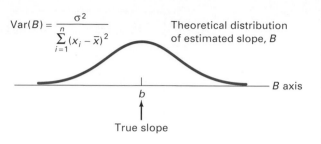

$$Var(B) = \frac{\sigma^2}{\sum\limits_{i=1}^{n}(x_i - \bar{x})^2}$$

Theoretical distribution
of estimated slope, $B$

$B$ axis

$b$

True slope

**FIG. 12.3**

(If the language here sounds familiar, it should: Theorem 12.1 is simply giving the properties of what Chapter 7 would have called the *sampling distribution of B*.)

**EXAMPLE 12.1**   A grass seed company is experimenting with a new hybrid fescue designed for golf courses. Still to be determined is the relationship between $x$, the density of seeds planted (in pounds per 500 ft$^2$), and $Y$, the quality of the resulting lawn. Eight presumably similiar fairways are selected to be test areas: each is sown according to the seed densities in Table 12.1. Four weeks later the grass growing on each fairway is subjectively rated on a scale from 0 to 100 (see column 3).

**TABLE 12.1**

| Fairway | Seed Density (lb/500 ft$^2$), $x$ | Lawn quality, $y$ |
|---------|----------------------------------|-------------------|
| 1 | 1.0 | 30 |
| 2 | 1.0 | 40 |
| 3 | 2.0 | 40 |
| 4 | 3.0 | 40 |
| 5 | 3.0 | 50 |
| 6 | 3.0 | 65 |
| 7 | 4.0 | 50 |
| 8 | 5.0 | 50 |

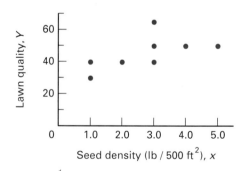

**FIG. 12.4**

Figure 12.4 shows the data plotted as a scatter diagram. Notice how the points reflect the assumptions stipulated by the linear model: (1) when $x$ is constant, $y$ still varies from sample to sample (an $x$ of 3.0, for example, produced $Y$ values of 40, 50, and 65) and (2) the actual means of the $Y$ distributions associated with each value of $x$ do not seem incompatible with the presumption that the *true* means are all collinear. Estimate the true regression line and compute the probability that $B$ is within 0.50 of the true slope. Assume $\sigma^2 = 80.0$.

**TABLE 12.2**

| $x_i$ | $Y_i$ | $x_iY_i$ | $x_i^2$ |
|-------|-------|----------|---------|
| 1.0 | 30 | 30.0 | 1.00 |
| 1.0 | 40 | 40.0 | 1.00 |
| 2.0 | 40 | 80.0 | 4.00 |
| 3.0 | 40 | 120.0 | 9.00 |
| 3.0 | 50 | 150.0 | 9.00 |
| 3.0 | 65 | 195.0 | 9.00 |
| 4.0 | 50 | 200.0 | 16.00 |
| 5.0 | 50 | 250.0 | 25.00 |
| 22.0 | 365 | 1065.0 | 74.00 |

$\bar{x} = 2.75$

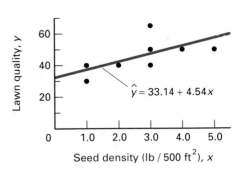

FIG. 12.5

**Solution**  Table 12.2 shows the intermediate computations for finding the values $\hat{a}$ and $\hat{b}$ of $A$ and $B$ for the given sample. From Theorem 12.1 the estimated slope is

$$\hat{b} = \frac{8(1065.0) - (22.0)(365)}{8(74.00) - (22.0)^2} = 4.54$$

and the estimated $y$ intercept is

$$\hat{a} = \frac{365 - (4.54)(22.0)}{8} = 33.14$$

The graph of the equation $y = 33.14 + 4.54x$ is shown in Figure 12.5.

Finding the probability that $B$ is within 0.50 of the true slope is an exercise in doing a Z transformation. The basic strategy follows the same pattern we have used so many times before. if $B$ is normally distributed with mean $b$ and variance $\sigma^2 \Big/ \sum\limits_{i=1}^{n} (x_i - \bar{x})^2$, then

$$\frac{B - b}{\sqrt{\dfrac{\sigma^2}{\sum\limits_{i=1}^{n} (x_i - \bar{x})^2}}}$$

$$\left( \frac{\text{Normal random variable} - E \text{ (normal random variable)}}{\text{Standard deviation of normal random variable}} \right)$$

is distributed like a *standard* normal, where $\sigma^2$, by assumption, is 80.0 and

$$\sum_{i=1}^{n} (x_i - \bar{x})^2 = (1.0 - 2.75)^2 + (1.0 - 2.75)^2 + \cdots + (5.0 - 275)^2 = 13.50$$

Therefore,

$$P(B \text{ is within } 0.50 \text{ of the true slope})$$

$$= P(-0.50 \leqslant B - b \leqslant 0.50)$$

$$= P\left(\frac{-0.50}{\sqrt{80.0/13.50}} \leqslant \frac{B - b}{\sqrt{80.0/13.50}} \leqslant \frac{0.50}{\sqrt{80.0/13.50}}\right)$$

$$= P\left(\frac{-0.50}{2.43} \leqslant Z \leqslant \frac{0.50}{2.43}\right) = P(-0.20 \leqslant Z \leqslant 0.20)$$

$$= P(Z \leqslant 0.20) - P(Z < -0.20) = 0.5793 - 0.4207$$

$$= 0.1586$$

If we collect $n = 8$ observations, in other words, when $\sigma^2 = 80.0$ and

$$\sum_{i=1}^{n} (x_i - \bar{x})^2 = 13.50$$

the probability is 0.16 that the estimated slope $B$ is within 0.50 of the true slope $b$.

What can we say about the particular slope estimate that we computed? Only this: as a member of the distribution of all possible $B$'s, 4.54 may be one of the 16% that lie "close" to $b$, or it may be one of the 84% that don't. The odds would suggest the latter, but there is no way of knowing for certain to which group it belongs.

---

QUESTIONS

**12.1** The final computation in Example 12.1 assigned a probability to the event "$-0.50 < B - b < 0.50$." Based on that result, what "name" could appropriately be given to the random interval $(B - 0.50, B + 0.50)$? Explain.

**12.2** For the data summarized in Table 12.1 is it surprising that the same value of $x$ would produce different values of $Y$? Why or why not?

**12.3** In case Study 2.9 the relationship between $x$, a county's index of radiation exposure, and $Y$, its cancer mortality rate, is described by the estimated regression line $y = 114.72 + 9.23x$. Suppose that $\sigma^2$, the variance for each of the $Y$ distributions, is 200. Compute the probability that the estimated $y$ intercept $A$ will be within 4.0 of the true $a$.
$$\left(\text{Hint: } \sum_{i=1}^{9} (x_i - \bar{x})^2 = 97.5.\right)$$

**12.4** Using the information in Question 12.3, construct a 95% confidence interval for the true slope $b$.

**12.5**   An experimenter collects the following set of regression data, which she presumes are following the linear model:

| $x_i$ | 5.0 | 5.0 | 10.0 | 15.0 | 15.0 | 15.0 |
|-------|-----|-----|------|------|------|------|
| $y_i$ | 6.0 | 8.0 | 5.0  | 4.0  | 5.0  | 2.0  |

(a) Plot the data as a scatter diagram.

(b) Compute the estimated regression line $y = \hat{a} + \hat{b}x$.

(c) Plot the regression line on the scatter diagram.

(d) Describe the sampling distributions of $A$ and $B$. Assume that $\sigma^2 = 1.5$.

**12.6**   Suppose an experimenter plans to take eight $(x_i, Y_i)$'s and intends for the $x_i$'s to be numbers somewhere in the interval from $-5$ to $+5$, inclusive. What values should be assigned to the $x_i$'s if the objective is to minimize the variance of $B$?

**12.7**   Suppose the relationship between a set of constants $(x_i)$ and a random variable $(Y_i)$ follows all three assumptions of the linear model. If $\sigma^2 = 12.0$ and $\sum_{i=1}^{n} (x_i - \bar{x})^2 = 6.5$, what is the probability that $B$ overestimates $b$ by at least 1.8?

### Computing the Variance

With Theorem 12.1 to tell us how to compute $\hat{a}$ and $\hat{b}$, only $\sigma^2$ is left to deal with. So far, our treatment of that particular parameter has been brief: when a value for the variance was needed in Example 12.1 we simply "assumed" one. In practice, doing that would not be acceptable. Like every other unknown parameter in every other statistical analysis, $\sigma^2$ needs to be estimated formally from the actual data.

Recall how $\sigma^2$ is estimated in nonregression situations. Figure 12.6 shows a set of one-sample data $(y_1, y_2, \ldots, y_n)$ and how the data might be scattered around their sample mean $\bar{y}$. The magnitude of that scatter is measured by first computing the sum of the squared deviations of the $y_i$'s from $\bar{y}$—that is, $\sum_{i=1}^{n} (y_i - \bar{y})^2$. For mathematical reasons we then divide that sum by $n - 1$, and the resulting quotient is our estimate; the sample variance

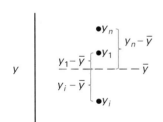

**FIG. 12.6**

$$s^2 = \frac{1}{n-1} \sum_{i=1}^{n} (y_i - \bar{y})^2 \tag{12.1}$$

(In this chapter's notation, $s^2$ would be written $\hat{\sigma}^2$).

Generalizing Equation 12.1 to include regression data is not difficult. Figure 12.7 shows the analog of Figure 12.6. The deviation of the $i$th point from its estimated expected value is $y_i - \hat{y}_i = y_i - (\hat{a} + \hat{b}x_i)$ rather than $y_i - \bar{y}$.

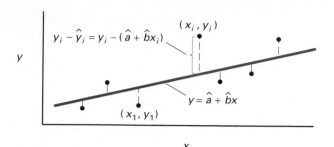

**FIG. 12.7**

The regression equivalent of $\sum_{i=1}^{n} (y_i - \bar{y})^2$, therefore, is

$$\sum_{i=1}^{n} (y_i - \hat{y}_i)^2 \quad \text{or} \quad \sum_{i=1}^{n} [y_i - (\hat{a} + \hat{b}x_i)]^2$$

Dividing the latter by $n - 2$ (instead of $n - 1$) gives the quotient we will use as our estimate for $\sigma^2$.

---

**DEFINITION 12.1** **Estimating the Variance**

Let $(x_1, y_1)$, $(x_2, y_2)$, . . . , $(x_n, y_n)$ be a set of regression data. Let $\sigma^2$ denote the true (unknown) variance of each $Y$ distribution, and let $\hat{y}_i = \hat{a} + \hat{b}x_i$ be the estimated regression line of $y_i$. Then $\sigma^2$ is estimated by $\hat{\sigma}^2$, where

$$\hat{\sigma}^2 = \frac{1}{n-2} \sum_{i=1}^{n} (y_i - \hat{y}_i)^2 = \frac{1}{n-2} \sum_{i=1}^{n} [y_i - (\hat{a} + \hat{b}x_i)]^2 \quad (12.2)$$

---

**COMMENT:** We call

$$\sum_{i=1}^{n} [y_i - (\hat{a} + \hat{b}x_i)]^2$$

the *sum of squares for error* (SSE), and

$$\frac{1}{n-2} \sum_{i=1}^{n} [y_i - (\hat{a} + \hat{b}x_i)]^2$$

the *mean sum of squares for error* (MSE). The estimated variance, then, can be written as $\hat{\sigma}^2$, MSE, or SSE/$(n - 2)$. The second and third forms are the ones most commonly used in commercial software packages.

Using Equation 12.2 to complete $\hat{\sigma}^2$ is tedious, even with the help of a calculator. An easier method is to use Equation 12.4 and then divide by

$n - 2$. Only the factor $\sum_{i=1}^{n} y_i^2$ entails any additional work: values for the other four factors are already known from finding the estimated regression line.

---

**THEOREM 12.2  A Computing Formula for $\hat{\sigma}^2$**

Let $(x_1, y_1), (x_2, y_2), \ldots, (x_n, y_n)$ be a set of data satisfying the linear model, where $y = \hat{a} + \hat{b}x$ is the estimated regression line and $\sigma^2$ is the unknown variance of the $Y$ distributions. Then.

$$\hat{\sigma}^2 = \frac{\text{SSE}}{n - 2} \tag{12.3}$$

where

$$\text{SSE} = \sum_{i=1}^{n} y_i^2 - \hat{a} \sum_{i=1}^{n} y_i - \hat{b} \sum_{i=1}^{n} x_i y_i \tag{12.4}$$

---

The quantity SSE is not only an estimate of the variance, but also the kernel of a method to measure how much of the variability in the data is explained by the linear model. Let $\hat{y}_i = \hat{a} + \hat{b}x_i$. We have already noted that in regression problems

$$\text{SSE} = \sum_{i=1}^{n} (y_i - \hat{y}_i)^2 \quad \text{replaces} \quad \sum_{i=1}^{n} (y_i - \bar{y})^2$$

as the key to estimating the variability of $Y$. Since $\sum_{i=1}^{y} (y_i - \bar{y})^2$ ignores the $xy$ relationship altogether, the difference

$$\sum_{i=1}^{n} (y_i - \bar{y})^2 - \sum_{i=1}^{n} (y_i \quad \hat{y}_i)^2 \tag{12.5}$$

measures the reduction in "error" due to using the linear model. A small value of this difference suggests that the linear model explains very little. However, we cannot say that the value is small in any absolute sense; there must be some scale of comparison. One way to put Equation 12.5 in perspective is to compare the reduction in error to the "total" error. That is, we will compute

$$r^2 = \frac{\sum_{i=1}^{n} (y_i - \bar{y})^2 - \sum_{i=1}^{n} (y_i - \hat{y}_i)^2}{\sum_{i=1}^{n} (y_i - \bar{y})^2} \tag{12.6}$$

This quantity is called the *coefficient of determination*: it measures the strength of the linear relationship between $x$ and $y$. Values of $r^2$ close to 1 indicate that the linear relationship is very strong.

Equivalently, if the denominator of the fraction in Equation 12.6 is divided into the numerator, then Equation 12.6 can be written

$$r^2 = 1 - \frac{\sum_{i=1}^{n} (y_i - \hat{y}_i)^2}{\sum_{i=1}^{n} (y_i - \bar{y})^2} = 1 - \frac{\text{SSE}}{\sum_{i=1}^{n} (y_i - \bar{y})^2}$$

The latter two expressions show why the coefficient of determination is denoted $r^2$. In the terminology of Definition 3.4,

$$1 - \frac{\sum_{i=1}^{n} (y_i - \hat{y}_i)^2}{\sum_{i=1}^{n} (y_i - \bar{y})^2} = 1 - \frac{\sum_{i=1}^{n} (y_i - (\hat{a} + \hat{b}x_i))^2}{\sum_{i=1}^{n} (y_i - \bar{y})^2}$$

is the square of the sample correlation coefficient.

---

**QUESTIONS**

**12.8** Compute $\hat{\sigma}^2$ for the grass seed data in Example 12.1. Is your answer close to the value assumed when we calculated $P(-0.50 < B - b < 0.50)$?

**12.9** What would the scatter diagram for a set of data look like if $\hat{\sigma}^2 = 0$?

**12.10** Use the formulas in Definition 12.1 and Theorem 12.2 to estimate $\sigma^2$ for the following set of regression data. Do your two answers agree?

| $x_i$ | 3 | 1 | 3 |
|-------|---|---|---|
| $y_i$ | 1 | 1 | 3 |

**12.11** Suppose $y = a + bx$ is the true relationship between the means of the $Y$ distributions and $x$. Is it possible for $\hat{\sigma}^2$ to be 0 if $\sigma^2 > 0$? Is it possible for $\sigma^2$ to be 0 if $\hat{\sigma}^2 > 0$?

### Hypothesis Testing

It was stated earlier in this section that generalizing from an estimated regression line $y = \hat{a} + \hat{b}x$ to a true regression line $y = a + bx$ is often an experimenter's primary statistical objective when collecting linear-model-type data. If hypothesis testing is the mode of inference to be used, three $H_0$'s immediately suggest themselves:

$$(1) \quad H_0: a = a_0, \qquad (2) \quad H_0: b = b_0, \qquad (3) \quad H_0: \sigma^2 = \sigma_0^2$$

We can handle the first two by setting up a $t$ ratio, using much the same format that we followed when testing means in Chapters 9, 10, and 11; the third requires a $\chi^2$ test, similar to what we did in Section 9.5. Theorem 12.3 shows how to carry out each test. Only the two-sided decision rules are given. The rejection regions can be modified to accommodate one-sided $H_1$'s in the usual way (see Example 12.2).

**THEOREM 12.3    Testing in the Linear Model**

Let $(x_1, y_1)$, $(x_2, y_2)$, . . . , $(x_n, y_n)$ be a set of regression data. Let $y = \hat{a} + \hat{b}x$ be the line estimating the true regression equation $y = a + bx$, and let

$$\hat{\sigma}^2 = \frac{1}{n-2}\left(\sum_{i=1}^n y_i^2 - \hat{a}\sum_{i=1}^n y_i - \hat{b}\sum_{i=1}^n x_i y_i\right)$$

be the estimate of the true variance $\sigma^2$ associated with each $Y$ distribution. Let $\alpha$ be the level of significance.

1. To test
$$H_0: \quad a = a_0$$
vs.
$$H_1: \quad a \neq a_0$$

compute
$$t = \frac{(\hat{a} - a_0)\sqrt{n\sum_{i=1}^n (x_i - \bar{x})^2}}{\sqrt{\hat{\sigma}^2 \sum_{i=1}^n x_i^2}}$$

If $t \leqslant -t_{\alpha/2, n-2}$ or $t \geqslant t_{\alpha/2, n-2}$, reject $H_0: a = a_0$.

2. To test
$$H_0: \quad b = b_0$$
vs.
$$H_1: \quad b \neq b_0$$

compute
$$t = \frac{(\hat{b} - b_0)\sqrt{\sum_{i=1}^n (x_i - \bar{x})^2}}{\sqrt{\hat{\sigma}^2}}$$

If $t \leqslant -t_{\alpha/2, n-2}$ or $t \geqslant t_{\alpha/2, n-2}$, reject $H_0: b = b_0$.

3. To test
$$H_0: \quad \sigma^2 = \sigma_0^2$$
vs.
$$H_1: \quad \sigma^2 \neq \sigma_0^2$$

compute
$$\chi^2 = \frac{(n-2)\hat{\sigma}^2}{\sigma_0^2}$$

If $\chi^2 \leqslant \chi_{\alpha/2, n-2}^2$ or $\chi^2 \geqslant \chi_{1-\alpha/2, n-2}^2$, reject $H_0: \sigma^2 = \sigma_0^2$.

In real-world problems the slope $b$ is often the most important linear model parameter. Its magnitude indicates the extent to which $Y$ is affected by a unit change in $x$, and its sign shows whether $Y$ varies *directly* or *inversely* with $x$. If the slope is 0, we can infer that there is no linear relationship. This last property leads to perhaps the most frequent application of Theorem 12.3—testing $H_0: b = 0$.

Arguably, the most fundamental question to be addressed in connection with linear model data is whether a presumed $xy$ relationship is, in fact, genuine. Just because we fit a regression line $y = \hat{a} + \hat{b}x$ to a set of $n$ points does not guarantee that $Y$ has any relationship to $x$ whatsoever. Whether a linear relationship is "real" depends *only* on the value of $b$. Specifically, if $b = 0$, we say that $Y$ is *not* linearly related to $x$, because if the true slope of a regression is zero, then changing the value for $x$ has no predictable effect on the magnitude of $Y$. In other words, $Y$ is "independent" of $x$ if $b = 0$.

Statistically, we deal with independence questions that arise in regression contexts by letting $b = 0$ be a null hypothesis. *Only if the estimated slope $\hat{b}$ is significantly different from zero will we conclude that the $xy$ relationship is real.*

---

*EXAMPLE 12.2*    The radiation data in Case Study 2.9 is a good example of a situation where establishing whether a relationship is genuine is an essential first step. How we interpret those data depends entirely on whether we believe that (1) $b = 0$ or (2) $b > 0$. In Table 2.10, each of nine Oregon counties was associated with an index of radiation exposure $x$ and a cancer mortality rate $y$. The latter can quite reasonably be viewed as the value of a random variable; even if two counties had identical values for $x$, we would not expect their $y$ values to be exactly the same.

A graph of the data (see Figure 2.10) shows an apparent linear relationship, the estimated regression line being $\hat{y} = 114.72 + 9.23x$. Can we infer that cancer mortality rates, in general, increase when radiation exposure increases?

*Solution*    By itself, $\hat{b} > 0$ does not imply that $b > 0$. Before we can legitimately conclude that Hanford's storage problems have produced an environmental hazard, we must eliminate the possibility that $b$ might be 0 despite the fact that $\hat{b}$ is 9.23. The only way to do that is by formally testing $H_0: b = 0$ versus $H_1: b > 0$ and rejecting $H_0$. (Notice that the alternative here is appropriately one-sided: there is no reason to believe that high levels of radiation would ever *lower* a person's risk of cancer.) Let $\alpha = 0.05$.

First, we need to estimate $\sigma^2$. For the data in Table 2.10.

$$\sum_{i=1}^{9} x_i = 41.56 \qquad \sum_{i=1}^{9} y_i = 1416.1$$

$$\sum_{i=1}^{9} x_i^2 = 289.4222 \qquad \sum_{i=1}^{9} y_i^2 = 232{,}498.97$$

$$\sum_{i=1}^{9} x_i y_i = 7{,}439.37$$

From Equation 12.4

$$SSE = \sum_{i=1}^{9} y_i^2 - \hat{a} \sum_{i=1}^{9} y_i - \hat{b} \sum_{i=1}^{9} x_i y_i$$

$$= 232{,}498.97 - (114.72)(1416.1) - (9.23)(7439.37)$$

$$= 1378.595$$

so that

$$\hat{\sigma}^2 = \frac{SSE}{n-2} = \frac{1378.595}{9-2} = 196.94$$

Also,

$$\sum_{i=1}^{9} (x_i - \bar{x})^2 = \left(2.49 - \frac{41.56}{9}\right)^2 + \left(2.57 - \frac{41.56}{9}\right)^2 + \cdots + \left(8.34 - \frac{41.56}{9}\right)^2$$

$$= 97.51$$

From part 2 of Theorem 12.3,

$$t = \frac{(\hat{b} - b_0)\sqrt{\sum_{i=1}^{n}(x_i - \bar{x})^2}}{\hat{\sigma}^2} = \frac{(9.23 - 0)\sqrt{97.51}}{\sqrt{196.94}} = 6.49$$

Figure 12.8 shows the appropriate rejection region (for a one-sided $H_1: b > 0$, we should reject $H_0$ if $t > t_{\alpha, n-2}$). Since the observed $t\ (=6.49)$ exceeds the "critical" $t\ (=1.8946)$, we reject the null hypothesis. Chance, the $t$ test is saying, can be effectively ruled out as the "cause" of the apparent linear dependency of $Y$ on $x$: an estimated slope of 9.23 is much too large to be compatible with a true slope of zero.

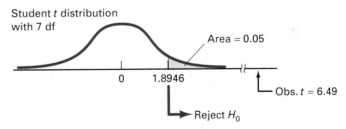

**FIG. 12.8**

*EXAMPLE 12.3*   When a straight line is to be fit to a set of data, the equation an experimenter usually has in mind is $y = a + bx$. In some applications, though, there are physical reasons for believing that the $y$ intercept $a$ must necessarily be zero, in which case the equation reduces to $y = bx$. The difference between the two models is more than cosmetic.

**Solution**   If $y = a + bx$, the estimate for $b$ is

$$\hat{b} = \frac{n \sum\limits_{i=1}^{n} x_i y_i - \left(\sum\limits_{i=1}^{n} x_i\right)\left(\sum\limits_{i=1}^{n} y_i\right)}{n \sum\limits_{i=1}^{n} x_i^2 - \left(\sum\limits_{i=1}^{n} x_i\right)^2}$$

On the other hand, the estimated slope when $y = bx$ is

$$\hat{b} = \frac{\sum\limits_{i=1}^{n} x_i y_i}{\sum\limits_{i=1}^{n} x_i^2}$$

In Case Study 2.10 we encountered a situation where the question of whether $a$ belonged in the model was a legitimate concern. The $x$ measurement being recorded was a galaxy's distance from the earth; the $y$ measurement was that same galaxy's velocity away from the earth (see Table 2.11). If the Big Bang hypothesis accurately explains the origin of the universe, then $y$ should equal 0 when $x$ equals 0. In analyzing the 11 data points, Hubble took precisely that position and fit the model $y = bx$. (His estimated regression line, $\hat{y} = 0.03544x$, is pictured in Figure 2.11.)

Are we absolutely certain though, that $a$ should be excluded from the original model? Not really. In matters concerning the origin of the universe, "certainty" is not a condition likely to occur very often. Fortunately, we can use Hubble's data to test the $a = 0$ hypothesis formally, by fitting the two-parameter model $y = a + bx$ and then using part 1 of Theorem 12.3 to test $H_0$: $a = 0$.

For the 11 $(x_i, y_i)$'s in Table 2.11,

$$\sum_{i=1}^{11} x_i = 4185 \qquad\qquad \sum_{i=1}^{11} y_i = 148.65$$

$$\sum_{i=1}^{11} x_i^2 = 2{,}685{,}141 \qquad\qquad \sum_{i=1}^{11} y_i^2 = 3376.313$$

and

$$\sum_{i=1}^{11} x_i y_i = 95{,}161.20$$

Also,

$$\sum_{i=1}^{11} (x_i - \bar{x})^2 = \sum_{i=1}^{11} \left(x_i - \frac{4185}{11}\right)^2 = 1{,}092{,}938.7$$

From Theorem 12.1 we find that

$$\hat{b} = \frac{11(95{,}161.20) - (4185)(148.65)}{11(2{,}685{,}141) - (4185)^2} = 0.03532$$

and

$$\hat{a} = \frac{148.65 - (0.03532)(4185)}{11} = 0.07598$$

Therefore, the estimated (two-parameter) regression line has the equation $\hat{y} = 0.07598 + 0.03532x$. (Notice that the estimate for $a$ is *not* zero and $\hat{b}$ is slightly different from the $\hat{b}$ estimated in Figure 2.10).

Our objective is to decide whether a "true $a$" of zero is likely to have produced an estimated $y$ intercept as large as 0.07598. If the two are compatible, then we should accept $H_0$: $a = 0$ (and conclude that the one-parameter model chosen in Case Study 2.10 was appropriate).

To begin, we estimate $\sigma^2$ by using Equations 12.3 and 12.4:

$$\text{SSE} = 3376.313 - (0.07598)(148.65) - (0.03532)(95,161.20)$$

$$= 3.92499$$

which makes

$$\hat{\sigma}^2 = \frac{3.92499}{11 - 2} = 0.43611$$

Write the hypotheses to be tested as

$$H_0: \quad a = 0$$

vs.

$$H_1: \quad a \neq 0$$

Suppose $\alpha$ is set at 0.05. By part 1 of Theorem 12.3, $H_0$ should be rejected if $t \leqslant -2.2622 \ (= -t_{0.025, 9})$ or $t \geqslant 2.2622 \ (= t_{0.025, 9})$, where

$$t = \frac{(\hat{a} - a_0)\sqrt{n \sum_{i=1}^{n} (x_i - \bar{x})^2}}{\sqrt{\hat{\sigma}^2 \sum_{i=1}^{n} x_i^2}}$$

In this case,

$$t = \frac{(0.07598 - 0)\sqrt{11(1,092,938.7)}}{\sqrt{(0.43611)(2,685,141)}} = 0.24$$

so our decision is clear-cut: we should accept $H_0$: $a = 0$. In other words, the model we chose in Case Study 2.10 is supported by the data.

---

QUESTIONS

**12.12**   For the grass seed data in Example 12.1, test $H_0$: $b = 0$ against a two-sided alternative. Let $\alpha = 0.05$. Note: Evaluating $t$ can be simplified by using the fact that

$$\sum_{i=1}^{n} (x_i - \bar{x})^2 = \sum_{i=1}^{n} x_i^2 - \frac{\left(\sum_{i=1}^{n} x_i\right)^2}{n}$$

**12.13**   Verify the computing formula given in Question 12.12 by showing that

$$\sum_{i=1}^{n} x_i^2 - \frac{\left(\sum_{i=1}^{n} x_i\right)^2}{n}$$

for the radiation data in Example 12.2 is equal to 97.51, the value for

$$\sum_{i=1}^{n} (x_i - \bar{x})^2$$

given on p. 519.

12.14    To answer the second question raised in Example 12.1, we assumed that $\sigma^2 = 80.0$. Test the reasonableness of that value by applying part 3 of Theorem 12.3 to the data in Table 12.2. Make $H_1$ two-sided and let $\alpha = 0.05$.

12.15    Speculation has persisted for years that the altitude of a ballpark and the number of home runs hit in that ballpark are linearly related. Shown below are the 1972 home run totals recorded at each American League stadium. Decide whether altitude $(x)$ and number of home runs $(y)$ are linearly related by testing $H_0: b = 0$ versus $H_1: b \neq 0$. Let $\alpha = 0.05$.

| Club | Altitude (ft) | Number of Home Runs |
|---|---|---|
| Cleveland | 660 | 138 |
| Milwaukee | 635 | 81 |
| Detroit | 585 | 135 |
| New York | 55 | 90 |
| Boston | 21 | 120 |
| Baltimore | 20 | 84 |
| Minnesota | 815 | 106 |
| Kansas City | 750 | 57 |
| Chicago | 595 | 109 |
| Texas | 435 | 74 |
| California | 340 | 61 |
| Oakland | 25 | 120 |

Note:

$$\sum_{i=1}^{12} x_i = 4936 \qquad \sum_{i=1}^{12} y_i = 1175$$

$$\sum_{i=1}^{12} x_i^2 = 3{,}071{,}116 \qquad \sum_{i=1}^{12} y_i^2 = 123{,}349$$

$$\sum_{i=1}^{12} x_i y_i = 480{,}565$$

12.16    By late 1971 all cigarette packs had to be labeled with the words, "Warning: The Surgeon General Has Determined That Cigarette Smoking Is Dangerous To Your Health." The case against smoking rested heavily on statistical, rather than laboratory, evidence. Extensive surveys of smokers and nonsmokers had revealed the former to have much higher risks of dying from a variety of causes, most notably lung cancer and heart disease. Other types of studies, some designed with a much broader focus, painted much the same picture. Typical are the data below: 21 countries were the subjects. Recorded for each country was $x$, its annual cigarette consumption, and $y$, its mortality rate due to coronary heart disease (108).

| Year | Country | Cigarette Consumption per Adult per Year, $x$ | CHD Mortality per 100,000 (ages 35–64), $y$ |
|------|---------|-----------------------------------------------|---------------------------------------------|
| 1962 | United States | 3900 | 256.9 |
| 1962 | Canada | 3350 | 211.6 |
| 1962 | Australia | 3220 | 238.1 |
| 1962 | New Zealand | 3220 | 211.8 |
| 1963 | United Kingdom | 2790 | 194.1 |
| 1962 | Switzerland | 2780 | 124.5 |
| 1962 | Ireland | 2770 | 187.3 |
| 1962 | Iceland | 2290 | 110.5 |
| 1962 | Finland | 2160 | 233.1 |
| 1963 | West Germany | 1890 | 150.3 |
| 1962 | Netherlands | 1810 | 124.7 |
| 1962 | Greece | 1800 | 41.2 |
| 1962 | Austria | 1770 | 182.1 |
| 1962 | Belgium | 1700 | 118.1 |
| 1962 | Mexico | 1680 | 31.9 |
| 1963 | Italy | 1510 | 114.3 |
| 1961 | Denmark | 1500 | 144.9 |
| 1962 | France | 1410 | 59.7 |
| 1962 | Sweden | 1270 | 126.9 |
| 1961 | Spain | 1200 | 43.9 |
| 1962 | Norway | 1090 | 136.3 |

(a) Plot the data. Do the assumptions of the linear model seem justified?

(b) Test $H_0$: $b = 0$ versus $H_1$: $b > 0$. Let $\alpha = 0.05$. Note:

$$\sum_{i=1}^{21} x_i = 45{,}110 \qquad \sum_{i=1}^{21} y_i = 3042.2$$

$$\sum_{i=1}^{21} x_i^2 = 109{,}957{,}100 \qquad \sum_{i=1}^{21} y_i^2 = 529{,}321.6$$

$$\sum_{i=1}^{21} x_i y_i = 7{,}319{,}602$$

12.17 For the tethered mosquito data in Question 1.8,

(a) Plot the data and use Theorem 12.1 to find the estimated regression line $y = \hat{a} + \hat{b}x$

(b) At the $\alpha = 0.05$ level of significance, test $H_0$: $b = 0$ versus $H_1$: $b \neq 0$.

12.18 Plot the nemesis data of Question 1.9. Fit the four points with a two-parameter linear model.

(a) What does the $y$ intercept represent in this situation?

(b) Is it likely that $a = -14.0$? Test an appropriate hypothesis. Let $\alpha = 0.05$.

12.19 A woman's chances of being rehired, as a function of how many years she had been away from the workforce, was discussed in Case Study 3.15. The estimated regression line was $y = 96.4 - 4.9x$ (see p. 136). What would we expect the *true* $y$ intercept to be in that situation? Set up and carry out the appropriate hypothesis test. Let $\alpha = 0.05$.

12.20    Part (**d**) of Question 3.62 asked for an informal conclusion about the relationship between two variables measured in a child behavior study. Answer that same question formally by doing the relevant hypothesis test. For the five children who participated,

$$\sum_{i=1}^{5} x_i = 29 \qquad \sum_{i=1}^{5} y_i = 26$$

$$\sum_{i=1}^{5} x_i^2 = 191 \qquad \sum_{i=1}^{5} y_i^2 = 156$$

$$\sum_{i=1}^{5} x_i y_i = 166$$

Let 0.01 be the level of significance.

12.21    Determining small quantities of calcium in the presence of magnesium has proven to be a difficult problem for analytical chemists. Direct precipitation, a standard technique often used in similar situations, is not feasible for these chemicals. One method proposed as a possible alternative involves the use of alcohol as a solvent. The data below show the results of applying the alcohol method to 10 mixtures containing known quantities of CaO. (70).

| Sample | CaO Present (mg), $x$ | CaO Recovered (mg), $Y$ |
|--------|----------------------|-------------------------|
| 1 | 4.0 | 3.7 |
| 2 | 8.0 | 7.8 |
| 3 | 12.5 | 12.2 |
| 4 | 16.0 | 15.6 |
| 5 | 20.0 | 19.8 |
| 6 | 25.0 | 24.5 |
| 7 | 31.0 | 31.1 |
| 8 | 36.0 | 35.5 |
| 9 | 40.0 | 39.4 |
| 10 | 40.0 | 39.5 |

If the method were perfect and the data were fit with a linear model, we would have $a = 0$, $y = x$, and $b = 1$. Analyze the procedure's efficiency by finding the estimated regression line $y = \hat{a} + \hat{b}x$ and testing $H_0: b = 1$ versus $H_1: b < 1$. Use 0.05 as the level of significance. Note: For the 10 $(x_i, y_i)$'s,

$$\sum_{i=1}^{10} x_i = 232.5 \qquad \sum_{i=1}^{10} y_i = 229.0$$

$$\sum_{i=1}^{10} x_i^2 = 6974.25 \qquad \sum_{i=1}^{10} y_i^2 = 6796.66$$

$$\sum_{i=1}^{10} x_i y_i = 6884.65$$

# 12.3

## THE LINEAR MODEL: ESTIMATION AND PREDICTION

As we have seen, hypothesis testing is not always the format an experimenter wishes to use, *or can use*, in phrasing an inference. Sometimes identifying the appropriate "standard" value for a parameter is physically impossible, so there is no way to formulate $H_0$ or $H_1$. In those situations we used the confidence interval as a backup. Researchers analyzing regression data often do the same.

For data satisfying the linear model, there are *five* interval-type estimates that may be of some use, depending on the situation. The three obvious ones are confidence intervals for the model's parameters $a$, $b$, and $\sigma^2$. Intervals can also be effective, though, in estimating the expected value of $Y$ for a given value of $x$ and the value of a future $Y$ (at a given value of $x$).

Intervals for the three parameters can be derived by "inverting" the three decision-rule functions in Theorem 12.3. It follows, for example, from the statement of part 2 that

$$\frac{(B - b)\sqrt{\sum_{i=1}^{n} (x_i - \bar{x})^2}}{\sqrt{\hat{\sigma}^2}}$$

has a Student $t$ distribution with $n - 2$ degrees of freedom. Therefore,

$$P\left(-t_{\alpha/2, n-2} \leqslant \frac{(B - b)\sqrt{\sum_{i=1}^{n} (x_i - \bar{x})^2}}{\sqrt{\hat{\sigma}^2}} \leqslant t_{\alpha/2, n-2}\right) = 1 - \alpha \quad (12.7)$$

in which case

$$P\left(-t_{\alpha/2, n-2} \frac{\sqrt{\hat{\sigma}^2}}{\sqrt{\sum_{i=1}^{n} (x_i - \bar{x})^2}} \leqslant B - b \leqslant t_{\alpha/2, n-2} \frac{\sqrt{\hat{\sigma}^2}}{\sqrt{\sum_{i=1}^{n} (x_i - \bar{x})^2}}\right) = 1 - \alpha \tag{12.8}$$

or, equivalently,

$$P\left(B - t_{\alpha/2, n-2} \frac{\sqrt{\hat{\sigma}^2}}{\sqrt{\sum_{i=1}^{n} (x_i - \bar{x})^2}} \leqslant b \leqslant B + t_{\alpha/2, n-2} \frac{\sqrt{\hat{\sigma}^2}}{\sqrt{\sum_{i=1}^{n} (x_i - \bar{x})^2}}\right) = 1 - \alpha \tag{12.9}$$

Equations 12.7, 12.8, and 12.9 parallel the algebraic manipulations in Section 9.3. We will also borrow that section's terminology and call

$$\left(\hat{b} - t_{\alpha/2, n-2} \frac{\sqrt{\hat{\sigma}^2}}{\sqrt{\sum_{i=1}^{n} (x_i - \bar{x})^2}}, \quad \hat{b} + t_{\alpha/2, n-2} \frac{\sqrt{\hat{\sigma}^2}}{\sqrt{\sum_{i=1}^{n} (x_i - \bar{x})^2}}\right)$$

a *100(1 − α)% confidence interval for b.* Theorem 12.4 gives confidence interval formulas for all three parameters. Each is obtained by following a series of steps similar to what we just did for *b*.

---

**THEOREM 12.4**   Confidence Interval Formulas for *a*, *b*, and $\sigma^2$

Let $(x_1, y_1)$, $(x_2, y_2)$, . . . , $(x_n, y_n)$ be a set of regression data. Let $y = \hat{a} + \hat{b}x$ be the line estimating the true regression equation $y = a + bx$, and let

$$\hat{\sigma}^2 = \frac{1}{n-2}\left(\sum_{i=1}^{n} y_i^2 - \hat{a}\sum_{i=1}^{n} y_i - \hat{b}\sum_{i=1}^{n} x_i y_i\right)$$

be the estimate of the true variance $\sigma^2$ associated with each *Y* distribution. Then

1. A *100(1 − α)% confidence interval for a* is $(\hat{a} - w, \hat{a} + w)$, where

$$w = t_{\alpha/2, n-2}\frac{\sqrt{\hat{\sigma}^2 \sum_{i=1}^{n} x_i^2}}{\sqrt{n \sum_{i=1}^{n} (x_i - \bar{x})^2}}$$

2. A *100(1 − α)% confidence interval for b* is $(\hat{b} - w, \hat{b} + w)$, where

$$w = t_{\alpha/2, n-2}\frac{\sqrt{\hat{\sigma}^2}}{\sqrt{\sum_{i=1}^{n} (x_i - \bar{x})^2}}$$

3. A *100(1 − α)% confidence interval for $\sigma^2$* is

$$\left(\frac{(n-2)\hat{\sigma}^2}{\chi_{1-\alpha/2, n-2}^2}, \frac{(n-2)\hat{\sigma}^2}{\chi_{\alpha/2, n-2}^2}\right)$$

---

**CASE STUDY 12.1**

*Airline Revenues*   For an airline hauling commercial cargo, freight revenues $(y)$ are a function of tonnage consigned $(x)$. Columns 2 and 3 in Table 12.3 show *x* and *y* values for the 10 largest airline freight companies in the United States, as reported in 1973 (87).

According to the data's scatter diagram in Figure 12.9, the relationship between *x* and *y* is clearly linear. The line estimating that relationship, shown superimposed over the $(x_i, y_i)$'s, has equation $y = 5.902760 +$

0.194811x, where

$$\hat{b} = \frac{10(494,774) - (4322)(901)}{10(2,408,810) - (4322)^2} = 0.194811$$

and
$$\hat{a} = \frac{901 - (0.194811)(4322)}{10} = 5.902760$$

To an entrepreneur the slope of this sort of $xy$ relationship is especially important. It represents *marginal revenue*, the change in dollars received that can be expected to accompany a "unit" increase in tonnage hauled. Here, the estimated slope is $\hat{b} = 0.194811$, implying that an increase of 1 million freight ton-miles should generate an additional $194,811 in revenues (=0.194811 millions of dollars).

Equally important, though, is the *precision* of the estimated slope. Is the $194,811 figure likely to be within, say, $5000 of the true $b$, or is it quite possible that $|\hat{b} - b|$ will be as large as $50,000? It makes a big difference: the greater the precision, the more "faith" the company can have in the numerical value of $\hat{b}$.

**TABLE 12.3**

| Airline | Freight Ton-Miles (in millions), $x_i$ | Freight Revenues (in millions), $y_i$ | $x_i^2$ | $y_i^2$ | $x_iy_i$ |
|---|---|---|---|---|---|
| Pan American | 860 | 188 | 739,600 | 35,344 | 161,680 |
| Flying Tiger | 681 | 120 | 463,761 | 14,400 | 81,720 |
| United | 645 | 135 | 416,025 | 18,225 | 87,075 |
| American | 529 | 114 | 279,841 | 12,996 | 60,306 |
| TWA | 475 | 98 | 225,625 | 9,604 | 46,550 |
| Seaboard | 359 | 53 | 128,881 | 2,809 | 19,027 |
| Northwest | 246 | 52 | 60,516 | 2,704 | 12,792 |
| Eastern | 207 | 56 | 42,849 | 3,136 | 11,592 |
| Delta | 176 | 56 | 30,976 | 3,136 | 9,856 |
| Continental | 144 | 29 | 20,736 | 841 | 4,176 |
| | 4,322 | 901 | 2,408,810 | 103,195 | 494,774 |

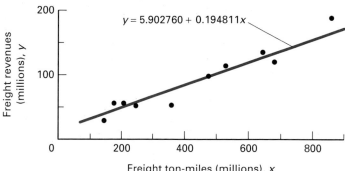

**FIG. 12.9**

Statistically, the standard way to make a statement about the credibility of an estimated slope is to construct a 95% confidence interval for $b$ and to let the width of that interval reflect $\hat{b}$'s precision. For the $(x_i, y_i)$'s in Table 12.3,

$$\hat{\sigma}^2 = \frac{103{,}195 - 5.902760(901) - 0.194811(494{,}774)}{10 - 2} = 186.1578$$

and

$$\sum_{i=1}^{10} (x_i - \bar{x})^2 = \sum_{i=1}^{10} x_i^2 - \frac{1}{10}\left(\sum_{i=1}^{10} x_i\right)^2$$

$$= 2{,}408{,}810 - \frac{(4322)^2}{10} = 540{,}841.6$$

Also,    $t_{\alpha/2,\, n-2} = t_{0.05/2,\, 10-2} = t_{0.025,\, 8} = 2.306$

so that from part 2 of Theorem 12.4,

$$w = 2.306\,\frac{\sqrt{186.1578}}{\sqrt{540{,}841.6}} = 0.042782$$

A 95% confidence interval for $b$, then, is the range of marginal revenues from \$152,000 to \$238,000:

$$(\hat{b} - w,\, \hat{b} + w) = (0.194811 - 0.042782,\, 0.194811 + 0.042782)$$

$$\doteq (0.152,\, 0.238) = (\$152{,}000,\, \$238{,}000)$$

---

**EXAMPLE 12.4**    Refer again to the radiation data in Example 12.2. For $x$, an index of radiation exposure, and $y$, a county's cancer mortality rate, the estimated regression line was found to have the equation $y = 114.72 + 9.23x$ (see Figure 2.10). The key to interpreting those data is the slope—the fact that 9.23 is significantly larger than 0 (at the $\alpha = 0.05$ level) allows us to conclude that cancer mortality levels, for whatever reason, are dependent on radiation exposure.

Still, it does not follow that $b$ is the *only* parameter in that problem warranting our attention. The $y$ intercept also has some special medical significance: it estimates cancer mortality *in the absence of* any abnormal radiation exposure. Whether $\hat{a}$ is actually useful, though, depends to some extent on its precision. So, for the same reason that motivated the analysis in Case Study 12.1, constructing a 95% confidence interval for $a$ makes good statistical sense.

*Solution*    We have already found for the nine $(x_i, y_i)$'s in Table 2.10 that

$$\sum_{i=1}^{9} x_i^2 = 289.4222 \qquad \sum_{i=1}^{9} (x_i - \bar{x})^2 = 97.51 \qquad \hat{\sigma}^2 = 196.94$$

Also, $t_{\alpha/2, n-2} = t_{0.05/2, 9-2} = t_{0.025, 7} = 2.3646$

From part 1 of Theorem 12.4, we get

$$w = 2.3646 \frac{\sqrt{(196.94)(289.4222)}}{\sqrt{9(97.51)}} = 19.06$$

A 95% confidence interval, then, for Oregon's "normal" cancer mortality rate ranges from 95.66 (deaths/100,000) to 133.78 (deaths/100,000):

$$(\hat{a} - w, \hat{a} + w) = (114.72 - 19.06, 114.72 + 19.06) = (95.66, 133.78)$$

---

**QUESTIONS**

**12.22** Derive the statement of part 3 in Theorem 12.4. Start with the fact that $(n-2)\hat{\sigma}^2/\sigma^2$ has a $\chi^2$ distribution with $n-2$ degrees of freedom.

**12.23** Construct a 95% confidence interval for the $\sigma^2$ in Case Study 12.1.

**12.24** For the cigarette data in Question 12.16 construct a 90% confidence interval for the slope $b$.

**12.25** The table below shows the monthly sales and the monthly sales expenses for a small manufacturing firm in Norman, Oklahoma (107).

| Month | Sales (1000s), $x$ | Expenses (1000s), $y$ |
|-------|--------------------|-----------------------|
| 4/78  | $187.1             | $25.4                 |
| 5/78  | 179.5              | 22.8                  |
| 6/78  | 157.0              | 20.6                  |
| 7/78  | 197.0              | 21.8                  |
| 8/78  | 239.4              | 32.4                  |
| 9/78  | 217.8              | 24.4                  |
| 10/78 | 227.1              | 29.3                  |
| 11/78 | 233.4              | 27.9                  |
| 12/78 | 242.0              | 27.8                  |
| 1/79  | 251.9              | 34.2                  |
| 2/79  | 190.0              | 29.2                  |
| 3/79  | 295.8              | 30.0                  |

(a) Given that

$$\sum_{i=1}^{12} x_i = 2618.0 \qquad \sum_{i=1}^{12} y_i = 325.8$$

$$\sum_{i=1}^{12} x_i^2 = 587{,}099.08 \qquad \sum_{i=1}^{12} y_i^2 = 9041.74$$

$$\sum_{i=1}^{12} x_i y_i = 72{,}375.09$$

find the estimated regression line.

(b) Construct a 95% confidence interval for the $y$ intercept. Note that $a$ in this context represents the company's fixed, or overhead cost (as opposed to $b$, which represents its marginal cost).

12.26   In Case Study 3.15 the estimated regression line has the equation $y = 96.4 - 4.9x$, where $y$ is the percentage of hospitals willing to hire a med tech who has been away from the workforce for $x$ years. Construct a 99% confidence interval for $b$.

12.27   Construct a 90% confidence interval for the change in a whisky's proof that one year's additional aging will produce. Use the data in Question 3.64.

12.28   Suppose a 95% confidence interval for $b$ does not contain the value 0. What does that imply about the outcome of testing $H_0$: $b = 0$ versus $H_1$: $b \neq 0$?

12.29   Construct a 95% confidence interval for the standard deviation associated with the nemesis data in Question 1.9. (Hint: If $(c, d)$ is a 95% confidence interval for $\sigma^2$, then $(\sqrt{c}, \sqrt{d})$ is a 95% confidence interval for $\sigma$.)

### *Estimating E(Y)*

Although $a$, $b$, and $\sigma^2$ are the only *parameters* in the linear model, they are not the only targets of a statistician's interest. Remember that high on the list of reasons why an experimenter computes an estimated regression line in the first place is prediction: What might reasonably be expected to happen to $y$ if $x$ is set equal to some specified value $x_0$?

   If we want to answer that question by identifying a single "most likely" $Y$ value, then we already know the answer: if $y = \hat{a} + \hat{b}x$ is the estimated regression line, the $Y$ value we "expect" to be associated with $x_0$ is simply $\hat{y}_0 = \hat{a} + \hat{b}x_0$.

   But $y = \hat{a} + \hat{b}x$ only *estimates* the $Y$ value that best corresponds to $x_0$. *On the average*, the "true" $Y$ value most likely to occur when $x = x_0$ is $E(Y)$, where

$$E(Y) = a + bx_0$$

(see Figure 12.10).

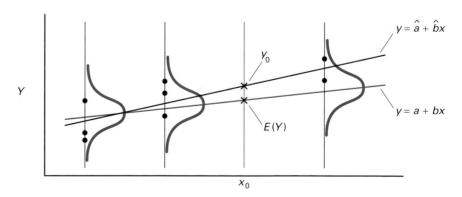

**FIG. 12.10**

Having computed $\hat{y}_0$, an experimenter needs to get some feeling for its precision. How likely is it, for example, that $\hat{y}_0$ will be within, say, 20 units of the quantity being estimated, $E(Y)$? Answering that leads us back to a familiar strategy: we construct a confidence interval.

---

**THEOREM 12.5   A Confidence Interval for $E(Y)$**

Let $(x_1, y_1)$, $(x_2, y_2)$, ..., $(x_n, y_n)$ be a set of regression data. Let $y = \hat{a} + \hat{b}x$ be the line estimating the true regression equation $y = a + bx$, and let

$$\hat{\sigma}^2 = \frac{1}{n-2}\left(\sum_{i=1}^{n} y_i^2 - \hat{a}\sum_{i=1}^{n} y_i - \hat{b}\sum_{i=1}^{n} x_i y_i\right)$$

be the estimate of the true variance $\sigma^2$ associated with each $Y$ distribution. Let $E(Y)$ be the mean of the $Y$ distribution when $x = x_0$. A $100(1-\alpha)\%$ confidence interval for $E(Y)$ is $(\hat{y}_0 - w, \hat{y}_0 + w)$, where

$$w = t_{\alpha/2, n-2}\sqrt{\hat{\sigma}^2\left\{\frac{1}{n} + \frac{(x_0 - \bar{x})^2}{\sum_{i=1}^{n}(x_i - \bar{x})^2}\right\}} \quad \text{and} \quad \hat{y}_0 = \hat{a} + \hat{b}x_0.$$

---

Figure 12.11 illustrates Theorem 12.5 graphically. Pictured is a $100(1-\alpha)\%$ confidence interval for $E(Y)$, one that actually contains $E(Y)$. We have no way of knowing, of course, whether $E(Y)$ lies inside the interval $(\hat{y}_0 - w, \hat{y}_0 + w)$ or outside. But we do have the assurance that $100(1-\alpha)\%$ of the intervals constructed in this fashion will, indeed, contain the true $E(Y)$ as an interior point.

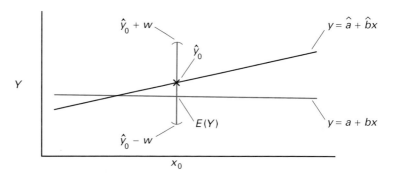

**FIG. 12.11**

## CASE STUDY 12.2

*Analysis of Connecting Rods*    A manufacturer of air conditioning units was having assembly problems caused by the failure of a connecting rod to meet weight specifications. Too many rods were going through the entire tooling process from rough casting to finished product only to be rejected as overweight. As a potential solution, the company's line supervisors recommended that a regression analysis be done on the weight of a rough casting and the weight of its finished rod. Using that relationship, they hoped it might be possible to identify early in the manufacturing process castings likely to end up as unsatisfactory. The rod could then be discarded before it wasted additional resources (130).

A total of 25 $(x_i, y_i)$ pairs were measured (see Table 12.4). Computing the usual sums and sums of squares gives

$$\sum_{i=1}^{25} x_i = 66.075 \qquad \sum_{i=1}^{25} y_i = 50.12$$

$$\sum_{i=1}^{25} x_i^2 = 174.673 \qquad \sum_{i=1}^{25} y_i^2 = 100.499$$

$$\sum_{i=1}^{25} x_i y_i = 132.491$$

Therefore,

$$\hat{b} = \frac{25(132.491) - (66.075)(50.12)}{25(174.673) - (66.075)^2} = 0.648$$

and

$$\hat{a} = \frac{50.12 - 0.648(66.075)}{25} = 0.292$$

**TABLE 12.4**

| Rod Number | Rough Weight | Finished Weight | Rod Number | Rough Weight | Finished Weight |
|---|---|---|---|---|---|
| 1 | 2.745 | 2.080 | 14 | 2.635 | 1.990 |
| 2 | 2.700 | 2.045 | 15 | 2.630 | 1.990 |
| 3 | 2.690 | 2.050 | 16 | 2.625 | 1.995 |
| 4 | 2.680 | 2.005 | 17 | 2.625 | 1.985 |
| 5 | 2.675 | 2.035 | 18 | 2.620 | 1.970 |
| 6 | 2.670 | 2.035 | 19 | 2.615 | 1.985 |
| 7 | 2.665 | 2.020 | 20 | 2.615 | 1.990 |
| 8 | 2.660 | 2.005 | 21 | 2.615 | 1.995 |
| 9 | 2.655 | 2.010 | 22 | 2.610 | 1.990 |
| 10 | 2.655 | 2.000 | 23 | 2.590 | 1.975 |
| 11 | 2.650 | 2.000 | 24 | 2.590 | 1.995 |
| 12 | 2.650 | 2.005 | 25 | 2.565 | 1.955 |
| 13 | 2.645 | 2.015 | | | |

Figure 12.12 shows the estimated regression line $y = 0.292 + 0.648x$ superimposed over the 25 data points.

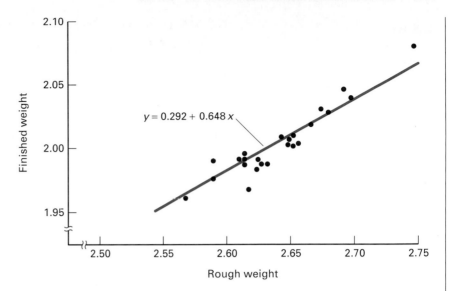

$y = 0.292 + 0.648\,x$

**FIG. 12.12**

Suppose the finished weight of a rod, if the unit is to work properly, needs to be between 1.95 and 2.03. Is it likely that castings whose rough weight is 2.61 will produce a satisfactory proportion of rods having an acceptable finished weight?

Based on cost considerations, the company's engineers decided, in general, that a rough weight $x_0$ will be acceptable if the corresponding 95% confidence interval for $E(Y)$ lies entirely within the 1.95 and 2.03 specifications. The key to the problem, therefore, is applying Theorem 12.5 when $x_0 = 2.61$.

Note, first of all, that

$$\hat{\sigma}^2 = \frac{100.499 - (0.292)(50.12) - (0.648)(132.491)}{25 - 2} = 0.000426$$

$$\bar{x} = \frac{(66.075)}{25} = 2.643$$

$$\sum_{i=1}^{25} (x_i - \bar{x})^2 = \sum_{i=1}^{25} x_i^2 - \frac{1}{25}\left(\sum_{i=1}^{25} x_i\right)^2 = 174.673 - \frac{(66.075)^2}{25} = 0.037$$

and

$$t_{\alpha/2,\,n-2} = t_{0.05/2,\,25-2} = t_{0.025,\,23} = 2.0687$$

Therefore,

$$w = 2.0687\sqrt{0.000426\left[\frac{1}{25} + \frac{(2.61 - 2.643)^2}{0.037}\right]} = 0.011$$

and

$$\hat{y}_0 = \hat{a} + \hat{b}x_0 = 0.292 + 0.648(2.61) = 1.983$$

The 95% confidence interval, then, for $E(Y)$ is

$$(\hat{y}_0 - w, \hat{y}_0 + w) = (1.983 - 0.011, 1.983 + 0.011) = (1.972, 1.994)$$

(see Figure 12.13).

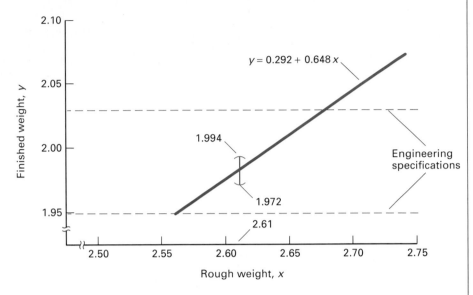

**FIG. 12.13**

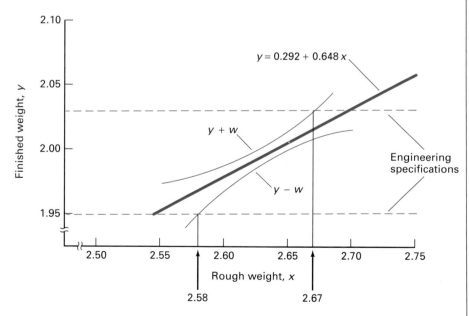

**FIG. 12.14**

Our answer, therefore, to the original question is "yes." Since $E(Y)$'s confidence interval, (1.972, 1.994), lies entirely inside the specification limits, (1.95, 2.03), we conclude, by the engineers' own criterion, that castings having a rough weight of 2.61 do show good promise and should not be discarded.

Notice from the formula for $w$ that the width of a 95% confidence interval for $E(Y)$ is a function of $x_0$; specifically, the interval is shortest when $x_0 = \bar{x}$ and gets wider as the difference between $x_0$ and $\bar{x}$ increases. Figure 12.14 illustrates that dependence: it shows $(\hat{y}_0 - w, \hat{y}_0 + w)$ computed for all values of $x_0$. The end points of the resulting intervals define two hyperbolas, one on either side of $y = \hat{a} + \hat{b}x$.

What the company wants to know in general—which castings should be discarded and which should be completed—can be deduced from Figure 12.14. By inspection, the lines $y = 1.95$ and $y = 2.03$ intersect the hyperbolas at $x = 2.58$ and $x = 2.67$. Any casting, therefore, having a rough weight outside either of those two limits should be discarded.

---

QUESTIONS

**12.30**  Does it make sense intuitively that a confidence interval for $E(Y)$ should get wider as $x_0$ gets further away from $\bar{x}$? Explain.

**12.31**  For the airline data in Case Study 12.1 use a 95% confidence interval to estimate the average revenue a commercial carrier would generate by hauling 600 million freight ton-miles.

**12.32**  One of the more useful applications of regression analysis occurs when two variables, $x$ and $y$, are related and one of the two (say $y$) is difficult to measure. Suppose, initially, the relationship between $x$ and $y$ can be quantified. The problem of directly measuring $y$ can then be routinely avoided. We need simply measure $x$ and use its value to estimate $y$ via the regression function. The table below lists the weights $x$ (in kilograms) and the volumes $y$ (in cubic decimeters) for 18 children between the ages of 5 and 8 (12).

| Weight, $x$ | Volume, $y$ | Weight, $x$ | Volume, $y$ |
|---|---|---|---|
| 17.1 | 16.7 | 15.8 | 15.2 |
| 10.5 | 10.4 | 15.1 | 14.8 |
| 13.8 | 13.5 | 12.1 | 11.9 |
| 15.7 | 15.7 | 18.4 | 18.3 |
| 11.9 | 11.6 | 17.1 | 16.7 |
| 10.4 | 10.2 | 16.7 | 16.6 |
| 15.0 | 14.5 | 16.5 | 15.9 |
| 16.0 | 15.8 | 15.1 | 15.1 |
| 17.8 | 17.6 | 15.1 | 14.5 |

(a) Given that

$$\sum_{i=1}^{18} x_i = 270.1 \qquad \sum_{i=1}^{18} y_i = 265.0$$

$$\sum_{i=1}^{18} x_i^2 = 4149.39 \qquad \sum_{i=1}^{18} y_i^2 = 3996.14$$

$$\sum_{i=1}^{18} x_i y_i = 4071.71$$

find the estimated regression line summarizing the relationship between $x$ and $y$. Are $x$ and $y$ equally easy to measure?

(b) Construct a 95% confidence interval for the average volume of children whose weight is 14.0 kg.

12.33 On the average, how far can a $3\frac{1}{2}$-week-old mosquito fly without stopping? Use the data in Question 1.8 and answer the question by constructing a 90% confidence interval for $E(Y)$, where $x_0 = 3.5$. (See Question 12.17.)

12.34 Refer to the sales data in Question 12.25. Based on the best marketing information available, management believes that the company's sales volume will stay in the \$275,000 neighborhood. Use Theorem 12.5 to construct a 95% confidence interval for the average expenses that sales at that level will incur.

12.35 The increase in the average weight of college football players from the turn of the century to the recent past was the focus of Question 2.33. For the University of Texas the relationship between weight $y$ and year $x$ was estimated to be $y = -1108.8 + 0.666x$. Is it likely that the *true* average weight of UT players in 1940 was less than 170 lb? Answer the question by finding a 95% confidence interval for $E(Y)$ and comparing the lower limit of that interval with 170. Note:

$$\sum_{i=1}^{6} x_i = 11{,}621 \qquad \sum_{i=1}^{6} y_i = 1093$$

$$\sum_{i=1}^{6} x_i^2 = 22{,}510{,}485 \qquad \sum_{i=1}^{6} y_i^2 = 200{,}327$$

$$\sum_{i=1}^{6} x_i y_i = 2{,}118{,}654$$

### Prediction Intervals

A variation on the preceding theme is the "prediction problem," where the objective is to use an interval to predict the location of a *single* future observation. Given some specified $x = x_0$, in other words, we want an interval that applies to $y$ rather than to $E(Y)$. By tradition, any such range of num-

**prediction interval**—In regression problems, a confidence interval for the value of a future $Y$ at a given value of $x$.

bers is called a *prediction interval* rather than a confidence interval. Conceptually, the two are the same.

---

**THEOREM 12.6  Prediction Interval**

Let $(x_1, y_1), (x_2, y_2), \ldots, (x_n, y_n)$ be a set of regression data. Let $y = \hat{a} + \hat{b}x$ be the line estimating the true regression equation $y = a + bx$, and let

$$\hat{\sigma}^2 = \frac{1}{n-2}\left(\sum_{i=1}^{n} y_i^2 - \hat{a}\sum_{i=1}^{n} y_i - \hat{b}\sum_{i=1}^{n} x_i y_i\right)$$

be the estimate of the true variance $\sigma^2$ associated with each $Y$ distribution. A $100(1-\alpha)\%$ *prediction interval for* $y_0$, when $x = x_0$, is $(\hat{y}_0 - w, \hat{y}_0 + w)$, where

$$w = t_{\alpha/2, n-2}\sqrt{\hat{\sigma}^2\left[1 + \frac{1}{n} + \frac{(x_0 - \bar{x})^2}{\sum_{i=1}^{n}(x_i - \bar{x})^2}\right]} \quad \text{and} \quad \hat{y}_0 = \hat{a} + \hat{b}x_0$$

---

## CASE STUDY 12.3

*Extinction of Life-Forms*

Geologists can document that mass extinctions of life-forms on our planet have occurred with alarming regularity, once every 26 or 27 million years. The cause of these "die-offs" remains a mystery, but more and more evidence is pointing in the direction of the *nemesis theory* as a plausible explanation. What the latter speculates is that some extraterrestrial mechanism periodically perturbs the orbits of the comets that surround our solar system, thereby increasing dramatically the probability of one or more of the comets striking the earth. When a collision does occur—approximately once every 26 million years—the vast amounts of dust raised block out the sun, thus creating a nuclear winter scenario.

Columns 1 and 2 of Table 12.5 reproduce the data from Question 1.9. Represented by $x$ are the four most recent extinctions, and by $y$ the number

**TABLE 12.5**

| Extinction Episode, $x_i$ | Millions of Years Before Present, $y_i$ | $x_i^2$ | $x_i y_i$ | $y_i^2$ |
|---|---|---|---|---|
| 1 | 11 | 1 | 11 | 121 |
| 2 | 38 | 4 | 76 | 1,444 |
| 3 | 65 | 9 | 195 | 4,225 |
| 4 | 91 | 16 | 364 | 8,281 |
| 10 | 205 | 30 | 646 | 14,071 |

of years before the present that each extinction occurred. Summarizing the $xy$ relationship is the estimated regression line $y = \hat{a} + \hat{b}x$, where

$$\hat{b} = \frac{4(646) - (10)(205)}{4(30) - (10)^2} = 26.7$$

and

$$\hat{a} = \frac{205 - 26.7(10)}{4} = -15.5$$

(see Figure 12.15).

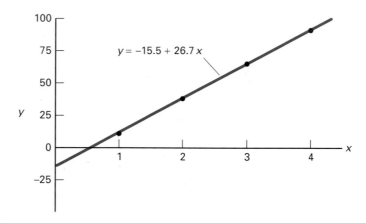

**FIG. 12.15**

    If the apparent periodicity of these mass extinctions continues, future generations need to worry about the value of $Y$ when $x = x_0 = 0$. (assuming they care whether they live or die) From the estimated regression line,

$$\hat{y}_0 = -15.5 + 26.7(0) = -15.5$$

implying that the next die-off will occur 15.5 million years from now. How much "error," though, is associated with that estimate? Should someone living, say, *14* million years from now be concerned?

    Here, the behavior of $Y$ is more relevant than the behavior of $E(Y)$ Why? Because the next die-off will be caused by a *single* $y$, not by an expected value. Statistically, therefore, we should use a *prediction* interval—not a confidence interval—to assess the precision of $\hat{y}_0$.

    From Table 12.5,

$$\hat{\sigma}^2 = \frac{14{,}071 - (15.5)(205) - 26.7(646)}{4 - 2} = 0.15$$

and

$$\sum_{i=1}^{4}(x_i - \bar{x})^2 = \sum_{i=1}^{4} x_i^2 - \frac{1}{4}\left(\sum_{i=1}^{4} x_i\right)^2 = 30 - \frac{(10)^2}{4} = 5$$

Let $\alpha = 0.01$. Then

$$t_{\alpha/2, n-2} = t_{0.01/2, 4-2} = t_{0.005, 2} = 9.9248$$

Therefore,   $w = 9.9248 \sqrt{0.15 \left[ 1 + \dfrac{1}{4} + \dfrac{(0 - 2.5)^2}{5} \right]} = 6.1$

and the 99% prediction interval for $y$ (when $x_0 = 0$) ranges from 9.4 million years in the future ($= \hat{y}_0 - w = -15.5 + 6.1$) to 21.6 million years in the future ($= \hat{y}_0 - w = -15.5 - 6.1$). So, is someone living 14 million years in the future a potential comet victim, based on the four observations in Table 12.5? Yes.

---

## QUESTIONS

**12.36**   As a function of $x$, the end points of a $100(1 - \alpha)\%$ confidence interval for $E(Y)$ trace two hyperbolas, one on either side of $y = \hat{a} + \hat{b}x$ (see Figure 12.14). What would the corresponding graph for prediction intervals look like?

**12.37**   (a)  Using the data in Case Study 12.1, construct a 95% prediction interval for the revenues that a single airline company can expect to earn by hauling 700 million ton-miles of freight.

   (b)  Is a company more likely to be interested in a prediction interval for $y$ or a confidence interval for $E(Y)$? Explain.

**12.38**   How variable are the individual weights for the connecting rods described in Case Study 12.2? Construct a 95% prediction interval for $y$ when the rough weight of a casting is 2.61. Compare your answer with the 95% confidence interval for $E(Y)$, as derived on p. 534.

**12.39**   For the grass seed data analyzed in Example 12.1, is it likely that planting a fairway with a seed density of 2.5 lb/500 ft$^2$ will produce a lawn with a quality rating as low as 35? Answer the question by constructing an appropriate 95% prediction interval.

**12.40**   In Case Study 2.9, cancer mortality ($y$) was found to be linearly related to a radiation index ($x$) according to the regression function $y = 114.72 + 9.23x$. What range of cancer mortalities might we expect to observe in a county whose radiation index is 7.0? Use the information in Example 12.2 to construct a 90% prediction interval for $y_0$.

**12.41**   The distance $y$ that a mosquito $x$-weeks-old can fly without stopping was the subject of a study described in Question 1.8.

   (a)  On the average, how far can a $1\frac{1}{2}$-week-old mosquito fly? (See Question 12.17.)

   (b)  Estimate the precision of the $\hat{y}_0$ determined in part ($a$) by constructing an appropriate 90% prediction interval.

12.42    If the limits of a 95% prediction interval for $y_0$ based on a given set of data and a given value for $x_0$ are 6.5 and 11.6, can we say that $\hat{y}_0$ will lie between those particular numbers 95% of the time (when $x = x_0$)? Why or why not?

# 12.4

## CORRELATION

The basic mechanics and rationale for finding regression lines were introduced much earlier. We pointed out in Section 2.5 that experimenters collecting quantitative information on two dissimilar traits have a very understandable desire to summarize the relationship between those traits with an equation. What we learned in Section 3.5—how to fit straight lines of the form $y = \hat{a} + \hat{b}x$—addressed precisely that objective.

Our intentions in this chapter have been more inferential. Fitting equations is a good way to *start* the analysis of an *xy* relationship, but additional questions of a more statistical nature invariably arise and need to be dealt with. How precise is the slope estimate? Is it reasonable to presume that $a = a_0$? If $x = x_0$, what can we expect for the value of $y$?

In general, the kinds of inferences capable of being formulated depend on the assumptions appropriate for $x$ and/or for $y$. Two scenarios are frequently encountered. If values for $x$ are fixed by the experimenter but $Y$ is a random variable, then the $(x_i, Y_i)$'s are *regression* data (a familiar example is the linear model in Sections 12.2 and 12.3). If *both* measurements are random variables, the $(X_i, Y_i)$'s are *correlation* data. This section analyzes correlation data.

Examples of correlation data are not difficult to imagine. Picture a team of anthropologists studying the relationship between height and weight for a newly discovered tribe of aborigines. How would they collect the necessary data? By choosing a sample of $n$ natives and measuring each native's height $(X)$ and weight $(Y)$. Approaching the problem in that fashion makes both $X$ and $Y$ random variables. Why? Because they are not fixing, or predetermining, either measurement (by looking specifically for a subject, say, 60 in tall or weighing 120 lb).

Table 12.6 shows, hypothetically, what the anthropologists' data might look like if the sample size was $n = 10$. The range of $X$ has been divided

**TABLE 12.6**

| Weight, $Y$ (lb) | Height, $X$ (in.) | | | | |
|---|---|---|---|---|---|
| | 55–59 | 60–64 | 65–69 | 70–74 | 75–79 |
| 90–99 | 1 | | | 1 | |
| 100–109 | | | 4 | | |
| 110–119 | | 1 | 2 | | |
| 120–129 | | | | | 1 |

into classes 55–59, . . . , 75–79 the range of $Y$ into classes 90–99, and so on. If the frequency associated with each height and weight class is represented by the height of a bar, the entries in the table produce the three-dimensional (or *bivariate*) histogram in Figure 12.16.

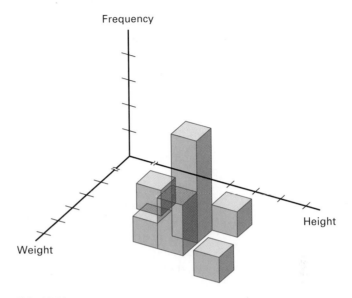

**FIG. 12.16**

Think about what would happen if *many* such measurements were taken. Most likely, a simple pattern would begin to emerge: short subjects tending to be light, tall subjects heavy, and most of the sample having both heights and weights near their respective averages. Approximating that large-sample pattern would be a surface, a shape similar, perhaps, to the configuration in Figure 12.17.

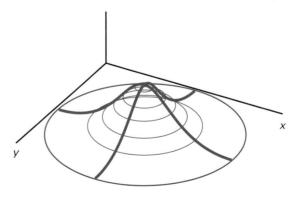

**FIG. 12.17**

Mathematically, surfaces describing the "joint" probabilistic behavior of two random variables are treated a bit differently than curves describing the individual behavior of single random variables. For the latter, *areas* represent probabilities. Figure 12.18, for example, is a familiar-looking diagram: it shows that the probability of X lying in the interval from a to b is

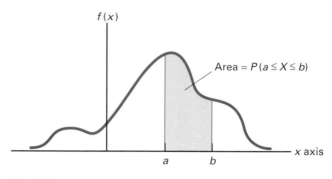

FIG. 12.18

the area under $f(x)$ that lies above $[a, b]$. When two random variables are being measured, events become regions (instead of intervals) and probabilities are represented by volumes (instead of areas). Look at Figure 12.19. The probability that X and Y lie in region A is numerically equal to the volume under the surface $f(x, y)$ that lies directly above A.

Since the end of Chapter 6, we have seen repeatedly that single random variables often have behaviors that can be well-approximated by specific curves known as normal distributions. Similarly, *joint* random variables

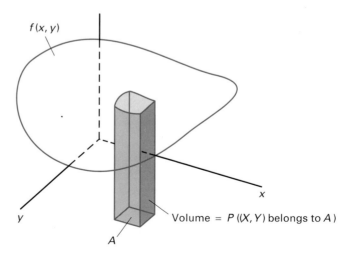

FIG. 12.19

**bivariate distribution**—A function $f(x, y)$ that defines a surface describing the joint probabilistic behavior of two random variables $X$ and $Y$. The volume under the surface lying above any region $A$ in the $xy$ plane represents the probability that the point $(X, Y)$ lies in the region $A$.

**true correlation coefficient** $(\rho)$—A parameter that measures the extent to which two bivariate normal random variables are linearly related.

frequently lead to outcomes that can be well-approximated by a family of surfaces known as *bivariate normal distributions*. An example is the surface in Figure 12.17. In general, bivariate normal distributions are surfaces whose shapes and locations are controlled by five parameters: $\mu_X$, $\sigma_X$, $\mu_Y$, $\sigma_Y$, and $\rho$. The first two represent the true mean and the true standard deviation characterizing the $X$ trait. The next two are defined similarly for the $Y$ trait. The fifth parameter is the *true correlation coefficient*. Numerically, $\rho$ varies between $-1$ and $+1$ and it represents the extent to which $X$ and $Y$ are not independent. (See the discussion of the *sample* correlation coefficient $r$ in Section 3.6. Those comments apply here; $r$ is simply a numerical estimate of $\rho$.)

---

**Definition**        **Bivariate Normal Distribution**

Let $X$ and $Y$ be a pair of random variables whose joint behavior is described by the surface $f(x, y)$. We say that $f(x, y)$ is a *bivariate normal distribution* if

$$f(x, y) = \frac{1}{2\pi\sigma_X\sigma_Y\sqrt{1 - \rho^2}}\, e^{-Q/2} \qquad \begin{array}{l} -\infty < x < \infty \\ -\infty < y < \infty \\ -1 \leqslant \rho \leqslant 1 \end{array} \quad (12.10)$$

where

$$Q = \frac{1}{1 - \rho^2}\left[\left(\frac{x - \mu_X}{\sigma_X}\right)^2 - 2\rho\left(\frac{x - \mu_X}{\sigma_X}\right)\left(\frac{y - \mu_Y}{\sigma_Y}\right) + \left(\frac{y - \mu_Y}{\sigma_Y}\right)^2\right]$$

---

Figure 12.20 shows some of the shapes that $f(x, y)$ can take, depending on the value of $\rho$. The other four parameters also have an effect: $\mu_X$ and $\mu_Y$ dictate the location of the surface, and $\sigma_X$ and $\sigma_Y$ reflect its

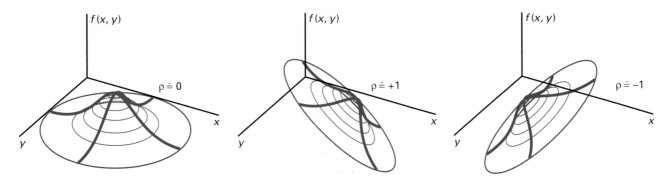

**FIG. 12.20**

dispersion relative to the $x$ axis and the $y$ axis, respectively. The key, though, to interpreting correlation data is $\rho$.

We will not need Equation 12.10 for computational purposes, but some of its properties will play important roles in the inference procedures discussed later in this section. Heading the list of what we need to know about $f(x, y)$ are estimates for its five parameters.

---

**THEOREM 12.7**  MLEs for $\mu_X$, $\sigma_X$, $\mu_Y$, $\sigma_Y$, and $\rho$

---

Let $(X_1, Y_1), (X_2, Y_2), \ldots, (X_n, Y_n)$ be a random sample from the correlation model. Assume the joint behavior of the $(X_i, Y_i)$'s is described by a bivariate normal distribution in accordance with Definition 12.2. The estimators for $\mu_X$, $\sigma_X$, $\mu_Y$, $\sigma_Y$, and $\rho$ based on the method of maximum likelihood are

$$\hat{\mu}_X = \bar{X} = \frac{1}{n} \sum_{i=1}^{n} X_i$$

$$\hat{\sigma}_X = S_X = \sqrt{\frac{n \sum_{i=1}^{n} X_i^2 - \left(\sum_{i=1}^{n} X_i\right)^2}{n(n-1)}}$$

$$\hat{\mu}_Y = \bar{Y} = \frac{1}{n} \sum_{i=1}^{n} Y_i$$

$$\hat{\sigma}_Y = S_Y = \sqrt{\frac{n \sum_{i=1}^{n} Y_i^2 - \left(\sum_{i=1}^{n} Y_i\right)^2}{n(n-1)}}$$

and $\quad \hat{\rho} = R = \dfrac{n \sum_{i=1}^{n} X_i Y_i - \left(\sum_{i=1}^{n} X_i\right)\left(\sum_{i=1}^{n} Y_i\right)}{\sqrt{n \sum_{i=1}^{n} X_i^2 - \left(\sum_{i=1}^{n} X_i\right)^2} \sqrt{n \sum_{i=1}^{n} Y_i^2 - \left(\sum_{i=1}^{n} Y_i\right)^2}}$

---

Among the consequences of both $X$ and $Y$ being random variables is that any set of correlation data has *two* "regression" lines. Since neither measurement is controlled by the experimenter, it may make just as much sense to think of $X$ as being a function of $y$ as it does to think of $Y$ as being a function of $x$. In practice, which of those two relationships is more appropriate depends entirely on what questions the data are being used to answer.

If $X$ and $Y$ have a bivariate normal distribution, it can be proved that both regression functions are linear. Traditionally, the slopes and $y$ intercepts of those two lines are expressed as functions of $r$, $s_X$, and $s_y$.

THEOREM 12.8 Regression Equations for Bivariate
Normal Distribution

Let $(x_1, y_1), (x_2, y_2), \ldots, (x_n, y_n)$ be a set of $n$ measurements taken on the random variables $X$ and $Y$ whose joint behavior is described by a bivariate normal distribution. The equation that best describes $Y$ as a function of $x$ is

$$y = \left( \bar{y} - \frac{rs_Y}{s_X} \bar{x} \right) + \frac{rs_Y}{s_X} x \qquad (12.11)$$

The equation that best describes $X$ as a function of $y$ is

$$x = \left( \bar{x} - \frac{rs_X}{s_Y} \bar{y} \right) + \frac{rs_X}{s_Y} y \qquad (12.12)$$

## CASE STUDY 12.4

*Cricket Thermometers*

Question 3.65 describes a linear relationship known to exist between $X$, the frequency of a cricket's chirps, and $Y$, the outdoor temperature. All 15 $(x_i, y_i)$'s shown in Table 12.7 qualify as correlation data, because neither variable at any time was controlled or predetermined by the experimenter.

**TABLE 12.7**

| Observation Number | Chirps per Second, $x_i$ | Temperature (°F), $y_i$ |
|---|---|---|
| 1 | 20.0 | 88.6 |
| 2 | 16.0 | 71.6 |
| 3 | 19.8 | 93.3 |
| 4 | 18.4 | 84.3 |
| 5 | 17.1 | 80.6 |
| 6 | 15.5 | 75.2 |
| 7 | 14.7 | 69.7 |
| 8 | 17.1 | 82.0 |
| 9 | 15.4 | 69.4 |
| 10 | 16.2 | 83.3 |
| 11 | 15.0 | 79.6 |
| 12 | 17.2 | 82.6 |
| 13 | 16.0 | 80.6 |
| 14 | 17.0 | 83.5 |
| 15 | 14.4 | 76.3 |

Assume the sample represents a random selection from a bivariate normal distribution. Find and graph the data's two regression lines.

Note, first of all, that

$$\sum_{i=1}^{15} x_i = 249.8 \qquad\qquad \sum_{i=1}^{15} x_i^2 = 4200.56$$

$$\sum_{i=1}^{15} y_i = 1200.6 \qquad\qquad \sum_{i=1}^{15} y_i^2 = 96{,}725.86$$

$$\sum_{i=1}^{15} x_i y_i = 20{,}127.47$$

From these five sums, we can compute estimates for $\mu_X$, $\sigma_X$, $\mu_Y$, $\sigma_Y$, and $\rho$, using Theorem 12.7:

$$\bar{x} = \hat{\mu}_X = \frac{249.8}{15} = 16.65$$

$$s_X = \hat{\sigma}_X = \sqrt{\frac{15(4200.56) - (249.8)^2}{15(14)}} = 1.70$$

$$\bar{y} = \hat{\mu}_Y = \frac{1200.6}{15} = 80.04$$

$$s_Y = \hat{\sigma}_Y = \sqrt{\frac{15(96{,}725.86) - (1200.6)^2}{15(14)}} = 6.71$$

$$r = \hat{\rho} = \frac{15(20{,}127.47) - (249.8)(1200.6)}{\sqrt{15(4200.56) - (249.8)^2}\,\sqrt{15(96{,}725.86)(1200.6)^2}}$$

$$= 0.84$$

From Equation 12.11 the regression equation of $Y$ on $x$ is

$$y = \left(80.04 - \frac{(0.84)(6.71)}{1.70}\,16.65\right) + \frac{(0.84)(6.71)}{1.70}\,x$$

$$= 24.8 + 3.3x$$

From Equation 12.12 the regression equation of $X$ on $y$ is

$$x = \left(16.65 - \frac{(0.84)(1.70)}{6.71}\,80.04\right) + \frac{(0.84)(1.70)}{6.71}\,y$$

$$= -0.4 + 0.21y$$

Figure 12.21 shows both regression lines graphed on a scatter diagram of the original data. Which do we use? It depends on what the experimenter wants to know. If the objective is to use a cricket as a thermometer, the more appropriate regression is

$$y = 24.8 + 3.3x$$

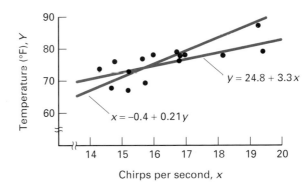

**FIG. 12.21**

Hearing, for example, 17 chirps per second would lead us to estimate that the outdoor temperature is 81°:

$$y = 24.8 + 3.3(17) = 81°$$

If, on the other hand, the purpose of the study is to see what effect temperature has on a cricket's chirping, the second regression makes more sense. We would estimate, for example, that a temperature of 75° should elicit a response of 15.4 chirps per second:

$$x = -0.4 + 0.21(75) = 15.4 \text{ chirps per second}$$

QUESTIONS

12.43   In Theorem 12.8 the slope of the regression line showing $y$ as a function of $x$ is given by the quotient $rs_Y/s_X$. Show that that expression reduces to the formula for $B$ given in Theorem 12.1.

12.44   The 13 observations in Question 2.32, all part of a study to characterize the relationship between plant diversity ($X$) and bird diversity ($Y$), are a clear-cut example of correlation data: neither variable was controlled by the experimenter. Part (**b**) claims that the regression of $Y$ on $x$ is given by $y = 0.26 + 1.70x$. Find the regression of $X$ on $y$. Note:

$$\sum_{i=1}^{13} x_i = 12.91 \qquad \sum_{i=1}^{13} x_i^2 = 15.6171$$

$$\sum_{i=1}^{13} y_i = 25.29 \qquad \sum_{i=1}^{13} y_i^2 = 57.8103$$

$$\sum_{i=1}^{13} x_i y_i = 29.8762$$

**12.45** In general, both regression lines in a correlation problem involving the bivariate normal distribution will go through the point $(\bar{x}, \bar{y})$. Verify the truth of that statement for the regression equations in Question 12.44.

**12.46** Graph the following set of six points and plot the two regression lines. Show that both lines go through the point $(\bar{x}, \bar{y})$.

| $x_i$ | 2 | 4 | 1 | 5 | 3 | 2 |
|-------|---|---|---|---|---|---|
| $y_i$ | 1 | 2 | 2 | 1 | 3 | 4 |

**12.47** An experimenter collects a set of data and draws the resulting scatter diagram shown below. Should the measurements be considered regression data or correlation data? Explain.

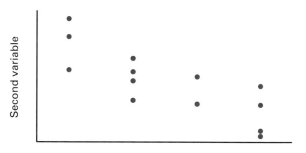

First variable

### Testing $H_0$: $\rho = 0$ versus $H_1$: $\rho \neq 0$

When a set of data is presumed to come from a bivariate normal distribution, the correlation coefficient $\rho$ is often the parameter that becomes the primary focus of interest. Independence is the reason. Whenever two dissimilar traits are being studied within the context of the regression–correlation structure, establishing whether a relationship genuinely exists is the first step that needs to be taken. If it can be shown, for example, that $Y$ is not influenced by $X$, there is no point is trying to describe the relationship between the two with an equation.

For regression data satisfying the linear model, we have already seen that the question of independence is attended to by testing $H_0: b = 0$ (recall Example 12.2). For correlation data coming from a bivariate normal distribution, that same question is handled in a slightly different way—by testing $H_0: \rho = 0$. If $H_0$ is rejected, we conclude that $X$ and $Y$ are *not* independent.

---

**THEOREM 12.9** Testing $H_0: \rho = 0$ versus $H_1: \rho \neq 0$

Let $(x_1, y_1), (x_2, y_2), \ldots, (x_n, y_n)$ be a set of $n$ measurements taken on the random variables $X$ and $Y$ whose joint behavior is described by a bivariate normal distribution. Let $r$ be the sample correlation coefficient estimating the true correlation coefficient $\rho$. To test

$$H_0: \quad \rho = 0$$
$$\text{vs.}$$
$$H_1: \quad \rho \neq 0$$

at the $\alpha$ level of significance, reject the null hypothesis if

$$\frac{\sqrt{n-2}\, r}{\sqrt{1-r^2}}$$

is either (1) $\leqslant -t_{\alpha/2,\, n-2}$ or (2) $\geqslant t_{\alpha/2,\, n-2}$. Rejecting $H_0$ is equivalent to concluding that $X$ and $Y$ are *not* independent.

---

## CASE STUDY 12.5

*Stock Market Behavior*

Being able to predict the future is everyone's fantasy, but no one is more obsessed with prophecy than the wheelers and dealers who play the stock market. With religious zeal they keep looking for anything that will give them an edge in predicting the market's behavior. One popular theory claims that the market will surge ahead if the National Conference wins the Super Bowl but will decline if the winner is a team from the American Conference. Somewhat less frivolous is the "early warning" system, which contends that the change in the market over the first five days in January foretells what it will do over the next 12 months.

We need not take such claims on faith. Theorem 12.9 is a perfectly objective mechanism for assessing the prognosticative merits of any alleged

**TABLE 12.8**

| Year | % Change for First 5 Days in Jan. | % Change for Year | Year | % Change for First 5 Days in Jan. | % Change for Year |
|------|------|------|------|------|------|
| 1950 | 2.0 | 21.8 | 1969 | −2.9 | −11.4 |
| 1951 | 2.3 | 16.5 | 1970 | 0.7 | 0.1 |
| 1952 | 0.6 | 11.8 | 1971 | 0.0 | 10.8 |
| 1953 | −0.9 | −6.6 | 1972 | 1.4 | 15.6 |
| 1954 | 0.5 | 45.0 | 1973 | 1.5 | −17.4 |
| 1955 | −1.8 | 26.4 | 1974 | −1.5 | −29.7 |
| 1956 | −2.1 | 2.6 | 1975 | 2.2 | 31.5 |
| 1957 | −0.9 | −14.3 | 1976 | 4.9 | 19.1 |
| 1958 | 2.5 | 38.1 | 1977 | −2.3 | −11.5 |
| 1959 | 0.3 | 8.5 | 1978 | −4.6 | 1.1 |
| 1960 | −0.7 | −3.0 | 1979 | 2.8 | 12.3 |
| 1961 | 1.2 | 23.1 | 1980 | 0.9 | 25.8 |
| 1962 | −3.4 | −11.8 | 1981 | −2.0 | −9.7 |
| 1963 | 2.6 | 18.9 | 1982 | −2.4 | 14.8 |
| 1964 | 1.3 | 13.0 | 1983 | 3.2 | 17.3 |
| 1965 | 0.7 | 9.1 | 1984 | 2.4 | 1.4 |
| 1966 | 0.8 | −13.1 | 1985 | −1.9 | 26.3 |
| 1967 | 3.1 | 20.1 | 1986 | −1.6 | 14.6 |
| 1968 | 0.2 | 7.7 | | | |

crystal ball. Table 12.8 lists $X$, the percentage change in the market for the first five days in January, and $Y$, the percentage change over the next 12 months. The period spanned is the 37 years from 1950 through 1986 (29). Is $X$ a legitimate predictor of $Y$?

To begin, we will assume that the $(x_i, y_i)$'s represent a random sample of size 37 from a bivariate normal distribution. Assessing the feasibility of using $X$ to predict $Y$ reduces to a test of

$$H_0: \quad \rho = 0$$

vs.

$$H_1: \quad \rho \neq 0$$

Let $\alpha = 0.05$. Computing the usual sums and sums of squares gives

$$\sum_{i=1}^{37} x_i = 9.1 \qquad \sum_{i=1}^{37} x_i^2 = 169.47$$

$$\sum_{i=1}^{37} y_i = 324.8 \qquad \sum_{i=1}^{37} y_i^2 = 12{,}814.64$$

$$\sum_{i=1}^{37} x_i y_i = 655.4$$

It follows from Theorem 12.7 that the sample correlation coefficient is

$$r = \frac{37(655.4) - (9.1)(324.8)}{\sqrt{37(169.47) - (9.1)^2}\sqrt{37(12{,}814.64) - (324.8)^2}} = 0.446$$

We need to decide (by doing a $t$ test) whether an $r$ as large as 0.446 is incompatible with a $\rho$ of 0. If it is, we should reject $H_0$ and conclude that the early warning system *does* have merit as a predictor of the market's year-long behavior.

For the $t$ distribution with 35 ($= n - 2$) df, $t_{0.025, 35} = 2.0301$. But

$$\frac{\sqrt{n-2}\,r}{\sqrt{1-r^2}} = \frac{\sqrt{37-2}\,(0.446)}{\sqrt{1-(0.446)^2}} = 2.95$$

By Theorem 12.9, therefore, we reject $H_0\colon \rho = 0$. It would appear that what happens to the market during the first five days in January does presage, to some extent, what will happen the rest of the year.

---

**QUESTIONS**

**12.48**  Test $H_0\colon \rho = 0$ versus $H_1\colon \rho \neq 0$ for the data in Question 12.46. Let $\alpha = 0.05$.

**12.49**  Are health problems precipitated by life-style changes? Attempting to answer that question, researchers surveyed a group of patients hospitalized for various chronic illnesses (187). Each subject was asked to fill out a Schedule of Recent Experience (SRE) questionnaire that measured the extent of his or her life-style changes in the past two years. At the same time, each participant's medical problem was graded according to the Seriousness of Illness Rating Scale (SIRS). The data are shown below.

| Admitting Diagnosis | Average SRE | SIRS |
|---|---|---|
| Dandruff | 26 | 21 |
| Varicose veins | 130 | 173 |
| Psoriasis | 317 | 174 |
| Eczema | 231 | 204 |
| Anemia | 325 | 312 |
| Hyperthyroidism | 816 | 393 |
| Gallstones | 563 | 454 |
| Arthritis | 312 | 468 |
| Peptic ulcer | 603 | 500 |
| High blood pressure | 405 | 520 |
| Diabetes | 599 | 621 |
| Emphysema | 357 | 636 |
| Alcoholism | 688 | 688 |
| Cirrhosis | 443 | 733 |
| Schizophrenia | 609 | 776 |
| Heart failure | 772 | 824 |
| Cancer | 777 | 1020 |

Set up and carry out an appropriate hypothesis test to decide whether SRE and SIRS scores can be considered independent. Let $\alpha = 0.01$. Note:

$$\sum_{i=1}^{17} x_i = 7973 \qquad \sum_{i=1}^{17} x_i^2 = 4{,}611{,}291$$

$$\sum_{i=1}^{17} y_i = 8517 \qquad \sum_{i=1}^{17} y_i^2 = 5{,}421{,}917$$

$$\sum_{i=1}^{17} x_i y_i = 4{,}759{,}470$$

**12.50** Both the Dow-Jones average and Standard and Poor's price index purport to measure the "state" of the stock market, implying the two should be highly correlated. The table below shows values recorded for both indices over a 13-week period.

| Week | Dow-Jones Composite | Standard and Poor's Composite |
|------|---------------------|-------------------------------|
| 1 | 276.61 | 91.62 |
| 2 | 271.26 | 89.69 |
| 3 | 272.47 | 89.89 |
| 4 | 268.35 | 88.58 |
| 5 | 271.44 | 89.62 |
| 6 | 272.35 | 90.08 |
| 7 | 263.86 | 87.96 |
| 8 | 265.07 | 88.49 |
| 9 | 262.15 | 87.45 |
| 10 | 265.38 | 88.88 |
| 11 | 269.41 | 90.20 |
| 12 | 267.01 | 89.36 |
| 13 | 266.94 | 89.21 |

Test $H_0: \rho = 0$ versus $H_1: \rho > 0$. Let $\alpha = 0.01$. (Be sure to adjust the rejection region to reflect the fact that $H_1$ is one-sided.) If $x_i$ and $y_i$ denote the $i$th week's Dow-Jones Composite and Standard and Poor's Composite, respectively, then

$$\sum_{i=1}^{13} x_i = 3492.30 \qquad \sum_{i=1}^{13} x_i^2 = 938{,}367.3224$$

$$\sum_{i=1}^{13} y_i = 1161.03 \qquad \sum_{i=1}^{13} y_i^2 = 103{,}705.5721$$

$$\sum_{i=1}^{13} x_i y_i = 311{,}946.5045$$

**12.51** The possible relationship between outdoor temperature $(X)$ and milk quality of dairy cows $(Y)$ led to the following 20 observations (129).

| April Date | Temperature (°F) | Butterfat (%) |
|---|---|---|
| 3 | 64 | 4.65 |
| 4 | 65 | 4.58 |
| 5 | 65 | 4.67 |
| 6 | 64 | 4.60 |
| 7 | 61 | 4.83 |
| 8 | 55 | 4.55 |
| 9 | 39 | 5.14 |
| 10 | 41 | 4.71 |
| 11 | 46 | 4.69 |
| 12 | 59 | 4.65 |
| 13 | 56 | 4.36 |
| 14 | 56 | 4.82 |
| 15 | 62 | 4.65 |
| 16 | 37 | 4.66 |
| 17 | 37 | 4.95 |
| 18 | 45 | 4.60 |
| 19 | 57 | 4.68 |
| 20 | 58 | 4.65 |
| 21 | 60 | 4.60 |
| 22 | 55 | 4.46 |

Can it be concluded that temperature and percent butterfat are related? Do the relevant hypothesis test at the 0.05 level of significance. Note:

$$\sum_{i=1}^{20} x_i = 1082 \qquad \sum_{i=1}^{20} x_i^2 = 60{,}304$$

$$\sum_{i=1}^{20} y_i = 93.50 \qquad \sum_{i=1}^{20} y_i^2 = 437.6406$$

$$\sum_{i=1}^{20} x_i y_i = 5044.50$$

**12.52** For Case Study 3.16 test whether the sample correlation coefficient $r$ computed from the plumage indices and behavioral indices is statistically significant. Let $\alpha = 0.01$ and make the alternative one-sided.

**12.53** Instead of using the $t$ distribution to test $H_0: \rho = 0$, experimenters sometimes take a slightly different approach that leads to an approximate Z test. It can be shown that the quotient

$$Z = \frac{\frac{1}{2}\ln\left(\frac{1+R}{1-R}\right)}{\sqrt{1/(n-3)}}$$

has approximately a standard normal distribution when $\rho = 0$, where ln is the natural logarithm. Use the statistic Z to retest the stock market data in Case Study 12.5. Keep the level of significance at 0.05.

# 12.5

## TESTING THE EQUALITY OF TWO SLOPES (Optional)

We have seen that the slope of a regression line is often its most important characteristic. To be sure, the $y$ intercept $a$ or the standard deviation $\sigma$ are sometimes the center of attention, but the slope is typically the parameter that illuminates most brightly an $xy$ relationship. That being the case, it should not be surprising that studies set up to compare two regression lines usually end up with the experimenter wanting to test whether the slope of one is equal to the slope of the other.

Figure 12.22 shows the problem graphically. A total of $n$ observations has been collected on a first $xy$ relationship, with the estimated regression line being the equation $y = \hat{a}_1 + \hat{b}_1 x$. An additional and independent set of $m$ observations has been taken on a second $xy$ relationship, resulting in a second estimated regression line $y = \hat{a}_2 + \hat{b}_2 x$. Let $b_1$ and $b_2$ denote the true slopes being estimated by $\hat{b}_1$ and $\hat{b}_2$, respectively. Are $b_1$ and $b_2$ equal?

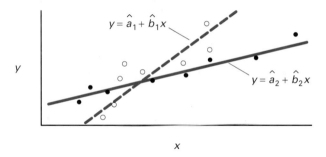

**FIG. 12.22**

Intuitively, if $\hat{b}_1$ and $\hat{b}_2$ are numerically similar, we would have no reason for rejecting the presumption that $b_1 = b_2$. As the difference between the estimated slopes gets further and further away from zero, though, the credibility of $b_1$ equaling $b_2$ diminishes. Written formally, what we are proposing to test becomes

$$H_0: \quad b_1 = b_2$$

vs.

$$H_1: \quad b_1 \neq b_2$$

Like the "one-sample" inference procedures involving slopes, this one also gets its critical values from the tail of a Student $t$ distribution.

---

**THEOREM 12.10**  Testing $H_0 : b_1 = b_2$ versus $H_1 : b_1 \neq b_2$

---

Let $y = \hat{a}_1 + \hat{b}_1 x$ be the estimated regression line based on a sample of $n$ observations representing a given $xy$ relationship. Let $y = \hat{a}_2 + \hat{b}_2 x$ be the estimated regression line based on another sample of $m$ observations representing a second $xy$ relationship. Let $b_1$ and $b_2$ denote the true slopes characterizing the two relationships.

Define $\sum_{(1)} (x_i - \bar{x}_1)^2$ to be the sum of the squared deviations of the $x_i$'s in the first sample from their own sample mean $(\bar{x}_1)$; define $\sum_{(2)} (x_i - \bar{x}_2)^2$ analogously for the second sample. Let

$$s = \sqrt{\frac{\sum_{(1)} [y_i - (\hat{a}_1 + \hat{b}_1 x_i)]^2 + \sum_{(2)} [y_i - (\hat{a}_2 + \hat{b}_2 x_i)]^2}{n + m - 4}} \qquad (12.13)$$

where $\sum_{(1)}$ indicates a summation over the $n$ observations in the first sample, and $\sum_{(2)}$ a summation over the $m$ observations in the second.

To test

$$H_0 : \quad b_1 = b_2$$

vs.

$$H_1 : \quad b_1 \neq b_2$$

at the $\alpha$ level of significance, reject $H_0$ if $t \leqslant -t_{\alpha/2, n+m-4}$ or $t \geqslant t_{\alpha/2, n+m-4}$, where

$$t = \frac{\hat{b}_1 - \hat{b}_2}{s \sqrt{\dfrac{1}{\sum_{(1)} (x_i - \bar{x}_i)^2} + \dfrac{1}{\sum_{(2)} (x_i - \bar{x}_2)^2}}}$$

---

COMMENT:   The calculations required by Theorem 12.10 can be simplified by using two of the computation formulas introduced in Section 12.2. For example,

$$\sum_{(1)} [y_i - (\hat{a}_1 + \hat{b}_1 x_i)]^2 = \sum_{(1)} y_i^2 - \hat{a}_1 \sum_{(1)} y_i - \hat{b}_1 \sum_{(1)} x_i y_i \qquad (12.14)$$

and

$$\sum_{(2)} [y_i - (\hat{a}_2 + \hat{b}_2 x_i)]^2 = \sum_{(2)} y_i^2 - \hat{a}_2 \sum_{(2)} y_i - \hat{b}_2 \sum_{(2)} x_i y_i \qquad (12.15)$$

Also

$$\sum_{(1)} (x_i - \bar{x}_1)^2 = \sum_{(1)} x_i^2 - \frac{1}{n} \left( \sum_{(1)} x_i \right)^2$$

and

$$\sum_{(2)} (x_i - \bar{x}_2)^2 = \sum_{(2)} x_i^2 - \frac{1}{m} \left( \sum_{(2)} x_i \right)^2$$

*EXAMPLE 12.5*  Natural selection, or survival of the fittest, was the overriding principle that Charles Darwin envisioned to be the driving force responsible for evolutionary change. According to Darwin, individuals possessing traits enabling them to cope particularly well with their environment will prosper. In turn, their success should somehow induce future generations to "evolve" in that same direction. As a sweeping overview of nature, Darwin's vision has come to be widely accepted among the scientific community; how it works, though, remains a hotly debated issue.

*Solution*  In recent years, evolutionists have begun to turn to molecular biology for clues that might help explain the actual mechanism of natural selection. One theory in vogue suggests that high genetic variability is the key. By the latter is meant the ability of a species to produce a wide assortment of genetically different offspring. It seems reasonable to suppose that species capable of greater genetic diversity have a higher probability of successfully adapting to environmental changes than do species whose genetic compositions limit their potential for modification.

Actually measuring the impact of genetic variability on a species' survival is not easy. To start with, subjects need to be chosen whose genetic makeups can be readily identified and whose lifetimes are relatively short. A suitably hostile environment then needs to be created, one that will make the subjects' continued existence a genuine struggle. In one experiment, fruit flies were the subjects, and a closed container with a limited food supply served as the hostile environment.

Two strains of fruit flies (*Drosophila serrata*) were raised: one was crossbred (strain A); the other was inbred (strain B). Fifty males and 50 females of each strain were put into identical, sealed glass containers. Food and space were kept at a minimum. Every 100 days, beginning on day 100 and ending on day 500, a count was made of the number of fruit flies living in each container. Since the life span of a fruit fly is about 20 days, a 500-day period corresponds to approximately 25 generations. The data [slightly modified from (4)] are summarized in Table 12.9.

**TABLE 12.9**

| Date | Day Number, $x$ | Strain A Population, $y$ | Strain B Population, $y$ |
|------|------|------|------|
| May 13 | 100 | 250 | 203 |
| Aug. 21 | 200 | 304 | 214 |
| Nov. 29 | 300 | 403 | 295 |
| March 8 | 400 | 446 | 330 |
| June 16 | 500 | 482 | 324 |

Figure 12.23 shows both sets of $(x_i, y_i)$'s graphed on the same scatter diagram. For the crossbred strain (A), the estimated regression line is $y = 195.2 + 0.606x$; for the inbred strain (B), $y = 165.8 + 0.358x$. Since it

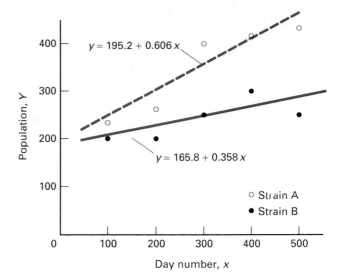

**FIG. 12.23**

measures a population's growth rate, the slope in this setting is unquestionably the key parameter: if genetic variability does enhance a strain's ability to survive, the true slope for the crossbred strain A (which has a high degree of genetic variability) should be greater than the true slope for the inbred strain B. Let $b_1$ and $b_2$ denote the true slopes for strain A and strain B, respectively. Translating the experiment's objectives into an $H_0$ and an $H_1$ leads to a test of

$$H_0: \quad b_1 = b_2$$

vs.

$$H_1: \quad b_1 > b_2$$

Let 0.05 be the level of significance.

For strain A,

$$\sum_{(1)} x_i = 1500 \qquad \sum_{(1)} x_i^2 = 550{,}000$$

$$\sum_{(1)} y_i = 1885 \qquad \sum_{(1)} y_i^2 = 748{,}565$$

$$\sum_{(1)} x_i y_i = 626{,}100$$

For strain B,

$$\sum_{(2)} x_i = 1500 \qquad \sum_{(2)} x_i^2 = 550{,}000$$

$$\sum_{(2)} y_i = 1366 \qquad \sum_{(2)} y_i^2 = 387{,}906$$

$$\sum_{(2)} x_i y_i = 445{,}600$$

Therefore, by Equations 12.14 and 12.15

$$\sum_{(1)} [y_i - (\hat{a}_1 - \hat{b}_1 x_i)]^2 = \sum_{(1)} y_i^2 - \hat{a}_1 \sum_{(1)} y_i - \hat{b}_1 \sum_{(1)} x_i y_i$$

$$= 748{,}565 - (195.2)(1885) - (0.606)(626{,}100)$$

$$= 1196.4$$

and

$$\sum_{(2)} [y_i - (\hat{a}_2 + \hat{b}_2 x_i)]^2 = \sum_{(2)} y_i^2 - \hat{a}_2 \sum_{(2)} y_i - \hat{b}_2 \sum_{(2)} x_i y_i$$

$$= 387{,}906 - (165.8)(1366) - (0.358)(445{,}600)$$

$$= 1898.4$$

From Equation 12.13, the pooled standard deviation $s$ is

$$s = \sqrt{\frac{1194.4 + 1898.4}{5 + 5 - 4}} = 22.71$$

Also,

$$\sum_{(1)} (x_i - \bar{x}_1)^2 = \sum_{(1)} x_i^2 - \frac{1}{n} \left( \sum_{(1)} x_i \right)^2$$

$$= 550{,}000 - \frac{(1500)^2}{5} = 100{,}000$$

which, in this case, is the same as the value for $\sum_{(2)} (x_i - \bar{x}_2)^2$, since both sets of $Y$ measurements were taken at the same set of $x_i$'s. From Theorem 12.10 the test statistic $t$, then, has a value of 2.44:

$$t = \frac{\hat{b}_1 - \hat{b}_2}{s \sqrt{\dfrac{1}{\displaystyle\sum_{(1)} (x_i - \bar{x}_1)^2} + \dfrac{1}{\displaystyle\sum_{(2)} (x_i - \bar{x}_2)^2}}}$$

$$= \frac{0.606 - 0.358}{22.71 \sqrt{\dfrac{1}{100{,}000} + \dfrac{1}{100{,}000}}} = 2.44$$

To finish the problem, we need to compare 2.44 with an appropriate critical value from the Student $t$ distribution with 6 ($=5 + 5 - 4$) df. Since $\alpha$ was set at 0.05 and $H_1$ is one-sided, the rejection region consists of all values $t \geq 1.9432$ (see Figure 12.24). Since $t = 2.44$ *does* exceed that cutoff, our conclusion is to reject $H_0$. The two estimated slopes, according to the $t$ test, are too far apart to be consistent with the null hypothesis that the two true slopes are equal. The data, in other words, are supporting the contention that genetic variability is a factor in the survival of a species.

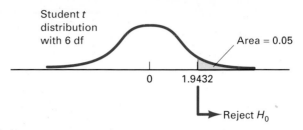

**FIG. 12.24**

---

QUESTIONS

**12.54** Which hypothesis test that we saw in an earlier chapter does Theorem 12.10 most closely resemble?

**12.55** In general, if two random variables $X$ and $Y$ are independent, Var $(X - Y) =$ Var $(X) +$ Var $(Y)$. Compare the denominator of $t$ in Theorem 12.10 with Var $(B)$ in Theorem 12.1. What does the denominator of $t$ represent?

**12.56** Based on your answer to Question 12.55, set up a decision rule for testing whether the difference between two estimated $y$ intercepts $(\hat{a}_1 - \hat{a}_2)$ is statistically significant; that is, test $H_0: a_1 = a_2$ versus $H_1: a_1 \neq a_2$.

**12.57** An experimenter collects the following two sets of regression data, reflecting conditions 1 and 2, respectively. Graph the two sets of data on the same scatter diagram and superimpose the two estimated regression lines. At the $\alpha = 0.05$ level of significance, test $H_0: b_1 = b_2$ against a two-sided alternative.

| CONDITION 1 | | CONDITION 2 | |
|---|---|---|---|
| $x_i$ | $y_i$ | $x_i$ | $y_i$ |
| 4 | 2 | 3 | 1 |
| 3 | 2 | 6 | 5 |
| 2 | 1 | 5 | 6 |
| 6 | 4 | 4 | 2 |
| 5 | 4 | 4 | 3 |
| | | 6 | 7 |

**12.58** Thirteen children of various ages, seven boys and six girls, are given a behavioral skills test that measures their social maturity. Scores on the test range from 0 to 100. Children making higher scores are presumably more capable of interacting with their peers in a responsible manner. Plot the data shown below on the same scatter diagram and graph the two estimated regression lines. Address the question of whether boys and girls develop socially at the same rate by testing $H_0: b_1 = b_2$ versus $H_1: b_1 \neq b_2$. Let $\alpha = 0.05$.

| | BOYS | | | GIRLS | |
|---|---|---|---|---|---|
| Subject | Age, $x$ | Score, $y$ | Subject | Age, $x$ | Score, $y$ |
| AL | 8 | 30 | DP | 5 | 20 |
| BD | 10 | 30 | CF | 12 | 80 |
| HM | 12 | 50 | SH | 9 | 80 |
| TT | 6 | 20 | EJ | 5 | 30 |
| GM | 6 | 30 | LF | 8 | 60 |
| BM | 9 | 50 | BB | 9 | 80 |
| JB | 12 | 40 | | | |

Note: For the boys,

$$\sum_{i=1}^{7} x_i = 63 \qquad \sum_{i=1}^{7} x_i^2 = 605$$

$$\sum_{i=1}^{7} y_i = 250 \qquad \sum_{i=1}^{7} y_i^2 = 9700$$

For the girls,

$$\sum_{i=1}^{6} x_i = 48 \qquad \sum_{i=1}^{6} x_i^2 = 420$$

$$\sum_{i=1}^{6} y_i = 350 \qquad \sum_{i=1}^{6} y_i^2 = 24{,}100$$

# 12.6

## FITTING NONLINEAR REGRESSIONS (Optional)

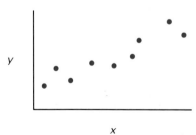

**FIG. 12.25**

Imagine yourself as an experimenter who has just collected $n$ data points as part of a regression study. You decide, quite properly, to draw the corresponding scatter diagram. Assume for the moment that neither measurement is considered a random variable. If the appearance of that scatter diagram looks like Figure 12.25, you know exactly what to do first: use Theorem 3.2 to find the estimated regression line $y = \hat{a} + \hat{b}x$, where

$$\hat{b} = \frac{n \sum_{i=1}^{n} x_i y_i - \left(\sum_{i=1}^{n} x_i\right)\left(\sum_{i=1}^{n} y_i\right)}{n \sum_{i=1}^{n} x_i^2 - \left(\sum_{i=1}^{n} x_i\right)^2} \quad \text{and} \quad \hat{a} = \frac{\sum_{i=1}^{n} y_i - \hat{b} \sum_{i=1}^{n} x_i}{n}$$

But suppose the $(x_i, y_i)$'s show a pattern that more closely resembles one of the scatter diagrams in Figure 12.26. How should you then proceed with the analysis? Clearly, in neither case would fitting a straight line make any sense: both relationships are distinctly *curvilinear* rather than linear. Nothing that we have done in this chapter would seem to be appropriate, at least at first glance. By doing a little algebra, though, we can make everything in this chapter appropriate.

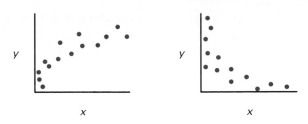

**FIG. 12.26**

The keys to coping with these more complicated *xy* relationships are two questions: (1) What mathematical functions produce scatter diagrams like Figure 12.26? (2) How do we fit an equation to a set of data when that equation does not have the form $y = a + bx$? Responding to either question in complete generality is impossible. Fortunately, partial answers are often sufficient.

In addressing the first question, our initial admission would have to be that the number of different functions leading to nonlinear scatter diagrams is infinite. That's the bad news. Still, by using only a handful of properly chosen special functions, we can adequately approximate a high proportion of the nonlinear scatter diagrams likely to be encountered in practice. That's the good news. Among the most useful of those special functions are

1.    $y = ax^b$ 	 4.    $y = \dfrac{1}{a + bx}$

2.    $y = ae^{bx}$ 	 5.    $y = \dfrac{x}{a + bx}$

3.    $y = a + \dfrac{b}{x}$

Figure 12.27, for example, shows some of the patterns associated with the equation $y = ax^b$. Quite clearly, by changing the values of *a* and *b*, scatter diagrams having quite different appearances can be approximated by that single function.

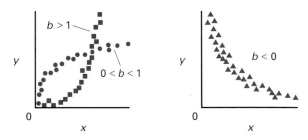

**FIG. 12.27**

Surprisingly, fitting any of these five nonlinear functions to a set of data is not especially difficult. The trick in each case is "transforming" the original function algebraically in such a way that a new equation is produced, one that *is* linear. For instance, suppose the appearance of a scatter diagram suggests that the *xy* relationship can be approximated by a function of the form

$$y = ax^b \tag{12.16}$$

Taking the log of both sides of Equation 12.16 gives

$$\log y = \log a + b \log x \tag{12.17}$$

Let $y' = \log y$, $a' = \log a$, $b' = b$, and $x' = \log x$. In terms of the primes, Equation 12.17 can be written in what is clearly a linear form:

$$y' = a' + b'x'$$

Moreover, the appearance of Equation 12.18 implies that we can estimate $a'$ and $b'$ by simply replacing each $x_i$ and $y_i$ in Theorem 3.2 with $x_i'$ and $y_i'$, respectively. That is,

$$\hat{b}' = \frac{n \sum\limits_{i=1}^{n} x_i' y_i' - \left( \sum\limits_{i=1}^{n} x_i' \right)\left( \sum\limits_{i=1}^{n} y_i' \right)}{n \sum\limits_{i=1}^{n} x_i'^2 - \left( \sum\limits_{i=1}^{n} x_i' \right)^2}$$

$$= \frac{n \sum\limits_{i=1}^{n} (\log x_i)(\log y_i) - \left( \sum\limits_{i=1}^{n} \log x_i \right)\left( \sum\limits_{i=1}^{n} \log y_i \right)}{n \sum\limits_{i=1}^{n} (\log x_i)^2 - \left( \sum\limits_{i=1}^{n} \log x_i \right)^2} \tag{12.19}$$

and

$$\hat{a}' = \frac{\sum\limits_{i=1}^{n} y_i' - \hat{b}' \sum\limits_{i=1}^{n} x_i'}{n}$$

$$= \frac{\sum\limits_{i=1}^{n} \log y_i - \hat{b}' \sum\limits_{i=1}^{n} \log x_i}{n} \tag{12.20}$$

If we know $\hat{a}'$ and $\hat{b}'$, though, we can easily compute $\hat{a}$ and $\hat{b}$, which are what we set out to find in the first place:

$$\hat{b} = \hat{b}' \quad \text{and} \quad \hat{a} = 10^{\hat{a}'} \quad \text{(since } a' = \log a\text{)}$$

---

*EXAMPLE 12.6*  The numbers described in Case Study 2.11 quantify the relationship between onset of locomotion $(x)$ and onset of play $(y)$ for 11 kinds of mammals including humans. The appearance of the data's scatter diagram (compare Figure 2.13 with Figure 12.27) suggests that a model of the form $y = ax^b$ might be a good choice for describing the *xy* relationship.

**TABLE 12.10**

| Species | Locomotion Begins, $x$ (days) | Play Begins, $y$ (days) | $x' = \log x$ | $y' = \log y$ |
|---|---|---|---|---|
| *Homo sapiens* | 360 | 90 | 2.55630 | 1.95424 |
| *Gorilla gorilla* | 165 | 105 | 2.21748 | 2.02119 |
| *Felis catus* | 21 | 21 | 1.32222 | 1.32222 |
| *Canis familiaris* | 23 | 26 | 1.36173 | 1.41497 |
| *Rattus norvegicus* | 11 | 14 | 1.04139 | 1.14613 |
| *Turdus merula* | 18 | 28 | 1.25527 | 1.44716 |
| *Macaca mulatta* | 18 | 21 | 1.25527 | 1.32222 |
| *Pan troglodytes* | 150 | 105 | 2.17609 | 2.02119 |
| *Saimiri sciurens* | 45 | 68 | 1.65321 | 1.83251 |
| *Cercocebus alb.* | 45 | 75 | 1.65321 | 1.87506 |
| *Tamiasciureus hud.* | 18 | 46 | 1.25527 | 1.66276 |

**Solution** Finding estimates for $a$ and $b$ requires that we first rewrite $y = ax^b$ in its equivalent linear form,

$$\log y = \log a + b \log x \qquad \text{or} \qquad y' = a' + b'x'.$$

Next to each of the original 11 data points $(x_i, y_i)$, Table 12.10 lists the "new" data point $(x_i', y_i')$, where $(x_i', y_i') = (\log x_i, \log y_i)$.

Computing the usual sums and sums of squares for the $x_i'$'s and the $y_i'$'s gives

$$\sum_{i=1}^{11} x_i' = \log x_1 + \log x_2 + \cdots + \log x_{11} = 17.74744$$

$$\sum_{i=1}^{11} y_i' = \log y_1 + \log y_2 + \cdots + \log y_{11} = 18.01965$$

$$\sum_{i=1}^{11} x_i'^2 = (\log x_1)^2 + (\log x_2)^2 + \cdots + (\log x_{11})^2 = 31.06764$$

and

$$\sum_{i=1}^{11} x_i'y_i' = (\log x_1)(\log y_1) + (\log x_2)(\log y_2) + \cdots + (\log x_{11})(\log y_{11})$$
$$= 30.43743$$

If we substitute all these values into Equations 12.19 and 12.20, the resulting slope and $y$ intercept are

$$\hat{b}' = \frac{11(30.43743) - (17.74744)(18.01965)}{11(31.06764) - (17.74744)^2} = 0.56$$

$$\hat{a}' = \frac{18.01965 - (0.56)(17.74744)}{11} = 0.73364$$

Mathematically, there are two ways to express what we just computed. *In terms of the transformed data,* the relationship between $\log x$ and $\log y$ is approximated by the equation

$$y' = \hat{a}' + \hat{b}'x' \qquad \text{or} \qquad \log y = 0.73364 + 0.56 \log x$$

*In terms of the original data,*

$$\hat{b} = \hat{b}' = 0.56 \qquad \text{and} \qquad \hat{a} = 10^{\hat{a}'} = 10^{0.73364} = 5.42$$

which implies that $\qquad y = \hat{a}x^{\hat{b}} = 5.42x^{0.56}$

Table 12.11 shows a set of points satisfying this equation. When plotted and connected, they produce the dashed curve in Figure 2.13.

**TABLE 12.11**

| $x$ | $y = 5.42x^{0.56}$ |
|---|---|
| 0 | 0 |
| 25 | 32.9 |
| 50 | 48.5 |
| 100 | 71.4 |
| 200 | 105.3 |
| 300 | 132.2 |

**QUESTIONS**

12.59 Reread Question 2.34 about the relationship between $x$, the foraging size of an ant colony, and $y$, the colony's total population. Plot the data given and fit them with a model of the form $y = ax^b$. Superimpose a sketch of the curve over the original scatter diagram. Note: If $x'_i = \log x_i$ and $y'_i = \log y_i$, then

$$\sum_{i=1}^{15} x'_i = 41.77440 \qquad\qquad \sum_{i=1}^{15} y'_i = 52.79857$$

$$\sum_{i=1}^{15} x'^2_i = 126.60443 \qquad\qquad \sum_{i=1}^{15} x'_i y'_i = 156.03807$$

12.60 Without actually plotting any points, describe what the scatter diagram for the data in Question 12.59 would look like if
(a) $\log x$ was plotted along the $x$ axis and $\log y$ along the $y$ axis?
(b) The original $(x_i, y_i)$'s were plotted on log-log paper?

12.61 An experimenter collects the following set of $n = 6$ data points:

| $x$ | 3.0 | 1.0 | 4.3 | 2.0 | 3.2 | 4.0 |
|---|---|---|---|---|---|---|
| $y$ | 1.5 | 1.0 | 4.1 | 1.0 | 2.4 | 2.5 |

(a) Plot the data.
(b) Fit the model $y' = a' + b'x'$, where $y' = \log y$, $a' = \log a$, $b' = b$, and $x' = \log x$.
(c) Use the answer to part (b) as a way of fitting the equation $y = ax^b$.
(d) Plot the equation $y = \hat{a}x^{\hat{b}}$ on the scatter diagram in part (a)

12.62 Listed below are the average prices of used cars in 1957 (25).

| Age, $x$ | Average Price, $y$ | Log $x$ | Log $y$ |
|---|---|---|---|
| 1 | $2,651 | 0.00000 | 3.42341 |
| 2 | 1,943 | 0.30103 | 3.28847 |
| 3 | 1,494 | 0.47712 | 3.17435 |
| 4 | 1,087 | 0.60206 | 3.03623 |
| 5 | 765 | 0.69897 | 2.88366 |
| 6 | 538 | 0.77815 | 2.73078 |
| 7 | 484 | 0.84510 | 2.68484 |
| 8 | 290 | 0.90309 | 2.46240 |
| 9 | 226 | 0.95424 | 2.35411 |
| 10 | 204 | 1.00000 | 2.30963 |

Let age be the $x$ variable, and average price the $y$ variable. Plot $x$ versus $y$. Fit the data with a function of the form

$$y = ax^b$$

Sketch $y = \hat{a}x^{\hat{b}}$ on the data's scatter diagram.

12.63 Suppose the curve calculated for the data in Question 12.62 remained valid for cars anywhere from 1 to 15 years old. On the average, what is the price of a 13-year-old car?

12.64 Over the years numerous efforts have been made to demonstrate that human brains are appreciably different in structure than the brains of lower-order primates. To the consternation of researchers everywhere, many of those efforts have proven futile. Listed below are the average areas of the striate cortex ($x$) and the prestriate cortex ($y$) recorded for humans and for three species of chimpanzees. Is it reasonable to conclude that $x$ and $y$ for all four primates can be modeled by the same function $y = ax^b$? Answer the question by plotting the data and graphing $y = \hat{a}x^{\hat{b}}$ (119).

| Primate | Striate Cortex Area, $x$ (mm$^2$) | Prestriate Cortex Area, $y$ (mm$^2$) |
|---|---|---|
| *Homo* | 2613 | 7838 |
| *Pongo* | 1876 | 2864 |
| *Cercopithecus* | 933 | 1334 |
| *Galago* | 78.9 | 40.8 |

In principle, fitting the other four nonlinear functions mentioned earlier in this section is accomplished by using the same basic strategy that was taken in dealing with $y = ax^b$. We first make whatever substitutions for $x$

and $y$ are necessary to change $y = f(x)$ into $y' = a' + b'x'$. Theorem 3.2 is then used to find $\hat{a}'$ and $\hat{b}'$. Once we know $\hat{a}'$ and $\hat{b}'$, we can transform "back" to find $\hat{a}$ and $\hat{b}$.

Suppose, for example, we have reason to believe that an $xy$ relationship can be approximated by the function

$$y = \frac{1}{a + bx}$$

Taking the reciprocal of both sides gives

$$\frac{1}{y} = a + bx \tag{12.21}$$

which implies that $1/y$ is linear with $x$. More formally, if we let $y' = 1/y$, $a' = a$, $b' = b$, and $x' = x$, Equation 12.21 can be written in the standard linear form,

$$y' = a' + b'x'$$

It follows, then, from Theorem 3.2 that

$$\hat{b}' = \frac{n \sum\limits_{i=1}^{n} x_i' y_i' - \left( \sum\limits_{i=1}^{n} x_i' \right)\left( \sum\limits_{i=1}^{n} y_i' \right)}{n \sum\limits_{i=1}^{n} x_i'^2 - \left( \sum\limits_{i=1}^{n} x_i' \right)^2}$$

$$= \frac{n \sum\limits_{i=1}^{n} x_i(1/y_i) - \left( \sum\limits_{i=1}^{n} x_i \right)\left( \sum\limits_{i=1}^{n} (1/y_i) \right)}{n \sum\limits_{i=1}^{n} x_i^2 - \left( \sum\limits_{i=1}^{n} x_i \right)^2} \tag{12.22}$$

and

$$\hat{a}' = \frac{\sum\limits_{i=1}^{n} y_i' - \hat{b}' \sum\limits_{i=1}^{n} x_i}{n}$$

$$= \frac{\sum\limits_{i=1}^{n} (1/y_i) - \hat{b}' \sum\limits_{i=1}^{n} x_i}{n} \tag{12.23}$$

The final step in a transformation problem—getting numerical values for $\hat{a}$ and $\hat{b}$—is easier with some functions than for others. Recall that for the model $y = ax^b$, the initial substitutions included the equation $a' = \log a$. Thus, recovering $\hat{a}$ required the additional computation $\hat{a} = 10^{\hat{a}'}$. For the model $y = 1/(a + bx)$, though, the initial substitutions keep the slope and the $y$ intercept the same; that is, $a' = a$ and $b' = b$ (which imply that $\hat{a} = \hat{a}'$ and $\hat{b} = \hat{b}'$). What we compute in Equations 12.22 and 12.23, therefore, are precisely the estimates we ultimately need.

12.65 Fit the dragonfly data in Question 1.13 with a regression function of the form

$$y = a + \frac{b}{x}$$

Let $y' = y$, $a' = a$, $b' = b$, and $x' = 1/x$. Draw a scatter diagram of the data and superimpose the curve $y = \hat{a} + \hat{b}/x$.

12.66 Find the best model of the form $y = 1/(a + bx)$ through the set of points

| $x$ | 0.4 | 0.5 | 4.0 | 1.0 | 3.0 |
|---|---|---|---|---|---|
| $y$ | 0.1 | 0.3 | 0.07 | 0.2 | 0.08 |

Use Equations 12.22 and 12.23. Plot the data and sketch the curve $y = 1/(\hat{a} + \hat{b}x)$.

12.67 What transformation linearizes the model $y = ae^{bx}$? (Hint: Take the natural log of both sides of the equation. What is the relationship between $b$ and $b'$? between $a$ and $a'$?)

12.68 One factor thought to cause skin cancer is ultraviolet (UV) radiation from the sun. It is well known that the amount of UV radiation a person receives is a function of the shielding thickness of the earth's ozone layer, and that, in turn, depends on the latitude of where the person lives. Listed blow for nine areas throughout the United States are the malignant skin cancer (melanoma) rates reported for white males. The location of each area is given in "degrees north latitude"; the rates refer to a three-year period (47). Fit these data with an exponential model $y = ae^{bx}$. Let $x$ denote "degrees north latitude," and $y$ "melanoma rate." Plot the data and sketch in the regression curve.

| Location Number | Degrees North Latitude | Melanoma Rate (per 100,000) |
|---|---|---|
| 1 | 32.8 | 9.0 |
| 2 | 33.9 | 5.9 |
| 3 | 34.1 | 6.6 |
| 4 | 37.9 | 5.8 |
| 5 | 40.0 | 5.5 |
| 6 | 40.8 | 3.0 |
| 7 | 41.7 | 3.4 |
| 8 | 42.2 | 3.1 |
| 9 | 45.0 | 3.8 |

The question asked at the beginning of this section (What mathematical functions lead to scatter diagrams having nonlinear patterns?) deserves a final comment. Being told that many curvilinear $xy$ relationships can be well-approximated by using only a handful of special functions is certainly reassuring, but it leaves unanswered what the experimenter may need to know most—*which* of those functions should be used in a particular situation? Given a set of data, for example, when is it reasonable to assume that $y = ax^b$ or that $y = x/(a + bx)$?

In practice, experimenters typically turn to computer software for help in choosing an appropriate regression function (if that choice is not obvious by simply looking at the scatter diagram). Given a set of data points, many of these computer programs will first fit the straight line $y = a + bx$ and then make the necessary transformations to write each of the five models cited on p. 561 in the form $y' = a' + b'x'$. Correlation coefficients are computed for all six regressions (the original straight line and the five linearized models). Of those six we choose as the "best" equation the one whose correlation coefficient (for $y' = a' + b'x'$) is closest to either $-1$ or $+1$.

# 12.7

## COMPUTER NOTES

Because of the many calculations required for regression problems, computers are especially useful. Most regression printouts provide many statistics and analyses. To make the printouts more understandable, we examine only the portion that we need.

Figure 12.28 gives part of the SAS printout involving the estimates of the parameters for the regression line for Example 12.2. The least squares estimates of the parameters $a$ and $b$ are located in the column headed PARAMETER ESTIMATE. The $t$ statistics that test the hypothesis that the parameter differs from zero appear in the next-to-last column. The last column gives the $P$ value for testing $H_0: b = 0$ versus $H_1: b \neq 0$. However, in Example 12.2 our alternative hypothesis was one-sided, so the $P$ value we want is

$$0.5(0.0003) = 0.00015$$

because of the symmetry of the $T$ distribution. For any $\alpha > 0.00015$ we can accept $H_0$.

The entry R-SQUARE gives the coefficient of determination which says that approximately 85% of the variability in the data is explained by the model.

Both SAS and Minitab have very convenient routines to calculate the correlation coefficient. The SAS procedure also contains the $P$ values for testing $H_0: \rho = 0$. Figure 12.29 is the portion of the SAS printout for Case Study 12.5. The coefficient 0.44563 is found on the X row and Y column and on the X column and Y row. Beneath that value is the $P$ value to test $H_0: \rho = 0$ versus $H_1: \rho \neq 0$. Since the $P$ value of $0.0057 < \alpha = 0.05$, we reject the null hypothesis.

```
DATA OBS;
   INPUT X Y @ @ ;
   CARDS;
```

(data from Case Study 2.9 here, in pairs)

```
PROC REG;
   MODEL Y = X;
```

DEP VARIABLE: Y

ANALYSIS OF VARIANCE

| SOURCE | DF | SUM OF SQUARES | MEAN SQUARE | F VALUE | PROB>F |
|--------|----|----------------|-------------|---------|--------|
| MODEL | 1 | 8309.55586 | 8309.55586 | 42.336 | 0.0003 |
| ERROR | 7 | 1373.94636 | 196.27805 | | |
| C TOTAL | 8 | 9683.50222 | | | |

| | | | |
|---|---|---|---|
| ROOT MSE | 14.00993 | R-SQUARE | 0.8581 |
| DEP MEAN | 157.3444 | ADJ R-SQ | 0.8378 |
| C.V. | 8.903986 | | |

PARAMETER ESTIMATES

| VARIABLE | DF | PARAMETER ESTIMATE | STANDARD ERROR | T FOR H0: PARAMETER=0 | PROB > |T| |
|----------|----|--------------------|-----------------|------------------------|-----------|
| INTERCEP | 1 | 114.71563 | 8.04566313 | 14.258 | 0.0001 |
| X | 1 | 9.23145627 | 1.41878693 | 6.507 | 0.0003 |

**FIG. 12.28**

```
DATA OBS;
   INPUT X Y @ @ ;
   CARDS;
```

(data from Table 12.8 entered here, in pairs)

```
PROC CORR;
```

PEARSON CORRELATION COEFFICIENTS / PROB > |R| UNDER H0 : RHO=0 / N = 37

| | X | Y |
|---|---|---|
| X | 1.00000 | 0.44563 |
| | 0.0000 | 0.0057 |
| Y | 0.44563 | 1.00000 |
| | 0.0057 | 0.0000 |

**FIG. 12.29**

## 12.7

### SUMMARY

The desire to draw inferences about $xy$ relationships was cited in the introduction as the motivation that would suggest ways in which regression–correlation data might best be analyzed. We saw numerous case studies where the obvious questions to be addressed involved hypothesis tests for the slope, confidence intervals for the $y$ intercept, prediction intervals for future $y$ values, and so on. *How* we carried out those procedures had a familiar ring. Theorems 12.3, 12.4, 12.5, 12.6, 12.9, and 12.10, for example, are all very much in the spirit of the $t$ tests, $\chi^2$ tests, and confidence intervals that we covered in Chapters 9, 10, and 11.

Assumptions have also played a major role in determining which topics should be covered in this chapter and why. What we even call a set of "relationship data" depends on certain assumptions. If only one of the measurements being recorded, say $Y$, can be considered a random variable, the $(x_i, Y_i)$'s are said to be *regression data* (an important special case being the *linear model* developed in Sections 12.2 and 12.3). If both measurements are random variables, the $(X_i, Y_i)$'s are classified as *correlation data*. Sometimes the distinction is important, sometimes not. If the experimenter's objective is to approximate $y$ using a linear function of $x$, the resulting equation is the same, regardless of whether the data satisfy the regression assumptions or the correlation assumptions. Testing $H_0$: $\rho = 0$, on the other hand, makes sense only if the measurements are correlation data.

The nature of the $xy$ relationship itself becomes a critical assumption. If we proceed as though $y = a + bx$ when, in fact, $y = ax^b$ or $y = x/(a + bx)$, any inferences we draw may be seriously in error. Therefore, scatter diagrams, are an absolutely essential first step in doing any regression–correlation problem. Until we see what the data look like, we have no way of knowing how to start the analysis mathematically.

Ostensibly, all the theorems appearing early in this chapter are conditional on the $xy$ relationship being linear. That isn't quite true. We saw in Section 12.6 that linearity is not nearly as restrictive an assumption as it might initially appear. By making certain transformations on either the $x_i$'s and/or the $y_i$'s, we can often rewrite nonlinear relationships in linear form and then analyze the latter.

The importance of its assumptions notwithstanding, what characterizes the analysis of regression–correlation data most distinctly is the sheer variety of statistical techniques it calls for routinely. Other data models we have seen can be mastered by learning two or three different procedures, but not so with regression–correlation data. Everything from descriptive statistics to curve fitting to confidence intervals to hypothesis tests may all need to be considered at the same time and all against the backdrop of some rather difficult probability theory.

### FORMULAS

$$\hat{b} = \frac{n \sum x_i y_i - (\sum x_i)(\sum y_i)}{n \sum x_i^2 - (\sum x_i)^2}$$

$$\boxed{\left( \sum = \sum_{i=1}^{n} \right)}$$

$$\hat{a} = \frac{\sum y_i - \hat{b} \sum x_i}{n}$$

$$\hat{\sigma}_Y = s_Y = \sqrt{\frac{n \sum y_i^2 - (\sum y_i)^2}{n(n-1)}}$$

$$\hat{\sigma}^2 = \frac{\text{SSE}}{n-2} = \frac{1}{n-2} \left( \sum y_i^2 - \hat{a} \sum y_i - \hat{b} \sum x_i y_i \right)$$

$$\hat{\rho} = r = \frac{n \sum x_i y_i - (\sum x_i)(\sum y_i)}{\sqrt{n \sum x_i^2 - (\sum x_i)^2} \sqrt{n \sum y_i^2 - (\sum y_i)^2}}$$

$$\hat{\mu}_X = \frac{1}{n} \sum x_i$$

$$y = \left( \bar{y} - \frac{r s_Y}{s_X} \bar{x} \right) + \frac{r s_Y}{s_X} x$$

$$\hat{\sigma}_X = s_X = \sqrt{\frac{n \sum x_i^2 - (\sum x_i)^2}{n(n-1)}}$$

$$x = \left( \bar{x} - \frac{r s_X}{s_Y} \bar{y} \right) + \frac{r s_X}{s_Y} y$$

$$\hat{\mu}_Y = \bar{y} = \frac{1}{n} \sum y_i$$

$$s = \sqrt{\frac{\sum_{(1)} [y_i - (\hat{a}_1 + \hat{b}_1 x_i)]^2 + \sum_{(2)} [y_i - (\hat{a}_2 + \hat{b}_2 x_i)]^2}{n + m - 4}}$$

### Hypothesis Tests:

| TEST | REJECTION REGION | STATISTIC |
|---|---|---|
| 1. $H_0: a = a_0$ vs. $H_1: a \neq a_0$ | $t \leqslant -t_{\alpha/2, n-2}$ or $t \geqslant t_{\alpha/2, n-2}$ | $t = \dfrac{(\hat{a} - a_0)\sqrt{n \sum (x_i - \bar{x})^2}}{\sqrt{\hat{\sigma}^2 \sum x_i^2}}$ |
| 2. $H_0: b = b_0$ vs. $H_1: b \neq b_0$ | $t \leqslant -t_{\alpha/2, n-2}$ or $t \geqslant t_{\alpha/2, n-2}$ | $t = \dfrac{(\hat{b} - b_0)\sqrt{\sum (x_i - \bar{x})^2}}{\sqrt{\hat{\sigma}^2}}$ |
| 3. $H_0: \sigma^2 = \sigma_0^2$ vs. $H_1: \sigma^2 \neq \sigma_0^2$ | $\chi^2 \leqslant \chi_{\alpha/2, n-2}^2$ or $\chi^2 \geqslant \chi_{1-\alpha/2, n-2}^2$ | $\chi = \dfrac{(n-2)\hat{\sigma}^2}{\sigma_0^2}$ |
| 4. $H_0: b_1 = b_2$ vs. $H_1: b_1 \neq b_2$ | $t \leqslant -t_{\alpha/2, n+m-4}$ or $t \geqslant t_{\alpha/2, n+m-4}$ | $t = \dfrac{\hat{b}_1 - \hat{b}_2}{s \sqrt{\dfrac{1}{\sum_{(1)} (x_i - \bar{x}_1)^2} + \dfrac{1}{\sum_{(2)} (x_i - \bar{x}_2)^2}}}$ |

### Confidence Intervals:

1. $(\hat{a} - w, \hat{a} + w)$  $\qquad w = t_{\alpha/2, n-2} \dfrac{\sqrt{\hat{\sigma}^2 \sum x_i^2}}{\sqrt{n \sum (x_i - \bar{x})^2}}$  $\qquad$ for $a$

2. $(\hat{b} - w, \hat{b} + w)$  $\qquad w = t_{\alpha/2, n-2} \dfrac{\sqrt{\hat{\sigma}^2}}{\sqrt{\sum (x_i - \bar{x})^2}}$  $\qquad$ for $b$

3. $\left( \dfrac{(n-2)\hat{\sigma}^2}{\chi_{1-\alpha/2, n-2}^2}, \dfrac{(n-2)\hat{\sigma}^2}{\chi_{\alpha/2, n-2}^2} \right)$  $\qquad$ for $\sigma^2$

4. $(\hat{y}_0 - w, \hat{y}_0 + w)$  $\qquad w = t_{\alpha/2, n-2} \sqrt{\hat{\sigma}^2 \left\{ \dfrac{1}{n} + \dfrac{(x_0 - \bar{x})^2}{\sum (x_i - \bar{x})^2} \right\}}$  $\qquad$ for $E(Y)$

*Prediction Interval:*

$$(\hat{y}_0 - w, \hat{y}_0 + w) \qquad\qquad w = t_{\alpha/2,\,n-2}\sqrt{\hat{\sigma}^2\left\{1 + \frac{1}{n} + \frac{(x_0 - \bar{x})^2}{\sum (x_i - \bar{x})^2}\right\}} \qquad \text{for } y_0$$

---

**REVIEW QUESTIONS**

**12.69**   To reduce drop out rates, school boards across the country have begun to implement a counseling program aimed at 10th graders. Nine schools serving similar populations were asked to help evaluate the program. For three schools, the program had been in effect for 2 years; for another three schools, the program had been in effect for 4 years; a final group of three had had the program for 6 years. Recorded for each school was last year's percentage of 11th and 12th graders who failed to graduate.

| School | Years Program in Effect, *x* | Drop Out %, *y* |
|--------|:----------------------------:|:---------------:|
| A60 | 2 | 19 |
| C14 | 2 | 12 |
| D19 | 2 | 14 |
| M08 | 4 | 8 |
| A22 | 4 | 11 |
| J16 | 4 | 17 |
| G21 | 6 | 12 |
| F04 | 6 | 4 |
| A16 | 6 | 10 |

(a) Plot the data.

(b) Compute the estimated regression line, $y = \hat{a} + \hat{b}x$.

(c) Graph your answer to part (b) on the scatter diagram in part (a).

(d) Estimate $\sigma^2$.

**12.70**   Decide whether time is a factor in the effectiveness of the program described in Question 12.69 by testing $H_0: b = 0$ against $H_1: b \neq 0$. Let $\alpha = 0.05$.

**12.71**   For the data in Question 12.69, construct a 90% confidence interval for the true *y*-intercept. Is is believable that $a = 30$?

**12.72**   Explain the difference between $b$, $\hat{b}$, and $B$. What is the difference between the $\hat{b}$ used in this chapter and the $\hat{b}$ that appeared in Section 3.5?

**12.73**   Six hypertensives participated in a 4 week clinical trial investigating the effectiveness of a new tranquilizer in reducing blood pressure. Listed below are the dosages administered and the blood pressure reductions observed.

| Subject | Dosage Received (mg), $x$ | Reduction in Blood Pressure (mmHg), $y$ |
|---|---|---|
| AF | 2.0 | 4 |
| LT | 2.0 | 8 |
| LC | 6.0 | 12 |
| SP | 6.0 | 10 |
| YC | 10.0 | 18 |
| LW | 10.0 | 14 |

(a) Plot the data.

(b) Find the estimated regression line, $y = \hat{a} + \hat{b}x$.

(c) Estimate $\sigma^2$.

(d) Are these regression data or correlation data?

**12.74**  (a) For the data in Question 12.73, construct 95% confidence intervals for $E(Y)$ when $x = 2.0$, $x = 6.0$, and $x = 10.0$.

(b) Graph the estimated regression line from Question 12.73 and plot the upper and lower limits for the confidence intervals asked for in part (a). Connect the three $\hat{y}_0 + w$ values with a curve. Do likewise for the $\hat{y}_0 - w$ values.

**12.75**  For the data in Question 12.73, compute the estimated variance of $B$. (Hint: A formula for the *true* variance of $B$ is given in Theorem 12.1. If $\sigma^2$ in the latter is replaced by an estimate of $\sigma^2$, then the formula in Theorem 12.1 gives the estimated variance of $B$.)

**12.76**  Is it likely that the variance of the blood pressure reductions described in Question 12.73 is as large as 10? Answer the question by testing $H_0: \sigma^2 = 10$ versus $H_1: \sigma^2 < 10$. Let $\alpha = 0.05$.

**12.77**  Using the data in Question 12.73, construct a 90% prediction interval to estimate the range in blood pressure reductions that might be expected from a subject given 8.0 mg of the tranquilizer.

**12.78**  Compute the coefficient of determination for the data in Question 12.73.

**12.79**  Which, if any, of the assumptions of the linear model appear to be violated in the following scatter diagram? Explain.

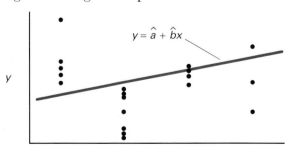

12.80   (a) Fit the model $y = a + b/x$ to the five $(x_i, y_i)$'s below.

| $x$ | 0.1 | 1 | 2 | 4 | 10 |
|---|---|---|---|---|---|
| $y$ | 14.2 | 2.8 | 2.4 | 2.1 | 1.9 |

(b) Graph the data and superimpose the equation $y = \hat{a} + \hat{b}/x$.

12.81   For graduating nurses, State Tech's Placement Office is studying the relationship between science GPA $(x)$ and number of job offers received $(y)$. A total of 12 students comprised the sample—four were selected because their GPA's rounded off to 2.5; another four had GPA's of 3.0; and the final four had GPA's of 3.5.

| Subject | GPA, $x$ | Number of Job Offers, $y$ |
|---|---|---|
| SH | 2.5 | 2 |
| CW | 2.5 | 3 |
| LM | 2.5 | 4 |
| BB | 2.5 | 3 |
| DF | 3.0 | 6 |
| ML | 3.0 | 4 |
| LN | 3.0 | 7 |
| NM | 3.0 | 5 |
| PF | 3.5 | 8 |
| KK | 3.5 | 8 |
| LD | 3.5 | 7 |
| JG | 3.5 | 10 |

(a) Graph the data.

(b) From the scatter diagram, does it look like the assumptions of the linear model are satisfied? Explain.

12.82   (a) Find the estimated regression line, $y = \hat{a} + \hat{b}x$, for the data in Question 12.81.

(b) Estimate $\sigma^2$ using Theorem 12.2.

(c) Construct an 80% confidence interval $E(Y)$ when $x = 3.5$.

(d) For what value of $x$ will a prediction interval for $y$ have the shortest width?

12.83   Suppose that $\sigma^2 = 1$ for the $Y$ distributions sampled in Question 12.81. Let $B$ denote the maximum likelihood estimator for $b$, the slope of the true regression line.

(a) Compute $P(-0.60 < B - b < 0.60)$.

(b) In the context of estimation, what does the probability computed in part (a) represent? Would an experimenter want.

$$P(-0.60 < B - b < 0.60)$$

to be large or small?

12.84   (a) Compute the coefficient of determination for the data in Question 12.81.

(b) In general, what must be true in order for the coefficient of determination to equal 1?

12.85   Fit the model $y = ax^b$ to the three points listed below.

| $x$ | 1 | 2 | 3 |
|---|---|---|---|
| $y$ | 4 | 50 | 235 |

What value of $y$ would we expect if $x = 6$?

12.86   Listed below are the number of marriages ($X$) and the number of children (or step-children) ($Y$) recorded for a sample of eight women recently retired.

| Subject | No. of Marriages, $x$ | No. of Children, $y$ |
|---|---|---|
| LL | 2 | 3 |
| HM | 1 | 1 |
| BP | 2 | 5 |
| LT | 3 | 4 |
| TK | 1 | 1 |
| KK | 4 | 5 |
| LC | 1 | 2 |
| AD | 2 | 2 |

(a) Find the regression equation of $Y$ on $x$.

(b) Find the regression equation of $X$ on $y$.

12.87   For the data in Question 12.86, test $H_0: \rho = 0$ versus $H_1: \rho \neq 0$. Let $\alpha = 0.05$. Can we conclude that $X$ and $Y$ are independent?

12.88   Board members of a large corporation believe that annual pay raises ($Y$) should be inversely proportional to employees' salaries ($x$). They are also aware that variation in $Y$ for a given value of $x$ needs to be carefully monitored, because employees perceive large differences in pay raises to be a sign of favoritism, whether those differences are justified or not. For the data below, construct a 95% confidence interval for $\sigma^2$.

| Employee | Salary (in 10's of thousands), $x$ | Pay Raise (%), $y$ |
|---|---|---|
| SG | 4 | 9.0 |
| ML | 4 | 8.0 |
| LB | 4 | 6.0 |
| PR | 6 | 6.0 |
| AW | 8 | 6.0 |
| KH | 8 | 5.0 |
| BG | 8 | 5.0 |
| EJ | 10 | 4.0 |

12.89 Based on the $xy$ relationship evidenced by the eight observations in Question 12.88, is it fair to conclude that an 8.5% pay raise is out of line for an employee making $94,000? Use a 90% prediction interval to support your decision.

12.90 Suppose that $(-0.25, 0.09)$ is the 95% confidence interval for $b$ based on a set of regression data. What conclusion would you draw? Explain.

12.91 Let $A$ and $B$ denote the maximum likelihood estimators for $a$ and $b$ in the linear model. Can Var $(A)$ ever be less than Var $(B)$? Explain.

12.92 An experimenter intends to collect two independent sets of regression data for the purpose of testing whether the difference between the two slope estimates, $\hat{b}_1$ and $\hat{b}_2$, is statistically significant. What factors affect the probability of committing a Type II error?

12.93 Four sets of six observations are taken on the same linear model. For each set, three of the $y_i$'s are observed when $x = -3$ and three are observed when $x = +3$. The four slope estimates are 2.6, 3.4, 4.2, and 1.6. Estimate $\sigma^2$.

# 13 Categorical Data

*Statistics means never having to say you're certain.*

*Anon.*

## 13.1

### INTRODUCTION

By their nature, certain data are *qualitative* rather than *quantitative*. We saw several examples in Sections 2.6 and 3.6, where "categorizing" an observation makes more sense than taking a numerical measurement (a patient lives or dies; a subject is Catholic or Protestant; a person's blood pressure is low, normal, or high). Situations also arise (see Case Study 3.18) where data that "begin" as numerical are deliberately reduced to a set of categories to simplify their interpretation. In short, qualitative data are frequently encountered in various contexts. How we analyze them formally is what this chapter is about.

Measurements classified according to a single criterion are said to be *one-dimensional*. The criterion itself can be essentially anything. An economist, for example, may have reason to divide industries into groups corresponding to the accounting methods they use. Public health officials, ever since the Bills of Mortality were introduced in the sixteenth century, have scrupulously associated every fatality with a cause of death. And on a more frivolous note, baseball statisticians summarize the outcome of each at-bat as a single, a double, a triple, a home run, a walk, an HBP, or an out.

If measurements are categorized according to a pair of criteria, the data are said to be *two-dimensional* or *cross-classified*. In Case Study 2.13 each of 1154 teenage girls was categorized according to (1) whether she had exhibited delinquent behavior and (2) which "birth order" she represented. Such data are typically displayed in a *contingency table*, where the rows correspond to categories associated with the first criterion, and the columns to categories associated with the second. Table 2.14 is an example.

Several inference questions arise in connection with qualitative measurements. For one-dimensional data experimenters often want to know whether a particular probability function adequately describes the way an observation varies from measurement to measurement. We answer that question by doing ***goodness-of-fit tests***. These are explained in detail in Sections 13.2 and 13.3.

For cross-classified data, *independence* is the key issue. Does an observation's location relative to one criterion give us an edge in predicting where it will fall relative to a second criterion? Does a teenage girl's birth order, for example, have any effect on her likelihood of becoming a delinquent? In general, if the answer to any such question is "yes," the two traits are *dependent*. As we will see, deciding whether two traits are dependent or independent can be critically important. Not surprisingly, the standard hypothesis test for making that judgment (a technique we take up in Section 13.4) is among the most widely used of all statistical procedures.

**goodness-of-fit statistic**—A function for quantifying the amount of disagreement between a set of observed frequencies and a set of expected frequencies.

## 13.2

### GOODNESS-OF-FIT TESTS: MODEL COMPLETELY SPECIFIED

Goodness-of-fit testing is different from anything we have seen so far. Its focus is more "global"—it draws inferences about the *form* of a pdf, not simply the value of a pdf's parameters. Also, the nature of the null hypothesis can vary considerably from problem to problem, in contrast to, say, a two-sample $t$ test, where every $H_0$ is the same. Finally, the structure of the test statistic has a form totally unlike the familiar $t$ ratios and Z ratios.

Motivating the definition of a test statistic for doing goodness-of-fit problems is best done with a specific example. The feathers of a frizzle chicken occur in three distinct variations (or *phenotypes*): extreme frizzle, mild frizzle, and normal (see Figure 13.1). It has been suggested that frizzle may be an example of a phenomenon known as *incomplete dominance*.

Extreme frizzle     Mild frizzle hybrid     Normal

**FIG. 13.1**

**TABLE 13.1**

| Phenotype | Observed Frequency | Expected Frequency |
|-----------|--------------------|--------------------|
| Extreme | 23 | 23.25 |
| Mild | 50 | 46.50 |
| Normal | 20 | 23.25 |
| | 93 | 93 |

If it is, genetic theory predicts that crossing two hybrid frizzles should produce twice as many milds as extremes or normals. The first two columns of Table 13.1 show the progeny resulting from 93 such crosses. Does the distribution of extremes, milds, and normals support the contention that frizzle is an example of incomplete dominance? (158)

Intuitively, we should answer "yes" if the *observed* frizzle distribution is "close" to the frequency distribution that we would *expect* to occur if our initial presumption were true. It was stated at the outset that incomplete dominance predicts that the ratio of extremes to milds to normals should be $1 : 2 : 1$. Written as a series of probability statements, that "model" reduces to three equations:

$$P \text{ (offspring has extreme frizzle)} = \tfrac{1}{4}$$

$$P \text{ (offspring has mild frizzle)} = \tfrac{1}{2}$$

$$P \text{ (offspring is normal)} = \tfrac{1}{4}$$

If each of 93 matings, though, has a *probability* of $\tfrac{1}{4}$ of producing an extreme frizzle offspring, it follows that the *expected number* of such progeny is

$(93)(\frac{1}{4}) = 23.25$ (recall Theorem 6.2). Similarly, the expected number of mild frizzles is $(93)(\frac{1}{2}) = 46.50$ and the expected number of normals is $(93)(\frac{1}{4}) = 23.25$. The last column of Table 13.1 shows the entire expected distribution.

How we proceed from this point is similar, in principle, to the path we followed in setting up hypothesis tests for parameters. In a one-sample $t$ test, for example, $H_0$ claims that $\mu = \mu_0$, which implies that the sample mean $\bar{y}$ should be "close" to $\mu_0$. We decided what "close" meant by computing $\dfrac{\bar{Y} - \mu_0}{S/\sqrt{n}}$ and using the fact that that ratio has a Student $t$ distribution with $n - 1$ degrees of freedom (when $H_0$ is true). Specifically, if $\dfrac{\bar{y} - \mu_0}{s/\sqrt{n}}$ is too far away from zero (in the direction of $H_1$), we reject $H_0$: $\mu = \mu_0$.

Extending that type of argument to Table 13.1 requires, first, that we identify a suitable test statistic—that is, a function capable of quantifying disagreement between observed distributions (like column 2 in Table 13.1) and expected distributions (like column 3). Second, we need to find the pdf describing the sampling distribution of that statistic. (If we know a test statistic and its pdf, we can find decision rules and compute critical values.)

The key to what we are looking for is the probability model known as the ***multinomial distribution***. Suppose an experiment consists of $n$ independent trials where the result of each trial is one of $k$ possible categories. Let the set of random variables $X_1, X_2, \ldots, X_k$ denote the numbers of outcomes belonging to categories $1, 2, \ldots, k$, respectively. (It must be true that $X_1 + X_2 + \cdots + X_k = n$). Also, let $p_j$ denote the probability that the outcome of a trial belongs to category $j, j = 1, 2, \ldots, k$. That is,

$$p_1 = P \text{ (outcome belongs to category 1)}$$

$$p_2 = P \text{ (outcome belongs to category 2)}$$

$$\vdots$$

$$p_k = P \text{ (outcome belongs to category } k)$$

(Since each outcome is necessarily a member of exactly one of the $k$ categories, $p_1 + p_2 + \cdots + p_k = 1$.) What we specifically need for data having this structure is a formula for the expected value of the $X_j$'s.

**multinomial distribution**—A generalization of the binomial distribution. The multinomial distribution describes the probabilistic behavior of the set of observed frequencies in a goodness-of-fit statistic.

---

**THEOREM 13.1**    Expected Value for a Multinomial Random Variable

Let $X_1, X_2, \ldots, X_k$ denote the numbers of observations falling into categories $1, 2, \ldots, k$, respectively, and let $p_1, p_2, \ldots, p_k$ denote the probabilities associated with each of those categories. If $n$ observations are recorded, the expected number falling into the $j$th category, $j = 1, 2, \ldots, k$, is

$$E(X_j) = np_j, \qquad j = 1, 2, \ldots, k \tag{13.1}$$

COMMENT: As a set, the random variables $X_1, X_2, \ldots, X_k$ have a *multinomial distribution*. The "joint" probability that

$$X_1 = x_1, X_2 = x_2, \ldots, \quad \text{and} \quad X_k = x_k$$

is

$$P(X_1 = x_1, X_2 = x_2, \ldots, X_k = x_k) = \frac{n!}{x_1! \, x_2! \cdots x_k!} \, p_1^{x_1} p_2^{x_2} \cdots p_k^{x_k}$$

When $k = 2$, the multinomial distribution reduces to the familiar *binomial distribution*:

$$P(X_1 = x_1, X_2 = x_2) = \frac{n!}{x_1! \, x_2!} \, p_1^{x_1} p_2^{x_2} = \frac{n!}{x_1! \, (n - x_1)!} \, p_1^{x_1} (1 - p_1)^{n - x_1}$$

$$= {}_nC_{x_1} p_1^{x_1} (1 - p_1)^{n - x_1}$$

since $x_1 + x_2 = n$ and $p_1 + p_2 = 1$.

Table 13.1 illustrates in particular what Theorem 13.1 describes in general. Summarized in its first two columns are the outcomes of 93 multinomial trials. The categories, into one of which each offspring must fall, are the three phenotypes: extreme frizzle, mild frizzle, and normal. If the random variables $X_1$, $X_2$, and $X_3$ denote the set of observed frequencies, then

$$X_1 = 23 \, (=x_1), \quad X_2 = 50 \, (=x_2), \quad \text{and} \quad X_3 = 20 \, (=x_3)$$

Also, by assumption, $p_1 = \frac{1}{4}$, $p_2 = \frac{1}{2}$, and $p_3 = \frac{1}{4}$. The entries in the third column are the corresponding expected frequencies, computed according to Equation 13.1:

$$E(X_1) = np_1 = (93)\left(\frac{1}{4}\right) = 23.25$$

$$E(X_2) = np_2 = (93)\left(\frac{1}{2}\right) = 46.50$$

$$E(X_3) = np_3 = (93)\left(\frac{1}{4}\right) = 23.25$$

Karl Pearson, one of the founding fathers of modern statistics, was the first to suggest a test statistic for comparing observed and expected distributions. In 1900 he proposed the function

$$v = \sum_{j=1}^{k} \frac{(x_j - np_j)^2}{np_j} \tag{13.2}$$

If the agreement or "fit" between the two distributions is perfect, $v = 0$: as the fit worsens—that is, as the row-by-row agreement between the $x_j$'s and the $np_j$'s deteriorates—$v$ increases. It follows that we should reject the proposed model if $v$ is too far *to the right* of zero.

---

### THEOREM 13.2    Testing Goodness of Fit

Let $X_1, X_2, \ldots, X_k$ be the numbers of outcomes falling into categories $1, 2, \ldots, k$, respectively, where the probabilities associated with those categories are $p_1, p_2, \ldots, p_k$.

1. If $np_j \geqslant 5$ for all $j$, then

$$V = \sum_{j=1}^{k} \frac{(X_j - np_j)^2}{np_j}$$

has approximately a $\chi^2$ distribution with $k - 1$ degrees of freedom.

2. To test

$$H_0: \quad p_1 = p_{1_0}, p_2 = p_{2_0}, \ldots, p_k = p_{k_0}$$

vs.

$$H_1: \quad \text{not all the } p_j\text{'s equal } p_{j_0}$$

at the $\alpha$ level of significance, reject $H_0$ if

$$v = \sum_{j=1}^{k} \frac{(x_j - np_{j_0})^2}{np_{j_0}} \geqslant \chi^2_{1-\alpha, k-1}$$

---

*EXAMPLE 13.1*   Consider the application of Theorem 13.2 to the frizzle data described earlier. Written formally, the hypotheses to be tested are

$$H_0: \quad p_1 = \tfrac{1}{4}, \quad p_2 = \tfrac{1}{2}, \quad p_3 = \tfrac{1}{4}$$

vs.

$$H_1: \quad \text{not all the } p_j\text{'s equal } p_{j_0}$$

Let $\alpha = 0.05$.

*Solution*   Substituting entries from the last two columns of Table 13.1 into Equation 13.2 gives

$$v = \frac{(23 - 23.25)^2}{23.25} + \frac{(50 - 46.5)^2}{46.50} + \frac{(20 - 23.25)^2}{23.25} = 0.72$$

According to part 2 of Theorem 13.2, we should reject $H_0$ only if

$$v \geqslant \chi^2_{1-\alpha, k-1} = \chi^2_{0.95, 2}$$

From Appendix A.3, $\chi^2_{0.95, 2} = 5.991$. Our conclusion, therefore, is clear: we should accept $H_0$. Observed frequencies of 23, 50, and 20 are not incompatible with expected frequencies of 23.25, 46.50, and 23.25. Or, phrased another way, these data are consistent with the speculation that frizzle inheritance is an example of incomplete dominance.

**CASE STUDY 13.1**

*Survival of Cerio-*
*Daphnia Cornuta*

Inhabiting many tropical waters is a small (<1 mm) crustacean, *Cerio-daphnia cornuta*, that occurs in two distinct morphological forms: one has a series of "horns" protruding from its exoskeleton, whereas the other is more rounded (Figure 13.2). An experiment was done (193) to test whether either variant is more conducive than the other to the survival of the species, in terms of its likelihood of not being eaten by predators.

Unhorned                          Horned

**FIG. 13.2**

Numerous *C. cornuta* were put into a holding tank in a 3 : 1 ratio: three of the unhorned variety for every one with horns. Also present in the tank was a small fish, *Melaniris chagresi*, whose diet routinely includes substantial numbers of *C. cornuta*. After approximately one hour, long enough for the predator to complete its feeding, the fish was sacrificed and the contents of its stomach examined. Among the 44 crustacean casualties, the unhorned-to-horned ratio was 40 : 4. In light of the 3 : 1 ratio initially present, is a 10 : 1 ratio of fatalities (based on a sample of size 44) enough evidence to conclude that one polymorph is more at risk than the other?

The two categories here for the 44 multinomial trials being recorded are unhorned and horned. Under the null hypothesis that morphology has no effect on survival, it would follow that the probability of either form's being eaten should be proportional to the numbers of each kind available. If $p_1 = P$ (unhorned *C. cornuta* is eaten) and $p_2 = P$ (horned *C. cornuta* is eaten), the question to be answered reduces to a test of

$$H_0: \quad p_1 = \tfrac{3}{4}, \quad p_2 = \tfrac{1}{4}$$

vs.

$$H_1: \quad p_1 \neq \tfrac{3}{4}, \quad p_2 \neq \tfrac{1}{4}$$

Let $\alpha = 0.05$.

Table 13.2 shows the observed and expected distributions. Since $k = 2$, the behavior of $V$ is approximated by a $\chi_1^2$ distribution, for which the 0.05 critical value is 3.84 (see Figure 13.3). Substituting the $x_j$'s and $p_{j_0}$'s into

**TABLE 13.2**

| Polymorph | Observed Frequency, $x_j$ | $p_{j_0}$ | $E(X_j) = 44p_{j_0}$ |
|---|---|---|---|
| Unhorned | 40 | $\frac{3}{4}$ | 33 |
| Horned | 4 | $\frac{1}{4}$ | 11 |
| | 44 | 1 | 44 |

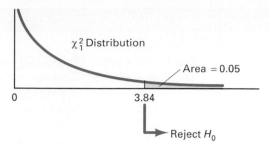

**FIG. 13.3**

Equation 13.2 gives

$$v = \frac{(40 - 33)^2}{33} + \frac{(4 - 11)^2}{11} = 5.93$$

Therefore, we reject $H_0$. It appears that morphology does affect *C. cornuta's* likelihood of being eaten. The unhorned variety is ending up as fish food significantly more often than it should (if chance were the only factor involved).

COMMENT:   Further experiments revealed that the absence of horns, despite what the goodness-of-fit test may have suggested, is *not* the underlying reason for the first polymorph having a higher probability of being eaten than the second. What actually causes the difference is the enlarged eyespot, which makes the unhorned crustacean the more visible of the two.

## CASE STUDY 13.2 (optional)

**TABLE 13.3**

| Series Length, $x$ | Observed Frequency |
|---|---|
| 4 | 9 |
| 5 | 11 |
| 6 | 8 |
| 7 | 22 |
| | 50 |

Table 13.3 summarizes the distribution of the lengths of the 50 World Series played from 1926 to 1975 (9 ended in 4 games, 11 lasted 5 games, and so on). Suppose someone who is not a fan says that baseball is a silly sport and claims that the winner of each World Series game could just as easily be determined by tossing a fair coin. Do the data in Table 13.3 support that contention? Is the distribution of World Series *lengths*, in other words, compatible with the presumption that World Series *games* are 50–50 propositions?

Finding an expected distribution in this case is a bit trickier than it was for the frizzle fowl data or for the *C. cornuta* data: the main thrust of the argument, though, has already been worked through in Example 6.10. Let the random variable $X$ denote the length of a World Series. We need to find $p_{j_0} = P(X = j)$, for $j = 4, 5, 6,$ and 7. Once we determine those values, we multiply each by 50 $(=n)$ to obtain the expected frequencies.

Note, first of all, that

$$P(X = 4) = p_{4_0} = P \text{ (World Series ends in 4 games)}$$

$$= P \text{ (American League wins in 4 games)}$$

$$+ P \text{ (National League wins in 4 games)}$$

$$= \left(\frac{1}{2}\right)^4 + \left(\frac{1}{2}\right)^4 = \frac{1}{8} \quad \text{(why?)}$$

If the outcome of each game, therefore, is assumed to be analogous to the flip of a fair coin, the expected number of four-game World Series, for a 50-year period, is $(50)(\frac{1}{8}) = 6.25$ (in contrast to the observed frequency of 9).

Computing the three remaining $p_{j_0}$'s is simplified if we realize that both teams have the same probability of winning the series in $j$ games (since each game is presumed to be a toss-up). That is,

$$P(X = j) = p_{j_0} = P \text{ (American League wins in } j \text{ games)}$$

$$+ P \text{ (National League wins in } j \text{ games)}$$

$$= 2P \text{ (American League wins in } j \text{ games)}$$

Also, recall from Example 6.10 that the only way in which the American League can win the series in $j$ games is to (1) win exactly three of the first $j - 1$ games and (2) win the $j$th game. Consider, for example, $p_{5_0}$:

$$p_{5_0} = P(X = 5) = P \text{ (World Series lasts 5 games)}$$

$$= 2P \text{ (American League wins in 5 games)}$$

$$= 2P \text{ (American League wins 3 of the first 4 games}$$
$$\text{and American League wins fifth game)}$$

$$= 2P \text{ (American League wins 3 of the first 4 games)}$$

$$\cdot P \text{ (American League wins fifth game)}$$

$$= 2 \,_4C_3 \left(\frac{1}{2}\right)^3 \left(1 - \frac{1}{2}\right)^1 \left(\frac{1}{2}\right) = \frac{1}{4}$$

Similar calculations will show that

$$p_{6_0} = P(X = 6) = \frac{5}{16} \quad \text{and} \quad p_{7_0} = P(X = 7) = \frac{5}{16}$$

(see Question 13.1).

The last two columns of Table 13.4 give the expected and observed frequencies for each possible value of $X$. Written formally, the hypotheses to be tested are

$$H_0: \quad p_4 = \tfrac{1}{8}, \quad p_5 = \tfrac{1}{4}, \quad p_6 = \tfrac{5}{16}, \quad p_7 = \tfrac{5}{16}$$

vs.

$$H_1: \quad \text{not all the } p_j\text{'s equal the } p_{j_0}\text{'s}$$

**TABLE 13.4**

| Series Length, $x$ | $p_{j_0} = P(X = j)$ | Expected Frequency ($= 50p_{j_0}$) | Observed Frequency |
|---|---|---|---|
| 4 | $\frac{1}{8}$ | 6.25 | 9 |
| 5 | $\frac{1}{4}$ | 12.50 | 11 |
| 6 | $\frac{5}{16}$ | 15.625 | 8 |
| 7 | $\frac{5}{16}$ | 15.625 | 22 |
| | 1 | 50 | 50 |

(How do the $p_j$'s differ from the $p_{j_0}$'s? Hint: What would $p_{4_0}$ equal if we assumed initially that the American League has a 60% chance of winning each game?)

Let $\alpha = 0.10$. Since $k = 4$, the null hypothesis should be rejected if $v \geqslant \chi^2_{0.90, 3} = 6.251$ (see Appendix A.3). But

$$v = \frac{(9 - 6.25)^2}{6.25} + \frac{(11 - 12.50)^2}{12.50} + \frac{(8 - 15.625)^2}{15.625} + \frac{(22 - 15.625)^2}{15.625}$$

$$= 7.71$$

so we *reject* $H_0$. It appears that World Series competition cannot be adequately modeled by successive tosses of a fair coin.

---

QUESTIONS

13.1   For the World Series problem described in Case Study 13.2, carry out the details to show that $p_{6_0} = p_{7_0} = \frac{5}{16}$.

13.2   Suppose that $\alpha$ for the World Series problem had been set at 0.05 instead of 0.10. Would our conclusion have changed?

13.3   Suppose we had access to the scores of all the games in the 50 World Series from 1926 to 1975. Let $p = P$ (American League wins any given game). How else might we test $H_0$: $p = \frac{1}{2}$, other than by using the goodness-of-fit procedure described in Case Study 13.2?

13.4   One hundred samples of size 2 are drawn without replacement from an urn presumed to contain four red chips and six white chips. Summarized below are the numbers of red chips in each of the 100 samples. Are the data compatible with our initial presumption of the urn's composition? Do an appropriate goodness-of-fit test at the $\alpha = 0.10$ level of significance.

| Number of Red Chips Drawn | Observed Frequency |
|---|---|
| 0 | 35 |
| 1 | 55 |
| 2 | 10 |
| | 100 |

13.5 In racing parlance, handicappers are people who try to quantify the effects that various factors might have on a horse's chances of winning. For example, does post position, the "placement" of a horse relative to the rail, influence its likelihood of finishing in the winner's circle? Shown below is the distribution of starting post positions for winners in 144 races, all run with a full field of eight horses.

| Starting Post | 1 | 2 | 3 | 4 | 5 | 6 | 7 | 8 |
|---|---|---|---|---|---|---|---|---|
| Number of Winners | 32 | 21 | 19 | 20 | 16 | 11 | 14 | 11 |

(a) Test an appropriate goodness-of-fit hypothesis. Let $\alpha = 0.05$.

(b) If you were a gambler, would your conclusion affect the way you placed a bet? Explain.

13.6 In one of Mendel's early genetics experiments, he observed two traits on each of 556 garden peas: *shape* (round or angular) and *color* (yellow or green). The results are listed below. According to his model, the four classes should occur in a $9:3:3:1$ ratio. Do the data support that presumption? Do a relevant hypothesis test at the $\alpha = 0.05$ level of significance.

| Shape | Color | Observed Frequency |
|---|---|---|
| Round | Yellow | 315 |
| Round | Green | 108 |
| Angular | Yellow | 101 |
| Angular | Green | 32 |

13.7 A widely held belief among expectant fathers is that babies are not born uniformly throughout the day. They choose to arrive instead at times purposely inconvenient. To test that notion statistically, a maternity hospital reviewed the records of 2650 births. Out of that number, some 494 were born at a conspicuously bad time, between midnight and 4:00 A.M (161). Define two classes and let $H_0$ reflect the presumption that births are random with respect to time of day. Do the corresponding goodness-of-fit test. Let $\alpha = 0.05$.

13.8 Medical researchers have raised the possibility that persons born early in a year may be more predisposed to becoming schizophrenic than will persons born later in a year. Out of 5139 patients in a British psychiatric clinic (69), 1383 were born between January 1 and April 30. Based on demographic considerations, the expected number that should have been born in that period (out of a random sample of 5139) was calculated to be 1292.1. Analyze the data. Let 0.05 be the level of significance.

13.9 Two traits that have been widely studied in tomato plants are *height* (tall versus dwarf) and *leaf type* (cut versus potato). Tall and cut are dominant. When a homozygous tall cut is crossed with a dwarf potato, the resulting progeny is called a dihybrid. (Its phenotype will be tall and cut.) When dihybrids are crossed, the phenotypes tall cut, tall potato, dwarf cut, and dwarf potato should appear in a $9:3:3:1$ ratio, provided the alleles governing the two traits segregate independently. Summarized below are the results of 1611 dihybrid crosses (158). Test the appropriateness of the $9:3:3:1$ model. Let $\alpha = 0.01$.

| Phenotype | Observed Frequency |
|---|---|
| Tall cut | 926 |
| Tall potato | 288 |
| Dwarf cut | 293 |
| Dwarf potato | 104 |
| | 1611 |

13.10   Computer software packages often include a random number generator—that is, a subroutine for producing numbers uniformly distributed over the interval from 0 to 1. Tabulated below is the distribution of 100 such numbers, all outputted by a program contained in a widely marketed statistical package. Set up and test an appropriate hypothesis. Let $\alpha = 0.05$.

| Interval | Observed Frequency |
|---|---|
| 0–0.249 | 23 |
| 0.250–0.499 | 23 |
| 0.500–0.749 | 28 |
| 0.750–1.000 | 26 |
| | 100 |

13.11   Suppose a goodness-of-fit problem involves $k = 2$ categories, in which case $p_2 = 1 - p_1$ and $p_{2_0} = 1 - p_{1_0}$. If we wish to test $H_0: p_1 = p_{1_0}$ versus $H_1: p_1 \neq p_{1_0}$ explain why the decision rule should *not* be

$$\text{``reject } H_0 \text{ if } v \leqslant \chi^2_{\alpha/2, 1} \quad \text{or} \quad v \geqslant \chi^2_{1-\alpha/2, 1} \text{''}$$

even though $H_1$ is two-sided.

13.12   A fair die is tossed 12 times. What is the probability that each face appears exactly twice? (Hint: Use the comment following Theorem 13.1. Let $X_j$ denote the number of times face $j$ appears, $j = 1, \ldots, 6$.)

13.13   A baseball player has appeared in 100 games, in each of which he batted exactly twice. Shown below are the numbers of games in which he got exactly $j$ hits, $j = 0, 1, 2$. Is the player's performance consistent with his claim that he's a .300 hitter? Answer the question by doing a goodness-of-fit test at the $\alpha = 0.10$ level of significance. Assume that each at-bat is an independent event.

| Number of Hits | Number of Games |
|:---:|:---:|
| 0 | 40 |
| 1 | 40 |
| 2 | 20 |

# 13.3

## GOODNESS-OF-FIT TESTS: PARAMETERS UNKNOWN

### Discrete Distributions

The null hypotheses tested in Section 13.2 were usually nothing more than a simple set of proportions (as in Case Study 13.1). If they did involve a formal pdf, all the parameters of that function were known (like the binomial in Case Study 13.2). Where the data were presumed to have originated, in other words, was completely specified. Not all goodness-of-fit problems, though, share that characteristic: sometimes the model to be tested is not entirely known, meaning its *form* is specified but it contains one or more parameters whose numerical values are open to speculation. Education officials, for example, may have good reason to believe that scores earned by next year's high school seniors on the SAT will be normally distributed, but until values for $\mu$ and $\sigma$ are determined, the model is not completely specified.

Many of the pdfs likely to show up as models in goodness-of-fit tests do have parameters that are typically unknown. The $p$ in a binomial, the $\lambda$ in a Poisson, and $\mu$ and $\sigma$ for a normal are all familiar examples. We must begin any problem involving a function of that sort by *estimating* the unknown parameter from the data. Sometimes the appropriate estimates are the same ones we encountered in earlier chapters; sometimes they are not. "What" to use "when" will be discussed a little later within the context of specific examples.

Once all the parameters have been estimated, we calculate expected frequencies for each of the data's $k$ categories, just as in Section 13.2, but the notation for designating those expected values is slightly different. According to Equation 13.1,

$$E(X_j) = np_j, \qquad j = 1, 2, \ldots, k$$

where $p_j$ is the *true* probability that a random observation falls into category $j$. For the problems in this section, $p_j$ is unknown, and so is the true expected value for $X_j$.

**expected frequency**—The number of observations, out of $n$, that "should" belong to a certain class (or cell) *if $H_0$ is true*. An expected frequency is said to be "estimated" if its computation is based on an $H_0$ model whose parameters, themselves, are estimated.

Technically, any probability computed from a model whose parameters have been estimated is, itself, an estimate. Thus, $p_j$ will now be written $\hat{p}_j$. Associated with each category, therefore, will be an ***estimated expected frequency*** $\widehat{E(X_j)}$, where

$$\widehat{E(X_j)} = \text{Estimated } E(X_j) = n\hat{p}_j, \qquad j = 1, 2, \ldots, k \qquad (13.3)$$

Whether or not we accept the presumed model will depend on how well the $X_j$'s match up with the $\widehat{E(X_j)}$'s.

---

**THEOREM 13.3**   Testing Goodness of Fit with Parameters Unknown

Let $f_Y(y)$ be a pdf having $q$ unknown parameters. To test the null hypothesis that a set of data $(y_1, y_2, \ldots, y_n)$ are a random sample from $f_Y(y)$:

1. Divide the range of the data into $k$ categories and let $X_j$ denote the number of observations in category $j$, $j = 1, 2, \ldots, k$.
2. Estimate the $q$ parameters in $f_Y(y)$ and compute $\hat{p}_j$, the estimated probability that a random observation belongs to category $j$.
3. If $n p_j < 5$ for any $j$, combine category $j$ with an adjacent category. The estimated expected frequency for each category appearing in the test statistic should be at least 5.
4. Define

$$V = \sum_{j=1}^{k} \frac{(X_j - n\hat{p}_j)^2}{n\hat{p}_j}$$

5. Reject $H_0$ if

$$v = \sum_{j=1}^{k} \frac{(x_j - n\hat{p}_j)^2}{n\hat{p}_j} \geqslant \chi^2_{1-\alpha,\, k-1-q} \qquad (13.4)$$

---

**EXAMPLE 13.2**   Sociologists often study characteristics or behavioral traits of persons who have achieved greatness for the purpose of testing whether those individuals differ from what might be expected in a more "normal" population. Suppose, for example, biographical records have been compiled on 120 prominent feminist scholars, all of whom have exactly two children. The first two columns of Table 13.5 summarize the distribution of boys in those 120 families. At the $\alpha = 0.05$ level can it be concluded that the number of boys in the families of feminist scholars is binomially distributed?

**TABLE 13.5**

| Number of Boys | Observed Frequency, $x_j$ | Estimated Expected Frequency, $120\hat{p}_j$ |
|:---:|:---:|:---:|
| 0 | 24 | 26.2 |
| 1 | 64 | 59.7 |
| 2 | 32 | 34.1 |
|  | 120 | 120.0 |

*Solution*    Let the random variable $Y$ denote the number of boys in a two-child family. We want to know whether the pdf for $Y$ has the form

$$f_Y(y) = P(Y = y) = {}_2C_y p^y (1 - p)^{2-y}, \qquad y = 0, 1, 2$$

Note that $p = P$ (child is a boy) is an unknown parameter.

This is not the first time that we have had to replace $p$ in a formula with a numerical estimate. Recall the confidence interval derivation in Section 9.3. In general,

$$\hat{p} = \text{Estimate for } p = \frac{\text{Total number of successes}}{\text{Total number of trials}}$$

which in this case reduces to

$$\hat{p} = \frac{\text{Total number of boys}}{\text{Total number of children}}$$

$$= \frac{0(24) + 1(64) + 2(32)}{2(120)} = 0.533$$

Let $X_j$ denote the number of families having $j$ boys, $j = 0, 1, 2$, and let $p_j = P$ (family has $j$ boys), $j = 0, 1, 2$. We obtain estimates for the $p_j$'s by evaluating $f_Y(y)$ after substituting $\hat{p}$ for $p$:

$$\hat{p}_0 = \text{estimated probability that family has 0 boys}$$

$$= {}_2C_0 (0.533)^0 (0.467)^2 = 0.218$$

$$\hat{p}_1 = \text{estimated probability that family has 1 boy}$$

$$= {}_2C_1 (0.533)^1 (0.467)^1 = 0.498$$

$$\hat{p}_2 = \text{estimated probability that family has 2 boys}$$

$$= {}_2C_2 (0.533)^2 (0.467)^0 = 0.284$$

Multiplying each $\hat{p}_j$ by 120 gives the set of estimated expected frequencies according to Equation 13.3:

$$\widehat{E(X_0)} = (120)(0.218) = 26.2$$

$$\widehat{E(X_1)} = (120)(0.498) = 59.7$$

$$\widehat{E(X_2)} = (120)(0.284) = 34.1$$

(see the last column of Table 13.5).

Since $k = 3$ for these data and $q = 1$, the number of degrees of freedom associated with the test statistic is $3 - 1 - 1 = 1$. By part 5 of Theorem 13.3, we should reject the binomial model if

$$v \geqslant \chi^2_{1-0.05,\, 3-1-1} = \chi^2_{0.95,\, 1} = 3.84$$

But, by Equation 13.4,

$$v = \frac{(24 - 26.2)^2}{26.2} + \frac{(64 - 59.7)^2}{59.7} + \frac{(32 - 34.1)^2}{34.1} = 0.62$$

so we accept $H_0$. Based on these data, it would *not* be unreasonable to assume that the number of boys in two-child families of prominent feminist scholars is binomially distributed.

---

**EXAMPLE 13.3**   The Poisson distribution was introduced in Section 6.5 as a probability model that is often useful in describing the occurrences of rare events. Not surprisingly, it frequently turns up as the null hypothesis in goodness-of-fit tests. (Recall that $Y$ is a Poisson random variable with parameter $\lambda$ if

$$f_Y(y) = \frac{e^{-\lambda}\lambda^y}{y!}, \qquad y = 0, 1, 2, \ldots$$

If $\lambda$ is not given, we can estimate it by $\hat{\lambda}$, where

$$\hat{\lambda} = \bar{y} = \frac{1}{n}\sum_{i=1}^{n} y_i$$

Each $y_i$, of course, is the number of occurrences recorded during the $i$th "unit," whatever the latter might be in the context of a particular problem).

Sporting events provide many opportunities to study phenomena where occurrences are likely to be Poisson events. The first two columns of Table 13.6, for example, give the distribution of the number of fumbles made by 110 college football teams over 55 games (156). Let $Y$ denote the number of fumbles made by a team during a game. Do the data in Table 13.6 support the contention at the 0.05 level of significance that $Y$ can be described by a Poisson pdf?

**TABLE 13.6**

| No. of Fumbles, $y$ | No. of Teams, $x_j$ | $\hat{p}_j$ | Estimated Expected No. |
|---|---|---|---|
| 0 | 8 | 0.078 | 8.6 |
| 1 | 24 | 0.199 | 21.9 |
| 2 | 27 | 0.254 | 27.9 |
| 3 | 20 | 0.215 | 23.7 |
| 4 | 17 | 0.137 | 15.1 |
| 5 | 10 | 0.070 | 7.7 |
| 6 | 3 | 0.030 | 3.3 |
| 7 | 1 | 0.017 | 1.8 |
|  | 110 | 1.00 | 110 |

*Solution*   The sample consists of 110 $y_i$'s, where 8 have the value 0, 24 have the value 1, and so on. If $Y$ is Poisson, $\lambda$ represents the *true* average number of fumbles per team. That number is unknown, but the *estimated* average number of fumbles per team is simply the sample mean of the 110 $y_i$'s:

$$\bar{y} = \hat{\lambda} = \frac{1}{n} \sum_{i=1}^{n} y_i = \frac{(8)(0) + (24)(1) + (27)(2) + \cdots + (1)(7)}{110}$$

$$= 2.55 \text{ fumbles per team}$$

Let $X_j$ denote the number of teams making exactly $j$ fumbles. Under the presumption that $Y$ is a Poisson random variable,

$$\hat{p}_j = \text{estimated probability that } X_j = j$$

$$= \frac{e^{-\lambda}\hat{\lambda}^j}{j!} = \frac{e^{-2.55}(2.55)^j}{j!}, \qquad j = 0, 1, 2, \ldots$$

The third column in Table 13.6 gives values for the $\hat{p}_j$'s. Note: For a goodness-of-fit test the $p_j$'s (or $\hat{p}_j$'s) must sum to 1. For a Poisson that will happen only if the last class is thought of as including the entire right-hand tail of the distribution, not just a single value. Here, that means the $y = 7$ class should be treated as though it were $y \geqslant 7$, in which case

$$\hat{p}_7 = P(Y \geqslant 7) = 1 - P(Y \leqslant 6)$$

$$= 1 - (0.078 + 0.199 + \cdots + 0.030) = 0.017$$

One final adjustment needs to be made before we compute a test statistic. Recall that part 4 of Theorem 13.3 requires that $n\hat{p}_j \geqslant 5$ for all $j$. These data violate that condition: the last two categories in Table 13.6 have expected frequencies *less than* 5:

$$n\hat{p}_6 = 3.3 \quad \text{and} \quad n\hat{p}_7 = 1.8$$

In general, if the $n\hat{p}_j \geqslant 5$ criterion is not satisfied, we need to combine adjacent classes until every category does have a large enough expected frequency. We can do that here by combining the $y = 6$ and $y = 7$ categories into the single class $y \geqslant 6$. Table 13.7 shows the revised data.

**TABLE 13.7**

| Number of Fumbles, $y$ | Number of Teams, $x_j$ | $\hat{p}_j$ | $n\hat{p}_j$ |
|---|---|---|---|
| 0 | 8 | 0.078 | 8.6 |
| 1 | 24 | 0.199 | 21.9 |
| 2 | 27 | 0.254 | 27.9 |
| 3 | 20 | 0.215 | 23.7 |
| 4 | 17 | 0.137 | 15.1 |
| 5 | 10 | 0.070 | 7.7 |
| 6+ | 4 | 0.047 | 5.1 |
| | 110 | 1.000 | 110 |

Substituting the $x_j$'s and $n\hat{p}_j$'s into Equation 13.4 gives

$$v = \frac{(8-8.6)^2}{8.6} + \frac{(24-21.9)^2}{21.9} + \cdots + \frac{(4-5.1)^2}{5.1} = 2.02$$

How many degrees of freedom are associated with $v$? From part 5 of Theorem 13.3,

$$\text{df} = k - 1 - q = 7 - 1 - 1 = 5$$

We should reject the Poisson model, therefore, if 2.02 ($=v$) exceeds the 95th percentile of the $\chi_5^2$ distribution. From Appendix A.3, $\chi^2_{0.95,\,5} = 11.070$, so $v$ is *not* significantly larger than zero and our conclusion is to accept $H_0$.

**QUESTIONS**

**13.14**  Fifty Vanderbilt students were asked how many broken bones they had sustained during their lifetimes (94). The results are summarized below. Can $Y$, the number of broken bones, be considered a Poisson random variable? Do a goodness-of-fit test at the $\alpha = 0.05$ level of significance.

| Number of Broken Bones, $y$ | Number of Students |
|---|---|
| 0 | 20 |
| 1 | 16 |
| 2 | 7 |
| 3 | 4 |
| 4 | 2 |
| 5 | 1 |
| | 50 |

**13.15**  Do a goodness-of-fit test on the radiation data summarized in Table 6.12. Multiply each $P(X = k)$ by 2608 to get the set of estimated expected frequencies. Let $\alpha = 0.01$. Would you conclude that radioactive decay can be described by a Poisson pdf?

**13.16**  For the Prussian cavalry data in Question 6.57, is it reasonable to argue that fatalities due to horse kicks are an example of a Poisson random variable? Do a goodness-of-fit test at the $\alpha = 0.05$ level of significance.

**13.17**  At the $\alpha = 0.05$ level of significance, test whether the war data in Question 6.6 have a Poisson distribution.

**13.18**  A gambler claims to have done 100 experiments where he flipped a coin until a head appeared for the first time. On each occasion the measurement recorded was the toss at which the first head appeared. His results are tabulated below. Do you believe that he did what he says he did? Note: For a geometric random variable (see Section 6.5), the parameter $p = P$ (success) is estimated by the reciprocal of the sample mean—that is, $\hat{p} = 1/\bar{y}$.

| Trial Number of First Head, $y$ | Observed Frequency |
|---|---|
| 1 | 20 |
| 2 | 25 |
| 3 | 20 |
| 4 | 20 |
| 5 | 13 |
| 6 | 0 |
| 7 | 2 |
|  | 100 |

**13.19**   A row of five automatic bank tellers is located outside a metropolitan office building. For 60 consecutive working days, a spot check is made of the number of tellers in use at 12:30 P.M. Do the results, as listed below, support management's contention that utilization of the machines can be modeled by a binomial distribution? Do the relevant goodness-of-fit test using the 0.05 level of significance.

| Number Answered Correctly, $y$ | Observed Frequency |
|---|---|
| 0 | 0 |
| 1 | 5 |
| 2 | 70 |
| 3 | 10 |
| 4 | 10 |
| 5 | 5 |
|  | 100 |

**13.20**   A class of 100 students takes a five-question true/false test. Shown below is the distribution of the total number of questions each student answered correctly.

(a) Can these scores be adequately described by a binomial distribution? Do an appropriate goodness-of-fit test at the 0.05 level of significance.

(b) In general, would we expect scores on a true/false test to have a binomial distribution? Why or why not?

| Number of Tellers in Use, $y$ | Observed Frequency |
|---|---|
| 0 | 22 |
| 1 | 18 |
| 2 | 14 |
| 3 | 4 |
| 4 | 1 |
| 5 | 1 |
|  | 60 |

**13.21** During the 96-year period from 1837 to 1932, the United States Supreme Court had 48 vacancies. The breakdown below shows the number of years in which exactly $y$ of those vacancies occurred. Can $Y$ be considered a Poisson random variable? Set up and carry out the relevant goodness-of-fit test at the $\alpha = 0.01$ level of significance (177).

| Number of Vacancies, $y$ | Observed Frequency |
|:---:|:---:|
| 0 | 59 |
| 1 | 27 |
| 2 | 9 |
| 3 | 1 |
| 4+ | 0 |
| | 96 |

### Continuous Distributions

Goodness-of-fit hypotheses are not limited to *discrete* probability models, like the binomial or Poisson. Sometimes an experimenter needs to test whether a set of data might reasonably have come from a certain *continuous* distribution. Example 13.4 is the prototype, verifying whether a set of $y_i$'s is *normally* distributed. Testing for normality is important because so many inference procedure (*t* tests, for example) are strictly valid only if the data are bell-shaped.

---

**EXAMPLE 13.4** Whether Etruscans were native to Italy was the focus of Case Study 9.1. The physical evidence on which a conclusion was to be based consisted of widths measured on 84 skulls that had been found over a period of years at a variety of archeological sites. Statistically, the analysis reduced to a one-sample $t$ test, the objective being to see whether the average size of the 84 Etruscan skulls was compatible with dimensions known to be characteristic of true Italians. Ultimately, the null hypothesis was rejected (see p. 418), but the credibility of that inference hinges on several factors, including an implicit assumption that all 84 observations represent a random sample from a normal distribution. Test that assumption.

*Solution* Whenever a model being fit is continuous, the first step to take when applying Theorem 13.3 is to construct a frequency distribution, thereby reducing the original $y_i$'s to a set of classes. The first and third columns of Table 13.8 categorize the 84 Etruscan skull widths, listed individually on p. 417, into seven classes, ranging from 125–129 to 155–159.

Recall that the bell-shaped curve being presumed by the $t$ test as the origin of the data,

$$f_Y(y) = \frac{1}{\sqrt{2\pi}\sigma} e^{-(1/2)((y-\mu)/\sigma)^2}$$

**TABLE 13.8**

| Width (mm) | Midpoint, $m_i$ | Obs. Freq., $f_i$ | $f_i m_i$ | $f_i m_i^2$ |
|---|---|---|---|---|
| 125–129 | 127 | 1 | 127 | 16,129 |
| 130–134 | 132 | 4 | 528 | 69,696 |
| 135–139 | 137 | 10 | 1,370 | 187,690 |
| 140–144 | 142 | 33 | 4,686 | 665,412 |
| 145–149 | 147 | 24 | 3,528 | 518,616 |
| 150–154 | 152 | 9 | 1,368 | 207,936 |
| 155–159 | 157 | 3 | 471 | 73,947 |
| | | 84 | 12,078 | 1,739,426 |

has two unknown parameters, $\mu$ and $\sigma$. It is best to estimate the first by the *grouped mean* $\bar{y}_g$, where

$$\bar{y}_g = \frac{1}{n} \sum_{i=1}^{k} f_i m_i$$

(see Question 3.29), and the second by the *grouped standard deviation* $s_g$, where

$$s_g = \sqrt{\frac{n \sum_{i=1}^{k} f_i m_i^2 - \left(\sum_{i=1}^{k} f_i m_i\right)^2}{n(n-1)}}$$

(see Question 3.29). All the midpoints $m_1, m_2, \ldots, m_k$ as well as the $f_i m_i$'s and $f_i m_i^2$'s are shown in Table 13.8. Substituting the necessary sums into the formulas for $\bar{y}_g$ and $s_g$ gives

$$\bar{y}_g = \frac{1(127) + 4(132) + \cdots + 3(157)}{84} = 143.8 \text{ mm}$$

and

$$s_g = \sqrt{\frac{84(1,739,426) - (12,078)^2}{84(83)}} = 5.79 \text{ mm}$$

We can compute estimated expected frequencies $\widehat{E(X_j)}, j = 1, 2, \ldots, k,$ for each category in Table 13.8 by using a Z transformation to find $\hat{p}_j$ and then multiplying $\hat{p}_j$ by $n$. Consider the third category, for example, which is designated 135–139 but really represents the set of values from 134.5 to 139.5. Let $\hat{p}_3$ denote the estimated probability that a randomly selected Etruscan skull has a width falling in that range. If we assume that skull widths have a normal distribution,

$$\hat{p}_3 = P(134.5 < Y < 139.5)$$

$$= P\left(\frac{134.5 - 143.8}{5.79} < \frac{Y - 143.8}{5.79} < \frac{139.5 - 143.8}{5.79}\right)$$

$$= P(-1.61 < Z < -0.74) = P(Z < -0.74) - P(Z < -1.61)$$

$$= 0.2297 - 0.0537 = 0.1760$$

**TABLE 13.9**

| Width (mm) | Obs. Freq. | $\hat{p}_j$ | $\widehat{E(X_j)} = n\hat{p}_j$ |
|:---:|:---:|:---:|:---:|
| $\leqslant 129$ | 1 | 0.0068 | 0.6 |
| 130–134 | 4 | 0.0469 | 3.9 |
| 135–139 | 10 | 0.1760 | 14.8 |
| 140–144 | 33 | 0.3181 | 26.7 |
| 145–149 | 24 | 0.2887 | 24.3 |
| 150–154 | 9 | 0.1313 | 11.0 |
| $\geqslant 155$ | 3 | 0.0322 | 2.7 |
| | 84 | 1 | 84 |

Thus,

$$\widehat{E(X_3)} = n\hat{p}_3 = 84(0.1760) = 14.8$$

Table 13.9 lists the $\hat{p}_j$'s and $\widehat{E(X_j)}$'s for each of the seven categories (along with the original $x_j$'s). In order to "force" the $\hat{p}_j$'s to sum to 1 and the $\widehat{E(X_j)}$'s to sum to $n$, we need to treat the first and last categories as open-ended. To compute $\hat{p}_1$, for example, we apply the $Z$ transformation to the event $Y \leqslant 129.5$ instead of $124.5 \leqslant Y \leqslant 129.5$. Similarly, $\hat{p}_7 = P(Y \geqslant 154.5)$.

The entries in columns 2 and 4 are almost what we need to compute the goodness-of-fit statistic $v$. Notice that not all categories have estimated expected frequencies of 5 or greater, as part 4 of Theorem 13.3 requires:

$$\widehat{E(X_1)} = 0.6, \qquad \widehat{E(X_2)} = 3.9, \quad \text{and} \quad \widehat{E(X_7)} = 2.7$$

Therefore we need to combine several of the "tail" classes into broader categories. Table 13.10 shows the data in an acceptable final form.

**TABLE 13.10**

| Width (mm) | Observed Frequency, $x_j$ | Estimated Expected Frequency, $n\hat{p}_j$ |
|:---:|:---:|:---:|
| $\leqslant 139$ | 15 | 19.3 |
| 140–144 | 33 | 26.7 |
| 145–149 | 24 | 24.3 |
| $\geqslant 150$ | 12 | 13.7 |
| | 84 | 84 |

The value of the test statistic $v$ for these data is

$$v = \frac{(15 - 19.3)^2}{19.3} + \frac{(33 - 26.7)^2}{26.7} + \frac{(24 - 24.3)^2}{24.3} + \frac{(12 - 13.7)^2}{13.7} = 2.66$$

and the critical value (at the $\alpha = 0.05$ level) is

$$\chi^2_{1-\alpha, k-1-q} = \chi^2_{1-0.05, 4-1-2} = \chi^2_{0.95, 1} = 3.84$$

Since $v$ does not exceed $\chi^2_{0.95,1}$, we *accept* the null hypothesis that skull widths are normally distributed. Therefore, doing a $t$ test on these data, as we did in Case Study 9.1 is justifiable.

<table>
<tr><td>QUESTIONS</td><td>13.22</td><td>Using the information given in Example 13.4, carry out the details to verify that $\widehat{E(X_5)} = 24.3$, as shown in Table 13.9.</td></tr>
</table>

**13.23**  Traffic fatality rates for each of the 50 states were listed in Question 6.109. Can those 50 numbers be treated as a random sample from a normal distribution? Start your analysis by constructing a frequency distribution. Let 2.0–2.9, 3.0–3.9, and so on, be the classes. Compute $\bar{y}_g$ and $s_g$. Use a Z transformation to estimate the probability associated with each of the classes and find the corresponding $\widehat{E(X_j)}$'s. Do any of the original classes need to be combined? Test the normality assumption by using the procedure in Theorem 13.3. Set the level of significance at 0.05.

**13.24**  An experiment for quantifying the reasons people have for making certain decisions is described in Case Study 3.2. Can the 106 observations, as summarized in Table 3.6, be considered a random sample from a normal distribution? Do an appropriate goodness-of-fit test at the $\alpha = 0.05$ level of significance.

**13.25**  (a)  Find $\bar{y}_g$ and $s_g$ for the 36 hurricane precipitation levels summarized in Table 3.10.

(b)  Compute the estimated expected frequency for the 12.00–15.99 class.

(c)  Does it seem appropriate to treat precipitation levels as random samples from a normal distribution?

**13.26**  For Question 3.13, group the 68 speeds recorded for drivers on the New Jersey Turnpike into classes ranging from 75–79 to 105–109 (ft/s). Carry out a goodness-of-fit test for normality. Let $\alpha = 0.10$.

# 13.4

## CONTINGENCY TABLES

What we learned in Sections 13.2 and 13.3 can be extended in a very natural way to provide an inference procedure for dealing with the most important kind of categorical data, those that are cross-classified. The latter arise (recall Section 2.6) when two dissimilar traits ($X$ and $Y$) are measured on each of $n$ subjects. Both measurements are recorded as categories rather than numbers along a continuous scale.

Case Study 3.17 describes an experiment typical of this sort of data: two measurements are recorded on each of 57 individuals being treated as outpatients. The $X$ trait refers to a subject's willingness to comply with doctors' orders and reduces to a set of two categories ("yes" or "no").

**$\chi^2$ test for independence**—The usual statistical procedure for analyzing categorical data. The null hypothesis that two traits, $X$ and $Y$, are independent is tested by rejecting $H_0$ if the observed and estimated expected frequencies in an $r \times c$ contingency table are too far apart.

Religion is the $Y$ trait, and those responses are similarly dichotomized, being listed as Catholic or Protestant.

Independence is the key issue that needs to be examined in analyzing categorical data. Does knowing a patient's religion, for example, help us predict whether that person will follow a doctor's orders? If the answer is "yes," then compliance and religion are *dependent*. Similar questions are raised in Case Studies 2.13 and 3.18. In the former we ask whether birth order and juvenile delinquency are related in a population of teenage girls. The latter looks at a possible link between transiency rates and suicide rates for residents of 25 major American cities.

Typically, we begin the analysis of cross-classified data by making *independence* the null hypothesis. Under that assumption, a set of estimated expected frequencies can easily be computed, one for each observed frequency appearing in the original contingency table. If the observed and expected frequencies are too far apart (as measured by a goodness-of-fit statistic) we reject what we initially presumed—that $X$ and $Y$ are independent.

Finding an appropriate set of estimated expected frequencies is an application of two results we saw much earlier, one from Chapter 5 and one from Chapter 6. Let $A_1, A_2, \ldots, A_r$ and $B_1, B_2, \ldots, B_c$ denote the categories associated with the $X$ trait and the $Y$ trait, respectively. To assume that $X$ and $Y$ are independent is to assume that knowing the $Y$ category of an observation is of no help in predicting where that data point will fall among the set of $X$ categories. Let $P(A_i|B_j)$ denote the conditional probability that an observation's $X$ trait belongs to $A_i$ *given that its $Y$ trait belongs to $B_j$*. Independence, then, implies that all the following equations must be true:

$$P(A_1|B_1) = P(A_1) \quad P(A_2|B_1) = P(A_2) \quad \cdots \quad P(A_r|B_1) = P(A_r)$$

$$P(A_1|B_2) = P(A_1) \quad P(A_2|B_2) = P(A_2) \quad \cdots \quad P(A_r|B_2) = P(A_r)$$

$$\vdots \qquad\qquad\qquad \vdots \qquad\qquad\qquad \vdots$$

$$P(A_1|B_c) = P(A_1) \quad P(A_2|B_c) = P(A_2) \quad \cdots \quad P(A_r|B_c) = P(A_r)$$

According to Definition 5.3, independence also implies that intersection probabilities factor. Therefore, in terms of the $A_i$'s and the $B_j$'s, we can write

$$P(A_1 \cap B_1) = P(A_1)P(B_1) \quad \cdots \quad P(A_r \cap B_1) = P(A_r)P(B_1)$$

$$P(A_1 \cap B_2) = P(A_1)P(B_2) \quad \cdots \quad P(A_r \cap B_2) = P(A_r)P(B_2)$$

$$\vdots \qquad\qquad\qquad\qquad \vdots$$

$$P(A_1 \cap B_c) = P(A_1)P(B_c) \quad \cdots \quad P(A_r \cap B_c) = P(A_r)P(B_c)$$

or, more concisely,

$$P(A_i \cap B_j) = P(A_i)P(B_j), \qquad 1 \leqslant i \leqslant r, 1 \leqslant j \leqslant c \qquad (13.5)$$

Equation 13.5 is the key result in setting up a statistic for testing that $X$ and $Y$ are independent.

**TABLE 13.11**

| | | Y TRAIT CATEGORIES | | |
|---|---|---|---|---|
| | | $B_1$ | $B_2$ | |
| X TRAIT CATEGORIES | $A_1$ | $f_{11}$ | $f_{12}$ | $R_1$ |
| | $A_2$ | $f_{21}$ | $f_{22}$ | $R_2$ |
| | | $C_1$ | $C_2$ | $n$ |

Consider, for the sake of simplicity, a $2 \times 2$ contingency table where $f_{ij}$ denotes the observed frequency in the $i$th row and $j$th column (see Table 13.11). Let $R_1, R_2, C_1,$ and $C_2$ denote the two row totals and the two column totals, respectively:

$$R_1 = f_{11} + f_{12} \qquad C_1 = f_{11} + f_{21}$$

$$R_2 = f_{21} + f_{22} \qquad C_2 = f_{12} + f_{22}$$

The overall total $n$ is the sum of either the $R_i$'s or the $C_j$'s (or the $f_{ij}$'s):

$$n = R_1 + R_2 = C_1 + C_2 = f_{11} + f_{12} + f_{21} + f_{22}$$

If we knew the numerical values of $P(A_i)$ and $P(B_j)$, it would follow from Theorem 6.2 that the expected frequency, assuming independence, for the $i$th row and $j$th column is $nP(A_i)P(B_j)$. In practice, though, we *don't* know the true probability associated with any of the categories, either the $A_i$'s or the $B_j$'s. Still, we can estimate them easily: it makes sense to approximate the probability of an $A_i$ or a $B_j$ by simply computing the proportion of observations that belong to that category. For a $2 \times 2$ table,

$$\widehat{P(A_1)} = \frac{R_1}{n} \qquad \widehat{P(B_1)} = \frac{C_1}{n}$$

$$\widehat{P(A_2)} = \frac{R_2}{n} \qquad \widehat{P(B_2)} = \frac{C_2}{n}$$

Multiplying each $\widehat{P(A_i)}\widehat{P(B_j)}$ by the total sample size $n$ gives us what we set out to find, an indication of what $f_{ij}$ "should" equal if $X$ and $Y$ were independent:

$$\widehat{E(f_{11})} = n\widehat{P(A_1)}\widehat{P(B_1)} = n\left(\frac{R_1}{n}\right)\left(\frac{C_1}{n}\right) = \frac{R_1 C_1}{n}$$

$$\widehat{E(f_{12})} = n\widehat{P(A_1)}\widehat{P(B_2)} = n\left(\frac{R_1}{n}\right)\left(\frac{C_2}{n}\right) = \frac{R_1 C_2}{n}$$

$$\widehat{E(f_{21})} = n\widehat{P(A_2)}\widehat{P(B_1)} = n\left(\frac{R_2}{n}\right)\left(\frac{C_1}{n}\right) = \frac{R_2 C_1}{n}$$

$$\widehat{E(f_{22})} = n\widehat{P(A_2)}\widehat{P(B_2)} = n\left(\frac{R_2}{n}\right)\left(\frac{C_2}{n}\right) = \frac{R_2 C_2}{n}$$

In general, for an $r \times c$ table,

$$\widehat{E(f_{ij})} = n\widehat{P(A_i)}\widehat{P(B_j)} = n\left(\frac{R_i}{n}\right)\left(\frac{C_j}{n}\right) = \frac{R_i C_j}{n}$$

Once we know the estimated expected frequencies, calculating a test statistic is straightforward. Following the precedent established in Theorem 13.3, let

$$V = \sum_{i,j} \frac{(f_{ij} - \widehat{E(f_{ij})})^2}{\widehat{E(f_{ij})}}$$

If $V$ is "close" to 0 (which occurs if the $\widehat{E(f_{ij})}$'s are approximately equal to the $f_{ij}$'s) we accept the null hypothesis that $X$ and $Y$ are independent.

Where does the rejection region begin? We will have to take on faith the assertion that $V$ has approximately a $\chi^2$ distribution. Furthermore, if the contingency table has $r$ rows and $c$ columns, the number of degrees of freedom associated with $V$ is $(r-1)(c-1)$ (see Question 13.41).

---

**THEOREM 13.4    Testing Cross-Classified Data**

Suppose $n$ observations are categorized according to an $X$ trait and a $Y$ trait and recorded as a set of frequencies in an $r \times c$ contingency table, where $R_1, R_2, \ldots, R_r$ and $C_1, C_2, \ldots, C_c$ are the row and column totals, respectively. Let $f_{ij}$ denote the observed frequency in the $i$th row and $j$th column, and let $\widehat{E(f_{ij})}$ denote the corresponding estimated expected frequency, where

$$\widehat{E(f_{ij})} = \frac{R_i C_j}{n}, \qquad 1 \leqslant i \leqslant r, 1 \leqslant j \leqslant c \qquad (13.6)$$

Define

$$V = \sum_{i,j} \frac{(f_{ij} - \widehat{E(f_{ij})})^2}{\widehat{E(f_{ij})}} \qquad (13.7)$$

Then

1. If $X$ and $Y$ are independent and $\widehat{E(f_{ij})} \geq 5$ for all $i$ and $j$, $V$ has approximately a $\chi^2$ distribution with $(r-1)(c-1)$ degrees of freedom.

2. To test

$$H_0: \quad X \text{ and } Y \text{ are independent}$$

vs.

$$H_1: \quad X \text{ and } Y \text{ are dependent}$$

at the $\alpha$ level of significance, reject $H_0$ if $v \geqslant \chi^2_{1-\alpha, (r-1)(c-1)}$.

*Magazine Marketing Strategies*

A trip to the nearest convenience store will bear ample witness to the claim that magazine publishing is a ruthlessly competitive business. Familiar favorites like *Time*, *Sports Illustrated*, and *Better Homes and Gardens* are still to be found, but their revenues are being threatened by the proliferation of specialty publications like *Omni* and *PC World*.

More than ever, a magazine's survival depends on effective marketing techniques and not just content. Setting prices as low as possible, for example, might seem to be a good strategy, but, surprisingly, the trend has been in the opposite direction; single-copy prices of new magazines have been increasing. Many analysts believe that a higher cover price definitely enhances a magazine's probability of success, the reason being that advertisers are more willing to pay for space if they perceive the targeted audience to be especially affluent. Others are not so sure.

In one study (76) two pieces of information were collected on each of 234 magazines: its cover price and whether it continues to be published. Table 13.12 shows the data summarized as a 4 × 2 contingency table. Notice how the last two rows seem to support the first position just stated, that higher prices are associated with greater likelihoods of survival. Is that association "real," or is it more reasonable to infer that a magazine's cover price and its long-term survival are independent?

**TABLE 13.12**

| | | Still Being Published | Has Ceased Publication | Totals |
|---|---|---|---|---|
| | < $1.99 | 70 | 65 | 135 |
| **COVER PRICE** | $2.00 – $2.49 | 39 | 28 | 67 |
| | $2.50 – $2.99 | 14 | 3 | 17 |
| | > $3.00 | 13 | 2 | 15 |
| | Totals | 136 | 98 | 234 |

We begin by computing the set of estimated expected frequencies. In the upper left-hand corner of the contingency table, for example, the observed frequency $f_{11}$ is 70. The corresponding estimated expected frequency $\widehat{E(f_{11})}$ is

$$\widehat{E(f_{11})} = \frac{R_1 C_1}{n} = \frac{(135)(136)}{234} = 78.5$$

All the other estimated expected frequencies are calculated similarly. Table 13.13 shows $\widehat{E(f_{ij})}$ in parentheses beneath $f_{ij}$ for all $i$ and $j$.

**TABLE 13.13**

|  | Still Being Published | Has Ceased Publication | Totals |
|---|---|---|---|
| ⩽ $1.99 | 70 (78.5) | 65 (56.5) | 135 |
| $2.00 – $2.49 | 39 (38.9) | 28 (28.1) | 67 |
| COVER PRICE   $2.50 – $2.99 | 14 (9.9) | 3 (7.1) | 17 |
| ⩾ $3.00 | 13 (8.7) | 2 (6.3) | 15 |
| Totals | 136 | 98 | 234 |

Substituting the $f_{ij}$'s and the $\widehat{E(f_{ij})}$'s into Equation 13.7 gives

$$v = \frac{(70 - 78.5)^2}{78.5} + \frac{(65 - 56.5)^2}{56.5} + \frac{(39 - 38.9)^2}{38.9} + \frac{(28 - 28.1)^2}{28.1}$$

$$+ \frac{(14 - 9.9)^2}{9.9} + \frac{(3 - 7.1)^2}{7.1} + \frac{(13 - 8.7)^2}{8.7} + \frac{(2 - 6.3)^2}{6.3} = 11.3$$

Since $r = 4$ and $c = 2$, the number of degrees of freedom associated with $v$ is $(4 - 1)(2 - 1) = 3$. Let $\alpha = 0.05$. To test

$H_0$:   cover price and survival are independent

vs.

$H_1$:   cover price and survival are dependent

we should reject $H_0$ if

$$v \geqslant \chi^2_{1 - \alpha, (r - 1)(c - 1)} = \chi^2_{0.95, 3} = 7.815$$

But $v \ (= 11.3)$ *does* exceed the critical value, so the $\chi^2$ test supports the conventional wisdom that price and survival are related. More specifically, magazines with higher cover prices seem to fare better financially than their lower priced competitors.

## CASE STUDY 13.4

*Methods of Accounting*   Evidence exists to suggest that *concentrated* industries, those where a relatively few firms control a major share of the market, are more profitable than *unconcentrated* industries. Not everyone agrees. Among the points of contention is the lack of a mutually agreed-upon definition of profitability. Since different accounting methods yield different profit figures, it becomes paramount to know whether "concentration" has any relationship to the way a company keeps its books.

Inventory evaluation is an area where accountants have some (legal) leeway in how they can manipulate a company's ledgers. Three methods are widely used. One is a system known as FIFO (first in, first out), where inventories are valued at current prices but costs of goods sold are calculated using earlier prices. During periods of inflation, FIFO paints a rosier profit picture than its conceptual opposite, LIFO (last in, first out). Also common is a third procedure, where inventories are valued by computing a weighted average that uses a range of prices, old as well as new.

Categorizing concentration is more subjective, although still relatively straightforward. First, the eight leading firms in each industry are identified. Then the percentage of an industry's total revenues accounted for by those eight top firms is computed, and that figure is assigned to one of three concentration classes: $\geq 90\%$, $50$–$89\%$, or $< 50\%$.

Table 13.14 gives a breakdown of 160 industries, each classified according to accounting system and level of concentration (62). Can we infer that those two traits are independent?

**TABLE 13.14**

|  | CONCENTRATION CLASS | | | |
|---|---|---|---|---|
|  | $\geq 90$ | $50-89$ | $<50$ | Totals |
| **LIFO** | 24 | 25 | 31 | 80 |
| **FIFO** | 9 | 19 | 9 | 37 |
| **Weighted Avg.** | 9 | 22 | 12 | 43 |
| **Totals** | 42 | 66 | 52 | 160 |

Table 13.15 shows the original data together with the frequencies (in parentheses) that would have been expected if a firm's method of accounting had had no relationship to that industry's level of concentration. Each $\widehat{E(f_{ij})}$ is derived from Equation 13.6: $\widehat{E(f_{ij})} = R_i C_j / n$.

**TABLE 13.15**

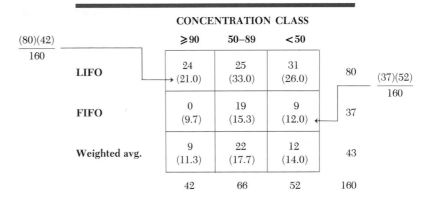

| | | CONCENTRATION CLASS | | | |
|---|---|---|---|---|---|
| | | $\geq 90$ | $50-89$ | $<50$ | |
| $\dfrac{(80)(42)}{160}$ | **LIFO** | 24 (21.0) | 25 (33.0) | 31 (26.0) | 80 |
| | **FIFO** | 0 (9.7) | 19 (15.3) | 9 (12.0) | 37 |
| | **Weighted avg.** | 9 (11.3) | 22 (17.7) | 12 (14.0) | 43 |
| | | 42 | 66 | 52 | 160 |

$\dfrac{(37)(52)}{160}$

Let $\alpha = 0.05$. To test

$$H_0: \quad \text{accounting and concentration are independent}$$
vs.
$$H_1: \quad \text{accounting and concentration are dependent}$$

we should reject the null hypothesis if

$$v \geqslant \chi^2_{1-\alpha,\,(r-1)(c-1)} = \chi^2_{1-0.05,\,(3-1)(3-1)} = \chi^2_{0.95,\,4} = 9.49$$

The value of the test statistic is

$$v = \frac{(24-21.0)^2}{21.0} + \frac{(25-33.0)^2}{33.0} + \cdots + \frac{(12-14.0)^2}{14.0} = 6.82$$

Thus we should accept $H_0$. These data do *not* support the contention that industries belonging to different concentration classes tend to measure their profitability in different ways.

---

**QUESTIONS**

**13.27**   A medical research team reviews the files of 51 male patients and 69 female patients, all of whom had had their carotid arteries examined by ultrasonics. They find that 26 of the males and 38 of the females have diseases related to cerebral arteriosclerosis.

(a) Summarize the data in a $2 \times 2$ contingency table. Use male and female to designate the two columns.

(b) Compute the proportions of men and women, respectively, having diseases related to arteriosclerosis. If sex and risk of arteriosclerosis are independent, what should we expect to be true of those two proportions?

(c) Assess the significance of the difference between the two proportions cited in part (b) by doing a $\chi^2$ test. Let $\alpha = 0.05$. Interpret the result.

**13.28**   Shown below is a cross-tabulation on 8605 full-time students attending a southern state university in the fall of 1986. Is the residency pattern for males different than it is for females? Do an appropriate hypothesis test at the $\alpha = 0.10$ level of significance.

|  | Male | Female |
|---|---|---|
| Resident | 3086 | 2915 |
| Nonresident | 1385 | 1219 |

**13.29**   A magazine's survival may depend on the frequency of its publication as well as on its single-issue cover price (recall Case Study 13.3). The first column in the following table shows the publication schedule for 134 new magazines still in business. The second column gives similar information on

96 magazines that have recently ceased publication. Set up and test the "independence" hypothesis. Let $\alpha = 0.05$.

| | Is Still Published | Has Ceased Publication |
|---|---|---|
| Monthly | 43 | 36 |
| Bimonthly | 52 | 30 |
| Quarterly | 26 | 10 |
| Other | 13 | 20 |
| Totals | 134 | 96 |

13.30  Do a $\chi^2$ test on Lister's data described in Case Study 2.12. Let $\alpha = 0.05$. Interpret your results.

13.31  (a) Using the data given in Case Study 2.13, test whether birth order and juvenile delinquency are independent. Let $\alpha = 0.01$.

(b) Is the conclusion reached in part (a) consistent with the fact that delinquency rates vary considerably from birth order to birth order?

13.32  (a) Do the data in Question 2.38 support the position that blood pressures of fathers and blood pressures of children are independent? Let $\alpha = 0.05$.

(b) What does your conclusion suggest about the feasibility of using one group to screen for high risk individuals in the other?

13.33  (a) Summarize the data described in Question 2.39 with a $2 \times 2$ contingency table.

(b) Do a goodness-of-fit test for independence. Let 0.05 be the level of significance.

13.34  Testing for independence in a $2 \times 2$ contingency table is equivalent to doing a two-sample Z test of $H_0: p_X = p_Y$ versus $H_1: p_X \neq p_Y$ (recall Section 10.5). Use Theorem 10.3 on the data being analyzed in Question 13.33. Keep the level of significance at 0.05. Show that (1) the observed $\chi^2$ statistic is the square of the observed Z statistic and (2) the $\chi^2$ critical value is the square of the Z critical value.

13.35  Using the data given in Question 2.40, test whether home ownership is a factor in whether a person's telephone number is listed or unlisted. Let $\alpha = 0.10$.

13.36  As part of the experiment described in Case Study 3.17, two traits (religion and compliance) were measured on a group of 57 patients. Do a formal analysis on those data. What hypotheses are being tested? Let 0.05 be the level of significance.

13.37  See Question 3.75. Are cheetahs, lions, and wild dogs equally adept at chasing down gazelles? Do an appropriate test for independence using the 0.01 level of significance.

**13.38** Records kept on 578 pregnancies, all involving mothers who had contracted a rubella infection, were the focus of Question 3.76. At the 0.01 level of significance, can we conclude that the risk of an abnormal birth is related to the time during the pregnancy that an infection first occurs?

**13.39** A market research study has investigated the relationship between an adult's self-perception and his or her attitude toward small cars. A total of 299 persons living in a metropolitan area were surveyed. On the basis of his or her responses to a questionnaire, each person was assigned to one of three distinct personality types: (1) cautious conservative, (2) middle-of-the-roader, and (3) confident explorer. At the same time, each person was asked to give an overall opinion of small cars. The results are listed in the table below (81). Test whether the three rows are independent of the three columns. Let $\alpha = 0.01$.

|  |  | SELF-PERCEPTION | | |
| --- | --- | --- | --- | --- |
|  |  | Cautious Conservative | Middle-of-the Roader | Confident Explorer |
|  | Favorable | 79 | 58 | 49 |
| OPINION OF SMALL CARS | Neutral | 10 | 8 | 9 |
|  | Unfavorable | 10 | 34 | 42 |

**13.40** The management of a garment factory is trying to decide with which of three yarn suppliers they should do business. To get an idea of each company's quality control, they order a set of samples from each of the three firms and count the number having zero defects, one defect, and so on. Do the results, as tabulated below, support management's suspicion that the quality levels of all three suppliers are essentially the same? Carry out an appropriate test at the $\alpha = 0.01$ level of significance.

|  |  | SUPPLIER | | |
| --- | --- | --- | --- | --- |
|  |  | A | B | C |
|  | 0 | 10 | 7 | 14 |
|  | 1 | 22 | 18 | 27 |
| Number of Defects | 2 | 25 | 23 | 27 |
|  | 3 | 19 | 19 | 18 |
|  | 4+ | 17 | 20 | 14 |

**13.41** In Section 13.3 the number of degrees of freedom associated with a goodness-of-fit statistic is given as $k - 1 - q$, where $k$ is the number of classes and $q$ is the number of estimated parameters. In Section 13.4 the formula used for the number of degrees of freedom in a test of independence is $(r - 1) \cdot (c - 1)$, where $r$ and $c$ are the numbers of rows and columns, respectively, in the data table. Show that the latter formula is nothing more than a special case of the former. (Hint: Recall that $\widehat{E(f_{ij})} = n\widehat{P(A_i)}\widehat{P(B_j)}$, *for all i and j*, suggesting that $r + c$ parameters are being estimated. However, both $\sum_{i=1}^{r} \widehat{P(A_i)}$ and $\sum_{j=1}^{c} \widehat{P(B_j)}$ *must* equal 1 (why?), so only $r - 1$ of the $\widehat{P(A_i)}$'s and $c - 1$ of the $\widehat{P(B_j)}$'s are actually computed from the data.)

# 13.5

## COMPUTER NOTES

There is no single routine to perform goodness-of-fit tests with commonly used statistical software. Such tests depend on the model, and the statistical packages vary in their ability to generate expected values. Once the expected values are available, the goodness-of-fit statistics can be readily calculated by any computer software with arithmetical routines.

The Minitab printout in Figure 13.4 shows how easy it is to calculate the chi-square statistic of Example 13.1.

```
SET C1
   23        50      20
SET C2
   23.25   46.50   23.25
NAME C1 = 'OBS' , C2 = 'EXP' , C3 = 'CHISQ'
LET 'CHISQ' = ('OBS' - 'EXP')**2/'EXP'
SUM 'CHISQ'
      SUM      =      0.72043
```

**FIG. 13.4**

SAS is capable of calculating a $P$ value to test goodness-of-fit. See Figure 13.5 for an application to Example 13.2.

```
DATA ONE;
  INPUT EXPT OBSV;
  CARDS;

    24  26.2
    64  59.7
    32  34.1

DATA;
  SET ONE END = FINAL;
  CHISQ = (EXPT - OBSV)**2/EXPT;
  TOT + CHISQ;
  P = 1 - PROBCHI (TOT,1);
  IF FINAL THEN PUT 'CHI SQUARE STATISTIC = ' TOT / 'P VALUE = ' P;


  CHI SQUARE STATISTIC = 0.6283854
  P VALUE = 0.4279482
```

**FIG. 13.5**

Producing contingency tables, with expected frequencies, is readily done in Minitab. Figure 13.6 demonstrates the Minitab treatment of Case Study 13.3.

The printout contains the value of the chi-square statistic, 11.30, but not the $P$ value for the test. Thus, we still need the chi-square table to find the critical value.

```
READ C1 C2
  70   65
  39   28
  14    3
  13    2
CHISQUARE C1 C2

CHISQUARE ANALYSIS ON TABLE IN C1, C2

EXPECTED FREQUENCIES ARE PRINTED BELOW OBSERVED FREQUENCIES

          I  C1   I  C2   ITOTALS
-------I-------I-------I-------
    1  I   70  I   65  I    135
       I  78.5I  56.5I
-------I-------I-------I-------
    2  I   39  I   28  I     67
       I  38.9I  28.1I
-------I-------I-------I-------
    3  I   14  I    3  I     17
       I   9.9I   6.3I
-------I-------I-------I-------
    4  I   13  I    2  I     15
       I   8.7I   6.3I
-------I-------I-------I-------
TOTALS I  136  I   98  I    234


TOTAL CHI SQUARE =
       .91+ 1.27+
       .00+  .00+
      1.72+ 2.38+
      2.10+ 2.92+

        = 11.30

DEGREES OF FREEDOM = ( 4-1 ) X ( 2-1 ) = 3
```

**FIG. 13.6**

# 13.6

## SUMMARY

Up until the previous chapter, parameters had been the primary focus of our attention. We learned how to analyze single means, paired means, and differences between means. On other occasions we dealt with variances, quotients of variances, proportions, and differences between proportions. Midway through Chapter 12 our objectives began to change. In the analysis of regression data, parameters were still important, but *relationships* between variables often took precedence. Here we continued that trend, to the extent that studying the relationship between two variables becomes our *only* objective.

Motivation for what we did in this chapter and the direction it took mathematically came from the nature of the data we were trying to analyze. Categorical data are *qualitative*, as opposed to *quantitative*, meaning that measurements are recorded as classes rather than as numbers. Thus, many of the more familiar statistical descriptors such as sample means and sample standard deviations have no relevance. Neither do test procedures based on $Z$ ratios, $t$ ratios, or $F$ ratios. We appealed instead to a goodness-of-fit argument, whereby the observed frequency associated with a class was compared to an *expected* frequency computed from the null hypothesis. If observed and expected frequencies were too far apart, $H_0$ was rejected.

Sections 13.2 and 13.3 provided the conceptual framework for the chapter's ultimate objective of doing a $\chi^2$ test for independence on an $r \times c$ contingency table. Learning how to compute expected frequencies, or estimated expected frequencies, was the first step. Equation 13.1 states the basic result: if the probability of an observation falling into category $j$ is $p_j$, and if a total of $n$ observations are recorded, then the expected number in that category $\widehat{E(X_j)}$, is the product, $np_j$.

The $p_j$'s come from the null hypothesis. In Sections 13.2 and 13.3 $H_0$ was a statement about which probability model might reasonably have produced a given set of data. We decide formally whether a probability model adequately describes a set of data by performing a *goodness-of-fit test*. Theorems 13.2 and 13.3 describe two variations. In both cases the test statistic has a $\chi^2$ distribution when $H_0$ is true.

With Sections 13.2 and 13.3 as background, motivating the statistical analysis of categorical data is not particularly difficult. Imagine $n$ subjects being "measured" with respect to two traits $X$ and $Y$. If the $r$ classes associated with $X$ are denoted $A_1, A_2, \ldots, A_r$ and the $c$ classes associated with $Y$ are written $B_1, B_2, \ldots, B_c$, then each subject belongs to exactly one of the $A_i \cap B_j$ intersections. The total number belonging to both $A_i$ and $B_j$ is the *observed frequency* for that "cell" and is denoted $f_{ij}$ (recall Table 13.11). The question of interest when data are categorical is whether a subject's $X$ trait is independent of its $Y$ trait. Is $P(A_i|B_j)$, in other words, different from $P(A_i)$? If it is, then $X$ and $Y$ are not independent.

Let $R_i$ denote the total number of observations in $A_i$, and $C_j$ the total number of observations in $B_j$. (For data displayed in a contingency table, $R_i$ is the $i$th row total and $C_j$ is the $j$th column total.) If traits $X$ and $Y$ are independent, the estimated expected frequency in $A_i \cap B_j$ is

$$\widehat{E(f_{ij})} = \frac{R_i C_j}{n}$$

Clearly, the credibility of the null hypothesis hinges on the similarity between the $f_{ij}$'s and the $\widehat{E(f_{ij})}$'s. To measure that similarity, we computed

$$V = \sum_{i,j} \frac{(f_{ij} - \widehat{E(f_{ij})})^2}{\widehat{E(f_{ij})}}$$

If $\widehat{E(f_{ij})} \geqslant 5$ for all $i$ and $j$ the sampling distribution of $V$ is well-approximated by a $\chi^2$ curve with $(r-1)(c-1)$ degrees of freedom. Moreover, the independence assumption assumption should be rejected (at the $\alpha$ level of significance) if

$$v \geqslant \chi^2_{1-\alpha,\,(r-1)(c-1)}$$

---

**REVIEW QUESTIONS**

**13.42** Forty-eight per cent of all registered voters in Fairview county are male. Of that number, 76% are white. The white to non-white ratio among women in Fairview is 71 to 39. Shown below is the race and sex breakdown of the 120 residents who served on the county's last ten grand juries. Has the makeup of the juries accurately reflected the demographic composition of Fairview county? Answer the question by doing an appropriate hypothesis test at the $\alpha = 0.01$ level.

| GRAND JURY MEMBERS | | | |
|---|---|---|---|
| White, Male | Non-white, Male | White, Female | Non-white, Female |
| 18 | 23 | 55 | 24 |

**13.43** A major lending institution conducts a survey of its clients every five years to see what proportion have overextended their credit. Do the data below support the company's contention that the proportion of borrowers experiencing credit problems remained constant from 1975 through 1985? Do a hypothesis test using $\alpha = 0.05$.

| | | YEAR | | |
|---|---|---|---|---|
| | | 1975 | 1980 | 1985 |
| Credit Overextended? | YES | 13 | 25 | 15 |
| | NO | 107 | 150 | 79 |

**13.44** Out of 1410 sorority women at State University, 296 had GPA's above 3.00 last semester. Similar GPA's were achieved by 461 of the 2012 women not in sororities. Can we conclude that academic performance and Greek affiliation are independent? Let $\alpha = 0.05$.

13.45 Listed below are the number of pieces of gum stuck underneath 60 desks in a high school study hall. Can the distribution of $Y$ be modeled adequately by a Poisson pdf? Do a goodness-of-fit test with $\alpha = 0.10$ (32).

| Number of Pieces of Gum, $y_i$ | Number of Desks |
|---|---|
| 0 | 11 |
| 1 | 20 |
| 2 | 10 |
| 3 | 7 |
| 4 | 6 |
| 5 | 4 |
| 6 | 2 |
| | 60 |

13.46 Three blossom colors are found in crosses of a new hybrid impatiens: red, variegated, and white. On the basis of the 212 plants categorized below, is it reasonable to expect that blossom colors of future crosses will show a $1:2:1$ ratio? Let $\alpha = 0.05$.

| Blossom Color | Number of Plants |
|---|---|
| Red | 49 |
| Variegated | 108 |
| White | 55 |
| | 212 |

13.47 Given the following sample of 70 observations, test at the $\alpha = 0.10$ level of significance,

$$H_0: \quad f_X(x) = \frac{9}{13}\left(\frac{1}{3}\right)^x, \qquad x = 0, 1, 2$$

vs.

$$H_1: \quad f_X(x) \neq \frac{9}{13}\left(\frac{1}{3}\right)^x, \qquad x = 0, 1, 2$$

```
0  0  2  0  1  0  2  2  2  1
1  1  0  2  0  1  0  1  0  1
0  0  1  0  0  0  0  1  0  2
0  2  0  0  1  0  1  0  1  0
2  0  0  1  0  0  2  0  0  1
1  0  0  0  0  1  0  0  0  2
0  1  0  1  0  2  0  1  1  2
```

13.48 Why is the critical region for a goodness-of-fit test only in the right-hand tail of a $\chi^2$ distribution?

13.49   A city planning study done several years ago analyzed eastbound traffic flow as it approached the Newberry and Ridgeview intersection. Sixty-five per cent of the cars continued straight ahead, 20% turned left, and 15% turned right. Since then a shopping mall has been opened nearby and traffic patterns may have changed. A recent follow up study released the figures summarized below. Do an appropriate goodness-of-fit test at the $\alpha = 0.05$ level of significance.

| Action | Number of Cars |
|---|---|
| Continue straight | 498 |
| Turn left | 172 |
| Turn right | 150 |
| | 820 |

13.50   An urn contains $w$ white chips and $r$ red chips. Two are drawn at random without replacement. Let the (hypergeometric) random variable $Y$ denote the number of red chips in the sample. Listed below are the $y$ values obtained in 100 repetitions of the experiment. Is it believable that $w = 3$ and $r = 5$? Carry out an appropriate hypothesis test at the 0.05 level of significance.

| Number of Red Chips, $y$ | Number of Samples |
|---|---|
| 0 | 8 |
| 1 | 59 |
| 2 | 33 |
| | 100 |

13.51   A Wall Street consulting firm claims that 41% of its middle managers are female. Unimpressed, a women's rights group contends that the corporation's salary policies are nevertheless discriminatory. Last year 732 middle managers received raises of 9% or better. Of that total, only 236 were women. Do the charges of discrimination have any merit? Draw an inference using the 0.05 level of significance.

13.52   Analyze the data in Question 13.51 by doing a one-sample binomial test of $H_0: p = p_0$. Keep $\alpha = 0.05$ and use a two-sided alternative. Let

$$p = P \text{ (woman receives raise of 9\% or better)}$$

13.53   Sixty-four replications of a physics experiment each produced an estimate of acceleration due to gravity. Subtracted from each estimate was the true value, 980 cm/sec/sec. Summarized on the next page is a frequency distribution of the resulting errors. Compute $\bar{y}_g$ and $s_g$.

| Errors in Gravity Estimates (cm/sec/sec) | Frequency |
|---|---|
| −31.5 to −22.5 | 6 |
| −22.5 to −13.5 | 8 |
| −13.5 to  4.5 | 11 |
| −4.5 to  4.5 | 14 |
| 4.5 to  13.5 | 10 |
| 13.5 to  22.5 | 9 |
| 22.5 to  31.5 | 6 |
|  | 64 |

13.54   At the $\alpha = 0.05$ level of significance, test whether the 64 estimation errors summarized in Question 13.53 can be considered a random sample from a normal distribution.

13.55   Analyze the data in Question 13.44 by setting up a two-sample Z test of $H_0$: $p_X = p_Y$. Use the same level of significance, $\alpha = 0.05$. Is your conclusion the same as it was in Question 13.44? Explain.

13.56   A sharpshooter fires three shots at a target and makes $y$ bullseyes, where $y = 0, 1, 2,$ or $3$. The $y$ values recorded for 80 volleys of three shots are listed below. Is the random variable $Y$ described adequately by a binomial distribution? Let $\alpha = 0.10$.

| Number of Bullseyes, $y$ | Number of Volleys |
|---|---|
| 0 | 13 |
| 1 | 15 |
| 2 | 35 |
| 3 | 17 |
|  | 80 |

13.57   As part of her research on the writing style of a major Elizabethan poet, a linguist has tallied the number of times the poet used the word *love* in passages of 50 words. Do the following data, representing 200 such passages, support the researcher's contention that the occurrences of *love* can be modeled by a Poisson pdf? Let $\alpha = 0.05$.

| Number of *loves*, $y$ | Number of Passages |
|---|---|
| 0 | 40 |
| 1 | 115 |
| 2 | 21 |
| 3 | 12 |
| 4 | 9 |
| 5 | 3 |
|  | 200 |

13.58   Scores on the first freshman chemistry exam tend to be normally distributed with $\mu = 72.0$ and $\sigma = 10.0$. If six exam papers are chosen at random, what is the probability that three will have scores less than or equal to 50, two will have scores between 50 and 90, and one will have a score of 90 or better? (Hint: Use the multinomial distribution on p. 581.)

13.59   A geneticist claims to have discovered a chromosome linkage phenomenon that produces a $9:6:1$ ratio of petal shapes (rounded, oval, pointed) in California sunflowers. She produces the following evidence based on 560 crosses to support her theory. Are you convinced that her model is correct? Explain.

| Petal Shape | Number of Plants |
|---|---|
| Rounded | 315 |
| Oval | 210 |
| Pointed | 35 |
| | 560 |

13.60   Why is $V$, the goodness-of-fit statistic, defined in terms of frequencies rather than probabilities? Intuitively, what would be wrong with using

$$\sum_{i,j} \frac{(p_{ij} - \hat{p}_{ij})^2}{\hat{p}_{ij}}$$

as a measure of agreement between observed and expected distributions?

13.61   Take 50 observations on a phenomenon you think might have a Poisson distribution. Use Section 6.5, Example 13.3 and Questions 13.14, 13.15, 13.16, 13.21, 13.45, 13.57 to suggest some possibilities. Summarize the data by constructing a histogram.

13.62   Use a goodness-of-fit test to decide whether the data you collected for Question 13.61 do follow a Poisson distribution. Let $\alpha = 0.05$.

13.63   Is it reasonable to presume that the 50 observations listed below are a random sample from

$$f_X(x) = \begin{cases} 1/8 & \text{for } x = 2 \\ 1/4 & \text{for } x = 4 \\ 1/8 & \text{for } x = 5 \\ 1/4 & \text{for } 6 \leqslant x \leqslant 8 \end{cases}$$

Do a goodness-of-fit test with $\alpha = 0.05$. (Hint: Divide the interval $6 \leqslant x \leqslant 8$ into two classes, $6 \leqslant x < 7$ and $7 \leqslant x \leqslant 8$. Remember that probabilities associated with a continuous random variable are represented by areas under $f_X(x)$).

| 4 | 5 | 7.4 | 7.3 | 7.6 | 6.3 |
| 6.2 | 2 | 6.3 | 4 | 6.7 | 4 |
| 2 | 6.9 | 5 | 6.5 | 2 | 7.2 |
| 6.8 | 5 | 2 | 2 | 4 | 5 |
| 5 | 6.1 | 6.0 | 5 | 2 | 6.8 |
| 4 | 6.8 | 5 | 4 | 5 | |
| 2 | 4 | 4 | 4 | 4 | |
| 7.1 | 5 | 7.4 | 7.0 | 6.8 | |
| 5 | 7.5 | 7.5 | 7.9 | 7.4 | |

13.64   Suppose the independence of treatments $X$ and $Y$ is to be tested by applying Theorem 13.4 to an $r \times c$ contingency table. Does it matter which treatment defines the rows and which defines the columns?

13.65   The formal way of testing the normality assumption in a linear model is to apply Theorem 13.3 to the $n$ residuals, $y_i - \hat{y}_i$, $i = 1, 2, \ldots, n$, where $\hat{y}_i = \hat{a} + \hat{b}x_i$. How can the normality assumption be checked "informally"?

13.66   Sixteen of 80 women who went on diet R were still following the regimen 90 days later. By way of comparison, 23 women of an initial 58 were still following diet T after the same length of time. Summarize the data with a $2 \times 2$ contingency table. State and test an appropriate null hypothesis. Let $\alpha = 0.05$.

# 14

# The k-Sample Problem

*Mathematicians are like Frenchmen;*
*Whatever you say to them,*
*They translate into their own language*
*And forthwith it is something entirely different.*

*Goethe*

## 14.1

### INTRODUCTION

If statistics chapters, like packs of cigarettes, were required to have consumer protection labels, the one emblazoned on Chapter 14 would probably say **Warning: What you are about to read may cause headaches, depression, and a feeling of uneasiness.** Of course, some would argue that a similar admonition would be appropriate for every chapter. But Chapter 14 *is* conceptually more demanding and computationally more unpleasant than anything we have seen up to this point.

The reason that things are about to take a turn for the worse is the complexity of $k$-sample data. What works for one-sample and two-sample data simply does not carry over for $k$-sample data. For example, we can define a one-sample $t$ test and a two-sample $t$ test, but there is no way to set up a $k$-sample $t$ test. Instead, data having more than two sets of independent samples are analyzed with an entirely different procedure, a technique known as the **analysis of variance** (ANOVA).

Actually, the analysis of variance is not one procedure but a collection of procedures, all based on the same mathematical principle. Chapter 14 introduces the simplest form of the analysis of variance, although even here—as our hypothetical warning would have cautioned—what we need to conceptualize is not simple.

> **ANOVA table**—A format for presenting (and helping to organize) the various sums of squares needed for doing an $F$ test on $H_0$: $\mu_1 = \mu_2 = \cdots = \mu_k$. In general, ANOVA tables show at a glance the exact model an experimenter is using and indicate the levels of significance that can be associated with the various hypotheses and subhypotheses being tested.

#### Historical Note

Much of the early development of the analysis of variance is credited to the great British statistician Sir Ronald A. Fisher. Shortly after the end of World War I, Fisher resigned a teaching position with which he was none too happy and accepted an appointment at the Rothamsted Statistical Laboratory, a facility heavily involved in agricultural research.

There he suddenly found himself entangled in problems where differences in the response variable (crop yields, for example) were constantly in danger of being obscured by the high level of uncontrollable heterogeneity in the experimental environment (different soil qualities, variation in the amount of rainfall, and so on). Seeing that traditional techniques were hopelessly inadequate under those conditions, Fisher began a search for alternatives that was destined to revolutionize the way scientists conduct research. In just a few years he succeeded in fashioning an entirely new statistical methodology, a panoply of data-collecting principles and mathematical tools that is today known as *experimental design*. The centerpiece of Fisher's creation is the analysis of variance.

# 14.2

## NOTATION

The top of Table 14.1 reminds us what $k$-sample data will look like. The columns might represent (1) $k$ different levels of the same treatment (all applied to $k$ independent sets of similar subjects) or (2) $k$ different kinds of subjects (all exposed to the same treatment). Examples of the first type are more common.

**TABLE 14.1**

|  | 1 | 2 | $\cdots$ | $k$ | Overall Totals |
|---|---|---|---|---|---|
|  | — | — |  | — |  |
|  | — | — |  | — |  |
|  | — | — |  | — |  |
|  | — |  |  |  |  |
| Sample Sizes | $n_1$ | $n_2$ | $\cdots$ | $n_k$ | $n$ |
| Sample Totals | $T_1$ | $T_2$ | $\cdots$ | $T_k$ | $T$ |
| Sample Means | $\bar{x}_1$ | $\bar{x}_2$ | $\cdots$ | $\bar{x}_k$ | $\bar{x}$ |
| Population Mean | $\mu_1$ | $\mu_2$ | $\cdots$ | $\mu_k$ |  |

The four rows at the bottom of the table show some of the additional notation we will encounter in dealing with $k$-sample data. The *sample sizes* associated with the $k$ sets of observations will be written $n_1, n_2, \ldots, n_k$. The *total sample size* is denoted by $n$, so

$$n = n_1 + n_2 + \cdots + n_k$$

The *sample total* for a given column is the sum of the data values listed in that column. The $k$ totals are denoted $T_1, T_2, \ldots, T_k$. Adding the $T_j$'s gives the *overall total* $T$:

$$T = \text{overall total} = T_1 + T_2 + \cdots + T_k$$

If a sample total is divided by the number of observations that went into that total, the resulting quotient is the *sample mean*. These are written $\bar{x}_1, \bar{x}_2, \ldots, \bar{x}_k$:

$$\bar{x}_1 = \frac{T_1}{n_1}, \qquad \bar{x}_2 = \frac{T_2}{n_2}, \qquad \cdots, \qquad \bar{x}_k = \frac{T_k}{n_k}$$

Dividing the overall total by the overall sample size gives the *overall mean* $\bar{x}$:

$$\bar{x} = \text{overall mean} = \frac{T}{n}$$

The $x_i$'s in a given column are presumed to be a random sample from a theoretical population (typically, the normal). The *true means* associated with those populations are denoted $\mu_1, \mu_2, \ldots, \mu_k$. These are parameters, so their exact numerical values are never known.

---

*EXAMPLE 14.1*   Recall the data described in Case Study 2.15 (see Table 14.2). The "subjects" are 12 slices of bacon, three representing each of four brands. Recorded for each slice is the amount of $N$-nitrosopyrrolidine (NPYR) recovered after frying. Compute $T$ and $\bar{x}$.

**TABLE 14.2 NPYR Recovered from Bacon (ppb)**

|        | Brand 1 | Brand 2 | Brand 3 | Brand 4 |
|--------|---------|---------|---------|---------|
|        | 20      | 75      | 15      | 25      |
|        | 40      | 25      | 30      | 30      |
|        | 18      | 21      | 21      | 31      |
| $n_j$  | 3       | 3       | 3       | 3       |
| $T_j$  | 78      | 121     | 66      | 86      |
| $\bar{x}_j$ | 26.0 | 40.3   | 22.0    | 28.7    |

*Solution*   Each sample size is 3, so $n_1 = n_2 = n_3 = n_4 = 3$ (and $n = 12$). For brand 1,

$$T_1 = 20 + 40 + 18 = 78 \quad \text{and} \quad \bar{x}_1 = \frac{78}{3} = 26.0$$

The other $T_j$'s and $\bar{x}_j$'s are computed similarly. The overall total is

$$T = 78 + 121 + 66 + 86 = 351,$$

and the overall average is

$$\bar{x} = \frac{351}{12} = 29.2.$$

Typically, the first objective in a $k$-sample problem is to test whether the populations being represented by the $k$ samples all have the same mean; that is, we need to choose between $H_0: \mu_1 = \mu_2 = \cdots = \mu_k$ and $H_1$: not all the $\mu_j$'s are equal. Here, for example, the initial question to be addressed is whether brands 1 through 4 all yield the same amounts of NPYR, *on the average*. Section 14.3 will show us how to make that decision.

---

QUESTIONS

14.1   Summarize the magnetic field declinations given in Table 2.15 by computing all the $T_j$'s and $\bar{x}_j$'s. Also, calculate $T$ and $\bar{x}$.

14.2   In general, what does

$$\sum_{j=1}^{k} n_j \bar{x}_j \quad \text{equal?}$$

14.3 For the octopus data in Question 2.43 compute the appropriate $T_j$'s and $\bar{x}_j$'s. Use the $T_j$'s to find $T$ and $\bar{x}$.

14.4 Give a formula for $\bar{x}$ in terms of the $T_j$'s.

14.5 Find the set of $T_j$'s and $\bar{x}_j$'s for the heart rate data in Question 3.81. Compute $\bar{x}$.

14.6 In general, does

$$\bar{x} = \frac{1}{k} \sum_{j=1}^{k} \bar{x}_j \quad ?$$

Explain.

# 14.3

## THE F TEST

**F test**—A procedure for testing $H_0$: $\mu_1 = \mu_2 = \cdots = \mu_k$. In general, F tests are based on the ratio of two different mean squares, each of which estimates a variance.

In the analysis of $k$-sample data, the assumptions made about the origin of the observations are analogous to what was presumed for the $X_i$'s and $Y_i$'s in the two-sample problem. Specifically, we require that the $n_j$ observations in column $j$ be a random sample from a normal distribution with mean $\mu_j$, $j = 1, 2, \ldots, k$, and variance $\sigma^2$ (see Figure 14.1). In addition, the observations in a column are assumed to be independent of all observations in all other columns.

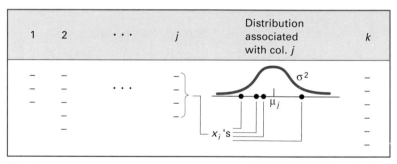

**FIG. 14.1**

For reasons we will better understand later, there is more convenient format for summarizing the conditions in Figure 14.1. If $X_i$ denotes the $i$th observation in the $j$th sample, we will write

$$X_i = \mu_j + \varepsilon_i, \qquad i = 1, 2, \ldots, n_j; \quad j = 1, 2, \ldots, k \qquad (14.1)$$

**model equation**—An equation that shows how each observation in a set of data can be "explained" as the sum of a random component added to one or more constants.

where $\varepsilon_i$ is an independent normal random variable with mean zero and variance $\sigma^2$, $i = 1, 2, \ldots, n$. The expression "$X_i = \mu_j + \varepsilon_i$" is called a ***model equation***. In general, model equations indicate the structural "components" of an observation; some of those components are *fixed*, and others are *random*. In this case, $\mu_j$ is fixed, and $\varepsilon_i$ is random.

**TABLE 14.3**

| 1 | 2 | j | ... | k |
|---|---|---|-----|---|
| – | – | 18 | | – |
| – | – | 19 | | – |
| – | – | 24 | | – |
| – | – | 16 | | – |
| – | – | | | – |

For example, suppose the data in column $j$ are the four numbers shown in Table 14.3. If $\mu_j = 20$, each observation can be represented by Equation 14.1 as follows:

$$X_1 = 18 = \mu_j + \varepsilon_1 = 20 + (-2)$$

$$X_2 = 19 = \mu_j + \varepsilon_2 = 20 + (-1)$$

$$X_3 = 24 = \mu_j + \varepsilon_3 = 20 + (+4)$$

$$X_4 = 16 = \mu_j + \varepsilon_4 = 20 + (-4)$$

The random components here are $\varepsilon_1 = -2$, $\varepsilon_2 = -1$, $\varepsilon_3 = +4$, and $\varepsilon_4 = -4$. By Equation 14.1, the $\varepsilon_i$'s are a random sample from a normal distribution with mean zero and variance $\sigma^2$ (see Figure 14.2).

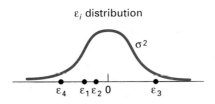

**FIG. 14.2**

**total sum of squares (SSTO)**—The sum of the squared deviations of each observation from the overall mean in the entire sample. It measures the total variability in a set of data—that is, the extent to which the observations are not all the same.

Are the assumptions imposed on the $X_i$'s and on the $\varepsilon_i$'s equivalent? Yes. If the $\varepsilon_i$'s have the distribution pictured in Figure 14.2, the $X_i$'s will have the distribution pictured in Figure 14.1, and vice versa.

Sums of squared deviations, terms similar to what we encountered in fitting least squares lines in Chapter 3, are the keys to setting up the analysis of variance. For $k$-sample data three sums of squares need to be defined. Each one tells us something specific about the data; together they lead to a formal procedure for testing $H_0$: $\mu_1 = \mu_2 = \cdots = \mu_k$.

The **total sum of squares** (SSTO) is the sum of the squared deviations of the observations from the overall mean:

$$\text{Total sum of squares} = \text{SSTO} = \sum\sum (x_i - \bar{x})^2 \qquad (14.2)$$

**treatment sum of squares (SSTR)**—The sum of the squared deviations of the *k* sample means from the overall mean. It measures the extent to which the $\mu_j$'s are not all equal.

The two summation signs in Equation 14.2 signify that the summation of $(x_i - \bar{x})^2$ includes all such terms in all *k* columns, the inner summation indicates that we should sum $(x_i - \bar{x})^2$ *within* a given column, and the outer summation tells us to continue the summation *from column to column*.

Another term to be computed is the **treatment sum of squares** (SSTR), which is defined to be the sum of the squared deviations between the sample means and the overall mean:

$$\text{Treatment sum of squares} = \text{SSTR} = \sum n_j(\bar{x}_j - \bar{x})^2 \qquad (14.3)$$

Here the summation is similar to what we have been doing since Chapter 1:

$$\text{SSTR} = n_1(\bar{x}_1 - \bar{x})^2 + n_2(\bar{x}_2 - \bar{x})^2 + \cdots + n_k(\bar{x}_k - \bar{x})^2$$

Finally, the **error sum of squares** (SSE) is the sum of the squared deviations of each observation from its own sample mean:

**error sum of squares (SSE)**—The sum of the squared deviations of each observation from its own sample mean.

$$\text{Error sum of squares} = \text{SSE} = \sum\sum (x_i - \bar{x}_j)^2 \qquad (14.4)$$

As for SSTO, the summation for the error sum of squares is first done *within* each sample, and then those *k* sums are combined to give the overall SSE.

---

**THEOREM 14.1    A Formula for Sums of Squares**

For *any* set of *k*-sample data,

$$\text{SSTO} = \text{SSTR} + \text{SSE} \qquad (14.5)$$

In summation notation,

$$\sum\sum (x_i - \bar{x})^2 = \sum n_j(\bar{x}_j - \bar{x})^2 + \sum\sum (x_i - \bar{x}_j)^2$$

---

A formal proof of Theorem 14.1 is unnecessary for our purposes, but we need to understand the theorem's implications. Specifically, we need to know what SSTO, SSTR, and SSE represent *conceptually* and how Equation 14.5 relates to the problem of testing $H_0$: $\mu_1 = \mu_2 = \cdots = \mu_k$.

First, consider SSTO. If all the $x_i$'s were numerically the same (that is, if the data showed absolutely no variability), each observation would equal $\bar{x}$ and SSTO would be zero. Conversely, as the numerical values of the $x_i$'s get further and further apart, $\sum\sum (x_i - \bar{x})^2$ gets larger and larger. What SSTO measures, therefore, is "total variability"—the extent to which the *n* observations are not all the same.

A similar argument suggests how SSTR should be interpreted. If the populations being sampled all have the same *true* means (that is, if $\mu_1 = \mu_2 = \cdots = \mu_k$, we would expect the corresponding *sample* means $\bar{x}_i$,

$\bar{x}_2, \ldots, \bar{x}_k$ to be approximately equal. But if the $\bar{x}_j$'s are approximately equal, the number to which they will be similar is $\bar{x}$, in which case $\sum n_j(\bar{x}_j - \bar{x})^2$ will be small. Conversely, if the true means were very different, we would expect those differences to be reflected in the sample means, the result being that $\sum n_j(\bar{x}_j - \bar{x})^2$ would inflate. In general, then, SSTR measures the credibility of $H_0: \mu_1 = \mu_2 = \cdots = \mu_k$: as SSTR *increases*, the likelihood of $H_0$ being true *decreases*.

What about SSE? If SSTR reflects variation associated with the treatments, then the error sum of squares reflects variation *not* associated with the treatments. Why? Look at the formula:

$$\text{SSE} = \sum\sum (x_i - \bar{x}_j)^2$$

If the only factor affecting the value of $x_i$ were the treatment level that that observation represents, then all the $x_i$'s in a given column would be the same and they would all equal $\bar{x}_j$ (and $\sum\sum (x_i - \bar{x}_j)^2$ would be zero). That they are *not* all the same shows that factors other than the treatment are influencing the measured response. The net effect of all these other factors is what SSE measures.

Now, look back at the statement of Theorem 14.1. In statistical terminology, what is being described there is the *partitioning* of the overall variability among the $x_i$'s (as measured by SSTO) into two components, SSTR and SSE: SSTR measures that portion of the total variability due to (or "caused by") the fact that the treatment levels are not all equivalent; SSE accumulates all the variability due to factors other than the one being controlled.

In principle, SSTR and SSE play roles analogous to the numerator and denominator in a two-sample $t$ statistic: SSE (like $s_p\sqrt{1/n + 1/m}$) reflects the data's inherent variability; SSTR (like $\bar{x} - \bar{y}$) quantifies the magnitude of the treatment effect.

If the analogy with a $t$ ratio is carried further, it suggests that we should base our decision rule for testing $H_0: \mu_1 = \mu_2 = \cdots = \mu_k$ on some function of the quotient SSTR/SSE. The function that proves to be the most convenient is

$$\frac{\text{SSTR}/(k-1)}{\text{SSE}/(n-k)}$$

because dividing SSTR by $k - 1$ and SSE by $n - k$ gives a ratio having an $F$ distribution (if $H_0$ is true), meaning we can use existing tables to find critical values. The numerator is called the *treatment mean square* (MSTR), and the denominator is called the *error mean square* (MSE). The quotient,

$$\frac{\text{MSTR}}{\text{MSE}} = \frac{\text{SSTR}/(k-1)}{\text{SSE}/(n-k)}$$

is the *observed $F$ ratio*.

---

**treatment mean square (MSTR)**— The treatment sum of squares (SSTR) divided by $k - 1$. If $H_0: \mu_1 = \mu_2 = \cdots = \mu_k$ is true, MSTR is an estimate of $\sigma^2$.

**error mean square (MSE)**—The error sum of squares (SSE) divided by $n - k$. Conceptually, MSE is analogous to the pooled variance $s_p^2$ in two-sample $t$ tests: it represents an "overall" estimate of the unknown variance $\sigma^2$ associated with each observation.

---

**THEOREM 14.2**

Let SSTR and SSE denote the treatment sum of squares and error sum of squares, respectively, for a set of $k$-sample data.

1. If $H_0: \mu_1 = \mu_2 = \cdots = \mu_k$ is true,

$$F = \frac{\text{MSTR}}{\text{MSE}} \tag{14.6}$$

has an $F$ distribution with $k - 1$ and $n - k$ degrees of freedom.

2. To test

$$H_0: \quad \mu_1 = \mu_2 = \cdots = \mu_k$$

vs.

$$H_1: \quad \text{not all the } \mu_j\text{'s are equal}$$

at the $\alpha$ level of significance, reject $H_0$ if[1]

$$F \geqslant F_{\alpha, k-1, n-k}.$$

---

**COMMENT:** It can be shown mathematically that if $H_0: \mu_1 = \mu_2 = \cdots = \mu_k$ is true, both $\text{SSTR}/(k - 1)$ and $\text{SSE}/(n - k)$ estimate $\sigma^2$, implying that their ratio should be close to 1. On the other hand, if $H_0$ is not true, $\text{SSTR}/(k - 1)$ will tend to be larger than $\text{SSE}/(n - k)$. Values of the observed $F$ ratio, therefore, that are too much larger than 1 are indications that $H_0$ is not true. The cutoff that tells us when the observed $F$ ratio has gotten "too much larger" is $F_{\alpha, k-1, n-k}$.

---

*EXAMPLE 14.2*  For the serum binding data in Case Study 3.19, Table 14.4 shows the four responses recorded for each of the five treatment levels. Let $\mu_1$ denote the true average serum binding percentage characteristic of penicillin G, $\mu_2$ the true average percentage characteristic of tetracycline, and so on. To compare the antibiotics, the experimenter would want to test

$$H_0: \quad \mu_1 = \mu_2 = \cdots = \mu_5$$

vs.

$$H_1: \quad \text{not all the } \mu_j\text{'s are equal}$$

Let $\alpha = 0.05$.

---

[1] Recall that $F_{\alpha, k-1, n-k}$ is the number that cuts off to its right an area of $\alpha$ under an $F$ distribution with $k - 1$ and $n - k$ degrees of freedom.

**TABLE 14.4 Serum Binding Percentages**

| | (1)<br>Penicillin G | (2)<br>Tetracycline | (3)<br>Streptomycin | (4)<br>Erythromycin | (5)<br>Chloramphenicol |
|---|---|---|---|---|---|
| | 29.6 | 27.3 | 5.8 | 21.6 | 29.2 |
| | 24.3 | 32.6 | 6.2 | 17.4 | 32.8 |
| | 28.5 | 30.8 | 11.0 | 18.3 | 25.0 |
| | 32.0 | 34.8 | 8.3 | 19.0 | 24.2 |
| $T_j$ | 114.4 | 125.5 | 31.3 | 76.3 | 111.2 |
| $\bar{x}_j$ | 28.6 | 31.4 | 7.8 | 19.1 | 27.8 |

**Solution**   Note that $k = 5$, $n = 20$, and

$$\bar{x} = \frac{T}{n} = \sum \frac{T_j}{n} = \frac{114.4 + 125.5 + \cdots + 111.2}{20}$$

$$= \frac{458.7}{20} = 22.935$$

Therefore,

$$\text{SSE} = \sum\sum (x_i - \bar{x})^2 = \sum\sum (x_i - 22.935)^2$$
$$= \underbrace{(29.6 - 22.935)^2 + (24.3 - 22.935)^2 + \cdots + (24.2 - 22.935)^2}_{20 \text{ terms}}$$

$$= 135.83$$

and

$$\text{SSTR} = \underbrace{4(28.6 - 22.935)^2 + 4(31.4 - 22.935)^2 + \cdots + 4(27.8 - 22.935)^2}_{5 \text{ terms}}$$

$$= 1480.83$$

The resulting test statistic is

$$F = \frac{1480.83/(5 - 1)}{135.83/(20 - 5)} = 40.9$$

Is 40.9 large enough for us to conclude that $H_0: \mu_1 = \mu_2 = \cdots = \mu_5$ is false? The *F* distribution will tell us. Figure 14.3 shows the *F* distribution with 4 $(=k-1)$ and 15 $(=n-k)$ degrees of freedom. According to Appendix A. 4c, the 95th percentile of that distribution $(=F_{0.05,4,15})$ is 3.06. Since the observed *F* ratio $(=40.9)$ exceeds that value, we should reject $H_0$, a decision that confirms what the nonoverlapping confidence intervals in Figure 3.35 strongly suggested.

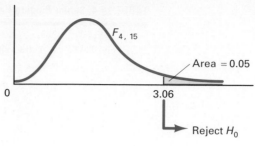

Area = 0.05

Reject $H_0$

**FIG. 14.3**

14.7   Compute SSTO, SSTR, and SSE for the following set of data:

| TREATMENTS | | |
| --- | --- | --- |
| 1 | 2 | 3 |
| 2 | 1 | 4 |
| 4 | 1 | 2 |
|   | 1 |   |
|   | 1 |   |

Also, show that the three sums satisfy Theorem 14.1.

14.8   For the data in Question 14.7, test $H_0: \mu_1 = \mu_2 = \mu_3$ versus $H_1$: not all the $\mu_j$'s are equal. Let $\alpha = 0.05$.

14.9   Three independent observations are taken on each of two treatments, as shown below:

| TREATMENTS | |
| --- | --- |
| 1 | 2 |
| 2 | 8 |
| 2 | 4 |
| 5 | 3 |

Compute SSTO, SSTR, and SSE. Show that the three sums satisfy Theorem 14.1.

14.10   Using Theorem 14.2, test $H_0: \mu_1 = \mu_2$ versus $H_1: \mu_1 \neq \mu_2$ for the data in Question 14.9. Let 0.05 be the level of significance.

14.11   It can be shown mathematically that an $F$ test done on $k$-sample data when $k = 2$ is equivalent to a two-sample $t$ test. In particular, the observed $F$ ratio is equal to the *square* of the observed $t$ ratio, and the critical value for the $F$ test is the *square* of each of the critical values for the $t$ test.

(a) Do a two-sample $t$ test on the data in Question 14.9.

(b) Verify the two relationships just cited by comparing the results of the $t$ test to the results of the $F$ test done in Question 14.10.

14.12 Give an example of a set of data satisfying each of the following conditions:

(a) SSE $= 0$ but SSTO $> 0$ and SSTR $> 0$

(b) SSTR $= 0$ but SSTO $> 0$ and SSE $> 0$

(c) SSE $=$ SSTR $=$ SSTO $= 0$

# 14.4
## COMPUTING FORMULAS

In practice, we never compute SSTR and SSE from Equations 14.3 and 14.4. Expressions like

$$\sum n_j(\bar{x}_j - \bar{x})^2 \quad \text{and} \quad \sum\sum (x_i - \bar{x})^2$$

are helpful in showing what the different sums of squares represent, but they are cumbersome to work with numerically. Equations 14.7, 14.8, and 14.9 may not look any nicer, but they are much easier to use.

---

**THEOREM 14.3** Computing Formulas for Sums of Squares

Let $n_1, n_2, \ldots, n_k$ be the sample sizes in a set of $k$-sample data, where the total sample size is

$$n = n_1 + n_2 + \cdots + n_k$$

Let $T_1, T_2, \ldots, T_k$ be the sums of the observations in each of the $k$ columns, and let $T = T_1 + T_2 + \cdots + T_k$ be the overall sum. Define $c = T^2/n$. Then

$$\text{SSTO} = \sum\sum x_i^2 - c \tag{14.7}$$

$$\text{SSTR} = \sum \frac{T_j^2}{n_j} - c \tag{14.8}$$

$$\text{SSE} = \text{SSTO} - \text{SSTR} \tag{14.9}$$

---

*EXAMPLE 14.3* For the serum binding data in Table 14.4, we find

$$\sum\sum x_i = T = 29.6 + 24.3 + \cdots + 24.2 = 458.7$$

and

$$\sum\sum x_i^2 = (29.6)^2 + (24.3)^2 + \cdots + (24.2)^2 = 12{,}136.93$$

Therefore,

$$c = \frac{T^2}{n} = \frac{(458.7)^2}{20} = 10{,}520.28$$

and, by Equation 14.7,

$$\text{SSTO} = \sum\sum x_i^2 - c = 12{,}136.93 - 10{,}520.28 = 1616.65$$

Squaring the column totals, dividing each by the column sample size, and subtracting $c$ gives the treatment sum of squares:

$$\text{SSTR} = \sum \frac{T_j^2}{n_j} - c = \frac{(114.4)^2}{4} + \frac{(125.5)^2}{4} + \cdots + \frac{(111.2)^2}{4} - 10{,}520.28$$

$$= 1480.83$$

Finally, by Equation 14.9,

$$\text{SSE} = \text{SSTO} - \text{SSR} = 1616.65 - 1480.83 = 135.83$$

Notice that the values computed here for SSTR and SSE are identical to what we found in Example 14.2.

---

QUESTIONS

14.13   Compute SSTO and SSTR for the data in Question 14.7. Compare your answers to the values obtained from Equations 14.2 and 14.3.

14.14   Use Theorem 14.3 to find SSTO and SSTR for the data in Question 14.9. Verify that Equations 14.2 and 14.3 give the same answers.

14.15   Compute SSTO, SSTR, and SSE for the methylmercury data in Question 2.42. Use Theorem 14.3.

14.16   For the octopus data in Question 2.43,

$$\sum\sum x_i = 524 \quad \text{and} \quad \sum\sum x_i^2 = 24{,}242$$

Find SSTO, SSTR, and SSE.

14.17   The sum and the sum of the squares for the 12 biofeedback measurements in Question 3.81 are given by

$$\sum\sum x_i = 62 \quad \text{and} \quad \sum\sum x_i^2 = 400$$

Use Theorem 14.3 to compute SSTO, SSTR, and SSE.

### The ANOVA Table

The factors, supporting calculations, and final conclusion for a *k*-sample *F* test are typically presented in a format known as an **ANOVA table** (see Table 14.5). The first column lists each of the sources of variation, as singled

**TABLE 14.5**

| Source of Variation | df | SS | MS | F |
|---|---|---|---|---|
| Treatments | $k-1$ | SSTR | MSTR | MSTR/MSE |
| Error | $n-k$ | SSE | MSE | |
| Total | $n-1$ | SSTO | | |

out by Theorem 14.1: treatments, error, and total. The second column shows degrees of freedom. The first two entries in the df column ($k - 1$ and $n - k$) are the scaling factors referred to in the paragraph preceding Theorem 14.2; the degrees of freedom for "Total" is always $n - 1$.

The third column lists the sums of squares (SS) associated with the sources of variation. The MS column contains the two mean squares, and the last column is the observed $F$ ratio, which we enter in the treatment row. If $F$ is *not significant at the $\alpha = 0.05$ level*, we put next to it the letters **NS**. If $H_0: \mu_1 = \mu_2 = \cdots = \mu_k$ can be *rejected at the $\alpha = 0.05$ level*, we put a $*$ next to MSTR/MSE; if $H_0$ can be *rejected at the $\alpha = 0.01$ level*, we write $**$ instead of $*$.

---

**EXAMPLE 14.4**  Table 14.6 reproduces the data from Table 2.15. The treatment levels are three different years in which Mount Etna erupted: 1669, 1780, and 1865. The measured variable is the direction of the magnetic field "trapped" in a block of lava. (Three blocks of lava were tested from each of the three eruptions.)

**TABLE 14.6 Declination of Magnetic Field**

| | 1669 | 1780 | 1865 |
|---|---|---|---|
| | 57.8 | 57.9 | 52.7 |
| | 60.2 | 55.2 | 53.0 |
| | 60.3 | 54.8 | 49.4 |
| $T_j$ | 178.3 | 167.9 | 155.1 |

When molten lava hardens, it retains the prevailing direction of the earth's magnetic field. With that property in mind, researchers collected the data in Table 14.6 to see if the orientation of the earth's magnetic field has shifted over the past several hundred years. Let $\mu_1, \mu_2$, and $\mu_3$ represent the true magnetic field declinations in 1669, 1780, and 1865, respectively. The hypotheses to be tested are

$$H_0: \quad \mu_1 = \mu_2 = \mu_3$$

vs.

$$H_1: \quad \text{not all the } \mu_j\text{'s are equal}$$

*Solution*   We start by computing $\sum\sum x_i$ and $\sum\sum x_i^2$:

$$\sum\sum x_i = T = 57.8 + 60.2 + \cdots + 49.4 = 501.3$$

and

$$\sum\sum x_i^2 = (57.8)^2 + (60.2)^2 + \cdots + (49.4)^2 = 28,030.11$$

From Theorem 14.3,

$$c = \frac{T^2}{n} = \frac{(501.3)^2}{9} = 27,922.41$$

$$\text{SSTO} = \sum\sum x_i^2 - c = 28,030.11 - 27,922.41 = 107.7$$

$$\text{SSTR} = \sum \frac{T_j^2}{n_j} - c = \frac{(178.3)^2}{3} + \frac{(167.9)^2}{3} + \frac{(155.1)^2}{3} - 27,922.41$$

$$= 90.026$$

$$\text{SSE} = \text{SSTO} - \text{SSTR} = 107.7 - 90.026 = 17.674$$

Table 14.7 shows the resulting ANOVA computations. (Recall that $k = 3$ and $n = 9$.) The observed $F$ ratio in the last column has $2\ (=k-1)$ and $6\ (=n-k)$ degrees of freedom.

**TABLE 14.7**

| Source of Variation | df | SS | MS | F | |
|---|---|---|---|---|---|
| Time | 2 | 90.026 | 45.013 | 15.28 | ** |
| Error | 6 | 17.674 | 2.9457 | | |
| Total | 8 | 107.7 | | | |

How do we assign a "significance" to the observed $F$ ratio? By comparing MSTR/MSE with upper percentiles of the $F$ distribution with 2 and 6 degrees of freedom. Figure 14.4 shows

$$F_{0.05, 2, 6}\ (=5.14) \quad \text{and} \quad F_{0.01, 2, 6}\ (=10.9)$$

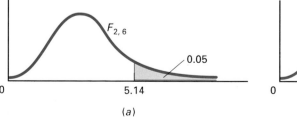

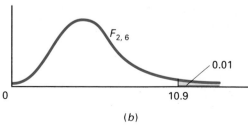

**FIG. 14.4**

Since 15.28 exceeds both of those values, we can reject $H_0$ at the $\alpha = 0.01$ level. It appears that the earth's magnetic field *did* shift in the less-than-200-year period from 1669 to 1865. Next to the observed $F$, therefore, we put **. (If the ratio MSTR/MSE had *not* exceeded $F_{0.05, 2, 6}$, we would have put *NS* to the right of the $F$ column.)

---

COMMENT: The procedure just described for assigning *'s to an ANOVA table is not strictly in keeping with the way in which we have done hypothesis tests up to this point. Technically, the level of significance of a test must be set by the experimenter *before looking at the data*. Modern computers, however, have encouraged the use of $P$ values rather than $\alpha$ values as a way of quantifying the significance of a result (see Question 10.31). The asterisk format is a widely used approximation to the notion of a $P$ value.

---

*EXAMPLE 14.5*  Can a teacher's opinion of a student's academic ability influence that child's performance on something as objective as an IQ test? There is evidence to suggest that the answer is "yes." The data described here are fictitious, but they are based on a study (139) set up to see whether teachers' opinions are, in fact, self-fulfilling prophecies.

Fifteen first-graders were all given a standard IQ test, but the children's teachers were told it was a special test for predicting whether a child would show sudden spurts of intellectual growth in the near future. The researchers divided the children into three groups of sizes six, five, and four *at random*, but they informed the teachers that, according to the test, the children in group I would not demonstrate any pronounced intellectual growth for the next year, those in group II would develop at a moderate rate, and those in group III could be expected to make exceptional progress.

A year later, the same 15 children were again given a standard IQ test. Table 14.8 shows the *differences* in the two scores (second test − first test) for each child. Can we conclude that the differences among the sample means (6.0 versus 8.4 versus 16.5) are statistically significant?

**TABLE 14.8  Changes in IQ (second test − first test)**

|  | Group I | Group II | Group III |
|---|---|---|---|
|  | 3 | 10 | 20 |
|  | 2 | 4 | 9 |
|  | 6 | 11 | 18 |
|  | 10 | 14 | 19 |
|  | 10 | 3 |  |
|  | 5 |  |  |
| $T_j$ | 36 | 42 | 66 |
| $n_j$ | 6 | 5 | 4 |
| $\bar{x}_j$ | 6.0 | 8.4 | 16.5 |

*Solution*   If a teacher's expectations are independent of a child's performance on the IQ test, the *true* average changes, $\mu_I$, $\mu_{II}$, and $\mu_{III}$, associated with groups I, II, and III should be the same. The hypotheses to be tested, then, are

$$H_0: \quad \mu_I = \mu_{II} = \mu_{III}$$

vs.

$$H_1: \quad \text{not all the } \mu_j\text{'s are equal}$$

The sum of the $x_i$'s and the sum of the $x_i^2$'s are

$$\sum\sum x_i = T = T_1 + T_2 + T_3 = 36 + 42 + 66 = 144$$

$$\sum\sum x_i^2 = 3^2 + 2^2 + \cdots + 19^2 = 1882$$

Therefore, by Theorem 14.3,

$$c = \frac{T^2}{n} = \frac{(144)^2}{15} = 1382.4$$

$$\text{SSTO} = \sum\sum x_i^2 - c = 1882 - 1382.4 = 499.6$$

$$\text{SSTR} = \sum \frac{T_j^2}{n_j} - c = \frac{(36)^2}{6} + \frac{(42)^2}{5} + \frac{(66)^2}{4} - 1382.4 = 275.4$$

$$\text{SSE} = \text{SSTO} - \text{SSTR} = 499.6 - 275.4 = 224.2$$

By Equation 14.6 the observed *F* ratio is

$$\frac{\text{MSTR}}{\text{MSE}} = \frac{275.4/2}{224.2/12} = 7.37$$

(with 2 and 12 degrees of freedom). Table 14.9 shows the details. The ** next to 7.37 means the observed *F* exceeds both $F_{0.05, 2, 12}$ and $F_{0.01, 2, 12}$ (see Figure 14.5).

**TABLE 14.9**

| Source of Variation | df | SS | MS | F | |
|---|---|---|---|---|---|
| Groups | 2 | 275.4 | 137.7 | 7.37 | ** |
| Error | 12 | 224.2 | 18.683 | | |
| Total | 14 | 499.6 | | | |

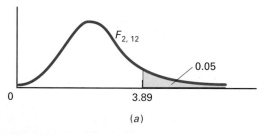

(a)

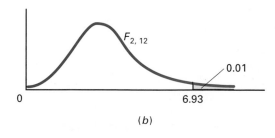

(b)

**FIG. 14.5**

Had $\alpha$ been set at 0.01, our conclusion would be to *reject* $H_0$. The *sample* means (6.0, 8.4, and 16.5) are too different to be consistent with the null hypothesis that the *true* means ($\mu_\text{I}$, $\mu_\text{II}$, and $\mu_\text{III}$) are all the same. What we suspected at the beginning, in other words, seems to have been borne out: for whatever reasons, students tend to perform on IQ tests at levels consistent with their teachers' expectations.

---

**QUESTIONS**

**14.18** (a) Do an $F$ test on the following set of $k$-sample data. Show the results in the ANOVA table format.

(b) Can $H_0$: $\mu_1 = \mu_2 = \mu_3$ be rejected at the 0.05 level of significances?

**TREATMENTS**

| 1 | 2 | 3 |
|---|---|---|
| 6 | 2 | 8 |
| 4 | 1 | 3 |
| 4 |   | 5 |

**14.19** The following gas mileages were recorded during a series of road tests on four new models of Japanese luxury sedans. Test the null hypothesis that all four models, *on the average*, give the same mileage. Let $\alpha = 0.05$. Use an ANOVA table to summarize the computations.

**MODEL**

| A | B | C | D |
|---|---|---|---|
| 22 | 28 | 31 | 23 |
| 26 | 24 | 32 | 24 |
|   | 29 | 28 |   |

**14.20** (a) Make an ANOVA table for the methylmercury data in Question 2.42. Use the values of SSTO, SSTR, and SSE computed in Question 14.15.

(b) Can the null hypothesis be rejected at the $\alpha = 0.05$ level of significance?

**14.21** Finish analyzing the octopus data in Question 2.43. Use the computations from Question 14.16.

**14.22** (a) Fill in the entries missing from the ANOVA table shown below:

| Source | df | SS | MS | F |
|---|---|---|---|---|
| Treatments |   | 421.5 |   |   |
| Error | 18 |   |   |   |
| Total | 21 | 1521.0 |   |   |

(b) What is the null hypothesis being tested? Can it be rejected at the $\alpha = 0.05$ level of significance?

14.23   Let $\mu_j, j = 1, 2, 3, 4$, denote the true average amounts of NPYR recovered from frying the four brands of bacon in Table 2.16. Test

$$H_0: \quad \mu_1 = \mu_2 = \mu_3 = \mu_4$$

vs.

$$H_1: \quad \text{not all the } \mu_j\text{'s are equal}$$

Indicate the significance of your conclusion by using the asterisk format.

14.24   (a) Complete the following ANOVA table:

| Source | df | SS | MS | F |
|---|---|---|---|---|
| Treatments | | | 69.7 | 6.58 |
| Error | | 117.0 | | |
| Total | | 256.4 | | |

(b) Should $H_0$ be rejected at the 0.01 level of significance?

14.25   Set up and carry out an appropriate hypothesis test for the data in Question 2.49. Define the parameters being compared. Let $\alpha = 0.05$. Note:

$$\sum\sum x_i = 3106 \quad \text{and} \quad \sum\sum x_i^2 = 460{,}228$$

14.26   Do an $F$ test on the catalytic converter data in Question 2.58. Use the ANOVA table format to summarize the computations. Note:

$$\sum\sum x_i = 278.0 \quad \text{and} \quad \sum\sum x_i^2 = 6467.10$$

14.27   Data from a study comparing the accuracy of four different air-to-surface missile launchers were described in Question 2.60. At the $\alpha = 0.01$ level, test

$$H_0: \quad \mu_1 = \mu_2 = \mu_3 = \mu_4$$

vs.

$$H_1: \quad \text{not all the } \mu_j\text{'s are equal}$$

Show the ANOVA table.

# 14.5

## MULTIPLE COMPARISONS: TUKEY'S TEST

The $F$ test described in Sections 14.3 and 14.4 is the usual beginning of the formal analysis for a set of $k$-sample data, but seldom the ending. Few researchers, after working months or even years to design and carry out an experiment, will be satisfied with a statistician whose conclusion is simply "Yes, the differences among the sample means are statistically significant," or "No, they are not." Often, what an experimenter most wants to know is *which* levels of the treatment are different? Are the responses to level 3, for example, significantly different from the responses to level 5?

Recall one more time the serum binding data in Example 14.2. Although five antibiotics were tested, it might be the case that a researcher is especially interested in one particular illness for which only the first two drugs are recommended. It would make sense under those circumstances to single out penicillin G and tetracycline and compare *their* binding percentages apart from all the others: that is, test the **subhypothesis**

$$H_0': \quad \mu_1 = \mu_2$$

Methods for following up the analysis of variance with inferences of a more restricted nature have greatly proliferated in the last 25 years. Entire books have been written on the subject. Some procedures require that we formulate the subhypotheses *before* we see the data. Others allow us to look at the $\bar{x}_j$'s first and then test whatever looks "suspicious." In this section we describe a well-known example of the latter type, a procedure credited to the American statistician John Tukey. It allows us to test all *pairwise* subhypotheses of the form

$$H_0': \quad \mu_i = \mu_j \qquad \text{versus} \qquad H_1': \quad \mu_i \neq \mu_j$$

First, a background result. Tukey's method derives from a sampling distribution we have not seen yet, the **studentized range**. In principle, a studentized range is the ratio of the range of a set of $k$ random variables (that is, the difference between the largest and the smallest) and their standard deviation. It has a two-parameter probability function that looks very much like an $F$ distribution.

> **subhypothesis**—A statement about a subset of the populations (and parameters) being studied.

> **studentized range**—A random variable derived from the range of a set of $k$ (normally distributed) observations or means.

---

**DEFINITION 14.1    Studentized Range**

Let $\bar{x}_1, \bar{x}_2, \ldots, \bar{x}_k$ be the column means in a set of $k$-sample data, where $n_1 = n_2 = \cdots = n_k$. Let $n = n_1 + n_2 + \cdots + n_k$. If $R$ denotes the difference between the largest sample mean and the smallest sample mean, then the ratio

$$Q_{k,\,n-k} = \frac{R}{\sqrt{\dfrac{\text{MSE}}{n/k}}}$$

has the *studentized range distribution with $k$ and $n - k$ degrees of freedom.*

---

Percentiles of studentized range distributions are given in Appendix A.5. For example, suppose a set of data consists of three treatment levels,

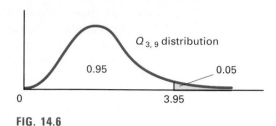

**FIG. 14.6**

each applied to four subjects. Then $k = 3$ and $n - k = 12 - 3 = 9$. According to Appendix A.5, the 95th percentile of the $Q_{3,9}$ distribution is 3.95 (see Figure 14.6). Written as an equation,

$$P(Q_{3,9} \leqslant 3.95) = 0.95$$

According to Definition 14.1, it also follows that

$$P\left(\frac{R}{\sqrt{MSE/4}} \leqslant 3.95\right) = 0.95$$

We will denote percentiles of the studentized range by the same format that we have used for $F$ distributions. The number 3.95, for example, would be referred to as $Q_{0.05,3,9}$.

To test subhypotheses of the form $H'_0: \mu_i = \mu_j$ versus $H'_1: \mu_i \neq \mu_j$, we construct a ***Tukey confidence interval***. The *center* of that interval is $\bar{x}_i - \bar{x}_j$; the *width* depends on $Q_{\alpha,k,n-k}$. If the Tukey interval does not contain zero, we reject $H'_0$.

**Tukey confidence interval**—An interval associated with a subhypothesis $H'_0: \mu_i = \mu_j$, $i < j$. If the interval fails to contain zero, we reject $H'_0$.

---

**Theorem 14.4   Constructing a Tukey Confidence Interval**

Let $\bar{x}_1, \bar{x}_2, \ldots, \bar{x}_k$ be the column means in a set of $k$-sample data where $n_1 = n_2 = \cdots = n_k = n/k$. To test

$$H'_0: \quad \mu_i = \mu_j$$
$$\text{vs.}$$
$$H'_1: \quad \mu_i \neq \mu_j$$

at the $\alpha$ level of significance, construct the interval

$$(\bar{x}_i - \bar{x}_j - D\sqrt{MSE}, \bar{x}_i - \bar{x}_j + D\sqrt{MSE})$$

where $D = Q_{\alpha,k,n-k}/\sqrt{n/k}$. If zero is not contained in the interval, reject $H'_0$.

---

**COMMENT:**   The Tukey intervals defined in Theorem 14.4 have the property that we can test *all* $_kC_2$ pairwise subhypotheses

$$H'_0: \mu_1 = \mu_2, \quad H'_0: \mu_1 = \mu_3, \ldots, \quad H'_0: \mu_{k-1} = \mu_k$$

and the probability is still $1 - \alpha$ that the entire set of $_kC_2$ conclusions will be simultaneously correct. Therefore we can use Tukey's procedure on sub-hypotheses formulated *after* looking at the data. No other procedure that we have seen thus far has had that property.

EXAMPLE 14.6   One method for measuring the purity of a fluid is to determine its concentration of particulate contaminants. Table 14.10 shows six such measurements made on intravenous fluids marketed by each of three pharmaceutical companies. The figures listed are the numbers of particles per liter that are greater than 5 microns in diameter. Use Tukey's method to compare (1) Cutter versus Abbott and (2) the best versus the worst. Let $\alpha = 0.05$ (168).

TABLE 14.10 **Number of Contaminant Particles**

|  | Cutter | Abbot | McGaw |
|---|---|---|---|
|  | 255 | 105 | 577 |
|  | 264 | 288 | 515 |
|  | 342 | 98 | 214 |
|  | 331 | 275 | 413 |
|  | 234 | 221 | 401 |
|  | 217 | 240 | 260 |
| $T_j$ | 1643 | 1227 | 2380 |
| $n_j$ | 6 | 6 | 6 |
| $\bar{x}_j$ | 273.83 | 204.50 | 396.67 |

Solution   First, we need to carry out the analysis of variance to find MSE. Listed below are the necessary computations in the order in which they would be performed. Table 14.11 on the next page summarizes the results.

1. $\sum\sum x_i = \sum T_j = 1643 + 1227 + 2380 = 5250$

2. $\sum\sum x_i^2 = (255)^2 + (264)^2 + \cdots + (260)^2 = 1{,}791{,}650$

3. $c = \dfrac{(5250)^2}{18} = 1{,}531{,}250$

4. $\text{SSTO} = \sum\sum x_i^2 - c = 1{,}791{,}650 - 1{,}531{,}250 = 260{,}400$

5. $\text{SSTR} = \sum \dfrac{T_j^2}{n_j} - c = \dfrac{(1643)^2}{6} + \dfrac{(1227)^2}{6} + \dfrac{(2380)^2}{6} - 1{,}531{,}250$

   $= 113{,}646.33$

6. $\text{SSE} = \text{SSTO} - \text{SSTR} = 260{,}400 - 113{,}646.33 = 146{,}753.67$

7. $\text{MSE} = \dfrac{\text{SSE}}{n - k} = \dfrac{146{,}753.67}{15} = 9783.58$

**TABLE 14.11**

| Source of Variation | df | SS | MS |
|---|---|---|---|
| Companies | 2 | 113,646.33 | |
| Error | 15 | 146,753.67 | 9783.58 |
| Total | 17 | 260,400 | |

Comparing Cutter versus Abbott means testing

$$H'_0: \quad \mu_1 = \mu_2$$

vs.

$$H'_1: \quad \mu_1 \neq \mu_2$$

From Appendix A.5, $Q_{\alpha, k, n-k} = Q_{0.05, 3, 15} = 3.67$. Since the sample means for Cutter and Abbott are 273.83 and 204.50, respectively, the corresponding Tukey interval from Theorem 14.4 is

$$\left( 273.83 - 204.50 - \frac{3.67}{\sqrt{6}} \sqrt{9783.58}, \ 273.83 - 204.50 + \frac{3.67}{\sqrt{6}} \sqrt{9783.58} \right)$$

$$= (69.33 - 148.20, \ 69.33 + 148.20)$$

$$= (-78.9, \ 217.5)$$

Should we reject $H'_0: \mu_1 = \mu_2$? No, because the Tukey interval contains zero as an interior point.

Comparing the best against the worst is done similarly. From Table 14.10 the fluids having the least contaminants and the most contaminants are Abbott and McGaw, respectively. Thus, testing the extremes becomes an inference about $\mu_2$ and $\mu_3$:

$$H'_0: \quad \mu_2 = \mu_3$$

vs.

$$H'_1: \quad \mu_2 \neq \mu_3$$

Values for MSE and $Q_{\alpha, k, n-k}$ remain the same, so the corresponding Tukey interval is simply repositioned around $\bar{x}_2 - \bar{x}_3$:

$$\left( 204.50 - 396.67 - \frac{3.67}{\sqrt{6}} \sqrt{9783.58}, \ 204.50 - 396.67 + \frac{3.67}{\sqrt{6}} \sqrt{9783.58} \right)$$

$$= (-192.17 - 148.20, \ -192.17 + 148.20)$$

$$= (-340.4, \ -44.0)$$

Here, the interval does *not* contain zero, so our conclusion is to reject $H'_0$. Fluids marketed by McGaw appear to have significantly higher numbers of particulate contaminants than do fluids sold by Abbott.

**QUESTIONS**

**14.28** Suppose three normally distributed observations are taken on each of five treatments. For the $R$, $k$, and $n$ given in Definition 14.1, find the 95th percentile of

$$\frac{R}{\sqrt{\dfrac{MSE}{n/k}}}$$

**14.29** Let $Q_{k,n-k}$ be a studentized range random variable with $k$ and $n-k$ degrees of freedom. Use Appendix A.5 to find the unknowns in the following equations:

(a) $P(Q_{4,6} \geqslant ?) = 0.05$      (b) $P(Q_{5,5} \leqslant ?) = 0.99$

(c) $P(Q_{4,?} \geqslant 5.77) = 0.01$      (d) $P(Q_{?,4} \leqslant 7.60) = 0.95$

(e) $P(Q_{11,13} \geqslant 5.43) = ?$      (f) $P(Q_{15,4} \leqslant 8.66) = ?$

**14.30** (a) Construct the third of the possible 95% Tukey confidence intervals for the particulate contaminant data in Table 14.10.

(b) What subhypothesis does it test? Is $H_0'$ accepted or rejected?

**14.31** (a) Construct a complete set of 95% Tukey confidence intervals for the magnetic field data in Table 2.15. As a value for the mean square for error, use the figure given in Table 14.7.

(b) Which means are significantly different from which other means?

**14.32** (a) Using 99% Tukey confidence intervals, do a brand-by-brand comparison of the NPYR data appearing in Table 2.16.

(b) Which brands are *not* significantly different from each other?

**14.33** Test each of the three pairwise subhypotheses for the methylmercury data in Question 2.42. Let 0.01 be the level of significance.

**14.34** (a) Compute the average mileages recorded for each of the three catalytic converters in Question 2.58.

(b) At the $\alpha = 0.01$ level, test whether the lowest average mileage and the highest average mileage are significantly different.

**14.35** For the data in Question 2.56, where $k = 2$, will a 95% confidence interval for $\mu_A - \mu_B$ based on Question 10.10 be the same as a 95% Tukey interval based on Theorem 14.4? Construct both intervals and compare the two lengths. (Both intervals will obviously be *centered* at the same point, $\bar{x}_A - \bar{x}_B$.)

# 14.6

## COMPUTER NOTES

Computer calculations for the ANOVA tables of Section 14.4 require little more than reading in the data and asking for the output.

Figure 14.7 is the Minitab version of Example 14.4. Note that the table corresponds exactly to Table 14.6. The Minitab program goes on to calculate confidence intervals for individual means. The standard deviation was estimated by pooling all of the data, much as was done for two-sample data. The Minitab command for this analysis is AOVONEWAY, because the technique of this chapter is often called a *one-way analysis of variance*.

```
READ C1 C2 C3
57.8  57.9  52.7
60.2  55.2  53.0
60.3  54.8  49.4

AOVONEWAY C1-C3

ANALYSIS OF VARIANCE

SOURCE    DF        SS        MS        F
FACTOR     2     90.03     45.01    15.28
ERROR      6     17.67      2.95
TOTAL      8    107.70

                                    INDIVIDUAL 95 PCT CI'S FOR MEAN
                                    BASED ON POOLED STDEV
LEVEL     N      MEAN     STDEV    -------+---------+---------+---------
1669      3     59.43      1.42                            (-----*-----)
1780      3     55.97      1.69                  (-----*-----)
1865      3     51.70      1.00     (-----*-----)
                                    -------+---------+---------+---------
POOL STDEV =     1.72               52.0      56.0      60.0
```

**FIG. 14.7**

SAS has an option in its ANOVA procedure to perform Tukey's test. Figure 14.8 demonstrates the usual SAS ANOVA printout for Example 14.6.

Figure 14.9 contains the output from exercising the option for Tukey's test. The radius of the Tukey interval, $D\sqrt{\text{MSE}}$ in Section 14.5, is

MINIMUM SIGNIFICANT DIFFERENCE = 148.33

The SAS routine does not calculate the intervals; instead, it summarizes the computations in Theorem 14.4 by using a convenient tabular format.

The SAS procedure places the variables into (possibly overlapping) groups, each group being coded by a letter. If two variables are in the same group, then we accept the null hypothesis that their means are equal. Since CUTTER and MCGAW are in the A group, we conclude that their means are equal. A similar conclusion holds for CUTTER and ABBOTT, because they are both in the B group. We reject $H'_0$ that the means for MCGAW and ABBOTT are equal, since MCGAW is in the A group but not the B, and ABBOTT is in the B group but not the A.

```
DATA;
  INPUT MAKER $ CONTAMS @ @ ;
  CARDS;
    CUTTER 255
    CUTTER 264

    (balance of data from Table 14.10 here)

PROC ANOVA;
  CLASS MAKER;
  MODEL CONTAMS = MAKER;
  MEANS MAKER / TUKEY;
```

ANALYSIS OF VARIANCE PROCEDURE

CLASS LEVEL INFORMATION

| CLASS | LEVELS | VALUES |
|-------|--------|--------|
| MAKER | 3 | ABBOTT CUTTER MCGAW |

NUMBER OF OBSERVATIONS IN DATA SET = 18

ANALYSIS OF VARIANCE PROCEDURE

DEPENDENT VARIABLE: CONTAMS

| SOURCE | DF | SUM OF SQUARES | MEAN SQUARE | F VALUE | PR > F | R-SQUARE | C.V. |
|--------|----|----|----|----|----|----|----|
| MODEL | 2 | 113646.33333333 | 56823.16666667 | 5.81 | 0.0136 | 0.436430 | 33.9127 |
| ERROR | 15 | 146753.66666667 | 9783.57777778 | | ROOT MSE | | CONTAMS MEAN |
| CORRECTED TOTAL | 17 | 260400.00000000 | | | 98.91196984 | | 291.66666667 |

| SOURCE | DF | ANOVA SS | F VALUE | PR > F |
|--------|----|----|----|----|
| MAKER | 2 | 113646.33333333 | 5.81 | 0.0136 |

**FIG. 14.8**

ANALYSIS OF VARIANCE PROCEDURE

TUKEY'S STUDENTIZED RANGE (HSD) TEST FOR VARIABLE: CONTAMS
NOTE: THIS TEST CONTROLS THE TYPE I EXPERIMENTWISE ERROR RATE
      BUT GENERALLY HAS A HIGHER TYPE II ERROR RATE THAN REGWQ

        ALPHA=0.05  DF=15  MSE=9783.58
        CRITICAL VALUE OF STUDENTIZED RANGE=3.673
        MINIMUM SIGNIFICANT DIFFERENCE=148.33

MEANS WITH THE SAME LETTER ARE NOT SIGNIFICANTLY DIFFERENT.

| TUKEY | GROUPING | | MEAN | N | MAKER |
|-------|----------|---|------|---|-------|
| | | A | 396.67 | 6 | MCGAW |
| | | A | | | |
| B | | A | 273.83 | 6 | CUTTER |
| B | | | | | |
| B | | | 204.50 | 6 | ABBOTT |

**FIG. 14.9**

# 14.7

## SUMMARY

Chapter 14 has two objectives. First and foremost is the analysis of $k$-sample data. Given a set of $k$ independent random samples (of sizes $n_1, n_2, \ldots, n_k$), we wanted a procedure that would tell us whether the true means represented by those samples ($\mu_1, \mu_2, \ldots, \mu_k$) are all the same. The hypotheses to be tested, in other words, are

$$H_0: \quad \mu_1 = \mu_2 = \cdots = \mu_k$$

vs.

$$H_1: \quad \text{not all the } \mu_j\text{'s are equal}$$

In spirit, choosing between $H_0$ and $H_1$ is equivalent to what motivated the *two-sample* problem in Chapter 10. *How* we test $H_0: \mu_1 = \mu_2 = \cdots = \mu_k$, though, is considerably different than how we test $H_0: \mu_X = \mu_Y$. Drawing that distinction is the second objective.

What we are seeing here for the first time is an inference procedure based on the *partitioning* of a sum of squared deviations. The specific decomposition that holds for $k$-sample data is

$$\text{SSTO} = \text{SSTR} + \text{SSE} \tag{14.10}$$

where
$$\text{SSTO} = \sum\sum (x_i - \bar{x})^2 = \text{Total sum of squares,}$$

$$\text{SSTR} = \sum n_j(\bar{x}_j - \bar{x})^2 = \text{Treatment sum of squares,}$$

$$\text{SSE} = \sum\sum (x_i - \bar{x}_j)^2 = \text{Error sum of squares.}$$

As a general technique in applied statistics, learning to understand and interpret expressions like Equation 14.10 is important. We will encounter a similar approach in Chapter 15 when we analyze randomized block data and again in Chapter 17 when the topic is multiple regression.

If $H_0: \mu_1 = \mu_2 = \cdots = \mu_k$ is true, it can be shown that, *on the average*, $\text{SSTR}/(k-1) = \text{SSE}/(n-k)$. On the other hand, if $H_0: \mu_1 = \mu_2 = \cdots = \mu_k$ is *not* true, then, on the average, $\text{SSTR}/(k-1) > \text{SSE}/(n-k)$. Put another way, if the null hypothesis is true, the ratio $\dfrac{\text{SSTR}/(k-1)}{\text{SSE}/(n-k)}$ should be fairly close to 1; if $H_0$ is *not* true, $\dfrac{\text{SSTR}/(k-1)}{\text{SSE}/(n-k)}$ will tend to be larger than 1. The magnitude of the ratio, therefore, can be used as a criterion for choosing between the two hypotheses.

Test statistics based on ratios of sums of squares always have $F$ distributions, only the numbers of degrees of freedom change. For $k$-sample data, the observed $F$ ratio MSTR/MSE (where $\text{MSTR} = \text{SSTR}/(k-1)$ and $\text{MSE} = \text{SSE}/(n-k)$) has an $F$ distribution with $k-1$ and $n-k$ degrees of freedom (when $H_0$ is true). Furthermore, since MSTR tends to be larger than MSE when $H_1$ is true, we should reject $H_0$ if the observed $F$ is too far to the *right*—specifically, $H_0$ will be rejected at the $\alpha$ level of significance if $\dfrac{\text{SSTR}/(k-1)}{\text{SSE}/(n-k)}$ exceeds $F_{\alpha,\,k-1,\,n-k}$.

Computations for an $F$ test are outlined in Table 14.12. ANOVA tables provide a format for summarizing the results

**TABLE 14.12.  Testing $H_0\colon \mu_1 = \mu_2 = \cdots = \mu_k$**

1. Compute $\sum\sum x_i$ and $\sum\sum x_i^2$.

2. Compute $c = T^2/n$, where $T = \sum\sum x_i$.

3. Compute $\text{SSTO} = \sum\sum x_i^2 - c$.

4. Compute $\text{SSTR} = \sum \dfrac{T_j^2}{n_j} - c$.

5. Compute $\text{SSE} = \text{SSTO} - \text{SSTR}$.

6. Fill in the first four columns of the ANOVA table as they appear in Table 14.5.

7. Compute the test statistic $F = \text{MSTR}/\text{MSE}$.

8. Compare the value of MSTR/MSE with the 95th and 99th percentiles of the $F$ distribution with $k-1$ and $n-k$ degrees of freedom. Next to $F$ write

    (a)   NS if $F < F_{0.05,\,k-1,\,n-k}$
    (b)   * if $F_{0.05,\,k-1,\,n-k} \leqslant F < F_{0.01,\,k-1,\,n-k}$
    (c)   ** if $F \geqslant F_{0.01,\,k-1,\,n-k}$.

Analyzing a set of $k$-sample data *begins* with the general $F$ test of $H_0\colon \mu_1 = \mu_2 = \cdots = \mu_k$ as described in Table 14.12, but it typically ends only after more specific questions have been formulated and addressed. There are many procedures for setting up inferences restricted to subsets of $\mu_j$'s—that is, to *subhypotheses*. Among the most frequently used is *Tukey's test*, the subject of Section 14.5. Based on the *studentized range*, $Q_{k,\,n-k}$, Tukey's test allows experimenters to make all possible *pairwise* comparisons of the form $H_0'\colon \mu_i = \mu_j$ versus $H_1'\colon \mu_i \neq \mu_j$. The Tukey confidence interval associated with $H_0'\colon \mu_i = \mu_j$ is

$$(\bar{x}_i - \bar{x}_j - D\sqrt{\text{MSE}},\ \bar{x}_i - \bar{x}_j + D\sqrt{\text{MSE}})$$

where $D = Q_{\alpha,\,k,\,n-k}/\sqrt{n/k}$. If the interval fails to contain zero, we reject $H_0'\colon \mu_i = \mu_j$.

---

**REVIEW QUESTIONS**

**14.36**  Nine political writers were asked to assess the United States' culpability in murders committed by revolutionary groups financed by the CIA. Scores were to be assigned using a scale of 0 to 100. Three of the writers were native Americans living in the U.S., three were native Americans living abroad, and three were foreign nationals. Formulate and carry out an appropriate $F$ test. Let $\alpha = 0.05$.

| Americans in U.S. | Americans Abroad | Foreign Nationals |
|:---:|:---:|:---:|
| 45 | 65 | 75 |
| 45 | 50 | 90 |
| 40 | 55 | 85 |

**14.37**   Construct a complete set of 95% Tukey confidence intervals for the data in Question 14.36. Which subhypotheses are rejected?

**14.38**   The same strain of bacteria was grown in each of nine Petri dishes. After all the cultures were well established, the nine dishes were randomly divided into three groups of three. Each group was inoculated with a different anti-bacterial agent. Two days later, diameters of zones of inhibition (areas showing no indication of bacterial growth) were measured (in cm). Test $H_0: \mu_1 = \mu_2 = \mu_3$. Use the ANOVA table format to display the results.

| ANTIBACTERIAL AGENT | | |
|:---:|:---:|:---:|
| (1)<br>M21z | (2)<br>ATC3 | (3)<br>B169 |
| 3.8 | 2.9 | 5.0 |
| 3.5 | 3.1 | 4.8 |
| 4.3 | 2.8 | 4.6 |

**14.39**   Construct 95% Tukey confidence intervals for the three pairwise comparisons possible with the data in Question 14.38. Interpret the results.

**14.40**   Fill in the entries missing from the ANOVA table shown below. What notation should be entered to the right of the *F* column?

| Source | df | SS | MS | F |
|:---|:---:|:---:|:---:|:---:|
| Treatments | 4 | | | 6.39 |
| Error | | | 10.57 | |
| Total | | 375.95 | | |

**14.41**   How many 95% Tukey confidence intervals could be constructed for the data represented by the ANOVA table in Question 14.40?

**14.42**   Responding to complaints that the response time of police to emergency calls is too long, the mayor of a city hires a consulting firm to recommend changes in the way patrol cars are dispatched. Three methods (A, B, and C) are proposed. They differ in the proportion of cars allocated to high crime rate areas. Each is implemented for several weeks. Listed below are response times (in minutes) to the first five emergency calls received. At the $\alpha = 0.05$ level, test whether differences among the three average response times are statistically significant.

| A | B | C |
|---|---|---|
| 15 | 6 | 8 |
| 14 | 12 | 10 |
| 19 | 8 | 12 |
| 18 | 9 | 11 |
| 12 | 10 | 10 |

**14.43**  Finish the analysis of the data in Question 14.42 by constructing 95% Tukey confidence intervals for the three pairwise differences, $\mu_i - \mu_j$, $i < j$.

**14.44**  Which of the following statements are correct:

(a) $\sum\sum x_i = \sum_{j=1}^{k} n_j \bar{x}_j$  (b) $\sum\sum x_i = n\bar{x}$

(c) $\sum_{j=1}^{k} T_j = \sum_{j=1}^{k} n_j \bar{x}_j$  (d) $\dfrac{T}{n} = \dfrac{1}{k}\sum_{j=1}^{k} \bar{x}_j$

**14.45**  Let $R$ be the range of a sample of size six from a normal distribution with $\sigma = 10$. Is $P(R > 39)$ greater than 0.05 or less than 0.05?

**14.46**  The owners of a music store are building a customer data file. They hope that knowing their clientele better will enable them to promote certain items more effectively. Listed below are the ages of seven people who bought the latest Bob Dylan album and five who bought the latest Bruce Springsteen album. Are the two average ages significantly different? Answer the question using the analysis of variance. Let $\alpha = 0.05$.

| Bob Dylan Album | Bruce Springsteen Album |
|---|---|
| 43 | 16 |
| 22 | 24 |
| 19 | 18 |
| 22 | 14 |
| 18 | 16 |
| 15 | |
| 21 | |

**14.47**  Analyze the data in Question 14.46 by using a two-sample $t$ test. Let $\alpha = 0.05$ and make $H_1$ two-sided. Show how the $F$ test and $t$ test are equivalent. (Hint: See Question 14.11.)

**14.48**  Weight losses were recorded for 15 men enrolled in a fitness program. Each participated in one of four exercise classes for a period of 3 months. Some classes were more strenuous than others. Let $\mu_j$ denote the true weight loss typical of the $j$th class. Test $H_0: \mu_1 = \mu_2 = \mu_3 = \mu_4$ at the $\alpha = 0.05$ level of significance. Show the ANOVA table.

| WEIGHT LOSS (lb) | | | |
|---|---|---|---|
| Class 1 | Class 2 | Class 3 | Class 4 |
| 25 | 22 | 10 | 20 |
| 17 | 19 | 14 | 17 |
| 19 | 18 | 13 | 19 |
| 16 | 21 | 13 | |

**14.49**  Justify each term on the right-hand-side of the following derivation. Summarize in words what the derivation has proved.

$$\sum_{i=1}^{n_j} (x_i - \bar{x}_j) = T_j - n_j \bar{x}_j = T_j - T_j = 0$$

**14.50**  Do the following data appear to violate any of the assumptions underlying the analysis of variance? Explain.

| TREATMENT | | | |
|---|---|---|---|
| A | B | C | D |
| 16 | 4 | 26 | 8 |
| 17 | 12 | 22 | 9 |
| 16 | 2 | 23 | 11 |
| 17 | 26 | 24 | 8 |

**14.51**  The notion of a model equation was introduced in Section 14.3. Given that

$$x_i = \mu_j + \varepsilon_i, \qquad i = 1, 2, \ldots, n_j \quad \text{and} \quad j = 1, 2, \ldots, k,$$

what is a reasonable estimate for $\mu_j$? for $\varepsilon_i$?

**14.52**  Three pottery shards from four widely scattered and now extinct western Indian tribes have been collected by a museum. Each of twelve archaeologists is sent one of the shards and asked to estimate its age. Based on the results shown below, is it conceivable that the four tribes were contemporaries of one another? Let $\alpha = 0.01$.

| ESTIMATED AGES OF SHARDS (years) | | | |
|---|---|---|---|
| Lakeside | Deep Gorge | Willow Ridge | Azalea Hill |
| 1200 | 850 | 1800 | 950 |
| 800 | 900 | 1450 | 1200 |
| 950 | 1100 | 1150 | 1150 |

Note: $\sum\sum x_i = 13{,}500$ and $\sum\sum x_i^2 = 16{,}055{,}000$.

**14.53**   For the data in Question 14.52, use a 99% Tukey confidence interval to compare the site having the smallest sample mean with the site having the largest sample mean.

**14.54**   Is the $P$ value for the $F$ test summarized in the following ANOVA table less than 0.05 or greater than 0.05?

| Source | df | SS | MS | F |
|---|---|---|---|---|
| Treatments | | | 30.45 | 4.98 |
| Error | | 36.69 | | |
| Total | | 97.59 | | |

**14.55**   Which of the following statements are true:

(a) $\sum\sum x_i^2 = T^2$       (b) $\sum\limits_{j=1}^{k} (n_j \bar{x}_j)^2 = \sum\sum x_i^2$

(c) $\dfrac{T^2}{n} = n\bar{x}^2$       (d) $\dfrac{1}{\sigma^2} \sum\limits_{j=1}^{k} n_j(\bar{x}_j - \bar{x})^2 = \sum\limits_{j=1}^{k} \left(\dfrac{\bar{x}_j - \bar{x}}{\sigma/\sqrt{n_j}}\right)^2$

**14.56**   Let $R_1$ denote the range of a random sample of size $k$ from a normal distribution with $\mu = 5$ and $\sigma = 10$. Let $R_2$ denote the range of a random sample of size $k$ from a normal distribution with $\mu = 50$ and $\sigma = 10$. Let $c > 0$. Which is larger, $P(R_1/\sigma > c)$ or $P(R_2/\sigma > c)$?

**14.57**   Let $\sum\limits_{j} x_i^2$ denote the sum of the squares of the $n_j$ observations in column $j$. An alternative formula for SSE, one that does not require that we know SSTO and SSTR, is

$$\text{SSE} = \left(\sum_{1} x_i^2 - \frac{T_1^2}{n_1}\right) + \left(\sum_{2} x_i^2 - \frac{T_2^2}{n_2}\right) + \cdots + \left(\sum_{k} x_i^2 - \frac{T_k^2}{n_k}\right)$$

Compute SSE using the above formula on the following data:

| 1 | 2 | 3 |
|---|---|---|
| 14 | 18 | 12 |
| 10 | 11 | 8 |
| 12 | 31 | 17 |

Compare your answer with the value of SSE computed from the formulas in Theorem 14.3.

**14.58**   (a) If 95% Tukey confidence intervals tell us to reject

$$H_0': \quad \mu_1 = \mu_2 \qquad \text{and} \qquad H_0': \quad \mu_1 = \mu_3$$

will we necessarily reject $H_0': \mu_2 = \mu_3$ (at $\alpha = 0.05$)?

(b) If 95% Tukey confidence intervals tell us to accept

$$H'_0: \quad \mu_1 = \mu_3 \quad \text{and} \quad H'_0: \quad \mu_2 = \mu_3$$

will we necessarily accept $H'_0: \mu_1 = \mu_2$ (at $\alpha = 0.05$)?

14.59 According to Theorem 14.4, the width of a Tukey confidence interval is

$$2\sqrt{\text{MSE}}\, Q_{\alpha, k, n-k} \Big/ \sqrt{\frac{n}{k}}.$$

If $k$ increases but $n/k$ and MSE stay the same, will the Tukey intervals get shorter or longer? Justify your answer intuitively.

14.60 Four companies have submitted bids to construct a noise-containment wall that will border an interstate as it passes through a residential area. Each has built a 50-foot-long segment for test purposes. Listed below are noise reductions (in decibels) recorded at various times during rush-hour traffic.

| NOISE REDUCTION (db) | | | |
|---|---|---|---|
| Company A | Company B | Company C | Company D |
| 14 | 13 | 20 | 30 |
| 16 | 12 | 18 | 26 |
| 19 | 15 | 26 | 22 |
| 17 | 16 | 22 | 25 |

(a) Do an appropriate $F$ test. Let $\alpha = 0.05$.

(b) Is Company D's wall significantly better than its closest competitor's? Answer the question by constructing a 95% Tukey confidence interval.

# 15 The Randomized Block Design

*Chance is always powerful.*
*Let your hook be always cast;*
*in the pool where you least expect it,*
*there will be a fish.*

*Ovid*

### INTRODUCTION

Somewhere in everyone's book of old jokes is the one about crazy Charlie, who is spending his first day in a mental institution. Dinner time comes, and he finds himself seated in a large cafeteria, surrounded by all his new companions. Before he can say so much as "Pass the chipped beef," someone yells "37!" and everyone breaks into hysterics. Minutes later, after a bemused calm has been restored, another number is shouted out—"52!" This time the pandemonium is even greater.

Bewildered, Charlie asks the elderly gentleman sitting to his left for an explanation. "It's really quite simple," he says. "We've all been here so long that we know each other's jokes by heart. To save time, we decided to number our favorites from 1 to 100. So when someone hollers 17, or 22, or 56, we know exactly what joke is being told, and we all laugh."

An interesting approach, Charlie thought, and maybe a good way to make some new friends. I'll give it a try. A little later, after sensing a lull in the conversation, he makes his move: "63!" Nothing happens—not even a snicker.

Charlie was crestfallen. How could this happen? It all seemed so easy when the other two did it. The old man sitting by his side tapped him gently on the arm. "Don't worry, son, it's not your fault. Some folks can tell jokes, and some folks can't."

There is a lesson to be learned from Charlie's joke-telling efforts that applies to statisticians as well: *what* we measure is sometimes not as important as *how* we measure. If that theme rings familiar, it should. Recall the distinction drawn earlier between two-sample data and paired data. Both are collected to compare the mean of one treatment with the mean of another. But *how* the observations are recorded is different, and that difference (independent samples versus dependent samples) can have a profound effect on the likelihood of $H_0$ being rejected.

In this chapter we encounter a similar situation, but the setting is slightly more complex. Our objective is the same as it was in Chapter 14—to compare the means of $k$ distributions, where $k > 2$. But here the samples are *dependent*.

Figure 15.1 shows the basic structure of *randomized block data*, as it was first presented in Section 2.8. The blocks play exactly the same role as the pairs do in paired data: they serve as a relatively homogeneous set of conditions under which the entire set of treatments can be compared. Mathematically, the observations recorded for, say, block 1 are dependent because their values all reflect whatever conditions characterize that particular block.

Many skills are involved in being a good researcher, not the least of which is knowing how to keep experimental error as small as possible. What we cover in this chapter addresses precisely that problem: blocking is one of the most effective yet cheapest ways to reduce $\sigma^2$. With that in mind,

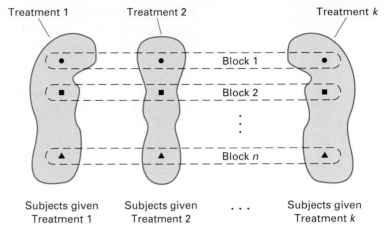

**FIG. 15.1**

pay particular attention to the case studies in the next several sections. Look at how the blocks are being used. Why were they defined the way they were? What was the experimenter trying to accomplish? Being able to answer those questions is the key to understanding the significance of the randomized block model.

## 15.2

### NOTATION

Randomized block data typically have the structure pictured in Table 15.1. All $k$ treatments are applied, once, within each of $b$ blocks. The notation for sample totals, sample means, and population means is the same as for $k$-sample data. Symbols for the overall sample size ($n$), overall sample total

**TABLE 15.1**

|  | TREATMENTS | | | | Block Totals |
|---|---|---|---|---|---|
|  | 1 | 2 | $\cdots$ | $k$ |  |
| **BLOCKS** $b$ | $-$ | $-$ |  | $-$ | $B_1$ |
|  | $-$ | $-$ |  | $-$ | $B_2$ |
|  | $-$ | $-$ |  | $-$ | $B_3$ |
|  | $\vdots$ |  | $\vdots$ |  | $\vdots$ |
|  | $-$ | $-$ |  | $-$ | $B_b$ |
| Sample sizes | $b$ | $b$ | $\cdots$ | $b$ | $n$ |
| Sample totals | $T_1$ | $T_2$ | $\cdots$ | $T_k$ | $T$ |
| Sample means | $\bar{x}_1$ | $\bar{x}_2$ | $\cdots$ | $\bar{x}_k$ | $\bar{x}$ |
| Population means | $\mu_1$ | $\mu_2$ | $\cdots$ | $\mu_k$ |  |

OVERALL TOTAL

$(T)$, and overall sample mean $(\bar{x})$ also remain the same. The only new notation is that $B_1, B_2, \ldots, B_b$ will represent the response totals recorded for the measurements in the first, second, . . . , $b$th blocks, respectively. Each $B_i$ is the sum of $k$ measurements (in the same way that each $T_j$ is the sum of $b$ measurements).

Table 15.2 lists some of the mathematical relationships associated with the terms in Table 15.1. Note, especially, the difference between $\bar{x}_i$ and $\bar{x}_j$.

**TABLE 15.2**

1. $n = bk$

2. $\bar{x}_j = \dfrac{T_j}{b}, \qquad j = 1, 2, \ldots, k$

3. $\bar{x}_i = \dfrac{B_i}{k}, \qquad i = 1, 2, \ldots, b$

4. $\bar{x} = \dfrac{T}{n}$

5. $T_1 + T_2 + \cdots + T_k = B_1 + B_2 + \cdots + B_b = T$

QUESTIONS

**15.1** Compute the entire set of $T_j$'s, $\bar{x}_j$'s, and $B_i$'s for the following set of randomized block data and calculate the overall mean $\bar{x}$.

|  | | TREATMENTS | |
|---|---|---|---|
|  | **1** | **2** | **3** |
| **1** | 10 | 8 | 8 |
| **2** | 6 | 7 | 10 |
| **BLOCKS  3** | 3 | 4 | 2 |
| **4** | 7 | 9 | 8 |

**15.2** For the Pressley strength index data in Question 2.46, compute $B_1$, $B_2$, $T_1$, and $T_5$.

**15.3** For randomized block data the overall mean is the average of the $k$ sample means:

$$\bar{x} = \frac{1}{k} \sum \bar{x}_j$$

Does a similar relationship hold, in general, for $k$-sample data? Explain.

**15.4** Is it possible for the $B_i$'s to be numerically very similar at the same time that the $T_j$'s are widely different? Is the reverse possible? Explain.

**15.5** For the affirmative action data in Question 2.52, compute the entire set of $T_j$'s, $\bar{x}_j$'s, and $B_i$'s as defined in Table 15.1.

15.6   Fill in the missing entries in the following table:

|  |  | TREATMENTS | | | | |
|---|---|---|---|---|---|---|
|  |  | 1 | 2 | 3 | 4 | $B_i$ |
|  | 1 |  |  | 2 | 4 | 20 |
|  | 2 | 2 |  | 2 | 0 |  |
| BLOCKS | 3 | 1 | 4 |  | 2 | 8 |
|  | $T_j$ |  | 15 | 5 |  |  |
|  | $\bar{x}_j$ | 3 |  |  |  |  |

# 15.3

## THE F TEST

In Chapter 14 the first step we took in motivating the analysis of variance was to describe a typical observation in terms of its model equation (recall the expression $X_i = \mu_j + \varepsilon_i$). We can also do that here. Figure 15.2 shows the "pedigree" of the response $X$ to treatment $j$ when the latter is applied in block $i$.

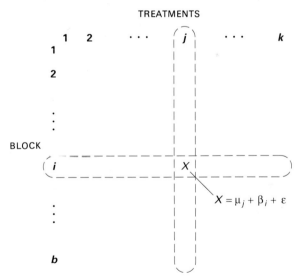

**FIG. 15.2**

The numerical value of the random variable $X$ can be thought of as the sum of three terms: $X = \mu_j + \beta_i + \varepsilon$. The first term, $\mu_j$, represents the true average effect of treatment $j$; $\beta_i$ denotes the *bias* introduced by block $i$. If the conditions characterizing block $i$ tend to *inflate* the values of the $X$'s recorded in that block, then $\beta_i > 0$. If block $i$ tends to produce *deflated* values, $\beta_i < 0$. If block $i$ is "average" relative to the rest of the blocks, $\beta_i = 0$.

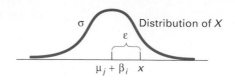

**FIG. 15.3**

The sum $\mu_j + \beta_i$ represents the *expected value* of X. We will assume that the *distribution* of X is normal with standard deviation $\sigma$. The $x$ that we measure will be some number located a distance $\varepsilon$ from the mean (see Figure 15.3). Therefore we can say that every observation in a set of randomized block data can be expressed in the form

$$X = \mu_j + \beta_i + \varepsilon \qquad (15.1)$$

where $\varepsilon$ is assumed to be normally distributed with mean zero and standard deviation $\sigma$.

The mathematical justification for any analysis of variance begins with a partitioning of the total sum of squares into a *sum* of sum of squares, where each individual sum of squares reflects on a particular term in the model equation. With $k$-sample data, for instance, where the model equation is $X_i = \mu_j + \varepsilon_i$, we showed in Section 14.3 that SSTO = SSTR + SSE. For randomized block data, an additional term ($\beta_i$) appears in the model equation, and that term gives rise to an additional sum of squares in the partitioning of SSTO. The details of the partitioning will have to be omitted, but Theorem 15.1 states the general result and gives computing formulas for each sum of squares.

---

**THEOREM 15.1    Computing Formulas for Sums of Squares**

For any set of randomized block data, the total sum of squares (SSTO) will always equal the sum of the treatment sum of squares (SSTR), the block sum of squares (SSB), and the error sum of squares (SSE):

$$SSTO = SSTR + SSB + SSE$$

Let $c = \dfrac{\left(\sum\sum x\right)^2}{n} = \dfrac{T^2}{n}$. Then

1. $SSTO = \sum\sum x^2 - c$

2. $SSTR = \sum \dfrac{T_j^2}{b} - c$

3. $SSB = \sum \dfrac{B_i^2}{k} - c$

4. $SSE = SSTO - SSTR - SSB$

---

**block sum of squares (SSB)**—The sum of the squared deviations of the *b* block means from the overall mean. The block sum of squares measures the extent to which the $\beta_i$'s are not all equal.

COMMENT: The *"block" sum of squares*, *SSB*, is defined to be the sum of the squared deviations of the block means from the overall mean. That is, if $\bar{x}_i$ denotes the average response of the *k* treatments in block *i*, $i = 1, 2, \ldots, b$, then

$$\text{SSB} = \sum\sum (\bar{x}_i - \bar{x})^2 = \sum k(\bar{x}_i - \bar{x})^2$$

Notice that SSB measures the "effect" of the blocks in the same sense that SSTR measures the effect of the treatments. If all the blocks were the same, all the $\beta_i$'s would be zero, all the $\bar{x}_i$'s would be similar, and SSB would be small. Conversely, if the conditions from block to block are highly variable, the $\bar{x}_i$'s will deviate markedly from $\bar{x}$, and $\sum k(\bar{x}_i - \bar{x})^2$ will be numerically large.

We use the sums of squares defined in Theorem 15.1 in the same way as we did for the sums of squares in Theorem 14.1 We divide the treatment *mean* square by the error *mean* square and use that quotient as our test statistic for choosing between

$$H_0: \quad \mu_1 = \mu_2 = \cdots = \mu_k \quad \text{and} \quad H_1: \quad \text{not all the } \mu_j\text{'s are equal}$$

---

**THEOREM 15.2 Testing Randomized Block Data**

Let SSTR and SSE denote the treatment sum of squares and error sum of squares, respectively, for a set of randomized block data.

1. If $H_0: \mu_1 = \mu_2 = \cdots = \mu_k$ is true, then

$$F = \frac{\text{SSTR}/(k - 1)}{\text{SSE}/((k-1)(b-1))} = \frac{\text{MSTR}}{\text{MSE}}$$

has an *F* distribution with $k - 1$ and $(k - 1)(b - 1)$ degrees of freedom.

2. To test

$$H_0: \quad \mu_1 = \mu_2 = \cdots = \mu_k$$
$$\text{vs.}$$
$$H_1: \quad \text{not all the } \mu_j\text{'s are equal}$$

at the $\alpha$ level of significance, reject $H_0$ if

$$F \geqslant F_{\alpha, k-1, (k-1)(b-1)}$$

where $F_{\alpha, k-1, (k-1)(b-1)}$ cuts off to its right an area of $\alpha$ under the *F* distribution with $k - 1$ and $(k - 1)(b - 1)$ degrees of freedom.

*EXAMPLE 15.1*  Case Study 2.16 described an attempt to quantify the "lunacy" brought on by the full moon—what writers of psychology papers (in deference to late-night werewolf movies) have dubbed the "Transylvania effect." Admissions to the emergency room of a mental health clinic were monitored for 12 months. Three measurements were recorded for each month: the admission rates before, during, and after the full moon. Table 15.3 summarizes the results.

The three different time periods—before, during, and after—are the treatment levels being compared; the 12 different months are serving as blocks. A look at the figures in Table 15.3 makes it clear that time *is* introducing biases into the recorded rates. April and July, for example, are showing much higher numbers than August or December, *regardless of the lunar cycle*. In the terminology introduced earlier, the $\beta$'s for April and July will be positive; those for August and December will be negative.

There is no pejorative connotation associated with the word *bias* as it applies to systematic variations from block to block. To say that the $x$'s are "biased" is not to say that their validity is compromised. The partitioning of SSTO as indicated in Theorem 15.1 and the $F$ ratio as described in Theorem 15.2 have the effect of removing the block biases before the treatment effect is addressed. Indeed, for reasons we will pursue in Section 15.4, we not only *want* the blocks to have biases, we want them to have *large* biases.

Let $\mu_1$, $\mu_2$, and $\mu_3$ denote the true average admission rates before the full moon, during the full moon, and after the full moon, respectively. The hypotheses to be tested are

$$H_0: \quad \mu_1 = \mu_2 = \mu_3$$

vs.

$$H_1: \quad \text{not all the } \mu_j\text{'s are equal}$$

Let $\alpha = 0.05$.

**TABLE 15.3 Admission Rates (patients per day)**

| Month | Before Full Moon | During Full Moon | After Full Moon | $B_i$ |
|-------|-----------------|-----------------|----------------|-------|
| Aug. | 6.4 | 5.0 | 5.8 | 17.2 |
| Sept. | 7.1 | 13.0 | 9.2 | 29.3 |
| Oct. | 6.5 | 14.0 | 7.9 | 28.4 |
| Nov. | 8.6 | 12.0 | 7.7 | 28.3 |
| Dec. | 8.1 | 6.0 | 11.0 | 25.1 |
| Jan. | 10.4 | 9.0 | 12.9 | 32.3 |
| Feb. | 11.5 | 13.0 | 13.5 | 38.0 |
| Mar. | 13.8 | 16.0 | 13.1 | 42.9 |
| Apr. | 15.4 | 25.0 | 15.8 | 56.2 |
| May | 15.7 | 13.0 | 13.3 | 42.0 |
| June | 11.7 | 14.0 | 12.8 | 38.5 |
| July | 15.8 | 20.0 | 14.5 | 50.3 |
| $T_j$ | 131.0 | 160.0 | 137.5 | |
| $\bar{x}_j$ | 10.9 | 13.3 | 11.5 | |

*Solution*   First, we find the sum and the sum of squares for the 36 observations:

$$\sum\sum x = 6.4 + 7.1 + \cdots + 14.5 = 428.5$$

$$\sum\sum x^2 = (6.4)^2 + (7.1)^2 + \cdots + (14.5)^2 = 5722.09$$

Then

$$c = \frac{\left(\sum\sum x\right)^2}{n} = \frac{(428.5)^2}{36} = 5100.34$$

and, from Theorem 15.1,

$$\text{SSTO} = \sum\sum x^2 - c = 5722.09 - 5100.34 = 621.75$$

The treatment sum of squares is computed by squaring each of the treatment totals, dividing by $b$, and subtracting $c$:

$$\text{SSTR} = \sum \frac{T_j^2}{b} - c = \frac{(131.0)^2}{12} + \frac{(160.0)^2}{12} + \frac{(137.5)^2}{12} - 5100.34 = 38.59$$

Similarly, the block sum of squares is computed by squaring each of the block totals, dividing by $k$, and subtracting $c$:

$$\text{SSB} = \sum \frac{B_i^2}{k} - c = \frac{(17.2)^2}{3} + \frac{(29.3)^2}{3} + \cdots + \frac{(50.3)^2}{3} - 5100.34 = 451.08$$

Finally, we find the error sum of squares by subtraction:

$$\text{SSE} = \text{SSTO} - \text{SSTR} - \text{SSB} = 621.75 - 38.59 - 451.08 = 132.08$$

By Theorem 15.2,

$$F = \frac{\text{MSTR}}{\text{MSE}} = \frac{38.59/(3 - 1)}{132.08/(3 - 1)(12 - 1)} = 3.22$$

and the appropriate critical value is the 95th percentile of the $F$ distribution with

$$k - 1 = 2 \quad \text{and} \quad (k - 1)(b - 1) = 22$$

degrees of freedom. From Appendix A.4c, $F_{0.05, 2, 22} \doteq 3.44$ (see Figure 15.4). Since $F < F_{0.05, 2, 22}$, we should accept $H_0$. At the 0.05 level of significance these data do not support the existence of a Transylvania effect.

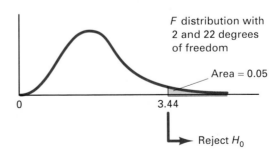

**FIG. 15.4**

15.7   Do an $F$ test on $H_0: \mu_1 = \mu_2 = \mu_3$ for the data in Question 15.1. Let $\alpha = 0.05$. Note:

$$\sum\sum x = 82 \qquad \text{and} \qquad \sum\sum x^2 = 636$$

15.8   A randomized block experiment comparing the effects of five different levels of potash on the strength of cotton fibers was summarized in Question 2.46. At the $\alpha = 0.01$ level, test the null hypothesis that the true average fiber strengths associated with the five different potash concentrations are all equal. Use the fact that

$$\sum\sum x = 115.83 \qquad \text{and} \qquad \sum\sum x^2 = 895.6183$$

15.9   For Question 2.48 let $\mu_j, j = 1, 2, 3$, denote the true average levels of cockroach hostility (encounters per minute) characterizing high-, intermediate-, and low-density environments, respectively. Use Theorem 15.2 to test $H_0: \mu_1 = \mu_2 = \mu_3$ at the $\alpha = 0.05$ level of significance. Note:

$$\sum\sum x = 6.32 \qquad \text{and} \qquad \sum\sum x^2 = 1.4810$$

15.10   For the affirmative action data of Question 2.52, test whether the true average salaries for the three races are all equal. Let $\alpha = 0.05$. What types of factors might account for the $B_i$'s not being all the same?

15.11   Let $\mu_1$, $\mu_2$, and $\mu_3$ denote the true average Nielsen ratings for the news programs broadcast by networks 1, 2, and 3, respectively. Using Theorem 15.2 on the data in Question 2.64, test $H_0: \mu_1 = \mu_2 = \mu_3$ at the $\alpha = 0.05$ level of significance. Note:

$$\sum\sum x = 191.5 \qquad \text{and} \qquad \sum\sum x^2 = 2510.37$$

15.12   When dependent observations are used to compare $k = 2$ treatments, testing $H_0: \mu_1 = \mu_2$ with an $F$ ratio is equivalent to testing $H_0: \mu_D = 0$ with a paired $t$. Analyze the color data in Question 2.66 both ways: first as a paired data problem and second as a randomized block problem. Carry out both procedures at the $\alpha = 0.05$ level of significance. (If all the computations are done correctly, the observed $F$ ratio should equal the *square* of the observed $t$ ratio. Also, the $F$ critical value should be the square of either of the $t$ critical values.)

### ANOVA Table

Computations for a randomized block analysis of variance are usually presented in a tabular format similar to Table 14.5. The same column designations are used, but an additional row for blocks is included. Table 15.4 shows the locations of the entries. The three "mean squares" listed are quotients of sums of squares divided by degrees of freedom:

$$\text{MSTR} = \frac{\text{SSTR}}{k-1}, \qquad \text{MSB} = \frac{\text{SSB}}{b-1}, \qquad \text{MSE} = \frac{\text{SSE}}{(b-1)(k-1)}$$

**TABLE 15.4**

| Source of Variation | df | SS | MS | F |
|---|---|---|---|---|
| Treatments | $k - 1$ | SSTR | MSTR | MSTR/MSE |
| Blocks | $b - 1$ | SSB | MSB | |
| Error | $(b - 1)(k - 1)$ | SSE | MSE | |
| Total | $n - 1$ | SSTO | | |

Table 15.5 is the ANOVA table for the Transylvania data. The NS appearing after the 3.22 in the *F* column is in keeping with the conventions introduced in Section 14.4. It means that the differences among the three sample means are *not significant* at the 0.05 level (since $3.22 < F_{0.05, 2, 22} = 3.44$).

**TABLE 15.5**

| Source of Variation | df | SS | MS | F | |
|---|---|---|---|---|---|
| Lunar cycle | 2 | 38.59 | 19.30 | 3.22 | NS |
| Months | 11 | 451.08 | 41.01 | | |
| Error | 22 | 132.08 | 6.00 | | |
| Total | 35 | 621.75 | | | |

---

*EXAMPLE 15.2*   A frequently used form of blocking is to apply all *k* treatments to each of *b* subjects—that is, make each subject into a block. An example of that strategy is the Ponzo illusion experiment described in Case Study 2.17.

Persons looking at the vertical lines in Figure 15.5 will tend to perceive the rightmost one as shorter, even though the two are equal. Moreover, the perceived difference in those two lengths—what psychologists call the "strength" of the illusion—has been shown to be a function of age.

Ponzo illusion

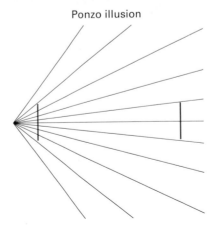

**FIG. 15.5**

Hypnosis entered the picture when Porter and his colleagues (126) wondered whether age regression could recapture the illusion strength/subject age relationship. If, for example, five-year-olds and nine-year-olds, on the average, perceive the illusion to have different strengths, will adults who are age-regressed under hypnosis to age 5 and to age 9 show a comparable difference in *their* perception of the illusion?

Table 15.6 repeats the data from Case Study 2.17. The three treatment levels are awake, regressed to age 9 and regressed to age 5. If age regression is able to recapture perceptual awareness, then the average illusion strengths associated with the three treatment levels should not be the same. The hypotheses to be tested, therefore, are $H_0: \mu_1 = \mu_2 = \mu_3$ and $H_1$: not all the $\mu_j$'s are equal. Let $\alpha = 0.05$.

**TABLE 15.6 Illusion Strengths**

| Subject | (1) Awake | (2) Regressed to Age 9 | (3) Regressed to Age 5 | $B_i$ |
|---|---|---|---|---|
| 1 | 0.81 | 0.69 | 0.56 | 2.06 |
| 2 | 0.44 | 0.31 | 0.44 | 1.19 |
| 3 | 0.44 | 0.44 | 0.44 | 1.32 |
| 4 | 0.56 | 0.44 | 0.44 | 1.44 |
| 5 | 0.19 | 0.19 | 0.31 | 0.69 |
| 6 | 0.94 | 0.44 | 0.69 | 2.07 |
| 7 | 0.44 | 0.44 | 0.44 | 1.32 |
| 8 | 0.06 | 0.19 | 0.19 | 0.44 |
| $T_j$ | 3.88 | 3.14 | 3.51 | |
| $\bar{x}_j$ | 0.48 | 0.39 | 0.44 | |
| True mean | $\mu_1$ | $\mu_2$ | $\mu_3$ | |

*Solution*   Following the sequence of steps in Example 15.1, we begin by finding

$$\sum\sum x = 10.53 \quad \text{and} \quad \sum\sum x^2 = 5.5889$$

Therefore,

$$c = \frac{(10.53)^2}{24} = 4.6200$$

and

$$\text{SSTO} = 5.5889 - 4.6200 = 0.9689$$

The sums of squares for treatments and for blocks are obtained from the $T_j$'s and $B_i$'s in Table 15.6:

$$\text{SSTR} = \frac{(3.88)^2}{8} + \frac{(3.14)^2}{8} + \frac{(3.51)^2}{8} - 4.6200 = 0.0343$$

$$\text{SSB} = \frac{(2.06)^2}{3} + \cdots + \frac{(0.44)^2}{3} - 4.6200 = 0.7709$$

By subtraction,

$$SSE = 0.9689 - 0.0343 - 0.7709 = 0.1637$$

Table 15.7 shows the ANOVA table for these data filled in according to the format prescribed in Table 15.4. The test statistic equals 1.47:

$$F = \frac{MSTR}{MSE} = \frac{0.0172}{0.0117} = 1.47$$

**TABLE 15.7**

| Source | df | SS | MS | F | |
|---|---|---|---|---|---|
| Ages | 2 | 0.0343 | 0.0172 | 1.47 | NS |
| Subjects | 7 | 0.7709 | 0.1101 | | |
| Error | 14 | 0.1637 | 0.0117 | | |
| Total | 23 | 0.9689 | | | |

By Theorem 15.2 the ratio MSTR/MSE has an $F$ distribution with 2 and 14 df. Since the 95th percentile of that distribution is 3.74 (see Figure 15.6), our conclusion is to accept $H_0$ at the $\alpha = 0.05$ level of significance. The differences among the *sample* means (0.48 versus 0.39 versus 0.44) are not sufficiently large to overturn our initial presumption that the *true* means are equal. The data do not demonstrate, in other words, that subjects age-regressed to an earlier age can recapture the perceptual patterns of that age.

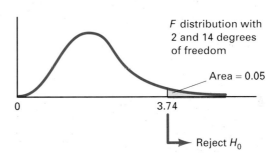

**FIG. 15.6**

QUESTIONS

15.13 Use the ANOVA table format to display the computations done for Question 15.7.

15.14 Do an $F$ test on the affirmative action data in Question 2.52. Put the results in the format of Table 15.4. Let $\alpha = 0.05$. Use the computations from Question 15.10.

15.15 Summarize the analysis done in Question 15.11 on the Nielsen ratings (from Question 2.64) by setting up the corresponding ANOVA table.

15.16 For Case Study 3.20 use the ANOVA table format to test whether rats prefer equally each of the four different artificial flavors. Note:

$$\sum\sum x = 460.6 \quad \text{and} \quad \sum\sum x^2 = 9191.30$$

15.17 Do a randomized block analysis on the blood-doping data in Question 3.84. Define the parameters being tested. Use the NS/*/** conventions to indicate whatever statistical significance the results may have. For the 18 observations,

$$\sum\sum x = 592.78 \quad \text{and} \quad \sum\sum x^2 = 19{,}538.7696$$

15.18 Heart rates were monitored (9) for six tree shrews (*Tupaia glis*) during three different stages of sleep: LSWS (light slow wave sleep), DSWS (deep slow wave sleep), and REM (rapid eye movement sleep). The results are shown below.

| HEART RATES (beats/5 s) | | | |
|---|---|---|---|
| Tree Shrew | LSWS | DSWS | REM |
| 1 | 14.1 | 11.7 | 15.7 |
| 2 | 26.0 | 21.1 | 21.5 |
| 3 | 20.9 | 19.7 | 18.3 |
| 4 | 19.0 | 18.2 | 17.0 |
| 5 | 26.1 | 23.2 | 22.5 |
| 6 | 20.5 | 20.7 | 18.9 |

(a) Do the appropriate analysis of variance to test the equality of the heart rates during these three phases of sleep.

(b) Show the ANOVA table.

(c) State your conclusion.

15.19 Four processes (1, 2, 3, and 4) related to the manufacture of penicillin are being investigated. Each is used one time in each of five nutrients (A through E). The numbers listed in the table are growth indices, with larger indices being preferable. Test $H_0: \mu_1 = \mu_2 = \mu_3 = \mu_4$, where $\mu_j$ is the true average growth index associated with process $j$, $j = 1, 2, 3, 4$.

| | | PROCESS | | | |
|---|---|---|---|---|---|
| | | 1 | 2 | 3 | 4 |
| | A | 29 | 28 | 37 | 34 |
| | B | 24 | 17 | 32 | 19 |
| NUTRIENT | C | 21 | 27 | 27 | 25 |
| | D | 27 | 32 | 29 | 24 |
| | E | 19 | 21 | 20 | 28 |

15.20   Safety officials are evaluating three types of workstations (A, B, and C) for air traffic controllers. The measurements recorded are scores on a stress test. A random sample of six controllers is chosen to participate. Each controller is tested three times, once after using each of the three stations. The data are shown below.

(a)  Do the appropriate analysis of variance.

(b)  Show the ANOVA table.

|  |  | STATION | | |
|---|---|---|---|---|
|  |  | A | B | C |
| | DF | 15 | 15 | 18 |
| | ML | 14 | 14 | 14 |
| CONTROLLER | SH | 10 | 11 | 15 |
| | DP | 13 | 12 | 17 |
| | JG | 16 | 13 | 16 |
| | CW | 13 | 13 | 13 |

# 15.4

## WHICH DESIGN IS BETTER?

Chapters 14 and 15 dealt with experiments whose objectives were the same; namely, to compare the means associated with $k$ different populations. Their abilities to make those judgements, though, may be quite different. Under certain circumstances a randomized block design is much more powerful than a completely randomized design. The former, in other words, has a higher probability of rejecting $H_0$ when $H_0$ is false. On the other hand, situations arise where the reverse is true, meaning the "completely randomized" is the better approach. The question that needs to be addressed, then, is obvious: *which design should we use when?*

The keys to comparing these two procedures are the partitioning statements made in Theorems 14.1 and 15.1. For a completely randomized design,

$$\text{SSTO} = \text{SSTR} + \text{SSE} \tag{15.2}$$

For a randomized block design,

$$\text{SSTO}' = \text{SSTR}' + \text{SSB}' + \text{SSE}' \tag{15.3}$$

where the primes here are being used simply to distinguish the terms in the two equations.

Imagine applying Equations 15.2 and 15.3 to two sets of data where the numbers of observations are the same (see Table 15.8). If we further assume for the sake of comparison that SSTO and SSTR remain the same

**TABLE 15.8**

| COMPLETELY RANDOMIZED | | | | | RANDOMIZED BLOCK | | | | |
| --- | --- | --- | --- | --- | --- | --- | --- | --- | --- |
| | Treatments | | | | | Treatments | | | |
| | 1 | 2 | $\cdots$ | k | Blocks | 1 | 2 | $\cdots$ | k |
| | – | – | | – | 1 | – | – | | – |
| | – | – | | – | 2 | – | – | | – |
| b replicates | – | – | | – | 3 | – | – | | – |
| | | | $\vdots$ | | $\vdots$ | | | $\vdots$ | |
| | – | – | | – | b | – | – | | – |

for both sets of data (that is, SSTO = SSTO′ and SSTR = SSTR′), then

$$SSE = SSB' + SSE' \qquad (15.4)$$

Figure 15.7 illustrates Equations 15.2 and 15.3 graphically. The areas of the two outer rectangles represent SSTO and SSTO′. The numerical partitioning of those sums of squares is equivalent to dividing the original rectangles into smaller areas representing SSTR, SSE, SSTR′, and so on, where the sizes of the smaller areas are proportional to the magnitudes of the sums of squares they represent. The numbers in parentheses are the degrees of freedom associated with each sum.

Completely randomized

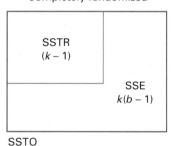

SSTO

Randomized block

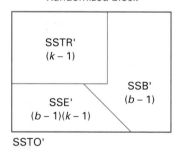

SSTO′

**FIG. 15.7**

Recall the quotients used for testing $H_0: \mu_1 = \mu_2 = \cdots = \mu_k$. For a completely randomized design,

$$F = \frac{SSTR/(k-1)}{SSE/k(b-1)} \quad \left( = \frac{SSTR/(k-1)}{SSE/(n-k)} \right) \qquad (15.5)$$

For the randomized block design,

$$F = \frac{SSTR'/(k-1)}{SSE'/(k-1)(b-1)} \qquad (15.6)$$

The numerators of these two $F$ ratios are the same (because of our presumption that SSTR = SSTR′). The denominator in Equation 15.6, however, is likely to differ from the denominator in Equation 15.5 for two reasons. First, we can see from Equation 15.4, that $SSE' \leqslant SSE$, which tends to make the denominator in Equation 15.6 *smaller* than the denominator in Equation 15.5. But $(k-1)(b-1) < k(b-1)$, which tends to make the denominator in Equation 15.6 *larger* than its counterpart in Equation 15.5.

Which denominator ends up being larger depends on the sizes of the differences between SSE and SSE' and between $(k-1)(b-1)$ and $k(b-1)$. If SSE' is much smaller than SSE, then $SSE'/(k-1)(b-1)$ will be smaller than $SSE/k(b-1)$. On the other hand, if SSE' is just a little smaller than SSE, then $SSE'/(k-1)(b-1)$ will be larger than $SSE/k(b-1)$. The size of the denominator, of course, is crucial because it directly affects the power of the test. For a fixed numerator a smaller denominator yields a larger quotient, meaning the observed $F$ ratio is more likely to be in the rejection region. (Note: The degrees of freedom for the two denominators are themselves different and that also has a bearing on the power of the two tests. However, compared with the effects of the differences between SSE and SSE' and between $k(b-1)$ and $(k-1)(b-1)$, the impact of the difference between the degrees of freedom for the two denominators is usually small. Ignoring this difference will help us focus better on the factors whose role is more critical.)

So what is the answer to our original question: which design should we use when? If we believe that the subjects are likely to show a good deal of variability, irrespective of which treatment they are given, then the preferred strategy is to look for some kind of blocking. In a situation of that sort, SSB' will be large, in which case SSE' will be much smaller than SSE and the observed $F$ for the randomized block design will be larger than the observed $F$ we would have gotten with a completely randomized design. If, on the other hand, the experimental environment is markedly homogeneous (that is, there is not much variability from subject to subject), then it would be better to collect the data in the format of a completely randomized design.

Figure 15.8 shows the two extremes just mentioned. In Figure 15.8a, the blocks have been able to account for a sizable proportion of the "non-treatment" sum of squares. As a result, SSE is relatively small. In contrast,

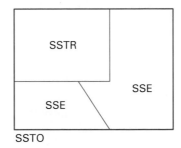
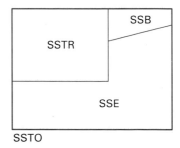

**FIG. 15.8**

the blocking for the experiment in Figure 15.8b was largely ineffective: SSE is still large, and the degrees of freedom expended for blocks were essentially wasted. If the experimenter had the opportunity to repeat these studies, the one in Figure 15.8a should be again done as a randomized block design. But for the data in Figure 15.8b, it appears that testing $H_0: \mu_1 = \mu_2 = \cdots = \mu_k$ might be accomplished more effectively with the methods of Chapter 14.

COMMENT:   One way of verifying *formally* whether a blocking scheme has been effective is to do an $F$ test on the differences among the block averages. Let $\beta_1, \beta_2, \ldots, \beta_b$ represent the true average responses characteristic of blocks $1, 2, \ldots, b$, respectively. If $H_0: \beta_1 = \beta_2 = \cdots = \beta_b$ is true, the ratio

$$\frac{\text{SSB}/(b-1)}{\text{SSE}/(b-1)(k-1)}$$

has an $F$ distribution with $b - 1$ and $(b - 1)(k - 1)$ degrees of freedom. *Rejecting* $H_0$ indicates that variability from block to block was substantial and confirms that the choice of a randomized block design was a prudent one. *Accepting* $H_0$ implies either that a different blocking criterion should be tried or that $H_0: \mu_1 = \mu_2 = \cdots = \mu_k$ should be tested by a completely randomized design.

---

QUESTIONS

**15.21**   Test the block effect for the Transylvania data in Case Study 2.16. Use the computations from Example 15.1. Let $\alpha = 0.05$. Has the experimenter benefitted by treating time as a blocking factor?

**15.22**   Test $H_0: \mu_1 = \mu_2 = \cdots = \mu_8$ for the Ponzo illusion data in Case Study 2.17. Base your analysis on the entries in Table 15.7. Let $\alpha = 0.05$.

**15.23**   By simply looking at the following two sets of responses, decide in each case if a future experiment should be conducted as a completely randomized design or as a randomized block design. Justify your choices.

(a)

|  |  | TREATMENTS | | |
|---|---|---|---|---|
|  |  | 1 | 2 | 3 |
|  | A | 20 | 25 | 32 |
| BLOCKS | B | 22 | 26 | 37 |
|  | C | 21 | 24 | 36 |

(b)

|  |  | TREATMENTS | | | |
|---|---|---|---|---|---|
|  |  | 1 | 2 | 3 | 4 |
|  | I | 10 | 12 | 11 | 13 |
| BLOCKS | II | 26 | 25 | 26 | 27 |
|  | III | 14 | 13 | 13 | 15 |

**15.24**   Test the block effect for the cockroach data in Question 2.48. Let $\alpha = 0.05$.

**15.25**   (a) In Case Study 3.20 can it be argued that environmental conditions remained relatively constant from survey to survey? Do the appropriate $F$ test. Let $\alpha = 0.05$.

(b) What does your answer imply about the way in which follow-up studies comparing bait preferences should be conducted?

15.26 Use the data in Question 15.20 to decide whether differences in average stress scores, from controller to controller, are statistically significant by putting NS, *, or ** next to the entry in the $F$ column.

# 15.5

## COMPUTER NOTES

Figure 15.9 is the SAS version of an analysis of randomized block design, using Example 15.1. Each admission rate is entered, along with the corresponding month and lunar cycle. Each entry in Table 15.3 is coded as a triple of values from the variables MONTH, LUNACY, and RATE. For example, the admission rate during August before the full moon was 6.4. This

```
DATA;
  INPUT MONTH $ LUNARCY $ RATES;
  CARDS;
    AUG   BEFORE  6.4
    SEP   BEFORE  7.1

   (Rest of data from Table 15.3.1 here)

PROC ANOVA;
  CLASS MONTH LUNARCY;
  MODEL RATES = MONTH LUNARCY;
```

```
                    ANALYSIS OF VARIANCE PROCEDURE
                      CLASS LEVEL INFORMATION

   CLASS     LEVELS    VALUES

   MONTH       12      APR AUG DEC FEB JAN JUL JUN MAR MAY NOV OCT SEP

   LUNARCY      3      AFTER BEFORE DURING

                NUMBER OF OBSERVATIONS IN DATA SET = 36

                    ANALYSIS OF VARIANCE PROCEDURE
```

DEPENDENT VARIABLE: RATES

| SOURCE | DF | SUM OF SQUARES | MEAN SQUARE | F VALUE | PR > F | R-SQUARE | C.V. |
|---|---|---|---|---|---|---|---|
| MODEL | 13 | 489.68027778 | 37.66771368 | 6.27 | 0.0001 | 0.787584 | 20.5846 |
| ERROR | 22 | 132.06944444 | 6.00315657 | | ROOT MSE | | RATES MEAN |
| CORRECTED TOTAL | 35 | 621.74972222 | | | 2.45013399 | | 11.90277778 |

| SOURCE | DF | ANOVA SS | F VALUE | PR > F |
|---|---|---|---|---|
| MONTH | 11 | 451.08305556 | 6.83 | 0.0001 |
| LUNARCY | 2 | 38.59722222 | 3.21 | 0.0596 |

FIG. 15.9

information was entered in the CARDS section as the data line AUG BEFORE 6.4.

Although the format of the printout differs from Table 15.3, it has all of that table's information. The $F$ statistic and the $P$ value, 0.0596, are on the last line of Figure 15.9 for testing the independence of means. For $\alpha = 0.05$ we accept the null hypothesis that the means for the three treatments are the same.

# 15.6

## SUMMARY

Chapter 15 can be summarized in a single word—*blocking*. To an experimenter, few concepts are more important. Why? Because replicating the levels of a treatment within a series of relatively homogeneous blocks can be a spectacularly efficient way of testing whether all $k$ treatment levels elicit the same average response.

Regardless of the specific nature of the data being collected, a major priority of every researcher is to reduce the amount of experimental error in the $y_i$'s. We measure that error by the magnitude of $\sigma^2$. The *direct* way of reducing $\sigma^2$ is to refine the measurement process itself: use better equipment, hire more competent technicians, and in general, be more careful. The *indirect* way is to collect data according to a randomized block design and to use the blocks to reduce SSE, which, in turn, reduces the estimate of $\sigma^2$. In many situations the indirect way is not only more effective but also more economical.

It is difficult to overstate the importance of blocking. A researcher's ability to know when and how an experiment should be blocked is analogous to a shortstop's ability to field ground balls: other skills are certainly necessary, but not having that one is a surefire way to guarantee failure.

No single definition adequately characterizes every possible type of block. In some experiments time is a block, as it was in the lunar cycle data of Example 15.1. Subjects themselves can serve as blocks if all treatment levels are applied to each person (recall Example 15.2). Other examples of blocks were described in Case Studies 11.1 through 11.4. In general, an experimenter's ingenuity is all that limits the way in which a set of blocks can be defined.

The mathematical formulation of a randomized block design is an extension of what we encountered in Chapter 14 for the $k$-sample problem. It will be assumed that treatment level $j$, when applied to a subject in block $i$, produces a response $X$, where,

$$X = \mu_j + \beta_i + \varepsilon$$

That is, we can write $X$ as the sum of a treatment effect ($\mu_j$), a block bias ($\beta_i$), and an error term ($\varepsilon$). Both $\mu_j$ and $\beta_i$ are unknown constants, whereas $\varepsilon$ is a random variable, assumed to be normally distributed with mean zero and variance $\sigma^2$. Our objective is to test $H_0: \mu_1 = \mu_2 = \cdots = \mu_k$.

Variability among the $X$'s, as measured by the total sum of squares, can be partitioned into the sum of three terms:

$$SSTO = SSTR + SSB + SSE$$

The basis for testing whether the treatment means are all equal is the ratio

$$F = \frac{MSTR}{MSE} = \frac{SSTR/(k-1)}{SSE/(b-1)(k-1)}$$

where $k$ is the number of treatment levels and $b$ is the number of blocks.

If $H_0$ is true, the observed $F$ ratio will have an $F$ distribution with $k-1$ and $(b-1)(k-1)$ degrees of freedom. Furthermore, if the null hypothesis is true, both

$$\frac{SSTR}{(k-1)} \quad \text{and} \quad \frac{SSE}{(b-1)(k-1)}$$

are estimates of $\sigma^2$, implying that their ratio should be close to 1. However, if $H_0$ is *not* true, MSTR will tend to be larger than MSE. Therefore, we should reject $H_0$ if the observed $F$ ratio is too far from 1 *in the right-hand tail* of the $F_{k-1,(b-1)(k-1)}$ distribution (see Figure 15.10).

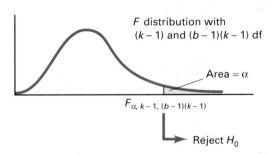

**FIG. 15.10**

Table 15.9 shows the sequence of steps to be followed in doing the analysis of variance for randomized block data representing $k$ treatments applied to $b$ blocks. Figure 15.11 reviews the necessary notation. Except for having to compute a sum of squares for blocks, the steps parallel the procedure outlined in Table 14.12 for the completely randomized design.

Although the primary focus of any randomized block design is the treatment effect, it may also be of interest to test the block effect. If

$$H_0: \quad \beta_1 = \beta_2 = \cdots = \beta_b = 0$$

is true, then

$$F = \frac{SSB/(b-1)}{SSE/(b-1)(k-1)}$$

| | | Treatments | | | | |
|---|---|---|---|---|---|---|
| | | 1 | 2 | ⋯ | $k$ | Totals |
| Blocks | 1 | - | - | | - | $B_1$ |
| | 2 | - | - | | - | $B_2$ |
| | ⋮ | | | ⋮ ⋯ | | |
| | $b$ | - | - | ⋯ | - | $B_b$ |
| Totals | | $T_1$ | $T_2$ | | $T_k$ | $T$ |

**FIG. 15.11**

**TABLE 15.9 Testing $H_0$: $\mu_1 = \mu_2 = \cdots = \mu_k$**

1. Compute $\sum\sum x$ and $\sum\sum x^2$.
2. Compute $c = \dfrac{T^2}{bk}$, where $T = \sum\sum x$.
3. Compute $\text{SSTO} = \sum\sum x^2 - c$.
4. Compute $\text{SSTR} = \sum \dfrac{T_j^2}{b} - c$.
5. Compute $\text{SSB} = \sum \dfrac{B_i^2}{k} - c$.
6. Compute $\text{SSE} = \text{SSTO} - \text{SSTR} - \text{SSB}$.
7. Fill in the first four columns of the ANOVA table as they appear in Table 15.4.
8. Compute the test statistic $F = \dfrac{\text{MSTR}}{\text{MSE}}$.
9. Compare the value of $\dfrac{\text{MSTR}}{\text{MSE}}$ with the 95th and 99th percentiles of the $F$ distribution with $k - 1$ and $(b - 1)(k - 1)$ degrees of freedom. Next to $\dfrac{\text{MSTR}}{\text{MSE}}$, write
   (a) NS if $\dfrac{\text{MSTR}}{\text{MSE}} < F_{0.05, k-1, (b-1)(k-1)}$
   (b) * if $F_{0.05, k-1, (b-1)(k-1)} \leqslant \dfrac{\text{MSTR}}{\text{MSE}} < F_{0.01, k-1, (b-1)(k-1)}$
   (c) ** if $\dfrac{\text{MSTR}}{\text{MSE}} \geqslant F_{0.01, k-1, (b-1)(k-1)}$

has an $F$ distribution with $b - 1$ and $(b - 1)(k - 1)$ degrees of freedom. A large value for $\dfrac{\text{MSB}}{\text{MSE}}$ indicates that the blocks were effective in accounting for a large portion of the total variability (and, therefore, effective in reducing the magnitude of SSE). A small value for the block $F$ ratio suggests that future studies should be done using the completely randomized design.

15.27    Ace Construction, the company renovating State Tech's basketball court, can use any of three patterns (A, B, or C) for the parquet floor. Coach Dribble is concerned that the appearance of the floor may have an effect on a team's shooting percentage. Players from four local high schools are invited to "test" the three floors. Listed below are the proportions of jump shots they made from a 15-ft range (out of 100 attempted).

|  |  | FLOOR PATTERN | | |
|  |  | A | B | C |
| --- | --- | --- | --- | --- |
| TEAMS | Vikings | 0.65 | 0.60 | 0.69 |
|  | Mud Hens | 0.42 | 0.46 | 0.49 |
|  | Redmen | 0.61 | 0.50 | 0.60 |
|  | Bobcats | 0.67 | 0.60 | 0.72 |

Use the analysis of variance to test whether floor patterns affect shooting percentage. Let $\alpha = 0.05$. Note:

$$\sum\sum x = 7.01 \quad \text{and} \quad \sum\sum x^2 = 4.1961$$

15.28    Listed below are starting salaries (in $1000s) reported by three employment agencies specializing in the placement of liberal arts graduates. At the 0.05 level, are the differences among average starting salaries for the four majors statistically significant? Show the ANOVA table and indicate your conclusion by using the NS/*/** notation.

|  |  | MAJORS | | | |
|  |  | Sociology | Psychology | Philosophy | History |
| --- | --- | --- | --- | --- | --- |
| PLACEMENT SERVICE | AAA | 19.0 | 20.0 | 22.5 | 21.0 |
|  | New Life | 22.3 | 23.5 | 24.0 | 24.5 |
|  | Upward | 24.1 | 28.1 | 27.0 | 26.5 |

Note:    $$\sum\sum x = 282.5 \quad \text{and} \quad \sum\sum x^2 = 6735.71$$

15.29    According to the Olympic diving-competition scores shown below, Australia won the gold medal, Norway the silver, and the United States the bronze. At the $\alpha = 0.05$ level, were the differences among the average scores statistically significant?

|  |  | Australia | Norway | United States |
| --- | --- | --- | --- | --- |
| JUDGE | 1 | 10 | 9 | 10 |
|  | 2 | 8 | 9 | 9 |
|  | 3 | 9 | 8 | 7 |
|  | 4 | 6 | 4 | 5 |
|  | 5 | 10 | 8 | 6 |

15.30   Two standard designs for moving customers through a cafeteria are random-access food stations and the single-server queue. In a study comparing the two, researchers recorded the following times (in minutes) required by diners to make a selection and pay the cashier. Each entry is an average computed from a sample of size 5. Use the analysis of variance to compare the two strategies. Let $\alpha = 0.05$.

|  | Single-Server | Food Stations |
|---|---|---|
| Breakfast, 8/20 | 4.3 | 3.0 |
| Lunch, 8/21 | 7.6 | 6.5 |
| Dinner, 8/22 | 4.5 | 4.2 |
| Lunch, 8/23 | 6.8 | 6.3 |
| Dinner, 8/24 | 4.8 | 5.0 |
| Lunch, 8/25 | 8.2 | 5.3 |

15.31   Fill in the entries missing from the following ANOVA table:

| Source | df | SS | MS | F |
|---|---|---|---|---|
| Treatments | 3 |  | 15.0 |  |
| Blocks |  |  |  | 2.0 |
| Error |  | 60.0 | 5.0 |  |
| Total |  |  |  |  |

15.32   A cosmetics firm has run three promotions (A, B, and C) advertising a new skin creme. Summarized below are the sales increases (in $1000s) reported by four major markets in response to the three campaigns. Are the differences in revenue attributed to A, B, and C statistically significant? Let $\alpha = 0.01$.

| | PROMOTION | | |
|---|---|---|---|
| | A | B | C |
| Atlanta | 50 | 45 | 60 |
| Cleveland | 35 | 40 | 45 |
| Dallas | 85 | 90 | 115 |
| Pittsburg | 65 | 60 | 70 |

15.33   Are the following statements true or false?

(a) MSB can never be negative.

(b) Either MSTR or MSB, or both, must be larger than MSE.

(c) $T^2/bk = \sum\sum x^2/n$.

**15.34** Propose some blocking criteria that might be appropriate for the following experiments:

(a) Comparing the bushels per acre harvested for five varieties of corn.

(b) Comparing gas mileages for six subcompacts.

(c) Comparing three antihistamines for treating allergy symptoms.

**15.35** Listed below are selected percentiles for the $F_{2,10}$ distribution. What can be said about the $P$ value associated with the following ANOVA table?

| | | | PERCENTILES ($F_{2,10}$) | | | | |
|------|------|------|------|------|------|------|------|
| 0.05 | 0.10 | 0.25 | 0.50 | 0.75 | 0.90 | 0.95 | 0.99 |
| 0.052 | 0.106 | 0.096 | 0.743 | 1.60 | 2.92 | 4.10 | 7.56 |

| Source | df | SS | MS | F |
|------------|----|------|------|---|
| Treatments | | 57.8 | | |
| Blocks | 5 | | | |
| Error | | | 17.2 | |
| Total | 17 | | | |

**15.36** Five television critics have been asked to rate (on a scale of 0 to 100) the violence content in prime-time police dramas broadcast by the three major networks. Are the differences among the three average ratings statistically significant? Show the ANOVA table and use the NS/*/** convention to indicate your conclusion.

| | | NETWORK | | |
|---|---|---|---|---|
| | | I | II | III |
| | BP | 60 | 80 | 75 |
| | NY | 90 | 95 | 90 |
| CRITIC | WW | 45 | 50 | 30 |
| | SG | 55 | 50 | 40 |
| | SQ | 65 | 90 | 75 |
| | Average: | 63.0 | 73.0 | 62.0 |

**15.37** Subhypotheses of the form $H'_0: \mu_i = \mu_j$ versus $H'_1: \mu_i \neq \mu_j$ can be tested for randomized block data by constructing Tukey confidence intervals for $\mu_i - \mu_j$. The appropriate formula is the expression given in Theorem 14.4, except that $D$ is redefined to be $Q_{\alpha, k, (b-1)(k-1)}/\sqrt{b}$. For the basketball data in Question 15.27, construct 95% Tukey confidence intervals to compare A versus B, A versus C, and B versus C.

15.38    The mathematics department at State University wants to compare the effectiveness of teaching calculus with and without computers. They intend to teach two sections, each containing five students. Shown below in ascending order are the quantitative SAT scores for the ten freshmen who have volunteered to participate.

| Name | Quantitative SAT |
|------|------------------|
| LT | 450 |
| WW | 480 |
| BM | 520 |
| TR | 520 |
| RL | 560 |
| SC | 600 |
| TM | 640 |
| BT | 710 |
| HW | 780 |
| RM | 800 |

(a) How would you assign students to the two sections?

(b) How would you set up the experiment if the ten SAT scores ranged from 590 to 620?

15.39    Analyze the prothrombin time data in Table 11.2 by using a randomized block ANOVA. Let $\alpha = 0.05$. Show that $F = t^2$. What relationship exists between the $F$ critical value and the $t$ critical value?

15.40    Four meat-packing companies each market bacon in 16 oz. packages. Once a month a consumer protection group weighs five packages representing each brand and records the difference (in oz.) between the actual weight and the stated weight. Can we conclude from the data below that the differences among the $\bar{x}_j$'s are statistically significant? Let $\alpha = 0.05$.

| | BRAND | | |
|---|---|---|---|
| | Oak Hill | Sunny Farm | Bryant |
| Apr. 17 | 0.5 | −0.4 | 0.1 |
| May 10 | −0.3 | −0.3 | 0.2 |
| June 15 | −0.4 | 0.1 | 0.4 |
| July 2 | −0.3 | 0.4 | 0.3 |

15.41    Which of the following statements are true?

(a) $\dfrac{T_1}{b} + \dfrac{T_2}{b} + \cdots + \dfrac{T_k}{b} = \dfrac{B_1}{k} + \dfrac{B_2}{k} + \cdots + \dfrac{B_b}{k}$

(b) $\dfrac{T_1}{bk} + \dfrac{T_2}{bk} + \cdots + \dfrac{T_k}{bk} = \dfrac{B_1}{n} + \dfrac{B_2}{n} + \cdots + \dfrac{B_b}{n}$

(c) $\text{SSE} = \sum\sum (x - \bar{x})^2$

(d) $\bar{x} = \left(\dfrac{1}{k}\right)(\bar{x}_1 + \bar{x}_2 + \cdots + \bar{x}_k)$

(e) $bk\bar{x} = T$

**15.42** For the randomized block data shown below, suppose $\mu_3 = 6$ and $\beta_2 = -1$. Find $\varepsilon$ for the circled observation.

|         |   | \multicolumn{4}{c}{TREATMENT} |
|---------|---|---|---|---|---|
|         |   | 1 | 2 | 3 | 4 |
|         | 1 | 8 | 4 | 7 | 9 |
| BLOCK   | 2 | 6 | 2 | ④ | 5 |
|         | 3 | 3 | 5 | 6 | 7 |

**15.43** Recall Question 15.27. Describe an experiment that would use the completely randomized design of Chapter 14 as the method for comparing the effects of floor patterns A, B, and C on shooting percentage.

**15.44** Based on the following ANOVA table, should we reject the null hypothesis that all the treatment means are equal?

| Source    | df | SS  | MS     | F |
|-----------|----|-----|--------|---|
| Treatment |    |     | 106.53 |   |
| Blocks    |    |     |        |   |
| Error     |    | 265 | 26.5   |   |
| Total     | 17 |     |        |   |

**15.45** Make up a set of responses for a randomized block experiment having $k = 3$ treatments and $b = 3$ blocks for which $\text{SSTR} = 0$, $\text{SSB} = 0$, but $\text{SSE} \neq 0$.

**15.46** Test the block effect for the data in Question 15.40.

**15.47** (a) Fill in the entries missing from the data table shown below.

|        |   | \multicolumn{4}{c}{TREATMENTS} | |
|--------|---|---|---|---|---|---|
|        |   | 1 | 2 | 3 | 4 | Totals |
|        | 1 | 4 |   |    |   |        |
| BLOCKS | 2 | 3 | 7 | 10 | 3 | 23     |
|        | 3 | 5 |   | 12 | 4 | 29     |
|        | Totals |   |   | 30 | 9 | 72 |

(b) Estimate $\mu_j$, $j = 1, 2, 3, 4$, and $\beta_i$, $i = 1, 2, 3$.

15.48   Suppose the $F$ ratio MSB/MSE implies that we should reject $H_0: \beta_1 = \beta_2 = \cdots = \beta_b = 0$.

(a) What does that tell us about the likelihood of rejecting $H_0: \mu_1 = \mu_2 = \cdots = \mu_k$?

(b) What does that tell us about the appropriateness of using the randomized block design?

15.49   Listed below are tire mileages (in 1000s of miles) reported by four drivers testing three brands of steel-belted radials. Are the differences among the averages shown at the bottom of the table statistically significant? Let $\alpha = 0.01$.

|  |  | TIRES | | |
|---|---|---|---|---|
|  |  | Regal | AK-400 | Roadway |
| DRIVER | MM | 42 | 36 | 31 |
|  | GF | 44 | 43 | 40 |
|  | PR | 39 | 32 | 33 |
|  | ME | 42 | 38 | 24 |
|  |  | 41.75 | 37.25 | 34.50 |

15.50   From Equation 15.1, the expected value for $X$, the observation representing the $j$th treatment and belonging to the $i$th block, is $\mu_j + \beta_i$. The *estimated* expected value for $X$ is $\bar{x}_j + (1/k)B_i - \bar{x}$. Suppose $\bar{x}_j + (1/k)B_i - \bar{x}$ is subtracted from each observation in a randomized block experiment. What would the distribution of differences look like?

15.51   A nursery is studying the effect of four fertilizers on the growth of redbud trees. Eight trees have been planted on each of four sites chosen at random. Each group of eight is divided into four pairs, and the two trees in a pair are given one of the fertilizers for a period of one year. Listed below are the estimated average annual growths (in inches) for each pair. Test whether the differences among the $\bar{x}_j$'s are statistically significant. Let $\alpha = 0.05$.

|  |  | FERTILIZER | | | |
|---|---|---|---|---|---|
|  |  | Medac | PNK-50 | Smith's | Biolite |
| SITES | 1 | 8.0 | 9.5 | 7.0 | 8.5 |
|  | 2 | 7.5 | 8.5 | 7.0 | 8.0 |
|  | 3 | 9.0 | 8.5 | 9.5 | 7.5 |
|  | 4 | 7.5 | 8.0 | 7.5 | 8.0 |

# 16 Nonparametric Statistics

*If your experiment needs statistics,
you ought to have done a better experiment.*

Lord Rutherford

# 16.1

## INTRODUCTION

Behind every hypothesis test and every confidence interval we have considered has been a specific set of assumptions about the population (or populations) from which the data were drawn. In comparing two proportions, $H_0: p_X = p_Y$ versus $H_1: p_X \neq p_Y$, we assume that one sample is derived from $n$ *independent Bernoulli trials* and the other from $m$ *independent Bernoulli trials*. The most common condition presumed for a set of data, of course, is that each observation comes from a *normal* population (see Theorems 8.2, 8.3, 9.1, 9.2, 10.1, 10.2, and 11.1).

But suppose our assumptions about the origin of the data prove to be incorrect? Do the computing formulas for the test statistic or the rules that tell us when to reject $H_0$ change? No. What *does* change is the sampling distribution of the test statistic. In the case of a one-sample $t$ test, for instance, if the $Y_i$'s are not a random sample from a normal distribution, the probabilistic behavior of $T = (\bar{Y} - \mu_0)/(S/\sqrt{n})$ is *not* described by a Student $t$ curve with $n - 1$ degrees of freedom. So what we *think* is the probability, $\alpha$, of committing a Type I error may be numerically quite different from the *actual* probability of the test statistic falling into the critical region (see Figure 16.1).

One approach taken by statisticians in response to the implications of Figure 16.1 has been to show that unless the $Y_i$'s are *markedly* nonnormal, the test statistic $T$ has still approximately a Student $t$ distribution with $n - 1$ degrees of freedom (in which case the presumed $\alpha$ will be close to the actual $\alpha$). The ability of sampling distributions to remain relatively stable in the face of violated assumptions is called *robustness* (see Section 9.2).

By and large, robustness studies done on $t$ tests, $\chi^2$ tests, and $F$ tests have been reassuring. Actual $\alpha$'s and presumed $\alpha$'s are often quite similar, even when assumptions underlying those tests are substantially violated. Nevertheless, when sample sizes are small and one or more of the assumptions are not even close to being satisfied, the effects on the sampling distribution of the test statistic may be too large to ignore.

To deal with problems where an appeal to robustness is not sufficient, statisticians have developed procedures for which the distribution of the test statistic is independent of the nature of the populations being sampled. The data, in other words, can be normal or nonnormal. Any procedure having that sort of latitude is called *nonparametric* or, more appropriately, *distribution-free*.

Although the use of rudimentary distribution-free techniques can be traced to the eighteenth century, what we now recognize as nonparametric statistics began in earnest in the 1930s. Since then, the number of such procedures has multiplied dramatically. Our objective here is not to cover nonparametric inference in any great depth; instead we will look at some of its approaches in the context of problems whose "parametric" solutions we have already discussed.

680

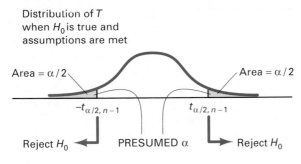

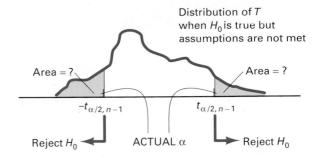

Distribution of $T$
when $H_0$ is true and
assumptions are met

Area $= \alpha/2$

Area $= \alpha/2$

$-t_{\alpha/2,\, n-1}$

$t_{\alpha/2,\, n-1}$

Reject $H_0$ ← PRESUMED $\alpha$ → Reject $H_0$

Distribution of $T$
when $H_0$ is true but
assumptions are not met

Area $= ?$

Area $= ?$

$-t_{\alpha/2,\, n-1}$

$t_{\alpha/2,\, n-1}$

Reject $H_0$ ← ACTUAL $\alpha$ → Reject $H_0$

**FIG. 16.1**

# 16.2

## THE SIGN TEST

**sign test**—A nonparametric procedure applicable to either one-sample data or paired data. The sign test is one of the most general of all nonparametric procedures—it assumes only that the pdf being sampled is continuous.

### One-Sample Problems

Among the simplest nonparametric procedures is the *sign test*. It can be used for one-sample tests of location as well as for the paired data situation described in Section 11.2.

Sign tests focus on a population's *median* instead of its *mean*. Recall that the median of a continuous random variable $Y$ is that value $\tilde{\mu}$ such that $P(Y < \tilde{\mu}) = P(Y > \tilde{\mu}) = 0.5$. If a random variable $Y$ has a symmetric pdf—that is, if $f_Y(\tilde{\mu} - y) = f_Y(\tilde{\mu} + y)$, for all $y$—then the mean and the median coincide. For asymmetric distributions, they may not. The mean, for example, of a chi-square random variable with $n$ degrees of freedom is $n$, but its median is less than $n$. For many one-sample problems the issue of prime importance is whether a shift in location has occurred. It matters very little whether that shift is measured in terms of a mean or a median. That being the case, testing $H_0: \tilde{\mu} = \tilde{\mu}_0$ is just as appropriate as testing $H_0: \mu = \mu_0$.

Suppose a random sample $Y_1, Y_2, \ldots, Y_n$ comes from a continuous pdf with a median that we think is $\tilde{\mu}_0$. If the $Y_i$'s are consistent with our presumption, then roughly half should lie above $\tilde{\mu}_0$ and half should lie below. We can use that idea to develop a formal procedure for testing whether the true median $\tilde{\mu}$ is equal to the presumed median $\tilde{\mu}_0$.

*Procedure for testing $H_0: \tilde{\mu} = \tilde{\mu}_0$*

1. Compute $y_i - \tilde{\mu}_0$, $i = 1, 2, \ldots, n$.
2. Replace each $y_i - \tilde{\mu}_0$ with its sign ($+$ or $-$). Discard the observation if $y_i - \tilde{\mu}_0 = 0$.
3. Let $X$ denote the *number* of $+$ signs. (If $H_0: \tilde{\mu} = \tilde{\mu}_0$ is true, $X$ is a binomial random variable with $p = \frac{1}{2}$, where $p = P(Y_i - \tilde{\mu}_0 > 0)$.)
4. Use Theorem 8.1 to test $H_0: p = \frac{1}{2}$.

*EXAMPLE 16.1*　In Case Study 9.3 we tested a theory claiming that certain aesthetic ideals are "universal" and not culture-dependent. Rectangles provided the context. Early Greek philosophers had argued that rectangles whose width-to-length $(w/l)$ ratios are approximately 0.618 are aesthetically more pleasing than other rectangles. The actual experiment consisted of measuring $w/l$ ratios of 20 rectangles sewn on leather goods by Shoshoni Indians, a northwest American tribe with no exposure to writings of ancient Greeks (recall Table 9.4). Letting $\mu$ denote the true average $w/l$ ratio preferred by the Shoshonis, we tested

$$H_0: \quad \mu = 0.618$$

vs.

$$H_1: \quad \mu \neq 0.618$$

and accepted $H_0$.

The observed $t$ ratio in that analysis ($= 2.05$) was disturbingly close to the right-hand critical value ($= 2.0930$). Suppose the 20 $w/l$ ratios making up the data are *not* samples from a normal distribution. The proximity of the observed $t$ to the critical $t$ suggests that even a moderate violation of the normality assumption might affect the sampling distribution of $T$ enough to produce a different conclusion. In light of that possibility, it makes sense to reexamine these data by using a nonparametric analysis.

*Solution*　Let $\tilde{\mu}$ be the true median $w/l$ ratio preferred by the Shoshonis. Table 16.1 shows steps 1 and 2 for testing $H_0: \tilde{\mu} = 0.618$. If $p$ is the probability that an observation lies to the right of 0.618, then $p$ should be 1/2 if the median is actually 0.618. Testing $H_0: p = 1/2$ versus $H_1: p \neq 1/2$ is therefore equivalent to testing $H_0: \tilde{\mu} = 0.618$ versus $H_1: \tilde{\mu} \neq 0.618$. Let $\alpha = 0.05$.

**TABLE 16.1**

| $y$ | $y - 0.618$ | Sign | $y$ | $y - 0.618$ | Sign |
|-------|-------------|------|-------|-------------|------|
| 0.693 | 0.075 | + | 0.654 | 0.036 | + |
| 0.662 | 0.044 | + | 0.615 | −0.003 | − |
| 0.690 | 0.072 | + | 0.668 | 0.050 | + |
| 0.606 | −0.012 | − | 0.601 | −0.017 | − |
| 0.570 | −0.048 | − | 0.576 | −0.042 | − |
| 0.749 | 0.131 | + | 0.670 | 0.052 | + |
| 0.672 | 0.054 | + | 0.606 | −0.012 | − |
| 0.628 | 0.010 | + | 0.611 | −0.007 | − |
| 0.609 | −0.009 | − | 0.553 | −0.065 | − |
| 0.844 | 0.226 | + | 0.933 | 0.315 | + |

The first part of Theorem 8.1 gives the appropriate decision rule. Since $\alpha = 0.05$ and $H_1$ is two-sided, we should reject $H_0$ if $z \leqslant -1.96$ or $z \geqslant 1.96$ (see Figure 16.2). From Table 16.1 the observed number of $+$'s ($=y$) is 11.

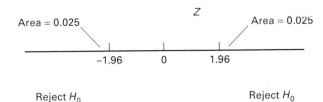

**FIG. 16.2**

Therefore,

$$z = \frac{y - np_0}{\sqrt{np_0(1 - p_0)}} = \frac{11 - 20(\frac{1}{2})}{\sqrt{20(\frac{1}{2})(1 - \frac{1}{2})}} = \frac{11 - 10}{\sqrt{5}} = 0.45$$

Both the $t$ test and the sign test, therefore, reach the same conclusion: accept $H_0$. The $w/l$ ratios in Table 16.1 are not inconsistent with the theory that Shoshoni Indians, like the Greeks, preferred the golden rectangle.

---

QUESTIONS

16.1   Analyze the airflow rates in Table 2.2 with a sign test. Let $\tilde{\mu}$ = true median $FEV_1/VC$ ratio for workers exposed to the *Bacillus subtilis* enzyme. Test $H_0$: $\tilde{\mu} = 0.80$ versus $H_1$: $\tilde{\mu} < 0.80$. Let $\alpha = 0.05$.

16.2   For the motivation study described in Case Study 3.2, is it reasonable to assume that the 106 observations in Table 3.5 represent a random sample from a population whose median is 13.7? Answer the question by setting up and carrying out an appropriate sign test. Let $\alpha = 0.05$.

16.3   Let $\tilde{\mu}$ denote the true median annual growth rate for major U.S. chemical companies. Use the data in Question 3.5 to test $H_0$: $\tilde{\mu} = 17.0\%$ versus $H_1$: $\tilde{\mu} < 17.0\%$. Choose between $H_0$ and $H_1$ on the basis of a sign test. Let $\alpha = 0.01$.

16.4   From 1832 to 1950, Hawaii's Mauna Loa volcano erupted 37 times. The data in Question 3.6 are the intervals (in months) between consecutive eruptions. At the 0.10 level is it reasonable to conclude that the true median interval between consecutive eruptions is 43.5 months? Do the appropriate sign test.

16.5   (a) Reexamine the pregnancy data in Case Study 8.2 by doing a sign test on $H_0$: $\tilde{\mu} = 266$ versus $H_1$: $\tilde{\mu} < 266$, where $\tilde{\mu}$ is the true median pregnancy length for women living in urban Nashville. Let $\alpha = 0.05$. Before doing the test, remove from the data all the $y_i$'s whose value *equals* the number specified for $\tilde{\mu}$ in $H_0$.

(b) Does your conclusion agree with the $z$ test?

16.6   (a) Test $H_0$: $\tilde{\mu} = 132.4$ versus $H_1$: $\tilde{\mu} \neq 132.4$ for the Etruscan skull data in Table 9.1. Let $\alpha = 0.05$.

(b) Does the sign test agree with the $t$ test?

### Paired Data

Applications of sign tests are not confined to one-sample problems—they can also be used with paired data (recall Section 11.3). Suppose $n$ sets of dependent observations,

$$(X_1, Y_1), (X_2, Y_2), \ldots, (X_n, Y_n)$$

are measured on two treatments, $X$ and $Y$, whose true means are $\mu_X$ and $\mu_Y$, respectively. As a measure of the relative location of $X$ and $Y$, define

$$p = P(Y_i > X_i)$$

If the $X$ and $Y$ populations have *not* shifted with respect to one another, $p$ should equal $\frac{1}{2}$. Testing $H_0\colon p = \frac{1}{2}$, therefore, is the nonparametric analog of testing $H_0\colon \mu_D = 0$, where $\mu_D = \mu_Y - \mu_X$.

*Procedure for Testing $H_0\colon \mu_D = 0$*

1. Compute $y_i - x_i$, $i = 1, 2, \ldots, n$.
2. Replace each $y_i - x_i$ with its sign $(+$ or $-)$. Discard $x_i$ and $y_i$ if $y_i - x_i = 0$.
3. Let $W$ denote the number of $+$ signs. (If $H_0\colon \mu_D = 0$ is true, $W$ is a binomial random variable with $p = \frac{1}{2}$, where $p = P(Y_i - X_i > 0)$.)
4. If $n$ is sufficiently large (see the footnote to Theorem 6.7), let $z = (w - n/2)/\sqrt{n/4}$ and use Theorem 8.1 to test $H_0\colon p = \frac{1}{2}$. For small $n$, test $H_0\colon p = \frac{1}{2}$ by using the binomial pdf,

$$P(W = w) = {}_nC_w \left(\frac{1}{2}\right)^w \left(1 - \frac{1}{2}\right)^{n-w}$$

to find an appropriate rejection region.

---

### CASE STUDY 16.1

*Drugs as Learning Facilitators*    Children with severe learning problems often have electroencephalograms and behavior patterns similar to those of children with petit mal, a mild form of epilepsy. Realizing that, scientists have speculated that drugs helpful in treating petit mal might also be useful as learning facilitators (157). To test that hypothesis, experimenters recruited 10 children, ranging in age from 8 to 14 and all having a history of learning and behavioral problems, to participate in a six-week study. A child was given a placebo for three weeks and ethosuximide, a widely prescribed anticonvulsant, for the other three weeks. After each three-week period all 10 children were given several parts of the standard Wechsler IQ test. Because a child might be expected to do better on the test the second time he or she took it, the order in which the placebo and the ethosuximide were administered was randomized; some children were given the placebo for the first three weeks while the others were given ethosuximide.

**TABLE 16.2**

| Child | IQ after Placebo, $x_i$ | IQ after Ethosuximide, $y_i$ | $y_i - x_i$ |
|-------|-------------------------|------------------------------|-------------|
| 1 | 97 | 113 | + |
| 2 | 106 | 113 | + |
| 3 | 106 | 101 | − |
| 4 | 95 | 119 | + |
| 5 | 102 | 111 | + |
| 6 | 111 | 122 | + |
| 7 | 115 | 121 | + |
| 8 | 104 | 106 | + |
| 9 | 90 | 110 | + |
| 10 | 96 | 126 | + |

Table 16.2 shows the two verbal IQ scores recorded for each subject. Entries in the fourth column are the +'s or −'s determined by the differences $y_i - x_i$.

Let $p = P(Y_i > X_i)$. If ethosuximide has no effect on a child's IQ, the probability of a $Y_i$ exceeding an $X_i$ is necessarily 1/2. One way of deciding, then, whether the drug should be pursued as a possible learning facilitator is to test

$$H_0: \quad p = \tfrac{1}{2}$$

vs.

$$H_1: \quad p \neq -\tfrac{1}{2}$$

(A two-sided alternative is being used to allow for the possibility that ethosuximide might actually have an *adverse* affect on IQ.) Let $\alpha = 0.10$.

Define $W$ to be the number of +'s among the response differences for the 10 children. If $H_0: p = \tfrac{1}{2}$ is true, then

$$f_W(w) = P(W = w) = {}_{10}C_w (\tfrac{1}{2})^w (1 - \tfrac{1}{2})^{10-w}, \qquad w = 0, 1, \dots, 10$$

Since $n$ is small, we should use the exact binomial, rather than a normal approximation, to construct our decision rule. Specifically, we should reject $H_0$ if $w \leqslant w_1^*$ or $w \geqslant w_2^*$, where $w_1^*$ and $w_2^*$ are chosen so that

$$P(W \leqslant w_1^* \,|\, H_0 \text{ is true}) = \sum_{w=0}^{w_1^*} {}_{10}C_w \left(\frac{1}{2}\right)^w \left(1 - \frac{1}{2}\right)^{10-w} \doteq 0.05$$

and

$$P(W \geqslant w_2^* \,|\, H_0 \text{ is true}) = \sum_{w=w_2^*}^{10} {}_{10}C_w \left(\frac{1}{2}\right)^w \left(1 - \frac{1}{2}\right)^{10-w} \doteq 0.05$$

By trial and error, $w_1^* = 2$ and $w_2^* = 8$. That is, we should reject $H_0$: $p = \tfrac{1}{2}$ if the observed number of +'s is either less than or equal to 2 or greater than or equal to 8. From the last column of Table 16.2, $w = 9$. It appears, therefore, that ethosuximide *does* affect IQ. The frequency with which the ethosuximide IQ exceeded the placebo IQ is too great to be written off to chance.

16.7    In a marketing research study 28 adult males were asked to shave one side of their faces with one brand of razor blade and the other side with a second brand. They were to use the blades for seven days and then decide which blade gave the smoother shave. Suppose that 19 of the subjects preferred brand A. Use a sign test to determine whether it can be claimed at the 0.05 level that both brands are equally effective.

16.8    In Case Study 16.1 what could be concluded if the after enthosuximide IQs were significantly higher than the after placebo IQs, but if all 10 children had been given the placebo first?

16.9    A new dairy is doing a survey on the popularity of its products. Milk purchases at each of 18 supermarkets are monitored over a 1-hr period. Shown below are the numbers of gallons purchased from the new dairy and its leading competitor. Test at the 0.05 level whether consumers are showing a brand preference.

| | GALLONS SOLD | |
|---|---|---|
| Store | New Brand | Established Brand |
| 1 | 5 | 7 |
| 2 | 3 | 6 |
| 3 | 8 | 6 |
| 4 | 9 | 11 |
| 5 | 12 | 9 |
| 6 | 8 | 9 |
| 7 | 3 | 12 |
| 8 | 7 | 5 |
| 9 | 8 | 14 |
| 10 | 7 | 12 |
| 11 | 9 | 11 |
| 12 | 6 | 9 |
| 13 | 10 | 9 |
| 14 | 6 | 11 |
| 15 | 7 | 8 |
| 16 | 17 | 15 |
| 17 | 14 | 18 |
| 18 | 6 | 8 |

16.10    Use a sign test to analyze the tensile strength data in Question 2.30. Define the parameter being tested. Let $\alpha = 0.05$ and make $H_1$ two-sided.

16.11    For Case Study 3.13 let $p = P(Y_i > X_i)$ and test $H_0$: $p = \frac{1}{2}$ versus $H_1$: $p < \frac{1}{2}$ at the 0.10 level of significance. Express the decision rule in terms of the exact binomial model (i.e., not a normal approximation).

16.12    Analyze the ESP data in Table 3.23 with a sign test. Make $H_1$ one-sided and let $\alpha = 0.05$. Phrase the decision rule in terms of the normal distribution.

16.13    Use the exact binomial model to find the 0.15 decision rule for the depth perception data in Question 3.59. Use a two-sided alternative.

16.14    In Case Study 11.1 seven of the eight chickens experienced a fall in pecking order following their unpleasant encounter with a rooster. Is seven out of eight statistically significant at the 0.05 level? Set up and carry out an appropriate sign test.

16.15    (a) Find the $P$ value for the tracheobronchial clearance data in Table 11.3. Use the exact binomial.

      (b) Is your $P$ value consistent with the conclusion reached by the $t$ test?

# 16.3

## THE WILCOXON SIGNED RANK TEST

### Small Samples ($n \le 12$)

**Wilcoxon signed rank test**—Like the sign test, a procedure for testing either $H_0: \tilde{\mu} = \tilde{\mu}_0$ or $H_0: \tilde{\mu}_D = 0$. The signed rank test is more powerful than the sign test but its assumptions are more restrictive—it requires that the pdf sampled be symmetric as well as continuous.

The **Wilcoxon signed rank test** was one of the first "modern" nonparametric procedures to be developed (it dates to 1945), and it remains one of the most widely used. Like the sign test, the Wilcoxon procedure can handle one-sample problems and paired data. The assumptions underlying a Wilcoxon test, though, are a bit more restrictive than what we encountered in Section 16.2. In particular, the pdf (or pdfs) being sampled is assumed to be *symmetric* as well as continuous. (A pdf $f_Y(y)$ is symmetric if $f_Y(y - \tilde{\mu}) = f_Y(y + \tilde{\mu})$ for all $y$. The point $\tilde{\mu}$ is necessarily both the mean and the median.)

Let $Y_1, Y_2, \ldots, Y_n$ be a random sample from an unspecified continuous and symmetric pdf $f_Y(y)$. We wish to test

$$H_0: \quad \tilde{\mu} = \tilde{\mu}_0$$

vs.

$$H_1: \quad \tilde{\mu} \ne \tilde{\mu}_0$$

where $\tilde{\mu}_0$ is some specified value of the median.

The Wilcoxon test is based on both the signs as well as the relative magnitudes of the differences $y_i - \tilde{\mu}_0$, $i = 1, 2, \ldots, n$.

*Procedure for Computing the Wilcoxon Test Statistic W*

1. Compute the differences $y_1 - \tilde{\mu}_0, y_2 - \tilde{\mu}_0, \ldots, y_n - \tilde{\mu}_0$.
2. Rank the differences from smallest to largest, *ignoring the signs*. (A "$-4$," in other words, would be considered larger than a "2"). Assign the smallest number a rank of 1, the second smallest a rank of 2, and so on.
3. Write the rank next to each difference.
4. Change the ranks associated with negative differences to zeros.
5. Sum the revised ranks; that is, add the ranks associated with the positive differences. We call that sum W, the *Wilcoxon signed rank statistic*.

Table 16.3 shows the computation of $w$ for a test of $H_0: \tilde{\mu} = 7$, where the data consist of a random sample of size four: 12, 6, 3, and 9. The value for $W$ is 6.

**TABLE 16.3**

| Data $(y_i)$ | Differences $(y_i - \tilde{\mu}_0)$ | Ordered Differences | Ranks | Revised Ranks |
|---|---|---|---|---|
| 12 | $12 - 7 = \ \ 5$ | $-1$ | 1 | 0 |
| 6 | $6 - 7 = -1$ | 2 | 2 | 2 |
| 3 | $3 - 7 = -4$ | $-4$ | 3 | 0 |
| 9 | $9 - 7 = \ \ 2$ | 5 | 4 | 4 |
| | | | $W = $ | 6 |

COMMENT: Notice that $W$ is based on the *ranks* of the deviations from $\tilde{\mu}_0$, and not on the deviations themselves. If the 12 in the first row of Table 16.3, for example, were 12.7 or 11.5, the value of $W$ would still be 6, since the ranks in columns 4 and 5 would not change. It is precisely that sort of "insensitivity" that is necessary if a procedure is to be nonparametric.

The smallest possible value of $W$ is zero, which would occur if all the original differences $(y_i - \tilde{\mu}_0)$ were negative. If all the differences were positive, on the other hand, $W$ would be as large as possible: its value in that case would be

$$1 + 2 + \cdots + n = \frac{n(n + 1)}{2}$$

Finding a decision rule based on $W$ is conceptually straightforward. If $H_0$ is true, we would expect the $Y_i$'s to be distributed at random above and below $\tilde{\mu}_0$, meaning that $W$ should tend to equal a number close to $n(n + 1)/4$, which is halfway between $W$'s smallest and largest values. If the observed $w$ is too far from $n(n + 1)/4$, in the direction of $H_1$, it makes sense to reject $H_0$.

Suppose, for example, the data consists of seven observations and the hypotheses to be tested are $H_0: \tilde{\mu} = \tilde{\mu}_0$ versus $H_1: \tilde{\mu} \neq \tilde{\mu}_0$. For $n = 7$, $W$ can range from 0 to $7(7 + 1)/2 = 28$. Depending on the desired $\alpha$, the appropriate critical region is a set of numbers in either tail of the $W$ distribution: for example, $(0, 28)$, $(0, 1, 27, 28)$, or $(0, 1, 2, 26, 27, 28)$. In general, the critical region $C$ is the set

$$C = \{w: \quad w \leqslant w_1^* \quad \text{or} \quad w \geqslant w_2^*\}$$

where

$$P(W \leqslant w_1^* \quad \text{or} \quad W \geqslant w_2^* \,|\, H_0 \text{ is true}) \leqslant \alpha$$

We should *reject* $H_0$ when $w \leqslant w_1^*$ or $w \geqslant w_2^*$.

Finding $w_1^*$ and $w_2^*$ directly is not easy because $f_W(w)$ is a complicated pdf. Fortunately, tables are available that simplify the problem enormously. Appendix A.6 gives values of $w_1^*$ and $w_2^*$ for a variety of $\alpha$ levels and for values of $n$ from 4 to 12. When $n = 7$, for example,

$$P(W \leqslant 2) + P(W \geqslant 26) = 0.023 + 0.023 = 0.046 \doteq 0.05$$

At approximately the 0.05 level, in other words, the null hypothesis $H_0$: $\tilde{\mu} = \tilde{\mu}_0$ will be rejected if the observed rank sum $w$ is either (1) less than or equal to 2 or (2) greater than or equal to 26.

---

**CASE STUDY 16.2**

*Swell Sharks*  Swell sharks (*Cephaloscyllium ventriosum*) are small reef-dwelling fish that inhabit the California coastal waters south of Monterey Bay. A second concentration is found near Catalina Island. Ichthyologists have hypothesized that the two populations never mix (6). Separating Santa Catalina from the mainland is a deep basin, and it may be that that basin keeps the coastal sharks away from the Catalina sharks.

One way of testing the "no-mixing" theory is to compare physical features of sharks caught in the two regions. If the populations have, indeed, remained separate, we would expect a certain number of morphological differences to have evolved. Table 16.4 lists the total length, the height of the first dorsal fin, and the length-to-fin ratio for 10 male swell sharks caught near Catalina ($13.32 = 906/68$, etc.).

Previous studies have suggested that the median length-to-fin ratio for *coastal* sharks is 14.60. Is the latter figure consistent with the data in Table 16.4? More formally, let $\tilde{\mu}$ denote the true median length-to-fin ratio for Catalina sharks. Can we reject $H_0$: $\tilde{\mu} = 14.60$, and thereby lend support to the no-mixing theory?

**TABLE 16.4**

| Total Length (mm) | Dorsal Fin Height (mm) | Ratio |
|---|---|---|
| 906 | 68 | 13.32 |
| 875 | 67 | 13.06 |
| 771 | 55 | 14.02 |
| 700 | 59 | 11.86 |
| 869 | 64 | 13.58 |
| 895 | 65 | 13.77 |
| 662 | 49 | 13.51 |
| 750 | 52 | 14.42 |
| 794 | 55 | 14.44 |
| 787 | 51 | 15.43 |

**TABLE 16.5**

| $y_i$ | $y_i - \tilde{\mu}_0$ | Ordered Diff. (ignoring signs) | Ranks | Revised Ranks |
|---|---|---|---|---|
| 13.32 | −1.28 | −0.16 | 1 | 0 |
| 13.06 | −1.54 | −0.18 | 2 | 0 |
| 14.02 | −0.58 | −0.58 | 3 | 0 |
| 11.86 | −2.74 | −0.83 | 4.5 | 0 |
| 13.58 | −1.02 | 0.83 | 4.5 | 4.5 |
| 13.77 | −0.83 | −1.02 | 6 | 0 |
| 13.51 | −1.09 | −1.09 | 7 | 0 |
| 14.42 | −0.18 | −1.28 | 8 | 0 |
| 14.44 | −0.16 | −1.54 | 9 | 0 |
| 15.43 | 0.83 | −2.74 | 10 | 0 |

Table 16.5 shows the signed rank procedure applied to the set of 10 observations in Table 16.4. Two of the entries in column 4—the 4.5's—deserve an explanation. In general, when two or more $(y_i - \tilde{\mu}_0)$'s have the same value, irrespective of sign, each is given the *average* of the ranks they otherwise would have received. Here,

$$|y_4 - 14.60| = |y_5 - 14.60| = 0.83$$

Furthermore, we can see from column 3 that 4 and 5 are the two positions for which those two deviations are competing. By convention, each is given a rank of $(4 + 5)/2$, or 4.5.

Should $H_1$ here be one-sided or two-sided? *Two-sided*, since $\tilde{\mu}$'s on either side of 14.60 will prompt the same ultimate conclusion. Specifically, we will test

$$H_0: \quad \tilde{\mu} = 14.60$$

vs.

$$H_1: \quad \tilde{\mu} \neq 14.60$$

at the $\alpha = 0.05$ level of significance.

Appropriate values for $w_1^*$ and $w_2^*$ can be found from Appendix A.6. With $n = 10$,

$$P(W \leqslant 8) + P(W \geqslant 47) = 0.024 + 0.024 = 0.048 \doteq 0.05$$

It follows that $H_0$ should be rejected if $w \leqslant 8$ or $w \geqslant 47$. Adding the entries in the last column of Table 16.5 gives $w = 4.5$, a rank sum lying to the left of $w_1^*$. Our conclusion, therefore, is to reject $H_0$. It appears that these two populations have *not* recently mixed.

---

**QUESTIONS**

16.16 One benefit of our space program has been the refinement in a variety of measurements pertaining to the solar system. Based on data received from early *Ranger* probes, scientists estimated the ratio $M$ of the mass of the earth to the mass of the moon to be approximately 81.3035. More sophisticated, techniques used on seven later flights have yielded a second set of estimates (2):

| Spacecraft | Estimate of $M$ |
|---|---|
| *Mariner 2* | 81.3001 |
| *Mariner 4* | 81.3015 |
| *Mariner 5* | 81.3006 |
| *Mariner 6* | 81.3011 |
| *Mariner 7* | 81.2997 |
| *Pioneer 6* | 81.3005 |
| *Pioneer 7* | 81.3021 |

Are these more recent data consistent with the original value of 81.3035? Test the appropriate hypotheses using the signed rank procedure. Let $\alpha$ be approximately 0.05.

16.17 For the invention data in Question 1.1, suppose a theory is advanced claiming that a typical conception-to-realization interval is longer than 10 years. Set up and carry out an appropriate hypothesis test. Let $\alpha \doteq 0.10$.

16.18   Let $\tilde{\mu}$ denote the true median age at which scientists make their greatest discoveries. Use the data in Case Study 2.2 to test $H_0$: $\tilde{\mu} = 30$ versus $H_1$: $\tilde{\mu} > 30$. Let $\alpha = 0.01$.

16.19   In the survey of commercially laundered napkins described in Question 2.9, the contamination level (in terms of bacteria per square inch) was the measurement recorded. Suppose $\tilde{\mu}$ denotes the true median bacteria count characteristic of commercially laundered napkins. Test $H_0$: $\tilde{\mu} = 300{,}000$ against a two-sided alternative. Let $\alpha \doteq 0.05$.

16.20   Survival times for 11 leukemia patients receiving bone marrow transplants were listed in Case Study 3.7. Can it be claimed at the 0.05 level of significance that the median survival time is longer than 50 days? Set up and carry out an appropriate hypothesis test.

16.21   (a) Analyze the $d$-hyoscyamine hydrobromide data in Table 9.2 with a signed rank test. Let $\alpha \doteq 0.10$.

   (b) Do a similar analysis on the $l$ isomer data in Table 9.3.

16.22   Use a signed rank test on the data in Question 9.12. Define an appropriate $H_0$ and $H_1$. Let $\alpha \doteq 0.05$.

16.23   To measure the effect of mild intoxication on coordination, researchers gave 13 subjects 15.7 mL of ethyl alcohol per square meter of body surface area, and asked them to write a certain phrase as many times as they could in the space of 1 min (110). The number of correctly written letters was then counted and scaled, with 0 representing the score that a subject *not* under the influence of alcohol would be expected to achieve. Negative scores indicate decreased writing speeds; positive scores, increased writing speeds. The results are listed below.

| Subject | Score |
|---------|-------|
| 1 | $-6$ |
| 2 | 10 |
| 3 | 9 |
| 4 | $-8$ |
| 5 | $-6$ |
| 6 | $-2$ |
| 7 | 20 |
| 8 | 0 |
| 9 | $-7$ |
| 10 | 5 |
| 11 | $-9$ |
| 12 | $-10$ |
| 13 | $-2$ |

Use a signed rank test to determine whether the level of alcohol provided in this study has any effect on writing speed. Let $\alpha = 0.05$. Delete subject 8 from your analysis.

### Large Samples ($n > 12$)

For sample sizes less than or equal to 12, critical values for signed rank tests are found in Appendix A.6. When $n > 12$, we use the normal curve to approximate the distribution of $W$. Critical values in that case come from Appendix A.1.

---

**THEOREM 16.1**   Computing the Wilcoxon Signed Rank Statistic for Large Samples

Let $W$ be the Wilcoxon signed rank statistic calculated from a random sample $Y_1, Y_2, \ldots, Y_n$. The mean and variance of $W$ are

$$E(W) = \frac{n(n+1)}{4} \quad \text{and} \quad \text{Var}(W) = \frac{n(n+1)(2n+1)}{24}$$

For $n \geqslant 12$, the distribution of

$$W' = \frac{W - E(W)}{\sqrt{\text{Var}(W)}}$$

can be adequately approximated by a standard normal curve.

---

*EXAMPLE 16.2*   The first column of Table 16.6 lists the Shoshoni rectangle data from Case Study 9.3. The last four columns show the steps required for computing the signed rank statistic $w$. If $\tilde{\mu}$ denotes the true median width-to-length ratio preferred by the Shoshonis, our objective is to test

$$H_0: \quad \tilde{\mu} = 0.618$$

vs.

$$H_1: \quad \tilde{\mu} \neq 0.618$$

Let $\alpha = 0.05$.

*Solution*   Here $n = 20$, so the large-sample approximation described in Theorem 16.1 is appropriate. Specifically,

$$E(W) = \frac{20(20+1)}{4} = 105$$

$$\text{Var}(W) = \frac{20(20+1)(40+1)}{24} = 717.5$$

and

$$W' = \frac{W - 105}{\sqrt{717.5}} \doteq Z$$

**TABLE 16.6**

| $y_i$ | $y_i - 0.618$ | Ordered Differences | Ranks | Revised Ranks |
|---|---|---|---|---|
| 0.693 | 0.075 | −0.003 | 1 | 0 |
| 0.662 | 0.044 | −0.007 | 2 | 0 |
| 0.690 | 0.072 | −0.009 | 3 | 0 |
| 0.606 | −0.012 | 0.010 | 4 | 4 |
| 0.570 | −0.048 | −0.012 | 5 | 0 |
| 0.749 | 0.131 | −0.012 | 6 | 0 |
| 0.672 | 0.054 | −0.017 | 7 | 0 |
| 0.628 | 0.010 | 0.036 | 8 | 8 |
| 0.609 | −0.009 | −0.042 | 9 | 0 |
| 0.844 | 0.226 | 0.044 | 10 | 10 |
| 0.654 | 0.036 | −0.048 | 11 | 0 |
| 0.615 | −0.003 | 0.050 | 12 | 12 |
| 0.668 | 0.050 | 0.052 | 13 | 13 |
| 0.601 | −0.017 | 0.054 | 14 | 14 |
| 0.576 | −0.042 | −0.065 | 15 | 0 |
| 0.670 | 0.052 | 0.072 | 16 | 16 |
| 0.606 | −0.012 | 0.075 | 17 | 17 |
| 0.611 | −0.007 | 0.131 | 18 | 18 |
| 0.553 | −0.065 | 0.226 | 19 | 19 |
| 0.933 | 0.315 | 0.315 | 20 | 20 |
| | | | $w =$ | 151 |

Since $H_1$ is two-sided, values of $w$ that are either too much smaller than 105 or too much larger than 105 will be grounds for rejecting $H_0$. With $\alpha$ set at 0.05, an equivalent decision rule, in terms of $w'$, is to reject $H_0$ if $w' \leqslant -1.96$ or $w' \geqslant 1.96$.

Note from the bottom of the last column in Table 16.6 that $w$ for these data is 151. Therefore,

$$w' = \frac{151 - 105}{\sqrt{717.5}} = 1.72$$

and our conclusion is to accept $H_0$. At the 0.05 level we cannot rule out the possibility that the Shoshoni Indians preferred exactly the same kind of rectangles as the ancient Greeks.

COMMENT: The signed rank test can also be extended to paired data problems in a very natural way. First, compute the within-pair response differences $d_i = y_i - x_i$, $i = 1, 2, \ldots, n$. Then apply the procedure on pp. 687–688 to the $d_i$'s (replace $\tilde{\mu}_0$ by 0). The null hypothesis being tested is $H_0$: $\tilde{\mu}_d = 0$, where $\tilde{\mu}_d$ is the true median of the response differences. As before, if $n \leqslant 12$, critical values are determined from Appendix A.6; if

$n > 12$, decision rules are phrased in terms of

$$w' = \frac{w - E(W)}{\sqrt{\text{Var}(W)}}$$

and critical values are obtained from Appendix A.1.

---

QUESTIONS

**16.24**   A study was done to evaluate the effectiveness of the drug cyclazocine for reducing a person's psychological dependence on heroin. Fourteen males, all chronic heroin addicts, were the subjects. After receiving the therapy for a prescribed period of time, each subject was asked a battery of questions that assessed his dependence on heroin. The responses were combined to form a $Q$ score. For addicts *not* treated with cyclazocine, the median $Q$ score is known from past experience to be 28. Scores recorded for the 14 subjects were 51, 53, 43, 36, 55, 55, 39, 43, 45, 27, 21, 26, 22, and 43. Do the data support the conclusion that cyclazocine is an effective treatment? Use the 0.05 level of significance. (Note: Higher $Q$ scores represent *less* psychological dependence (134).)

**16.25**   In Case Study 2.1 let $\tilde{\mu}$ denote the true median $\text{FEV}_1/\text{VC}$ ratio characteristic of workers exposed to the *Bacillus subtilis* enzyme.

   (a) Use the normal approximation to the signed rank statistic to test $H_0$: $\tilde{\mu} = 0.80$ versus $H_1$: $\tilde{\mu} < 0.80$. Let $\alpha = 0.05$.

   (b) Does your conclusion agree with the one-sample $t$ test (see Question 9.5)?

**16.26**   The lengths of 15 release chirps were listed in Case Study 3.3. Do those data support the hypothesis that the population being sampled has a median of 0.12 seconds? Set up and carry out an appropriate signed rank test at the $\alpha = 0.05$ level of significance.

**16.27**   Is it reasonable to assume that the platelet counts in Question 3.7 represent a random sample from a population whose median is 210,000? Do an appropriate test at the $\alpha = 0.10$ level of significance.

**16.28**   Let $\tilde{\mu}$ denote the true median cockpit noise level (in decibels) characteristic of commerical aircraft. Use the data in Question 3.8 to test $H_0$: $\tilde{\mu} = 76$ versus $H_1$: $\tilde{\mu} > 76$. Indicate the significance of your conclusion by computing a $P$ value.

**16.29**   The average energy expenditures for eight elderly women were estimated on the basis of information received from a battery-powered heart rate monitor worn by each subject. Two overall averages were calculated for each woman, one for summer months and one for winter months (145). The results (in kilocalories) are shown in the table. Let $\tilde{\mu}_d$ denote the true median difference between the summer and winter energy expenditures. Compute $y_i - x_i$, $i = 1, 2, \ldots, 8$, and use the Wilcoxon signed rank procedure to test $H_0$: $\tilde{\mu}_d = 0$ versus $H_1$: $\tilde{\mu}_d \neq 0$. Let $\alpha = 0.15$.

**AVERAGE DAILY ENERGY EXPENDITURES (kcal)**

| Subject | Summer, $x_i$ | Winter, $y_i$ |
|---------|---------------|---------------|
| 1 | 1458 | 1424 |
| 2 | 1353 | 1501 |
| 3 | 2209 | 1495 |
| 4 | 1804 | 1739 |
| 5 | 1912 | 2031 |
| 6 | 1366 | 934 |
| 7 | 1598 | 1401 |
| 8 | 1406 | 1339 |

16.30   Recall the tensile strength data in Question 2.30. Let $x_i$ and $y_i$ represent tensile strengths when annealing is done at moderate temperatures and high temperatures, respectively. Define $d_i = y_i - x_i$, $i = 1, 2, \ldots, 15$, and let $\tilde{\mu}_d$ denote the median of the population from which the $d_i$'s are a sample. Test $H_0: \tilde{\mu}_d = 0$ versus $H_1: \tilde{\mu}_d \neq 0$ by using the normal approximation to the signed rank statistic. Let $\alpha = 0.10$.

16.31   Do a signed rank test on the ESP data in Case Study 3.14. Omit the scores for student 14. Let $\alpha = 0.05$. Make the alternative hypothesis two-sided.

16.32   An attempt to alter a chicken's place in its pecking order was described in Case Study 11.1.

   Analyze the data in Table 11.1 with a signed rank test. Let $\alpha = 0.05$.

   Does your conclusion agree with the paired $t$ test?

16.33   Redo the statistical analysis in Case Study 11.3 by using a signed rank test. Keep $\alpha$ the same.

   Does your conclusion agree with the paired $t$ test?

# 16.4

## THE WILCOXON RANK SUM TEST

### Small Samples $(n + m \leqslant 12)$

We have seen that it is not uncommon for the same basic problem to have several potentially different solutions, with each solution being based on slightly different assumptions about the populations being sampled. The one-sample $t$ test, the sign test, and the signed rank test, for instance, are all aimed at the one-sample location problem. The first assumes normality, the second assumes that the data's pdf is continuous, and the third assumes that the pdf is continuous and symmetric. In much the same spirit, comparing two sets of dependent observations is sometimes best done *parametrically* (with a paired $t$ test) and sometimes *nonparametrically* (with a signed rank test).

**Wilcoxon rank sum test—A nonparametric procedure for the two-sample location problem.**

Section 16.4 continues this theme of exploring alternative solutions to familiar problems. Our objective here is the **Wilcoxon rank sum test**, a procedure widely used for comparing the locations of population $X$ and population $Y$ when the $X_i$'s and $Y_i$'s are independent but not normally distributed. As such, it provides a nonparametric analysis for data that would otherwise be handled with a two-sample $t$ test.

Suppose that $f_X$ and $f_Y$ are continuous pdfs having the same shape but possibly different medians. The graph of $f_X$, in other words, is a shifted version of the graph of $f_Y$. Using function notation, we would express the relationship between $f_X$ and $f_Y$ by writing $f_X(x) = f_Y(x - \theta)$, where $\theta$ is an unknown parameter. It follows that testing whether the two pdfs are identical is equivalent to deciding whether $\theta = 0$.

Assume that random samples $X_1, X_2, \ldots, X_n$ from $f_X$ and $Y_1, Y_2, \ldots, Y_m$ from $f_Y$ are given, where the $X_i$'s and $Y_i$'s are independent. *We will assume the $X_i$'s represent the smaller of the two samples (in terms of sample size).* If $f_X$ is identical to $f_Y$( and both equal, say, $f$), then combining the $X_i$'s and $Y_i$'s forms a random sample of size $n + m$ from $f$. If the pooled set of observations is ordered, the $X_i$'s and $Y_i$'s should be well intermixed. On the other hand, if $f_X$ has shifted with respect to $f_Y$, we would expect the $X_i$'s to be either relatively small or relative large with respect to the $Y_i$'s.

Finding a procedure for testing that $f_X = f_Y$ requires that we find a statistic for measuring the extent to which the $X_i$'s and $Y_i$'s, when combined, are *not* randomly intermixed. Wilcoxon's solution was straight-forward:

*Procedure for Computing the Wilcoxon Rank Sum Statistic ($n + m \leqslant 12$)*

1. Pool the two samples.
2. Rank the entire set of $x_i$'s and $y_i$'s.
3. Sum the ranks of the $x_i$'s.
4. Use the magnitude of the sum of the ranks of the $x_i$'s as the basis for a decision rule.

For instance, suppose the two $X$'s are 4 and 9 and the $Y$ observations are 3, 5, and 7. Ordering the data gives $3 < 4 < 5 < 7 < 9$. Notice from Table 16.7 that the $X$ observations in the pooled sample have ranks 2 and 5, giving a sum of 7. (In general, we will use $V$ to denote the sum of the ranks associated with the set of $X$'s.) If $V$ is too small or too large, it seems reasonable that we should reject $H_0$: $\theta = 0$ in favor of $H_1$: $\theta \neq 0$. As always,

**TABLE 16.7**

| Ordered Data | 3 | 4 | 5 | 7 | 9 |
|---|---|---|---|---|---|
| Origin | Y | X | Y | Y | X |
| Rank | 1 | 2 | 3 | 4 | 5 |

the precise definition of what constitutes "too small" and "too large" requires that we know the pdf of the test statistic. Unfortunately, there is no easy-to-use formula for $f_V(v)$, but for small values of $n$ and $m$, cutoff values for $V$ are given in Appendix A.7. These values can be used to set up decision rules in the same way that entries in Appendix A.6 define the possible rejection regions for a signed rank test.

---

*EXAMPLE 16.3*    A contracting firm is negotiating the construction site for a major housing development. Since the ease with which homes can be sold depends on a favorable money market, the firm would like to build in a community where interest rates are relatively low. Pleasant Hill and Northside are two possible locations. Seven financial institutions in Pleasant Hill are randomly selected and asked about their interest rates for a conventional 30-year $75,000 mortgage. Similar information is collected from five lenders in Northside. The data are summarized in Table 16.8. Is there a difference in the "typical" mortgage rate prevailing in these two communities?

*Solution*    Let $\theta$ denote the shift between the theoretical distributions of Pleasant Hill and Northside mortgage rates, respectively. The hypotheses to be tested are

$$H_0: \quad \theta = 0$$

vs.

$$H_1: \quad \theta \neq 0$$

Let $\alpha = 0.05$.

Table 16.9 shows the steps necessary to calculate the rank sum statistic V. As before, data that are tied are assigned ranks equal to the average of the positions for which they are competing. Notice how the ranks

**TABLE 16.8  Interest Rates**

| Northside, $x_i$ | Pleasant Hill, $y_i$ |
|---|---|
| 9.50 | 9.75 |
| 9.70 | 10.25 |
| 9.60 | 10.00 |
| 10.00 | 9.90 |
| 9.65 | 10.00 |
| | 9.80 |
| | 9.60 |

**TABLE 16.9**

| Pooled Data | Ordered Data | Sample | Rank | Revised Rank |
|---|---|---|---|---|
| 9.75 | 9.50 | X | 1 | 1 |
| 10.25 | 9.60 | X | 2.5 | 2.5 |
| 10.00 | 9.60 | Y | 2.5 | 0 |
| 9.90 | 9.65 | X | 4 | 4 |
| 10.00 | 9.70 | X | 5 | 5 |
| 9.80 | 9.75 | Y | 6 | 0 |
| 9.60 | 9.80 | Y | 7 | 0 |
| 9.50 | 9.90 | Y | 8 | 0 |
| 9.70 | 10.00 | X | 10 | 10 |
| 9.60 | 10.00 | Y | 10 | 0 |
| 10.00 | 10.00 | Y | 10 | 0 |
| 9.65 | 10.25 | Y | 12 | 0 |
| | | | $v =$ | 22.5 |

corresponding to the $Y$ observations are eventually replaced by zeros. Doing that makes the sum of the entries in the last column equal to the desired test statistic $v$.

According to Appendix Table A.7, the 0.025 probability cutoffs for $V$ when $n = 5$ and $m = 7$ are 20 and 45. That is, under the null hypothesis

$$P(V \leqslant 20) \doteq 0.025 \quad \text{and} \quad P(V \geqslant 45) \doteq 0.025$$

Since $v = 22.5$, we accept $H_0$. These two samples do not constitute statistical proof that interest rates in Northside are any different, on the average, from interest rates in Pleasant Hill.

---

**QUESTIONS**

**16.34**  A trucking company is choosing between two routes for moving equipment from a factory to a warehouse. Times were recorded for five trips made along route 1 and six along route 2. The results (in minutes) are shown below. Is there a significant difference between the two routes? Set up and carry out an appropriate rank sum test. Let $\alpha = 0.05$.

| Route 1 | Route 2 |
|---------|---------|
| 39 | 46 |
| 52 | 51 |
| 45 | 40 |
| 65 | 50 |
| 49 | 45 |
|    | 43 |

**16.35**  Do a rank sum test on the SAT data of Question 2.57. Make the alternative hypothesis reflect the belief that review courses, *if* they have any effect at all, will be beneficial. Let $\alpha = 0.05$.

**16.36**  Use the data in Question 2.19 to compare the caffeine contents of spray-dried and freeze-dried coffee. Do the appropriate rank sum test, letting $\alpha = 0.05$. Should the alternative hypothesis be one-sided or two-sided? Justify your choice.

**16.37**  (a) Using a rank sum test, find the $P$ value that would be associated with the paramecium data in Question 2.18. Assume the alternative hypothesis is one-sided.

(b) What would the $P$ value be if $H_1$ were two-sided?

**16.38**  It was argued in Case Study 2.5 that special exercises given to infants might enable them to walk at an earlier age. Use a rank sum test on the data in Table 2.5 to test that claim at the $\alpha = 0.05$ level of significance.

**16.39**  Recall the discussion of thermoluminescent dating given in Case Study 10.2. Use the rank sum procedure to test whether Terral Lewis and Jaketown evolved technologically at the same rate (as measured by their ability to make kiln-fired ornaments). Let $\alpha = 0.05$.

16.40   Suppose two symmetric distributions have the same mean but possibly different variances. Random samples $x_1, x_2, \ldots, x_n$ and $y_1, y_2, \ldots, y_m$ are drawn from the $X$ distribution and $Y$ distribution, respectively. The experimenter wishes to test $H_0: \sigma_X^2 = \sigma_Y^2$ versus $H_1: \sigma_X^2 \neq \sigma_Y^2$. How might a rank sum be computed that would be capable of distinguishing between $H_0$ and $H_1$?

### Large Samples ($n + m > 12$)

As was true for the signed rank test in Section 16.3, the way we do a rank sum test depends on the sample size. If the total number of observations drawn $(n + m)$ is less than or equal to 12, decision rules for rank sum tests are based directly on $V$. meaning that critical values ($v_1^*$ and/or $v_2^*$) are obtained from Appendix Table A.7. For $n + m > 12$ we phrase decision rules in terms of

$$V' = \frac{V - E(V)}{\sqrt{\mathrm{Var}\,(V)}}$$

which has approximately a standard normal distribution.

---

**THEOREM 16.2   The Rank Sum Statistic ($n + m > 12$)**

Let $V$ be the Wilcoxon rank sum statistic computed from random samples $X_1, X_2, \ldots, X_n$ and $Y_1, Y_2, \ldots, Y_m$. Let the $X$'s be assigned to the smaller sample. The mean and variance of $V$ are

$$E(V) = \frac{n(n + m + 1)}{2} \quad \text{and} \quad \mathrm{Var}\,(V) = \frac{nm(n + m + 1)}{12}$$

For $n + m \geqslant 12$, the distribution of

$$V' = \frac{V - E(V)}{\sqrt{\mathrm{Var}\,(V)}}$$

can be adequately approximated by a standard normal curve.

---

## CASE STUDY 16.3

*Alcohol Consumption versus Life-Span*

Considerable work has been done to document the harmful effects of immoderation in the consumption of alcoholic beverages. One study compared the life-spans of certain twentieth-century American authors known to be heavy drinkers with the life-spans of authors who presumably were not (36).

Table 16.10 lists the ages at death for 9 writers in the former group and 12 in the latter.

**TABLE 16.10**

| AUTHORS NOTED FOR ALCOHOL ABUSE | | AUTHORS NOT NOTED FOR ALCOHOL ABUSE | |
|---|---|---|---|
| **Name** | **Age at Death** | **Name** | **Age at Death** |
| Ring Lardner | 48 | Carl Van Doren | 65 |
| Sinclair Lewis | 66 | Ezra Pound | 87 |
| Raymond Chandler | 71 | Randolph Bourne | 32 |
| Eugene O'Neill | 65 | Van Wyck Brooks | 77 |
| Robert Benchley | 56 | Samuel Eliot Morrison | 89 |
| J. P. Marquand | 67 | John Crowe Ransom | 86 |
| Dashiell Hammett | 67 | T. S. Eliot | 77 |
| e. e. cummings | 70 | Conrad Aiken | 84 |
| Edmund Wilson | 77 | Ben Ames Williams | 64 |
| Median = $\tilde{x}$ = | 67.0 | Henry Miller | 88 |
| | | Archibald MacLeish | 90 |
| | | James Thurber | 67 |
| | | Median = $\tilde{y}$ = | 80.5 |

Let $\tilde{\mu}_X$ and $\tilde{\mu}_Y$ denote the true median life-spans for heavy drinkers and not heavy drinkers, respectively. Define

$$\theta = \tilde{\mu}_Y - \tilde{\mu}_X$$

Since past experience suggests strongly that alcohol abuse will not prolong a person's life, the hypotheses to be tested are

$$H_0: \quad \theta = 0$$

vs.

$$H_1: \quad \theta > 0$$

Let $\alpha = 0.05$.

Table 16.11 shows the steps necessary to compute the rank sum statistic $v$. The combined sample size here is fairly large ($n + m = 9 + 12 = 21$), so it becomes appropriate to formulate the decision rule in terms of an approximate Z ratio. From Theorem 16.2,

$$E(V) = \frac{9(9 + 12 + 1)}{2} = 99.0,$$

$$\text{Var}(V) = \frac{9(12)(9 + 12 + 1)}{12} = 198.0$$

and

$$V' = \frac{V - 99.0}{\sqrt{198.0}}$$

**TABLE 16.11**

| Pooled Data | Ordered Data | Sample | Ranks | Revised Ranks |
|---|---|---|---|---|
| 48 | 32 | Y | 1 | 0 |
| 66 | 48 | X | 2 | 2 |
| 71 | 56 | X | 3 | 3 |
| 65 | 64 | Y | 4 | 0 |
| 56 | 65 | X | 5.5 | 5.5 |
| 67 | 65 | Y | 5.5 | 0 |
| 67 | 66 | X | 7 | 7 |
| 70 | 67 | X | 9 | 9 |
| 77 | 67 | X | 9 | 9 |
| 65 | 67 | Y | 9 | 0 |
| 87 | 70 | X | 11 | 11 |
| 32 | 71 | X | 12 | 12 |
| 77 | 77 | X | 14 | 14 |
| 89 | 77 | Y | 14 | 0 |
| 86 | 77 | Y | 14 | 0 |
| 77 | 84 | Y | 16 | 0 |
| 84 | 86 | Y | 17 | 0 |
| 64 | 87 | Y | 18 | 0 |
| 88 | 88 | Y | 19 | 0 |
| 90 | 89 | Y | 20 | 0 |
| 67 | 90 | Y | 21 | 0 |
| | | | | $v = $ 72.5 |

Where should we locate the rejection region? In the *left-hand* tail of the standard normal curve. If, on the average, alcohol impacts negatively on longevity, it follows that the $x$'s will tend to be smaller than the $y$'s, $v$ will tend to be smaller than $E(V)$ (why?), and $V'$ will tend to be negative. More specifically, we should reject $H_0$ if $v' \leqslant -1.64$, since $P(Z \leqslant -1.64) = 0.05$, the latter being the level chosen for $\alpha$.

Notice from the last column of Table 16.11 that $v = 72.5$. Therefore,

$$v' = \frac{72.5 - 99.0}{\sqrt{198.0}} = -1.88$$

and our conclusion is to reject $H_0$. The 12.5-year gap between the two sample median life-spans ($\tilde{y} - \tilde{x} = 80.5 - 67.0 = 12.5$) is statistically significant.

QUESTIONS

16.41 Use a rank sum test to compare the silver contents in Table 2.4. Let $\alpha = 0.05$ and make the alternative hypothesis two-sided. (Notice that the variables need to be relabeled: observations in the late coinage sample must be called $x$'s; those in the early coinage sample, $y$'s.)

**16.42**  For Question 2.20 let $\theta$ denote the difference between the true median spectropenetration gradients for males and females; that is,

$$\theta = \tilde{\mu}_{\text{female}} - \tilde{\mu}_{\text{male}}$$

Test $H_0\colon \theta = 0$ versus $H_1\colon \theta \neq 0$. Let $\alpha = 0.05$.

**16.43**  Compare the true proportions of three-letter words written by Mark Twain and Quintus Curtius Snodgrass by using the data in Table 10.1.

(a)  Do a rank sum test with a two-sided alternative. Let $\alpha = 0.01$.

(b)  Does your conclusion agree with the inference reached in Case Study 10.1?

**16.44**  Analyze the half-life data in Table 10.3 nonparametrically. Define the parameter being tested. Let $\alpha = 0.05$.

**16.45**  Suppose an experimenter collects two independent random samples

$$X_1, X_2, \ldots, X_n \quad \text{and} \quad Y_1, Y_2, \ldots, Y_m$$

and wishes to compare the locations of the two populations from which the data came. On what basis is the decision made to use a rank sum test rather than a two-sample $t$ test?

**16.46**  Set up and carry out an appropriate one-sided rank sum test on the presidential life-span data in Table 3.21. Let $\alpha = 0.05$.

**16.47**  The use of carpeting in hospitals, while having obvious esthetic merits, raises an important medical question: Are carpeted floors sanitary? Among the ways of addressing that concern is to compare bacterial levels in carpeted and uncarpeted rooms. Airborne bacteria can be counted by passing air at a known rate over a growth medium, incubating that medium, and then counting the number of bacterial colonies that form. In one such study done in a Montana hospital (179), room air was pumped over a Petri dish at the rate of 1 ft$^3$/min. The procedure was repeated in 16 patient rooms, 8 carpeted and 8 uncarpeted. The results, expressed in terms of bacteria per cubic foot of air are listed below. Do an appropriate nonparametric test. Let $\alpha = 0.05$. Should the alternative be one-sided or two-sided?

| Carpeted Rooms | Bacteria/ft$^3$ | Uncarpeted Rooms | Bacteria/ft$^3$ |
|---|---|---|---|
| 212 | 11.8 | 210 | 12.1 |
| 216 | 8.2 | 214 | 8.3 |
| 220 | 7.1 | 215 | 3.8 |
| 223 | 13.0 | 217 | 7.2 |
| 225 | 10.8 | 221 | 12.0 |
| 226 | 10.1 | 222 | 11.1 |
| 227 | 14.6 | 224 | 10.1 |
| 228 | 14.0 | 229 | 13.7 |

## 16.5

### THE KRUSKAL-WALLIS TEST

**Kruskal-Wallis test—A** nonparametric alternative to the analysis of variance. Based on rank sums associated with observations representing each of $k$ treatment levels, the Kruskal-Wallis test is a procedure for choosing between $H_0$: $\tilde{\mu}_1 = \tilde{\mu}_2 = \cdots = \tilde{\mu}_k$ and $H_1$: not all the $\tilde{\mu}_j$'s are equal.

This section and the following one present two widely used nonparametric techniques for the ANOVA models described in Chapters 14 and 15. The first to be considered, the **Kruskal-Wallis test**, applies to the $k$-sample problem. A fairly recent procedure, it was first proposed in 1952.

Suppose that $k\,(\geqslant 2)$ independent random samples of sizes $n_1$, $n_2, \ldots, n_k$, respectively, are drawn from $k$ populations whose pdfs all have the same shape but possibly different locations. We wish to test whether the latter—as measured by the $k$ population medians—are equal; that is,

$$H_0: \quad \tilde{\mu}_1 = \tilde{\mu}_2 = \cdots = \tilde{\mu}_k$$

vs.

$$H_1: \quad \text{not all the } \tilde{\mu}_j\text{'s are equal}$$

(Recall that if the populations being sampled were *normal*, $H_0$: $\tilde{\mu}_1 = \tilde{\mu}_2 = \cdots = \tilde{\mu}_k$ would be the same as $H_0$: $\mu_1 = \mu_2 = \cdots = \mu_k$, and the best way to analyze the data would be the $F$ test given in Theorem 14.2.)

*Procedure for the Kruskal-Wallis Test*

1. Pool and rank the entire set of $n = n_1 + n_2 + \cdots + n_k$ observations (the smallest observation is assigned a rank of 1; the largest a rank of $n$).

2. Compute the sums of the ranks associated with the observations in each of the $k$ samples and denote them by $r_1, r_2, \ldots, r_k$, respectively (see Table 16.12).

3. Compute the Kruskal-Wallis statistic $B$ (see Equation 16.2).

4. Test the null hypothesis according to Theorem 16.3.

If the null hypothesis is true, the $n$ observations represent a random sample from a single distribution, and we would expect the corresponding $n$ ranks to be randomly distributed among the $k$ columns. Similarly, the overall *sum* of the ranks $(1 + 2 + \cdots + n = n(n + 1)/2)$ should also be divided proportionately among the $k$ columns, with the proportions being based on the various sample sizes. In the $j$th column, for example, where

**TABLE 16.12**

|  | SAMPLE 1 | | SAMPLE 2 | | SAMPLE 3 | |
|---|---|---|---|---|---|---|
|  | Data | Rank | Data | Rank | Data | Rank |
|  | 15.0 | 8 | 13.1 | 6 | 11.9 | 5 |
|  | 8.2 | 2 | 14.2 | 7 | 10.4 | 4 |
|  | 9.6 | 3 |  |  | 6.2 | 1 |
| Sample sizes: | $n_1 = 3$ | | $n_2 = 2$ | | $n_3 = 3$ | |
| Rank totals: | | $r_1 = 13$ | | $r_2 = 13$ | | $r_3 = 10$ |

there are $n_j$ observations, the sum of the ranks should equal, on the average,

$$\frac{n_j}{n}\left[\frac{n(n+1)}{2}\right] = \frac{n_j(n+1)}{2}$$

*if the population medians are all the same.*

The squared difference between an observed rank sum $R_j$ and its expected value $n_j(n+1)/2$ is clearly a measure of the extent to which the $j$th sample deviates from what $H_0$ predicts. Therefore, an intuitively reasonable test statistic could be defined by summing such differences over all $j$'s; that is, by computing

$$\sum_{j=1}^{k}\left[R_j - \frac{n_j(n+1)}{2}\right]^2$$

If this sum were zero, for example, we would surely accept $H_0$. Unfortunately, this statistic does not have a convenient expression for its pdf. We will use instead the related statistic $B$, where

$$B = \frac{12}{n(n+1)}\sum_{j=1}^{k}\frac{1}{n_j}\left[R_j - \frac{n_j(n+1)}{2}\right]^2 \tag{16.1}$$

or, equivalently,

$$B = \frac{12}{n(n+1)}\sum_{j=1}^{k}\frac{R_j^2}{n_j} - 3(n+1) \tag{16.2}$$

It can be shown that $B$ has approximately a $\chi^2$ distribution.

---

### THEOREM 16.3    Kruskal-Wallis Test

If $H_0; \tilde{\mu}_1 = \tilde{\mu}_2 = \cdots = \tilde{\mu}_k$ is true and either $k = 3$ with each $n_j \geqslant 5$ or $k \geqslant 4$ with each $n_j \geqslant 4$, then

$$B = \frac{12}{n(n+1)}\sum_{j=1}^{k}\frac{R_j^2}{n_j} - 3(n+1)$$

has approximately a chi-square distribution with $k-1$ degrees of freedom. Furthermore, $H_0$ should be rejected at the $\alpha$ level of significance if $b \geqslant \chi^2_{1-\alpha, k-1}$.

---

**CASE STUDY 16.4**

*Was the 1969 Draft Lottery Fair?*    On December 1, 1969, a lottery was held in Selective Service headquarters in Washington, D.C., to determine the draft status of all 19-year-old males. It was the first time such a procedure had been used since World War II. Priorities were established according to a person's birthday. Each of the 366 possible birthdates was written on a slip of paper and placed inside a small capsule. All 366 capsules were then thrown into a large urn, mixed, and drawn out one by one. By agreement, men whose birthday corresponded to the first capsule drawn would have the dubious distinction of being placed at the top of their local draft board's "eligible" list; those whose birthday corresponded to the second capsule drawn were to have the second highest draft priority, and so on. Table 16.13 shows the actual order in which the 366 days were selected, September 14 being first and June 8 last (150).

**TABLE 16.13  1969 Draft Lottery**

| Date | Jan. | Feb. | Mar. | Apr. | May | June | July | Aug. | Sept. | Oct. | Nov. | Dec. |
|------|------|------|------|------|------|------|------|------|------|------|------|------|
| 1 | 305 | 086 | 108 | 032 | 330 | 249 | 093 | 111 | 225 | 359 | 019 | 129 |
| 2 | 159 | 144 | 029 | 271 | 298 | 228 | 350 | 045 | 161 | 125 | 034 | 328 |
| 3 | 251 | 297 | 267 | 083 | 040 | 301 | 115 | 261 | 049 | 244 | 348 | 157 |
| 4 | 215 | 210 | 275 | 081 | 276 | 020 | 279 | 145 | 232 | 202 | 266 | 165 |
| 5 | 101 | 214 | 293 | 269 | 364 | 028 | 188 | 054 | 082 | 024 | 310 | 056 |
| 6 | 224 | 347 | 139 | 253 | 155 | 110 | 327 | 114 | 006 | 087 | 076 | 010 |
| 7 | 306 | 091 | 122 | 147 | 035 | 085 | 050 | 168 | 008 | 234 | 051 | 012 |
| 8 | 199 | 181 | 213 | 312 | 321 | 366 | 013 | 048 | 184 | 283 | 097 | 105 |
| 9 | 194 | 338 | 317 | 219 | 197 | 335 | 277 | 106 | 263 | 342 | 080 | 043 |
| 10 | 325 | 216 | 323 | 218 | 065 | 206 | 284 | 021 | 071 | 220 | 282 | 041 |
| 11 | 329 | 150 | 136 | 014 | 037 | 134 | 248 | 324 | 158 | 237 | 046 | 039 |
| 12 | 221 | 068 | 300 | 346 | 133 | 272 | 015 | 142 | 242 | 072 | 066 | 314 |
| 13 | 318 | 152 | 259 | 124 | 295 | 069 | 042 | 307 | 175 | 138 | 126 | 163 |
| 14 | 238 | 004 | 354 | 231 | 178 | 356 | 331 | 198 | 001 | 294 | 127 | 026 |
| 15 | 017 | 089 | 169 | 273 | 130 | 180 | 322 | 102 | 113 | 171 | 131 | 320 |
| 16 | 121 | 212 | 166 | 148 | 055 | 274 | 120 | 044 | 207 | 254 | 107 | 096 |
| 17 | 235 | 189 | 033 | 260 | 112 | 073 | 098 | 154 | 255 | 288 | 143 | 304 |
| 18 | 140 | 292 | 332 | 090 | 278 | 341 | 190 | 141 | 246 | 005 | 146 | 128 |
| 19 | 058 | 025 | 200 | 336 | 075 | 104 | 227 | 311 | 177 | 241 | 203 | 240 |
| 20 | 280 | 302 | 239 | 345 | 183 | 360 | 187 | 344 | 063 | 192 | 185 | 135 |
| 21 | 186 | 363 | 334 | 062 | 250 | 060 | 027 | 291 | 204 | 243 | 156 | 070 |
| 22 | 337 | 290 | 265 | 316 | 326 | 247 | 153 | 339 | 160 | 117 | 009 | 053 |
| 23 | 118 | 057 | 256 | 252 | 319 | 109 | 172 | 116 | 119 | 201 | 182 | 162 |
| 24 | 059 | 236 | 258 | 002 | 031 | 358 | 023 | 036 | 195 | 196 | 230 | 095 |
| 25 | 052 | 179 | 343 | 351 | 361 | 137 | 067 | 286 | 149 | 176 | 132 | 084 |
| 26 | 092 | 365 | 170 | 340 | 357 | 022 | 303 | 245 | 018 | 007 | 309 | 173 |
| 27 | 355 | 205 | 268 | 074 | 296 | 064 | 289 | 352 | 233 | 264 | 047 | 078 |
| 28 | 077 | 299 | 223 | 262 | 308 | 222 | 088 | 167 | 257 | 094 | 281 | 123 |
| 29 | 349 | 285 | 362 | 191 | 226 | 353 | 270 | 061 | 151 | 229 | 099 | 016 |
| 30 | 164 | | 217 | 208 | 103 | 209 | 287 | 333 | 315 | 038 | 174 | 003 |
| 31 | 211 | | 030 | | 313 | | 193 | 011 | | 079 | | 100 |
| Totals | 6236 | 5886 | 7000 | 6110 | 6447 | 5872 | 5628 | 5377 | 4719 | 5656 | 4462 | 3768 |

Was the lottery fair? No! As statisticians across the country were quick to point out, the sequence of dates drawn was not even close to being random. Twenty-six of the 31 days in December, for example, appeared in the first half of the draft (ranks 001 through 183); in contrast, only 10 of the 31 days in March were chosen that early. Probably the best way to get a feeling for these data, though, is to apply the Kruskal-Wallis test.

If the lottery was random, the distribution of ranks assigned to each month should be comparable. Specifically, the true median rank for January ($\tilde{\mu}_{Jan}$) should equal the true median rank for February ($\tilde{\mu}_{Feb}$), and so on. Testing that equivalence becomes an exercise in the use of Theorem 16.3 when $k = 12$.

If we substitute the $r_j$'s at the bottom of Table 16.13 into Equation 16.2, we get

$$b = \frac{12}{366(367)} \left[ \frac{(6236)^2}{31} + \cdots + \frac{(3768)^2}{31} \right] - 3(367) = 25.95$$

Under the null hypothesis that $\tilde{\mu}_{Jan} = \tilde{\mu}_{Feb} = \cdots = \tilde{\mu}_{Dec}$, $B$ has approximately a $\chi^2$ distribution with 11 df. Let $\alpha = 0.01$. From Appendix A.3, $\chi^2_{1-\alpha, k-1} = \chi^2_{0.99, 11} = 24.7$. Since $b > 24.7$, we should reject

$$H_0: \quad \tilde{\mu}_{Jan} = \tilde{\mu}_{Feb} = \cdots = \tilde{\mu}_{Dec}$$

The lottery was *not* fair.

We can obtain an even more resounding rejection of the randomness hypothesis by dividing the 12 months into two half-years. Under that formulation the hypotheses to be tested are

$$H_0: \quad \tilde{\mu}_{Jan-June} = \tilde{\mu}_{July-Dec}$$

vs.

$$H_1: \quad \tilde{\mu}_{Jan-June} \neq \tilde{\mu}_{July-Dec}$$

Table 16.14, derived from Table 16.13, summarizes what we need to know about the two half-years. The sum of the ranks for days in the first six months was 37,551 ($= r_1$); for days in the last six months, 29,610 ($= r_2$). Putting those totals into Equation 16.2 gives

$$b = \frac{12}{366(367)} \left[ \frac{(37,551)^2}{182} + \frac{(29,610)^2}{184} \right] - 3(367) = 16.85$$

**TABLE 16.14**

|       | January–June (1) | July–December (2) |
|-------|------------------|-------------------|
| $r_j$ | 37,551           | 29,610            |
| $n_j$ | 182              | 184               |

How significant is 16.85? *Very* significant. Notice from Appendix A.3 that the critical value for this second test (at the $\alpha = 0.01$ level) is "only" 6.635. If we had set $\alpha$ as small as 0.0005, the critical value (see (38)) would increase to 12.116, which is still considerably smaller than the observed $b$. There can be no doubt, in other words, that the lottery failed miserably to achieve its stated objective—the generation of a random sequence of draft priorities.

What happened? Why were days late in the year so much more likely to appear early in the lottery? An investigation revealed that two serious errors were made in preparing the urn. To begin with, the 366 capsules were placed in the urn in a decidedly non-random fashion: the January capsules were put in first, then the February capsules, and so on, with the December capsules being put in last. If the capsules had then been thoroughly mixed, no problems of the kind encountered would have occurred. Unfortunately, they were not. Despite efforts to the contrary, the capsules were *not* mixed well, and days late in the year tended to remain near the top of the urn.

---

**QUESTIONS**

**16.48**  Suppose you were in charge of a future draft lottery and had prepared the 366 capsules described in Case Study 16.4. What procedures would you follow to prevent a reoccurrence of what happened in 1969? Be specific.

**16.49**  For the following data, test the null hypothesis that the three population medians are all equal. Let $\alpha = 0.05$.

| Sample 1 | Sample 2 | Sample 3 |
|----------|----------|----------|
| 13 | 42 | 36 |
| 42 | 16 | 18 |
| 23 | 5 | 4 |
| 32 | 3 | 30 |
| 22 | | |

**16.50**  A franchise food company is comparing the effectiveness of three different advertising campaigns (A, B, and C). Each is the focus of a series of television commercials run in three widely separated cities. One campaign emphasizes convenience, another stresses quality, and the third promotes the restaurant's low cost. Listed below are average monthly sales increases (in thousands of dollars) for several outlets doing business in each of the three cities. Test at the 0.05 level of significance the null hypothesis that the true median sales increases characteristic of the three advertising approaches are all the same.

| CAMPAIGN | | |
|---|---|---|
| **A** | **B** | **C** |
| 16 | 23 | 17 |
| 27 | 4 | 17 |
| 13 | 76 | 11 |
| 6 | 10 | 34 |
| | | 11 |

16.51    Do a Kruskal-Wallis test on the ammonium chloride data in Question 2.49. Let $\alpha = 0.01$.

16.52    Compare the accuracies of the four missile launchers described in Question 2.60. Base your conclusion on a Kruskal-Wallis analysis with $\alpha = 0.05$. In assigning ranks to tied observations, use the same convention followed in Sections 16.3 and 16.4.

16.53    (a) Recall the serum binding percentages given in Case Study 3.19. Set up and carry out an appropriate Kruskal-Wallis test to compare the "locations" characteristic of the five antibiotics. Define the parameters being tested. Let $\alpha = 0.10$.

    (b) Does your answer agree with the parametric analysis detailed in Example 14.2?

16.54    For the data in Example 14.5, let $\tilde{\mu}_j$, $j = 1, 2, 3$, denote the true median IQ change for groups I, II, and III, respectively. Use a Kruskal-Wallis analysis to test $H_0: \tilde{\mu}_1 = \tilde{\mu}_2 = \tilde{\mu}_3$. Let $\alpha = 0.01$.

16.55    Using the data in Example 14.6, test whether the Cutter, Abbott, and McGaw pharmaceutical companies all produce fluids having the same median contaminant size. Let $\alpha = 0.05$.

16.56    Show that Equations 16.1 and 16.2 are equivalent. Use the fact that

$$\sum_{j=1}^{k} R_j = 1 + 2 + \cdots + n = \frac{n(n + 1)}{2}$$

# 16.6

## FRIEDMAN'S TEST

Friedman's test—A way of analyzing randomized block data without making the usual normality assumption.

We conclude this chapter by turning our attention to the randomized block problem. The standard nonparametric analysis for data having that structure is **Friedman's test**. In principle, Friedman's test is quite similar to the Kruskal–Wallis test, the only real difference between the two being the way in which the ranks are assigned.

We saw in Chapters 3 and 15 that a typical set of randomized block data can be summarized in a table having $b$ rows and $k$ columns (see Section 15.2). Each row represents a block; the observations *within* a block are dependent, and the observations *between* blocks are independent. Each column corresponds to a treatment, meaning the observation in the $i$th row and $j$th column is the result of applying treatment $j$ to block $i$. Our objective is to test the null hypothesis that the population medians associated with the $k$ treatments $\tilde{\mu}_1, \tilde{\mu}_2, \ldots, \tilde{\mu}_k$ are all equal.

We begin by ranking the $k$ observations *within each block* (that is, within each row) from smallest to largest. Each row of ranks will necessarily be a permutation of the integers from 1 to $k$. Each row *total*, therefore, will be the sum of the first $k$ integers, and we have seen before that

$$1 + 2 + \cdots k = k(k + 1)/2$$

Altogether, the table has $b$ rows, so the total of the ranks for the entire set of observations is $bk(k + 1)/2$.

Now, let $R_j$ denote the total of the ranks associated with the $b$ observations in the $j$th column. If the $k$ treatments all have the same effect, each $R_j$ should not deviate much from its expected value, $E(R_j)$, where

$$E(R_j) = \frac{bk(k + 1)/2}{k} = \frac{b(k + 1)}{2} \quad \text{(Why?)}$$

From the argument used to justify the Kruskal-Wallis procedure, it makes sense to base a test statistic for $H_0: \tilde{\mu}_1 = \tilde{\mu}_2 = \cdots = \tilde{\mu}_k$ on

$$\sum_{j=1}^{k} \left[ R_j - \frac{b(k + 1)}{2} \right]^2$$

Only if the latter is "large" should we reject $H_0$. In practice, through, a slightly different formulation is used, one whose pdf is approximated by a familiar distribution. Specifically, we test $H_0$ by computing the Friedman statistic $G$, where

$$G = \frac{12}{bk(k + 1)} \sum_{j=1}^{k} R_j^2 - 3b(k + 1) \tag{16.3}$$

Like the Kruskal-Wallis $B$, $G$ has a $\chi^2$ distribution.

---

**THEOREM 16.4   Friedman's Test**

If either $n > 5$ or $k > 5$, and if $H_0: \tilde{\mu}_1 = \tilde{\mu}_2 = \cdots = \tilde{\mu}_k$ is true, then

$$G = \frac{12}{bk(k + 1)} \sum_{j=1}^{k} R_j^2 - 3b(k + 1)$$

has approximately a chi-square distribution with $k - 1$ degrees of freedom. We should reject $H_0$ at the $\alpha$ level of significance if $g \geqslant \chi^2_{1-\alpha, k-1}$.

---

**CASE STUDY 16.5**

*Running the Bases*   Baseball rules allow a batter considerable leeway in how he is permitted to round first base when running from home to second. Two possibilities are the narrow-angle and wide-angle paths diagrammed in Figure 16.3. If you were the coach, which would you recommend?

As a way of comparing these two options, time trials were held involving 22 players (185). Each person ran both paths (thus acting as his own block). Recorded on each occasion was the time (in seconds) it took to go from a point 35 ft from home plate to a point 15 ft from second base. These are listed in the second and fourth columns of Table 16.15. In the third and

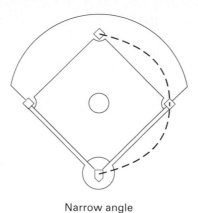
Narrow angle

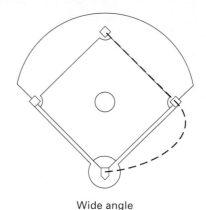
Wide angle

**FIG. 16.3**

**TABLE 16.15**

| Player | Narrow Angle | Rank | Wide Angle | Rank |
|--------|--------------|------|------------|------|
| 1 | 5.50 | 1 | 5.55 | 2 |
| 2 | 5.70 | 1 | 5.75 | 2 |
| 3 | 5.60 | 2 | 5.50 | 1 |
| 4 | 5.50 | 2 | 5.40 | 1 |
| 5 | 5.85 | 2 | 5.70 | 1 |
| 6 | 5.55 | 1 | 5.60 | 2 |
| 7 | 5.40 | 2 | 5.35 | 1 |
| 8 | 5.50 | 2 | 5.35 | 1 |
| 9 | 5.15 | 2 | 5.00 | 1 |
| 10 | 5.80 | 2 | 5.70 | 1 |
| 11 | 5.20 | 2 | 5.10 | 1 |
| 12 | 5.55 | 2 | 5.45 | 1 |
| 13 | 5.35 | 1 | 5.45 | 2 |
| 14 | 5.00 | 2 | 4.95 | 1 |
| 15 | 5.50 | 2 | 5.40 | 1 |
| 16 | 5.55 | 2 | 5.50 | 1 |
| 17 | 5.55 | 2 | 5.35 | 1 |
| 18 | 5.50 | 1 | 5.55 | 2 |
| 19 | 5.45 | 2 | 5.25 | 1 |
| 20 | 5.60 | 2 | 5.40 | 1 |
| 21 | 5.65 | 2 | 5.55 | 1 |
| 22 | 6.30 | 2 | 6.25 | 1 |
| | | $\overline{39}$ | | $\overline{27}$ |

fifth columns are the within-block ranks for each path. A rank of 1 was assigned to the better path—this is, the path enabling the runner to round first base faster.

If $\tilde{\mu}_1$ and $\tilde{\mu}_2$ denote the true median rounding times associated with the narrow-angle and wide-angle options, respectively, the hypotheses to be tested are

$$H_0: \quad \tilde{\mu}_1 = \tilde{\mu}_2$$

vs.

$$H_1: \quad \tilde{\mu}_1 \neq \tilde{\mu}_2$$

Let $\alpha = 0.05$. According to Theorem 16.4, the Friedman statistic (under $H_0$) will have approximately a $\chi^2$ distribution with 1 df and the decision rule will be

reject $H_0$ if $g \geqslant 3.84$

(since $\chi^2_{0.95,1} = 3.84$). But

$$g = \frac{12}{22(2)(3)} \left[ (39)^2 + (27)^2 \right] - 3(22)(3) = 6.54$$

implying that the two paths are *not* equivalent. (Which is better?)

---

*EXAMPLE 16.4*   Case Study 2.16 described an experiment whose objective was to test whether lunar cycles have an effect on human behavior, as measured by admission rates to a mental hospital before, during, and after a full moon (10). If it is reasonable to presume that all three admission rates are essentially normal random variables with the same variance, then the appropriate way to analyze the data is with an *F* test, which is what we did in Example 15.1. If, on the other hand, we have reason to suspect that the normality assumption might *not* be true, the Friedman test would be a better approach to take.

Table 16.16 reproduces the data from Case Study 2.16. The numbers in parentheses are the within-block ranks. Let $\tilde{\mu}_{\text{before}}$, $\tilde{\mu}_{\text{during}}$, and $\tilde{\mu}_{\text{after}}$ denote the true median admission rates associated with the three phases of the moon. We wish to test

$$H_0: \quad \tilde{\mu}_{\text{before}} = \tilde{\mu}_{\text{during}} = \tilde{\mu}_{\text{after}}$$

vs.

$$H_1: \quad \text{not all the } \tilde{\mu}\text{'s are equal}$$

Let $\alpha = 0.05$.

*Solution*   Substituting $b = 12$, $k = 3$, and the three treatment rank sums (21, 27, and 24) into Equation 16.3 gives

$$g = \frac{12}{12(3)(4)} \left[ (21)^2 + (27)^2 + (24)^2 \right] - 3(12)(4) = 1.5$$

**TABLE 16.16**

| Month | Before Full Moon | During Full Moon | After Full Moon |
|---|---|---|---|
| | ADMISSION RATES | | |
| Aug. | 6.4 (3) | 5.0 (1) | 5.8 (2) |
| Sept. | 7.1 (1) | 13.0 (3) | 9.2 (2) |
| Oct. | 6.5 (1) | 14.0 (3) | 7.9 (2) |
| Nov. | 8.6 (2) | 12.0 (3) | 7.7 (1) |
| Dec. | 8.1 (2) | 6.0 (1) | 11.0 (3) |
| Jan. | 10.4 (2) | 9.0 (1) | 12.9 (3) |
| Feb. | 11.5 (1) | 13.0 (2) | 13.5 (3) |
| Mar. | 13.8 (2) | 16.0 (3) | 13.1 (1) |
| Apr. | 15.4 (1) | 25.0 (3) | 15.8 (2) |
| May | 15.7 (3) | 13.0 (1) | 13.3 (2) |
| June | 11.7 (1) | 14.0 (3) | 12.8 (2) |
| July | 15.8 (2) | 20.0 (3) | 13.5 (1) |
| Rank sums: | 21 | 27 | 24 |

Since $k = 3$, the sampling distribution of $G$ is approximated by a $\chi^2$ curve with 2 df, which makes $\chi^2_{1-\alpha, k-1} = \chi^2_{0.95, 2} = 5.991$. Our conclusion, therefore, is to accept $H_0$. These data, when analyzed with a Friedman test, do *not* support the existence of a Transylvania effect.

**QUESTIONS**

**16.57** Let $\tilde{\mu}_i$, $i = 1, 2, 3, 4$, denote the population medians associated with treatments 1, 2, 3, and 4, respectively. Use Friedman's method on the data below to test $H_0$: $\tilde{\mu}_1 = \tilde{\mu}_2 = \tilde{\mu}_3 = \tilde{\mu}_4$. Let $\alpha = 0.05$.

| | | TREATMENT | | | |
| | | 1 | 2 | 3 | 4 |
|---|---|---|---|---|---|
| | A | 44 | 161 | 55 | 18 |
| | B | 612 | 20 | 175 | 59 |
| BLOCK | C | 103 | 127 | 111 | 7 |
| | D | 5 | 44 | 14 | 104 |
| | E | 76 | 114 | 82 | 43 |

**16.58** Another way to analyze the base-running data in Table 16.15 is with a paired $t$ test. Does a paired $t$ test seem more reasonable than a Friedman test? Explain.

**16.59** Set up and carry out a Friedman test for the Ponzo data in Case Study 2.17 to see whether illusion strength is independent of regressed age. Define the parameters that appear in $H_0$. Let $\alpha = 0.05$. Compare your answer with the $F$ test done in Example 15.2.

16.60   The data in Question 2.47 compare three laboratory methods for measuring the amount of solids in uncreamed cottage cheese. Can it be claimed that the median responses characteristic of the three methods are not all the same? Do an appropriate Friedman test at the $\alpha = 0.05$ level of significance.

16.61   An experiment was described in Question 2.48 where the objective was to see whether colony density has any effect on the hostility of cockroaches (to other cockroaches). Analyze the data with a Friedman test. Approximate the $P$ value of the test statistic.

16.62   Set up and carry out an appropriate Friedman test on the Nielsen ratings in Question 2.64. Let $\alpha = 0.10$. Define the medians being compared.

16.63   (a) Analyze nonparametrically the rat bait data in Case Study 3.20. Let $\alpha = 0.05$.

   (b) Does your answer agree with the $F$ test asked for in Question 15.7?

16.64   Define the medians associated with the blood doping data in Question 3.84. Use a Friedman test to decide whether it would be reasonable to argue that all three medians are the same. Let $\alpha = 0.05$.

## 16.7

### COMPUTER NOTES

Nonparametric hypothesis tests require little more than entering the data and calling for the test. Figure 16.4 gives the Minitab version of the Wilcoxon signed rank test performed in Example 16.2. The median is called the *center* in this printout.

```
SET C1
  .693 .662 .690 .606 .570 .749 .672 .628 .609 .844
  .654 .615 .668 .601 .576 .670 .606 .611 .553 .933
WTEST CENTER .618 ON C1

 TEST OF CENTER = 0.6180 VERSUS CENTER N.E. 0.6180

                      N FOR   WILCOXON              ESTIMATED
                 N    TEST   STATISTIC  P-VALUE       CENTER
     C1         20     20      151.0     0.089        0.6420
```

**FIG. 16.4**

The rank sum test is sometimes called the Mann-Whitney test, and that name is used by Minitab. Figure 16.5 is the Minitab output for the rank sum test of Example 16.3. The printout provides a confidence interval for the difference of the medians as well as the $P$ value for the test, 0.1229.

SAS also has nonparametric routines. The Kruskal-Wallis test illustrates the SAS procedure for this chapter. Figure 16.6 gives the output for the Kruskal-Wallis test applied to the data of Example 14.6.

```
SET C1
   9.50   9.70   9.60  10.00   9.65
SET C2
   9.75  10.25  10.00   9.90  10.00   9.80   9.60
MANN-WHITNEY ON C1 AND C2
    C1          N =   5         MEDIAN =     9.6500
    C2          N =   7         MEDIAN =     9.9000
    A POINT ESTIMATE FOR ETA1-ETA2 IS        -0.250
    A 96.5 PERCENT C.I. FOR ETA1-ETA2 IS    (   -0.50,    0.10)

    W=    22.5
    TEST OF ETA1 = ETA2 VS. ETA1 N.E. ETA2
    THE TEST IS SIGNIFICANT AT .1229

CANNOT REJECT AT ALPHA = 0.05
```

**FIG. 16.5**

```
DATA;
  INPUT MAKER $ CONTAM;
  CARDS;

PROC NPAR1WAY WILCOXON;
  CLASS MAKER;
  VAR CONTAM;
```

ANALYSIS FOR VARIABLE CONTAM CLASSIFIED BY VARIABLE MAKER

WILCOXON SCORES (RANK SUMS)

| LEVEL | N | SUM OF SCORES | EXPECTED UNDER H0 | STD DEV UNDER H0 | MEAN SCORE |
|-------|---|---------------|-------------------|------------------|------------|
| CUTTER | 6 | 55.00 | 57.00 | 10.68 | 9.17 |
| ABBOTT | 6 | 38.00 | 57.00 | 10.68 | 6.33 |
| MCGAW  | 6 | 78.00 | 57.00 | 10.68 | 13.00 |

KRUSKAL-WALLIS TEST (CHI-SQUARE APPROXIMATION)
CHISQ=   4.71   DF= 2     PROB > CHISQ=0.0947

**FIG. 16.6**

# 16.8
## SUMMARY

In practice, collecting and analyzing data is an exercise in asking questions and making choices. How many treatments should be included and what needs to be measured? In which data structure should the experiment be cast? How large a sample size is necessary? Should the inference be framed as a confidence interval or as a hypothesis test? How should $H_0$ and $H_1$ be defined? At what level should $\alpha$ be set?

Certainly not the least important issue of this sort is the choice of a statistical procedure itself. Suppose, for example, our objective is to do a one-sample hypothesis test on a location parameter $\mu$. Should the analysis be carried out as a one-sample $t$ test, a sign test, or a Wilcoxon signed rank test? The choice is important. What we ultimately conclude may be entirely different, depending on which method has been selected, and some of the methods may not be valid.

Chapter 16 introduced an entire set of alternative procedures known as *nonparametric statistics*. Each of the five methods described—the *sign test*, the *signed rank test*, the *rank sum test*, the *Kruskal–Wallis test*, and the *Friedman test*—parallels one of the standard parametric procedures introduced in Chapters 9, 10, 11, 14, and 15.

How do nonparametric procedures differ from parametric procedures? Despite what the terminology might suggest, the distinction has nothing to do with parameters. What *does* separate the two sets of procedures is the probabilistic behavior of their test statistics. In general, nonparametric methods are based on test statistics whose sampling distributions remain the same even when the data come from markedly different pdfs. Parametric techniques, on the other hand, assume the observations come from a specific pdf. The one-sample $t$ test, for instance, is strictly valid only when the population being sampled is a bell-shaped curve. Its nonparametric counterpart, the signed rank test, requires only that the data's pdf be symmetric.

Once a statistical objective has been articulated, an experimenter decides whether to seek a parametric or a nonparametric solution by looking at the assumptions underlying the tests being considered. Parametric procedures, *if their assumptions are met*, yield inferences that are more precise than do nonparametric procedures (in terms of shorter confidence intervals and hypothesis tests with smaller $\beta$'s). Presented with a set of data, then, a statistician's first objective is to use the descriptive techniques of Chapter 3 to get a feeling for whether the assumptions of a parametric test are likely to be satisfied. Only if it seems as if they are not should a nonparametric procedure be considered.

Figure 16.7 shows two sets of one-sample data plotted in the format first introduced in Case Study 3.1. Assuming our objective is to choose between $H_0$: $\mu = \mu_0$ and $H_1$: $\mu \neq \mu_0$, should we do the analysis as a one-sample $t$ test or as a signed rank test? Notice in (a) that the density of points is greatest in the center of the distribution and falls off sharply in both tails. Such a pattern would be entirely consistent with the assumption that the population being sampled is a normal distribution. The proper analysis to use on those data, then, would be the one-sample $t$ test.

In contrast, the points in (**b**) are symmetric, but their heaviest density is in the two tails, which suggests the observations may have come from a U-shaped distribution (see Fig. 16.8). Since the latter is decidedly nonnormal, a $t$ test would be inappropriate; a signed rank test is what we should use.

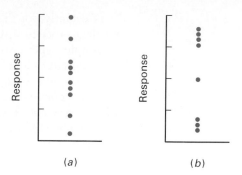

(a)                    (b)

**FIG. 16.7**

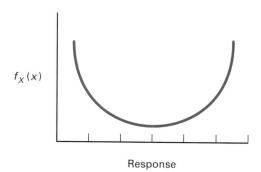

$f_X(x)$

Response

**FIG. 16.8**

The material in this chapter has served two purposes. On the one hand, it describes a set of very useful techniques for analyzing data. The rank sum test and the Kruskal-Wallis test, in particular, deal with experimental situations that are widely encountered in practice. At the same time, it forced us to think more critically about assumptions and how they can be verified. Indeed, the more we learn about nonparametric statistics, the better we are at doing parametric statistics.

---

**REVIEW QUESTIONS**        **16.65**   Two companies that sell research papers are competing for business among upperclassmen taking a required ethics course. Shown below are the grades earned by eight papers bought from Research, Inc. and by seven papers bought from Know It All. Use a nonparametric procedure to test an appropriate $H_0$. Let $\alpha = 0.01$.

| Research, Inc. | | Know It All | |
|---|---|---|---|
| 87 | 85 | 76 | 65 |
| 92 | 84 | 65 | 83 |
| 76 | 93 | 82 | 79 |
| 90 | 91 | 74 | |

16.66 The 28 most recent claims paid by an automobile insurance company ranged from \$350 to \$15,000:

| | | | | | | |
|---|---|---|---|---|---|---|
| \$2,450 | 10,350 | 475 | 750 | 11,500 | 650 | 875 |
| 550 | 1,500 | 620 | 350 | 1,000 | 3,450 | 4,300 |
| 8,175 | 480 | 540 | 825 | 840 | 7,500 | 2,650 |
| 1,050 | 9,650 | 15,000 | 12,500 | 630 | 1,500 | 975 |

If this sample is typical, can we conclude that the median claim is significantly different from \$700? Use a sign test and let $\alpha = 0.05$.

16.67 Two herbicides (TD-X and Smith's Weedkill) were each applied to different halves of 20 one-acre fields planted with soybeans. Two weeks later, a visual comparison of the 20 fields was made. Analyze the resulting data with a sign test. Use a two-sided alternative and let $\alpha = 0.05$.

| Field | TD-X Better Than Weedkill? | Field | TD-X Better Than Weedkill? |
|---|---|---|---|
| 1 | yes | 11 | no |
| 2 | yes | 12 | yes |
| 3 | yes | 13 | yes |
| 4 | no | 14 | yes |
| 5 | yes | 15 | yes |
| 6 | yes | 16 | yes |
| 7 | yes | 17 | yes |
| 8 | no | 18 | yes |
| 9 | yes | 19 | no |
| 10 | yes | 20 | yes |

16.68 Listed below are the numbers of personal fouls called by four referees while officiating a total of 21 Division I basketball games. Define what $\tilde{\mu}_i$ represents in this situation and test $H_0: \tilde{\mu}_1 = \tilde{\mu}_2 = \tilde{\mu}_3 = \tilde{\mu}_4$. Let $\alpha = 0.05$.

| Referee 1 | Referee 2 | Referee 3 | Referee 4 |
|---|---|---|---|
| 13 | 11 | 15 | 8 |
| 12 | 8 | 13 | 10 |
| 15 | 14 | 10 | 10 |
| 9 | 6 | 16 | 7 |
| 13 | 7 | 16 | |
| 12 | 8 | | |

16.69 Before buying a new photocopier, a company buyer decides to compare the reliability of three similarly priced machines. Each is used in six offices for a period of one month. A tally is kept of the number of times any sort of maintenance service is required. Listed below are the average number of repairs per week. Use a Friedman test to decide whether the differences among the machines are statistically significant. Let $\alpha = 0.05$.

|  | | MACHINE | |
|---|---|---|---|
| | **A** | **B** | **C** |
| 1 | 8.5 | 10.6 | 7.6 |
| 2 | 7.4 | 8.8 | 7.5 |
| **OFFICE**  3 | 15.3 | 19.2 | 14.4 |
| 4 | 4.2 | 4.1 | 3.0 |
| 5 | 9.2 | 11.1 | 10.3 |
| 6 | 8.3 | 9.9 | 8.6 |

**16.70**  In Review Exercise 16.69, the offices served as blocks. Is there anything about those data to suggest that we would not want to compare machines A, B, and C by using independent samples? Explain.

**16.71**  Is it believable that the following sample of 30 observations came from a population whose median is 65? Answer the question by doing a sign test at the $\alpha = 0.05$ level of significance.

| | | | | | |
|---|---|---|---|---|---|
| 67.6 | 61.2 | 58.6 | • 66.4 | 66.3 | 69.4 |
| 68.2 | 66.8 | 59.8 | 68.1 | 64.1 | 68.5 |
| 64.9 | 65.1 | 67.3 | 63.2 | 69.4 | 66.3 |
| 65.3 | 66.2 | 65.1 | 69.4 | 62.6 | 57.4 |
| 72.1 | 68.3 | 71.2 | 70.4 | 70.4 | 67.5 |

**16.72**  What $\alpha$ level is associated with a test of $H_0: \tilde{\mu} = \tilde{\mu}_0$ versus $H_1: \tilde{\mu} \neq \tilde{\mu}_0$ if $n = 5$ and we use the Wilcoxon rank sum statistic by rejecting $H_0$ if $w \leqslant 1$ or $w \geqslant 13$?

**16.73**  Movietime and Show Palace are two competitors in the video rental market. Listed below are percentage increases in profits reported by 8 franchises owned by Movietime and 12 owned by Show Palace. Set up and carry out an appropriate inference using the Wilcoxon rank sum test. Let $\alpha = 0.05$.

| Movietime | | Show Palace | |
|---|---|---|---|
| 6.2 | 7.5 | 5.9 | 8.6 |
| 9.6 | 11.2 | 8.9 | 11.2 |
| 4.8 | 8.0 | 14.2 | 13.4 |
| 8.6 | 7.3 | 12.1 | 7.4 |
| | | 10.9 | 8.6 |

**16.74**  Would it be better to analyze the following data by using a sign test or a Wilcoxon signed rank test? Explain.

| | | | | | |
|---|---|---|---|---|---|
| 51 | 40 | 45 | 35 | 8 | 41 |
| 18 | 58 | 32 | 55 | 52 | 12 |
| 43 | 4 | 56 | 30 | 49 | 50 |
| 22 | 53 | 54 | 49 | 51 | 47 |
| 58 | 29 | 42 | 57 | 20 | 50 |

16.75 Three faculty members of State University's English department have been nominated for a teaching award. Summarized below are the average ratings on student evaluations that each professor received during six recent semesters. Are the differences among the ratings statistically significant? Do an appropriate nonparametric test at the $\alpha = 0.05$ level of significance.

| Prof. C | Prof. T | Prof. W |
|---|---|---|
| 4.6 | 4.5 | 4.4 |
| 4.5 | 4.1 | 4.7 |
| 4.7 | 4.0 | 4.8 |
| 4.1 | 4.1 | 4.7 |
| 4.2 | 4.2 | 4.9 |
| 4.8 | 4.0 | 4.4 |

16.76 Let $W$ denote the Wilcoxon signed rank statistic as described in Theorem 16.1. What is the variance of $W$ if $E(W) = 60$?

16.77 Analyze the shear strength data in Question 2.56 with a Wilcoxon rank sum test. Let $\alpha = 0.10$.

16.78 In the recent past, the length of time required by a child psychology journal to review a submitted manuscript averaged 35 days. Beginning in January, the journal has been using a new refereeing system designed to expedite the reviewing process. Do the 30 turnaround times listed below suggest that the new procedure has been successful? Answer the question by doing a sign test at the $\alpha = 0.05$ level of significance.

| | | | | | |
|---|---|---|---|---|---|
| 38 | 28 | 32 | 25 | 36 | 34 |
| 32 | 36 | 40 | 28 | 28 | 38 |
| 31 | 29 | 36 | 27 | 34 | 37 |
| 25 | 30 | 32 | 37 | 33 | 30 |
| 29 | 31 | 36 | 29 | 26 | 33 |

16.79 Use a signed rank test to analyze the data in Review Question 16.78. Keep the level of significance at 0.05.

16.80 Shown below is a partially completed ranking of six treatments in four blocks. Fill in the missing entries.

| | | TREATMENT | | | | | |
|---|---|---|---|---|---|---|---|
| | | A | B | C | D | E | F |
| BLOCKS | I | 3 | 1 | 4 | 6 | | |
| | II | 1 | 2 | 4 | 3 | 6 | |
| | III | 2 | 4 | 1 | | 3 | 6 |
| | IV | | | | | 4 | |
| | $R_j$: | 8 | 8 | 15 | | 15 | 19 |

16.81   In general, what can an experimenter do to decide whether a one-sample problem should be analyzed with a *t* test, a sign test, or a signed rank test?

16.82   Do the following weight losses (in pounds) reported by 18 participants in a three-month fitness program enable us to conclude that, at the $\alpha = 0.01$ level of significance, diets T, X, and B60 are not equally effective? Use a nonparametric analysis.

| Subject | Diet | Weight Loss (lb) |
|---------|------|------------------|
| CF | X | 10 |
| DF | T | 6 |
| DP | X | 12 |
| BB | X | 14 |
| ML | T | 9 |
| JH | B60 | 6 |
| LM | X | 11 |
| LF | B60 | 8 |
| ME | B60 | 9 |
| LB | T | 14 |
| SH | X | 9 |
| CW | B60 | 4 |
| AW | B60 | 5 |
| CT | T | 7 |
| KH | B60 | 10 |
| JG | T | 8 |
| PR | T | 6 |
| EJ | B60 | 3 |

16.83   Theorem 16.1 states that $W' = (W - E(W))/\sqrt{\text{Var }(W)}$ can be adequately approximated by a standard normal curve if $n > 12$. Intuitively, the approximation should still be fairly good for $n = 12$. According to Appendix Table A.6, $P(W \leqslant 18 | n = 12) = 0.055$. Compute $W'$ for $n = 12$. Is the number close to what you would expect? Explain.

16.84   Analyze the data in Question 2.49 using a nonparametric procedure. Let $\alpha = 0.05$.

16.85   Six television critics were asked to indicate their personal preferences among three of the most famous programs from the 1960's: *The Honeymooners*, *Twilight Zone*, and *Gunsmoke*. Set up and test an appropriate hypothesis. Let $\alpha = 0.10$.

|    | The Honeymooners | Twilight Zone | Gunsmoke |
|----|------------------|---------------|----------|
| JL | 3 | 1 | 2 |
| GS | 1 | 3 | 2 |
| BS | 1 | 2 | 3 |
| GD | 3 | 2 | 1 |
| JB | 1 | 3 | 2 |
| BL | 1 | 3 | 2 |

16.86 Recall the data in Example 2.1. Use the Wilcoxon rank sum statistic to test an appropriate $H_0$ against a one-sided $H_1$. Let $\alpha = 0.05$.

16.87 For the data in Question 16.66, test $H_0$: $\tilde{\mu} = 700$ versus $H_1$: $\tilde{\mu} \neq 700$ with a signed rank test. Let $\alpha = 0.05$.

16.88 Attrition is a major problem for community colleges. The percentage of students completing a degree in six years or less is discouragingly low. Some educators believe that a graduation rate of 40% is a realistic goal. Listed below are the six-year graduation rates reported by a state's 10 community colleges. Eight of the 10 are below 40%. At the 0.01 level of significance, can we rule out the possibility that the true median graduation rate might still be 40%? Use a signed rank test and let $H_1$ be one-sided.

| 29 | 24 | 36 | 38 | 25 |
| 32 | 42 | 38 | 44 | 29 |

16.89 Use a signed rank test on the data in Review Question 11.30 to decide whether daytime batting averages tend to be higher than nighttime batting averages. Define the decision rule so that the probability of committing a Type I error comes as close as possible to 0.10 without exceeding 0.10.

# 17

# Multiple Regression

'*T*is fate that flings the dice, and as she flings
of Kings makes peasants, and of peasants Kings.

*Dryden*

## 17.1

### INTRODUCTION

The financial section of many metropolitan newspapers features a regular column explaining the behavior of yesterday's stock market. "Foreign Interest Rate Cuts Spur Market Rally" is a typical headline. The prime rate, balance of trade, tax rates, inflation, and similar economic indicators are frequently mentioned as causes of market declines and advances. Even factors that might seem irrelevant to the casual observer are cited: U.S.–Soviet relations, wars in distant places, changes in political leadership, and so on. While the selection of the particular variables needed to predict market behavior is best left to the financial "wizards," there is no doubt that the Dow Jones average *is* heavily influenced by a variety of measurable quantities. We encountered in Chapter 12 the beginnings of a method to build models for such relationships: recall our discussion of the models $y = a + bx$ and $y = ax^b$. The stock market, though, is one of many situations that demand more complexity. Specifically, we need techniques capable of modeling random variables that are functions of more than one factor.

The applications considered in this chapter are a generalization of the regression–correlation data structure introduced in Section 2.5 and analyzed in Chapter 12. If a stock market example had appeared in Chapter 12 the data would have been in the form (8.75, 1730), where the first entry $(x)$ represents, say, the prime rate on a certain day and the second entry $(y)$ is the Dow Jones average for the following day. In contrast, the kind of data we learn to analyze in this chapter will have the form

$$(8.75, 6.00, 9.10, 483, 1911)$$

where    8.75 = Tuesday's prime rate

              6.00 = Tuesday's discount rate

              9.10 = Tuesday's rate for 30-year Treasury bonds

             483 = Tuesday's price of gold (per ounce)

           1911 = Wednesday's Dow Jones average of 30 industrial stocks

Increasing the number of factors in a regression model necessitates a notation change. In Chapter 12 the $i$th data point is denoted $(x_i, y_i)$. Putting more factors into a regression equation might suggest that we write the data as $(w_i, x_i, y_i)$ or $(v_i, w_i, x_i, y_i)$. Notation of that sort, though, is never used, primarily because it does not lend itself to summation techniques and other mathematical devices. Instead, we will denote the $k$ factors (also known as *independent variables* or *regressors*) as $x_1, x_2, \ldots, x_k$. The $i$th value recorded for factor $x_j$ is written $x_{i,j}$. The $i$th data point, therefore, has $k + 1$ components—a set of values for the regressors $x_1, x_2, \ldots, x_k$ and a value for the response variable $y$. We write it

$$(x_{i,1}, x_{i,2}, \ldots, x_{i,k}, y_i)$$

EXAMPLE 17.1    Colleges sometimes use the average MCAT score, $Y$, earned by their students as a measure of the quality of their pre-med program. Among the factors known to influence $Y$ are (1) the average SAT of incoming freshmen $(x_1)$ and (2) the percentage of students who consider themselves pre-med $(x_2)$. Table 17.1 shows measurements on $x_1$, $x_2$, and $Y$ taken at five different schools. For School 1, the values recorded for factors $x_1$ and $x_2$ are 1210 and 3.0, respectively. Written in the *double* subscript notation just described,

$$x_{1,1} = 1210 \quad \text{and} \quad x_{1,2} = 3.0$$

Similarly, $x_{2,1} = 1120$, $x_{2,2} = 2.1$, and so on. There is no need to use two subscripts on the response variable $y$, so we write $y_1 = 10.2$, $y_2 = 10.0$, . . . , $y_5 = 8.1$.

**TABLE 17.1**

| School, $i$ | Freshman SAT, $x_1$ | Percent pre-med, $x_2$ | Average MCAT, $y$ |
|---|---|---|---|
| 1 | 1210 | 3.0 | 10.2 |
| 2 | 1120 | 2.1 | 10.0 |
| 3 | 1040 | 3.0 | 8.4 |
| 4 | 1260 | 4.3 | 10.6 |
| 5 | 980 | 1.6 | 8.1 |

QUESTIONS

17.1    A baseball enthusiast is trying to show that the number of games a team wins during a season can be predicted quite well by three factors—its batting average, earned run average, and fielding percentage. Listed below is a portion of her data.

| Team | BA | ERA | FP | Number of Wins, $y$ |
|---|---|---|---|---|
| 1 | .260 | 4.82 | .962 | 86 |
| 2 | .262 | 4.12 | .956 | 90 |
| 3 | .220 | 4.90 | .940 | 65 |
| 4 | .281 | 3.96 | .971 | 101 |
| 5 | .253 | 3.74 | .962 | 98 |

If BA, ERA, and FP are denoted as factors $x_1$, $x_2$, and $x_3$, respectively, what is

(a) $x_{3,1}$    (b) $x_{1,3}$    (c) $x_{2,3}$    (d) $x_{5,2}$    (e) $y_4$?

17.2     Suppose a study similar to the one proposed in Example 17.1 claims that the relationship between $x_1$, $x_2$, and $y$ can be described by the equation

$$y = 0.01x_1 - 0.8x_2$$

What average MCAT score would we expect from a school where 3.1% of the students are pre-meds and the average freshman SAT is 1090?

In Chapter 12, we drew a distinction between *regression* data, where $Y$ is a random variable and $x$ is a constant, and *correlation* data, where both $X$ and $Y$ are random variables. A similar dichotomy exists here, where $k$ factors are involved. The mathematics involved, though, in extending the notion of correlation to $k$ factors is prohibitively difficult, so we will restrict our attention to the regression format.

Written formally, the particular regression model that we will be studying for much of this chapter satisfies three assumptions:

**general linear model**—A particular set of assumptions defining the most common formulation of multiple regression. If $Y$ is a random variable and $x_1, x_2, \ldots, x_k$ are fixed variables, the general linear model assumes that

$$Y = a + b_1x_1 + b_2x_2 + \cdots + b_kx_k + \varepsilon$$

where $\varepsilon$ is normally distributed with mean zero and variance $\sigma^2$.

1. $Y = a + b_1x_1 + b_2x_2 + \cdots + b_kx_k + \varepsilon$, where the $x_i$'s are constants set by the experimenter and $\varepsilon$ is a random error term.

2. The random variable $\varepsilon$ is normally distributed with mean 0 and variance $\sigma^2$.

3. $\varepsilon$ is independent of the values set for $x_1, x_2, \ldots,$ and $x_k$.

Borrowing from the nomenclature introduced in Chapter 12, we call any equation of the form $Y = a + b_1x_1 + b_2x_2 + \cdots + b_kx_k + \varepsilon$ a **general linear model** (or **multiple regression**). The primary objective of this chapter is to show how the coefficients $a, b_1, b_2, \ldots, b_k$ are estimated and what they tell us about the impact of regressors $x_1, x_2, \ldots,$ and $x_k$ on the response variable, $Y$.

## *17.2*

### *LEAST SQUARES ESTIMATION IN MULTIPLE REGRESSION*

In Section 3.5, we estimated the slope and $y$-intercept in the model

$$y = a + bx$$

by choosing for $a$ and $b$ values that minimize

$$\sum_{i=1}^{n} [y_i - (a + bx_i)]^2$$

The resulting $\hat{a}$ and $\hat{b}$ (recall Theorem 3.2) were called *least squares estimates*. The analogous criterion can be used to find estimates for $a, b_1, \ldots,$ and $b_k$ in the general linear model.

---

**DEFINITION 17.1**

Let

$$(x_{1,1}, x_{1,2}, \ldots, x_{1,k}, y_1), \quad (x_{2,1}, x_{2,2}, \ldots, x_{2,k}, y_2), \quad \ldots,$$

$$(x_{n,1}, x_{n,2}, \ldots, x_{n,k}, y_n)$$

be a set of $n$ observations on the linear model

$$y = a + b_1 x_1 + b_2 x_2 + \cdots + b_k x_k$$

The values $\hat{a}, \hat{b}_1, \hat{b}_2, \ldots$ and $\hat{b}_k$ that minimize

$$\sum_{i=1}^{n} [y_i - (a + b_1 x_{i,1} + b_2 x_{i,2} + \cdots + b_k x_{i,k})]^2$$

are called the *least squares estimates* of $a, b_1, b_2, \ldots$, and $b_k$, respectively.

---

Finding least squares estimates for a general linear model is difficult. Even writing formulas for $\hat{a}, \hat{b}_1, \hat{b}_2, \ldots$, and $\hat{b}_k$ requires substantially more notation than we can develop here. Only when $k = 2$ is it feasible to implement Definition 17.1 without the aid of a computer. For the remainder of this section, then, we will focus on the model

$$y = a + b_1 x_1 + b_2 x_2$$

Let

$$\sum x_j = x_{1,j} + x_{2,j} + \cdots + x_{n,j} \quad \text{and} \quad \sum x_j^2 = x_{1,j}^2 + x_{2,j}^2 + \cdots + x_{n,j}^2$$

be the sum and the sum of squares, respectively, of the $n$ values recorded for the $j$th independent variable. Also, let

$$\sum x_j x_m = x_{1,j} x_{1,m} + x_{2,j} x_{2,m} + \cdots + x_{n,j} x_{n,m}$$

be the sum of the products of the $n$ values recorded for regressors $j$ and $m$. Finally, let

$$S_{ii} = n \sum x_i^2 - \left(\sum x_i\right)^2$$

$$S_{ij} = n \sum x_i x_j - \sum x_i \sum x_j$$

and

$$S_{iy} = n \sum x_i y - \sum x_i \sum y$$

where

$$\sum y = y_1 + y_2 + \cdots + y_n$$

and

$$\sum x_i y = x_{1,i} y_1 + x_{2,i} y_2 + \cdots + x_{n,i} y_n$$

---

THEOREM 17.1

Let
$$(x_{1,1}, x_{1,2}, y_1), (x_{2,1}, x_{2,2}, y_2), \quad \ldots, \quad (x_{n,1}, x_{n,2}, y_n)$$
be a set of $n$ observations on the linear model
$$y = a + b_1 x_1 + b_2 x_2$$
The least squares estimates for $a$, $b_1$, and $b_2$ are given by

$$\hat{b}_1 = \frac{S_{1y}S_{22} - S_{2y}S_{12}}{S_{11}S_{22} - S_{12}^2}, \qquad \hat{b}_2 = \frac{S_{2y}S_{11} - S_{1y}S_{12}}{S_{11}S_{22} - S_{12}^2}$$

$$\hat{a} = \frac{1}{n}\sum y - \left(\frac{1}{n}\sum x_1\right)\hat{b}_1 - \left(\frac{1}{n}\sum x_2\right)\hat{b}_2$$

---

**EXAMPLE 17.2**    The first three columns of Table 17.2 show seven data points collected on the model $y = a + b_1 x_1 + b_2 x_2$. Find $\hat{a}$, $\hat{b}_1$, and $\hat{b}_2$.

**TABLE 17.2**

| $x_{i,1}$ | $x_{i,2}$ | $y_i$ | $x_{i,1}^2$ | $x_{i,2}^2$ | $x_{i,1}x_{i,2}$ | $x_{i,1}y_i$ | $x_{i,2}y_i$ |
|---|---|---|---|---|---|---|---|
| 3.1 | 2.4 | 22.1 | 9.61 | 5.76 | 7.44 | 68.51 | 53.04 |
| 2.5 | 5.7 | 36.2 | 6.25 | 32.49 | 14.25 | 90.50 | 206.34 |
| 3.1 | 2.7 | 23.5 | 9.61 | 7.29 | 8.37 | 72.85 | 63.45 |
| 6.8 | 5.7 | 46.0 | 46.24 | 32.49 | 38.76 | 312.80 | 262.20 |
| 4.4 | 8.0 | 50.6 | 19.36 | 64.00 | 35.20 | 222.64 | 404.80 |
| 3.1 | 4.7 | 32.5 | 9.61 | 22.09 | 14.57 | 100.75 | 152.75 |
| 5.1 | 3.4 | 29.0 | 26.01 | 11.56 | 17.34 | 147.90 | 98.60 |
| 28.1 | 32.6 | 239.9 | 126.69 | 175.68 | 135.93 | 1015.95 | 1241.18 |

**Solution**    Look at the sum of the eight columns. In the notation we just introduced,

$$28.1 = \sum x_1 \qquad\qquad 175.68 = \sum x_2^2$$
$$32.6 = \sum x_2 \qquad\qquad 135.93 = \sum x_1 x_2$$
$$239.9 = \sum y \qquad\qquad 1015.95 = \sum x_1 y$$
$$126.69 = \sum x_1^2 \qquad\qquad 1241.18 = \sum x_2 y$$

Also,

$$S_{11} = n\sum x_1^2 - \left(\sum x_1\right)^2 = 7(126.69) - (28.1)^2 = 97.22$$
$$S_{22} = n\sum x_2^2 - \left(\sum x_2\right)^2 = 7(175.68) - (32.6)^2 = 167.00$$

$$S_{12} = n \sum x_1 x_2 - \sum x_1 \sum x_2 = 7(135.93) - (28.1)(32.6) = 35.45$$

$$S_{1y} = n \sum x_1 y - \sum x_1 \sum y = 7(1015.95) - (28.1)(239.9) = 370.46$$

and

$$S_{2y} = n \sum x_2 y - \sum x_2 \sum y = 7(1241.18) - (32.6)(239.9) = 867.52$$

Substituting, then, into the formulas given in Theorem 17.1, we find

$$\hat{b}_1 = \frac{(370.46)(167.00) - (867.52)(35.45)}{(97.22)(167.00) - (35.45)^2} = 2.07718506$$

$$\hat{b}_2 = \frac{(867.52)(97.22) - (370.46)(35.45)}{(97.22)(167.00) - (35.45)^2} = 4.753809275$$

and

$$\hat{a} = \frac{1}{7}(239.9) - \left(\frac{1}{7} \cdot 28.1\right)2.07718506 - \left(\frac{1}{7} \cdot 32.6\right)4.753809275$$

$$= 3.794112517$$

The equation

$$y = 3.794112517 + 2.07718506 x_1 + 4.753809275 x_2 \qquad (17.1)$$

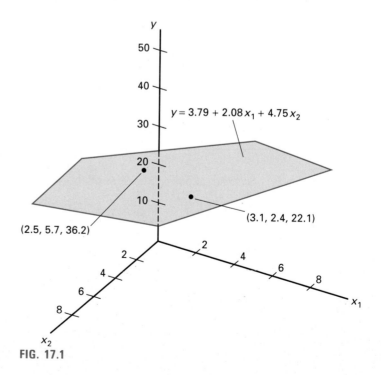

**FIG. 17.1**

is the linear model, therefore, that best fits the seven data points. Geometrically, Equation 17.1 defines the plane shown in Figure 17.1. Also shown are the first two data points, (3.1, 2.4, 22.1) and (2.5, 5.7, 36.2).

---

QUESTIONS

**17.3** Are the two data points shown in Figure 17.1 above or below the plane described by the equation $y = 3.79 + 2.08x_1 + 4.75x_2$?

**17.4** Find the equation of the form $y = a + b_1x_1 + b_2x_2$ that best describes the five points listed below.

| $x_1$ | $x_2$ | $y$ |
|---|---|---|
| 1 | 1 | 10 |
| 0 | 1 | 9 |
| 1 | 0 | 4 |
| 2 | 1 | 16 |
| 1 | 2 | 20 |

**17.5** Find $\hat{a}$, $\hat{b}_1$, and $\hat{b}_2$ for the MCAT data in Table 17.1. Suppose a group of students come from a school where 2.1% of the students are pre-meds and the average SAT of incoming freshmen is 1300. What MCAT score would we expect them to earn?

$$\sum x_1 = 5610 \qquad \sum x_2 = 14.0$$
$$\sum y = 47.3 \qquad \sum x_1^2 = 6,348,100$$
$$\sum x_2^2 = 43.46 \qquad \sum x_1x_2 = 16,088$$

**17.6** Skein strength $(y)$ is a measure of the quality of cotton yarn. It may be a function of fiber length $(x_1)$ and fiber tensile strength $(x_2)$. Listed below are test results on six samples of cotton yarn examined by the Department of Agriculture (169). Assume that $y$ can be described by a regression equation of the form $y = a + b_1x_1 + b_2x_2$. Find $\hat{a}$, $\hat{b}_1$, and $\hat{b}_2$.

| Fiber length, $x_1$ (hundredths of an inch) | Tensile Strength, $x_2$ (1000s of lb/sq in) | Skein Strength, $y$ (lb) |
|---|---|---|
| 85 | 76 | 99 |
| 74 | 72 | 97 |
| 70 | 73 | 88 |
| 76 | 76 | 91 |
| 64 | 79 | 80 |
| 74 | 78 | 100 |

**17.7** Fundraisers for State University's scholarship program suspect that the amount of money alumni donate $(y)$ may be a linear function of their

(1) GPA ($x_1$) and (2) number of years since graduation ($x_2$). Listed below is information on $x_1$, $x_2$, and $y$ for five recent donors. Fit these data with a function of the form $y = a + b_1 x_1 + b_2 x_2$.

| Subject | GPA, $x_1$ | Years Since Graduation, $x_2$ | Donation, $y$ |
|---------|-----------|-------------------------------|---------------|
| DP | 3.00 | 3 | $100 |
| CF | 3.20 | 1 | 50 |
| EJ | 3.50 | 4 | 200 |
| LM | 2.80 | 6 | 20 |
| ME | 2.60 | 3 | 30 |

If an experimenter's sole objective is fitting the equation

$$y = a + b_1 x_1 + b_2 x_2$$

to a set of $n$ points, then using Theorem 17.1 to find $\hat{a}$, $\hat{b}_1$, and $\hat{b}_2$ completes the analysis. We saw in Chapter 12, though, that curve-fitting is more likely to be the *start* of a statistical analysis rather than the end. The same is true here: estimating $a$, $b_1$, $b_2$, ..., and $b_k$ almost inevitably raises a variety of inference questions. But before any of those can be addressed, we need to estimate $\sigma^2$.

We take our cue, once again, from Chapter 12. In Definition 12.1, $\sigma^2$ is estimated by averaging the squared deviations of the observed $Y_i$'s from the predicted $Y_i$'s. Definition 17.2 extends that approach in the obvious way. (Note: Any inference procedure involving regression data requires that $Y$ be considered a random variable. Accordingly, we will resume at this point our practice of using lowercase letters for constants and uppercase letters for random variables.

---

**DEFINITION 17.2**

Let
$$(x_{1,1}, x_{1,2}, \ldots, x_{1,k}, Y_1), \quad (x_{2,1}, x_{2,2}, \ldots, x_{2,k}, Y_2), \quad \ldots,$$
$$(x_{n,1}, x_{n,2}, \ldots, x_{n,k}, Y_n)$$

be a set of $n$ data points taken on the general linear model

$$Y = a + b_1 x_1 + b_2 x_2 + \cdots + b_k x_k + \varepsilon$$

where $\varepsilon$ is normally distributed with mean 0 and variance $\sigma^2$. Let $\hat{a}$, $\hat{b}_1$, $\hat{b}_2$, ..., $\hat{b}_k$ be the estimates for $a$, $b_1$, $b_2$, ..., $b_k$ and let

$$\hat{y}_i = \hat{a} + \hat{b}_1 x_{i,1} + \hat{b}_2 x_{i,2} + \cdots + \hat{b}_k x_{i,k}$$

Then

1. The *error sum of squares*, SSE, is given by

$$SSE = \sum_{i=1}^{n} (y_i - \hat{y}_i)^2$$

2. The *error mean square*, MSE, is given by

$$MSE = \frac{1}{n - (k + 1)} \cdot SSE$$

We will use MSE as the estimate for $\sigma^2$.

---

**EXAMPLE 17.3**   The first three columns of Table 17.3 repeat the seven data points in Example 17.2. The fourth column lists the predicted $Y$ value for each observed $(x_{i,1}, x_{i,2})$. For example,

$$\hat{y}_1 = \hat{a} + \hat{b}_1 x_{1,1} + \hat{b}_2 x_{1,2}$$
$$= 3.79 + 2.07718506(3.1) + 4.753809275(2.4)$$
$$= 21.64232215$$

Find $\hat{\sigma}^2$.

**TABLE 17.3**

| $x_{i,1}$ | $x_{i,2}$ | $y_i$ | $\hat{y}_i$ | $(y_i - \hat{y}_i)^2$ |
|-----------|-----------|-------|-------------|-----------------------|
| 3.1 | 2.4 | 22.1 | 21.64232215 | 0.20946902 |
| 2.5 | 5.7 | 36.2 | 36.08362165 | 0.01354392 |
| 3.1 | 2.7 | 23.5 | 23.06846493 | 0.18622252 |
| 6.8 | 5.7 | 46.0 | 45.01523123 | 0.96976954 |
| 4.4 | 8.0 | 50.6 | 50.96390814 | 0.13242914 |
| 3.1 | 4.7 | 32.5 | 32.57608348 | 0.00578870 |
| 5.1 | 3.4 | 29.0 | 30.55036843 | 2.40364228 |
| | | | SSE = | 3.92086512 |

*Solution*   The last column lists the squared deviation of each observed $y_i$ from its predicted value, $\hat{y}_i$. Adding the entries in that column gives the error sum of squares:

$$SSE = \sum_{i=1}^{7} (y_i - \hat{y}_i)^2 = 3.92086512$$

Therefore,   $$\hat{\sigma}^2 = MSE = \frac{1}{7 - (2 + 1)} \cdot 3.92086512 = 0.98$$

COMMENT:    The data in Table 17.3 were computer-generated by assuming that the true regression equation is

$$Y = 3 + 2x_1 + 5x_2 + \varepsilon$$

where $\varepsilon$ is normally distributed with $\mu = 0$ and $\sigma^2 = 1$. Notice that the least squares estimates for the model's four parameters are all quite close to their true values, despite the small sample size:

$$\hat{a} = 3.79 \qquad \hat{b}_1 = 2.08 \qquad \hat{b}_2 = 4.75 \qquad \hat{\sigma}^2 = 0.98$$

In practice, SSE is calculated more easily by using a formula similar to the one that appeared in Theorem 12.2.

---

**THEOREM 17.2**

Let $\hat{a}, \hat{b}_1, \hat{b}_2, \ldots, \hat{b}_k$ be parameter estimates based on $n$ data points taken from the model

$$Y = a + b_1x_1 + b_2x_2 + \cdots + b_kx_k + \varepsilon$$

Let $S_{YY} = n \sum y^2 - (\sum y)^2$. Then

$$\text{SSE} = \frac{1}{n} \{S_{yy} - \hat{b}_1 S_{1y} - \hat{b}_2 S_{2y} - \cdots - \hat{b}_k S_{ky}\}$$

---

*EXAMPLE 17.4*   We conclude our discussion of the data in Example 17.2 by using Theorem 17.2 to find SSE. From Table 17.2,

$$\sum y = 239.9 \qquad \text{and} \qquad \sum y^2 = 8924.71$$

Therefore,       $S_{yy} = 7(8924.71) - (239.9)^2 = 4920.96$

Substituting the values already found for $\hat{b}_1$, $S_{1y}$, $\hat{b}_2$, and $S_{2y}$ into Theorem 17.2 gives

$$\text{SSE} = \frac{1}{7}\{4920.96 - (2.07718506)(370.46) - (4.753809275)(867.52)\}$$

$$= 3.92086515$$

Except for some rounding error in the eighth decimal place, this is the same value for SSE that we found in Example 17.3 using Definition 17.2.

---

QUESTIONS          17.8   Suppose $\hat{\sigma}^2 = 0$ for a set of $n$ data points taken on the model

$$Y = a + b_1x_1 + b_2x_2 + \varepsilon$$

What can be said, geometrically, about the $n$ points?

17.9   If a set of regression data satisfy the assumptions on p. 725 that define a general linear model, then each difference, $y_i - \hat{y}_i$, represents a random sample from a normal distribution with mean 0 and variance $\sigma^2$. What procedure in Chapter 13 could be used to test the normality assumption made for $\varepsilon$? Explain.

17.10   For the data in Question 17.4,

(a) use Definition 17.2 to estimate $\sigma^2$;

(b) use Theorem 17.2 to find SSE and then use SSE to find $\hat{\sigma}^2$.

17.11   Use Theorem 17.2 to estimate $\sigma^2$ for the MCAT data in Question 17.5. For what proportion of schools will students exceed their predicted MCAT scores by more than 0.3?

17.12   Estimate the standard deviation of skein strengths using the data in Question 17.6.

17.13   Suppose a set of regression data do *not* follow a general linear model. Will $\dfrac{1}{n - (k + 1)} \cdot \text{SSE}$ still be an appropriate estimate for $\sigma^2$? Explain. (Hint: Think of what happens to $\hat{\sigma}^2$ in a one-variable regression if we fit a straight line to a set of points that have a distinctly curvilinear relationship.)

# 17.3

## HYPOTHESIS TESTING IN THE GENERAL LINEAR MODEL

When an experimenter presumes that a random variable $Y$ can be written in the form

$$Y = a + b_1 x_1 + b_2 x_2 + \cdots + b_k x_k + \varepsilon$$

two questions immediately arise. First, how do we know that the expression

$$a + b_1 x_1 + b_2 x_2 + \cdots + b_k x_k$$

has any relationship whatsoever to the value of $Y$? And second, how can we be sure that a given factor—say, $x_j$—is worth keeping in the model? In both cases, we will look to hypothesis testing for guidance. To determine whether the model *as a whole* has any validity we test

$$H_0: \quad b_1 = b_2 = \cdots = b_k = 0$$

To test whether factor $x_j$ contributes significantly to the prediction of $Y$, we test $H_0: b_j = 0$.

We begin by describing a procedure for testing

$$H_0: \quad b_1 = b_2 = \cdots = b_k = 0$$

In practice, that's what an experimenter would do first. Given a set of $n$ data points, the total variation among the $y_i$'s can be measured by the quantity

$$\sum_{i=1}^{n} (y_i - \bar{y})^2 \quad \text{(Why?)}$$

Also, the variation "left over" or "unaccounted for" after fitting the model $\hat{y} = \hat{a} + \hat{b}_1 x_1 + \cdots + \hat{b}_k x_k$ is SSE. It follows, then, that the difference,

$$\sum_{i=1}^{n} (y_1 - \bar{y})^2 - \text{SSE}$$

is the amount of variability "explained" by the regression. Equivalently,

$$\frac{\sum_{i=1}^{n} (y_i - \bar{y})^2 - \text{SSE}}{\sum_{i=1}^{n} (y_i - \bar{y})^2} = 1 - \frac{\text{SSE}}{\sum_{i=1}^{n} (y_i - \bar{y})^2} = 1 - \frac{n \cdot \text{SSE}}{S_{yy}} \quad (17.2)$$

is the *proportion* of the total variability explained by the regression.

---

**DEFINITION 17.3**

Let $\hat{y} = \hat{a} + \hat{b}_1 x_1 + \cdots + \hat{b}_k x_k$ be the estimated regression equation fit to a set of $n$ points. The *coefficient of determination*, $r^2$, where

$$r^2 = 1 - \frac{n \cdot \text{SSE}}{S_{yy}}$$

is the proportion of the total variation in the $y_i$'s that is explained by the linear model.

---

The numerical value of $r^2$ indicates the strength of the relationship between the observed $y_i$'s and the predicted $\hat{y}_i$'s. If $r^2 = 0$, SSE must equal $\sum_{i=1}^{n} (y_i - \bar{y})^2$, which means that the regression is not accounting for any of the variation in the data. But if $a + b_1 x_1 + \cdots + b_k x_k$ is incapable of explaining the variation in the $y_i$'s, we may as well set all the $b_j$'s $= 0$—that is, we should accept $H_0$: $b_1 = b_2 = \cdots = b_k = 0$. At the other extreme, if $r^2 = 1$ then SSE must be zero, which implies that the model is perfect: every $\hat{y}_i$ equals $y_i$. In that case, of course, we should reject $H_0$: $b_1 = b_2 = \cdots = b_k = 0$. Theorem 17.3 tells us how to interpret values of $r^2$ *between* 0 and 1—that is, it gives the $\alpha$ decision rule for testing $H_0$: $b_1 = b_2 = \cdots = b_k = 0$.

---

**THEOREM 17.3**

Let $r^2$ be the coefficient of determination based on $n$ observations fit to a linear model,

$$Y = a + b_1 x_1 + \cdots + b_k x_k + \varepsilon$$

To test

$$H_0: \quad b_1 = b_2 = \cdots = b_k = 0$$

vs.

$$H_1: \quad \text{not all the } b_j\text{'s} = 0$$

at the $\alpha$ level of significance, reject $H_0$ if

$$\frac{r^2/k}{(1 - r^2)/[n - (k + 1)]} \geq F_{\alpha, k, n - (k + 1)}$$

---

*EXAMPLE 17.5*  In economics, a graph showing the relationship between the price of a product and the amount purchased is called a *demand curve*. Figure 17.2 shows a demand curve for chicken, covering the 15-year period from 1948 to 1962 (97). The original data are listed in Table 17.4. (To correct for the changing dollar, values of $x$ have been adjusted by the consumer price index.)

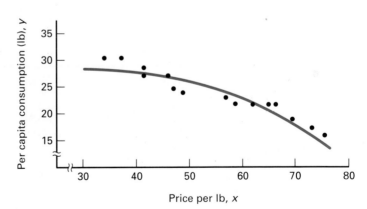

**FIG. 17.2**

The shape of the scatter diagram in Figure 17.2 suggests that the relationship between $x$ and $Y$ might be quadratic. That is,

$$Y = a + b_1 x + b_2 x^2 + \varepsilon \tag{17.3}$$

**TABLE 17.4**

| Year | Price per lb $(x)$ | Lb per person $(y)$ |
|------|------|------|
| 1948 | 75.4 | 18.3 |
| 1949 | 71.8 | 19.6 |
| 1950 | 68.0 | 20.6 |
| 1951 | 66.0 | 21.7 |
| 1952 | 65.0 | 22.1 |
| 1953 | 62.8 | 21.9 |
| 1954 | 56.4 | 22.8 |
| 1955 | 58.7 | 21.3 |
| 1956 | 50.4 | 24.4 |
| 1957 | 47.6 | 25.5 |
| 1958 | 45.8 | 28.2 |
| 1959 | 41.4 | 28.9 |
| 1960 | 41.4 | 28.2 |
| 1961 | 37.0 | 30.3 |
| 1962 | 38.6 | 30.2 |

Whether Equation 17.3 is a reasonable description of Figure 17.2 can be ascertained by testing $H_0$: $b_1 = b_2 = 0$. If $H_0$ is accepted, we reject the model; if $H_0$ is rejected, we accept the model.

*Solution*   Table 17.5 shows the sums necessary to estimate $a$, $b_1$, $b_2$ and SSE. Using the formulas introduced in Section 17.2,

$$S_{22} = 15(47806.33) - (826.3)^2 = 34323.26$$

$$S_{22} = 15(180,605,075) - (47806.33)^2 = 423,630,936.9$$

**TABLE 17.5**

| Year | Price, $x_1$ | $x_2 = x_1^2$ | $y$ | $x_1^2$ | $x_2^2$ |
|------|------|------|------|------|------|
| 1948 | 75.4 | 5685.16 | 18.3 | 5685.16 | 32,321,044 |
| 1949 | 71.8 | 5155.24 | 19.6 | 5155.24 | 26,576,499 |
| 1950 | 68.0 | 4624.00 | 20.6 | 4624.00 | 21,381,376 |
| ⋮ | ⋮ | ⋮ | ⋮ | ⋮ | ⋮ |
| 1962 | 38.6 | 1489.96 | 30.2 | 1489.96 | 2,219,980 |
|      | 826.3 | 47806.33 | 364.0 | 47806.33 | 180,605,075 |

| $x_1 x_2$ | $x_1 y$ | $x_2 y$ | $y^2$ |
|------|------|------|------|
| 428661.06 | 1379.82 | 104038.43 | 334.89 |
| 370146.23 | 1407.28 | 101042.70 | 384.16 |
| 314432.00 | 1400.80 | 95254.40 | 424.36 |
| ⋮ | ⋮ | ⋮ | ⋮ |
| 57512.46 | 1165.72 | 44996.79 | 912.04 |
| 2886729.09 | 19354.03 | 1083777.25 | 9057.28 |

$$S_{12} = 15(2,866,729.091) - (826.3)(47806.33) = 3,798,565.89$$

$$S_{1y} = 15(19354.03) - (826.3)(364) = -10462.75$$

$$S_{2y} = 15(1,083,777.25) - (47806.33)(364) = -1,144,845.32$$

$$S_{yy} = 15(9057.28) - (364)^2 = 3363.2$$

Therefore,

$$\hat{b}_1 = \frac{(-10462.75)(423,630,936.9) - (-1,144,845.32)(3,798,565.89)}{(34323.26)(423,630,936.9) - (3,798,565.89)^2}$$

$$= -0.750945153$$

$$\hat{b}_2 = \frac{(-1,144,845.32)(34323.26) - (-10462.75)(3,798,565.89)}{(34323.26)(423,630,936.9) - (3,798,565.89)^2}$$

$$= 0.00403103$$

and

$$\hat{a} = \frac{1}{15}[364 - (826.3)(-0.750945153) - (47806.33)(0.00403103)]$$

$$= 52.78648197$$

Finally, from Theorem 17.2,

$$\text{SSE} = \frac{1}{15}[3363.2 - (-0.750945153)(-10462.75)$$

$$- (0.00403103)(-1,144,845.32)]$$

$$= 8.076963392$$

Now, let $\alpha = 0.05$. Given that $n = 15$ and $k = 2$, we should reject $H_0: b_1 = b_2 = 0$ if

$$\frac{r^2/2}{(1 - r^2)/[15 - (2 + 1)]} \geqslant F_{0.05, 2, 15-(2+1)} = F_{0.05, 2, 12} = 3.89$$

But $\quad\quad r^2 = 1 - \dfrac{n \cdot \text{SSE}}{S_{yy}} = 1 - \dfrac{15(8.077)}{3363.2} = 0.96$

so the observed $F$ ratio is

$$\frac{0.96/2}{(1 - 0.96)/12} = 144.0$$

Since $144.0 > 3.89$, our conclusion is to reject $H_0$—the model $a + b_1x + b_2x^2$ *does* help predict the per capita consumption of chicken.

**17.14** What inference procedure that we studied in an earlier chapter is motivated by an argument similar to the discussion that precedes Equation 17.2?

**17.15** Suppose $r^2 = 0.34$ for a set of 16 data points collected on the model

$$Y = a + b_1x_1 + b_2x_2 + b_3x_3 + \varepsilon$$

At the $\alpha = 0.05$ level, test $H_0$: $b_1 = b_2 = b_3 = 0$. Should we conclude that the model is effective in predicting the behavior of $Y$?

**17.16** Does the statistical significance we associate with $r^2$ depend on the sample size $n$? (Hint: Look at the observed $F$ ratio given in Theorem 17.3. What happens if $r^2$ and $k$ remain fixed but $n$ increases?)

**17.17** Test $H_0$: $b_1 = b_2 = 0$ for the data in Question 17.4. Let $\alpha = 0.05$.

**17.18** Use the data in Table 17.1 to test whether MCAT scores can be predicted by a linear combination of freshman SATs and the per cent of a school's students that are pre-med. Let $\alpha = 0.05$. (Some of the sums you will need are given in Question 17.5)

**17.19** At the $\alpha = 0.01$ level, test $H_0$: $b_1 = b_2 = 0$ for the skein strength data in Question 17.6

**17.20** Recall Question 17.7. Is it reasonable to presume that donations are a linear function of GPA and years since graduation? Set up and carry out an appropriate hypothesis test. Let $\alpha = 0.01$.

Suppose a linear regression is found to be useful—that is, the data allow us to reject $H_0$: $b_1 = b_2 = \cdots = b_k = 0$. What do we do next? Usually an experimenter at that point tries to simplify the problem by identifying a smaller subset of the original $k$ factors that would, themselves, constitute an effective model. In multiple regression, eliminating unnecessary factors is very important because of the enormous number of additional computations that each extra factor requires. General techniques for identifying a "best" model are quite complicated and lay beyond the scope of our intentions. What we can do instead to illustrate the ideas involved is to treat the particular case where $Y$ is initially presumed to be a function of just two factors, $x_1$ and $x_2$. The question to be addressed is whether the model can be reduced to *one* factor.

Conceptually, we assess the significance of a factor in a model by testing whether its coefficient can be set equal to zero. We have encountered that idea before, in the context of the simple linear model $Y = a + bx + \varepsilon$. Recall Theorem 12.3. Testing whether $Y$ is independent of $x$ is equivalent to testing $H_0$: $b = 0$. If $H_0$ is accepted we conclude that $x$ is not effective in predicting the behavior of $Y$. Here, just as in Section 12.2, testing whether the coefficient of a regression factor is zero is accomplished by setting up a $t$ ratio.

THEOREM 17.4

Assume that $Y = a + b_1 x_1 + b_2 x_2 + \varepsilon$, where $\varepsilon$ is normally distributed with mean zero and variance $\sigma^2$. Let $\hat{a}$, $\hat{b}_1$, and $\hat{b}_2$ be the estimates of $a$, $b_1$, and $b_2$, respectively, based on $n$ observations. If $S_1$ and $S_2$ denote the standard deviations of $B_1$ and $B_2$, respectively, then

$$S_1 = \sqrt{\frac{n \cdot \text{MSE} \cdot S_{22}}{(S_{11}S_{22} - S_{12}^2)}} \quad \text{and} \quad S_2 = \sqrt{\frac{n \cdot \text{MSE} \cdot S_{11}}{(S_{11}S_{22} - S_{12}^2)}}$$

Also, if $H_0$: $b_i = 0$ is true, the ratio $B_i/S_i$ has a Student $t$ distribution with $n - 3$ degrees of freedom (for $i = 1$ or 2).

---

**EXAMPLE 17.6** Recall the data in Example 17.2. Does factor $x_1$ contribute any useful information to the model? Can we replace the expression $Y = a + b_1 x_1 + b_2 x_2 + \varepsilon$, in other words, with the simpler equation $Y = a + b_2 x_2 + \varepsilon$? To answer that requires that we test

$$H_0: \quad b_1 = 0$$

vs.

$$H_1: \quad b_1 \neq 0$$

**Solution** Let $\alpha = 0.05$. Since $n = 7$, the ratio $B_1/S_1$ has a Student $t$ distribution with $4 \, (=7 - 3)$ degrees of freedom when $H_0$ is true. The critical values for rejecting the null hypothesis are $\pm 2.7764$ (see Figure 17.3).

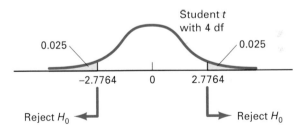

**FIG. 17.3**

From p. 728, $B_1 = 2.077$. Also, using the values already computed for $S_{11}$, $S_{22}$, $S_{12}$, and MSE, we find that the estimated standard deviation of $B_1$ is 0.277:

$$S_1 = \sqrt{\frac{7(0.980)(167.00)}{(97.22)(167.00) - (35.45)^2}} = 0.277$$

The test statistic, therefore, is

$$B_1/S_1 = \frac{2.077}{0.277} = 7.5$$

and our conclusion at the $\alpha = 0.05$ level is to reject $H_0$—factor $x_1$ *does* contribute significantly to the prediction of $Y$.

---

**EXAMPLE 17.7**    Example 17.5 proposed that consumer demand for chicken ($Y$) can be written as a function of price ($x$) and price squared ($x^2$). Is that particular model reasonable? Yes. By rejecting $H_0$: $b_1 = b_2 = 0$, as we did on p. 737, we showed that a function of the form $a + b_1x + b_2x^2$ *is* effective in describing the behavior of $Y$. Is there a simpler model that might also be reasonable? Maybe yes and maybe no. The two possibilities are

$$Y = a + b_2x^2 + \varepsilon \tag{17.4}$$

or $$Y = a + b_1x + \varepsilon \tag{17.5}$$

*Solution*    Assessing the appropriateness of Equations 17.4 and 17.5 requires that we test $H_0$: $b_1 = 0$ and $H_0$: $b_2 = 0$.

Suppose, first, that we test

$$H_0: \quad b_1 = 0$$

vs.

$$H_1: \quad b_1 \neq 0$$

Recall that $B_1 = -0.751$, $S_{11} = 34323.26$, $S_{22} = 423{,}630{,}936.9$, $S_{12} = 3{,}798{,}565.89$, and MSE $= 0.67308028$. From Theorem 17.4,

$$S_1 = \sqrt{\frac{15(0.67308028)(423{,}630{,}936.9)}{(34323.26)(423{,}630{,}936.9) - (3{,}798{,}565.89)^2}} = 0.196$$

and the observed $t$ statistic is

$$\frac{B_1}{S_1} = \frac{-0.751}{0.196} = -3.83$$

Let $\alpha = 0.05$. The numbers that cut off areas of 0.025 in either tail of the $t$ distribution with $15 - 3 = 12$ degrees of freedom are $\pm 2.1788$. Therefore, we should reject $H_0$: $b_1 = 0$ and conclude that the "$x$" term does belong in the model.

Similarly, $B_2 = 0.004$ and

$$S_2 = \sqrt{\frac{15(0.67308028)(34323.26)}{(34323.26)(423{,}630{,}936.9) - (3{,}798{,}565.89)^2}} = 0.0018$$

The decision rule for testing $H_0: b_2 = 0$ is the same as the decision rule for testing $H_0: b_1 = 0$. The test statistic, though, becomes

$$\frac{B_2}{S_2} = \frac{0.004}{0.0018} = 2.28$$

Since $2.28 > 2.1788$, we should reject $H_0: b_2 = 0$ (and keep $x^2$ in the model).

## CASE STUDY 17.1

Estimating the market value of a house is a familiar example of using a set of fixed variables to predict the value of a random variable. Ultimately, the worth of a house is the amount, $Y$, that someone is willing to pay for it. Real estate agents, though, are able to approximate that figure quite well by looking at factors such as square footage, number of bedrooms, age, lot size, and so on.

Listed in Table 17.6 is information describing 34 homes sold in 1983 in Knox County, Tennessee (191). As a first question to explore, suppose we wish to see whether a combination of floor area (SQFT) and age (AGE) can be used to predict selling price (PR). Is it worthwhile, in other words, to write

$$PR = a + b_1\text{SQFT} + b_2\text{AGE} + \varepsilon$$

We begin by finding estimates for $a$, $b_1$, and $b_2$. Let $x_1$ and $x_2$ denote the square foot and age factors, respectively. Then

$$\sum x_1 = 64026 \qquad \sum x_2 = 741 \qquad \sum y = 2{,}163{,}360$$
$$\sum x_1^2 = 131{,}603{,}500 \quad \sum x_2^2 = 19959 \qquad \sum y^2 = 149{,}239{,}000{,}000$$
$$\sum x_1 x_2 = 1{,}408{,}389 \quad \sum x_1 y = 4{,}364{,}214{,}000 \quad \sum x_2 y = 48{,}043{,}410$$

Also,

$$S_{11} = 34(131{,}603{,}500) - (64026)^2 = 375{,}190{,}324$$
$$S_{22} = 34(19959) - (741)^2 = 129{,}525$$
$$S_{12} = 34(1{,}408{,}389) - (64026)(741) = 441{,}960$$
$$S_{1y} = 34(4{,}364{,}214{,}000) - (64026)(2{,}163{,}360) = 9{,}871{,}988{,}640$$
$$S_{2y} = 34(48{,}043{,}410) - (741)(2{,}163{,}360) = 30{,}426{,}180$$
$$S_{yy} = 34(149{,}239{,}000{,}000) - (2{,}163{,}360)^2 = 393{,}999{,}510{,}411$$

**TABLE 17.6**

| Selling Price ($) (PR) | Number of Bedrooms (BR) | Area (sq ft) (SQFT) | Age (yrs) (AGE) | Lot Size (sq yd) (LS) | Days on Market (DOM) |
|---|---|---|---|---|---|
| 38,500 | 3 | 1040 | 2 | 2911 | 78 |
| 40,000 | 3 | 1103 | 26 | 2244 | 19 |
| 40,000 | 2 | 1000 | 25 | 833 | 95 |
| 43,321 | 3 | 912 | 16 | 3940 | 44 |
| 47,500 | 3 | 1300 | 28 | 1274 | 84 |
| 49,000 | 3 | 1539 | 26 | 2053 | 307 |
| 42,800 | 3 | 1095 | 26 | 1417 | 159 |
| 47,000 | 3 | 1535 | 10 | 2222 | 91 |
| 47,000 | 3 | 1885 | 17 | 1667 | 169 |
| 51,000 | 2 | 1954 | 35 | 1783 | 184 |
| 52,500 | 3 | 1332 | 23 | 1667 | 76 |
| 48,671 | 3 | 1675 | 22 | 3583 | 105 |
| 55,000 | 3 | 1463 | 20 | 4308 | 44 |
| 60,500 | 3 | 1900 | 16 | 2750 | 67 |
| 55,000 | 3 | 1903 | 20 | 4103 | 120 |
| 58,000 | 3 | 2190 | 7 | 2173 | 76 |
| 58,093 | 3 | 1906 | 9 | 1291 | 164 |
| 63,500 | 3 | 1760 | 27 | 3093 | 101 |
| 66,750 | 3 | 1596 | 16 | 1833 | 274 |
| 63,500 | 3 | 1800 | 21 | 5808 | 75 |
| 65,000 | 3 | 1800 | 23 | 2667 | 130 |
| 67,000 | 3 | 2100 | 19 | 2222 | 63 |
| 69,900 | 3 | 2363 | 43 | 4033 | 240 |
| 64,000 | 5 | 2888 | 18 | 3513 | 129 |
| 65,000 | 3 | 1350 | 35 | 4156 | 55 |
| 70,950 | 4 | 2452 | 21 | 2133 | 43 |
| 74,800 | 5 | 2348 | 20 | 1878 | 91 |
| 79,900 | 3 | 2173 | 44 | 2108 | 191 |
| 89,500 | 3 | 2000 | 3 | 873 | 76 |
| 86,925 | 4 | 2937 | 25 | 2933 | 130 |
| 99,000 | 3 | 2440 | 50 | 1007 | 152 |
| 98,750 | 4 | 2504 | 18 | 7051 | 45 |
| 105,000 | 3 | 2829 | 20 | 1956 | 1 |
| 100,000 | 4 | 2954 | 10 | 5266 | 520 |

Therefore,

$$B_1 = \frac{(9{,}871{,}988{,}640)(129525) - (30{,}426{,}180)(441960)}{(375{,}190{,}324)(129525) - (441960)^2}$$

$$= 26.14030694$$

$$B_2 = \frac{(30{,}426{,}180)(375{,}190{,}324) - (9{,}871{,}988{,}640)(441960)}{(375{,}190{,}324)(129525) - (441960)^2}$$

$$= 145.7109434$$

$$A = \frac{1}{34} [2{,}163{,}360 - (64026)(26.14030694) - (741)(145.7109434)]$$

$$= 11127.32055$$

and by Theorem 17.2,

$$SSE = \frac{1}{34} [393{,}999{,}510{,}411 - (26.14030694)(9{,}871{,}988{,}640)$$

$$- (145.7109434)(30{,}426{,}180)]$$

$$= 3{,}867{,}919{,}701$$

The coefficient of determination for these 34 points is 0.66621971:

$$r^2 = 1 = \frac{n \cdot SSE}{S_{yy}} = 1 - \frac{34(3{,}867{,}919{,}701)}{393{,}999{,}510{,}411} = 0.66621971$$

By Theorem 17.3, we should reject $H_0: b_1 = b_2 = 0$ if

$$\frac{r^2/k}{(1 - r^2)/[n - (k + 1)]} \geq F_{0.05, k, n-(k+1)} = F_{0.05, 2, 31} = 3.31$$

But

$$\frac{0.66621971/2}{(1 - 0.66621971)/31} = 30.94$$

To answer our first question, then, the model $a + b_1 SQFT + b_2 AGE$ *is* helpful in estimating a home's selling price.

Next, we turn our attention to each of the factors separately. SQFT belongs in the model only if we can reject $H_0: b_1 = 0$. But $B_1 = 26.140$ and

$$S_1 = \sqrt{\frac{34(3{,}867{,}919{,}701/31)(129{,}525)}{(375{,}190{,}324)(129{,}525) - (441{,}960)^2}} = 3.369$$

so $B_1/S_1 = 26.140/3.369 = 7.76$. For $\alpha = 0.05$, the numbers defining the critical region are $\pm 2.0395$. It follows, then, that we should reject $H_0: b_1 = 0$. Similarly,

$$S_2 = \sqrt{\frac{34(3{,}867{,}919{,}701/31)(375{,}190{,}324)}{(375{,}190{,}324)(129{,}525) - (441{,}960)^2}} = 181.341$$

and

$$\frac{B_2}{S_2} = \frac{145.711}{181.341} = 0.80$$

Since $-2.0395 < 0.80 < 2.0395$, we should accept $H_0: b_2 = 0$. Taken together, these last two hypothesis tests suggest that we should keep SQFT in the model but drop AGE.

**17.21**  Test $H_0: b_1 = 0$ and $H_0: b_2 = 0$ for the data in Question 17.4. Let $\alpha = 0.05$.

**17.22**  What effect, if any, does the magnitude of $\sigma^2$ have on a test of $H_0: b_1 = 0$? Explain.

**17.23**  Recall the fundraising data in Question 17.7. Do an appropriate hypothesis test to determine whether GPA is a useful predictor of the amount a graduate donates. Let $\alpha = 0.05$.

**17.24**  Listed below are weights (WEIGHT), post positions (PP), and completion times (TIME) for 10 dogs who competed in a recent week's festivities at the Pensacola Greyhound Track (120).

| Name | WEIGHT (lb) | PP | TIME (sec) |
|------|------------|-----|-----------|
| Sure Fire Winner | 59.0 | 2 | 31.47 |
| JE's Estelle | 54.0 | 1 | 31.44 |
| Retrospeed | 71.5 | 7 | 31.08 |
| Forward Late | 76.0 | 4 | 32.08 |
| Anne Lemon | 64.0 | 3 | 31.97 |
| Waco Fats | 63.5 | 3 | 31.30 |
| Handy Zipper | 56.0 | 5 | 31.68 |
| Ninja X | 71.5 | 3 | 31.32 |
| Sickem Dana | 65.0 | 5 | 31.71 |
| C Me Fly | 60.0 | 8 | 32.07 |

Estimate $a$, $b_1$, and $b_2$ for the model

$$\text{TIME} = a + b_1\text{WEIGHT} + b_2\text{PP} + \varepsilon$$

Test $H_0: b_1 = b_2 = 0$ at the $\alpha = 0.01$ level. If $H_0$ is rejected, decide whether the model can be simplified.

Even though only one of the two variables considered in Case Study 17.1 contributes significantly to the prediction of a home's selling price, there remains the possibility that other linear combinations of the full set of factors in Table 17.6 might yield a better model. Hypothesizing situations where a random variable $Y$ depends on more than two regressors is not difficult—doing the necessary computations, though, is quite another story. Fortunately, computers have come to the rescue. Most mainframes and many microcomputers have software that can do multiple regression.

Figure 17.4 shows a printout from SAS, one of the more popular software packages. The data entered were the PR, SQFT, and AGE figures from Table 17.6. Those portions of the printout that relate specifically to what we did in Case Study 17.1 are labeled. There are slight differences in some of the numbers, due to round-off errors. The top block of entries,

DEP VARIABLE: PR

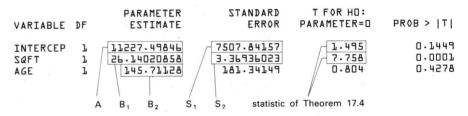

ANALYSIS OF VARIANCE

| SOURCE | DF | SUM OF SQUARES | MEAN SQUARE | F VALUE | PROB>F |
|--------|----|----------------|-------------|---------|--------|
| | | | | statistic of Theorem 17.3 | |
| MODEL | 2 | 7720269275 | 3860134638 | 30.937 | 0.0001 |
| ERROR | 31 | 3867956675 | 124762796 | | |
| C TOTAL | 33 | 11588225950 | | | |
| | | | SSE    MSE | | |

| | | | | | |
|--|--|--|--|--|--|
| ROOT MSE | 11170.17 | R-SQUARE | 0.6662 | $r^2$ | |
| DEP MEAN | 63628.24 | ADJ R-SQ | 0.6447 | | |
| C.V. | 17.55537 | | | | |

PARAMETER ESTIMATES

| VARIABLE | DF | PARAMETER ESTIMATE | STANDARD ERROR | T FOR HO: PARAMETER=0 | PROB > |T| |
|----------|----|--------------------|----------------|-----------------------|-----------|
| INTERCEP | 1 | 11227.49846 | 7507.84157 | 1.495 | 0.1449 |
| SQFT | 1 | 26.14020858 | 3.36936023 | 7.758 | 0.0001 |
| AGE | 1 | 145.71128 | 181.34149 | 0.804 | 0.4278 |

A   $B_1$   $B_2$   $S_1$   $S_2$   statistic of Theorem 17.4

**FIG. 17.4**

labeled ANALYSIS OF VARIANCE, refer to the overall model,

$$Y = a + b_1\text{SQFT} + b_2\text{AGE} + \varepsilon$$

and how well it fits the data. In the lower block of numbers, labeled PARAM-ETER ESTIMATES, are the values for $A$, $B_1$, and $B_2$ and the results of $t$ tests indicating the contribution of each individual term in the original model.

As we have seen before, computer printouts do not summarize hypothesis tests by writing "accept $H_0$" or "reject $H_0$." Instead, they list a $P$ value—that is, the probability that the test statistic, by chance, would be as extreme or more extreme than what was actually observed. The entry, for example, in the PROB > F column is 0.0001. That number refers to the overall model and means that the probability of

$$\frac{r^2/k}{(1-r^2)/[n-(k+1)]} = \frac{r^2/2}{(1-r^2)/31}$$

exceeding 30.937 (the observed $F$ ratio) is less than or equal to 0.0001. We can reject $H_0$: $b_1 = b_2 = 0$, in other words, at any $\alpha$ greater than 0.0001.

Similar inferences can be drawn about individual terms in the model by looking at the PROB > |T| column. Specifically, we can (1) reject $H_0$: $a = 0$ at any $\alpha > 0.1449$, (2) reject $H_0$: $b_1 = 0$ at any $\alpha > 0.0001$, and (3) reject $H_0$: $b_2 = 0$ at any $\alpha > 0.4278$. If $\alpha$ is set at 0.05, we should (1) accept $H_0$: $a = 0$, (2) reject $H_0$: $b_1 = 0$, and (3) accept $H_0$: $b_2 = 0$. According to the program, then, a suitable model is

$$PR = b_1 SQFT + \varepsilon \tag{17.6}$$

(Note: The coefficient of SQFT in Equation 17.6 is not equal to 26.14020858, which was the value assigned to $b_1$ in the original regression. If we reduce a model by eliminating factors, estimates for the remaining coefficients need to be recomputed. We will come back to that point shortly.)

Figure 17.5 shows the SAS printout for all the data in Table 17.6. The model being fit is

$$PR = a + b_1 SQFT + b_2 AGE + b_3 BR + b_4 LS + b_5 DOM + \varepsilon \tag{17.7}$$

DEP VARIABLE: PR

ANALYSIS OF VARIANCE

| SOURCE | DF | SUM OF SQUARES | MEAN SQUARE | F VALUE | PROB>F |
|--------|----|----|----|----|----|
| MODEL | 5 | 7846749728 | 1569349946 | 11.745 | 0.0001 |
| ERROR | 28 | 3741476222 | 133624151 | | |
| C TOTAL | 33 | 11588225950 | | | |

| | | | | |
|--|--|--|--|--|
| ROOT MSE | 11559.59 | R-SQUARE | 0.6771 | |
| DEP MEAN | 63628.24 | ADJ R-SQ | 0.6195 | |
| C.V. | 18.16739 | | | |

PARAMETER ESTIMATES

| VARIABLE | DF | PARAMETER ESTIMATE | STANDARD ERROR | T FOR H0: PARAMETER=0 | PROB > |T| |
|----------|----|----|----|----|----|
| INTERCEP | 1 | 16126.61488 | 12153.99083 | 1.327 | 0.1953 |
| BR | 1 | -3029.84010 | 4262.56302 | -0.711 | 0.4831 |
| SQFT | 1 | 27.94915421 | 4.58051123 | 6.102 | 0.0001 |
| AGE | 1 | 124.35683 | 195.45826 | 0.636 | 0.5298 |
| LS | 1 | 1.04666557 | 1.46271065 | 0.716 | 0.4802 |
| DOM | 1 | -8.67563603 | 21.33722894 | -0.407 | 0.6874 |

**FIG. 17.5**

Look at the PROB > |T| entries. Only one of the regressors (SQFT) has a small $P$-value. Equation 17.7, therefore, can be simplified to

$$PR = b_1 SQFT + \varepsilon \tag{17.8}$$

To find the $B_1$ appropriate for the $b_1$ in Equation 17.8, we again run the SAS program, this time entering only the PR values and the SQFT values. We add the instruction that the fit be made restricting the intercept to zero. Figure 17.6 shows the results. The second entry in the PARAMETER ESTIMATE column is the number we need—that is,

$$\text{PR} = 33.16182827 \cdot \text{SQFT} \tag{17.9}$$

```
DEP VARIABLE: PR

                          ANALYSIS OF VARIANCE

                              SUM OF          MEAN
          SOURCE      DF      SQUARES        SQUARE      F VALUE     PROB>F

          MODEL        0    7074531435                    0.000     1.0000
          ERROR       33    4513694516     136778622
          C TOTAL     33   11588225950

                ROOT MSE       11695.24     R-SQUARE       0.6105
                DEP MEAN       63628.24     ADJ R-SQ       0.6105
                C.V.           18.38058

                          PARAMETER ESTIMATES

                        PARAMETER       STANDARD     T FOR H0:
      VARIABLE  DF       ESTIMATE          ERROR    PARAMETER=0    PROB > |T|

      INTERCEP   1     1.81899E-12    0.000075462        0.000       1.0000
      SQFT       1      33.16182827    1.01947207       32.528       0.0001
      RESTRICT  -1      40140.78346   19747.03316        2.033       0.0502
```

**FIG. 17.6**

Notice that going from a six-parameter model (Equation 17.7) to a one-parameter model (Equation 17.9) has not reduced by very much our ability to explain the variation in the $y_i$'s. Figure 17.5 shows that $r^2$ for the six-parameter model is 0.6771; for the one-parameter model, the $r^2$ shown in Figure 17.6 is almost as large, 0.6105.

**CASE STUDY 17.2**

The total personal income of a county is an indication of its general economic health. Public planners and potential investors often collect a variety of information to help them understand the variables on which that income depends. Table 17.7 shows a set of economic statistics for 27 counties in the Panhandle and Big Bend regions of Florida (51). Figure 17.7 is the SAS printout for the corresponding six-parameter regression,

$$\text{INCOME} = a + b_1\text{TOP} + b_2\text{MIG} + b_3\text{LABOR}$$

$$+ b_4\text{UNEMP} + b_5\text{SALES} + \varepsilon$$

**TABLE 17.7**

| County | POP | MIG | LABOR | UNEMP | SALES | INCOME |
|---|---|---|---|---|---|---|
| Bay | 129.7 | 6844 | 57762 | 4536 | 782796 | 1562.7 |
| Escambia | 278.4 | 14231 | 122648 | 7196 | 1616914 | 3382.5 |
| Holmes | 16.3 | 657 | 6548 | 442 | 36187 | 124.1 |
| Okaloosa | 149.0 | 9522 | 63443 | 3693 | 941399 | 1801.4 |
| Santa Rosa | 66.2 | 3053 | 29474 | 1613 | 243955 | 841.6 |
| Walton | 27.5 | 1036 | 11887 | 765 | 84575 | 244.4 |
| Washington | 15.4 | 535 | 66 | 487 | 47614 | 153.3 |
| Calhoun | 9.7 | 221 | 3236 | 333 | 34667 | 83.8 |
| Franklin | 8.5 | 254 | 4038 | 198 | 26628 | 69.2 |
| Gadsden | 46.2 | 638 | 18716 | 884 | 111877 | 375.3 |
| Gulf | 12.0 | 365 | 4035 | 357 | 30072 | 119.7 |
| Jackson | 43.7 | 1768 | 17346 | 1202 | 160503 | 410.6 |
| Jefferson | 11.9 | 314 | 4734 | 206 | 37951 | 105.2 |
| Leon | 176.5 | 5933 | 102057 | 3577 | 1132152 | 2362.6 |
| Liberty | 5.0 | 81 | 2306 | 117 | 7645 | 42.6 |
| Wakulla | 13.7 | 303 | 6348 | 224 | 19221 | 133.4 |
| Alachua | 179.7 | 5017 | 97177 | 2829 | 1116430 | 2120.3 |
| Bradford | 24.1 | 521 | 10188 | 492 | 96777 | 218.3 |
| Columbia | 41.5 | 882 | 19950 | 1689 | 246512 | 425.6 |
| Dixie | 9.9 | 259 | 3538 | 237 | 34867 | 73.6 |
| Gilchrist | 7.1 | 236 | 2790 | 122 | 13494 | 79.3 |
| Hamilton | 9.4 | 253 | 3544 | 310 | 36964 | 88.0 |
| Lafayette | 5.1 | 68 | 2169 | 104 | 7767 | 68.9 |
| Madison | 15.9 | 387 | 6827 | 377 | 60631 | 157.3 |
| Suwannee | 26.2 | 712 | 11176 | 653 | 115984 | 258.4 |
| Taylor | 18.8 | 367 | 9494 | 471 | 72352 | 199.3 |
| Union | 10.7 | 162 | 3956 | 83 | 17888 | 70.0 |

POP = population in thousands
MIG = new registration of out-of-town vehicles (migration measure)
LABOR = size of labor force
UNEMP = number unemployed
SALES = total retail sales in thousands
INCOME = total personal income in millions

Note, first, that the model as a whole is effective—$r^2$ is large (0.9981) and PROB > F is small (0.0001). On the other hand, most of the variability is being explained by just two of the regressors, MIG and LABOR; all the others have large values for PROB > |T|. A two-parameter model, in other words, would appear to be quite sufficient.

Figure 17.8 shows the final step in the analysis. Values for MIG, LABOR, and INCOME are the only variables to be considered by the SAS program. According to the printout, the three are related by the equation

$$INCOME = 0.07510642 \cdot MIG + 0.01841024 \cdot LABOR$$

DEP VARIABLE: INCOME

ANALYSIS OF VARIANCE

| SOURCE | DF | SUM OF SQUARES | MEAN SQUARE | F VALUE | PROB>F |
|--------|-----|----------------|-------------|---------|--------|
| MODEL | 5 | 19750863.72 | 3950172.74 | 2238.617 | 0.0001 |
| ERROR | 21 | 37055.74307 | 1764.55919 | | |
| C TOTAL | 26 | 19787919.46 | | | |

| | | | | |
|---|---|---|---|---|
| ROOT MSE | 42.00666 | R-SQUARE | 0.9981 | |
| DEP MEAN | 576.7185 | ADJ R-SQ | 0.9977 | |
| C.V. | 7.283736 | | | |

PARAMETER ESTIMATES

| VARIABLE | DF | PARAMETER ESTIMATE | STANDARD ERROR | T FOR H0: PARAMETER=0 | PROB > \|T\| |
|----------|-----|--------------------|----------------|-----------------------|--------------|
| INTERCEP | 1 | -13.82177433 | 13.89100051 | -0.995 | 0.3311 |
| POP | 1 | 1.37921054 | 1.98237050 | 0.696 | 0.4942 |
| MIG | 1 | 0.06260899 | 0.01661780 | 3.768 | 0.0012 |
| LABOR | 1 | 0.01712948 | 0.003680622 | 4.654 | 0.0001 |
| UNEMP | 1 | 0.01687852 | 0.02497052 | 0.676 | 0.5065 |
| SALES | 1 | -0.000087213 | 0.000231213 | -0.377 | 0.7098 |

FIG. 17.7


DEP VARIABLE: INCOME

ANALYSIS OF VARIANCE

| SOURCE | DF | SUM OF SQUARES | MEAN SQUARE | F VALUE | PROB>F |
|--------|-----|----------------|-------------|---------|--------|
| MODEL | 1 | 19747907.61 | 19747907.61 | 12338.787 | 0.0001 |
| ERROR | 25 | 40011.84872 | 1600.47395 | | |
| C TOTAL | 26 | 19787919.46 | | | |

| | | | | |
|---|---|---|---|---|
| ROOT MSE | 40.00592 | R-SQUARE | 0.9980 | |
| DEP MEAN | 576.7185 | ADJ R-SQ | 0.9979 | |
| C.V. | 6.93682 | | | |

PARAMETER ESTIMATES

| VARIABLE | DF | PARAMETER ESTIMATE | STANDARD ERROR | T FOR H0: PARAMETER=0 | PROB > \|T\| |
|----------|-----|--------------------|----------------|-----------------------|--------------|
| INTERCEP | 1 | -2.22045E-16 | 3.72584E-08 | -0.000 | 1.0000 |
| MIG | 1 | 0.07510642 | 0.005392271 | 13.929 | 0.0001 |
| LABOR | 1 | 0.01841024 | 0.000520028 | 35.402 | 0.0001 |
| RESTRICT | -1 | -45.58015919 | 170.86301 | -0.267 | 0.7918 |

FIG. 17.8

17.25    Listed below are construction costs for 31 ranch-style homes built in the Fort Walton Beach, Florida area (80).

| Floor Area (sq ft) | Cost (1000 $) | Number of Baths | Number of rooms |
|---|---|---|---|
| 900 | 25 | 1 | 5 |
| 950 | 28 | 1.5 | 5 |
| 950 | 31 | 1.5 | 6 |
| 1000 | 29 | 2 | 6 |
| 1000 | 25 | 1.5 | 5 |
| 1150 | 36 | 2 | 6 |
| 1200 | 33 | 2 | 6 |
| 1200 | 28 | 2 | 5 |
| 1300 | 39 | 2 | 7 |
| 1450 | 43 | 2 | 7 |
| 1500 | 45 | 2 | 6 |
| 1500 | 49 | 2 | 6 |
| 1600 | 46 | 2 | 6 |
| 1600 | 42 | 2 | 6 |
| 1750 | 57 | 2.5 | 7 |
| 1800 | 49 | 2 | 6 |
| 1900 | 65 | 2 | 6 |
| 1900 | 68 | 3 | 8 |
| 1900 | 52 | 2 | 6 |
| 1950 | 58 | 2.5 | 7 |
| 2000 | 60 | 3 | 7 |
| 2000 | 57 | 2.5 | 7 |
| 2200 | 63 | 2 | 7 |
| 2250 | 61 | 3 | 7 |
| 2400 | 57 | 2.5 | 7 |
| 2500 | 75 | 3 | 9 |
| 2600 | 72 | 3.5 | 9 |
| 2600 | 63 | 3 | 8 |
| 2950 | 69 | 3.5 | 9 |
| 3000 | 81 | 4 | 9 |
| 3000 | 72 | 3.5 | 9 |

Use the SAS printout below to test whether COST can be predicted by using a linear function of the other three factors. Also, test whether each factor by itself contributes significantly to the prediction of the dependent variable. Let $\alpha = 0.05$.

DEP VARIABLE: COST

ANALYSIS OF VARIANCE

| SOURCE | DF | SUM OF SQUARES | MEAN SQUARE | F VALUE | PROB>F |
|---|---|---|---|---|---|
| MODEL | 3 | 7094.38686 | 2364.79572 | 70.918 | 0.0001 |
| ERROR | 27 | 900.32281 | 33.34528938 | | |
| C TOTAL | 30 | 7994.70968 | | | |

```
        ROOT MSE      5.774538      R-SQUARE      0.8874
        DEP MEAN      50.90323      ADJ R-SQ      0.8749
        C.V.          11.34415
```

PARAMETER ESTIMATES

| VARIABLE | DF | PARAMETER ESTIMATE | STANDARD ERROR | T FOR H0: PARAMETER=0 | PROB > |T| |
|----------|----|--------------------|--------------  |-----------------------|-----------|
| INTERCEP | 1  | -2.52963923        | 7.44703743     | -0.340                | 0.7367    |
| AREA     | 1  | 0.01901251         | 0.003933489    | 4.833                 | 0.0001    |
| BATHS    | 1  | 0.34499241         | 4.19028197     | 0.082                 | 0.9350    |
| ROOMS    | 1  | 2.69778010         | 2.16661512     | 1.245                 | 0.2238    |

17.26  Six factors have been identified as having a possible effect on the monthly number of HOURS needed to staff a Navy hospital:

1. VISITS (number of patient visits)
2. XRAY (number of X-ray exposures)
3. ADM (number of admissions)
4. STAY (average length of stay, in days)
5. POP (eligible population, in thousands)
6. MIL (military population, in thousands)

Measurements on all seven quantities taken at 17 hospitals are summarized in the following table (111).

| HOURS | VISITS | XRAY | ADM | STAY | POP | MIL |
|-------|--------|------|-----|------|-----|-----|
| 566.5 | 6,026 | 2,463 | 114 | 4.45 | 18.0 | 3.5 |
| 696.8 | 5,272 | 2,048 | 178 | 6.92 | 9.5 | 2.8 |
| 1,033.1 | 11,327 | 3,940 | 150 | 4.28 | 12.8 | 5.6 |
| 1,603.6 | 12,690 | 6,505 | 159 | 3.90 | 36.7 | 8.6 |
| 1,611.6 | 10,598 | 5,723 | 264 | 5.50 | 35.7 | 9.9 |
| 1,613.3 | 25,459 | 11,520 | 301 | 4.60 | 24.0 | 11.4 |
| 1,854.2 | 15,641 | 5,779 | 302 | 5.62 | 43.3 | 8.0 |
| 2,160.5 | 14,811 | 5,969 | 316 | 5.15 | 46.7 | 11.2 |
| 2,305.6 | 29,535 | 8,461 | 552 | 6.18 | 78.7 | 12.0 |
| 3,503.9 | 38,458 | 20,106 | 612 | 6.15 | 180.5 | 24.2 |
| 3,571.9 | 24,393 | 13,313 | 504 | 5.88 | 60.9 | 12.3 |
| 3,741.4 | 30,734 | 10,771 | 818 | 4.88 | 103.7 | 25.9 |
| 4,026.5 | 33,742 | 15,543 | 681 | 5.50 | 126.8 | 30.7 |
| 10,343.8 | 61,402 | 36,194 | 1,124 | 7.00 | 157.7 | 27.8 |
| 11,732.2 | 41,024 | 34,703 | 1,106 | 10.80 | 169.4 | 34.3 |
| 15,414.9 | 82,168 | 39,204 | 2,099 | 7.05 | 331.4 | 92.8 |
| 18,854.4 | 11,564 | 86,533 | 2,505 | 6.35 | 371.6 | 96.4 |

Use the SAS printout below to discuss the regression model,

$$HOURS = a + b_1 VISITS + b_2 XRAY + b_3 ADM + b_4 STAY$$
$$+ b_5 POP + b_6 MIL + \varepsilon$$

DEP VARIABLE: HOURS

## ANALYSIS OF VARIANCE

| SOURCE | DF | SUM OF SQUARES | MEAN SQUARE | F VALUE | PROB>F |
|--------|-----|--------------|-------------|---------|--------|
| MODEL | 6 | 487778798 | 81296466.29 | 117.293 | 0.0001 |
| ERROR | 10 | 6931030.54 | 693103.05 | | |
| C TOTAL | 16 | 494709828 | | | |

| | | | | |
|---|---|---|---|---|
| ROOT MSE | 832.5281 | R-SQUARE | 0.9860 | |
| DEP MEAN | 4978.482 | ADJ R-SQ | 0.9776 | |
| C.V. | 16.72253 | | | |

## PARAMETER ESTIMATES

| VARIABLE | DF | PARAMETER ESTIMATE | STANDARD ERROR | T FOR H0: PARAMETER=0 | PROB > \|T\| |
|----------|-----|--------------------|-----------------|------------------------|--------------|
| INTERCEP | 1 | -3545.33729 | 873.12877 | -4.060 | 0.0023 |
| VISITS | 1 | 0.04315552 | 0.01816528 | 2.376 | 0.0389 |
| XRAY | 1 | 0.15070373 | 0.04163377 | 3.620 | 0.0047 |
| ADM | 1 | 2.08964981 | 2.48608847 | 0.841 | 0.4203 |
| STAY | 1 | 580.60715 | 176.66801 | 3.286 | 0.0082 |
| POP | 1 | -23.13670478 | 10.62269838 | -2.178 | 0.0544 |
| MIL | 1 | 90.44196384 | 50.72751750 | 1.783 | 0.1049 |

17.27    In forestry, regression is used to help predict hard-to-measure characteristics of a tree. Listed below is information collected on 20 stands of loblolly pines: its AGE, the NUMBER of trees it contains, the average tree DIAMETER, measured 4.5 ft above the ground, and the average HEIGHT of dominant trees.

| AGE (yr) | NUMBER | DIAM. (ft) | HEIGHT (ft) |
|----------|--------|------------|-------------|
| 19 | 500 | 7.0 | 51.5 |
| 14 | 900 | 5.0 | 41.3 |
| 11 | 650 | 6.2 | 36.7 |
| 13 | 480 | 5.2 | 32.2 |
| 13 | 520 | 6.2 | 39.0 |
| 12 | 610 | 5.2 | 29.8 |
| 18 | 700 | 6.2 | 51.2 |
| 14 | 760 | 6.4 | 46.8 |
| 20 | 930 | 6.4 | 61.8 |
| 17 | 690 | 6.4 | 55.8 |
| 13 | 800 | 5.4 | 37.3 |
| 21 | 650 | 6.4 | 54.2 |
| 11 | 530 | 5.4 | 32.5 |
| 19 | 680 | 6.7 | 56.3 |
| 17 | 620 | 6.7 | 52.8 |
| 15 | 900 | 5.9 | 47.0 |
| 16 | 620 | 6.9 | 53.0 |
| 16 | 730 | 6.9 | 50.3 |
| 14 | 680 | 6.9 | 50.5 |
| 22 | 480 | 7.9 | 57.7 |

Can HEIGHT be predicted by the other three factors? Use the following SAS printout to discuss the model

$$HEIGHT = a + b_1 AGE + b_2 NUMBER + b_3 DIAMETER + \varepsilon$$

DEP VARIABLE: HT

ANALYSIS OF VARIANCE

| SOURCE | DF | SUM OF SQUARES | MEAN SQUARE | F VALUE | PROB>F |
|--------|----|----|----|----|----|
| MODEL | 3 | 1587.06739 | 529.02246 | 73.939 | 0.0001 |
| ERROR | 16 | 114.47811 | 7.15488185 | | |
| C TOTAL | 19 | 1701.54550 | | | |

| | | | |
|--|--|--|--|
| ROOT MSE | 2.674861 | R-SQUARE | 0.9327 |
| DEP MEAN | 46.885 | ADJ R-SQ | 0.9201 |
| C.V. | 5.705153 | | |

PARAMETER ESTIMATES

| VARIABLE | DF | PARAMETER ESTIMATE | STANDARD ERROR | T FOR H0: PARAMETER=0 | PROB > \|T\| |
|--------|----|----|----|----|----|
| INTERCEP | 1 | -33.46739740 | 7.16707136 | -4.670 | 0.0003 |
| AGE | 1 | 1.53025256 | 0.26195898 | 5.842 | 0.0001 |
| NO | 1 | 0.02395589 | 0.004797754 | 4.993 | 0.0001 |
| DIAM | 1 | 6.41093998 | 1.20459690 | 5.322 | 0.0001 |

17.28  A cable TV company has customers in 11 suburbs of a large metropolitan area. Management would like to have a linear function relating operating COST to several variables measuring a suburb's current business capacity and its potential for expansion. Four factors are being considered as possible regressors:

1. SUBS (the number of subscribers, in 1000s)
2. CABLE (the number of miles of cable)
3. NEW (the number of new subscribers, in 1000s)
4. POTL (the potential market, in 1000s)

Data from the eleven locations are summarized in the following table. Use the accompanying SAS printout to propose a model for COST. What would you do to finish the analysis?

| District | SUBS | CABLE | NEW | POTL | COST |
|--------|----|----|----|----|----|
| 1 | 66.0 | 763 | 3.1 | 93.2 | 3465.1 |
| 2 | 43.8 | 322 | 1.4 | 57.3 | 2089.8 |
| 3 | 15.9 | 238 | 0.6 | 23.2 | 810.0 |
| 4 | 6.5 | 86 | 0.5 | 8.7 | 379.5 |
| 5 | 19.9 | 267 | 1.5 | 39.9 | 1072.2 |
| 6 | 7.6 | 95 | 0.4 | 13.1 | 505.3 |
| 7 | 27.3 | 177 | 0.6 | 37.5 | 910.5 |
| 8 | 91.5 | 675 | 17.6 | 141.8 | 4579.6 |
| 9 | 10.7 | 97 | 0.8 | 13.9 | 774.2 |
| 10 | 9.9 | 127 | 5.1 | 25.2 | 662.0 |
| 11 | 21.3 | 280 | 1.7 | 43.2 | 1084.7 |

DEP VARIABLE: COST

### ANALYSIS OF VARIANCE

| SOURCE | DF | SUM OF SQUARES | MEAN SQUARE | F VALUE | PROB>F |
|--------|-----|----------------|-------------|---------|--------|
| MODEL   | 4  | 18179899.86 | 4544974.96 | 166.224 | 0.0001 |
| ERROR   | 6  | 164055.11 | 27342.51888 | | |
| C TOTAL | 10 | 18343954.97 | | | |

| | | | | |
|--|--|--|--|--|
| ROOT MSE | 165.3557 | R-SQUARE | 0.9911 |
| DEP MEAN | 1484.809 | ADJ R-SQ | 0.9851 |
| C.V. | 11.1365 | | |

### PARAMETER ESTIMATES

| VARIABLE | DF | PARAMETER ESTIMATE | STANDARD ERROR | T FOR H0: PARAMETER=0 | PROB > \|T\| |
|----------|-----|--------------------|-----------------|------------------------|-------------|
| INTERCEP | 1 | 31.66322932 | 99.69899865 | 0.318 | 0.7616 |
| SUBS     | 1 | 35.77737317 | 5.92217817 | 6.041 | 0.0009 |
| CABLE    | 1 | 1.82307863 | 0.87347482 | 2.087 | 0.0819 |
| NEW      | 1 | 36.18631460 | 34.71939702 | 1.042 | 0.3375 |
| POTL     | 1 | -5.09414081 | 4.52674230 | -1.125 | 0.3034 |

# 17.4

## MODELS WITH PRODUCT TERMS

Up to this point, the regressors we have used have all been quite simple in structure—each has been either a factor raised to the first power $(x)$ or a factor raised to the second power $(x^2)$. The general linear model, though, does allow more complicated expressions to be used as dependent variables. In practice, the complexity of the mathematical model we choose is limited primarily by our ability to manage the calculations. Including terms that are *products* of variables is a good way to make linear models substantially more useful without unduly complicating their interpretation.

**general second-order linear model—**
A particular regression function involving two regressors that includes all possible first-order and second-order terms. If $x_1$ and $x_2$ are the two regressors, the general second-order linear model for $Y$ has the form

$$Y = a + b_1x_1 + b_2x_2 + b_3x_1^2 + b_4x_2^2 + b_5x_1x_2 + \varepsilon$$

where $\varepsilon$ is normally distributed with mean zero and variance $\sigma^2$.

---

### DEFINITION 17.4

Suppose a random variable $Y$ is presumed to be a function of two factors, $x_1$ and $x_2$. The **general second-order linear model** is written

$$Y = a + b_1x_1 + b_2x_2 + b_3x_1^2 + b_4x_2^2 + b_5x_1x_2 + \varepsilon$$

where $\varepsilon$ is normally distributed with mean zero and variance $\sigma^2$. If $H_0: b_5 = 0$ can be rejected, we say that factors $x_1$ and $x_2$ *interact*.

**EXAMPLE 17.8**   In the logging industry, estimates need to be made of the amount of lumber that can be sawn from a tree. A variety of rules exist for that purpose—typically, they relate the yield, $Y$, to a log's length and to its radius. Intuitively, we would expect any such rule to involve products, since the formula for the volume of a cylinder depends on the square of the radius times the length.

The unit of volume for lumber is the *board foot*. By definition, a board foot is the volume of a piece of lumber 1 in thick by 12 in wide by 12 in long. Table 17.8 gives the number of board feet, denoted BDFT, sawn from 24 logs. DIAM is the diameter of a log in inches at its small end. LENGTH is measured in feet.

**TABLE 17.8**

| DIAM | LENGTH | BDFT | DIAM | LENGTH | BDFT |
|------|--------|------|------|--------|------|
| 10 | 10 | 27.77 | 24 | 12 | 305.97 |
| 12 | 14 | 62.05 | 26 | 15 | 462.56 |
| 12 | 15 | 51.59 | 27 | 14 | 478.83 |
| 12 | 16 | 52.20 | 28 | 7 | 246.30 |
| 12 | 16 | 72.11 | 28 | 16 | 577.92 |
| 16 | 14 | 112.00 | 30 | 16 | 668.32 |
| 16 | 16 | 141.22 | 31 | 15 | 671.15 |
| 16 | 20 | 188.10 | 33 | 12 | 640.01 |
| 20 | 16 | 242.15 | 33 | 18 | 944.44 |
| 20 | 20 | 321.31 | 34 | 12 | 682.67 |
| 24 | 14 | 353.24 | 35 | 19 | 1147.02 |
| 24 | 16 | 403.53 | 37 | 16 | 1094.22 |

**Solution**   According to Definition 17.4, the general second-order linear model relating BDFT to DIAM and LENGTH has the equation

$$\text{BDFT} = a + b_1\text{DIAM} + b_2\text{LENGTH} + b_3\text{DIAM}^2 + b_4\text{LENGTH}^2$$
$$+ b_5\text{DIAM} \cdot \text{LENGTH} + \varepsilon \tag{17.9}$$

Figure 17.9 shows the SAS printout for Equation 17.9. Judging from the entries in the PROB > |T| column, every term except $b_4\text{LENGTH}^2$ should be kept in the model.

COMMENT:   The high value of $r^2$ (0.9988) in Figure 17.9 indicates that the right-hand side of Equation 17.9 predicts very effectively the random variable BDFT. That $r^2$ is so large, though, may be a bit of a surprise since the volume of a cylinder is actually a *third-order* effect (volume $= \pi \cdot$ radius$^2 \cdot$ length). Not much is being sacrificed, apparently, by ignoring third-order terms and fitting the data with a second-order model.

DEP VARIABLE: BDFT

ANALYSIS OF VARIANCE

| SOURCE | DF | SUM OF SQUARES | MEAN SQUARE | F VALUE | PROB>F |
|--------|-----|----------------|-------------|---------|--------|
| MODEL | 5 | 2516375.08 | 503275.02 | 2979.836 | 0.0001 |
| ERROR | 18 | 3040.08339 | 168.89352 | | |
| C TOTAL | 23 | 2519415.16 | | | |

| | | | | |
|---|---|---|---|---|
| ROOT MSE | 12.9959 | R-SQUARE | 0.9988 | |
| DEP MEAN | 414.4454 | ADJ R-SQ | 0.9985 | |
| C.V. | 3.135734 | | | |

PARAMETER ESTIMATES

| VARIABLE | DF | PARAMETER ESTIMATE | STANDARD ERROR | T FOR H0: PARAMETER=0 | PROB > |T| |
|----------|-----|--------------------|----------------|------------------------|------------|
| INTERCEP | 1 | 413.41204 | 63.93332492 | 6.466 | 0.0001 |
| DIAM | 1 | -41.86134639 | 2.33755215 | -17.908 | 0.0001 |
| LENGTH | 1 | -28.32042457 | 7.02078921 | -4.034 | 0.0008 |
| DIAM*DIAM | 1 | 0.94331715 | 0.05037410 | 18.726 | 0.0001 |
| LENGTH* LENGTH | 1 | 0.05860128 | 0.20425405 | 0.287 | 0.7775 |
| DIAM* LENGTH | 1 | 2.31829513 | 0.12681002 | 18.282 | 0.0001 |

**FIG. 17.9**

---

QUESTIONS

**17.29**  How many board feet might we expect from a 17-ft tree with a 29-in diameter? Use the original model.

qualitative ("dummy") variable—A regressor whose values refer to categories rather than a numerical scale.

All the variables in multiple regression equations need not be *quantitative*, like DIAM and LENGTH in Example 17.8. Some may represent categorical information such as male/female, Democrat/Republican, or white/black/other. Any factor whose possible values are limited to traits (as opposed to a range of numbers) is said to be a *qualitative* (or "dummy") *variable*. The next several examples show the contribution that qualitative variables can make in the construction and interpretation of a regression function.

---

EXAMPLE 17.9

Manufacturers of electronic equipment sometimes buy circuit boards from subcontractors. One such subcontractor has two plants—one in Florida, the other in California. The cost of materials and labor to produce a single board is the same at both plants ($30). Fixed expenses are $2000 in Florida and $2400 in California. Producing $x_1$ boards, then, costs $Y$ dollars, where $Y = 2000 + 30x_1$ in Florida and $Y = 2400 + 30x_1$ in California. (To simplify notation, we will omit writing the $\varepsilon$ term in most of the regression equations in this section.) Figure 17.10 is a graph of the two cost functions: the lines have the same slope but different $Y$-intercepts.

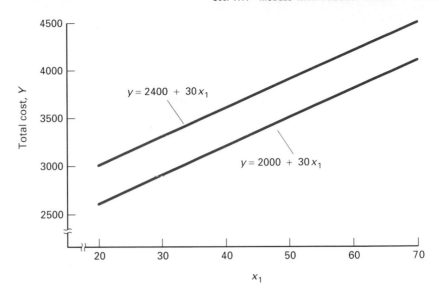

**FIG. 17.10**

*Solution*  As a way of simplifying certain problems, it sometimes helps to combine a system of equations into a single expression. That can be accomplished by introducing a dummy variable, $x_2$. Here, let $x_2 = 0$ if a data point $(x_1, Y)$ comes from Florida, and let $x_2 = 1$ if $(x_1, Y)$ comes from California. Our objective is to write both cost functions in the form

$$y = a + b_1 x_1 + b_2 x_2 \qquad (17.10)$$

which means that Equation 17.10 must reduce to $2000 + 30x_1$ when $x_2 = 0$ and to $2400 + 30x_1$ when $x_2 = 1$. That is,

$$2000 + 30x_1 = a + b_1 x_1 + b_2 \cdot 0 \qquad (17.11)$$

and $$2400 + 30x_1 = a + b_1 x_1 + b_2 \cdot 1 \qquad (17.12)$$

Look at Equation 17.11. By equating coefficients, $a = 2000$ and $b_1 = 30$. Substituting those values into Equation 17.12 gives

$$2400 + 30x_1 = 2000 + 30x_1 + b_2$$

implying that $b_2 = 400$. The single equation, then, that describes both cost functions is

$$y = 2000 + 30x_1 + 400x_2$$

By going one step further and adding a product term involving the dummy variable, we can use a single equation to model even more complicated situations. Suppose, for example, the direct cost to produce a circuit board in California increases to \$35, but remains at \$30 in Florida. Given that scenario, the two cost functions have different $Y$-intercepts *and* different slopes (Figure 17.11). Still, the two can be combined. Consider the

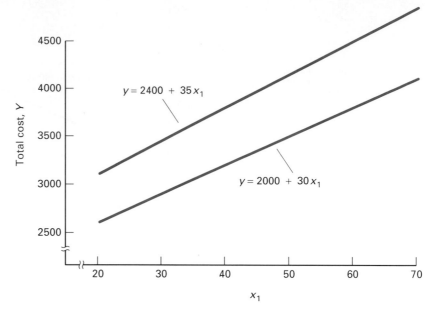

**FIG. 17.11**

expression

$$y = a + b_1x_1 + b_2x_2 + b_3x_1x_2$$

If $x_2 = 0$,

$$2000 + 30x_1 = a + b_1x_1 + b_2 \cdot 0 + b_3 \cdot x_1 \cdot 0$$

which implies, as before, that $a = 2000$ and $b_1 = 30$. Then, if $x_2 = 1$,

$$2400 + 35x_1 = 2000 + 30x_1 + b_2 \cdot 1 + b_3 \cdot x_1 \cdot 1$$

or

$$400 + 5x_1 = b_2 + b_3x_1$$

Therefore, $b_2 = 400$ and $b_3 = 5$ so the unifying equation becomes

$$y = 2000 + 30x_1 + 400x_2 + 5x_1x_2$$

---

COMMENT: The two cases just described illustrate the difference between two factors that *don't* interact and two factors that *do* interact (recall Definition 17.4). If the graphs of $Y$ versus $x_1$ are parallel for all the different levels of $x_2$ (as in Figure 17.10), we say that factors $x_1$ and $x_2$ do not interact. When that happens, the coefficient of $x_1x_2$ in the regression model will be zero. On the other hand, if the graphs of $Y$ versus $x_1$ are not parallel for all the different levels of $x_2$ (as in Figure 17.11), then $x_1$ and $x_2$ do interact—

meaning the product $x_1x_2$ contributes meaningfully to the prediction of $Y$. The $x_1x_2$ term in those cases should be kept in the regression model.

---

*EXAMPLE 17.10*    A bottler in Pensacola, Florida, has studied the time it takes to deliver cases of soft drinks. If $x_1$ represents the number of cases on a delivery truck, his data show that the number of minutes required to unload the truck can be described by the equation

$$y = 5.97 + 0.50x_1 \tag{17.13}$$

The bottler has a second plant in Tallahassee. There the relationship between $x_1$ and $y$ is estimated to be

$$y = 5.36 + 0.57x_1 \tag{17.14}$$

Combine Equations 17.13 and 17.14.

*Solution*    We could, of course, use the technique in Example 17.9 to produce a single equation. Another approach, though, is to apply SAS to the model

$$Y = a + b_1x_1 + b_2x_2 + b_3x_1x_2 + \varepsilon \tag{17.15}$$

where $\qquad x_2 = \begin{cases} 0 & \text{if } (x_1, Y) \text{ comes from Pensacola} \\ 1 & \text{if } (x_1, Y) \text{ comes from Tallahassee} \end{cases}$

The $x_1$ and $Y$ columns in Table 17.9 are the data from which Equations 17.13 and 17.14 were derived. Figure 17.12 is the SAS printout for Equation

**TABLE 17.9**

| PENSACOLA | | | TALLAHASSE | | |
|---|---|---|---|---|---|
| Cases $(x_1)$ | Time $(y)$ | Plant $(x_2)$ | Cases $(x_1)$ | Time $(y)$ | Plant $(x_2)$ |
| 200 | 105.14 | 0 | 247 | 144.49 | 1 |
| 225 | 118.21 | 0 | 150 | 93.59 | 1 |
| 60 | 34.64 | 0 | 322 | 190.11 | 1 |
| 241 | 125.41 | 0 | 70 | 46.93 | 1 |
| 175 | 95.64 | 0 | 55 | 33.23 | 1 |
| 170 | 90.25 | 0 | 310 | 177.01 | 1 |
| 185 | 96.64 | 0 | 165 | 98.44 | 1 |
| 163 | 83.51 | 0 | 75 | 44.99 | 1 |
| 110 | 65.18 | 0 | 82 | 53.00 | 1 |
| 205 | 110.13 | 0 | 255 | 147.42 | 1 |
| 241 | 126.25 | 0 | 300 | 179.81 | 1 |
| 80 | 43.60 | 0 | 144 | 85.95 | 1 |
| 175 | 89.96 | 0 | 60 | 42.38 | 1 |
| 330 | 168.31 | 0 | 75 | 48.09 | 1 |
| 170 | 90.37 | 0 | 210 | 122.93 | 1 |

DEP VARIABLE: TIME

ANALYSIS OF VARIANCE

| SOURCE | DF | SUM OF SQUARES | MEAN SQUARE | F VALUE | PROB>F |
|--------|-----|----------------|-------------|---------|--------|
| MODEL | 3 | 59419.95314 | 19806.65105 | 3239.889 | 0.0001 |
| ERROR | 26 | 158.94769 | 6.11337279 | | |
| C TOTAL | 29 | 59578.90083 | | | |

| | | | | |
|--|--|--|--|--|
| ROOT MSE | 2.472524 | R-SQUARE | 0.9973 |
| DEP MEAN | 98.387 | ADJ R-SQ | 0.9970 |
| C.V. | 2.513059 | | |

PARAMETER ESTIMATES

| VARIABLE | DF | PARAMETER ESTIMATE | STANDARD ERROR | T FOR H0: PARAMETER=0 | PROB > \|T\| |
|----------|-----|--------------------|----------------|------------------------|-------------|
| INTERCEP | 1 | 5.97328995 | 1.90585301 | 3.134 | 0.0042 |
| CASES | 1 | 0.49583907 | 0.009866757 | 50.253 | 0.0001 |
| PLANT | 1 | -0.61780039 | 2.30298167 | -0.262 | 0.7906 |
| PROD | 1 | 0.07084254 | 0.01192198 | 5.942 | 0.0001 |

**FIG. 17.12**

17.15. Rounded to two decimal places, the estimated regression function for both sets of data, combined, is

$$y = 5.97 + 0.50x_1 - 0.62x_2 + 0.07x_1x_2$$

COMMENT:   In every multiple regression example we have seen thus far, PROB > F has been close to zero and, at the same time, $r^2$ has been close to 1. That relationship will not always be true. In situations where the sample size is large, PROB > F can be very small even when $r^2$ is only modest. If that happens, it means (1) that the model does enhance our ability to predict the behavior of $Y$ and (2) that a large portion of the variability in the data remains unexplained. Case Study 17.2, for example, describes a set of data where an $r^2$ of 0.263 produces a P-value of 0.0001.

## CASE STUDY 17.2

Multiple regression models are frequently used to analyze salary structures for the purpose of identifying possible instances of sex or race discrimination. One such study is the 1988 report from the National Center for Education Statistics, entitled *Employment Outcomes of Recent Master's and Bachelor's Degree Recipients*. Based on a 1985 survey, the report compiled salary, work, education, and demographic information on 10,311 recent recipients of either a bachelor's degree or a master's degree. Seven variables were included, five qualitative and two quantitative:

1. SALARY (1985 annual salary)
2. DEGREE (1 if recipient has master's degree; 0 otherwise)
3. SEX (1 for males; 0 for females)
4. WORK (years of prior work experience)
5. MAJORPR (1 for a degree in a professional field; 0 otherwise)
6. MAJORAS (1 for a degree in arts and sciences; 0 otherwise)
7. RACE (1 if recipient is white non-Hispanic; 0 otherwise)

Figure 17.13 is the SAS printout for the model

$$\text{SALARY} = a + b_1\text{DEGREE} + b_2\text{SEX} + b_3\text{WORK} + b_4\text{MAJORPR}$$
$$+ b_5\text{MAJORAS} + b_6\text{RACE} + \varepsilon$$

```
DEP VARIABLE: SALARY

                              ANALYSIS OF VARIANCE

                            SUM OF          MEAN
          SOURCE      DF    SQUARES        SQUARE      F VALUE     PROB>F

          MODEL        ·       ·             ·        267.423     0.0001
          ERROR        ·       ·             ·
          C TOTAL      ·       ·

              ROOT MSE          ·        R-SQUARE      0.263
              DEP MEAN          ·        ADJ R-SQ        ·
              C.V.              ·

                          PARAMETER ESTIMATES

                      PARAMETER     STANDARD     T FOR H0:
       VARIABLE  DF    ESTIMATE       ERROR     PARAMETER=0    PROB > |T|

       INTERCEP   1    12311.97         ·            ·           0.0001
       DEGREE     1     5912.99         ·            ·           0.0001
       SEX        1     4545.19         ·            ·           0.0001
       WORK       1      416.67         ·            ·           0.0001
       MAJORPR    1     3240.64         ·            ·           0.0001
       MAJORAS    1     1409.58         ·            ·           0.0001
       RACE       1      580.29         ·            ·           0.0667
```

**FIG. 17.13**

Entries not needed for our discussion have been deleted. Note that $r^2$ is only 0.263, yet the $P$-value for $H_0$: $b_1 = b_2 = \cdots = b_6 = 0$ is 0.0001. The large sample size ($n = 10{,}311$) is responsible—as $n$ increases, the "power" of the test also increases, meaning that smaller values of $r^2$ are sufficient to overturn $H_0$.

Individually, each regressor has a small $P$-value, suggesting that the original model should be left unchanged. In general, parameter estimates for qualitative variables are worth examining—they indicate the amount that $Y$ is expected to increase when that particular dummy variable goes from zero to 1. For example, $B_1 = \$5912.99$, implying that a master's degree recipient can expect to earn \$5912.99 more a year than a bachelor's degree recipient, *all other factors being equal*. Similarly, a man can expect to earn \$4545.19 more a year than a woman, all other factors being equal.

## QUESTIONS

**17.30**  Use a dummy variable to combine the two equations

$$y = 7 + 5x_1 \qquad \text{and} \qquad y = 12 + 5x_1$$

**17.31**  With the help of a second variable, write

$$y = 15 - 3x_1 \qquad \text{and} \qquad y = 10 + 8x_1$$

as a single equation.

**17.32**  As the first step taken in a university salary study, data are collected on 20 support staff having the same job classification. The results are summarized below. SERVICE is recorded in years, SEX is 0 for females and 1 for males, AGE is in years, TEMP is 0 for permanent employees and 1 for temporary employees, and SALARY is in annual dollars.

| SERVICE | SEX | AGE | TEMP | SALARY |
|---------|-----|-----|------|--------|
| 5 | 1 | 27 | 0 | 11,653 |
| 7 | 0 | 39 | 1 | 10,555 |
| 4 | 0 | 23 | 0 | 11,130 |
| 15 | 1 | 33 | 1 | 11,715 |
| 9 | 1 | 41 | 0 | 11,445 |
| 8 | 1 | 53 | 1 | 11,523 |
| 5 | 0 | 36 | 0 | 11,360 |
| 16 | 1 | 35 | 0 | 11,930 |
| 10 | 1 | 40 | 0 | 11,755 |
| 12 | 1 | 30 | 1 | 11,596 |
| 7 | 0 | 32 | 0 | 11,468 |
| 18 | 1 | 36 | 0 | 12,459 |
| 4 | 1 | 27 | 0 | 12,117 |
| 4 | 0 | 26 | 0 | 11,421 |
| 9 | 1 | 43 | 1 | 11,546 |
| 10 | 1 | 42 | 0 | 11,773 |
| 7 | 1 | 34 | 1 | 11,356 |
| 11 | 1 | 41 | 0 | 12,091 |
| 8 | 1 | 26 | 0 | 11,658 |
| 14 | 1 | 35 | 0 | 11,917 |

Do these data suggest a bias for sex? for age? Answer the questions by using the following SAS printout. Let $\alpha = 0.05$.

DEP VARIABLE: SALARY

ANALYSIS OF VARIANCE

| SOURCE | DF | SUM OF SQUARES | MEAN SQUARE | F VALUE | PROB>F |
|--------|-----|-------------|-------------|---------|--------|
| MODEL | 4 | 2239244.46 | 559811.12 | 10.613 | 0.0003 |
| ERROR | 15 | 791208.34 | 52747.22250 | | |
| C TOTAL | 19 | 3030452.80 | | | |

| | | | | |
|--------|-------------|--------|---------|--------|
| ROOT MSE | 229.6676 | R-SQUARE | 0.7389 | |
| DEP MEAN | 11623.4 | ADJ R-SQ | 0.6693 | |
| C.V. | 1.975908 | | | |

PARAMETER ESTIMATES

| VARIABLE | DF | PARAMETER ESTIMATE | STANDARD ERROR | T FOR H0: PARAMETER=0 | PROB > \|T\| |
|----------|-----|---------------|-------------|-------------|----------|
| INTERCEP | 1 | 11232.06261 | 265.43481 | 42.316 | 0.0001 |
| SERVICE | 1 | 23.12197088 | 15.26885267 | 2.104 | 0.0527 |
| SEX | 1 | 497.55222 | 142.89742 | 3.482 | 0.0033 |
| AGE | 1 | -4.41679909 | 8.05419588 | -0.548 | 0.5915 |
| TEMP | 1 | -404.58562 | 119.27455 | -3.392 | 0.0040 |

# 17.5

## SUMMARY

As a data model, regression problems fall into one of two broad categories. If a random variable $Y$ is presumed to be a function of a single factor $x$, the $(x, Y)$ relationship is said to be a *simple regression*. All the case studies in Sections 2.5, 3.5 and 12.1, 12.2, 12.3, for example, are simple regressions. In contrast, *multiple regression* refers to situations where $Y$ is a function of $k$ factors ($k \geq 2$).

Theoretically, any function of $k$ factors qualifies as a multiple regression. In practice, though, the most common formulation—by far—is a special case known as the *general linear model*, which assumes that $Y$ can be written in the form

$$Y = a + b_1 x_1 + b_2 x_2 + \cdots + b_k x_k + \varepsilon \qquad (17.16)$$

where $\varepsilon$ is normally distributed with mean zero and variance $\sigma^2$. Values of $x_1, x_2, \ldots, x_k$ are presumed not to be random and are often controlled, or fixed, by the experimenter. The $i$th data point collected on a $k$-factor regression is denoted $(x_{i,1}, x_{i,2}, \ldots, x_{i,k}, y_i)$.

How we fit Equation 17.16 to a set of data depends to some extent on the value of $k$. If $k = 2$, it is not prohibitively difficult to find least squares estimates for $a$, $b_1$, and $b_2$ using only a hand calculator. The necessary formulas are given in Theorem 17.1. If more than two regressors are involved, the calculations become so lengthy that for all practical purposes we can do the analysis only by using a computer. SAS and MINITAB are two of the many statistical packages that are widely available.

Every application of multiple regression leads to at least two inference questions: (1) Is the proposed model, as a whole, effective in predicting the behavior of $Y$? and (2) Is it possible to eliminate one or more of the regressors and not significantly diminish the model's usefulness? The first question is resolved by doing an $F$ test on the hypothesis $H_0\colon b_1 = b_2 = \cdots b_k = 0$. Rejecting $H_0$ validates the model. Conceptually, we reject $H_0$ if the coefficient of determination, $r^2$, is significantly larger than zero. Theorem 17.3 gives the appropriate decision rule.

If $H_0\colon b_1 = b_2 = \cdots = b_k = 0$ is accepted, the experimenter needs to discard the model and look for some other function or some other set of regressors that will better explain the behavior of $Y$. If $H_0\colon b_1 = b_2 = \cdots = b_k = 0$ is rejected, the model has been demonstrated to be effective but the possibility still remains that a simpler model—one with fewer regressors—might also be effective. As a general rule of thumb, we want to eliminate as many regressors as we can without significantly reducing the model's ability to predict $Y$. Testing $H_0\colon b_j = 0$ is one way to see whether regressor $x_j$ is needed. If $H_0\colon b_j = 0$ is accepted, $x_j$ can be discarded. Case Studies 17.1 and 17.2 show how $P$-values associated with $t$ tests of $H_0\colon b_j = 0$ are used to reduce the original set of regressors to a more manageable size.

Qualitative, or dummy, variables are introduced in Section 17.4. These are used for measurements that initially refer to categories rather than numerical responses.

---

**REVIEW QUESTIONS**

17.33　Which equation, $y = 2 + x_1 + 3x_2$ or $y = 1 + 2x_1 + 3x_2$, better fits the following set of points?

| $x_1$ | $x_2$ | $y$ |
|-------|-------|-----|
| 3 | 2 | 10 |
| 1 | 4 | 15 |
| 2 | 0 | 8 |
| 2 | 5 | 11 |

17.34　Use Theorem 17.1 to find the equation of the form $y = a + b_1 x_1 + b_2 x_2$ that best describes the data in Question 17.33.

17.35　Estimate the variance associated with the linear model found in Question 17.34. Use the method illustrated in Example 17.3.

17.36　A large scientific laboratory stores expensive equipment in a central facility. The time it takes technicians to walk to the storage area, check out what they need, and return to their workstations presumably depends on their distance from the storage facility ($x_1$) and the number of items they need

($x_2$). Using the data below, test whether time is, in fact, a linear function of $x_1$ and $x_2$. Let $\alpha = 0.05$.

| Distance (ft) | Number of Instruments | Time (min) |
|---|---|---|
| 400 | 4 | 18.5 |
| 100 | 4 | 15.7 |
| 560 | 10 | 35.8 |
| 300 | 7 | 24 8 |
| 500 | 7 | 33.9 |
| 560 | 6 | 26.2 |
| 100 | 5 | 15.3 |
| 230 | 11 | 28.9 |
| 280 | 2 | 21.6 |
| 300 | 5 | 24.9 |

Questions 17.37–17.40 refer to the following incomplete SAS printout.

DEP VARIABLE: Y

ANALYSIS OF VARIANCE

| SOURCE | DF | SUM OF SQUARES | MEAN SQUARE | F VALUE | PROB>F |
|---|---|---|---|---|---|
| MODEL | 3 | 0.18255987 | 0.06085329 | . | . |
| ERROR | 13 | . | . | | |
| C TOTAL | 16 | 1.46437647 | | | |

| | | | | |
|---|---|---|---|---|
| ROOT MSE | 0.3140084 | R-SQUARE | 0.1247 |
| DEP MEAN | 31.61118 | ADJ R-SQ | . |
| C.V. | 0.9933461 | | |

PARAMETER ESTIMATES

| VARIABLE | DF | PARAMETER ESTIMATE | STANDARD ERROR | T FOR H0: PARAMETER=0 | PROB > \|T\| |
|---|---|---|---|---|---|
| INTERCEP | 1 | 28.37061353 | 9.16212282 | 3.097 | 0.0085 |
| X1 | 1 | 0.09547388 | 0.28703830 | . | . |
| X2 | 1 | 0.000665571 | 0.01140596 | 0.058 | 0.9544 |
| X3 | 1 | 0.05026119 | 0.03973992 | 1.265 | 0.2282 |

17.37  Find SSE and MSE.

17.38  Compute the $F$ statistic. At the 0.05 level of significance, should we conclude that a linear model is helpful in explaining the behavior of $Y$?

17.39  Calculate the $t$ statistic for X1. Does X1 contribute significantly to the prediction of $Y$? Answer the question by doing an appropriate hypothesis test at the 0.05 level of significance.

17.40    How many points are in the data set?

17.41    Suppose the relationship between a dependent variable $y$ and two independent variables $x_1$ and $x_2$ is given by

$$y = 3.5 + 7.2x_1 + 5.6x_2$$

(a)  Find $y$ when $x_1 = 10$ and $x_2 = 4$.

(b)  Find $y$ when $x_1 = 8$ and $x_2 = -2.1$.

(c)  If $x_1 = 4$ and $y = 7$, what is $x_2$?

17.42    Let $y = 3.5 + 7.2x_1 + 5.6x_2$, as in Question 17.41. Set $x_2 = 1$ and graph $y$ as a function of $x_1$. On the same graph, plot $y$ as a function of $x_1$ when $x_2 = 2$ and when $x_2 = 3$.

17.43    Consider a linear regression of the form $y = 4 + b_1x_1 + b_2x_2$. For which values of $b_1$ and $b_2$ will the points $(2, 3, 7)$ and $(5, 8, 9)$ satisfy the equation?

17.44    (a)  Suppose a set of multiple regression data consists of $n = 20$ points collected on $k = 4$ variables. If $F = 10.232$, what is $r^2$?

(b)  Suppose that $F = 3.269$ and $r^2 = 0.43$ for a set of multiple regression data for which $k = 3$. Find $n$.

17.45    The table below is a summary of economic indicators for eleven districts in the state of Florida. The variable SALES is the total taxable sales measured in millions of dollars, INDEX is a measurement of the level of retail activity, and EMPL is total employment in 1000's. Are the variables SALES and INDEX useful in predicting EMPL? Test the two-variable model and the individual coefficients at the 0.05 level of significance (50).

| District | SALES | INDEX | EMPL |
|---|---|---|---|
| West | 307.2 | 263.4 | 265.5 |
| Apalachee | 146.2 | 271.1 | 161.2 |
| North Central | 150.6 | 250.6 | 159.1 |
| Northeast | 591.9 | 268.0 | 461.9 |
| Withlacoochee | 193.2 | 360.3 | 152.6 |
| East Central | 1542.4 | 348.1 | 898.1 |
| Central | 303.2 | 258.6 | 208.8 |
| Tampa Bay | 1476.4 | 288.3 | 966.3 |
| Southwest | 788.9 | 334.8 | 355.9 |
| Treasure Coast | 949.7 | 358.6 | 518.6 |
| South | 2464.9 | 262.1 | 1513.1 |

17.46    An $(x_1, x_2, y)$ data set yielded the following values for the S terms: $S_{11} = 116$, $S_{22} = 777$, $S_{12} = -258$, $S_{1y} = -1568.125$, $S_{2y} = 5352.015625$, and $S_{yy} = 38380.25$. Calculate $\hat{b}_1$ and $\hat{b}_2$.

17.47   For the six measurements summarized below, assume that $y$ is linearly related to $x_1$ and $x_2$. Use Theorem 17.1 to estimate $a$, $b_1$, and $b_2$.

| $x_1$ | $x_2$ | $y$ |
|------|------|------|
| 6.5 | 11.8 | 62.7 |
| 6.7 | 10.2 | 110.2 |
| 5.8 | 10.6 | 68.2 |
| 5.7 | 9.8 | 62.5 |
| 4.1 | 13.0 | 56.9 |
| 8.8 | 7.0 | 54.7 |

17.48   For each data point in Question 17.47, calculate $\hat{y}_i$ and $y_i - \hat{y}_i$. Compute

$$\sum_{i=1}^{6} (y_i - \hat{y}_i)^2$$

and use it to estimate $\sigma$.

17.49   Use Theorem 17.2 to calculate SSE for the data in Question 17.47.

17.50   To construct a model for the salary levels of corporate executives, information was collected on salary, sales, profits, and employment. A SAS run on 15 observations produced the following printout, where SALARY is executive annual salary in thousands of dollars, SALES is measured in millions of dollars, PROFIT is annual profit in millions, and EMPL is employment of the firm in hundreds. At the 0.10 level, test

(a) the significance of the model and

(b) the individual variables.

DEP VARIABLE: SALARY

ANALYSIS OF VARIANCE

| SOURCE | DF | SUM OF SQUARES | MEAN SQUARE | F VALUE | PROB>F |
|--------|----|----------------|-------------|---------|--------|
| MODEL | 3 | 15294.59626 | 5098.19875 | 0.454 | 0.7200 |
| ERROR | 11 | 123659.80 | 11241.80034 | | |
| C TOTAL | 14 | 138954.40 | | | |

| | | | | |
|---|---|---|---|---|
| ROOT MSE | 106.0274 | R-SQUARE | 0.1101 | |
| DEP MEAN | 440.8 | ADJ R-SQ | -0.1326 | |
| C.V. | 24.05339 | | | |

PARAMETER ESTIMATES

| VARIABLE | DF | PARAMETER ESTIMATE | STANDARD ERROR | T FOR H0: PARAMETER=0 | PROB > \|T\| |
|----------|----|--------------------|----------------|-----------------------|-------------|
| INTERCEP | 1 | 347.25743 | 87.73808717 | 3.958 | 0.0022 |
| SALES | 1 | -0.006414957 | 0.02346350 | -0.273 | 0.7896 |
| PROFIT | 1 | 0.04149417 | 0.28152471 | 0.147 | 0.8855 |
| EMPL | 1 | 0.33610851 | 0.36853847 | 0.912 | 0.3813 |

**17.51**  Use dummy variables to combine the following regressions into a single equation.

(a) $y = 9x$  and  $y = 4x$

(b) $y = 5 - 9x$  and  $y = 1 + 4x$

**17.52**  Shown below are price and consumption figures for beef reported for the years 1948 through 1957. Fit the data with a quadratic regression, $y = a + b_1 x + b_2 x^2$. Graph the points and superimpose the estimated regression function (97).

| Year | Price per lb, $x$ (constant \$) | Lb per Capita, $y$ |
|---|---|---|
| 1948 | 82.9 | 63.1 |
| 1949 | 76.3 | 63.9 |
| 1950 | 88.3 | 63.4 |
| 1951 | 90.0 | 56.1 |
| 1952 | 85.4 | 62.2 |
| 1953 | 66.2 | 77.6 |
| 1954 | 64.1 | 80.1 |
| 1955 | 63.2 | 82.0 |
| 1956 | 60.9 | 85.4 |
| 1957 | 63.1 | 84.6 |

**17.53**  For the data in Question 17.52, test

(a) $H_0: b_1 = b_2 = 0$

(b) $H_0: b_1 = 0$

(c) $H_0: b_2 = 0$.

In each case, let $\alpha = 0.05$.

**17.54**  Four factors thought to affect fuel consumption were measured on 33 models of 1988 automobiles: WEIGHT is the weight in pounds, HP is horsepower, TANK is the capacity of the fuel tank in gallons, and CYL is the number of cylinders. The dependent variable FUEL was the amount of gasoline consumed over a 195-mile test drive. Interpret the accompanying SAS printout.

```
DEP VARIABLE: FUEL

                              ANALYSIS OF VARIANCE

                          SUM OF        MEAN
        SOURCE     DF     SQUARES       SQUARE       F VALUE    PROB>F

        MODEL       4    601.87126    150.46782      39.675     0.0001
        ERROR      28    106.18935    3.79247663
        C TOTAL    32    708.06061

               ROOT MSE    1.947428    R-SQUARE     0.8500
               DEP MEAN   29.57576     ADJ R-SQ     0.8286
               C.V.        6.584542
```

PARAMETER ESTIMATES

| VARIABLE | DF | PARAMETER ESTIMATE | STANDARD ERROR | T FOR H0: PARAMETER=0 | PROB > \|T\| |
|---|---|---|---|---|---|
| INTERCEP | 1 | 50.08408487 | 2.34926358 | 21.319 | 0.0001 |
| WEIGHT | 1 | -0.007675111 | 0.001766261 | -4.345 | 0.0002 |
| HP | 1 | -0.03222449 | 0.02474035 | -1.303 | 0.2034 |
| TANK | 1 | -0.01481602 | 0.21908355 | -0.068 | 0.9466 |
| CYL | 1 | 1.01296533 | 0.57892636 | 1.750 | 0.0911 |

17.55   Suppose $y$ is related to factors $x_1$ and $x_2$ according to the nonlinear equation

$$y = ax_1^{b_1}x_2^{b_2}$$

How might the data be transformed so as to produce a linear regression. (Hint: See Section 12.6.)

17.56   What can we infer about the sample size in a multiple regression problem if both R-SQUARE and PROB > F in the SAS printout are small?

# Bibliography

1. Allen, R.L., and Doubleday, L.W. "The Deterrence of Unauthorized Intrusion in Residences, a Systems Analysis." *First International Electronic Crime Countermeasures Conference*. Lexington, Ky.: ORES Publications, 1973.

2. Anderson, J.D., Efron, L., and Wong, S.K. "Martian Mass and Earth—Moon Mass Ratio from Coherent S-Band Tracking of Mariners 6 and 7." *Science*, **167** (1970), 277–9.

3. Asimov, I. *Asimov on Astronomy*. New York: Bonanza Books, 1979, p. 31.

4. Ayala, F.J. "The Mechanisms of Evolution." *Evolution, A Scientific American Book*. San Francisco: W.H. Freeman, 1978, pp. 14–27.

5. Ball, J.A.C., and Taylor, A.R. "The Effect of Cyclandelate on Mental Function and Cerebral Blood Flow in Elderly Patients." *Research on the Cerebral Circulation*. Edited by Meyer, J.S., Lechner, H., and Eichhorn, O. Springfield, Ill.: Thomas, 1969.

6. Barnicot, N.A., and Brothwell, D.R. "The Evaluation of Metrical Data in the Comparison of Ancient and Modern Bones." *Medical Biology and Etruscan Origins*. Edited by Wolstenholme, G.E.W. and O'Connor, C.M. Boston: Little, Brown, 1959.

7. Bellany, I. "Strategic Arms Competition and the Logistic Curve." *Survival*, **16** (1974), 228–30.

8. Bennett, W.R., Jr. "How Artificial Is Intelligence?" *American Scientist*, **65**, No. 6 (1977), 694–702.

9. Berger, R.J., and Walker, J.M. "A Polygraphic Study of Sleep in the Tree Shrew." *Brain, Behavior and Evolution*, **5** (1972), 62.

10. Blackman, S., and Catalina, D. "The Moon and the Emergency Room." *Perceptual and Motor Skills*, **37** (1973), 624–26.

11. Bortkiewicz, L. *Das Gesetz der Kleinen Zahlen*. Leipzig: Teubner, 1898.

12. Boyd, E. "The Specific Gravity of the Human Body." *Human Biology*, **5** (1933), 651–52.

13. Breed, M. D., and Byers, J.A. "The Effect of Population Density on Spacing Patterns and Behavioral Interactions in the Cockroach, *Byrsotria fumigata* (Guérin)." *Behavioral and Neural Biology*, **27** (1979), 526.

771

14. Breed, M.D., Smith, S.K., and Gall, B.G. "Systems of Mate Selection in a Cockroach Species with Male Dominance Heirarchies." *Animal Behaviour*, **28** (1980), 131.

15. Brien, A.J., and Simon, T.L. "The Effects of Red Blood Cell Infusion on 10-km Race Time." *Journal of the American Medical Association*, May 22 (1987), 2764.

16. Brinegar, C.S. "Mark Twain and the Quintus Curtius Snodgrass Letters: A Statistical Test of Authorship." *Journal of the American Statistical Association*, **58** (1963), 85–96.

17. Brown, J.L. *The Evolution of Behavior*. New York: Norton, 1975, p. 111.

18. Brown, L.E., and Littlejohn, M.J. "Male Release Call in the *Bufo americanus* Group." *Evolution in the Genus Bufo*. Edited by Blair, W.F. Austin: University of Texas Press, 1972, p. 316.

19. Buchanan, T.M., Brooks, G.F., and Brachman, P.S. "The Tularemia Skin Test." *Annals of Internal Medicine*, **74** (1971), 336–43.

20. Bullard, R.W., and Shumake, S.A. "Food Temperature Preference Response of *Desmodus rotundus*." *Journal of Mammalogy*, **54** (1973), 299–302.

21. Burnham, E. Personal communication.

22. Camner, P., and Philipson, K. "Urban Factor and Tracheobronchial Clearance." *Archives of Environmental Health*, **27** (1973), 82.

23. Campbell, C., and Joiner, B.L. "How to Get the Answer Without Being Sure You've Asked the Question." *American Statistician*, **27** (1973), 229–31.

24. Casler, L. "The Effects of Hynosis on GESP." *Journal of Parapsychology*, **28** (1964), 126–34.

25. Chow, G.C. "Statistical Demand Functions and Their Use in Forecasting." *The Demand for Durable Goods*. Edited by Harberger, A.C. Chicago: University of Chicago Press, 1960.

26. Clason, C.B. *Exploring the Distant Stars*. New York: G.P. Putnam, 1958, p. 337.

27. Cochran, W.G., and Cox, G.M. *Experimental Designs*, 2nd ed. New York: John Wiley, 1957, p. 108.

28. *Commercial Appeal*, Memphis. Nov. 4, 1985.

29. ———, Jan. 12, 1987.

30. Connolly, K. "The Social Facilitation of Preening Behaviour in *Drosophila Melanogaster*." *Animal Behavior*, **16** (1968), 385–91.

31. Coulson, J.C. "The Significance of the Pair-Bond in the Kittiwake." *Parental Behavior in Birds*. Edited by Silver, R. Stroudsburg, Pa.: Dowden, Hutchinson, & Ross, 1977.

32. Crowder, J. and Shalash, A. Personal communication.

33. Cummins, H. and Midlo, C. *Finger Prints, Palms, and Soles*. Philadelphia: Blakiston, 1943.

34. Cushny, A.R., and Peebles, A.R. "The Action of Optical Isomers." *Journal of Physiology*, **32** (1904–05), 501–10.

35. David, F. N. *Games, Gods and Gambling*. New York: Hafner, 1962, p. 16.

36. Davis, M. "Premature Mortality Among Prominent American Authors Noted for Alcohol Abuse." *Drug and Alcohol Dependence*, **18** (1986), 133–38.

37. Dewey, G. *Relative Frequency of English Spellings*. New York: Teachers College Press, Columbia University, 1970.

38. *Documenta Geigy Scientific Tables*, 6th ed. Edited by Diem, K. Ardsley, N.Y.: Geigy Pharmaceuticals, 1962.

39. Dubois, C. *Lowie's Selected Papers in Anthropology*. Berkeley: University of California Press, 1960, pp. 137–42.

40. Edie, L.C., Herman, R., and Lam, T.N. "Observed Multilane Speed Distribution and the Kinetic Theory of Vehicular Traffic." *Transportation Science*, **14** (1980), 69.

41. Edmunds, M. *Defence in Animals*. New York: Longman, 1974, p. 276.

42. Emmons, D.B., Larmond, E., and Beckett, D.C. "Determination of Total Solids in Heterogeneous Heat-Sensitive Foods." *Journal of the Association of Official Analytical Chemists*, **54** (1971), 1403–05.

43. Evans, L., and Herman, R. "Automobile Fuel Economy on Fixed Urban Driving Schedules." *Transportation Science*, **12** (1978), 137–51.

44. Fadeley, R.C. "Oregon Malignancy Pattern Physiographically Related to Hanford, Washington, Radioisotope Storage." *Journal of Environmental Health*, **27** (1965), 883–97.

45. Fagen, R.M. "Exercise, Play, and Physical Training in Animals." *Perspectives in Ethology*. Edited by Bateson, P.P.G. and Klopfer, P.H. New York: Plenum Press, 1976.

46. Fairley, W.B. "Evaluating the 'Small' Probability of a Catastrophic Accident from the Marine Transportation of Liquefied Natural Gas." *Statistics and Public Policy*. Edited by Fairley, W.B. and Mosteller, F. Reading, Mass.: Addison-Wesley, 1977.

47. Fears, T., Scotts, J., and Scheiderman, M.A. "Skin Cancer, Melanoma, and Sunlight." *American Journal of Public Health*, **66** (1976), 461–64.

48. Feller, W. "Statistical Aspects of ESP." *Journal of Parapsychology*, **4** (1940), 271–98.

49. Fishbein, M. *Birth Defects*. Philadelphia: Lippincott, 1962, p. 177.

50. *Florida Monthly Economic Report*, **8**, No. 9. Tallahassee: The Florida Legislature, 1988.

51. *Florida Trend*, **30**, No. 30. St. Petersburg: Trend Magazines, Inc., 1988.

52. Free, J.B. "The Stimuli Releasing the Stinging Response of Honeybees." *Animal Behavior*, **9** (1961), 193–96.

53. ———. "Effect of Flower Shapes and Nectar Guides on the Behavior of Foraging Honeybees." *Behavior*, **37** (1970), 269–85.

54. Furuhata, T., and Yamamoto, K. *Forensic Odontology*. Springfield, Ill.: Thomas, 1967, p. 84.

55. Gendreau, Paul, et al. "Changes in EEG Alpha Frequency and Evoked Response Latency During Solitary Confinement." *Journal of Abnormal Psychology*, **79** (1972), 54–59.

56. Geotis, S. "Thunderstorm Water Contents and Rain Fluxes Deduced from Radar." *Journal of Applied Meteorology*, **10** (1971), 1234.

57. Glover, J.D. *The Revolutionary Corporations: Engines of Plenty, Engines of Growth, Engines of Change*. Homewood, Ill.: Dow Jones–Irwin, 1980, pp. 412–17.

58. Graw, R.G., Jr., and Santos, G.W. "Bone Marrow Transplantation in Patients with Leukemia." *Transplantation*, **11** (1971), 198.

59. Griffin, D.R., Webster, F.A., and Michael, C.R. "The Echolocation of Flying Insects by Bats." *Animal Behavior*, **8** (1960), 148.

60. Gross, N., Mason, W.S., and McEachern, A.W. *Explorations in Role Analysis*. New York: John Wiley, 1958, p. 297.

61. Grover, C.A. "Population Differences in the Swell Shark *Cephaloscyllium ventriosum*." *California Fish and Game*, **58** (1972), 191–97.

62. Hagerman, R.L., and Senbet, L.W. "A Test of Accounting Bias and Marketing Structure." *Journal of Business*, **49** (1976), 509–14.

63. Haggard, W.H., Bilton, T.H., and Crutcher, H.L. "Maximum Rainfall from Tropical Cyclone Systems Which Cross the Appalachians." *Journal of Applied Meteorology*, **12** (1973), 50–61.

64. Hammer, M., and Salzinger, K. "Some Formal Characteristics of Schizophrenic Speech as a Measure of Social Deviance." *Annals of the New York Academy of Sciences*, **15** (1964), 865–68.

65. Handson, P.D. "Lead and Arsenic Levels in Wines Produced from Vineyards Where Lead Arsenate Sprays Are Used for Caterpillar Control." *Journal of the Science of Food and Agriculture*, **35** (1984), 216.

66. Hankins, F.H. "Adolphe Quetelet As Statistician." *Studies in History, Economics, and Public Law*, **xxxi**, No. 4. New York: Longman, Green, 1908, p. 497.

67. Hansel, C.E.M. *ESP: A Scientific Evaluation*. New York: Scribner, 1966, pp. 86–89.

68. Hansen, W.R., and Eckel, E.B. "The Alaska Earthquake, March 27, 1964: Field Investigations and Reconstruction Effort." *Focus on Environmental Geology*, 2nd ed. Edited by Tank, R.W. New York: Oxford University Press, 1976, pp. 73–74.

69. Hare, E., Price, J., and Slater, E. "Mental Disorder and Season of Birth: A National Sample Compared with the General Population." *British Journal of Psychiatry*, **124** (1974), 81–86.

70. Hazel, W.M., and Eglof, W.K. "Determination of Calcium in Magnesite and Fused Magnesia." *Industrial and Engineering Chemistry*, Analytical Edition, **18** (1946), 759–60.

71. Hendy, M.F., and Charles, J.A. "The Production Techniques, Silver Content and Circulation History of the Twelfth-Century Byzantine Trachy." *Archaeometry*, **12** (1970), 13–21.

72. Hersen, M. "Personality Characteristics of Nightmare Sufferers." *Journal of Nervous and Mental Diseases*, **153** (1971), 29–31.

73. Horvath, F.S., and Reid, J.E. "The Reliability of Polygraph Examiner Diagnosis of Truth and Deception." *Journal of Criminal Law, Criminology, and Police Science*, **62** (1971), 276–81.

74. Hudgens, G.A., Denenberg, V.H., and Zarrow, M.X. "Mice Reared with Rats: Effects of Preweaning and Postweaning Social Interactions Upon Adult Behavior." *Behaviour*, **30** (1968), 259–74.

75. Hulbert, R.H., and Krumbiegel, E.R. "Synthetic Flavors Improve Acceptance of Anticoagulant-Type Rodenticides." *Journal of Environmental Health*, **34** (1972), 407–11.

76. Husni, S.A. "The Typical American Consumer Magazine of the 1980's." Presentation to the Association for Education in Journalism and Mass Communication, Annual Convention. Gainesville, Fla., 1984.

77. Huxtable, J., Aitken, M.J., and Weber, J.C. "Thermoluminescent Dating of Baked Clay Balls of the Poverty Point Culture." *Archaeometry*, 14 (1972), 269–75.

78. Hynek, J.A. *The UFO Experience: A Scientific Inquiry*. Chicago: Rognery, 1972.

79. Ibrahim, M.A., et al. "Coronary Heart Disease: Screening by Familial Aggregation." *Archives of Environmental Health*, **16** (1968), 235–40.

80. Institute for Statistical and Mathematical Modelling. University of West Florida, 1988.

81. Jacobson, E., and Kossoff, J. "Self-percept and Consumer Attitudes Toward Small Cars." *Consumer Behavior in Theory and in Action*. Edited by Britt, S.H. New York: John Wiley, 1970.

82. Jamcs, Λ., and Moncada, R. "Many Sct Color TV Lounges Show Highest Radiation." *Journal of Environmental Health*, **31** (1969), 359–60.

83. Jones, J.C., and Pilitt, D.R. "Blood-feeding Behavior of Adult *Aedes Aegypti* Mosquitoes." *Biological Bulletin*, **145** (1973), 127–39.

84. Jones, N.G.B. "An Ethological Study of Some Aspects of Social Behaviour of Children in Nursery School." *Evolution of Play Behavior*. Edited by Müller-Schwarze, D. Stroudsburg, Pa.: Dowden, Hutchinson & Ross, 1978, p. 361.

85. Kahn, D. *The Codebreakers*. New York: Macmillan, 1967.

86. Kendall, M.G. "The Beginnings of a Probability Calculus." *Studies in the History of Statistics and Probability*. Edited by Pearson, E.S., and Kendall, M.G. Darien, Conn.: Hafner, 1970, pp. 8–11.

87. Kneafsey, J.T. *Transportation Economic Analysis*. Lexington, Mass.: Heath, 1975.

88. Kozlowski, T.T. *Growth and Development of Trees*, vol. II. New York: Academic Press, 1971, p. 231.

89. Kronoveter, K.J., and Somerville, G.W. "Airplane Cockpit Noise Levels and Pilot Hearing Sensitivity." *Archives of Environmental Health*, **20** (1970), 498.

90. Kruk-De Bruin, M., Röst, Luc C.M., and Draisma, Fons G.A.M. "Estimates of the Number of Foraging Ants with the Lincoln-Index Method in Relation to the Colony Size of *Formica Polyctena*." *Journal of Animal Ecology*, **46** (1977), 463–65.

91. Laurie–Ahlberg, C.C., and McKinney, F. "The Nod-Swim Display of Male Green-Winged Teal (*Anas Crecca*)." *Animal Behaviour*, **27** (1979), 170.

92. Lemmon, W.B., and Patterson, G.H. "Depth Perception in Sheep: Effects of Interrupting the Mother-Neonate Bond." *Comparative Psychology: Research in Animal Behavior*. Edited by Denny M.R., and Ratner, S. Homewood, Ill.: Dorsey Press, 1970, p. 403.

93. Li, F.P. "Suicide Among Chemists." *Archives of Environmental Health*, **19** (1969), 519.

94. Lockwood, L. Personal communication.

95. Ludd, E.C., Jr., and Lipset, S.M. "Politics of Academic Natural Scientists and Engineers." *Science*, **176** (1972), 1091–100.

96. MacDonald, G.A. and Abbott, A.T. *Volcanoes in the Sea*. Honolulu: University of Hawaii Press, 1970, pp. 56–57.

97. Maddala, G.S. *Econometrics*. New York: McGraw-Hill, 1977.

98. Malina, R.M. "Comparison of the Increase in Body Size between 1899 and 1970 in a Specially Selected Group with That in the General Population." *Physical Anthropology*, **37** (1972), 135–41.

99. Mares, M.A., et al. "The Strategies and Community Patterns of Desert Animals." *Convergent Evolution in Warm Deserts*. Edited by Orians, G.H., and Solbrig, O.T. Stroudsburg, Pa.: Dowden, Hutchinson & Ross, 1977, p. 141.

100. Marx, M.L. Personal communication.

101. McConnell, T.R., Jr. "Suggestibility in Children As a Function of Chronological Age." *Journal of Abnormal and Social Psychology*, **67** (1963), 288.

102. McIntyre, D.B.. "Precision and Resolution in Geochronometry." *The Fabric of Geology*. Edited by Albritton, C.C., Jr. Stanford, Calif.: Freeman, Cooper, 1963.

103. "Medical News." *Journal of the American Medical Association*, **219** (1972), 981.

104. Merchant, L. *The National Football Lottery*. New York: Holt, Rinehart, and Winston, 1973.

105. Miettinen, J.K. "The Accumulation and Excretion of Heavy Metals in Organisms." *Heavy Metals in the Aquatic Environment.* Edited by Krenkel, P.A. Oxford: Pergamon Press, 1975.

106. Minkoff, E.C. "A Fossil Baboon from Angola, with a Note on *Australopithecus.*" *Journal of Paleontology,* **46** (1972), 836–44.

107. Moriarty, S. Personal communication.

108. Mulcahy, R., McGilvray, J.W., and Hickey, N. "Cigarette Smoking Related to Geographic Variations in Coronary Heart Disease Mortality and to Expectation of Life in the Two Sexes." *American Journal of Public Health,* **60** (1970), 1516.

109. Nader, R. "The Case for Federal Chartering." *Corporate Power in America.* Edited by Nader, R. and Green, M.J. New York: Grossman Publishers, 1973, pp. 90–91.

110. Nash, H. *Alcohol and Caffeine.* Springfield, Ill.: Thomas, 1962, p. 96.

111. Navy Manpower and Material Analysis Center. *Procedures and Analysis for Staffing Standards Development: Data/Regression Analysis Handbook.* San Diego, 1979.

112. *Newsweek.* June 9, 1975, p. 73.

113. ———. March 6, 1978, p. 78.

114. ———. June 2, 1986, p. 1.

115. Nicholes, P.S. "Bacteria in Laundered Fabrics." *American Journal of Public Health,* **60** (1970), 2177.

116. Nye, F.I. *Family Relationships and Delinquent Behavior.* New York: John Wiley, 1958, p. 37.

117. Ore, O. *Cardano, The Gambling Scholar.* Princeton, N.J.: Princeton University Press, 1953, pp. 25–26.

118. Pascal, G., and Suttell, B. "Testing the Claims of a Graphologist." *Journal of Personality,* **16** (1947), 192–97.

119. Passingham, R.E. "Anatomical Differences between the Neocortex of Man and Other Primates." *Brain, Behavior, and Evolution,* **7** (1973), 337–59.

120. Pensacola Greyhound Track. *Official Program.* Pensacola, Fla., June 22, 1988.

121. Petersen, W. *Population,* 2nd ed. New York: Macmillan, 1969, p. 66.

122. Phillips, D.P. "Deathday and Birthday: An Unexpected Connection." *Statistics: A Guide to the Unknown.* Edited by Tanur, J.M. et al. San Francisco: Holden–Day, 1972.

123. Pickering, Sir George. "The Quantitative Approach to Disease." *Ciba Foundation: Significant Trends in Medical Research.* Boston: Little, Brown, 1960, pp. 278–91.

124. Pierce, G.W. *The Songs of Insects.* Cambridge, Mass.: Harvard University Press, 1949, pp. 12–21.

125. Pillay, K.K.S., et al. "Mercury Pollution of Lake Erie Ecosphere." *Environmental Research,* **5** (1972), 172–81.

126. Porter, J.W., et al. "Effect of Hypnotic Age Regression on the Magnitude of the Ponzo Illusion." *Journal of Abnormal Psychology,* **79** (1972), 189–94.

127. Premack, D. "Language in Chimpanzee?" *Science,* **172** (1971), 808–22.

128. Quetelet, L.A.J. *Lettres sur la Theorie des Probabilites, Appliquee aux Sciences Morales et Politiques.* Bruxelles: Hayez, M., Imprimeur de L'Academie Royal des Sciences, des Lettres et des Beaux-Arts de Belgique, 1846, p. 400.

129. Ragsdale, A.C., and Brody, S. *Journal of Dairy Science,* **5** (1922), 214.

130. Rahman, N.A. *Practical Exercises in Probability and Statistics*. New York: Hafner, 1972.

131. Ratner, S.C. "Effect of Learning to Be Submissive on Status in the Peck Order of Domestic Fowl." *Comparative Psychology: Research in Animal Behavior*. Edited by Denny, M., and Ratner, S. Homewood, Ill.: Dorsey Press, 1970, pp. 436–41.

132. Reichler, J.L. *The Great All-Time Baseball Record Book*. New York: Macmillan, 1981.

133. Reid, D.H., Doerr, J. E., and Buckman, J.A. "Determination of Parachute Ripcord Pull Forces During Free-Fall: Physiological Studies of Military Parachutists Via FM/FM Telemetry-IV." *Aerospace Medicine*, **44** (1973), 1164–68.

134. Resnick, R.B., Fink, M., and Freedman, A.M. "A Cyclazocine Typology in Opiate Dependence." *American Journal of Psychiatry*, **126** (1970), 1256–60.

135. Rich, C.L. "Is Random Digit Dialing Really Necessary?" *Journal of Marketing Research*, **14** (1977), 300–05.

136. Richardson, L.F. "The Distribution of Wars in Time." *Journal of the Royal Statistical Society*, **107** (1944), 242–50.

137. Ritter, B. "The Use of Contact Desensitization, Demonstration–Plus–Participation and Demonstration–Alone in the Treatment of Acrophobia." *Behaviour Research and Therapy*, **7** (1969), 157–64.

138. Roberts, C.A. "Retraining of Inactive Medical Technologists—Whose Responsibility?" *American Journal of Medical Technology*, **42** (1976), 115–23.

139. Rosenthal, R., and Jacobson, L.F. "Teacher Expectations for the Disadvantaged." *Scientific American*, **218** (1968), 19–23.

140. Rowley, W.A. "Laboratory Flight Ability of the Mosquito, *Culex Tarsalis Coq.*" *Journal of Medical Entomology*, **7** (1970), 713–16.

141. Roy, R.H. *The Cultures of Management*. Baltimore: Johns Hopkins University Press, 1977, p. 261.

142. Ruckman, J.E., Zscheile, F.P., Jr., and Qualset, C.O. "Protein, Lysine, and Grain Yields of Triticale and Wheat as Influenced by Genotype and Location." *Journal of Agricultural and Food Chemistry*, **21** (1973), 697–700.

143. Rutherford, Sir Ernest, Chadwick, J., and Ellis, C.D. *Radiations from Radioactive Substances*. London: Cambridge University Press, 1951, p. 172.

144. Ryzl, M. "Precognition Scoring and Attitude Toward ESP." *Journal of Parapsychology*, **32** (1968), 1–8.

145. Salvosa, C.B., Payne, P.R., and Wheeler, E.F. "Energy Expenditure of Elderly People Living Alone or in Local Authority Homes." *American Journal of Clinical Nutrition*, **24** (1971), 1468.

146. Samaras, T.T. "That Song Put Down Short People, But . . . " *Science Digest*, **84** (1978), 76–79.

147. Schaller, G.B. "The Behavior of the Mountain Gorilla." *Primate Patterns*. Edited by Dolhinow, P. New York: Holt, Rinehart and Winston, 1972, p. 95.

148. Schell, E.D. "Samuel Pepys, Isaac Newton, and Probability." *The American Statistician*, **14** (1960), 27–30.

149. Schoeneman, R.L., Dyer, R.H.; and Earl, E.M. "Analytic Profile of Straight Bourbon Whiskies." *Journal of the Association of Official Analytical Chemists,* **54** (1971), 1247–61.

150. Selective Service System. Office of the Director. Washington, D.C., 1969.

151. Sen, Nrisinha, et al. "Effect of Sodium Nitrite Concentration on the Formation of Nitrosopyrrolidine and Dimethylnitrosamine in Fried Bacon." *Journal of Agricultural and Food Chemistry*, **22** (1974), 540–41.

152. Shahidi, S.A., et al. "Celery Implicated in High Bacteria Count Salads." *Journal of Environmental Health*, **32** (1970), 669.

153. Sharpe, R.S., and Johnsgard, P.A. "Inheritance of Behavioral Characters in $F_2$ Mallard x Pintail (*Anas Platyrhynchos L.* x *Anas Acuta L.*) Hybrids." *Behaviour*, **27** (1966), 259–72.

154. Shore, N.S., Greene, R., and Kazemi, H. "Lung Dysfunction in Workers Exposed to *Bacillus subtilis* Enzyme." *Environmental Research*, **4** (1971), 512–19.

155. Siddiqui, S.H., and Parizek, R.R. "Application of Nonparametric Statistical Tests in Hydrogeology." *Ground Water*, **10** (1972), 26–31.

156. Smith, P. Personal communication.

157. Smith, W.L. "Facilitating Verbal-Symbolic Functions in Children with Learning Problems and 14-6 Positive Spike EEG Patterns with Ethosuximide (Zarontin)." *Drugs and Cerebral Function*. Edited by Smith, W. Springfield, Ill.: Thomas, 1970.

158. Srb, A.M., Owen, R.D., and Edgar, R.S. *General Genetics*, 2nd ed. San Francisco: W.H. Freeman, 1965.

159. Statkiewicz, W.R., and Schein, M.W. "Variability and Periodicity of Dustbathing Behaviour in Japanese Quail (*Coturnix Coturnix Japonica*)." *Animal Behaviour*, **28** (1980), 464.

160. Student. "The Probable Error of a Mean." *Biometrika*, **6**, No. 6 (1908), 1–25.

161. Sutton, D.H. "Gestation Period." *Medical Journal of Australia*, **1** (1945), 611–13.

162. Szalontai, S., and Timaffy, M. "Involutional Thrombopathy." *Age with a Future*. Edited by Hansen, P.F. Philadelphia: F.A. Davis, 1964.

163. Tanguy, J.C. "An Archaeomagnetic Study of Mount Etna: The Magnetic Direction Recorded in Lava Flows Subsequent to the Twelfth Century." *Archaeometry*, **12** (1970), 115–28.

164. *Tennessean* (Nashville). Jan. 20, 1973.

165. *Time*. May 6, 1985, p. 74.

166. Treuhaft, P.S., and McCarty, D.J. "Synovial Fluid pH, Lactate, Oxygen and Carbon Dioxide Partial Pressure in Various Joint Diseases." *Arthritis and Rheumatism*, **14** (1971), 476–77.

167. Trugo, L.C., Macrae, R., and Dick, J. "Determination of Purine Alkaloids and Trigonelline in Instant Coffee and Other Beverages Using High Performance Liquid Chromatography." *Journal of the Science of Food and Agriculture*, **34** (1983), 300–06.

168. Turco, S., and Davis, N. "Particulate Matter in Intravenous Infusion Fluids—Phase 3." *American Journal of Hospital Pharmacy*, **30** (1973), 612.

169. U.S. Department of Agriculture. "Results of Fiber and Spinning Tests for Some Varieties of Upland Cotton Grown in the United States." Washington, D.C., 1945.

170. Van Twyver, H., and Allison, T. "Sleep in the Armadillo *Dasypus novemcinctus* at Moderate and Low Ambient Temperatures." *Brain, Behavior and Evolution*, **9** (1974), 115.

171. Vilenkin, N.Y. *Combinatorics*. New York: Academic Press, 1971, p. 25.

172. Vincent, P. "Factors Influencing Patient Noncompliance: A Theoretical Approach." *Nursing Research*, **20** (1971), 514.

173. Vogel, J.H.K., Horgan, J.A., and Strahl, C.L. "Left Ventricular Dysfunction in Chronic Constrictive Pericarditis." *Chest*, **59** (1971), 489.

174. Vogt, E.Z., and Hyman, R. *Water Witching U.S.A.* Chicago: University of Chicago Press, 1959, p. 55.

175. Walker, H. *Studies in the History of Statistical Method.* Baltimore: Williams and Wilkins, 1929.

176. Wallechinsky, D., Wallace, I., and Wallace, A. *The Book of Lists.* New York: Bantam Books, 1978.

177. Wallis, W.A. "The Poisson Distribution and the Supreme Court." *Journal of the American Statistical Association,* **31** (1936), 376–80.

178. Walsberg, G.E. *Ecology and Energetics of Contrasting Social Systems in Phainopepla Nitens (Aves: Ptilogonatidae).* Berkeley: University of California Press, 1977, p. 14.

179. Walter, W.G., and Stober, A. "Microbial Air Sampling in a Carpeted Hospital." *Journal of Environmental Health,* **30** (1968), 405.

180. Weiss, W. "Cigarette Smoke Gas Phase and *Paramecium* Survival." *Archives of Environmental Health,* **17** (1968), 63.

181. Werner, M., Stabenau, J.R., and Pollin, W. "Thematic Apperception Test Method for the Differentiation of Families of Schizophrenics, Delinquents, and 'Normals'." *Journal of Abnormal Psychology,* **75** (1970), 139–145.

182. Winslow, C. *The Conquest of Epidemic Disease.* Princeton, NJ: Princeton University Press, 1943, p. 303.

183. Wolf, S. *The Artery and the Process of Arteriosclerosis: Measurement and Modification.* Proceedings of an Interdisciplinary Conference on Fundamental Data on Reactions of Vascular Tissue in Man, April 19–25, 1970, Lindau, West Germany. New York: Plenum Press, 1972, p. 116.

184. Wood, R.M. "Giant Discoveries of Future Science." *Virginia Journal of Science,* **21** (1970), 169–77.

185. Woodward, W.F. "A Comparison of Base Running Methods in Baseball." M.Sc. Thesis, Florida State University, 1970.

186. Wrightman, L.S. "Wallace Supporters and Adherence to 'Law and Order'." *Human Social Behavior.* Edited by Baron, R.A., and Liebert, R.M. Homewood, Ill.: Dorsey Press, 1971.

187. Wyler, A.R., Minoru, M., and Holmes, T.H. "Magnitude of Life Events and Seriousness of Illness." *Psychosomatic Medicine,* **33** (1971), 115–22.

188. Yarnall, J.L. "Aspects of the Behaviour of *Octopus Cyanea Gray.*" *Animal Behaviour,* **17** (1969), 751.

189. Yochem, D., and Roach, D. "Aspirin: Effect on Thrombus Formation Time and Prothrombin Time of Human Subjects." *Angiology,* **22** (1971), 72.

190. Young, P.V., and Schmid, C. *Scientific Social Surveys and Research.* Englewood Cliffs, NJ: Prentice Hall, 1966, p. 319.

191. Younger, M.S. *A First Course in Linear Analysis,* 2nd ed. Boston: Duxbury Press, 1985.

192. Zaleznik, A., Christensen, C.R., and Roethlisberger, F.J. *The Motivation, Productivity, and Satisfaction of Workers.* Boston: Harvard University Press, 1958, p. 260.

193. Zaret, T.M. "Predators, Invisible Prey, and the Nature of Polymorphism in the *Cladocera* (Class *Crustacea*)." *Limnology and Oceanography,* **17** (1972), 171–84.

194. Zelazo, P.R., Zelazo, N.A., and Kolb, S. "'Walking' in the Newborn." *Science,* **176** (1972), 314–15.

195. Ziv, G., and Sulman, F.G. "Binding of Antibiotics to Bovine and Ovine Serum." *Antimicrobial Agents and Chemotherapy,* **2** (1972), 206–13.

196. Zucker, N. "The Role of Hood-Building in Defining Territories and Limiting Combat in Fiddler Crabs." *Animal Behaviour,* **29** (1981), 391.

## APPENDIX TABLE A.1  Cumulative Areas Under the Standard Normal Distribution

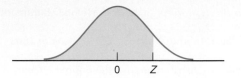

| Z | 0 | 1 | 2 | 3 | 4 | 5 | 6 | 7 | 8 | 9 |
|------|--------|--------|--------|--------|--------|--------|--------|--------|--------|--------|
| −3. | 0.0013 | 0.0010 | 0.0007 | 0.0005 | 0.0003 | 0.0002 | 0.0002 | 0.0001 | 0.0001 | 0.0000 |
| −2.9 | 0.0019 | 0.0018 | 0.0017 | 0.0017 | 0.0016 | 0.0016 | 0.0015 | 0.0015 | 0.0014 | 0.0014 |
| −2.8 | 0.0026 | 0.0025 | 0.0024 | 0.0023 | 0.0023 | 0.0022 | 0.0021 | 0.0021 | 0.0020 | 0.0019 |
| −2.7 | 0.0035 | 0.0034 | 0.0033 | 0.0032 | 0.0031 | 0.0030 | 0.0029 | 0.0028 | 0.0027 | 0.0026 |
| −2.6 | 0.0047 | 0.0045 | 0.0044 | 0.0043 | 0.0041 | 0.0040 | 0.0039 | 0.0038 | 0.0037 | 0.0036 |
| −2.5 | 0.0062 | 0.0060 | 0.0059 | 0.0057 | 0.0055 | 0.0054 | 0.0052 | 0.0051 | 0.0049 | 0.0048 |
| −2.4 | 0.0082 | 0.0080 | 0.0078 | 0.0075 | 0.0073 | 0.0071 | 0.0069 | 0.0068 | 0.0066 | 0.0064 |
| −2.3 | 0.0107 | 0.0104 | 0.0102 | 0.0099 | 0.0096 | 0.0094 | 0.0091 | 0.0089 | 0.0087 | 0.0084 |
| −2.2 | 0.0139 | 0.0136 | 0.0132 | 0.0129 | 0.0126 | 0.0122 | 0.0119 | 0.0116 | 0.0113 | 0.0110 |
| −2.1 | 0.0179 | 0.0174 | 0.0170 | 0.0166 | 0.0162 | 0.0158 | 0.0154 | 0.0150 | 0.0146 | 0.0143 |
| −2.0 | 0.0228 | 0.0222 | 0.0217 | 0.0212 | 0.0207 | 0.0202 | 0.0197 | 0.0192 | 0.0188 | 0.0183 |
| −1.9 | 0.0287 | 0.0281 | 0.0274 | 0.0268 | 0.0262 | 0.0256 | 0.0250 | 0.0244 | 0.0238 | 0.0233 |
| −1.8 | 0.0359 | 0.0352 | 0.0344 | 0.0336 | 0.0329 | 0.0322 | 0.0314 | 0.0307 | 0.0300 | 0.0294 |
| −1.7 | 0.0446 | 0.0436 | 0.0427 | 0.0418 | 0.0409 | 0.0401 | 0.0392 | 0.0384 | 0.0375 | 0.0367 |
| −1.6 | 0.0548 | 0.0537 | 0.0526 | 0.0516 | 0.0505 | 0.0495 | 0.0485 | 0.0475 | 0.0465 | 0.0455 |
| −1.5 | 0.0668 | 0.0655 | 0.0643 | 0.0630 | 0.0618 | 0.0606 | 0.0594 | 0.0582 | 0.0570 | 0.0559 |
| −1.4 | 0.0808 | 0.0793 | 0.0778 | 0.0764 | 0.0749 | 0.0735 | 0.0722 | 0.0708 | 0.0694 | 0.0681 |
| −1.3 | 0.0968 | 0.0951 | 0.0934 | 0.0918 | 0.0901 | 0.0885 | 0.0869 | 0.0853 | 0.0838 | 0.0823 |
| −1.2 | 0.1151 | 0.1131 | 0.1112 | 0.1093 | 0.1075 | 0.1056 | 0.1038 | 0.1020 | 0.1003 | 0.0985 |
| −1.1 | 0.1357 | 0.1335 | 0.1314 | 0.1292 | 0.1271 | 0.1251 | 0.1230 | 0.1210 | 0.1190 | 0.1170 |
| −1.0 | 0.1587 | 0.1562 | 0.1539 | 0.1515 | 0.1492 | 0.1469 | 0.1446 | 0.1423 | 0.1401 | 0.1379 |
| −0.9 | 0.1841 | 0.1814 | 0.1788 | 0.1762 | 0.1736 | 0.1711 | 0.1685 | 0.1660 | 0.1635 | 0.1611 |
| −0.8 | 0.2119 | 0.2090 | 0.2061 | 0.2033 | 0.2005 | 0.1977 | 0.1949 | 0.1922 | 0.1894 | 0.1867 |
| −0.7 | 0.2420 | 0.2389 | 0.2358 | 0.2327 | 0.2297 | 0.2266 | 0.2236 | 0.2206 | 0.2177 | 0.2148 |
| −0.6 | 0.2743 | 0.2709 | 0.2676 | 0.2643 | 0.2611 | 0.2578 | 0.2546 | 0.2514 | 0.2483 | 0.2451 |
| −0.5 | 0.3085 | 0.3050 | 0.3015 | 0.2981 | 0.2946 | 0.2912 | 0.2877 | 0.2843 | 0.2810 | 0.2776 |
| −0.4 | 0.3446 | 0.3409 | 0.3372 | 0.3336 | 0.3300 | 0.3264 | 0.3228 | 0.3192 | 0.3156 | 0.3121 |
| −0.3 | 0.3821 | 0.3783 | 0.3745 | 0.3707 | 0.3669 | 0.3632 | 0.3594 | 0.3557 | 0.3520 | 0.3483 |
| −0.2 | 0.4207 | 0.4168 | 0.4129 | 0.4090 | 0.4052 | 0.4013 | 0.3974 | 0.3936 | 0.3897 | 0.3859 |
| −0.1 | 0.4602 | 0.4562 | 0.4522 | 0.4483 | 0.4443 | 0.4404 | 0.4364 | 0.4325 | 0.4286 | 0.4247 |
| −0.0 | 0.5000 | 0.4960 | 0.4920 | 0.4880 | 0.4840 | 0.4801 | 0.4761 | 0.4721 | 0.4681 | 0.4641 |

| Z | 0 | 1 | 2 | 3 | 4 | 5 | 6 | 7 | 8 | 9 |
|---|---|---|---|---|---|---|---|---|---|---|
| 0.0 | 0.5000 | 0.5040 | 0.5080 | 0.5120 | 0.5160 | 0.5199 | 0.5239 | 0.5279 | 0.5319 | 0.5359 |
| 0.1 | 0.5398 | 0.5438 | 0.5478 | 0.5517 | 0.5557 | 0.5596 | 0.5636 | 0.5675 | 0.5714 | 0.5753 |
| 0.2 | 0.5793 | 0.5832 | 0.5871 | 0.5910 | 0.5948 | 0.5987 | 0.6026 | 0.6064 | 0.6103 | 0.6141 |
| 0.3 | 0.6179 | 0.6217 | 0.6255 | 0.6293 | 0.6331 | 0.6368 | 0.6406 | 0.6443 | 0.6480 | 0.6517 |
| 0.4 | 0.6554 | 0.6591 | 0.6628 | 0.6664 | 0.6700 | 0.6736 | 0.6772 | 0.6808 | 0.6844 | 0.6879 |
| 0.5 | 0.6915 | 0.6950 | 0.6985 | 0.7019 | 0.7054 | 0.7088 | 0.7123 | 0.7157 | 0.7190 | 0.7224 |
| 0.6 | 0.7257 | 0.7291 | 0.7324 | 0.7357 | 0.7389 | 0.7422 | 0.7454 | 0.7486 | 0.7517 | 0.7549 |
| 0.7 | 0.7580 | 0.7611 | 0.7642 | 0.7673 | 0.7703 | 0.7734 | 0.7764 | 0.7794 | 0.7823 | 0.7852 |
| 0.8 | 0.7881 | 0.7910 | 0.7939 | 0.7967 | 0.7995 | 0.8023 | 0.8051 | 0.8078 | 0.8106 | 0.8133 |
| 0.9 | 0.8159 | 0.8186 | 0.8212 | 0.8238 | 0.8264 | 0.8289 | 0.8315 | 0.8340 | 0.8365 | 0.8389 |
| 1.0 | 0.8413 | 0.8438 | 0.8461 | 0.8485 | 0.8508 | 0.8531 | 0.8554 | 0.8577 | 0.8599 | 0.8621 |
| 1.1 | 0.8643 | 0.8665 | 0.8686 | 0.8708 | 0.8729 | 0.8749 | 0.8770 | 0.8790 | 0.8810 | 0.8830 |
| 1.2 | 0.8849 | 0.8869 | 0.8888 | 0.8907 | 0.8925 | 0.8944 | 0.8962 | 0.8980 | 0.8997 | 0.9015 |
| 1.3 | 0.9032 | 0.9049 | 0.9066 | 0.9082 | 0.9099 | 0.9115 | 0.9131 | 0.9147 | 0.9162 | 0.9177 |
| 1.4 | 0.9192 | 0.9207 | 0.9222 | 0.9236 | 0.9251 | 0.9265 | 0.9278 | 0.9292 | 0.9306 | 0.9319 |
| 1.5 | 0.9332 | 0.9345 | 0.9357 | 0.9370 | 0.9382 | 0.9394 | 0.9406 | 0.9418 | 0.9430 | 0.9441 |
| 1.6 | 0.9452 | 0.9463 | 0.9474 | 0.9484 | 0.9495 | 0.9505 | 0.9515 | 0.9525 | 0.9535 | 0.9545 |
| 1.7 | 0.9554 | 0.9564 | 0.9573 | 0.9582 | 0.9591 | 0.9599 | 0.9608 | 0.9616 | 0.9625 | 0.9633 |
| 1.8 | 0.9641 | 0.9648 | 0.9656 | 0.9664 | 0.9671 | 0.9678 | 0.9686 | 0.9693 | 0.9700 | 0.9706 |
| 1.9 | 0.9713 | 0.9719 | 0.9726 | 0.9732 | 0.9738 | 0.9744 | 0.9750 | 0.9756 | 0.9762 | 0.9767 |
| 2.0 | 0.9772 | 0.9778 | 0.9783 | 0.9788 | 0.9793 | 0.9798 | 0.9803 | 0.9808 | 0.9812 | 0.9817 |
| 2.1 | 0.9821 | 0.9826 | 0.9830 | 0.9834 | 0.9838 | 0.9842 | 0.9846 | 0.9850 | 0.9854 | 0.9857 |
| 2.2 | 0.9861 | 0.9864 | 0.9868 | 0.9871 | 0.9874 | 0.9878 | 0.9881 | 0.9884 | 0.9887 | 0.9890 |
| 2.3 | 0.9893 | 0.9896 | 0.9898 | 0.9901 | 0.9904 | 0.9906 | 0.9909 | 0.9911 | 0.9913 | 0.9916 |
| 2.4 | 0.9918 | 0.9920 | 0.9922 | 0.9925 | 0.9927 | 0.9929 | 0.9931 | 0.9932 | 0.9934 | 0.9936 |
| 2.5 | 0.9938 | 0.9940 | 0.9941 | 0.9943 | 0.9945 | 0.9946 | 0.9948 | 0.9949 | 0.9951 | 0.9952 |
| 2.6 | 0.9953 | 0.9955 | 0.9956 | 0.9957 | 0.9959 | 0.9960 | 0.9961 | 0.9962 | 0.9963 | 0.9964 |
| 2.7 | 0.9965 | 0.9966 | 0.9967 | 0.9968 | 0.9969 | 0.9970 | 0.9971 | 0.9972 | 0.9973 | 0.9974 |
| 2.8 | 0.9974 | 0.9975 | 0.9976 | 0.9977 | 0.9977 | 0.9978 | 0.9979 | 0.9979 | 0.9980 | 0.9981 |
| 2.9 | 0.9981 | 0.9982 | 0.9982 | 0.9983 | 0.9984 | 0.9984 | 0.9985 | 0.9985 | 0.9986 | 0.9986 |
| 3. | 0.9987 | 0.9990 | 0.9993 | 0.9995 | 0.9997 | 0.9998 | 0.9998 | 0.9999 | 0.9999 | 1.0000 |

SOURCE: B. W. Lindgren, *Statistical Theory* (New York: Macmillan, 1962), pp. 392–393.

## APPENDIX TABLE A.2  Upper Percentiles of Student $t$ Distributions

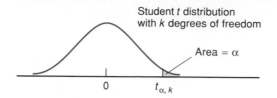

Student $t$ distribution
with $k$ degrees of freedom

Area = $\alpha$

0    $t_{\alpha, k}$

| df | $\alpha$ 0.20 | 0.15 | 0.10 | 0.05 | 0.025 | 0.01 | 0.005 |
|----|------|------|------|------|-------|------|-------|
| 1  | 1.376 | 1.963 | 3.078 | 6.3138 | 12.706 | 31.821 | 63.657 |
| 2  | 1.061 | 1.386 | 1.886 | 2.9200 | 4.3027 | 6.965 | 9.9248 |
| 3  | 0.978 | 1.250 | 1.638 | 2.3534 | 3.1825 | 4.541 | 5.8409 |
| 4  | 0.941 | 1.190 | 1.533 | 2.1318 | 2.7764 | 3.747 | 4.6041 |
| 5  | 0.920 | 1.156 | 1.476 | 2.0150 | 2.5706 | 3.365 | 4.0321 |
| 6  | 0.906 | 1.134 | 1.440 | 1.9432 | 2.4469 | 3.143 | 3.7074 |
| 7  | 0.896 | 1.119 | 1.415 | 1.8946 | 2.3646 | 2.998 | 3.4995 |
| 8  | 0.889 | 1.108 | 1.397 | 1.8595 | 2.3060 | 2.896 | 3.3554 |
| 9  | 0.883 | 1.100 | 1.383 | 1.8331 | 2.2622 | 2.821 | 3.2498 |
| 10 | 0.879 | 1.093 | 1.372 | 1.8125 | 2.2281 | 2.764 | 3.1693 |
| 11 | 0.876 | 1.088 | 1.363 | 1.7959 | 2.2010 | 2.718 | 3.1058 |
| 12 | 0.873 | 1.083 | 1.356 | 1.7823 | 2.1788 | 2.681 | 3.0545 |
| 13 | 0.870 | 1.079 | 1.350 | 1.7709 | 2.1604 | 2.650 | 3.0123 |
| 14 | 0.868 | 1.076 | 1.345 | 1.7613 | 2.1448 | 2.624 | 2.9768 |
| 15 | 0.866 | 1.074 | 1.341 | 1.7530 | 2.1315 | 2.602 | 2.9467 |
| 16 | 0.865 | 1.071 | 1.337 | 1.7459 | 2.1199 | 2.583 | 2.9208 |
| 17 | 0.863 | 1.069 | 1.333 | 1.7396 | 2.1098 | 2.567 | 2.8982 |
| 18 | 0.862 | 1.067 | 1.330 | 1.7341 | 2.1009 | 2.552 | 2.8784 |
| 19 | 0.861 | 1.066 | 1.328 | 1.7291 | 2.0930 | 2.539 | 2.8609 |
| 20 | 0.860 | 1.064 | 1.325 | 1.7247 | 2.0860 | 2.528 | 2.8453 |
| 21 | 0.859 | 1.063 | 1.323 | 1.7207 | 2.0796 | 2.518 | 2.8314 |
| 22 | 0.858 | 1.061 | 1.321 | 1.7171 | 2.0739 | 2.508 | 2.8188 |
| 23 | 0.858 | 1.060 | 1.319 | 1.7139 | 2.0687 | 2.500 | 2.8073 |
| 24 | 0.857 | 1.059 | 1.318 | 1.7109 | 2.0639 | 2.492 | 2.7969 |
| 25 | 0.856 | 1.058 | 1.316 | 1.7081 | 2.0595 | 2.485 | 2.7874 |
| 26 | 0.856 | 1.058 | 1.315 | 1.7056 | 2.0555 | 2.479 | 2.7787 |
| 27 | 0.855 | 1.057 | 1.314 | 1.7033 | 2.0518 | 2.473 | 2.7707 |
| 28 | 0.855 | 1.056 | 1.313 | 1.7011 | 2.0484 | 2.467 | 2.7633 |
| 29 | 0.854 | 1.055 | 1.311 | 1.6991 | 2.0452 | 2.462 | 2.7564 |
| 30 | 0.854 | 1.055 | 1.310 | 1.6973 | 2.0423 | 2.457 | 2.7500 |
| 31 | 0.8535 | 1.0541 | 1.3095 | 1.6955 | 2.0395 | 2.453 | 2.7441 |
| 32 | 0.8531 | 1.0536 | 1.3086 | 1.6939 | 2.0370 | 2.449 | 2.7385 |
| 33 | 0.8527 | 1.0531 | 1.3078 | 1.6924 | 2.0345 | 2.445 | 2.7333 |
| 34 | 0.8524 | 1.0526 | 1.3070 | 1.6909 | 2.0323 | 2.441 | 2.7284 |
| 35 | 0.8521 | 1.0521 | 1.3062 | 1.6896 | 2.0301 | 2.438 | 2.7239 |
| 36 | 0.8518 | 1.0516 | 1.3055 | 1.6883 | 2.0281 | 2.434 | 2.7195 |
| 37 | 0.8515 | 1.0512 | 1.3049 | 1.6871 | 2.0262 | 2.431 | 2.7155 |
| 38 | 0.8512 | 1.0508 | 1.3042 | 1.6860 | 2.0244 | 2.428 | 2.7116 |
| 39 | 0.8510 | 1.0504 | 1.3037 | 1.6849 | 2.0227 | 2.426 | 2.7079 |
| 40 | 0.8507 | 1.0501 | 1.3031 | 1.6839 | 2.0211 | 2.423 | 2.7045 |
| 41 | 0.8505 | 1.0498 | 1.3026 | 1.6829 | 2.0196 | 2.421 | 2.7012 |
| 42 | 0.8503 | 1.0494 | 1.3020 | 1.6820 | 2.0181 | 2.418 | 2.6981 |
| 43 | 0.8501 | 1.0491 | 1.3016 | 1.6811 | 2.0167 | 2.416 | 2.6952 |
| 44 | 0.8499 | 1.0488 | 1.3011 | 1.6802 | 2.0154 | 2.414 | 2.6923 |

**APPENDIX TABLE A.2   Upper Percentiles of Student $t$ Distributions (*Continued*)**

| df | $\alpha$ 0.20 | 0.15 | 0.10 | 0.05 | 0.025 | 0.01 | 0.005 |
|----|------|------|------|------|-------|------|-------|
| 45 | 0.8497 | 1.0485 | 1.3007 | 1.6794 | 2.0141 | 2.412 | 2.6896 |
| 46 | 0.8495 | 1.0483 | 1.3002 | 1.6787 | 2.0129 | 2.410 | 2.6870 |
| 47 | 0.8494 | 1.0480 | 1.2998 | 1.6779 | 2.0118 | 2.408 | 2.6846 |
| 48 | 0.8492 | 1.0478 | 1.2994 | 1.6772 | 2.0106 | 2.406 | 2.6822 |
| 49 | 0.8490 | 1.0476 | 1.2991 | 1.6766 | 2.0096 | 2.405 | 2.6800 |
| 50 | 0.8489 | 1.0473 | 1.2987 | 1.6759 | 2.0086 | 2.403 | 2.6778 |
| 51 | 0.8448 | 1.0471 | 1.2984 | 1.6753 | 2.0077 | 2.402 | 2.6758 |
| 52 | 0.8486 | 1.0469 | 1.2981 | 1.6747 | 2.0067 | 2.400 | 2.6738 |
| 53 | 0.8485 | 1.0467 | 1.2978 | 1.6742 | 2.0058 | 2.399 | 2.6719 |
| 54 | 0.8484 | 1.0465 | 1.2975 | 1.6736 | 2.0049 | 2.397 | 2.6700 |
| 55 | 0.8483 | 1.0463 | 1.2972 | 1.6731 | 2.0041 | 2.396 | 2.6683 |
| 56 | 0.8481 | 1.0461 | 1.2969 | 1.6725 | 2.0033 | 2.395 | 2.6666 |
| 57 | 0.8480 | 1.0460 | 1.2967 | 1.6721 | 2.0025 | 2.393 | 2.6650 |
| 58 | 0.8479 | 1.0458 | 1.2964 | 1.6716 | 2.0017 | 2.392 | 2.6633 |
| 59 | 0.8478 | 1.0457 | 1.2962 | 1.6712 | 2.0010 | 2.391 | 2.6618 |
| 60 | 0.8477 | 1.0455 | 1.2959 | 1.6707 | 2.0003 | 2.390 | 2.6603 |
| 61 | 0.8476 | 1.0454 | 1.2957 | 1.6703 | 1.9997 | 2.389 | 2.6590 |
| 62 | 0.8475 | 1.0452 | 1.2954 | 1.6698 | 1.9990 | 2.388 | 2.6576 |
| 63 | 0.8474 | 1.0451 | 1.2952 | 1.6694 | 1.9984 | 2.387 | 2.6563 |
| 64 | 0.8473 | 1.0449 | 1.2950 | 1.6690 | 1.9977 | 2.386 | 2.6549 |
| 65 | 0.8472 | 1.0448 | 1.2948 | 1.6687 | 1.9972 | 2.385 | 2.6537 |
| 66 | 0.8471 | 1.0447 | 1.2945 | 1.6683 | 1.9966 | 2.384 | 2.6525 |
| 67 | 0.8471 | 1.0446 | 1.2944 | 1.6680 | 1.9961 | 2.383 | 2.6513 |
| 68 | 0.8470 | 1.0444 | 1.2942 | 1.6676 | 1.9955 | 2.382 | 2.6501 |
| 69 | 0.8469 | 1.0443 | 1.2940 | 1.6673 | 1.9950 | 2.381 | 2.6491 |
| 70 | 0.8468 | 1.0442 | 1.2938 | 1.6669 | 1.9945 | 2.381 | 2.6480 |
| 71 | 0.8468 | 1.0441 | 1.2936 | 1.6666 | 1.9940 | 2.380 | 2.6470 |
| 72 | 0.8467 | 1.0440 | 1.2934 | 1.6663 | 1.9935 | 2.379 | 2.6459 |
| 73 | 0.8466 | 1.0439 | 1.2933 | 1.6660 | 1.9931 | 2.378 | 2.6450 |
| 74 | 0.8465 | 1.0438 | 1.2931 | 1.6657 | 1.9926 | 2.378 | 2.6640 |
| 75 | 0.8465 | 1.0437 | 1.2930 | 1.6655 | 1.9922 | 2.377 | 2.6431 |
| 76 | 0.8464 | 1.0436 | 1.2928 | 1.6652 | 1.9917 | 2.376 | 2.6421 |
| 77 | 0.8464 | 1.0435 | 1.2927 | 1.6649 | 1.9913 | 2.376 | 2.6413 |
| 78 | 0.8463 | 1.0434 | 1.2925 | 1.6646 | 1.9909 | 2.375 | 2.6406 |
| 79 | 0.8463 | 1.0433 | 1.2924 | 1.6644 | 1.9905 | 2.374 | 2.6396 |
| 80 | 0.8462 | 1.0432 | 1.2922 | 1.6641 | 1.9901 | 2.374 | 2.6388 |
| 81 | 0.8461 | 1.0431 | 1.2921 | 1.6639 | 1.9897 | 2.373 | 2.6380 |
| 82 | 0.8460 | 1.0430 | 1.2920 | 1.6637 | 1.9893 | 2.372 | 2.6372 |
| 83 | 0.8460 | 1.0430 | 1.2919 | 1.6635 | 1.9890 | 2.372 | 2.6365 |
| 84 | 0.8459 | 1.0429 | 1.2917 | 1.6632 | 1.9886 | 2.371 | 2.6357 |
| 85 | 0.8459 | 1.0428 | 1.2916 | 1.6630 | 1.9883 | 2.371 | 2.6350 |
| 86 | 0.8458 | 1.0427 | 1.2915 | 1.6628 | 1.9880 | 2.370 | 2.6343 |
| 87 | 0.8458 | 1.0427 | 1.2914 | 1.6626 | 1.9877 | 2.370 | 2.6336 |
| 88 | 0.8457 | 1.0426 | 1.2913 | 1.6624 | 1.9873 | 2.369 | 2.6329 |
| 89 | 0.8457 | 1.0426 | 1.2912 | 1.6622 | 1.9870 | 2.369 | 2.6323 |
| 90 | 0.8457 | 1.0425 | 1.2910 | 1.6620 | 1.9867 | 2.368 | 2.6316 |
| 91 | 0.8457 | 1.0424 | 1.2909 | 1.6618 | 1.9864 | 2.368 | 2.6310 |
| 92 | 0.8456 | 1.0423 | 1.2908 | 1.6616 | 1.9861 | 2.367 | 2.6303 |
| 93 | 0.8456 | 1.0423 | 1.2907 | 1.6614 | 1.9859 | 2.367 | 2.6298 |
| 94 | 0.8455 | 1.0422 | 1.2906 | 1.6612 | 1.9856 | 2.366 | 2.6292 |
| 95 | 0.8455 | 1.0422 | 1.2905 | 1.6611 | 1.9853 | 2.366 | 2.6286 |
| 96 | 0.8454 | 1.0421 | 1.2904 | 1.6609 | 1.9850 | 2.366 | 2.6280 |
| 97 | 0.8454 | 1.0421 | 1.2904 | 1.6608 | 1.9848 | 2.365 | 2.6275 |
| 98 | 0.8453 | 1.0420 | 1.2903 | 1.6606 | 1.9845 | 2.365 | 2.6270 |
| 99 | 0.8453 | 1.0419 | 1.2902 | 1.6604 | 1.9843 | 2.364 | 2.6265 |
| 100 | 0.8452 | 1.0418 | 1.2901 | 1.6602 | 1.9840 | 2.364 | 2.6260 |
| $\infty$ | 0.84 | 1.04 | 1.28 | 1.64 | 1.96 | 2.33 | 2.58 |

SOURCE: *Scientific Tables*, 6th ed. (Basel, Switzerland: J. R. Geigy, 1962), pp. 32–33.

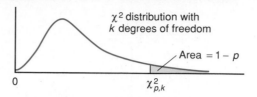

| $p$<br>df | 0.010 | 0.025 | 0.050 | 0.10 | 0.90 | 0.95 | 0.975 | 0.99 |
|---|---|---|---|---|---|---|---|---|
| 1 | 0.000157 | 0.000982 | 0.00393 | 0.0158 | 2.706 | 3.841 | 5.024 | 6.635 |
| 2 | 0.0201 | 0.0506 | 0.103 | 0.211 | 4.605 | 5.991 | 7.378 | 9.210 |
| 3 | 0.115 | 0.216 | 0.352 | 0.584 | 6.251 | 7.815 | 9.348 | 11.345 |
| 4 | 0.297 | 0.484 | 0.711 | 1.064 | 7.779 | 9.488 | 11.143 | 13.277 |
| 5 | 0.554 | 0.831 | 1.145 | 1.610 | 9.236 | 11.070 | 12.832 | 15.086 |
| 6 | 0.872 | 1.237 | 1.635 | 2.204 | 10.645 | 12.592 | 14.449 | 16.812 |
| 7 | 1.239 | 1.690 | 2.167 | 2.833 | 12.017 | 14.067 | 16.013 | 18.475 |
| 8 | 1.646 | 2.180 | 2.733 | 3.490 | 13.362 | 15.507 | 17.535 | 20.090 |
| 9 | 2.088 | 2.700 | 3.325 | 4.168 | 14.684 | 16.919 | 19.023 | 21.666 |
| 10 | 2.558 | 3.247 | 3.940 | 4.865 | 15.987 | 18.307 | 20.483 | 23.209 |
| 11 | 3.053 | 3.816 | 4.575 | 5.578 | 17.275 | 19.675 | 21.920 | 24.725 |
| 12 | 3.571 | 4.404 | 5.226 | 6.304 | 18.549 | 21.026 | 23.336 | 26.217 |
| 13 | 4.107 | 5.009 | 5.892 | 7.042 | 19.812 | 22.362 | 24.736 | 27.688 |
| 14 | 4.660 | 5.629 | 6.571 | 7.790 | 21.064 | 23.685 | 26.119 | 29.141 |
| 15 | 5.229 | 6.262 | 7.261 | 8.547 | 22.307 | 24.996 | 27.488 | 30.578 |
| 16 | 5.812 | 6.908 | 7.962 | 9.312 | 23.542 | 26.296 | 28.845 | 32.000 |
| 17 | 6.408 | 7.564 | 8.672 | 10.085 | 24.769 | 27.587 | 30.191 | 33.409 |
| 18 | 7.015 | 8.231 | 9.390 | 10.865 | 25.989 | 28.869 | 31.526 | 34.805 |
| 19 | 7.633 | 8.907 | 10.117 | 11.651 | 27.204 | 30.144 | 32.852 | 36.191 |
| 20 | 8.260 | 9.591 | 10.851 | 12.443 | 28.412 | 31.410 | 34.170 | 37.566 |
| 21 | 8.897 | 10.283 | 11.591 | 13.240 | 29.615 | 32.671 | 35.479 | 38.932 |
| 22 | 9.542 | 10.982 | 12.338 | 14.041 | 30.813 | 33.924 | 36.781 | 40.289 |
| 23 | 10.196 | 11.688 | 13.091 | 14.848 | 32.007 | 35.172 | 38.076 | 41.638 |
| 24 | 10.856 | 12.401 | 13.848 | 15.659 | 33.196 | 36.415 | 39.364 | 42.980 |
| 25 | 11.524 | 13.120 | 14.611 | 16.473 | 34.382 | 37.652 | 40.646 | 44.314 |
| 26 | 12.198 | 13.844 | 15.379 | 17.292 | 35.563 | 38.885 | 41.923 | 45.642 |
| 27 | 12.879 | 14.573 | 16.151 | 18.114 | 36.741 | 40.113 | 43.194 | 46.963 |
| 28 | 13.565 | 15.308 | 16.928 | 18.939 | 37.916 | 41.337 | 44.461 | 48.278 |
| 29 | 14.256 | 16.047 | 17.708 | 19.768 | 39.087 | 42.557 | 45.722 | 49.588 |
| 30 | 14.953 | 16.791 | 18.493 | 20.599 | 40.256 | 43.773 | 46.979 | 50.892 |
| 31 | 15.655 | 17.539 | 19.281 | 21.434 | 41.422 | 44.985 | 48.232 | 52.191 |
| 32 | 16.362 | 18.291 | 20.072 | 22.271 | 42.585 | 46.194 | 49.480 | 53.486 |
| 33 | 17.073 | 19.047 | 20.867 | 23.110 | 43.745 | 47.400 | 50.725 | 54.776 |
| 34 | 17.789 | 19.806 | 21.664 | 23.952 | 44.903 | 48.602 | 51.966 | 56.061 |
| 35 | 18.509 | 20.569 | 22.465 | 24.797 | 46.059 | 49.802 | 53.203 | 57.342 |
| 36 | 19.233 | 21.336 | 23.269 | 25.643 | 47.212 | 50.998 | 54.437 | 58.619 |
| 37 | 19.960 | 22.106 | 24.075 | 26.492 | 48.363 | 52.192 | 55.668 | 59.892 |
| 38 | 20.691 | 22.878 | 24.884 | 27.343 | 49.513 | 53.384 | 56.895 | 61.162 |
| 39 | 21.426 | 23.654 | 25.695 | 28.196 | 50.660 | 54.572 | 58.120 | 62.428 |
| 40 | 22.164 | 24.433 | 26.509 | 29.051 | 51.805 | 55.758 | 59.342 | 63.691 |
| 41 | 22.906 | 25.215 | 27.326 | 29.907 | 52.949 | 56.942 | 60.561 | 64.950 |
| 42 | 23.650 | 25.999 | 28.144 | 30.765 | 54.090 | 58.124 | 61.777 | 66.206 |
| 43 | 24.398 | 26.785 | 28.965 | 31.625 | 55.230 | 59.304 | 62.990 | 67.459 |
| 44 | 25.148 | 27.575 | 29.787 | 32.487 | 56.369 | 60.481 | 64.201 | 68.709 |
| 45 | 25.901 | 28.366 | 30.612 | 33.350 | 57.505 | 61.656 | 65.410 | 69.957 |
| 46 | 26.657 | 29.160 | 31.439 | 34.215 | 58.641 | 62.830 | 66.617 | 71.201 |
| 47 | 27.416 | 29.956 | 32.268 | 35.081 | 59.774 | 64.001 | 67.821 | 72.443 |
| 48 | 28.177 | 30.755 | 33.098 | 35.949 | 60.907 | 65.171 | 69.023 | 73.683 |
| 49 | 28.941 | 31.555 | 33.930 | 36.818 | 62.038 | 66.339 | 70.222 | 74.919 |
| 50 | 29.707 | 32.357 | 34.764 | 37.689 | 63.167 | 67.505 | 71.420 | 76.154 |

SOURCE: *Scientific Tables*, 6th ed. (Basel, Switzerland: J. R. Geigy, 1962), p. 36.

### TABLE A.4 Percentiles of the *F*-Distribution with *m* and *n* Degrees of Freedom

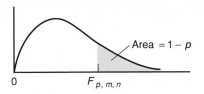

$p = 0.90$

| | | NUMERATOR DEGREES OF FREEDOM | | | | | | | |
|---|---|---|---|---|---|---|---|---|---|
| **m** **n** | 1 | 2 | 3 | 4 | 5 | 6 | 7 | 8 | 9 |
| 1 | 39.86 | 49.50 | 53.59 | 55.83 | 57.24 | 58.20 | 58.91 | 59.44 | 59.86 |
| 2 | 8.53 | 9.00 | 9.16 | 9.24 | 9.29 | 9.33 | 9.35 | 9.37 | 9.38 |
| 3 | 5.54 | 5.46 | 5.39 | 5.34 | 5.31 | 5.28 | 5.27 | 5.25 | 5.24 |
| 4 | 4.54 | 4.32 | 4.19 | 4.11 | 4.05 | 4.01 | 3.98 | 3.95 | 3.94 |
| 5 | 4.06 | 3.78 | 3.62 | 3.52 | 3.45 | 3.40 | 3.37 | 3.34 | 3.32 |
| 6 | 3.78 | 3.46 | 3.29 | 3.18 | 3.11 | 3.05 | 3.01 | 2.98 | 2.96 |
| 7 | 3.59 | 3.26 | 3.07 | 2.96 | 2.88 | 2.83 | 2.78 | 2.75 | 2.72 |
| 8 | 3.46 | 3.11 | 2.92 | 2.81 | 2.73 | 2.67 | 2.62 | 2.59 | 2.56 |
| 9 | 3.36 | 3.01 | 2.81 | 2.69 | 2.61 | 2.55 | 2.51 | 2.47 | 2.44 |
| 10 | 3.29 | 2.92 | 2.73 | 2.61 | 2.52 | 2.46 | 2.41 | 2.38 | 2.35 |
| 11 | 3.23 | 2.86 | 2.66 | 2.54 | 2.45 | 2.39 | 2.34 | 2.30 | 2.27 |
| 12 | 3.18 | 2.81 | 2.61 | 2.48 | 2.39 | 2.33 | 2.28 | 2.24 | 2.21 |
| 13 | 3.14 | 2.76 | 2.56 | 2.43 | 2.35 | 2.28 | 2.23 | 2.20 | 2.16 |
| 14 | 3.10 | 2.73 | 2.52 | 2.39 | 2.31 | 2.24 | 2.19 | 2.15 | 2.12 |
| 15 | 3.07 | 2.70 | 2.49 | 2.36 | 2.27 | 2.21 | 2.16 | 2.12 | 2.09 |
| 16 | 3.05 | 2.67 | 2.46 | 2.33 | 2.24 | 2.18 | 2.13 | 2.09 | 2.06 |
| 17 | 3.03 | 2.64 | 2.44 | 2.31 | 2.22 | 2.15 | 2.10 | 2.06 | 2.03 |
| 18 | 3.01 | 2.62 | 2.42 | 2.29 | 2.20 | 2.13 | 2.08 | 2.04 | 2.00 |
| 19 | 2.99 | 2.61 | 2.40 | 2.27 | 2.18 | 2.11 | 2.06 | 2.02 | 1.98 |
| 20 | 2.97 | 2.59 | 2.38 | 2.25 | 2.16 | 2.09 | 2.04 | 2.00 | 1.96 |
| 21 | 2.96 | 2.57 | 2.36 | 2.23 | 2.14 | 2.08 | 2.02 | 1.98 | 1.95 |
| 22 | 2.95 | 2.56 | 2.35 | 2.22 | 2.13 | 2.06 | 2.01 | 1.97 | 1.93 |
| 23 | 2.94 | 2.55 | 2.34 | 2.21 | 2.11 | 2.05 | 1.99 | 1.95 | 1.92 |
| 24 | 2.93 | 2.54 | 2.33 | 2.19 | 2.10 | 2.04 | 1.98 | 1.94 | 1.91 |
| 25 | 2.92 | 2.53 | 2.32 | 2.18 | 2.09 | 2.02 | 1.97 | 1.93 | 1.89 |
| 26 | 2.91 | 2.52 | 2.31 | 2.17 | 2.08 | 2.01 | 1.96 | 1.92 | 1.88 |
| 27 | 2.90 | 2.51 | 2.30 | 2.17 | 2.07 | 2.00 | 1.95 | 1.91 | 1.87 |
| 28 | 2.89 | 2.50 | 2.29 | 2.16 | 2.06 | 2.00 | 1.94 | 1.90 | 1.87 |
| 29 | 2.89 | 2.50 | 2.28 | 2.15 | 2.06 | 1.99 | 1.93 | 1.89 | 1.86 |
| 30 | 2.88 | 2.49 | 2.28 | 2.14 | 2.05 | 1.98 | 1.93 | 1.88 | 1.85 |
| 40 | 2.84 | 2.44 | 2.23 | 2.09 | 2.00 | 1.93 | 1.87 | 1.83 | 1.79 |
| 60 | 2.79 | 2.39 | 2.18 | 2.04 | 1.95 | 1.87 | 1.82 | 1.77 | 1.74 |
| 120 | 2.75 | 2.35 | 2.13 | 1.99 | 1.90 | 1.82 | 1.77 | 1.72 | 1.68 |
| ∞ | 2.71 | 2.30 | 2.08 | 1.94 | 1.85 | 1.77 | 1.72 | 1.67 | 1.63 |

DENOMINATOR DEGREES OF FREEDOM

$p = 0.90$

| | | | | NUMERATOR DEGREES OF FREEDOM | | | | | | |
|---|---|---|---|---|---|---|---|---|---|---|
| *n* | 10 | 12 | 15 | 20 | 24 | 30 | 40 | 60 | 120 | ∞ |
| 1 | 60.19 | 60.71 | 61.22 | 61.74 | 62.00 | 62.26 | 62.53 | 62.79 | 63.06 | 63.33 |
| 2 | 9.39 | 9.41 | 9.42 | 9.44 | 9.45 | 9.46 | 9.47 | 9.47 | 9.48 | 9.49 |
| 3 | 5.23 | 5.22 | 5.20 | 5.18 | 5.18 | 5.17 | 5.16 | 5.15 | 5.14 | 5.13 |
| 4 | 3.92 | 3.90 | 3.87 | 3.84 | 3.83 | 3.82 | 3.80 | 3.79 | 3.78 | 3.76 |
| 5 | 3.30 | 3.27 | 3.24 | 3.21 | 3.19 | 3.17 | 3.16 | 3.14 | 3.12 | 3.10 |
| 6 | 2.94 | 2.90 | 2.87 | 2.84 | 2.82 | 2.80 | 2.78 | 2.76 | 2.74 | 2.72 |
| 7 | 2.70 | 2.67 | 2.63 | 2.59 | 2.58 | 2.56 | 2.54 | 2.51 | 2.49 | 2.47 |
| 8 | 2.54 | 2.50 | 2.46 | 2.42 | 2.40 | 2.38 | 2.36 | 2.34 | 2.32 | 2.29 |
| 9 | 2.42 | 2.38 | 2.34 | 2.30 | 2.28 | 2.25 | 2.23 | 2.21 | 2.18 | 2.16 |
| 10 | 2.32 | 2.28 | 2.24 | 2.20 | 2.18 | 2.16 | 2.13 | 2.11 | 2.08 | 2.06 |
| 11 | 2.25 | 2.21 | 2.17 | 2.12 | 2.10 | 2.08 | 2.05 | 2.03 | 2.00 | 1.97 |
| 12 | 2.19 | 2.15 | 2.10 | 2.06 | 2.04 | 2.01 | 1.99 | 1.96 | 1.93 | 1.90 |
| 13 | 2.14 | 2.10 | 2.05 | 2.01 | 1.98 | 1.96 | 1.93 | 1.90 | 1.88 | 1.85 |
| 14 | 2.10 | 2.05 | 2.01 | 1.96 | 1.94 | 1.91 | 1.89 | 1.86 | 1.83 | 1.80 |
| 15 | 2.06 | 2.02 | 1.97 | 1.92 | 1.90 | 1.87 | 1.85 | 1.82 | 1.79 | 1.76 |
| 16 | 2.03 | 1.99 | 1.94 | 1.89 | 1.87 | 1.84 | 1.81 | 1.78 | 1.75 | 1.72 |
| 17 | 2.00 | 1.96 | 1.91 | 1.86 | 1.84 | 1.81 | 1.78 | 1.75 | 1.72 | 1.69 |
| 18 | 1.98 | 1.93 | 1.89 | 1.84 | 1.81 | 1.78 | 1.75 | 1.72 | 1.69 | 1.66 |
| 19 | 1.96 | 1.91 | 1.86 | 1.81 | 1.79 | 1.76 | 1.73 | 1.70 | 1.67 | 1.63 |
| 20 | 1.94 | 1.89 | 1.84 | 1.79 | 1.77 | 1.74 | 1.71 | 1.68 | 1.64 | 1.61 |
| 21 | 1.92 | 1.87 | 1.83 | 1.78 | 1.75 | 1.72 | 1.69 | 1.66 | 1.62 | 1.59 |
| 22 | 1.90 | 1.86 | 1.81 | 1.76 | 1.73 | 1.70 | 1.67 | 1.64 | 1.60 | 1.57 |
| 23 | 1.89 | 1.84 | 1.80 | 1.74 | 1.72 | 1.69 | 1.66 | 1.62 | 1.59 | 1.55 |
| 24 | 1.88 | 1.83 | 1.78 | 1.73 | 1.70 | 1.67 | 1.64 | 1.61 | 1.57 | 1.53 |
| 25 | 1.87 | 1.82 | 1.77 | 1.72 | 1.69 | 1.66 | 1.63 | 1.59 | 1.56 | 1.52 |
| 26 | 1.86 | 1.81 | 1.76 | 1.71 | 1.68 | 1.65 | 1.61 | 1.58 | 1.54 | 1.50 |
| 27 | 1.85 | 1.80 | 1.75 | 1.70 | 1.67 | 1.64 | 1.60 | 1.57 | 1.53 | 1.49 |
| 28 | 1.84 | 1.79 | 1.74 | 1.69 | 1.66 | 1.63 | 1.59 | 1.56 | 1.52 | 1.48 |
| 29 | 1.83 | 1.78 | 1.73 | 1.68 | 1.65 | 1.62 | 1.58 | 1.55 | 1.51 | 1.47 |
| 30 | 1.82 | 1.77 | 1.72 | 1.67 | 1.64 | 1.61 | 1.57 | 1.54 | 1.50 | 1.46 |
| 40 | 1.76 | 1.71 | 1.66 | 1.61 | 1.57 | 1.54 | 1.51 | 1.47 | 1.42 | 1.38 |
| 60 | 1.71 | 1.66 | 1.60 | 1.54 | 1.51 | 1.48 | 1.44 | 1.40 | 1.35 | 1.29 |
| 120 | 1.65 | 1.60 | 1.55 | 1.48 | 1.45 | 1.41 | 1.37 | 1.32 | 1.26 | 1.19 |
| ∞ | 1.60 | 1.55 | 1.49 | 1.42 | 1.38 | 1.34 | 1.30 | 1.24 | 1.17 | 1.00 |

DENOMINATOR DEGREES OF FREEDOM

SOURCE: From M. Merrington and C. M. Thompson, "Tables of Percentage Points of the Inverted Beta (*F*)-Distribution," *Biomertrika*, 1943, 33, 73–88. Reproduced by permission of the *Biometrika* trustees.

TABLE A.4 Percentiles of the *F*-Distribution with *m* and *n* Degrees of Freedom (*Continued*)

$p = 0.95$

|  | NUMERATOR DEGREES OF FREEDOM | | | | | | | | |
|---|---|---|---|---|---|---|---|---|---|
| *n* \ *m* | 1 | 2 | 3 | 4 | 5 | 6 | 7 | 8 | 9 |
| 1 | 161.4 | 199.5 | 215.7 | 224.6 | 230.2 | 234.0 | 236.8 | 238.9 | 240.5 |
| 2 | 18.51 | 19.00 | 19.16 | 19.25 | 19.30 | 19.33 | 19.35 | 19.37 | 19.38 |
| 3 | 10.13 | 9.55 | 9.28 | 9.12 | 9.01 | 8.94 | 8.89 | 8.85 | 8.81 |
| 4 | 7.71 | 6.94 | 6.59 | 6.39 | 6.26 | 6.16 | 6.09 | 6.04 | 6.00 |
| 5 | 6.61 | 5.79 | 5.41 | 5.19 | 5.05 | 4.95 | 4.88 | 4.82 | 4.77 |
| 6 | 5.99 | 5.14 | 4.76 | 4.53 | 4.39 | 4.28 | 4.21 | 4.15 | 4.10 |
| 7 | 5.59 | 4.74 | 4.35 | 4.12 | 3.97 | 3.87 | 3.79 | 3.73 | 3.68 |
| 8 | 5.32 | 4.46 | 4.07 | 3.84 | 3.69 | 3.58 | 3.50 | 3.44 | 3.39 |
| 9 | 5.12 | 4.26 | 3.86 | 3.63 | 3.48 | 3.37 | 3.29 | 3.23 | 3.18 |
| 10 | 4.96 | 4.10 | 3.71 | 3.48 | 3.33 | 3.22 | 3.14 | 3.07 | 3.02 |
| 11 | 4.84 | 3.98 | 3.59 | 3.36 | 3.20 | 3.09 | 3.01 | 2.95 | 2.90 |
| 12 | 4.75 | 3.89 | 3.49 | 3.26 | 3.11 | 3.00 | 2.91 | 2.85 | 2.80 |
| 13 | 4.67 | 3.81 | 3.41 | 3.18 | 3.03 | 2.92 | 2.83 | 2.77 | 2.71 |
| 14 | 4.60 | 3.74 | 3.34 | 3.11 | 2.96 | 2.85 | 2.76 | 2.70 | 2.65 |
| 15 | 4.54 | 3.68 | 3.29 | 3.06 | 2.90 | 2.79 | 2.71 | 2.64 | 2.59 |
| 16 | 4.49 | 3.63 | 3.24 | 3.01 | 2.85 | 2.74 | 2.66 | 2.59 | 2.54 |
| 17 | 4.45 | 3.59 | 3.20 | 2.96 | 2.81 | 2.70 | 2.61 | 2.55 | 2.49 |
| 18 | 4.41 | 3.55 | 3.16 | 2.93 | 2.77 | 2.66 | 2.58 | 2.51 | 2.46 |
| 19 | 4.38 | 3.52 | 3.13 | 2.90 | 2.74 | 2.63 | 2.54 | 2.48 | 2.42 |
| 20 | 4.35 | 3.49 | 3.10 | 2.87 | 2.71 | 2.60 | 2.51 | 2.45 | 2.39 |
| 21 | 4.32 | 3.47 | 3.07 | 2.84 | 2.68 | 2.57 | 2.49 | 2.42 | 2.37 |
| 22 | 4.30 | 3.44 | 3.05 | 2.82 | 2.66 | 2.55 | 2.46 | 2.40 | 2.34 |
| 23 | 4.28 | 3.42 | 3.03 | 2.80 | 2.64 | 2.53 | 2.44 | 2.37 | 2.32 |
| 24 | 4.26 | 3.40 | 3.01 | 2.78 | 2.62 | 2.51 | 2.42 | 2.36 | 2.30 |
| 25 | 4.24 | 3.39 | 2.99 | 2.76 | 2.60 | 2.49 | 2.40 | 2.34 | 2.28 |
| 26 | 4.23 | 3.37 | 2.98 | 2.74 | 2.59 | 2.47 | 2.39 | 2.32 | 2.27 |
| 27 | 4.21 | 3.35 | 2.96 | 2.73 | 2.57 | 2.46 | 2.37 | 2.31 | 2.25 |
| 28 | 4.20 | 3.34 | 2.95 | 2.71 | 2.56 | 2.45 | 2.36 | 2.29 | 2.24 |
| 29 | 4.18 | 3.33 | 2.93 | 2.70 | 2.55 | 2.43 | 2.35 | 2.28 | 2.22 |
| 30 | 4.17 | 3.32 | 2.92 | 2.69 | 2.53 | 2.42 | 2.33 | 2.27 | 2.21 |
| 40 | 4.08 | 3.23 | 2.84 | 2.61 | 2.45 | 2.34 | 2.25 | 2.18 | 2.12 |
| 60 | 4.00 | 3.15 | 2.76 | 2.53 | 2.37 | 2.25 | 2.17 | 2.10 | 2.04 |
| 120 | 3.92 | 3.07 | 2.68 | 2.45 | 2.29 | 2.17 | 2.09 | 2.02 | 1.96 |
| ∞ | 3.84 | 3.00 | 2.60 | 2.37 | 2.21 | 2.10 | 2.01 | 1.94 | 1.88 |

DENOMINATOR DEGREES OF FREEDOM

**TABLE A.4  Percentiles of the *F*-Distribution with *m* and *n* Degrees of Freedom (*Continued*)**

$p = 0.95$

| | | | | | | | | | | | |
|---|---|---|---|---|---|---|---|---|---|---|---|
| | *m* | NUMERATOR DEGREES OF FREEDOM | | | | | | | | | |
| *n* | | 10 | 12 | 15 | 20 | 24 | 30 | 40 | 60 | 120 | ∞ |
| 1 | | 241.9 | 243.9 | 245.9 | 248.0 | 249.1 | 250.1 | 251.1 | 252.2 | 253.3 | 254.3 |
| 2 | | 19.40 | 19.41 | 19.43 | 19.45 | 19.45 | 19.46 | 19.47 | 19.48 | 19.49 | 19.50 |
| 3 | | 8.79 | 8.74 | 8.70 | 8.66 | 8.64 | 8.62 | 8.59 | 8.57 | 8.55 | 8.53 |
| 4 | | 5.96 | 5.91 | 5.86 | 5.80 | 5.77 | 5.75 | 5.72 | 5.69 | 5.66 | 5.63 |
| 5 | | 4.74 | 4.68 | 4.62 | 4.56 | 4.53 | 4.50 | 4.46 | 4.43 | 4.40 | 4.36 |
| 6 | | 4.06 | 4.00 | 3.94 | 3.87 | 3.84 | 3.81 | 3.77 | 3.74 | 3.70 | 3.67 |
| 7 | | 3.64 | 3.57 | 3.51 | 3.44 | 3.41 | 3.38 | 3.34 | 3.30 | 3.27 | 3.23 |
| 8 | | 3.35 | 3.28 | 3.22 | 3.15 | 3.12 | 3.08 | 3.04 | 3.01 | 2.97 | 2.93 |
| 9 | | 3.14 | 3.07 | 3.01 | 2.94 | 2.90 | 2.86 | 2.83 | 2.79 | 2.75 | 2.71 |
| 10 | | 2.98 | 2.91 | 2.85 | 2.77 | 2.74 | 2.70 | 2.66 | 2.62 | 2.58 | 2.54 |
| 11 | | 2.85 | 2.79 | 2.72 | 2.65 | 2.61 | 2.57 | 2.53 | 2.49 | 2.45 | 2.40 |
| 12 | | 2.75 | 2.69 | 2.62 | 2.54 | 2.51 | 2.47 | 2.43 | 2.38 | 2.34 | 2.30 |
| 13 | | 2.67 | 2.60 | 2.53 | 2.46 | 2.42 | 2.38 | 2.34 | 2.30 | 2.25 | 2.21 |
| 14 | | 2.60 | 2.53 | 2.46 | 2.39 | 2.35 | 2.31 | 2.27 | 2.22 | 2.18 | 2.13 |
| 15 | | 2.54 | 2.48 | 2.40 | 2.33 | 2.29 | 2.25 | 2.20 | 2.16 | 2.11 | 2.07 |
| 16 | | 2.49 | 2.42 | 2.35 | 2.28 | 2.24 | 2.19 | 2.15 | 2.11 | 2.06 | 2.01 |
| 17 | | 2.45 | 2.38 | 2.31 | 2.23 | 2.19 | 2.15 | 2.10 | 2.06 | 2.01 | 1.96 |
| 18 | | 2.41 | 2.34 | 2.27 | 2.19 | 2.15 | 2.11 | 2.06 | 2.02 | 1.97 | 1.92 |
| 19 | | 2.38 | 2.31 | 2.23 | 2.16 | 2.11 | 2.07 | 2.03 | 1.98 | 1.93 | 1.88 |
| 20 | | 2.35 | 2.28 | 2.20 | 2.12 | 2.08 | 2.04 | 1.99 | 1.95 | 1.90 | 1.84 |
| 21 | | 2.32 | 2.25 | 2.18 | 2.10 | 2.05 | 2.01 | 1.96 | 1.92 | 1.87 | 1.81 |
| 22 | | 2.30 | 2.23 | 2.15 | 2.07 | 2.03 | 1.98 | 1.94 | 1.89 | 1.84 | 1.78 |
| 23 | | 2.27 | 2.20 | 2.13 | 2.05 | 2.01 | 1.96 | 1.91 | 1.86 | 1.81 | 1.76 |
| 24 | | 2.25 | 2.18 | 2.11 | 2.03 | 1.98 | 1.94 | 1.89 | 1.84 | 1.79 | 1.73 |
| 25 | | 2.24 | 2.16 | 2.09 | 2.01 | 1.96 | 1.92 | 1.87 | 1.82 | 1.77 | 1.71 |
| 26 | | 2.22 | 2.15 | 2.07 | 1.99 | 1.95 | 1.90 | 1.85 | 1.80 | 1.75 | 1.69 |
| 27 | | 2.20 | 2.13 | 2.06 | 1.97 | 1.93 | 1.88 | 1.84 | 1.79 | 1.73 | 1.67 |
| 28 | | 2.19 | 2.12 | 2.04 | 1.96 | 1.91 | 1.87 | 1.82 | 1.77 | 1.71 | 1.65 |
| 29 | | 2.18 | 2.10 | 2.03 | 1.94 | 1.90 | 1.85 | 1.81 | 1.75 | 1.70 | 1.64 |
| 30 | | 2.16 | 2.09 | 2.01 | 1.93 | 1.89 | 1.84 | 1.79 | 1.74 | 1.68 | 1.62 |
| 40 | | 2.08 | 2.00 | 1.92 | 1.84 | 1.79 | 1.74 | 1.69 | 1.64 | 1.58 | 1.51 |
| 60 | | 1.99 | 1.92 | 1.84 | 1.75 | 1.70 | 1.65 | 1.59 | 1.53 | 1.47 | 1.39 |
| 120 | | 1.91 | 1.83 | 1.75 | 1.66 | 1.61 | 1.55 | 1.50 | 1.43 | 1.35 | 1.25 |
| ∞ | | 1.83 | 1.75 | 1.67 | 1.57 | 1.52 | 1.46 | 1.39 | 1.32 | 1.22 | 1.00 |

*(Left vertical label: DENOMINATOR DEGREES OF FREEDOM)*

SOURCE: From M. Merrington and C. M. Thompson, "Tables of Percentage Points of the Inverted Beta (*F*)-Distribution," *Biometrika*, 1943, 33, 73—88. Reproduced by permission of the *Biometrika* trustees.

$p = 0.975$

| | | NUMERATOR DEGREES OF FREEDOM | | | | | | | |
|---|---|---|---|---|---|---|---|---|---|
| *n* \ *m* | 1 | 2 | 3 | 4 | 5 | 6 | 7 | 8 | 9 |
| 1 | 647.8 | 799.5 | 864.2 | 899.6 | 921.8 | 937.1 | 948.2 | 956.7 | 963.3 |
| 2 | 38.51 | 39.00 | 39.17 | 39.25 | 39.30 | 39.33 | 39.36 | 39.37 | 39.39 |
| 3 | 17.44 | 16.04 | 15.44 | 15.10 | 14.88 | 14.73 | 14.62 | 14.54 | 14.47 |
| 4 | 12.22 | 10.65 | 9.98 | 9.60 | 9.36 | 9.20 | 9.07 | 8.98 | 8.90 |
| 5 | 10.01 | 8.43 | 7.76 | 7.39 | 7.15 | 6.98 | 6.85 | 6.76 | 6.68 |
| 6 | 8.81 | 7.26 | 6.60 | 6.23 | 5.99 | 5.82 | 5.70 | 5.60 | 5.52 |
| 7 | 8.07 | 6.54 | 5.89 | 5.52 | 5.29 | 5.12 | 4.99 | 4.90 | 4.82 |
| 8 | 7.57 | 6.06 | 5.42 | 5.05 | 4.82 | 4.65 | 4.53 | 4.43 | 4.36 |
| 9 | 7.21 | 5.71 | 5.08 | 4.72 | 4.48 | 4.32 | 4.20 | 4.10 | 4.03 |
| 10 | 6.94 | 5.46 | 4.83 | 4.47 | 4.24 | 4.07 | 3.95 | 3.85 | 3.78 |
| 11 | 6.72 | 5.26 | 4.63 | 4.28 | 4.04 | 3.88 | 3.76 | 3.66 | 3.59 |
| 12 | 6.55 | 5.10 | 4.47 | 4.12 | 3.89 | 3.73 | 3.61 | 3.51 | 3.44 |
| 13 | 6.41 | 4.97 | 4.35 | 4.00 | 3.77 | 3.60 | 3.48 | 3.39 | 3.31 |
| 14 | 6.30 | 4.86 | 4.24 | 3.89 | 3.66 | 3.50 | 3.38 | 3.29 | 3.21 |
| 15 | 6.20 | 4.77 | 4.15 | 3.80 | 3.58 | 3.41 | 3.29 | 3.20 | 3.12 |
| 16 | 6.12 | 4.69 | 4.08 | 3.73 | 3.50 | 3.34 | 3.22 | 3.12 | 3.05 |
| 17 | 6.04 | 4.62 | 4.01 | 3.66 | 3.44 | 3.28 | 3.16 | 3.06 | 2.98 |
| 18 | 5.98 | 4.56 | 3.95 | 3.61 | 3.38 | 3.22 | 3.10 | 3.01 | 2.93 |
| 19 | 5.92 | 4.51 | 3.90 | 3.56 | 3.33 | 3.17 | 3.05 | 2.96 | 2.88 |
| 20 | 5.87 | 4.46 | 3.86 | 3.51 | 3.29 | 3.13 | 3.01 | 2.91 | 2.84 |
| 21 | 5.83 | 4.42 | 3.82 | 3.48 | 3.25 | 3.09 | 2.97 | 2.87 | 2.80 |
| 22 | 5.79 | 4.38 | 3.78 | 3.44 | 3.22 | 3.05 | 2.93 | 2.84 | 2.76 |
| 23 | 5.75 | 4.35 | 3.75 | 3.41 | 3.18 | 3.02 | 2.90 | 2.81 | 2.73 |
| 24 | 5.72 | 4.32 | 3.72 | 3.38 | 3.15 | 2.99 | 2.87 | 2.78 | 2.70 |
| 25 | 5.69 | 4.29 | 3.69 | 3.35 | 3.13 | 2.97 | 2.85 | 2.75 | 2.68 |
| 26 | 5.66 | 4.27 | 3.67 | 3.33 | 3.10 | 2.94 | 2.82 | 2.73 | 2.65 |
| 27 | 5.63 | 4.24 | 3.65 | 3.31 | 3.08 | 2.92 | 2.80 | 2.71 | 2.63 |
| 28 | 5.61 | 4.22 | 3.63 | 3.29 | 3.06 | 2.90 | 2.78 | 2.69 | 2.61 |
| 29 | 5.59 | 4.20 | 3.61 | 3.27 | 3.04 | 2.88 | 2.76 | 2.67 | 2.59 |
| 30 | 5.57 | 4.18 | 3.59 | 3.25 | 3.03 | 2.87 | 2.75 | 2.65 | 2.57 |
| 40 | 5.42 | 4.05 | 3.46 | 3.13 | 2.90 | 2.74 | 2.62 | 2.53 | 2.45 |
| 60 | 5.29 | 3.93 | 3.34 | 3.01 | 2.79 | 2.63 | 2.51 | 2.41 | 2.33 |
| 120 | 5.15 | 3.80 | 3.23 | 2.89 | 2.67 | 2.52 | 2.39 | 2.30 | 2.22 |
| ∞ | 5.02 | 3.69 | 3.12 | 2.79 | 2.57 | 2.41 | 2.29 | 2.19 | 2.11 |

*DENOMINATOR DEGREES OF FREEDOM*

**TABLE A.4  Percentiles of the *F*-Distribution with *m* and *n* Degrees of Freedom (*Continued*)**

$p = 0.975$

| $n$ | \ $m$ | 10 | 12 | 15 | 20 | 24 | 30 | 40 | 60 | 120 | $\infty$ |
|---|---|---|---|---|---|---|---|---|---|---|---|
| | | | | | | NUMERATOR DEGREES OF FREEDOM | | | | | |

| $n$ | 10 | 12 | 15 | 20 | 24 | 30 | 40 | 60 | 120 | $\infty$ |
|---|---|---|---|---|---|---|---|---|---|---|
| 1 | 968.6 | 976.7 | 984.9 | 993.1 | 997.2 | 1001 | 1006 | 1010 | 1014 | 1018 |
| 2 | 39.40 | 39.41 | 39.43 | 39.45 | 39.46 | 39.46 | 39.47 | 39.48 | 39.49 | 39.50 |
| 3 | 14.42 | 14.34 | 14.25 | 14.17 | 14.12 | 14.08 | 14.04 | 13.99 | 13.95 | 13.90 |
| 4 | 8.84 | 8.75 | 8.66 | 8.56 | 8.51 | 8.46 | 8.41 | 8.36 | 8.31 | 8.26 |
| 5 | 6.62 | 6.52 | 6.43 | 6.33 | 6.28 | 6.23 | 6.18 | 6.12 | 6.07 | 6.02 |
| 6 | 5.46 | 5.37 | 5.27 | 5.17 | 5.12 | 5.07 | 5.01 | 4.96 | 4.90 | 4.85 |
| 7 | 4.76 | 4.67 | 4.57 | 4.47 | 4.42 | 4.36 | 4.31 | 4.25 | 4.20 | 4.14 |
| 8 | 4.30 | 4.20 | 4.10 | 4.00 | 3.95 | 3.89 | 3.84 | 3.78 | 3.73 | 3.67 |
| 9 | 3.96 | 3.87 | 3.77 | 3.67 | 3.61 | 3.56 | 3.51 | 3.45 | 3.39 | 3.33 |
| 10 | 3.72 | 3.62 | 3.52 | 3.42 | 3.37 | 3.31 | 3.26 | 3.20 | 3.14 | 3.08 |
| 11 | 3.53 | 3.43 | 3.33 | 3.23 | 3.17 | 3.12 | 3.06 | 3.00 | 2.94 | 2.88 |
| 12 | 3.37 | 3.28 | 3.18 | 3.07 | 3.02 | 2.96 | 2.91 | 2.85 | 2.79 | 2.72 |
| 13 | 3.25 | 3.15 | 3.05 | 2.95 | 2.89 | 2.84 | 2.78 | 2.72 | 2.66 | 2.60 |
| 14 | 3.15 | 3.05 | 2.95 | 2.84 | 2.79 | 2.73 | 2.67 | 2.61 | 2.55 | 2.49 |
| 15 | 3.06 | 2.96 | 2.86 | 2.76 | 2.70 | 2.64 | 2.59 | 2.52 | 2.46 | 2.40 |
| 16 | 2.99 | 2.89 | 2.79 | 2.68 | 2.63 | 2.57 | 2.51 | 2.45 | 2.38 | 2.32 |
| 17 | 2.92 | 2.82 | 2.72 | 2.62 | 2.56 | 2.50 | 2.44 | 2.38 | 2.32 | 2.25 |
| 18 | 2.87 | 2.77 | 2.67 | 2.56 | 2.50 | 2.44 | 2.38 | 2.32 | 2.26 | 2.19 |
| 19 | 2.82 | 2.72 | 2.62 | 2.51 | 2.45 | 2.39 | 2.33 | 2.27 | 2.20 | 2.13 |
| 20 | 2.77 | 2.68 | 2.57 | 2.46 | 2.41 | 2.35 | 2.29 | 2.22 | 2.16 | 2.09 |
| 21 | 2.73 | 2.64 | 2.53 | 2.42 | 2.37 | 2.31 | 2.25 | 2.18 | 2.11 | 2.04 |
| 22 | 2.70 | 2.60 | 2.50 | 2.39 | 2.33 | 2.27 | 2.21 | 2.14 | 2.08 | 2.00 |
| 23 | 2.67 | 2.57 | 2.47 | 2.36 | 2.30 | 2.24 | 2.18 | 2.11 | 2.04 | 1.97 |
| 24 | 2.64 | 2.54 | 2.44 | 2.33 | 2.27 | 2.21 | 2.15 | 2.08 | 2.01 | 1.94 |
| 25 | 2.61 | 2.51 | 2.41 | 2.30 | 2.24 | 2.18 | 2.12 | 2.05 | 1.98 | 1.91 |
| 26 | 2.59 | 2.49 | 2.39 | 2.28 | 2.22 | 2.16 | 2.09 | 2.03 | 1.95 | 1.88 |
| 27 | 2.57 | 2.47 | 2.36 | 2.25 | 2.19 | 2.13 | 2.07 | 2.00 | 1.93 | 1.85 |
| 28 | 2.55 | 2.45 | 2.34 | 2.23 | 2.17 | 2.11 | 2.05 | 1.98 | 1.91 | 1.83 |
| 29 | 2.53 | 2.43 | 2.32 | 2.21 | 2.15 | 2.09 | 2.03 | 1.96 | 1.89 | 1.81 |
| 30 | 2.51 | 2.41 | 2.31 | 2.20 | 2.14 | 2.07 | 2.01 | 1.94 | 1.87 | 1.79 |
| 40 | 2.39 | 2.29 | 2.18 | 2.07 | 2.01 | 1.94 | 1.88 | 1.80 | 1.72 | 1.64 |
| 60 | 2.27 | 2.17 | 2.06 | 1.94 | 1.88 | 1.82 | 1.74 | 1.67 | 1.58 | 1.48 |
| 120 | 2.16 | 2.05 | 1.94 | 1.82 | 1.76 | 1.69 | 1.61 | 1.53 | 1.43 | 1.31 |
| $\infty$ | 2.05 | 1.94 | 1.83 | 1.71 | 1.64 | 1.57 | 1.48 | 1.39 | 1.27 | 1.00 |

DENOMINATOR DEGREES OF FREEDOM

$p = 0.99$

| $n$ | \multicolumn{9}{c}{NUMERATOR DEGREES OF FREEDOM} |
| | 1 | 2 | 3 | 4 | 5 | 6 | 7 | 8 | 9 |
|---|---|---|---|---|---|---|---|---|---|
| 1 | 4,052 | 4,999.5 | 5,403 | 5,625 | 5,764 | 5,859 | 5,928 | 5,982 | 6,022 |
| 2 | 98.50 | 99.00 | 99.17 | 99.25 | 99.30 | 99.33 | 99.36 | 99.37 | 99.39 |
| 3 | 34.12 | 30.82 | 29.46 | 28.71 | 28.24 | 27.91 | 27.67 | 27.49 | 27.35 |
| 4 | 21.20 | 18.00 | 16.69 | 15.98 | 15.52 | 15.21 | 14.98 | 14.80 | 14.66 |
| 5 | 16.26 | 13.27 | 12.06 | 11.39 | 10.97 | 10.67 | 10.46 | 10.29 | 10.16 |
| 6 | 13.75 | 10.92 | 9.78 | 9.15 | 8.75 | 8.47 | 8.26 | 8.10 | 7.98 |
| 7 | 12.25 | 9.55 | 8.45 | 7.85 | 7.46 | 7.19 | 6.99 | 6.84 | 6.72 |
| 8 | 11.26 | 8.65 | 7.59 | 7.01 | 6.63 | 6.37 | 6.18 | 6.03 | 5.91 |
| 9 | 10.56 | 8.02 | 6.99 | 6.42 | 6.06 | 5.80 | 5.61 | 5.47 | 5.35 |
| 10 | 10.04 | 7.56 | 6.55 | 5.99 | 5.64 | 5.39 | 5.20 | 5.06 | 4.94 |
| 11 | 9.65 | 7.21 | 6.22 | 5.67 | 5.32 | 5.07 | 4.89 | 4.74 | 4.63 |
| 12 | 9.33 | 6.93 | 5.95 | 5.41 | 5.06 | 4.82 | 4.64 | 4.50 | 4.39 |
| 13 | 9.07 | 6.70 | 5.74 | 5.21 | 4.86 | 4.62 | 4.44 | 4.30 | 4.19 |
| 14 | 8.86 | 6.51 | 5.56 | 5.04 | 4.69 | 4.46 | 4.28 | 4.14 | 4.03 |
| 15 | 8.68 | 6.36 | 5.42 | 4.89 | 4.56 | 4.32 | 4.14 | 4.00 | 3.89 |
| 16 | 8.53 | 6.23 | 5.29 | 4.77 | 4.44 | 4.20 | 4.03 | 3.89 | 3.78 |
| 17 | 8.40 | 6.11 | 5.18 | 4.67 | 4.34 | 4.10 | 3.93 | 3.79 | 3.68 |
| 18 | 8.29 | 6.01 | 5.09 | 4.58 | 4.25 | 4.01 | 3.84 | 3.71 | 3.60 |
| 19 | 8.18 | 5.93 | 5.01 | 4.50 | 4.17 | 3.94 | 3.77 | 3.63 | 3.52 |
| 20 | 8.10 | 5.85 | 4.94 | 4.43 | 4.10 | 3.87 | 3.70 | 3.56 | 3.46 |
| 21 | 8.02 | 5.78 | 4.87 | 4.37 | 4.04 | 3.81 | 3.64 | 3.51 | 3.40 |
| 22 | 7.95 | 5.72 | 4.82 | 4.31 | 3.99 | 3.76 | 3.59 | 3.45 | 3.35 |
| 23 | 7.88 | 5.66 | 4.76 | 4.26 | 3.94 | 3.71 | 3.54 | 3.41 | 3.30 |
| 24 | 7.82 | 5.61 | 4.72 | 4.22 | 3.90 | 3.67 | 3.50 | 3.36 | 3.26 |
| 25 | 7.77 | 5.57 | 4.68 | 4.18 | 3.85 | 3.63 | 3.46 | 3.32 | 3.22 |
| 26 | 7.72 | 5.53 | 4.64 | 4.14 | 3.82 | 3.59 | 3.42 | 3.29 | 3.18 |
| 27 | 7.68 | 5.49 | 4.60 | 4.11 | 3.78 | 3.56 | 3.39 | 3.26 | 3.15 |
| 28 | 7.64 | 5.45 | 4.57 | 4.07 | 3.75 | 3.53 | 3.36 | 3.23 | 3.12 |
| 29 | 7.60 | 5.42 | 4.54 | 4.04 | 3.73 | 3.50 | 3.33 | 3.20 | 3.09 |
| 30 | 7.56 | 5.39 | 4.51 | 4.02 | 3.70 | 3.47 | 3.30 | 3.17 | 3.07 |
| 40 | 7.31 | 5.18 | 4.31 | 3.83 | 3.51 | 3.29 | 3.12 | 2.99 | 2.89 |
| 60 | 7.08 | 4.98 | 4.13 | 3.65 | 3.34 | 3.12 | 2.95 | 2.82 | 2.72 |
| 120 | 6.85 | 4.79 | 3.95 | 3.48 | 3.17 | 2.96 | 2.79 | 2.66 | 2.56 |
| ∞ | 6.63 | 4.61 | 3.78 | 3.32 | 3.02 | 2.80 | 2.64 | 2.51 | 2.41 |

DENOMINATOR DEGREES OF FREEDOM

$p = 0.99$

| $n$ | NUMERATOR DEGREES OF FREEDOM | | | | | | | | | |
|---|---|---|---|---|---|---|---|---|---|---|
| | 10 | 12 | 15 | 20 | 24 | 30 | 40 | 60 | 120 | ∞ |
| 1 | 6,056 | 6,106 | 6,157 | 6,209 | 6,235 | 6,261 | 6,287 | 6,313 | 6,339 | 6,366 |
| 2 | 99.40 | 99.42 | 99.43 | 99.45 | 99.46 | 99.47 | 99.47 | 99.48 | 99.49 | 99.50 |
| 3 | 27.23 | 27.05 | 26.87 | 26.69 | 26.60 | 26.50 | 26.41 | 26.32 | 26.22 | 26.13 |
| 4 | 14.55 | 14.37 | 14.20 | 14.02 | 13.93 | 13.84 | 13.75 | 13.65 | 13.56 | 13.46 |
| 5 | 10.05 | 9.89 | 9.72 | 9.55 | 9.47 | 9.38 | 9.29 | 9.20 | 9.11 | 9.02 |
| 6 | 7.87 | 7.72 | 7.56 | 7.40 | 7.31 | 7.23 | 7.14 | 7.06 | 6.97 | 6.88 |
| 7 | 6.62 | 6.47 | 6.31 | 6.16 | 6.07 | 5.99 | 5.91 | 5.82 | 5.74 | 5.65 |
| 8 | 5.81 | 5.67 | 5.52 | 5.36 | 5.28 | 5.20 | 5.12 | 5.03 | 4.95 | 4.86 |
| 9 | 5.26 | 5.11 | 4.96 | 4.81 | 4.73 | 4.65 | 4.57 | 4.48 | 4.40 | 4.31 |
| 10 | 4.85 | 4.71 | 4.56 | 4.41 | 4.33 | 4.25 | 4.17 | 4.08 | 4.00 | 3.91 |
| 11 | 4.54 | 4.40 | 4.25 | 4.10 | 4.02 | 3.94 | 3.86 | 3.78 | 3.69 | 3.60 |
| 12 | 4.30 | 4.16 | 4.01 | 3.86 | 3.78 | 3.70 | 3.62 | 3.54 | 3.45 | 3.36 |
| 13 | 4.10 | 3.96 | 3.82 | 3.66 | 3.59 | 3.51 | 3.43 | 3.34 | 3.25 | 3.17 |
| 14 | 3.94 | 3.80 | 3.66 | 3.51 | 3.43 | 3.35 | 3.27 | 3.18 | 3.09 | 3.00 |
| 15 | 3.80 | 3.67 | 3.52 | 3.37 | 3.29 | 3.21 | 3.13 | 3.05 | 2.96 | 2.87 |
| 16 | 3.69 | 3.55 | 3.41 | 3.26 | 3.18 | 3.10 | 3.02 | 2.93 | 2.84 | 2.75 |
| 17 | 3.59 | 3.46 | 3.31 | 3.16 | 3.08 | 3.00 | 2.92 | 2.83 | 2.75 | 2.65 |
| 18 | 3.51 | 3.37 | 3.23 | 3.08 | 3.00 | 2.92 | 2.84 | 2.75 | 2.66 | 2.57 |
| 19 | 3.43 | 3.30 | 3.15 | 3.00 | 2.92 | 2.84 | 2.76 | 2.67 | 2.58 | 2.49 |
| 20 | 3.37 | 3.23 | 3.09 | 2.94 | 2.86 | 2.78 | 2.69 | 2.61 | 2.52 | 2.42 |
| 21 | 3.31 | 3.17 | 3.03 | 2.88 | 2.80 | 2.72 | 2.64 | 2.55 | 2.46 | 2.36 |
| 22 | 3.26 | 3.12 | 2.98 | 2.83 | 2.75 | 2.67 | 2.58 | 2.50 | 2.40 | 2.31 |
| 23 | 3.21 | 3.07 | 2.93 | 2.78 | 2.70 | 2.62 | 2.54 | 2.45 | 2.35 | 2.26 |
| 24 | 3.17 | 3.03 | 2.89 | 2.74 | 2.66 | 2.58 | 2.49 | 2.40 | 2.31 | 2.21 |
| 25 | 3.13 | 2.99 | 2.85 | 2.70 | 2.62 | 2.54 | 2.45 | 2.36 | 2.27 | 2.17 |
| 26 | 3.09 | 2.96 | 2.81 | 2.66 | 2.58 | 2.50 | 2.42 | 2.33 | 2.23 | 2.13 |
| 27 | 3.06 | 2.93 | 2.78 | 2.63 | 2.55 | 2.47 | 2.38 | 2.29 | 2.20 | 2.10 |
| 28 | 3.03 | 2.90 | 2.75 | 2.60 | 2.52 | 2.44 | 2.35 | 2.26 | 2.17 | 2.06 |
| 29 | 3.00 | 2.87 | 2.73 | 2.57 | 2.49 | 2.41 | 2.33 | 2.23 | 2.14 | 2.03 |
| 30 | 2.98 | 2.84 | 2.70 | 2.55 | 2.47 | 2.39 | 2.30 | 2.21 | 2.11 | 2.01 |
| 40 | 2.80 | 2.66 | 2.52 | 2.37 | 2.29 | 2.20 | 2.11 | 2.02 | 1.92 | 1.80 |
| 60 | 2.63 | 2.50 | 2.35 | 2.20 | 2.12 | 2.03 | 1.94 | 1.84 | 1.73 | 1.60 |
| 120 | 2.47 | 2.34 | 2.19 | 2.03 | 1.95 | 1.86 | 1.76 | 1.66 | 1.53 | 1.38 |
| ∞ | 2.32 | 2.18 | 2.04 | 1.88 | 1.79 | 1.70 | 1.59 | 1.47 | 1.32 | 1.00 |

DENOMINATOR DEGREES OF FREEDOM

# APPENDIX TABLE A.5 Upper Percentiles of Studentized Range Distributions

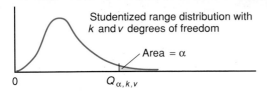

Studentized range distribution with $k$ and $v$ degrees of freedom

Area = $\alpha$

0    $Q_{\alpha,k,v}$

| $v$ | $1-\alpha$   $k$ | 2 | 3 | 4 | 5 | 6 | 7 | 8 | 9 | 10 | 11 | 12 | 13 | 14 | 15 | 16 |
|---|---|---|---|---|---|---|---|---|---|---|---|---|---|---|---|---|
| 1 | 0.95 | 18.0 | 27.0 | 32.8 | 37.1 | 40.4 | 43.1 | 45.4 | 47.4 | 49.1 | 50.6 | 52.0 | 53.2 | 54.3 | 55.4 | 56.3 |
|   | 0.99 | 90.0 | 135 | 164 | 186 | 202 | 216 | 227 | 237 | 240 | 253 | 260 | 266 | 272 | 277 | 282 |
| 2 | 0.95 | 6.09 | 8.3 | 9.8 | 10.9 | 11.7 | 12.4 | 13.0 | 13.5 | 14.0 | 14.4 | 14.7 | 15.1 | 15.4 | 15.7 | 15.9 |
|   | 0.99 | 14.0 | 19.0 | 22.3 | 24.7 | 26.6 | 28.2 | 29.5 | 30.7 | 31.7 | 32.6 | 33.4 | 34.1 | 34.8 | 35.4 | 36.0 |
| 3 | 0.95 | 4.50 | 5.91 | 6.82 | 7.50 | 8.04 | 8.48 | 8.85 | 9.18 | 9.46 | 9.72 | 9.95 | 10.2 | 10.4 | 10.5 | 10.7 |
|   | 0.99 | 8.26 | 10.6 | 12.2 | 13.3 | 14.2 | 15.0 | 15.6 | 16.2 | 16.7 | 17.1 | 17.5 | 17.9 | 18.2 | 18.5 | 18.8 |
| 4 | 0.95 | 3.93 | 5.04 | 5.76 | 6.29 | 6.71 | 7.05 | 7.35 | 7.60 | 7.83 | 8.03 | 8.21 | 8.37 | 8.52 | 8.66 | 8.79 |
|   | 0.99 | 6.51 | 8.12 | 9.17 | 9.96 | 10.6 | 11.1 | 11.5 | 11.9 | 12.3 | 12.6 | 12.8 | 13.1 | 13.3 | 13.5 | 13.7 |
| 5 | 0.95 | 3.64 | 4.60 | 5.22 | 5.67 | 6.03 | 6.33 | 6.58 | 6.80 | 6.99 | 7.17 | 7.32 | 7.47 | 7.60 | 7.72 | 7.83 |
|   | 0.99 | 5.70 | 6.97 | 7.80 | 8.42 | 8.91 | 9.32 | 9.67 | 9.97 | 10.2 | 10.5 | 10.7 | 10.9 | 11.1 | 11.2 | 11.4 |
| 6 | 0.95 | 3.46 | 4.34 | 4.90 | 5.31 | 5.63 | 5.89 | 6.12 | 6.32 | 6.49 | 6.65 | 6.79 | 6.92 | 7.03 | 7.14 | 7.24 |
|   | 0.99 | 5.24 | 6.33 | 7.03 | 7.56 | 7.97 | 8.32 | 8.61 | 8.87 | 9.10 | 9.30 | 9.49 | 9.65 | 9.81 | 9.95 | 10.1 |
| 7 | 0.95 | 3.34 | 4.16 | 4.68 | 5.06 | 5.36 | 5.61 | 5.82 | 6.00 | 6.16 | 6.30 | 6.43 | 6.55 | 6.66 | 6.76 | 6.85 |
|   | 0.99 | 4.95 | 5.92 | 6.54 | 7.01 | 7.37 | 7.68 | 7.94 | 8.17 | 8.37 | 8.55 | 8.71 | 8.86 | 9.00 | 9.12 | 9.24 |
| 8 | 0.95 | 3.26 | 4.04 | 4.53 | 4.89 | 5.17 | 5.40 | 5.60 | 5.77 | 5.92 | 6.05 | 6.18 | 6.29 | 6.39 | 6.48 | 6.57 |
|   | 0.99 | 4.74 | 5.63 | 6.20 | 6.63 | 6.96 | 7.24 | 7.47 | 7.68 | 7.87 | 8.03 | 8.18 | 8.31 | 8.44 | 8.55 | 8.66 |
| 9 | 0.95 | 3.20 | 3.95 | 4.42 | 4.76 | 5.02 | 5.24 | 5.43 | 5.60 | 5.74 | 5.87 | 5.98 | 6.09 | 6.19 | 6.28 | 6.36 |
|   | 0.99 | 4.60 | 5.43 | 5.96 | 6.35 | 6.66 | 6.91 | 7.13 | 7.32 | 7.49 | 7.65 | 7.78 | 7.91 | 8.03 | 8.13 | 8.23 |
| 10 | 0.95 | 3.15 | 3.88 | 4.33 | 4.65 | 4.91 | 5.12 | 5.30 | 5.46 | 5.60 | 5.72 | 5.83 | 5.93 | 6.03 | 6.11 | 6.20 |
|   | 0.99 | 4.48 | 5.27 | 5.77 | 6.14 | 6.43 | 6.67 | 6.87 | 7.05 | 7.21 | 7.36 | 7.48 | 7.60 | 7.71 | 7.81 | 7.91 |
| 11 | 0.95 | 3.11 | 3.82 | 4.26 | 4.57 | 4.82 | 5.03 | 5.20 | 5.35 | 5.49 | 5.61 | 5.71 | 5.81 | 5.90 | 5.99 | 6.06 |
|   | 0.99 | 4.39 | 5.14 | 5.62 | 5.97 | 6.25 | 6.48 | 6.67 | 6.84 | 6.99 | 7.13 | 7.25 | 7.36 | 7.46 | 7.56 | 7.65 |
| 12 | 0.95 | 3.08 | 3.77 | 4.20 | 4.51 | 4.75 | 4.95 | 5.12 | 5.27 | 5.40 | 5.51 | 5.62 | 5.71 | 5.80 | 5.88 | 5.95 |
|   | 0.99 | 4.32 | 5.04 | 5.50 | 5.84 | 6.10 | 6.32 | 6.51 | 6.67 | 6.81 | 6.94 | 7.06 | 7.17 | 7.26 | 7.36 | 7.44 |
| 13 | 0.95 | 3.06 | 3.73 | 4.15 | 4.45 | 4.69 | 4.88 | 5.05 | 5.19 | 5.32 | 5.43 | 5.53 | 5.63 | 5.71 | 5.79 | 5.86 |
|   | 0.99 | 4.26 | 4.96 | 5.40 | 5.73 | 5.98 | 6.19 | 6.37 | 6.53 | 6.67 | 6.79 | 6.90 | 7.01 | 7.10 | 7.19 | 7.27 |
| 14 | 0.95 | 3.03 | 3.70 | 4.11 | 4.41 | 4.64 | 4.83 | 4.99 | 5.13 | 5.25 | 5.36 | 5.46 | 5.55 | 5.64 | 5.72 | 5.79 |
|   | 0.99 | 4.21 | 4.89 | 5.32 | 5.63 | 5.88 | 6.08 | 6.20 | 6.41 | 6.54 | 6.66 | 6.77 | 6.87 | 6.96 | 7.05 | 7.12 |
| 15 | 0.95 | 3.01 | 3.67 | 4.08 | 4.37 | 4.60 | 4.78 | 4.94 | 5.08 | 5.20 | 5.31 | 5.40 | 5.49 | 5.58 | 5.65 | 5.72 |
|   | 0.99 | 4.17 | 4.83 | 5.25 | 5.56 | 5.80 | 5.99 | 6.16 | 6.31 | 6.44 | 6.55 | 6.66 | 6.76 | 6.84 | 6.93 | 7.00 |
| 16 | 0.95 | 3.00 | 3.65 | 4.05 | 4.33 | 4.56 | 4.74 | 4.90 | 5.03 | 5.15 | 5.26 | 5.35 | 5.44 | 5.52 | 5.59 | 5.66 |
|   | 0.99 | 4.13 | 4.78 | 5.19 | 5.49 | 5.72 | 5.92 | 6.08 | 6.22 | 6.35 | 6.46 | 6.56 | 6.66 | 6.74 | 6.82 | 6.90 |
| 17 | 0.95 | 2.98 | 3.63 | 4.02 | 4.30 | 4.52 | 4.71 | 4.86 | 4.99 | 5.11 | 5.21 | 5.31 | 5.39 | 5.47 | 5.55 | 5.61 |
|   | 0.99 | 4.10 | 4.74 | 5.14 | 5.43 | 5.66 | 5.85 | 6.01 | 6.15 | 6.27 | 6.38 | 6.48 | 6.57 | 6.66 | 6.73 | 6.80 |
| 18 | 0.95 | 2.97 | 3.61 | 4.00 | 4.28 | 4.49 | 4.67 | 4.82 | 4.96 | 5.07 | 5.17 | 5.27 | 5.35 | 5.43 | 5.50 | 5.57 |
|   | 0.99 | 4.07 | 4.70 | 5.09 | 5.38 | 5.60 | 5.79 | 5.94 | 6.08 | 6.20 | 6.31 | 6.41 | 6.50 | 6.58 | 6.65 | 6.72 |
| 19 | 0.95 | 2.96 | 3.59 | 3.98 | 4.25 | 4.47 | 4.65 | 4.79 | 4.92 | 5.04 | 5.14 | 5.23 | 5.32 | 5.39 | 5.46 | 5.53 |
|   | 0.99 | 4.05 | 4.67 | 5.05 | 5.33 | 5.55 | 5.73 | 5.89 | 6.02 | 6.14 | 6.25 | 6.34 | 6.43 | 6.51 | 6.58 | 6.65 |
| 20 | 0.95 | 2.95 | 3.58 | 3.96 | 4.23 | 4.45 | 4.62 | 4.77 | 4.90 | 5.01 | 5.11 | 5.20 | 5.28 | 5.36 | 5.43 | 5.49 |
|   | 0.99 | 4.02 | 4.64 | 5.02 | 5.29 | 5.51 | 5.69 | 5.84 | 5.97 | 6.09 | 6.19 | 6.29 | 6.37 | 6.45 | 6.52 | 6.59 |
| 24 | 0.95 | 2.92 | 3.53 | 3.90 | 4.17 | 4.37 | 4.54 | 4.68 | 4.81 | 4.92 | 5.01 | 5.10 | 5.18 | 5.25 | 5.32 | 5.38 |
|   | 0.99 | 3.96 | 4.54 | 4.91 | 5.17 | 5.37 | 5.54 | 5.69 | 5.81 | 5.92 | 6.02 | 6.11 | 6.19 | 6.26 | 6.33 | 6.39 |
| 30 | 0.95 | 2.89 | 3.49 | 3.84 | 4.10 | 4.30 | 4.46 | 4.60 | 4.72 | 4.83 | 4.92 | 5.00 | 5.08 | 5.15 | 5.21 | 5.27 |
|   | 0.99 | 3.89 | 4.45 | 4.80 | 5.05 | 5.24 | 5.40 | 5.54 | 5.65 | 5.76 | 5.85 | 5.93 | 6.01 | 6.08 | 6.14 | 6.20 |
| 40 | 0.95 | 2.86 | 3.44 | 3.79 | 4.04 | 4.23 | 4.39 | 4.52 | 4.63 | 4.74 | 4.82 | 4.91 | 4.98 | 5.05 | 5.11 | 5.16 |
|   | 0.99 | 3.82 | 4.37 | 4.70 | 4.93 | 5.11 | 5.27 | 5.39 | 5.50 | 5.60 | 5.69 | 5.77 | 5.84 | 5.90 | 5.96 | 6.02 |
| 60 | 0.95 | 2.83 | 3.40 | 3.74 | 3.98 | 4.16 | 4.31 | 4.44 | 4.55 | 4.65 | 4.73 | 4.81 | 4.88 | 4.94 | 5.00 | 5.06 |
|   | 0.99 | 3.76 | 4.28 | 4.60 | 4.82 | 4.99 | 5.13 | 5.25 | 5.36 | 5.45 | 5.53 | 5.60 | 5.67 | 5.73 | 5.79 | 5.84 |
| 120 | 0.95 | 2.80 | 3.36 | 3.69 | 3.92 | 4.10 | 4.24 | 4.36 | 4.48 | 4.56 | 4.64 | 4.72 | 4.78 | 4.84 | 4.90 | 4.95 |
|   | 0.99 | 3.70 | 4.20 | 4.50 | 4.71 | 4.87 | 5.01 | 5.12 | 5.21 | 5.30 | 5.38 | 5.44 | 5.51 | 5.56 | 5.61 | 5.66 |
| ∞ | 0.95 | 2.77 | 3.31 | 3.63 | 3.86 | 4.03 | 4.17 | 4.29 | 4.39 | 4.47 | 4.55 | 4.62 | 4.68 | 4.74 | 4.80 | 4.85 |
|   | 0.99 | 3.64 | 4.12 | 4.40 | 4.60 | 4.76 | 4.88 | 4.99 | 5.08 | 5.16 | 5.23 | 5.29 | 5.35 | 5.40 | 5.45 | 5.49 |

SOURCE: Olive Jean Dunn and Virginia A. Clark, *Applied Statistics: Analysis of Variance and Regression* (New York: Wiley, 1974), pp. 371–372.

| | $w_1^*$ | $w_2^*$ | $P(W \leqslant w_1^*) = P(W \geqslant w_2^*)$ |
|---|---|---|---|
| $n = 4$ | 0 | 10 | 0.062 |
| | 1 | 9 | 0.125 |
| $n = 5$ | 0 | 15 | 0.031 |
| | 1 | 14 | 0.062 |
| | 2 | 13 | 0.094 |
| | 3 | 12 | 0.156 |
| $n = 6$ | 0 | 21 | 0.016 |
| | 1 | 20 | 0.031 |
| | 2 | 19 | 0.047 |
| | 3 | 18 | 0.078 |
| | 4 | 17 | 0.109 |
| | 5 | 16 | 0.156 |
| $n = 7$ | 0 | 28 | 0.008 |
| | 1 | 27 | 0.016 |
| | 2 | 26 | 0.023 |
| | 3 | 25 | 0.039 |
| | 4 | 24 | 0.055 |
| | 5 | 23 | 0.078 |
| | 6 | 22 | 0.109 |
| | 7 | 21 | 0.148 |
| $n = 8$ | 0 | 36 | 0.004 |
| | 1 | 35 | 0.008 |
| | 2 | 34 | 0.012 |
| | 3 | 33 | 0.020 |
| | 4 | 32 | 0.027 |
| | 5 | 31 | 0.039 |
| | 6 | 30 | 0.055 |
| | 7 | 29 | 0.074 |
| | 8 | 28 | 0.098 |
| | 9 | 27 | 0.125 |
| $n = 9$ | 1 | 44 | 0.004 |
| | 2 | 43 | 0.006 |
| | 3 | 42 | 0.010 |
| | 4 | 41 | 0.014 |
| | 5 | 40 | 0.020 |
| | 6 | 39 | 0.027 |
| | 7 | 38 | 0.037 |
| | 8 | 37 | 0.049 |
| | 9 | 36 | 0.064 |
| | 10 | 35 | 0.082 |
| | 11 | 34 | 0.102 |
| | 12 | 33 | 0.125 |

| | $w_1^*$ | $w_2^*$ | $P(W \leqslant w_1^*) = P(W \geqslant w_2^*)$ |
|---|---|---|---|
| $n = 10$ | 3 | 52 | 0.005 |
| | 4 | 51 | 0.007 |
| | 5 | 50 | 0.010 |
| | 6 | 49 | 0.014 |
| | 7 | 48 | 0.019 |
| | 8 | 47 | 0.024 |
| | 9 | 46 | 0.032 |
| | 10 | 45 | 0.042 |
| | 11 | 44 | 0.053 |
| | 12 | 43 | 0.065 |
| | 13 | 42 | 0.080 |
| | 14 | 41 | 0.097 |
| | 15 | 40 | 0.116 |
| | 16 | 39 | 0.138 |
| $n = 11$ | 5 | 61 | 0.005 |
| | 6 | 60 | 0.007 |
| | 7 | 59 | 0.009 |
| | 8 | 58 | 0.012 |
| | 9 | 57 | 0.016 |
| | 10 | 56 | 0.021 |
| | 11 | 55 | 0.027 |
| | 12 | 54 | 0.034 |
| | 13 | 53 | 0.042 |
| | 14 | 52 | 0.051 |
| | 15 | 51 | 0.062 |
| | 16 | 50 | 0.074 |
| | 17 | 49 | 0.087 |
| | 18 | 48 | 0.103 |
| | 19 | 47 | 0.120 |
| | 20 | 46 | 0.139 |
| $n = 12$ | 7 | 71 | 0.005 |
| | 8 | 70 | 0.006 |
| | 9 | 69 | 0.008 |
| | 10 | 68 | 0.010 |
| | 11 | 67 | 0.013 |
| | 12 | 66 | 0.017 |
| | 13 | 65 | 0.021 |
| | 14 | 64 | 0.026 |
| | 15 | 63 | 0.032 |
| | 16 | 62 | 0.039 |
| | 17 | 61 | 0.046 |
| | 18 | 60 | 0.055 |
| | 19 | 59 | 0.065 |
| | 20 | 58 | 0.076 |
| | 21 | 57 | 0.088 |
| | 22 | 56 | 0.102 |
| | 23 | 55 | 0.117 |
| | 24 | 54 | 0.133 |

SOURCE: Wilfrid J. Dixon and Frank J. Massey, Jr., *Introduction to Statistical Analysis*, 2nd ed. (New York: McGraw-Hill, 1957), pp. 443–444.

### TABLE A.7 Critical Values $v_1^*$ and $v_2^*$ for the Wilcoxon Rank Sum Statistic, $V$

*Test statistic is rank sum associated with smaller sample (if equal sample sizes, either rank sum can be used).*

**a.** $\alpha = .025$ one-tailed;  $\alpha = .05$ two-tailed

| $m$ \ $n$ | 3 | | 4 | | 5 | | 6 | | 7 | | 8 | | 9 | | 10 | |
|---|---|---|---|---|---|---|---|---|---|---|---|---|---|---|---|---|
| | $v_1^*$ | $v_2^*$ | $v_1^*$ | $v_2^*$ | $v_1^*$ | $v_2^*$ | $v_1^*$ | $v_2^*$ | $v_1^*$ | $v_2^*$ | $v_1^*$ | $v_2^*$ | $v_1^*$ | $v_2^*$ | $v_1^*$ | $v_2^*$ |
| 3 | 5 | 16 | 6 | 18 | 6 | 21 | 7 | 23 | 7 | 26 | 8 | 28 | 8 | 31 | 9 | 33 |
| 4 | 6 | 18 | 11 | 25 | 12 | 28 | 12 | 32 | 13 | 35 | 14 | 38 | 15 | 41 | 16 | 44 |
| 5 | 6 | 21 | 12 | 28 | 18 | 37 | 19 | 41 | 20 | 45 | 21 | 49 | 22 | 53 | 24 | 56 |
| 6 | 7 | 23 | 12 | 32 | 19 | 41 | 26 | 52 | 28 | 56 | 29 | 61 | 31 | 65 | 32 | 70 |
| 7 | 7 | 26 | 13 | 35 | 20 | 45 | 28 | 56 | 37 | 68 | 39 | 73 | 41 | 78 | 43 | 83 |
| 8 | 8 | 28 | 14 | 38 | 21 | 49 | 29 | 61 | 39 | 73 | 49 | 87 | 51 | 93 | 54 | 98 |
| 9 | 8 | 31 | 15 | 41 | 22 | 53 | 31 | 65 | 41 | 78 | 51 | 93 | 63 | 108 | 66 | 114 |
| 10 | 9 | 33 | 16 | 44 | 24 | 56 | 32 | 70 | 43 | 83 | 54 | 98 | 66 | 114 | 79 | 131 |

**b.** $\alpha = .05$ one-tailed;  $\alpha = .10$ two-tailed

| $m$ \ $n$ | 3 | | 4 | | 5 | | 6 | | 7 | | 8 | | 9 | | 10 | |
|---|---|---|---|---|---|---|---|---|---|---|---|---|---|---|---|---|
| | $v_1^*$ | $v_2^*$ | $v_1^*$ | $v_2^*$ | $v_1^*$ | $v_2^*$ | $v_1^*$ | $v_2^*$ | $v_1^*$ | $v_2^*$ | $v_1^*$ | $v_2^*$ | $v_1^*$ | $v_2^*$ | $v_1^*$ | $v_2^*$ |
| 3 | 6 | 15 | 7 | 17 | 7 | 20 | 8 | 22 | 9 | 24 | 9 | 27 | 10 | 29 | 11 | 31 |
| 4 | 7 | 17 | 12 | 24 | 13 | 27 | 14 | 30 | 15 | 33 | 16 | 36 | 17 | 39 | 18 | 42 |
| 5 | 7 | 20 | 13 | 27 | 19 | 36 | 20 | 40 | 22 | 43 | 24 | 46 | 25 | 50 | 26 | 54 |
| 6 | 8 | 22 | 14 | 30 | 20 | 40 | 28 | 50 | 30 | 54 | 32 | 58 | 33 | 63 | 35 | 67 |
| 7 | 9 | 24 | 15 | 33 | 22 | 43 | 30 | 54 | 39 | 66 | 41 | 71 | 43 | 76 | 46 | 80 |
| 8 | 9 | 27 | 16 | 36 | 24 | 46 | 32 | 58 | 41 | 71 | 52 | 84 | 54 | 90 | 57 | 95 |
| 9 | 10 | 29 | 17 | 39 | 25 | 50 | 33 | 63 | 43 | 76 | 54 | 90 | 66 | 105 | 69 | 111 |
| 10 | 11 | 31 | 18 | 42 | 26 | 54 | 35 | 67 | 46 | 80 | 57 | 95 | 69 | 111 | 83 | 127 |

SOURCE: From F. Wilcoxon and R. A. Wilcox, "Some Rapid Approximate Statistical Procedures," 1964, 20–23. Reproduced with the permission of American Cyanamid Company.

# Glossary

**ANOVA table**—A format for presenting (and helping to organize) the various sums of squares needed for doing an $F$ test on $H_0$: $\mu_1 = \mu_2 = \cdots = \mu_k$. In general, ANOVA tables show at a glance the exact model an experimenter is using and indicate the levels of significance that can be associated with the various hypotheses and subhypotheses being tested.

**Bayes' rule**—A formula that "reverses" conditional probabilities. If events $A_1, A_2, \ldots, A_k$ partition a sample space $S$ and if $B$ is any event defined on $S$ such that $P(B) > 0$, then Bayes' rule gives $P(A_j|B)$, for any $j$, in terms of the conditional probabilities $P(B|A_i)$, $i = 1, 2, \ldots, k$.

**bell-shaped histogram**—A histogram in which class frequencies are largest in the center and taper off symmetrical in the two tails.

**Bernoulli trial**—An experiment with only two possible outcomes: success and failure.

**Bernoulli variable**—A variable that *for each subject* can have only one of two possible values, 0 or 1.

**binomial distribution**—The pdf that describes the number of successes occurring in a sequence of $n$ independent Bernoulli trials.

**bivariate distribution**—A function $f(x, y)$ that defines a surface describing the joint probabilistic behavior of two random variables $X$ and $Y$. The volume under the surface lying above any region $A$ in the $xy$ plane represents the probability that the point $(X, Y)$ lies in the region $A$.

**block**—A group of similar subjects or similar experimental conditions to which the entire set of treatments being compared is applied.

**block mean square (MSB)**—The block sum of squares (SSB) divided by $b - 1$. If $H_0$: $\beta_1 = \beta_2 = \cdots = \beta_b$ is true, MSB (like MSTR and MSE) is an estimate of $\sigma^2$.

**block sum of squares (SSB)**—The sum of the squared deviations of the $b$ block means from the overall mean. The block sum of squares measures the extent to which the $\beta_i$'s are not all equal.

**categorical data**—Two dissimilar variables, $x$ and $y$, are recorded on each subject. The range of possible $x$ responses is reduced to a set of $r$ categories; the range of possible $y$ responses is reduced to a set of $c$ categories. The number of observations belonging to each ($x$ category, $y$ category) combination is reported.

**central limit theorem**—A powerful result showing that sample means are approximately normal, regardless of the pdf from which the individual $Y_i$'s are drawn.

**chi-square distribution**—The pdf that describes the behavior of the statistic $(n - 1)S^2/\sigma^2$, where $S^2$ is computed from a random sample of size $n$ taken from a normal distribution with variance $\sigma^2$.

**combination**—An unordered collection of objects. The total number of combinations of size $r$ that can be formed from a set of $n$ distinct objects is

$$_nC_r \quad \text{or} \quad \binom{n}{r},$$

where

$$_nC_r = \frac{n!}{r!\,(n - r)!}$$

**combinatorics**—The branch of mathematics concerned with the selection, arrangement, and enumeration of ordered and unordered sequences.

**complement**—If $A$ is any event defined on a sample space S, the *complement* of $A$, written $A^c$, is the event consisting of all the outcomes in S that are not in $A$.

**conditional probability**—A probability computed on the presumption that some other event has already occurred.

**contingency table**—A grid of $r$ rows and $c$ columns for summarizing categorical data. Entries are the numbers of observations belonging to each ($x$ category, $y$ category) combination.

**continuous random variable**—A random variable that can take on any value in a certain interval—that is, its possible values are uncountably infinite.

**data models**—All data have certain fundamental properties relating to the nature, number, and type of variables recorded. Most data sets fall into one of seven categories or models.

**decile**—Percentiles that are multiples of 10.

**degrees of freedom**—Constants related to sample size that appear in the pdfs for Student $t$, $\chi^2$, and $F$ random variables.

**dependent data**—Two variables, $x$ and $y$, are dependent if the conditions under which $x_i$ is measured are deliberately "matched" to be similar to the conditions under which $y_i$ is measured.

**descriptive statistics**—Procedures for summarizing and displaying data. They may be graphical (e.g., a histogram) or numerical (e.g., a standard deviation).

**discrete random variable**—A random variable whose set of possible values is either finite or countably infinite.

**dispersion**—The amount of scatter in a set of $y_i$'s.

**distribution**—See *probability density function*.

**error mean square (MSE)**—The error sum of squares (SSE) divided by $n - k$. Conceptually, MSE is analogous to the pooled variance $s_p^2$ in two-sample $t$ tests: it represents an "overall" estimate of the unknown variance $\sigma^2$ associated with each observation.

**error sum of squares (SSE)**—The sum of the squared deviations of each observation from its own sample mean.

**estimated regression line**—A function that approximates the "true" relationship between two measurements.

**event**—A collection of outcomes in the sample space that share or satisfy some condition.

**expected frequency**—The number of observations, out of $n$, that "should" belong to a certain class (or cell) *if $H_0$ is true*. An expected frequency is said to be "estimated" if its computation is based on an $H_0$ model whose parameters, themselves, are estimated.

**expected value**—A number that measures the "center" of a pdf. $E(Y)$ does for a probability density function what $\bar{y}$ does for a sample.

**experiment**—Any procedure capable of producing a well-defined set of possible outcomes.

***F* distribution**—The pdf that describes the behavior of the quotient of two independent chi-square random variables, each divided by its degrees of freedom.

***F* test**—A procedure for testing $H_0: \sigma_X^2 = \sigma_Y^2$. If both populations sampled are normal and have the same variance, then $S_Y^2/S_X^2$ has an $F$ distribution (with $m - 1$ and $n - 1$ degrees of freedom). Values of $s_Y^2/s_X^2$ too far from 1 are evidence that $H_0: \sigma_X^2 = \sigma_Y^2$ is not true. $F$ tests are also used to test $H_0: \mu_1 = \mu_2 = \cdots = \mu_k$. In general, $F$ tests are based on the ratio of two different mean squares, each of which estimates a variance.

**frequency distribution**—A listing that shows the numbers of observations falling into each of several classes.

**frequency polygon**—A descriptive technique that presents graphically the information contained in a frequency distribution. The frequency of a class is represented by the height of a point drawn above the midpoint of that class. Successive points are connected with straight lines. The vertical axis can be scaled in terms of either frequency, relative frequency, or percentage frequency.

**Friedman's test**—A way of analyzing randomized block data without making the usual normality assumption.

**function**—A formula that associates a number $x$ with a possibly different number $f(x)$.

**general linear model**—A particular set of assumptions defining the most common formulation of multiple regression. If $Y$ is a random variable and $x_1, x_2, \ldots, x_k$ are fixed variables, the general linear model assumes that

$$Y = a + b_1 x_1 + b_2 x_2 + \cdots + b_k x_k + \varepsilon$$

where $\varepsilon$ is normally distributed with mean zero and variance $\sigma^2$.

**general second-order linear model**—A particular regression function involving two regressors that includes all possible first-order and second-order terms. If $x_1$ and $x_2$ are the two regressors, the general second-order linear model for $Y$ has the form

$$Y = a + b_1 x_1 + b_2 x_2 + b_3 x_1^2 + b_4 x_2^2 + b_5 x_1 x_2 + \varepsilon$$

where $\varepsilon$ is normally distributed with mean zero and variance $\sigma^2$.

**geometric distribution**—The pdf that describes the occurrence of the first success in a series of independent Bernoulli trials.

**goodness-of-fit statistic**—A function for quantifying the amount of disagreement between a set of observed frequencies and a set of expected frequencies.

**histogram**—A graphical format for grouped data: the frequency of a class is represented by the height of a bar. The vertical axis of a histogram can be scaled in terms of either frequency, relative frequency, or percentage frequency.

**hypergeometric distribution**—The pdf that models the experiment of drawing samples without replacement from an urn.

**independent data**—Two variables, $x$ and $y$, are independent if the value of one does not help to predict the value of the others.

**independent events**—Events $A$ and $B$ are independent if the occurrence of one has no effect on the likelihood of the other—that is, if $P(A|B) = P(A)$.

**inferential statistics**—Procedures for making generalizations from samples to populations.

**intersection**—If $A$ and $B$ are two events defined on the sample space S, the *intersection* of $A$ and $B$, written $A \cap B$, is the event whose outcomes are in *both* $A$ and $B$.

**$k$-sample data**—Samples of sizes $n_1, n_2, \ldots, n_k$ $(k > 2)$ are chosen to represent treatment 1, treatment 2, $\ldots$, treatment $k$, respectively. All observations are independent.

**Kruskal–Wallis test**—A nonparametric alternative to the analysis of variance. Based on rank sums associated with observations representing each of $k$ treatment levels, the Kruskal–Wallis test is a procedure for choosing between $H_0: \tilde{\mu}_1 = \tilde{\mu}_2 = \cdots = \tilde{\mu}_k$ and $H_1$: not all the $\tilde{\mu}_j$'s are equal.

**least squares criterion**—A rule for deciding which straight line $y = a + bx$ best fits a set of regression-correlation data. The slope and $y$ intercept should minimize the function $L$, where

$$L = \sum_{i=1}^{n} (y_i - (a + bx_i))^2$$

**least squares line**—The "best" straight line through a set of $n$ points. Also called the *regression line*.

**linear model**—A set of assumptions frequently invoked when doing a regression analysis. For each given value of $x$, the corresponding distributions of $Y$ are assumed to be normal with the same variance $\sigma^2$ and with means all lying on the same straight line $y = a + bx$.

**linear regression**—A kind of regression–correlation data where the relationship between the $x$ and $y$ measurements can be satisfactorily approximated by a function of the form $y = a + bx$.

**location**—The position or center of a distribution. Location is measured by the sample mean. If the distribution is markedly skewed, location is more appropriately measured by the sample median.

**mean sum of squares for error (MSE)**—A term often used to denote the estimate of the parameter $\sigma^2$ that appears in the linear model.

**model equation**—An equation that shows how each observation in a set of data can be "explained" as the sum of a random component added to one or more constants.

**multinomial distribution**—A generalization of the binomial distribution. The multinomial distribution describes the probabilistic behavior of the set of observed frequencies in a goodness-of-fit statistic.

**multiplication rule**—If an ordered sequence consists of $k$ steps and if $n_i$ options are available at the $i$th step, $i = 1, 2, \ldots, k$, then the number of possible sequences is $n_1 n_2 \cdots n_k$.

**mutually exclusive events**—Events that have no outcomes in common.

**nonlinear regression**—Regression–correlation data that can *not* be adequately approximated by a function of the form $y = a + bx$.

**normal distribution**—The bell-shaped curve. There are an infinite number of normal distributions, each having a different value for $\mu$ and/or $\sigma$. All of them can be transformed to a *standard normal distribution*.

**NS**—An abbreviation for Not Signficant (at the $\alpha = 0.05$ level). When NS appears to the right of the observed $F$ ratio in an ANOVA table, it means that $H_0: \mu_1 = \mu_2 = \cdots = \mu_k$ would be accepted at the 0.05 level of significance.

**null event** $(\varnothing)$—The event that has no outcomes.

**one-sample data**—A single sample of size $n$ is observed, $y_1, y_2, \ldots, y_n$. The $y_i$'s measure (1) a single set of conditions or (2) the effects of one particular treatment.

**one-sample $t$ test**—A procedure for testing $H_0: \mu = \mu_0$ when the data consist of a random sample of size $n$ presumed to have come from a normal distribution having mean $\mu$ and variance $\sigma^2$, both unknown.

**$P$ value**—The probability under $H_0$ associated with values of a test statistic as extreme or more extreme than what actually occurred. It is the smallest value of $\alpha$ for which the given data could have rejected $H_0$.

**paired data**—Two dependent observations, representing treatment X and treatment Y, are observed on each of $n$ pairs. The set of within-pair response differences, $d_1 = y_1 - x_1, \quad d_2 = y_2 - x_2, \quad \ldots, \quad d_n = y_n - x_n$ becomes the basis for comparing the treatments.

**paired $t$ test**—A procedure for testing whether the average difference in $n$ paired observations is significantly different from zero. If $\mu_D$ denotes the true average difference, we test $H_0: \mu_D = 0$ by computing the statistic, $\bar{d}/(s_D/\sqrt{n})$. If $H_0$ is true and the $d_i$'s are normally distributed, $\bar{D}/(S_D/\sqrt{n})$ has a Student $t$ distribution with $n - 1$ degrees of freedom.

**pdf**—See *probability density function*.

**percentage frequency**—One hundred times the frequency of a class divided by the total number of observations in the sample.

**percentile**—A number having the property that it equals or exceeds a certain percentage of all the observations in a distribution.

**plot**—Another name for a pair. Historically, many of the early applications of the paired $t$ test involved agricultural research. The treatments being compared were applied within small, geographically contiguous pieces of land. The latter were called *plots*. Conditions within a plot relative to the response being measured were kept as homogeneous as possible.

**point estimate**—A single number that represents our best guess for the value of an unknown parameter.

**Poisson distribution**—A pdf used for (1) approximating binomial probabilities when $n$ is large and $p$ is small and (2) describing the occurrences of rare events.

**pooled standard deviation** $(s_p)$—The square root of a weighted average of the sample variances of the $x$'s and $y$'s. It is an estimate of the unknown $\sigma$ that is presumed to be the same for the $X$ and $Y$ distributions.

**population**—The (usually) hypothetical set of measurements from which the observed sample is presumed to have come.

**power**—The probability that $H_0$ is rejected when $H_1$ is true. Numerically, it equals $1 - \beta$, where $\beta$ is the probability of committing a Type II error. A graph of $1 - \beta$ versus the possible values of the parameter being tested is called a *power curve*.

**prediction interval**—In regression problems, a confidence interval for the value of a future $Y$ at a given value of $x$.

**probability density function (pdf)**—A function that describes the probability structure or "behavior" of a random variable.

**qualitative variable**—A variable whose possible values are not numerical, but descriptive.

**quantitative variable**—A variable whose possible values are numerical.

**quartile**—Percentiles that are multiples of 25.

**radius (of a confidence interval)**—The distance from the center of a confidence interval to either end point. The radius is a measure of the precision with which a parameter is being estimated.

**random sample of size $n$**—A set of $n$ independent observations taken from the same pdf under presumably identical conditions.

**random variable**—A function that assigns a number to each outcome in an experiment's original sample space.

**randomized block data**—$k$ ($>2$) dependent observations, representing treatment 1, treatment 2, . . . , treatment $k$, are observed in each of $n$ blocks. The objective is to compare the $k$ treatments.

**regression–correlation data**—Two (dissimilar) measurements, $x$ and $y$, are taken on each of $n$ subjects. The objective is to quantify the relationship evidenced by the resulting set of $n$ points,

$$(x_1, y_1), \quad (x_2, y_2), \quad \ldots, \quad (x_n, y_n)$$

**relative frequency**—The frequency of a class divided by the total number of observations in the sample.

**robustness**—The ability of the sampling distribution of a test statistic to be relatively unaffected by violations of its underlying assumptions.

**sample**—A set of measurements chosen (or selected or drawn) from a population of all possible such measurements.

**sample correlation coefficient ($r$)**—A measure of the strength of a linear relationship.

**sample mean ($\bar{y}$)**—The average of all observations in a sample.

**sample median**—The "middle" observation in a sample, *in terms of magnitude* (if the number of observations is even, the sample median is the average of the middle two).

**sample mode** $(y_m)$—The most frequently occurring value in a set of data.

**sample standard deviation** $(s)$—A measure of the dispersion in a set of $y_i$'s.

**sampling distribution**—The pdf describing the probabilistic behavior of a statistic.

**scatter diagram**—The basic graphical descriptive technique for regression-correlation data. The $n$ observations are plotted as points in the $xy$ plane.

**sigma notation** $(\Sigma)$—A mathematical shorthand for representing the sum of a set of observations. Given $x_1, x_2, \ldots, x_n$,

$$\sum_{i=1}^{n} x_i = x_1 + x_2 + \cdots + x_n$$

The notation applies to *functions* of the $x_i$'s as well. For example,

$$\sum_{i=1}^{n} x_i^2 = x_1^2 + x_2^2 + \cdots + x_n^2$$

**sign test**—A nonparametric procedure applicable to either one-sample data or paired data. The sign test is one of the most general of all nonparametric procedures—it assumes only that the pdf being sampled is continuous.

**standard deviation**—A number that measures the spread or dispersion associated with the values of a random variable. The standard deviation $\sigma$ does for a pdf what the standard deviation $s$ does for a sample.

**standard error of the mean**—A number that reflects the dispersion that would be observed among a set of sample means if each mean was based on a different random sample.

**standard normal distribution**—The particular normal distribution having $\mu = 0$ and $\sigma = 1$.

**statistic**—A function computed from a random sample.

**statistical inference**—The science of making generalizations from samples to populations.

**statistics**—The numbers recorded or information collected. In other contexts, it refers to all the procedures and mathematical properties associated with the collection, analysis, and interpretation of data.

**Student $t$ distribution**—The pdf that describes the behavior of the statistic

$$\frac{\bar{Y} - \mu}{S/\sqrt{n}}$$

where $\bar{Y}$ and $S$ are calculated from a random sample of size $n$ drawn from a normal distribution with mean $\mu$.

**studentized range**—A random variable derived from the range of a set of $k$ normally distributed observations or means.

**subhypothesis**—A statement about a subset of the populations (and parameters) being studied.

**sum of squares for error (SSE)**—In the linear model, the sum of the squared deviations of the $y_i$'s from the estimated regression line $y = \hat{a} + \hat{b}x$.

**test statistic**—A function of the data on whose value we base the decision between $H_0$ and $H_1$. If the test statistic falls into the critical region, we reject $H_0$. For many of the most important problems in statistical inference, the test statistic will have a pdf described by either the Z, the $t$, the $\chi^2$, or the $F$ distribution.

**total sum of squares (SSTO)**—The sum of the squared deviations of each observation from the overall mean in the entire sample. It measures the total variability in a set of data—that is, the extent to which the observations are not all the same.

**treatment**—Any identifiable condition or trait whose effect on the measured variable is being studied.

**treatment mean square (MSTR)**—The treatment sum of squares (SSTR) divided by $k - 1$. If $H_0: \mu_1 = \mu_2 = \cdots = \mu_k$ is true, MSTR is an estimate of $\sigma^2$.

**treatment sum of squares (SSTR)**—The sum of the squared deviations of the $k$ sample means from the overall mean. It measures the extent to which the $\mu_j$'s are not all equal.

**true correlation coefficient ($\rho$)**—A parameter that measures the extent to which two bivariate normal random variables are linearly related.

**true regression line**—The line that connects the means of the $Y$ distributions in a linear model.

**Tukey confidence interval**—An interval associated with a subhypothesis $H_0': \mu_i = \mu_j$, $i < j$. If the interval fails to contain zero, we reject $H_0'$.

**two-sample data**—Two sets of independent samples,

$$x_1, x_2, \ldots, x_n \quad \text{and} \quad y_1, y_2, \ldots, y_m$$

representing treatment X and treatment Y are observed. The objective is to use the $x_i$'s and $y_i$'s to compare the effects of the two treatments.

**two-sample $t$ test**—A procedure for testing $H_0: \mu_X = \mu_Y$. Data for a two-sample $t$ test should consist of independent random samples from two normal distributions, each having the same variance.

**two-sided alternative**—An $H_1$ that allows for the possibility that the parameter being tested may be larger or smaller than the value specified in $H_0$.

**Type I error**—Rejecting $H_0$ when $H_0$ is true. The *probability* of committing a Type I error is denoted $\alpha$ and referred to as the *level of significance*.

**Type II error**—Accepting $H_0$ when $H_1$ is true. The *probability* of committing a Type II error is denoted by $\beta$.

**union**—If $A$ and $B$ are any two events defined on the sample space $S$, the *union* of $A$ and $B$, written $A \cup B$, is the event whose outcomes are in either $A$ or $B$ or both.

**variance**—The square of the standard deviation.

**Venn diagram**—A graphical representation of events defined on a sample space.

**Wilcoxon rank sum text**—A nonparametric procedure for the two-sample location problem.

**Wilcoxon signed rank test**—Like the sign test, a procedure for testing either $H_0$: $\tilde{\mu} = \tilde{\mu}_0$ or $H_0$: $\tilde{\mu}_D = 0$. The signed rank test is more powerful than the sign test but its assumptions are more restrictive—it requires that the pdf sampled be symmetric as well as continuous.

**$\chi^2$ test for independence**—The usual statistical procedure for analyzing categorical data. The null hypothesis that two traits, $X$ and $Y$, are independent is tested by rejecting $H_0$ if the observed and expected frequencies in an $r \times c$ contingency table are too far apart.

# Answers to Odd-Numbered Questions

**1.1** (a) 216   (b) 27

**1.3** (a) 11   (b) 37   (c) 121
     (d) 9   (e) 9   (f) 81

**1.5** (a) $\sum_{i=1}^{3} x_i^2$
     (b) $\left(\sum_{i=1}^{3} x_i\right)^2$
     (c) $\sum_{i=1}^{3} (x_i - 1)^2$
     (d) $\sum_{i=1}^{3} (1/4)x_i$
     (e) $\sum_{i=1}^{3} (1/2)(x_i - 4)^2$

**1.7** (a) 126.55
     (b) The regional averages must be weighted by the sample size.

**1.9** (a) 10   (b) 30   (c) 205
     (d) 14,071   (e) 2050
     (f) 646

**1.11** (a) $\left(\sum_{i=1}^{3} x_i\right)\left(\sum_{i=1}^{3} y_i\right)$
     (b) $\sum_{i=1}^{3} x_i y_i$
     (c) $\left(\sum_{i=1}^{3} x_i\right)^2\left(\sum_{i=1}^{3} y_i^2\right)$
     (d) $\left(\sum_{i=1}^{3} x_i^2 y_i^2\right)^2$

**1.13** (a) 64.7   (b) 976.85
     (c) 74.7   (d) 1055.83
     (e) 4833.09   (f) 668.44
     (g) 25.854   (h) 285.494
     (i) 668.44

**1.17** Sum of the first n odd numbers

**1.19** (a) $\bar{x} = (1/n) \sum_{i=1}^{5} x_i$
     (b) 62.4

**1.21** (a) $\sum_{i=1}^{3} x_i y_i^2$
     (b) $\sum_{i=1}^{4} i \cdot x_i$
     (c) $\sum_{i=1}^{n} (x_i + y_i)^2$

**1.23** (a) 2.0   (b) does not exist
     (c) 6.5   (d) 7.9

**1.25** $x_{22}/\left(\sum_{i=1}^{62} x_i\right)$

**1.27** (a) 3   (b) 3   (c) 9

**1.29** (a) 74.4   (b) Tom

**1.31** (a) $\sum_{i=1}^{n} (x_i + 1)$
     (b) $\sum_{i=1}^{n} x_i^{i-1}$

**1.33** (a) 18   (b) 70   (c) 324

**1.35** 21.5

**1.37** (a) $\prod_{i=1}^{n} x_i$
     (b) $\prod_{i=1}^{n} (x_i - 2)$
     (c) $\prod_{i=1}^{n} i$

**3.21** 35.4

**3.23** 3.647

**3.25** 5.3

**3.27** 315.0

**3.29** 165.2;  166;  160

**3.33** 0.084

**3.35** 0.240

**3.37** 7.2

**3.39** 6.6

**3.41** Yes

**3.43** (a) 68%    (b) 95%
(c) 81.5%    (d) 84%
(e) 0%

**3.45** (a) $\bar{y} = 6.625$, $s = 4.53$
(b) 45%
(c) data not bell-shaped

**3.47** Third quartile

**3.49** (a) 11.5    (b) 17.5
(c) 26.8

**3.51** 63.2; 11.7; 44.7; 18.4

**3.55** 6.74, 0.54; 5.61; 0.36

**3.59** 3.23; 2.59

**3.61** Yes; no

**3.65** (a) $y = 25.2 + 3.3x$
(c) 84.6°F

**3.69** $r = 1$

**3.73** neither

**3.79** (a) 59.43; 1.42
(b) 55.97; 1.69
(c) 51.7; 2.00
(d) yes

**3.81** Procedure S:
$\bar{y} = 7.75$,   $s = 1.7$
Procedure L:
$\bar{y} = 4.4$,   $s = 1.5$
Procedure N:
$\bar{y} = 3.0$,   $s = 3.0$

**3.83** Treatment averages: 7.85,
8.053, 7.743, 7.513, 7.45
Block averages: 7.63, 7.826,
7.71

**3.85** Observations for blocks are
not independent.

**3.87** Yes, the line given is the least
squares line.

**3.89** (a) Mean is 12.45; grouped
mean is 12.9.
(b) For example, when the
observations in each class
are equal.

**3.91** No, better students may take
the more difficult test.

**3.93** 4.57; 4; 2.64

**3.95** The first, since the small
standard deviations in the
second place more significance
on the observed differences.

**3.97** (b) $y = -288.8 + 1567.7x$

**3.99** The first may be bimodal,
whereas the second seems
unimodal or bell-shaped.

**3.101** (b) $y = 3.795 - 0.886x$

**3.103** 96

**3.107** The observations should be
paired.

**4.1** $20 \cdot 20 \cdot 9 \cdot 6 = 21,600$

**4.3** $3 \cdot 4 = 12$

**4.5** (a) $5 \cdot 6 = 30$
(b) $30 \cdot 30 = 900$

**4.7** $90 \cdot 3! = 540$

**4.9** $2 + 2^2 + 2^3 = 14$

**4.11** $2 \cdot 3 \cdot 2 \cdot 2 = 24$

**4.13** $26 + 26^2 + \cdots + 26^9$
$= 5.647 \times 10^{12}$

**4.15** $6 \cdot 5 \cdot 4 = 120$

**4.17** $15 \cdot 14 \cdot 13 \cdot 12 = 32,760$

**4.19** $6! = 720$

**4.21** (a) $26 \cdot 25 \cdot 10 \cdot 9 \cdot 8 \cdot 7$
$= 3,276,000$
(b) $26 \cdot 25 \cdot 10^4$
$= 6,500,000$
(c) $26^2 \cdot 10^4 = 6,760,000$

**4.23** (a) $4! \cdot 5!$
(b) $6! \times 4! \times 5! = 17,280$

**4.25** $4 \cdot 3 \cdot 2 - 1 = 23$

**4.27** $2 \cdot 2 \cdot 3! = 24$

**4.29** $_4C_2 = 6$

**4.31** $\dfrac{10!}{3!\, 2!\, 4!} = 12,600$

**4.33** $\dfrac{5 \cdot 6!}{2!\, 3!} = 300$

**4.35** $\dfrac{6!}{3!\, 3!} = 20$

**4.37** (a) 21    (b) $21^2 = 441$

**4.39** $_5C_3 = 10$

**4.41** $_{52}C_5 = 2,598,960$

**4.43** $_8C_3 \cdot {_4C_2} = 336$

**4.45** (a) $_6C_2 = 15$    (b) 5
(c) 5/15

**4.47** 5

**4.51** $_{10}C_5 = 252$

**4.53** (a) 216 (b) 6/216

**4.55** (a) 3/9 (b) 5/9 (c) 5/9

**4.57** (a) 4/52 (b) 13/52
(c) 1/52 (d) 26/52

**4.59** (a) $_{52}C_2 = 1326$
(b) $\dfrac{_4C_2}{_{52}C_2} = \dfrac{6}{1326}$

**4.61** 126, 136, 146, 156, 236, 246,
256, 346, 356, 456

**4.63** $\dfrac{_5C_3 \; _{10}C_3}{_{15}C_6} = \dfrac{240}{1001}$

**4.65** $1 - \dfrac{_{365}P_4}{365^4} = 0.016$

**4.67** (a) $7!/7^7 = 0.006$
(b) $7/7^7 = 1/117{,}649$

**4.69** (a) $\dfrac{13}{_{15}C_3} = \dfrac{1}{35}$ (b) yes

**4.71** $60/12! = 0.000000125$

**4.73** (a) yes (b) 5

**4.75** (a) $3! \cdot 5! \cdot 4! = 17{,}280$
(b) $3!9!/12! = 1/220$
(c) $\dfrac{_3C_1 \cdot _5C_1 \cdot _4C_1}{_{12}C_3} = \dfrac{60}{220}$

**4.77** 10/1296

**4.79** $\dfrac{_3C_2 \cdot _7C_2}{_{10}C_4} = \dfrac{3}{10}$

**4.81** $\dfrac{_{11}C_3}{_{47}C_3} - 0.01$

**4.83** $2.645 \cdot 10^{32}$

**4.85** $\dfrac{_{2n}C_n}{2}$

**4.89** (a) $6! = 720$
(b) $6!/(3! \cdot 3!) = 20$
(c) $2 \cdot 5 \cdot 4!/6! = 1/3$

**4.91** 986,409

**4.93** Placement doesn't matter.

**4.95** $_{19}C_2 = 171$

**4.97** (a) $6! \cdot 7!/13! = 1/1716$
(b) No, husbands and wives
are apt to be adjacent.

**4.99** Choice of suit times choice of
cards in suit minus number of
straight flushes.

**4.101** $\dfrac{(_2C_1 \cdot _2C_1)^4 \cdot _{32}C_4}{_{48}C_{12}} = 0.000132$

**4.103** (a) $\dfrac{2}{_{47}C_2} = 0.00185$
(b) $\dfrac{_{10}C_2}{_{47}C_2} = 0.0416$

**5.1** jack, queen, king of diamonds
and hearts

**5.3** (a) {4, 5, 7, 9, 10}
(b) {1, 2, 3, 6, 7, 8, 9, 10}
(c) {2, 3} (d) {1, 6, 8}
(e) {1, 4, 5, 6, 7, 8, 9, 10}
(f) {1, 4, 5, 6, 7, 8, 9, 10}

**5.7** (a) {(6, 1), (5, 2), (4, 3), (3, 4),
(2, 5), (1, 6)}, {(6, 5), (5, 6)}
(b) 8 (c) 8/36

**5.9** $A = \{$Jack the Ripper,
Marquis de Sade$\}$
$B = \{$Nero$\}$
$C = \{$Jack the Ripper, Lizzie
Borden$\}$
(a) Yes (b) No

**5.13** (a) Female math professors
without tenure
(b) Not a female math
professor with tenure
(c) Tenured male professor
not in mathematics

**5.15** (a) U, V, X, Y
(b) None
(c) W
(d) U, W, Y, Z

**5.17** (a) 24 (b) 22

**5.19** 1/5

**5.21** 0.985

**5.23** 6/8

**5.25** (a) 0.15 (b) 0.60
(c) 0.45 (d) 0.15

**5.27** 0.33

**5.29** 4/6

**5.31** (a) 0.5 (b) 0.8

**5.33** 2/3

**5.35** 1/3

**5.37** (a) 0.40 (b) 0.30
(c) 10/44

**5.39** 0.366

**5.41** 94/182

**5.43** 0.51

**5.45**  0.43

**5.47**  2/3

**5.49**  Yes

**5.51**  0.49

**5.53**  0.4375

**5.55**  (a) 0.275    (b) 0.1636

**5.57**  Margo; $0.54 > 0.45$

**5.59**  Yes

**5.61**  0.52

**5.63**  Yes, $1/4 = (1/2) \cdot (1/2)$

**5.65**  No, $1/18 \neq (1/2) \cdot (1/6)$

**5.67**  24

**5.69**  8/81

**5.71**  29/90

**5.73**  6/36

**5.75**  0.952

**5.77**  0.625

**5.79**  0.632

**5.81**  1, 2, 3, 4, 5, 7, 8, 9, 10, 11

**5.83**  1/3

**5.85**  0.3

**5.87**  0.1152

**5.89**  47/70

**5.91**  0.15

**5.93**  (a) [0, 1]    (b) [0, 1/k]

**5.95**  32/33

**5.97**  1/4

**5.99**  0.944

**5.101**  5/18

**5.103**  Independence cannot be assumed.

**6.1**  0.50

**6.3**  0.38

**6.7**  0.218

**6.9**  (a)

| $s$ | $(1, 1)$ | $(1, 2)$ | $(1, 3)$ | $(1, 4)$ |
|---|---|---|---|---|
| $p(s)$ | $\frac{4}{49}$ | $\frac{6}{49}$ | $\frac{2}{49}$ | $\frac{2}{49}$ |
| $s$ | $(2, 1)$ | $(2, 2)$ | $(2, 3)$ | $(2, 4)$ |
| $p(s)$ | $\frac{6}{49}$ | $\frac{9}{49}$ | $\frac{3}{49}$ | $\frac{3}{49}$ |
| $s$ | $(3, 1)$ | $(3, 2)$ | $(3, 3)$ | $(3, 4)$ |
| $p(s)$ | $\frac{2}{49}$ | $\frac{3}{49}$ | $\frac{1}{49}$ | $\frac{1}{49}$ |
| $s$ | $(4, 1)$ | $(4, 2)$ | $(4, 3)$ | $(4, 4)$ |
| $p(s)$ | $\frac{2}{49}$ | $\frac{3}{49}$ | $\frac{1}{49}$ | $\frac{1}{49}$ |

(b) 25/49

**6.11**  $S = \{2, 3, 4, 5, 6, 8, 10, 12, 15, 20\}$;
$p(s) = 1/10$, all $s$

**6.13**

| $y$ | 0 | 1 | 2 |
|---|---|---|---|
| $f_Y(y)$ | 0.1 | 0.6 | 0.3 |

**6.15**  (a) $f_Y(7) = 3/16$
(b) Ace-six flats: $1/6 < 3/16$

**6.17**  $f_Y(y) = \dfrac{{}_4C_{Y} \, {}_{48}C_{5-Y}}{{}_{52}C_5}$,
$y = 0, 1, 2, 3, 4$

**6.19**  (a) 0.0625
(b) 0.6517
(c) 0.096256
(d) 0.419904

**6.21**  0.127

**6.23**  (a) 0.201    (b) 0.1672

**6.25**  0.00995

**6.27**  (a) 0.6651    (b) 0.6187
(c) 0.5974

**6.29**  0.040

**6.31**  (a) 30/84    (b) 0
(c) 34/84

**6.33**  $f_Y(y) = \dfrac{{}_5C_k \cdot {}_3C_{4-k}}{{}_8C_4}$, $1 \leq k \leq 4$

**6.35**  (a) 0.0128    (b) 0.0821
(c) 0.2163

**6.37**  0.019

**6.39**  11/42

**6.41**  $y$ = number of defectives
$$\dfrac{{}_{100-y}C_5 \cdot {}_yC_0 + {}_{100-y}C_4 \cdot {}_yC_1}{{}_{100}C_5}$$

**6.43**  (a) 0    (b) 2/9    (c) 5/9
(d) 16/81

**6.45**  125/1296

**6.47**  0.81

**6.49**  (a) 75/1296
(b) $f_Y(k)$
$$= (k-1)\left(\frac{5}{6}\right)^{k-2}\left(\frac{1}{6}\right)^2, k \geq 2$$

**6.51**  (a) 0.497    (b) 0.875

**6.53**  0.602

**6.55** 0.472

**6.57** (a) Expected numbers: 108.7, 66.3, 20.2, 4.1, 0.6

(b) Yes

**6.59** 0.264

**6.61** −$0.26

**6.63** 0

**6.65** 1.6

**6.67** (a) 1   (b) 1

**6.69** (a) $107p$   (b) 53.5

**6.71** 0.271

**6.73** 2

**6.75** 4.805

**6.77** (a) 0.3375   (b) 0.5809

**6.79** 5/9

**6.81** 1.09

**6.83** 1.45

**6.85** 2.1

**6.87** 0.36

**6.89** (a) 0.2514   (b) 0.5
(c) 0.1093   (d) 0.8264
(e) 0.2978   (f) 0.2978

**6.91** 0.8461

**6.93** 0.0495

**6.95** 0.9162

**6.97** 0.1335

**6.99** (a) 0.5   (b) 0.4013
(c) 0.1809

**6.101** No, since the probability is 0.8413.

**6.103** 16%

**6.105** 0.0062

**6.107** (a) 0.1292   (b) 0.1922

**6.109** (a) Relative frequencies: 0.02, 0.08, 0.32, 0.30, 0.14, 0.08, 0.06
(b) 0.2444

**6.111** (a) 0.95   (b) 83rd

**6.113** 58.4

**6.115** (a) 83   (b) 75

**6.117** $\dfrac{{}_{10}C_5 \cdot {}_{15}C_7 + {}_{10}C_6 \cdot {}_{15}C_6}{{}_{25}C_{12}}$
$= 0.514$

**6.119** 12

**6.121** 0.5457

**6.125** Greater than

**6.127** 0.22

**6.129** Yes, the random variable is binomial.

**6.131** They are equal.

**6.133** 0.57

**6.135** Mathematicians are like Frenchmen, etc.

**6.137** 1/2

**6.139** (a) ${}_{39}C_{18}\left(\dfrac{1}{2}\right)^{39}$

(b) $39\left(\dfrac{1}{2}\right) = 19.5$

(c) 0.7389; this quantity is the probability of the observed amount or greater, assuming that dominance is not a factor.

**7.1** (b) $E(Y) = E(\bar{Y}) = 15/8$,
Var $(Y) = 23/64$,
Var $(\bar{Y}) = 23/128$

**7.3** (a) 15, $\sqrt{245}$   (b) 3, $\sqrt{49/5}$

**7.5** (a) 0.3192   (b) 0.1904

**7.7** 0.0465

**7.9** 0.8966

**7.11** 1.533, 2.1318, 2.7764, 3.747

**7.13** 17

**7.15** −1.415

**7.17** 2.4469

**7.19** 1.8331

**7.21** (a) 1   (b) 0   (c) 0

**7.23** (a) 0.1112   (b) 0.0071
(c) 0.1813

**7.25** 0.0017

**7.27** 0.0409

**7.29** 0.5878

**7.31** 17.535

**7.33** 9.236

**7.35** 7

**7.37** (a) 2.159   (b) 9.014

**7.39** 3.37

**7.41** 3.29

**7.43** 2.98

**7.45** 0.206

**7.47** 0.265

**7.49** The new machine will not be used.

**7.51** 0.3446

**7.53** $-\$10$

**7.55** 0.2514

**7.57** 0.7852

**7.61** 16

**7.65** 30.625

**7.67** The former denotes a right tail; the latter, a left tail.

**7.69** 210.4

**7.71** Yes

**7.73** 0.2119

**7.75** The latter

**8.1** $y^* = 13; \alpha = 0.048$

**8.3** (a) Reject $H_0$ if $Y \geqslant 8$; $\alpha = 0.055$    (b) No

**8.5** Test statistic is $(\bar{y} - 37.5)/4.84$.
(a) $-2.58, 2.58$    (b) $-2.33$
(c) 2.33

**8.7** (a) Accept $H_0$ since $-1.96 < 1.90 < 1.96$.
(b) Accept $H_0$ since $-2.58 < 1.90 < 2.58$.
(c) Accept $H_0$ since $1.90 < 2.33$.
(d) Accept $H_0$ since $-2.33 < -1.90$.

**8.9** Reject $H_0$ since $3.03 > 2.58$.

**8.11** 62

**8.13** Yes, because $1.83 > 1.64$.

**8.15** (a) Accept $H_0$ since $-2.58 < -2.46 < 2.58$.
(b) Reject $H_0$ since $-2.46 < -2.33$.
(c) Reject $H_0$ since $-2.46 < -1.96$.
(d) Reject $H_0$ since $-2.46 < -1.64$.

**8.17** 0.03

**8.19** Reject $H_0$ since $2.03 > 1.96$.

**8.21** Accept $H_0$ since $-2.33 < -2.24$.

**8.25** 0.8023

**8.27** (a) 0.5    (b) 0.333
(c) 0.183

**8.29** (a) Reject if $\bar{y} < 48.56$, $\bar{y} < 48.16$, $\bar{y} < 47.38$.
(b) 0.44

**8.31** (a) 0.0808    (b) 0.8413
(c) Reject if $y < 0.84$.
(d) 0.9726

**8.33** (a) 0.61, 0.93, 0.986, 0.997

**8.37** (45.72, 47.68)

**8.39** (0.380, 0.588)

**8.41** (a) (12.4, 16.2)
(b) (34.4, 37.0)
(c) (13.2, 14.6)
(d) (12.4, 17.6)

**8.43** 85%

**8.45** Accept $H_1: p > 0.25$ since $4.90 > 2.33$.

**8.47** No, since $1.47 < 2.33$.

**8.49** $H_0: p = 1/2;$   $H_1: p > 1/2;$ $\alpha = 0.14$

**8.51** (1.067, 1.167)

**8.53** Reject $H_0$ since $2.71 > 1.28$.

**8.55** Reject $H_0$ since $1.82 > z_{0.185} = 0.90$

**8.57** The former measures one standard deviation of $Y$ from the sample mean; the latter measures one standard deviation of $\bar{Y}$ from the sample mean.

**8.59** $n$

**8.61** The former, since $0.41\sigma < 0.52\sigma$

**8.63** $\alpha$

**8.65** 0.262

**8.67** No, the data is not bell-shaped.

**8.69** Reject $H_0$ if the $P$ value is $\leqslant \alpha$.

**9.1** (a) 1.7109    (b) 2.492
(c) $-2.0639$
(d) $-2.0639, 2.0639$
(e) $-1.7109, 1.7109$

**9.3** Accept $H_0$ since
$-2.0930 < -0.53 < 2.0930$.

**9.5** (a) Accept $H_0$
since $-1.7341 < -1.71$.
(b) It is the value for a healthy
adult.

**9.7** Accept $H_0$
since $-2.7764 < 2.60 < 2.7764$.

**9.11** Accept $H_0$
since $-1.8331 < -1.46$.

**9.13** (a) $(5.42, 6.14)$
(b) $(5.34, 6.22)$

**9.15** (a) $(5.63, 6.29)$
(b) $(5.15, 6.77)$

**9.17** $(31.7, 39.1)$

**9.19** $(157.2, 173.2)$

**9.21** (a) 95%      (b) 80%
(c) 94.5%    (d) 95%

**9.23** (a) $(0.33, 0.53)$
(b) $(0.35, 0.51)$

**9.25** $(0.17, 0.25)$

**9.27** $(0.76, 0.86)$

**9.29** $(0.53, 0.72)$

**9.31** 190

**9.33** (a) 1557     (b) 1076
(c) 2663

**9.35** 97

**9.37** 482

**9.39** (a) 1068     (b) 1849
(c) 125

**9.41** 543

**9.43** (a) $(99.47, 700.83)$
(b) $(41.79, 154.12)$

**9.45** $(0.0050, 0.0184)$

**9.47** $\sigma$, since it preserves units of
measurement

**9.49** $(-\infty, 0.0141), (0.0046, \infty)$

**9.51** 92.5%

**9.53** Reject $H_0$
since $16.812 < 25.078$.

**9.55** Accept $H_0$ since $3.325 < 6.41$.

**9.57** 9

**9.59** $(0.554, 0.646)$

**9.61** 1681

**9.63** $(0.51, 0.75)$

**9.65** Reject $H_0$ since $16.919 < 40.83$.

**9.67** (a) $t$ with $n$ degrees of
freedom vs. normal.
(b) As $n$ increases, the $t$
statistic approaches the
normal.
(c) Whether $\sigma$ is known.

**9.69** Yes, since $-2.30 < -2.0150$.

**9.71** To measure the accuracy of the
estimate.

**9.73** The first is a "random" interval;
the second, an interval of
numbers.

**9.75** (a) It depends on whether the
experimenter wants a
"yes–no" answer or an
estimate of the parameter.
(b) Yes, reject $H_0$ if its
parameter value does not
fall in the confidence
interval.

**9.77** 1530

**9.79** Accept the hypothesis that
$3500 is the average since
$-1.9432 < -1.35 < 1.9432$.

**9.81** 1.32

**10.1** (a) Reject $H_0$ since $1.833$
$< 1.93$.
(b) Accept $H_0$ since $2.120$
$1.54 < 2.120$.
(c) Reject $H_0$ since $2.704$
$< 5.40$.

**10.3** Reject $H_0$ since $2.1448 < 4.77$.

**10.5** Accept $H_0$ since $-3.1693$
$< 2.01 < 3.1693$.

**10.7** Reject $H_0$ since $2.0244 < 2.66$.

**10.9** Accept $H_0$ since $-2.58$
$< -0.21 < 2.58$.

**10.11** $(5.89, 25.79)$

**10.13** (a) 0.125, 5.80
(b) 0.181, 4.32
(c) 0.269, 2.98

**10.15** (a) 0.257, 6.52     (b) 4.68
(c) 0.198

**10.17** Reject $H_0$ since $7.15 < 32.8$.

**10.19** Accept equality of the variances since $0.299 < 1.18 < 8.50$.

**10.21** $\left(\dfrac{S_Y^2}{S_X^2}\dfrac{1}{F_{\alpha/2,\,m-1,\,n-1}},\dfrac{S_Y^2}{S_X^2}\right.$

$\left. F_{\alpha/2,\,n-1,\,m-1}\right)$

**10.23** $(0.27, 1.73)$

**10.25** Accept $H_0$ since $-1.96 < 1.88 < 1.96$.

**10.27** Reject $H_0$ since $-5.024 < -1.64$.

**10.29** Accept $H_0$ since $-2.58 < -0.07 < 2.58$.

**10.31** $0.5222$

**10.33** $\left(\dfrac{x}{n} - \dfrac{y}{m}\right.$

$-Z_{\alpha/2}\sqrt{\dfrac{x/n(1-x/n)}{n} + \dfrac{y/m(1-y/m)}{m}},$

$\dfrac{x}{n} - \dfrac{y}{m}$

$\left. +Z_{\alpha/2}\sqrt{\dfrac{x/n(1-x/n)}{n} + \dfrac{y/m(1-y/m)}{m}}\right)$

**10.35** Accept $H_0$ since $0.156 < 4.21 < 6.39$.

**10.37** The times are different since $2.228 < 2.24$.

**10.39** $\mu_X =$ mean teacher salaries in Tennessee.
$\mu_Y =$ mean teacher salaries in North Carolina.
Accept $H_0$ since $-1.397 < -1.392 < 1.397$.

**10.41** No, since $-1.96 < -0.39 < 1.96$.

**10.43** $(25{,}237.4, 294{,}810.8)$

**10.45** No, it is a measure of variability, not location.

**10.47** No

**10.49** $0.2843$

**10.51** That definition ignores the weights due to relative sample size.

**10.53** Let $p_x = p_y$ be the probability that an observation falls in a certain range of values.

**10.55** The first hypothesis test assumes that the data is pooled; the second is for a two-sample scheme. Having established equality of means, one might want to then test for the value of the common mean.

**10.57** Usually, the test for equality of the means and not accuracy.

**10.59** It is best for $S\sqrt{1/n + 1/m}$ to be small so choose 20,20.

**11.1** (a) Accept $H_0$ since $-2.160 < 1.66 < 2.160$.
 (b) Reject $H_0$ since $2.492 < 16.25$.
 (c) Accept $H_0$ since $-1.476 < -0.95$.

**11.3** (a) Accept $H_0$ since $0.37 < 2.074$.
 (b) Paired, since within-pair data points are dependent.

**11.5** (a) Reject $H_0$ since $2.861 < 6.19$.

**11.7** $\mu =$ average protein content of triticale crop. $H_1: \mu_D \neq 0$. Accept $H_0$ since $-3.182 < 2.12 < 3.182$.

**11.9** $H_0: \mu_D = 0$;  $H_1: \mu_D \neq 0$. Reject $H_0$ since $2.353 < 2.42$

**11.11** Reject $H_0$ since $1.782 < 4.50$.

**11.13** $H_0: \mu_D = 0$;  $H_1: \mu_D > 0$. Accept $H_0$ since $1.57 < 2.132$.

**11.15** $H_0: \mu_D = 0$;  $H_1: \mu_D \neq 0$. Accept $H_0$ since $-2.015 < 0.69 < 2.015$.

**11.17** $\left(\bar{d} - \dfrac{t_{\alpha/2,\,n-1}s_D}{\sqrt{n}}, \bar{d} + \dfrac{t_{\alpha/2,\,n-1}s_D}{\sqrt{n}}\right)$

**11.19** Accept $H_0$ since 0 is in the interval.

**11.21** Yes, $\bar{d}$ will be normally distributed via the central limit theorem.

**11.23** (BG, BP), (WW, NW), (JS, SG), (LL, JM)

**11.25** (a) Paired data columns are necessarily the same length.
(b) Within-pair data points are connected by a line.

**11.29** Good; the use of paired data is confirmed.

**11.31** No, within-pair independence is still required.

**11.33** Increase $n$.

**11.35** Reject $H_0$ since $-7.61 < -1.476$.

**12.1** A 15.86% confidence interval.

**12.3** 0.3758

**12.5** $Y = 8.586 - 0.331x$
$A$ is normal with mean $a$ and variance 1.7069.
$B$ is normal with mean $b$ and variance 0.0124.

**12.7** 0.0926

**12.9** The data points will lie on the regression line.

**12.11** $\hat{\sigma}^2$ can be zero with $\sigma^2 > 0$ if all data points lie on the regression line. If $\sigma^2 > 0$, then $\hat{\sigma}^2 > 0$.

**12.13** 97.51 for both expressions

**12.15** Accept $H_0$ since $-2.2281 < -0.0922 < 2.2281 = t_{0.025, 10}$.

**12.17** (a) $y = 13.8 - 1.5x$
(b) Accept $H_0$ since $-4.3027 < -1.5865 < 4.3027 = t_{0.025, 2}$.

**12.19** (a) The true intercept is 100.
(b) Accept $H_0$ since $-t_{0.05, 4} = -2.1318 < -0.617$.

**12.21** (a) $y = -0.2022 + 0.9941x$
(b) Accept $H_0$ since $-t_{0.05, 8} = -1.8595 < -1.3354$.

**12.23** (84.9028, 683.1172)

**12.25** (a) $y = 9.4052 + 0.0813x$
(b) $(-2.359, 21.169)$

**12.27** (0.8095, 1.1011)

**12.29** (0.2016, 2.425)

**12.31** (110.52, 135.06)

**12.33** (4.4086, 12.6914)

**12.35** (177.5285, 190.9315) It is unlikely that the average weight of the players was less than 170 pounds.

**12.37** (a) (107.34, 177.20)
(b) The prediction interval, which provides an estimate of the revenue when hauling 700 million ton-miles.

**12.39** (20.896, 68.084) The rating could be as low as 35.

**12.41** (a) 11.55
(b) (4.116, 18.984)

**12.45** $\bar{y} = 0.26 + 1.70(0.99) = 1.94$
$\bar{x} = -0.825 + 0.553(1.94) = 0.99$

**12.47** Regression data. There are several $y$-values for each $x$, which suggests that the $x$-values were set by the experimenter.

**12.49** Reject $H_0$ since $t_{0.005, 15} = 2.9467 < 4.5616$.

**12.51** Reject $H_0$ since $-2.1576 < -2.1009 = -t_{0.025, 18}$.

**12.53** Reject $H_0$ since $z_{0.025} = 1.96 < 2.7971$.

**12.55** The denominator of the $t$ statistic in Theorem 12.10 is the standard deviation of $B_1 - B_2$.

**12.57** Condition 1:
$y = -0.6 + 0.8x$
Condition 2:
$y = -4.2727 + 1.7727x$
Accept $H_0$ since $-2.3646 < 2.2061 < 2.3646 = t_{0.025, 7}$.

**12.59** $y = 11.995x^{0.8764}$

**12.61** $y = 0.78x^{0.8793}$

**12.63** $197.56

**12.65** $y = 5.0585 + 58.1308/x$

**12.67** $\ln(y) = \ln(a) + bx$;
$a' = \ln(a)$, $b' = b$

**12.69** (b) $y = 18.222 - 1.5833x$
(d) 14.6746

**12.71** (11.82, 24.62)
It is unlikely that $a = 30$

**12.73** (b) $y = 3.50 + 1.25x$
(c) 4.5
(d) Regression

**12.75** 0.0703

**12.77** (8.49, 18.51)

**12.79** The $y$'s are not normally
distributed, and the variance
seems to depend on $x$.

**12.81** (a) For each $x$ value, the data
seems to have a similar
distribution.
(b) The $y$'s seem normally
distributed.

**12.83** (a) 0.7699
(b) The probability represents
the precision of the
estimator.
(c) Large

**12.85** $y = 3.9552x^{3.7022}$

**12.87** Reject $H_0$ since $t_{0.025,6}$
$= 2.4469 < 3.43$
The variables are not
independent.

**12.89** No, since the 90% prediction
interval is (2.24, 6.58).

**12.91** If $(\sum x^2)/n < 1$, then Var $(A)$
could be less than Var $(b)$.

**12.93** 66.78

**13.3** One-sample binomial test

**13.5** (a) Observed $X^2 = 17.6$.
$X^2_{0.95,7} = 14.067$. Reject
$H_0$.
(b) Yes, there is a significant
bias toward the number
1 post.

**13.7** Observed $X^2 = 7.43$.
$X^2_{0.95,1} = 3.841$. Reject $H_0$.

**13.9** Observed $X^2 = 1.47$.
$X^2_{0.99,3} = 11.345$. Accept $H_0$.

**13.11** Small values of $d$ imply good
agreement with the null
hypothesis.

**13.13** No. Observed $X^2 = 15.19$.
$X^2_{0.90,2} = 4.605$. Reject $H_0$.

**13.15** Observed $X^2 = 30.38$.
$X^2_{0.99,8} = 20.090$. Reject $H_0$.

**13.17** Observed $X^2 = 3.83$.
$X^2_{0.95,2} = 5.991$. Accept $H_0$.

**13.19** Observed $X^2 = 3.11$.
$X^2_{0.95,1} = 3.841$. Accept $H_0$.

**13.21** Observed $X^2 = 0.34$.
$X^2_{0.99,1} = 6.635$. Accept $H_0$.

**13.23** Observed $X^2 = 3.89$.
$X^2_{0.95,2} = 5.991$. Accept $H_0$.

**13.25** (a) 5.0
(b) No, histogram
is skewed to the left.

**13.27** (b) 0.51, 0.55. We would
expect these two to be
the same.
(c) Observed $X^2 = 0.20$.
$X^2_{0.95,1} = 3.841$. Accept $H_0$.

**13.29** Observed $X^2 = 9.90$.
$X^2_{0.95,3} = 7.815$. Reject $H_0$.

**13.31** (a) Observed $X^2 = 42.24$.
$X^2_{0.99,3} = 11.345$. Reject
$H_0$.
(b) Yes

**13.33** Observed $X^2 = 12.61$.
$X^2_{0.95,1} = 3.841$. Reject $H_0$.

**13.35** Observed $X^2 = 2.77$.
$X^2_{0.90,1} = 2.706$. Reject $H_0$.

**13.37** Observed $X^2 = 27.58$.
$X^2_{0.99,2} = 9.210$. Reject $H_0$.

**13.39** Observed $X^2 = 27.29$.
$X^2_{0.99,4} = 13.277$. Reject $H_0$.

**13.43** Observed $X^2 = 1.29$.
$X^2_{0.95,2} = 5.991$. Accept $H_0$.

**13.45** Reject the Poisson
model since the observed
$X^2 = 6.73$, which exceeds
$X^2_{0.90,3} = 6.251$.

**13.47** Observed $X^2 = 11.28$.
$X^2_{0.90,2} = 4.605$. Reject $H_0$.

**13.49** Observed $X^2 = 8.62$.
$X^2_{0.95,2} = 5.991$. Reject the model.

**13.53** $\bar{y}_g = 0.14$, $s_g = 15.91$.

**13.55** Yes, recall that a $Z$ statistic squared is $X^2_1$.

**13.57** Reject the Poisson model since the observed $X^2$ $= 48.89$, which exceeds $X^2_{0.95,3} = 7.815$.

**13.63** Observed $X^2 = 3.04$. $X^2_{0.95,4} = 9.488$. Accept the model.

**13.65** Use a histogram of the residuals.

**14.1** $T_1 = 178.3$, $T_2 = 167.9$, $T_3 = 155.1$
$\bar{x}_1 = 59.43$, $\bar{x}_2 = 55.96$, $\bar{x}_3 = 51.7$
$T = 501.3$, $\bar{x} = 55.7$

**14.3** $T_1 = 34$, $T_2 = 39$, $T_3 = 127$, $T_4 = 99$, $T_5 = 136$, $T_6 = 89$
$\bar{x}_1 = 11.33$, $\bar{x}_2 = 13$, $\bar{x}_3 = 42.33$, $\bar{x}_4 = 33$, $\bar{x}_5 = 45.33$, $\bar{x}_6 = 29.67$
$T = 524$, $\bar{x} = 29.11$

**14.5** $T_1 = 200.82$, $T_2 = 197.98$, $T_3 = 193.98$
$\bar{x}_1 = 33.47$, $\bar{x}_2 = 32.9767$, $\bar{x}_3 = 32.33$
$\bar{x} = 32.9322$

**14.7** SSTO $= 12 = 8 + 4$
$=$ SSTR $+$ SSE

**14.9** SSTO $= 26 = 6 + 20$
$=$ SSTR $+$ SSE

**14.11** Accept $H_0$ since $-2.7764$ $< -1.0945 < 2.7764$
$= t_{0.025,4}$
$(2.7764)^2 = 7.7084$,
$(-1.0954)^2 = 1.200$

**14.13** SSTO $= 12$, SSTR $= 8$

**14.15** SSTO $= 0.3465$,
SSTR $= 0.2050$, SSE $= 0.1415$

**14.17** SSTO $= 79.6667$,
SSTR $= 43.7167$,
SSE $= 35.9500$

**14.19** (a)

| Source | df | SS | MS | F |
|---|---|---|---|---|
| Treatment | 3 | 74.9333 | 24.9788 | 4.81 |
| Error | 6 | 31.1667 | 5.1945 | |
| Total | 9 | 106.1000 | | |

(**b**) Reject $H_0$ since
$F_{0.05,3,6} = 4.76 < 4.81$.

**14.21** (a)

| Source | df | SS | MS | F |
|---|---|---|---|---|
| Treatment | 5 | 3087.11 | 617.42 | 1.26 |
| Error | 12 | 5900.67 | 491.72 | |
| Total | 17 | 8987.78 | | |

(**b**) Accept $H_0$ since
$1.26 < 3.11 = F_{0.05,5,12}$.

**14.23**

| Source | df | SS | MS | F |
|---|---|---|---|---|
| Treatment | 3 | 558.92 | 186.31 | 0.67 NS |
| Error | 8 | 2241.33 | 280.17 | |
| Total | 11 | 2800.25 | | |

**14.25** Reject $H_0$ since
$F_{0.05,2,18} < 13.99$.

**14.27** (a)

| Source | df | SS | MS | F |
|---|---|---|---|---|
| Treatment | 3 | 109 | 36.33 | 16.80 |
| Error | 12 | 26 | 2.17 | |
| Total | 15 | 135 | | |

(**b**) Reject $H_0$ since
$F_{0.01,3,12} = 5.95 < 16.80$.

**14.29** (a) 4.90   (b) 8.42
(c) 10   (d) 9   (e) 0.05
(f) 0.95

**14.31** (a) $(-0.84, 7.76)$,
$(3.43, 12.03)$,
$(-0.03, 8.57)$
(**b**) $\mu_{1669}$ and $\mu_{1865}$ are significantly different.

**14.33** Accept all subhypotheses since 0 belongs to each interval:
$(-0.603, 0.078)$,
$(-0.630, 0.050)$,
$(-0.368, 0.313)$.

**14.35** Both intervals are $(-27.9, 7.6)$.

**14.37** $(-29.51, 2.84)$,
$(-42.84, -10.49)$,
$(-56.17, -23.83)$
$\mu_{\text{U.S.}}$ and $\mu_{\text{Abroad}}$ are not significantly different; the other pairs are.

**14.39** $(0.24, 1.62), (-1.62, -0.24)$,
$(-2.56, -1.18)$
All pairs are significantly different.

**14.41** $_5C_2 = 10$

**14.43** $(2.77, 10.43), (1.57, 9.23)$,
$(-5.03, 2.63)$

**14.45** Greater

**14.47** Accept $H_0$ since $-2.2281$
$< 1.1891 < 2.2281 = t_{0.025, 10}$.
$(1.1891)^2 = 1.414$,
$(2.2281)^2 = 4.96$

**14.49** The sum of the deviations of a sample from the mean is zero.

**14.51** Estimate $\mu_j$ by $\bar{x}_j$ and $\varepsilon_i$ by $\sqrt{\text{MSE}}$

**14.53** $-246.18, 1279.52)$

**14.55** (c) and (d)

**14.57** 254.6667

**14.59** The intervals will be longer.

**15.1** $B_1 = 26, B_2 = 23, B_3 = 9$,
$B_4 = 24$
$T_1 = 26, T_2 = 28, T_3 = 28$
$\bar{x}_1 = 6.5, \bar{x}_2 = 7, \bar{x}_3 = 7$
$T = 82, \bar{x} = 6.8333$

**15.3** Only if each of the $k$ samples is of the same size.

**15.5** $B_1 = 59.5, B_2 = 58.8$,
$B_3 = 58.1$
$T_1 = 60.5, T_2 = 58.8$,
$T_3 = 57.5$
$\bar{x}_1 = 20.1667, \bar{x}_2 = 19.6$,
$\bar{x}_3 = 19.1667$
$T = 176.8, \bar{x} = 19.6444$

**15.7** Accept $H_0$ since $0.14 < 5.14$
$= F_{0.05, 2, 6}$.

**15.9** Reject $H_0$ since $F_{0.05, 2, 18}$
$= 3.57 < 3.59$.

**15.11** Accept $H_0$ since $0.14 < 4.46$
$= F_{0.05, 2, 8}$.

**15.13**

| Source | df | SS | MS | F |
|---|---|---|---|---|
| Treatment | 2 | 0.6667 | 0.3333 | 0.14 NS |
| Blocks | 3 | 60.3333 | 20.1111 | |
| Error | 6 | 14.6667 | 2.4444 | |
| Total | 11 | 75.6667 | | |

**15.15**

| Source | df | SS | MS | F |
|---|---|---|---|---|
| Treatment | 2 | 0.9613 | 0.4807 | 0.14 NS |
| Blocks | 4 | 37.8067 | 9.4517 | |
| Error | 8 | 26.7853 | 3.3482 | |
| Total | 14 | 65.5533 | | |

**15.17**

| Source | df | SS | MS | F |
|---|---|---|---|---|
| Treatment | 2 | 3.9362 | 1.9681 | 10.36** |
| Blocks | 5 | 9.3690 | 1.8738 | |
| Error | 10 | 1.9017 | 0.1902 | |
| Total | 17 | 15.2069 | | |

**15.19**

| Source | df | SS | MS | F |
|---|---|---|---|---|
| Treatment | 3 | 70 | 23.33 | 1.24 NS |
| Blocks | 4 | 264 | 66.00 | |
| Error | 12 | 226 | 18.83 | |
| Total | 19 | 560 | | |

**15.21** Reject $H_0$ since
$F_{0.05, 11, 22} = 2.26 < 6.83$.
The experimenter benefited from using time as a blocking factor.

**15.23** (a) Completely randomized design
(b) Randomized block design

**15.25** (a) Reject $H_0$ since
$F_{0.05, 4, 12} = 3.26 < 49.93$.
(b) Retain the randomized block design

**15.27** Reject $H_0$ since
$F_{0.05, 2, 6} = 5.14 < 6.53$.

**15.29** Accept $H_0$ since
$1.80 < 4.46 = F_{0.05, 2, 8}$.

**15.31**

| Source | df | SS | MS | F |
|---|---|---|---|---|
| Treatment | 3 | 45.0 | 15.0 | 3.0 |
| Blocks | 4 | 40.0 | 10.0 | 2.0 |
| Error | 12 | 60.0 | 5.0 | |
| Total | 19 | 145.0 | | |

**15.33** (a) True  (b) False
 (c) False

**15.35** $0.10 < P$ value $< 0.25$

**15.37** $(-0.025, 0.120)$,
 $(-0.110, 0.035)$
 $(-0.157, -0.013)$

**15.39** (a) Accept $H_0$ since $0.5476$
 $< 4.84 = F_{0.05, 1, 11}$.
 (b) $(-0.74)^2 = 0.5476$,
 $(2.20)^2 = 4.84$

**15.41** (a) False  (b) True
 (c) False  (d) True
 (e) True

**15.43** Randomly assign one-third of
 the pool of players to each
 coach. The coach could group
 them by ability in threes and
 assign them randomly to
 courts by group.

**15.45**

**TREATMENTS**

| | 1 | 2 | 3 |
|---|---|---|---|
| **BLOCKS** | 2 | 3 | 1 |
| | 3 | 1 | 2 |

**15.47** (a)

**TREATMENTS**

| | 4 | 6 | 8 | 2 | 20 |
|---|---|---|---|---|---|
| **BLOCKS** | 3 | 7 | 10 | 3 | 23 |
| | 5 | 8 | 12 | 4 | 29 |
| | 12 | 21 | 30 | 9 | 72 |

 (b) $\hat{\mu}_1 = 4, \hat{\mu}_2 = 7, \hat{\mu}_3 = 10,$
 $\hat{\mu}_4 = 3$
 $\hat{\beta}_1 = 5, \hat{\beta}_2 = 5.75, \hat{\beta}_3 = 7.25$

**15.49** Accept $H_0$ since $6.63 < 10.9$
 $= F_{0.01, 2, 6}$.

**15.51** Accept $H_0$ since $0.95 < 3.86$
 $= F_{0.05, 3, 9}$.

**16.1** Accept $H_0$. Test statistic
 value $= -0.23$. Critical value:
 1.64.

**16.3** Reject $H_0$. Test statistic
 value $= -3.13$. Critical value:
 $-2.33$.

**16.5** (a) Reject $H_0$. Test statistic
 value $= -3.05$. Critical
 value: $-1.64$.
 (b) Yes

**16.7** Accept $H_0$. Test statistic
 value $= 1.89$. Critical values:
 $-1.96, 1.96$.

**16.9** Accept $H_0$. Test statistic
 value $= 1.89$. Critical values:
 $-1.96, 1.96$.

**16.11** Reject $H_0$. Test statistic
 value $= 2$. Critical value: 3.
 $\alpha = 0.1133$.

**16.13** Reject $H_0$. Test statistic
 value $= 13$. Critical values: 4,
 11. $\alpha = 0.144$.

**16.15** (a) 0.0625  (b) Yes

**16.17** Reject $H_0$. Test statistic
 value $= 33$. Critical value: 28.
 $\alpha - 0.098$.

**16.19** Accept $H_0$. Test statistic
 value $= 21$. Critical values: 8,
 47. $\alpha = 0.048$.

**16.21** (a) Accept $H_0$. Test statistic
 value $= 36$. Critical value:
 41.
 $\alpha = 0.097$.
 (b) Reject $H_0$. Test statistic
 value $= 53.3$. Critical
 value: 41. $\alpha = 0.097$.

**16.23** Accept $H_0$. Test statistic
 value $= 34$. Critical values:
 14, 64. $\alpha = 0.052$.

**16.25** (a) Accept $H_0$. Test statistic
 value $= -1.37$. Critical
 value: $-1.64$.
 (b) Yes

**16.27** Accept $H_0$. Test statistic value $= -1.40$. Critical values: $-1.64$, $1.64$.

**16.29** Accept $H_0$. Test statistic value $= 27$. Critical values: 7, 29. $\alpha = 0.148$.

**16.31** Reject $H_0$. Test statistic value $= 2.32$. Critical values: $-1.97$, $1.97$.

**16.33** (a) Accept $H_0$. Test statistic value $= 22$. Critical value: 24. $\alpha = 0.055$.
(b) Yes

**16.35** Reject $H_0$. Test statistic value $= 36$. Critical value: 36. $\alpha = 0.048$.

**16.37** (a) 0.001    (b) 0.002

**16.39** Accept $H_0$. Test statistic value $= 15$. Critical values: 11, 25. $\alpha = 0.058$

**16.41** Reject $H_0$. Test statistic value $= -3.18$. Critical values: $-1.96$, $1.96$.

**16.43** (a) Reject $H_0$. Test statistic value $= 3.02$. Critical values: $-2.58$, $2.58$.
(b) Yes

**16.45** If the normality assumption is not appropriate, the rank sum test should be used.

**16.47** Accept $H_0$. Test statistic value $= 0.68$. Critical values: $-1.96$, $1.96$. A two-sided alternative is appropriate; carpeted rooms could be more sanitary.

**16.49** Accept $H_0$. Test statistic value $= 1.25$. Critical value: 5.991.

**16.51** Reject $H_0$. Test statistic value $= 12.48$. Critical value: 9.210.

**16.53** Reject $H_0$. Test statistic value $= 15.33$. Critical value: 7.779.

**16.55** Accept $H_0$. Test statistic value $= 4.71$. Critical value: 5.991.

**16.57** Accept $H_0$. Test statistic value $= 3.48$. Critical value: 7.815.

**16.59** Accept $H_0$. Test statistic value $= 1.31$. Critical value: 5.991. This result agrees with the $F$ test done previously.

**16.61** The $P$ value is slightly less than 0.05.

**16.63** (a) Reject $H_0$. Test statistic value $= 8.28$. Critical value: 7.815.
(b) Yes

**16.65** Reject $H_0$. Test statistic value $= -2.84$. Critical values: $-2.58$, $2.58$.

**16.67** Reject $H_0$. Test statistic value $= 16$. Critical values: 5, 15. $\alpha = 0.0414$.

**16.69** Reject $H_0$. Test statistic value $= 6.33$. Critical value: 5.991.

**16.71** Reject $H_0$. Test statistic value $= 2.56$. Critical values: $-1.96$, $1.96$.

**16.73** Accept $H_0$. Test statistic value $= -1.73$. Critical values: $-1.96$, $1.96$.

**16.75** Reject $H_0$. Test statistic value $= 8.13$. Critical value: 5.991.

**16.77** Accept $H_0$. Test statistic value $= 30$. Critical values: 28, 50. $\alpha = 0.094$.

**16.79** Reject $H_0$. Test statistic value $= -3.22$. Critical value: $-1.64$.

**16.81** Bell shaped data: $t$ test Symmetric data: signed rank test Asymmetric data: sign data

**16.83** $w' = -1.6474$. $P(Z < -1.65) = 0.0495$.

**16.85** Accept $H_0$. Test statistic value $= 1.33$. Critical value: 4.605.

**16.87** Reject $H_0$. Test statistic value $= 3.28$. Critical values:

$-1.96, 1.96.$

**16.89** Reject $H_0$. Test statistic value $= 32.5$. Critical value: $28 \; \alpha = 0.098$.

**17.1** (a) $0.220$ (b) $0.962$
(c) $0.956$ (d) $3.74$
(e) $101$

**17.3** Both points are above the plane.

**17.5** $y = -3.1473 + 0.0123x_1 - 0.4136x_2$
Expected MCAT score $= 11.9$

**17.7** $y = 503.122 + 184.585x_1 - 7.551x_2$

**17.9** The $y - \hat{y}$ differences could be tested by a goodness of fit test with unknown parameter $\sigma$.

**17.11** $0.080, 0.1446$

**17.13** If the linear model does not hold, then $y - \hat{y}$ is not necessarily related to $\sigma$.

**17.15** The model appears not to be effective in predicting $Y$ since $2.06 < 3.49 = F_{0.05, 3, 12}$.

**17.17** The model appears to be effective in predicting $Y$ since $F_{0.05, 2, 2} = 19.0 < 35.47$.

**17.19** Accept $H_0$ since $2.61 < 30.8 = F_{0.01, 2, 3}$.

**17.21** Accept $H_0$: $b_1 = 0$ since $3.38 < 4.3027 = t_{0.025, 2}$.
Reject $H_0$: $b_2 = 0$ since $t_{0.025, 2} \; 4.3027 < 7.72$.

**17.23** GPA does not contribute to the model since $2.31 < 4.3027 = t_{0.025, 2}$.

**17.25** The model is appropriate since the $P$ value is $\leq 0.0001 < 0.05 = \alpha$.
$b_1$ is not zero since the $P$ value is $< 0.05 = \alpha$.
For the other two parameters the $P$ values—$0.9350$ and $0.2238$—exceed $0.05$.

**17.27** The model is significant, and all parameters are significantly different from zero.

**17.29** $669.38$

**17.31** $y = 15 - 3x_1 - 5x_2 + 11x_1x_2$, $x_2 = 0, 1$

**17.33** The first, since its sum of squared deviations is $81$; the second, $99$.

**17.35** MSE $= 7.147$

**17.37** SSE $= 1.2818$, MSE $= 0.986$

**17.39** $0.33$. Not significant since $0.33 < 2.1448 = t_{0.025, 14}$.

**17.41** (a) $97.9$ (b) $49.34$
(c) $-4.5179$

**17.43** $b_1 = 9$, $b_2 = -5$

**17.45** The model is useful since the test statistic has value $320.25$.
Reject $H_0$: $b_1 = 0$ since $t_{0.025, 8} = 2.3060 < 25.31$
Reject $H_0$: $b_2 = 0$ since $-2.63 < -2.3060 = -t_{0.025, 8}$.

**17.47** $\hat{a} = 5.2165$, $\hat{b}_1 = 4.7108$, $\hat{b}_2 = 3.3137$

**17.49** SSE $= 2068.82$

**17.51** (a) $y = 4x_1 + 5x_1x_2$, $x_2 = 0, 1$
(b) $y = 1 + 4x_1 + 4x_2 - 13x_1x_2$, $x_2 = 0, 1$

**17.53** (a) Reject $H_0$: $b_1 = b_2 = 0$ since the test statistic has value $93.06$.
(b) Reject $H_0$: $b_1 = 0$ since $-2.83 < -2.3646 = -t_{0.025, 7}$.
(c) Accept $H_0$: $b_2 = 0$ since $2.32 < 2.3646 = t_{0.025, 7}$.

**17.55** Regress $\log y$ on $\log x_1$ and $\log x_2$.

# Index